A FIRST COURSE

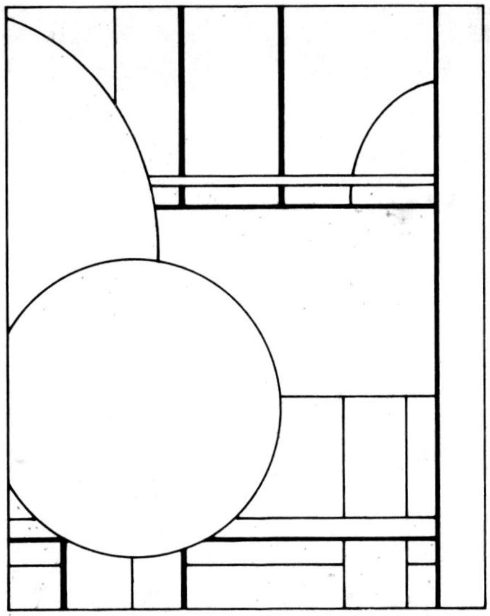

IN

LINEAR REGRESSION

A FIRST COURSE IN LINEAR REGRESSION

SECOND EDITION

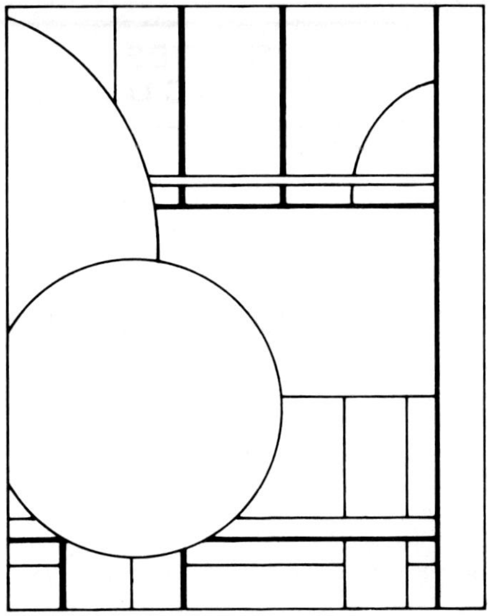

MARY SUE YOUNGER

THE UNIVERSITY OF TENNESSEE

DUXBURY PRESS ▪ BOSTON

PWS PUBLISHERS

Prindle, Weber & Schmidt • Duxbury Press • PWS Engineering •
Statler Office Building • 20 Park Plaza • Boston, Massachusetts 02116

PWS Publishers is a division of Wadsworth, Inc.

LIBRARY OF CONGRESS CATALOGING IN PUBLICATION DATA

Younger, Mary Sue.
A First Course in Linear Regression.
Rev. ed. of: Handbook for linear regression, c1979.
Includes index.
1. Regression analysis. 2. Regression analysis—
Data processing. I. Title.
AQ278.2Y68 1985 519.5′36 84-20592
ISBN 0-87150-865-6

ISBN 0-87150-865-6

Printed in the United States of America
85 86 87 88 89 — 10 9 8 7 6 5 4 3 2 1

Manuscript Editor: Alice Cheyer
Design and Coordinating: Susan London
Interior Illustration: Deborah Schneck
Typesetting: Monotype Composition Company
Cover Printing: New England Book Components, Inc.
Printing and Binding: Halliday Lithograph, Arcata Graphics

P R E F A C E

A First Course in Linear Regression introduces the techniques and applications of linear regression and provides a sound basis for more advanced study. It can also be used as a manual for those who use regression techniques without the benefit of a formal course.

It is expected that some readers may already have had an introductory survey course in regression analysis; nevertheless, this book begins at an elementary level. Building step-by-step on these elementary concepts, it offers a thorough—yet readable—explanation of regression. The text is "user-friendly," favoring verbal explanations rather than mathematical ones. Statistical theory and advanced mathematical techniques are kept to a minimum. (Theoretical assumptions and the derivations of formulas are explained intuitively in relation to commonsense considerations.) The proper application of statistical results is contrasted with cases of their incorrect use. The clear and descriptive notation conforms to that most widely used.

- The text has an early introduction to the matrix formulation of the regression problem (expanded in this revision). This serves two very important purposes:

  (1) It provides a unified framework for simple, polynomial, and multiple regression and thus saves the reader from being overwhelmed by special formulas;

and

  (2) It provides a basis from which to advance to the analysis of variance and covariance and to multivariate statistics.

- New material has been added on exponential functions and on dummy variables and interactions.

• *A First Course in Linear Regression* provides step-by-step instruction in modern regression and also provides step-by-step analysis techniques using four widely available computer programs: BMDP, Minitab, SAS, and $SPSS^X$. Many printouts and plots are examined in detail to guide the reader in applying this software to regression analysis. Since it is expected that the reader will use only one of these four packages, each computer section stands on its own, and exercise problems may be solved using any one of the four programs. Although the matrix algebra formulation and the instructions for computer use provide many benefits to the reader, coverage of these topics can be minimized without loss of understanding if course length requires it.

• The first nine chapters discuss the simple linear regression model and introduce the matrix algebra formulation and the computer programs with regression diagnostics. Integrated with this discussion are topics presumably introduced, but seldom well-understood, in previous courses: measuring variability, constructing confidence intervals, and testing hypotheses. A single example is continued throughout the first nine chapters so that attention can be focused on new concepts rather than on new data. Also, results obtained algebraically, by matrix algebra, or from the computer, become familiar through repeated use, reinforcing the equivalence of these procedures.

• Chapter 10 serves as a bridge between single-predictor and multiple-predictor models. Fitting exponential models is a simple alteration to fitting straight-line models. Polynomial models are treated as an extension of simple linear models. Both exponential and polynomial models are reviewed. In connection with polynomial models, simultaneous, sequential, and partial tests of hypotheses are introduced.

• Chapters 11–16 discuss multiple regression. An extension of the earlier example helps illustrate concepts when more than one predictor is used. The multiple regression model is introduced using algebraic formulas, and the matrix formulation and computer programs are illustrated in detail. Simultaneous, sequential, and partial tests are developed, and the concepts of collinearity and partial correlation are explained. Dummy variables are discussed. A data set from an actual marketing study is used to illustrate various variable-selection and model-selection techniques.

• An ample number of exercises are provided at the end of each section. Many exercises introduced early are referred to later, so that the reader may utilize results already obtained.

• A large data set is included in the Appendix. Though the reader may not care to analyze the entire set, even by computer, these data may be useful in illustrating variability by drawing a random sample of cases. Also in the Appendix are "Analyses of Data Sets," which include complete numerical analyses for all exercise data sets, "Answers to Selected Exercises," and supplementary topics such as the binary number system and calculus-based derivations. *A First Course in Linear Regression* will be a valuable learning and reference source for the reader.

ACKNOWLEDGMENTS

There are many people to thank for their help in the preparation of this edition, in addition to all those listed in the preface of the first edition. Judy Lovely and Nancy Tipton did a great job of typing. Barbara Lawson attended to thousands of details, such as copying, mailing, and typing correspondence. Susan London of PWS and copy editor Alice Cheyer did an excellent job of preparing my manuscript for publication.

The University of Tennessee Computing Center has always cheerfully provided the most valuable and expert help. *A First Course in Linear Regression* has benefited especially from the advice of Paul Wright, Don Broach, and Bob Muenchen of the University of Tennessee Computing Center.

Professor John C. Rennie, Department of Forestry, Wildlife, and Fisheries, University of Tennessee, Knoxville, was kind enough to provide real data concerning the application of regression to the study of forestry. Professor David W. Cravens, formerly of the Department of Marketing at University of Tennessee, Knoxville but now at Texas Christian University, was kind enough to allow me to continue to use his data, which are so marvelously effective in illustrating variable selection procedures.

Dean C. Warren Neel of the College of Business Administration at University of Tennessee, Knoxville, and Professors Charles C. Thigpen and David L. Sylwester, former and present Chairman of the Statistics Department, respectively, have provided the facilities and resources to make this text possible. Finally, thanks to those of my colleagues and students who have been kind enough to say nice things about the first edition.

Mary Sue Younger

CONTENTS

ONE
SIMPLE LINEAR REGRESSION 1

TWO
THE LEAST SQUARES REGRESSION EQUATION 24

THREE
THE MATRIX APPROACH 60

FOUR
THE USE OF COMPUTER PROGRAMS IN SIMPLE REGRESSION 104

FIVE

THE USE AND INTERPRETATION OF THE REGRESSION EQUATION 152

SIX

MEASURING ERROR IN ESTIMATION 177

SEVEN
INFERENCES CONCERNING THE REGRESSION COEFFICIENTS 211

EIGHT
CORRELATION 247

NINE
APPROPRIATENESS OF THE MODEL 264

TEN

EXPONENTIAL AND POLYNOMIAL MODELS 306

ELEVEN

MULTIPLE REGRESSION 372

TWELVE
CORRELATION IN MULTIPLE REGRESSION 425

THIRTEEN
TESTS OF SIGNIFICANCE IN MULTIPLE REGRESSION 435

FOURTEEN
DUMMY VARIABLES 458

FIFTEEN
SOME VARIABLE-SELECTION PROCEDURES 479

SIXTEEN
ANALYSIS OF RESIDUALS IN MULTIPLE REGRESSION 574

Scatter Diagram

5. It is hard to pinpoint the relationship between size and price just by looking at the raw data. A graph of the data will let us view all 20 pairs of values at once. Such a graph, called a **scatter diagram,** consists of a set of coordinate axes on which the 20 (X, Y)-values are plotted as points. The scatter diagram for the data in Table 1.1 is shown in Figure 1.2.

FIGURE 1.2 Scatter Diagram for Sizes and Prices of 20 Houses

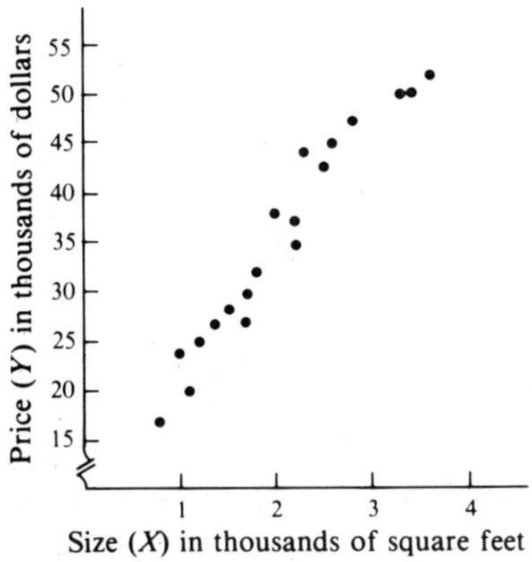

The construction of a scatter diagram is somewhat a matter of personal taste, but there are some general rules to observe. Since the purpose of a scatter diagram is to give an overall view of the problem, one should take care that it does convey information and that it does not distort things. Attention should be given to (1) labeling, (2) scaling, (3) the points, and (4) the quadrant.

L A B E L I N G It is useful to have an informative title. The axes should be labeled with the names of the variables and their units. It is generally customary, as in algebra, to plot the predictor variable (X) on the horizontal axis and the response (Y) on the vertical axis.

S C A L I N G The units on both the horizontal axis and the vertical axis should be marked off and labeled. Scales on the two axes need not be the same; for example, one inch may indicate 100 units on the vertical axis and 1,000 units on the horizontal axis, depending upon the magnitudes of the values of X and Y to be plotted. Similarly, it is not necessary for the axes to intersect at the origin, especially if there are not values of X or Y near zero. To show a true zero on a scatter diagram will often result in crowding the points into one corner of the picture instead of centering them in the middle. Consider, for example, the ten pairs of points (X, Y) shown in Table 1.3 on page 10. These points are plotted in Figure 1.3. In Figure 1.3(a) and (b) the scale on the vertical axis is one-tenth that on the horizontal axis, since values of Y are roughly one-tenth as large as values of X. In Figure 1.3(a) the axes have been labeled so as to intersect on the origin (0, 0). This results in the points being crowded together in the upper right-hand portion of the graph, with a lot of wasted space taking up most of the picture. With a little advance planning, a much more attractive scatter diagram can be constructed, as in Figure 1.3(b).

The choice of scale can greatly affect the impression given by the scatter diagram. Figure 1.3(c) is another graph of the ten pairs of values in Table 1.3. It looks very different from the graph in Figure 1.3(b), at least at first glance, although all three graphs are really the same and none of them is actually incorrect. Since it is impossible to interpret the visual picture presented by a scatter diagram without reference to the units on the vertical and horizontal axes, these units must always be given.

TABLE 1.3
Some Points to Graph

X	Y
5,000	250
5,200	270
5,400	280
5,800	320
4,100	220
4,500	210
4,700	250
4,000	210
5,000	260
5,500	300

FIGURE 1.3

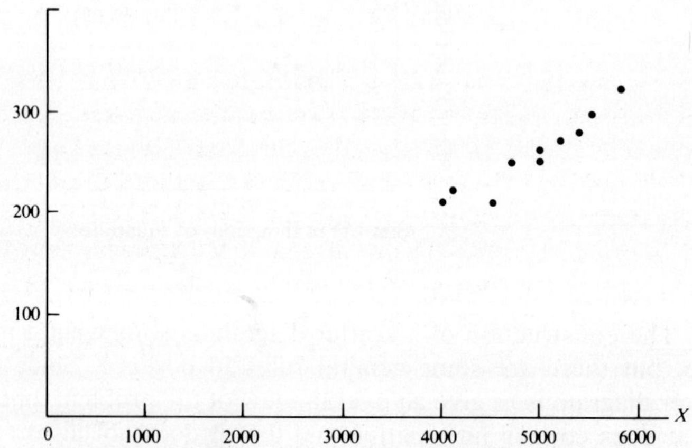

(a) Axes intersect at (0, 0), resulting in points being clustered in a corner of the graph

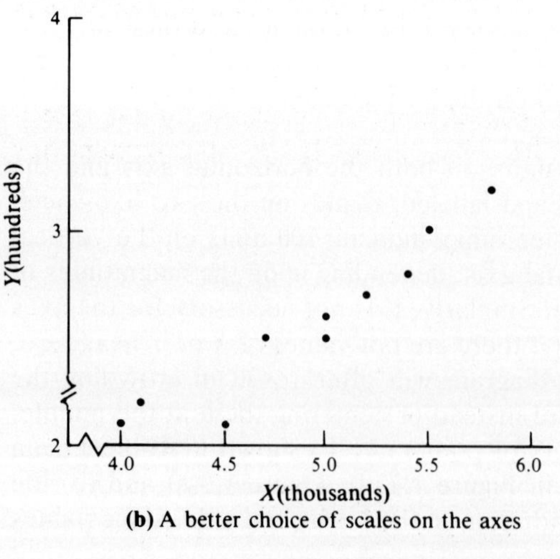

(b) A better choice of scales on the axes

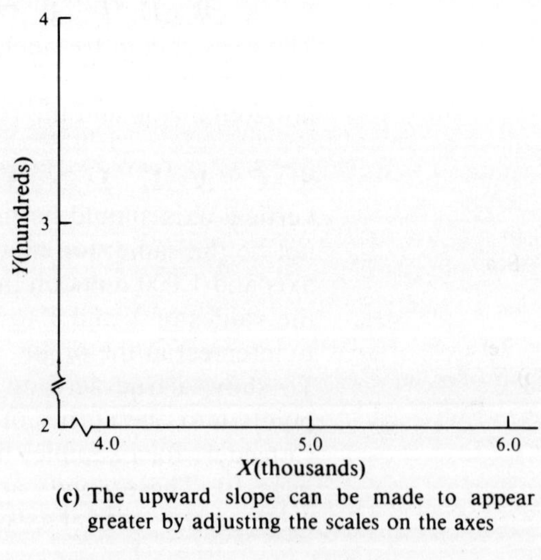

(c) The upward slope can be made to appear greater by adjusting the scales on the axes

THE POINTS The points stand alone. Labeling them would only clutter up the picture. Neither are the points connected with lines. Such lines would suggest a *time series,* a change in a single variable over time. Recall that the typical regression problem concerns a fixed point in time and not change in a variable over time.

THE QUADRANT Since only positive X- and Y-values are usually observed, most scatter diagrams lie completely in the first quadrant. However, they can extend into other quadrants if either variable assumes negative values. For example, one can observe negative temperatures in the wintertime.

THE DIRECTION OF THE RELATIONSHIP

What can we tell from a scatter diagram? First, is there a relationship? By *relationship* we mean (1) the larger the X-value, the generally larger the Y-value; or (2) the larger the X-value, the generally smaller the Y-value. if large X-values and large Y-values tend to occur together (large sizes and large prices) so that the mass of points tends to rise from left to right, then we say that X and Y have a **direct relationship.** Indeed, there appears to be a direct relationship between sizes and prices of houses in Figure 1.2, as we expected. On the other hand, if large X-values and small Y-values tend to occur together, so that the mass of points falls from left to right, then the relationship between X and Y is **inverse.**

Direct Relationship

Inverse Relationship

Inverse relationships are as common as direct ones. For example, one expects an inverse relationship between average winter temperature and consumption of heating oil: the warmer the temperature, generally the less oil used for heating. Similarly, for many commodities studied by economists, the higher the price of an item, the lower the demand. An inverse relationship is shown in the scatter diagram of Figure 1.4(a).

Figure 1.4(b) shows a typical scatter diagram indicating no relationship. During the oil shortage of 1974 the price of gasoline increased sharply, but it was found that this price increase has almost no effect on the demand for

FIGURE 1.4 Scatter Diagrams Indicating (a) an Inverse Relationship and (b) No Relationship Between Two Variables

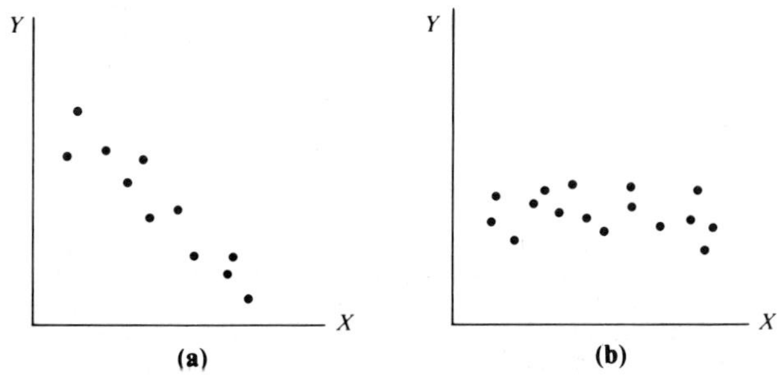

gasoline. To an economist, this was a classic case of *inelastic demand.* A mathematician would say that quantity demanded was a *constant function* of the price. To a statistician it meant that the two variables were not related: regardless of the value of X, the value of Y was generally constant. Note on Figure 1.4(b) that the points lie more or less parallel to the horizontal axis.

LINEAR AND CURVILINEAR RELATIONSHIPS

Linear Relationship

Curvilinear Relationship

If Y increases or decreases steadily with X so that the points tend to lie within a band formed by two parallel straight lines, then we say that there is a **linear relationship** between the two variables. If, however, the rate of change of Y with X is not generally constant, then the relationship is **curvilinear,** and the points on the scatter diagram lie within segments of concentric curves. Figure 1.5 shows examples of some linear and curvilinear relationships. The curvilinear relationships in Figure 1.5(b) and (c) show Y increasing and decreasing with X at *decreasing* rates. Of course, Y could also increase or decrease with X at an increasing rate. (See Exercise 1.14.) Figure 1.5(d) shows Y increasing and then decreasing with X.

FIGURE 1.5
Examples of Linear and Curvilinear Relationships

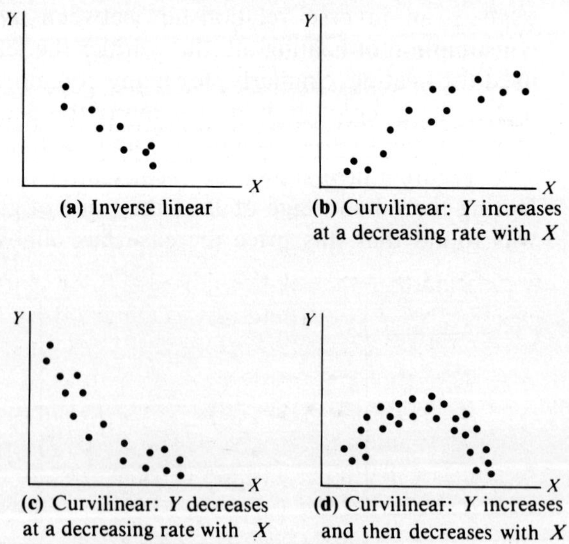

(a) Inverse linear

(b) Curvilinear: Y increases at a decreasing rate with X

(c) Curvilinear: Y decreases at a decreasing rate with X

(d) Curvilinear: Y increases and then decreases with X

The characteristic of a linear relationship is that Y increases by a constant amount for each one-unit increase in X, regardless of the size of X. For example, suppose it takes a typist 2 minutes to type a page of manuscript. Then it will take 2 minutes longer to type four pages than to type three pages, and 2 minutes longer to type twenty-five pages than to type twenty-four pages. For each additional page, the additional time is 2 minutes: this constant amount of increase of Y (time) with X (number of pages) is what makes the relationship linear.

Suppose, however, that after the typist has typed a certain number of pages, fatigue begins to set in. Suppose that after thirty-five pages the typist starts to slow down, and each subsequent page takes just a little longer to type than the one before it. Then we will see that Y (time) increases more and more with each one-page increase in X. Since the amount of change in Y is not constant for each one-unit increase in X, this relationship would not be linear; rather, it would be curvilinear. (Would a scatter diagram of Y with X for this example look like any of the pictures in Figure 1.5? If so, which one? If not, what would it look like? See Exercise 1.14.)

Some thought should be given to the form of relationship you would expect to see between a pair of variables before data are collected. For example, if X (number of miles a person lives from work) and Y (time required to drive to work) are to be measured for a group of people who work in the same office building, would you expect to see a linear or a curvilinear relationship between these two variables? Suppose X = age of a male employee, and Y = number of days missed from work per year—would you expect this relationship to be linear or curvilinear, and why? When data are collected, they should be graphed in a scatter diagram; then you can see whether your hypothesis about the form of the relationship is supported by the data or not. If your data do not support the sort of relationship you expected to see, you would need to rethink your hypothesis about the relationship, or you would need to check that you are measuring what you thought you were (that the items are a sample from the true population of interest). In many instances, if data do not look like you thought they should, you might have an error in your hypothesis or in your research design.

We will study both linear and curvilinear relationships in this text. At first, in order to keep the mathematics simple, our analyses will emphasize relationships that are considered strictly linear. In the absence of theory indicating that the relationship should be curvilinear, we can ignore any very slight curvature, attributing it to the chance variability inherent in the sampling process. For example, in Figure 1.2 you might see curvature in the range X = 3,000 sq ft to X = 4,000 sq ft. But it is slight and may only indicate that we sampled some relatively low-priced houses in that size range. Thus, we begin by assuming that the relationship is linear, although it will be important for us to test this assumption later. Even if a curvilinear relationship is seen in a given problem, the analysis is still considered part of linear regression analysis, as will be explained in Chapter 2.

THE STRENGTH OF THE RELATIONSHIP

Figure 1.2 seems to indicate a relatively strong relationship between size and price; for any small range of sizes the range of prices is also rather small. That is, the vertical scatter of points over a small horizontal range is not large. This means, of course, that if two houses are close to the same size, then their prices are also relatively close. Thus, size should be a strong predictor of price. Suppose that we have two predictors of Y—X_1 and X_2—and that their scatter diagrams are as shown in Figure 1.6(a) and (b). Since for any small range of X-values, the vertical scatter of Y-values is smaller in (a) than in (b), X_1 is a stronger predictor of Y than is X_2. One can predict Y from X_1 more accurately than from X_2.

FIGURE 1.6
Scatter Diagrams for Two Different Predictors of Y, Showing Relative Strengths of the Predictors

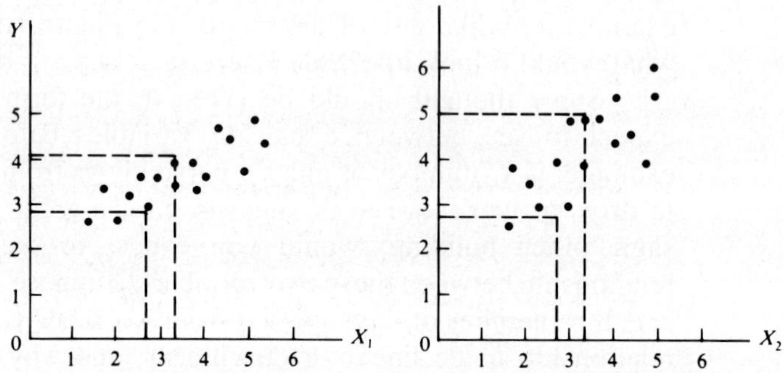

(a) Small vertical scatter

(b) Large vertical scatter

IS THE SCATTER UNIFORM?

Homogeneous Scatter

Heterogeneous Scatter

The strength of a relationship is measured not only by the amount of scatter over a small range but also by the constancy of the vertical scatter throughout the range of X-values. The scatter diagram, again, helps us determine whether or not the scatter remains constant. As long as the points lie within parallel straight lines, we have evidence of constant, or **homogeneous, scatter,**† depicted in Figure 1.7(a). Two types of **heterogeneous** (nonconstant) vertical **scatter** are shown in Figure 1.7(b) and (c).

† For curvilinear relationships, as long as the points lie within arcs of concentric curves, we can consider that the scatter is homogeneous.

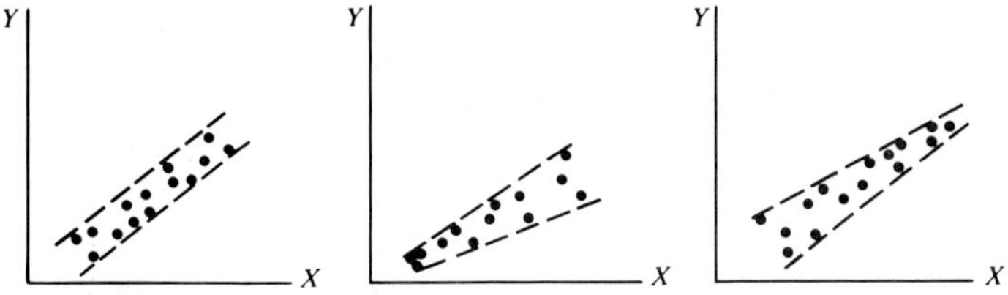

(a) Homogeneous scatter (b) Scatter increases with X (c) Scatter decreases as X increases

FIGURE 1.7 Scatter Diagrams Showing Homogeneous and Heterogeneous Vertical Scatter

EXERCISES

Data used in the exercises will be called a data set *and given a short descriptive title. In most instances, further analyses will be performed using the same data in later chapters. Appendix L, entitled "Analyses of Data Sets," contains complete analyses of these data, and you should refer to that appendix to check your numerical answers.*

1.5 The manager of the soft drink concession for a football stadium needs to estimate the demand for soft drinks on game days. Attendance at the football games is rather variable, depending on the home team's won-lost record in previous weeks, who the opponent is, weather conditions, and other factors. As a beginning, the manager decides to collect information on the number of tickets sold by Wednesday of the week preceding each game and the soft drink sales per game for the past seven home games. These data are shown in Table 1.4.

TABLE 1.4 Concession Data

	Game						
	1	2	3	4	5	6	7
Advance ticket sales (thousands)	60.5	70.9	72.0	74.9	62.8	68.7	54.6
Gallons of syrup used	89.2	105.6	106.4	108.6	90.7	100.6	80.7

a. Draw a scatter diagram.
b. What kind of relationship appears to exist between ticket sales and syrup used?
c. Does the relationship appear to be linear?
d. How strong does the relationship appear to be?
e. Does the scatter appear to be uniform?

1.6 The number of employees (X) and the average weekly retail sales in thousands of dollars (Y) of 15 randomly selected retail outlets of a large retail chain are shown in Table 1.5.

TABLE 1.5 Sales Data

X	17	39	32	17	25	43	25	32
Y	7	17	10	5	7	15	11	13

X	48	10	48	42	36	30	9
Y	19	3	17	15	14	12	8

a. Draw a scatter diagram.
b. Is the relationship direct or inverse?
c. Is the relationship linear?
d. How strong does the relationship appear to be?
e. Does the scatter appear to be uniform?

1.7 Suppose a survey of cities of comparable sizes, randomly selected from cities in the United States, yields the data shown in Table 1.6 for a given year.

TABLE 1.6
Police Data

City	Number of Police Officers (X)	Number of Robberies (Y)	City	Number of Police Officers (X)	Number of Robberies (Y)
1	64	625	8	90	510
2	53	750	9	50	800
3	67	560	10	77	550
4	52	690	11	88	550
5	82	515	12	71	525
6	59	680	13	58	625
7	67	630			

a. Graph the data on a scatter diagram.
b. Is the relationship direct or inverse?
c. Is the relationship linear or curvilinear?
d. How strong does the relationship appear to be?
e. Is the scatter uniform?

1.8 A homeowner in a small northwestern city wishes to estimate the daily winter heating oil bill from the average daily temperature. On 15 randomly selected days during December, January, and February, the homeowner records the average daily temperature and the cost of oil burned that day. The data are shown in Table 1.7.

TABLE 1.7
Heating Data

Day	Average Daily Temperature (° F)	Daily Oil Bill (dollars)	Day	Average Daily Temperature (° F)	Daily Oil Bill (dollars)
1	50	0.30	9	27	0.50
2	4	0.70	10	−2	0.80
3	5	0.65	11	6	0.70
4	19	0.65	12	26	0.55
5	33	0.50	13	49	0.40
6	48	0.30	14	0	0.70
7	6	0.65	15	2	0.75
8	11	0.60			

a. Graph the data.
b. Is the relationship direct or inverse?
c. Is the relationship linear?
d. How strong is the relationship?
e. Is the scatter uniform?

1.9 The information shown in Table 1.8 was collected on a sample of employees of a large company. (Maximum number of days of paid sick leave = 14.)

TABLE 1.8
Smokers Data

	Employee							
	1	**2**	**3**	**4**	**5**	**6**	**7**	**8**
Avg. no. of cigarettes smoked daily	5	0	10	30	3	25	20	0
No. sick days taken last year	3	1	11	13	11	4	1	7

	Employee							
	9	**10**	**11**	**12**	**13**	**14**	**15**	**16**
Cigarettes	25	45	0	30	12	45	25	0
Sick days	1	10	2	11	5	3	7	11

	Employee								
	17	**18**	**19**	**20**	**21**	**22**	**23**	**24**	**25**
Cigarettes	50	20	30	15	30	0	40	2	40
Sick days	8	14	0	12	11	13	12	0	14

a. Graph the data.
b. Does it appear that the number of sick days is related to the number of cigarettes smoked daily?
c. Is the scatter uniform?

1.10 Advanced-level engineering textbooks range from very short to very long and also range in price from quite reasonable to extremely expensive. These books also vary as to the number of illustrations and formulas, the quality of the paper, and so on. A random sampling of such textbooks gave the information shown in Table 1.9 for suggested retail price and number of pages.

TABLE 1.9
Textbook Data

Number of Pages (hundreds)	Price (dollars)	Number of Pages (hundreds)	Price (dollars)
2.8	29	1.1	17
5.5	44	4.3	48
1.5	10	5.4	54
3.1	42	4.1	24
5.5	26	4.9	59
1.8	18	2.4	18
5.0	35	3.3	31
5.8	70		

a. Construct a scatter diagram showing how price is related to number of pages.
b. Does the relationship appear to be direct or inverse?
c. Is the relationship linear?
d. Is the scatter uniform?
e. Given your answer to (d), what can you say about the strength of the relationship?

1.11 A professional association for a group of people working in the same profession reported the results shown in Table 1.10 from a simple random sampling of its members.

TABLE 1.10
Professionals Data

Years of Professional Experience (X)	Average Income From Profession (Y) (thousands of dollars)
5	30.7
10	37.2
15	42.5
20	47.6
25	49.8
30	52.4
35	54.0
40	54.9

a. Graph the data.
b. Does income appear to be related to years of experience? How?
c. Comment on the strength of the relationship.

1.12 Plot the data of Table 1.1 on a scatter diagram in which the vertical and horizontal axes intersect at zero.

1.13 Use nontechnical terms to describe the following relationships. For example, a direct relationship between the size of a house and its price means that larger houses tend to cost more than smaller houses.
a. A direct relationship between the amount of fertilizer applied per plot and the number of bushels of tomatoes harvested.
b. An inverse relationship among college students between the average number of hours spent per week in intramural sports and grade average.
c. No relationship between the price per jar of baby food and the number of jars sold.
d. An inverse relationship between the time it takes to cook a vegetable and the percentage of vitamin content retained.
e. A direct relationship between consumption of electricity and consumption of beer among families in a given location.

1.14 Sketch a scatter diagram showing
a. Y increasing with X at an increasing rate.
b. Y decreasing with X at an increasing rate.

1.15 Suppose that after a certain number of pages have been typed, a typist begins to slow down because of fatigue. Each subsequent page takes a bit longer to type than the one before. Sketch a scatter diagram showing your prediction of the relationship between X = number of pages typed, and Y = number of minutes required.

1.16 For each of the pairs of variables listed below, tell whether you would expect to see a linear or curvilinear relationship, and why.
a. X = number of typewriters given routine cleaning in an office;
Y = number of minutes the service person spends at the office.

b. X = number of cases to be shelved in a warehouse;
Y = time required to shelve a shipment of cases.
c. X = weight of a truckload of oranges;
Y = volume of orange juice per truckload.
d. X = age of males;
Y = percent of employed males in age group.
e. X = annual income;
Y = annual savings.

1 . 4

THE MODEL

Now that we have collected and plotted the data on houses and determined that there is a linear relationship between size and price, how do we describe that relationship? We know that price increases with size, but how much, and starting where? We would like to have a mathematical equation, or function, or **model,** which specifies what the relationship is. Since the relationship is linear, we want a **linear equation** stating Y as a function of X.

Model

Linear Equation

LINEAR EQUATIONS

Parameters

At this point, it might be well to digress and review linear equations. The equation $Y = \alpha + \beta X$ is the equation of a straight line. The Greek letters α (alpha) and β (beta) are **parameters** of the line: once they are specified, the line is fixed. The first parameter, α, is the Y-intercept, the value of Y when $X = 0$. The parameter β is the slope of the line, the number of units and direction that Y changes for each one-unit increase in X.†

For example, suppose that $\alpha = 4$ and $\beta = 2$, so that the equation is

$$Y = 4 + 2X.$$

It is easy to see that if $X = 0$,

$$Y = 4 + 2(0) = 4,$$

and $\alpha = 4$ is the value of Y when $X = 0$. Notice that as the value of X is increased by 1, the value of Y increases by $\beta = +2$:

$$\text{If } X = 1, \quad Y = 4 + 2(1) = 4 + 2 = 6;$$

$$\text{If } X = 2, \quad Y = 4 + 2(2) = 4 + 4 = 8;$$

$$\text{If } X = 3, \quad Y = 4 + 2(3) = 4 + 6 = 10;$$

and so on.

To plot the line $Y = 4 + 2X$ we could first plot the Y-intercept ($X = 0$, $Y = 4$). Knowing that a straight line is determined by two points, we could

† Some writers prefer to denote the intercept by β_0 and the slope by β_1.

then choose another X-value, say $X = 5$, for which $Y = 4 + 2(5) = 4 + 10 = 14$. After plotting the point (5, 14), we could then connect the two points with a straightedge, as shown in Figure 1.8. Note that we would not ordinarily label the coordinates of the points we have chosen to plot, since they are completely arbitrary; labeling draws attention to the labeled points.

FIGURE 1.8 Plotting the Equation $Y = 4 + 2X$

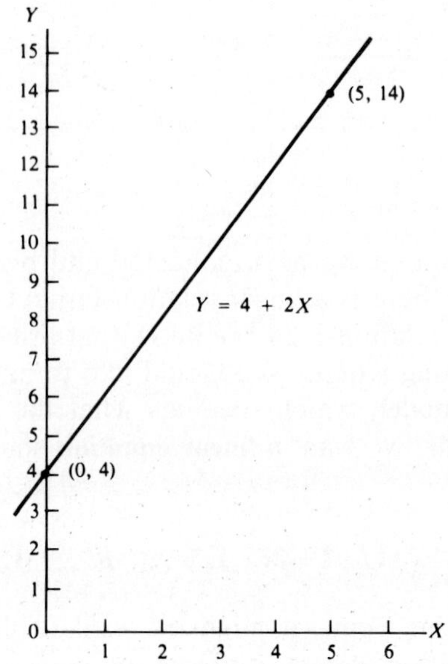

Although the points chosen to plot a straight line are arbitrarily chosen, one usually plots the Y-intercept for convenience. However if, as in Figure 1.2, the axes do not intersect at zero, the Y-intercept cannot be plotted. Another X-value, $X \neq 0$, must be selected. In addition, a straight line can usually be graphed more accurately if the two points are chosen at opposite ends of the range of X-values. If the two points are quite close together, small and often unavoidable errors in locating the points can result in rather serious errors in plotting the line. Such a situation is illustrated in Figure 1.9, in which the dashed line indicates the accurate plot, with the line passing through the points A, B, and C. If a small error is made in locating the point B and the line passes instead through the points A and B′, note how much the plot of the line is affected. If the same amount of error had been made in locating the point C, however, the graph of the line would not have been affected so seriously.

Here are a few further guidelines to remember when graphing a straight line. The Y-intercept can be any real number. If it is positive, then the line crosses the Y-axis above the origin; if it is negative, the line crosses the Y-axis below the origin; and if it is zero, the line passes through the origin. As for the slope, it can also be any real number. Note that the maximum possible value for the slope is *not* 1 and that a vertical line has slope equal to infinity. A line with positive slope rises from left to right; one with negative slope falls from left to right; and a horizontal line is one with slope zero.

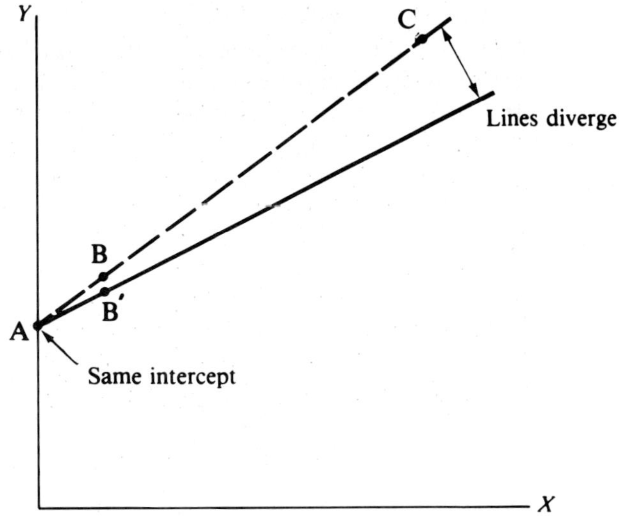

FIGURE 1.9
Large Divergence in Lines Because of Small Error in Plotting Second Point (B) When Second X-Value (B') Is Chosen Too Close to First One (A)

DETERMINISTIC AND PROBABILISTIC MODELS

In an equation of the form

$$Y = \alpha + \beta X$$

the points all lie on a straight line. But, in a scatter diagram, all points do not lie on a straight line. How, then, can the relationship between house size and price be described using the equation for a straight line?

Deterministic Model

The equation $Y = \alpha + \beta X$ represents a **deterministic,** or mathematical, **model**—once a value of X is chosen, the value of Y is automatically determined by the specific values of α and β and the rules of arithmetic. Deterministic models are often encountered in such conversion formulas as

$$F = 32 + \frac{9}{5}C,$$

which converts Celsius to Fahrenheit temperature, or

$$I = 12F,$$

which converts feet to inches. However, no such exact relationship exists between the size of a house and its price. Many factors other than size help determine the price of a house. When some factors are disregarded, two houses of the same size tend to sell for somewhat different prices.

Furthermore, even if we considered every conceivable factor affecting price and found two houses with all the same characteristics, we would still expect their prices to differ. Why? We can only attribute the difference to some unpredictable whim of the buyer or the seller, or to chance. This uncontrollable, unmeasurable variation is what distinguishes one identical twin from another, or one laboratory rat from its littermate; it's what gives a chemist slightly different results on any two runs of the same experiment. By the same token, our statistical predictions will always be subject to **random error:** They will always be, to some degree, imperfect when applied to any specific situation.

Random Error

Recognizing that we can never predict anything exactly, we can describe a relationship by means of the probabilistic or **statistical, model**

Statistical Model

$$Y = \alpha + \beta X + \varepsilon.$$

Here, ε, the Greek letter epsilon, represents the error in the predictions. In this model $Y = \alpha + \beta X$ says that there is a linear relationship, that generally Y increases or decreases at a constant rate with X, and ε says that this relationship is not exact for every individual pair of observations. The error term ε accounts for variables that affect Y but are not included as predictors. It accounts for chance, or random, variability as well as imprecision in the specified model, which might be almost, but not exactly, linear.

We can thus say that the error term is composed of two general kinds of error:

1. *Model error,* or *lack of fit,* meaning that all relevant predictors are not specified or that the form of the relationship is not correctly specified; and
2. *Random error,* which is unpredictable and uncontrollable.

ESTIMATING THE UNDERLYING RELATIONSHIP

The model $Y = \alpha + \beta X + \varepsilon$ merely states the conceptual framework of the problem. It is a shorthand way of saying that we are trying to investigate a problem in which there is an imperfect linear relationship between two variables. The exact values of α, β, and ε can never be known. Our primary goal, then, is to get a handle on this underlying relationship so that we can estimate Y from X. Thus, we shall use the data to get numerical estimates, which we will call a and b of α and β, so that the **estimated price** $\hat{Y}$ of a given house can be obtained by simply substituting its size (X) into the equation

Estimated Value

$$\hat{Y} = a + bX.$$

Then, since the estimated price can be expected to differ somewhat from the actual price in any individual instance, our second concern is to estimate how far off one can expect the prediction to be. That is, we want to estimate how much error is involved in our predictions. If this error is quite large, this might be an indication that the relationship is not strong enough to bother with.

Finally, we recognize that because our information comes from a randomly chosen sample, there is a chance that it is giving us a distorted picture of how X and Y are related. It might indicate a relationship when in fact none exists. The question of whether or not a relationship exists is central to the whole study of regression: We certainly would not spend time trying to describe a nonexistent relationship and trying to make predictions and measure error from it. However, we will investigate this question last because of the many steps leading up to it. In Chapter 2 we begin work on our primary goal: obtaining numerical estimates of the slope and the intercept of the underlying linear relationship.

EXERCISES

1.17 Explain the difference between a deterministic and a probabilistic model.

1.18 **a.** Refer to the Textbook Data in Exercise 1.10. There are two textbooks that are both 550 pages long. How do you explain the fact that their prices are different?

b. In the Police Data (Exercise 1.7) cities 3 and 7 both have the same number of police officers, but more robberies are committed in city 7 than in city 3. How do you explain this?

c. Suppose that two check-out lanes in a supermarket each have three customers in line. Do you expect both lanes to service their three customers equally quickly? Why, or why not?

d. A physicist rolls a metal sphere down an incline twice and on each roll measures the time elapsed with an electronic timer. Will the physicist see identical times elapsed both times the experiment is performed?

e. How might the physicist in (d) formulate a physical "law" relating the degree of incline to the time it takes the ball to roll down it?

1.19 Graph the following linear equation:

a. $Y = 4 - 2X$; **b.** $Y = -4 - 2X$; **c.** $R = -5 + 3S$,

d. $T = 3 + 1.1U$; **e.** $Q = 160 - 2.7P$.

1 . 5

SUMMARY

We have made a good start at analyzing a simple regression problem. Several concepts and terms have been defined, both generally and with reference to a specific data problem. We have plotted some data and noticed some important features of the example problem. Having decided that a simple linear model will most likely provide an adequate representation for the house size-price problem, in the next three chapters we will study methods by which to describe the underlying relationship present in these data.

TWO

THE LEAST SQUARES REGRESSION EQUATION

2.1 INTRODUCTION

In this chapter we will look at the least squares method for fitting a straight line to bivariate data and at the rationale for use of the least squares criterion. Alternative, labor-saving methods of calculation and data transformation will also be considered.

2.2 CRITERIA FOR THE "BEST" LINE

Recall that the model for a simple linear regression prediction problem,

$$Y = \alpha + \beta X + \varepsilon,$$

indicates that X and Y are linearly related, but imperfectly; for a given problem we need to estimate α and β in order to obtain the **regression,** or prediction, **equation**

Regression Equation

$$\hat{Y} = a + bX,$$

where a and b are numerical values and $\hat{Y}$ is the predicted value of Y for a given value of X. Since the linear relationship is imperfect, we must also estimate ε, the size of the error of estimation.

THE LINEARITY ASSUMPTION

Suppose for a moment that the whole population of houses within the desired size range is at our disposal. There would be many different sizes, and for each size there would be not one but a variety of prices. Thus, there would be a subpopulation of prices for each different size house, as shown in Figure 2.1. (Note the uniform scatter of the subpopulations.) Under these circumstances, the (population) mean price for a house of a given size could be

FIGURE 2.1
Scatter Diagram for the Population of Houses

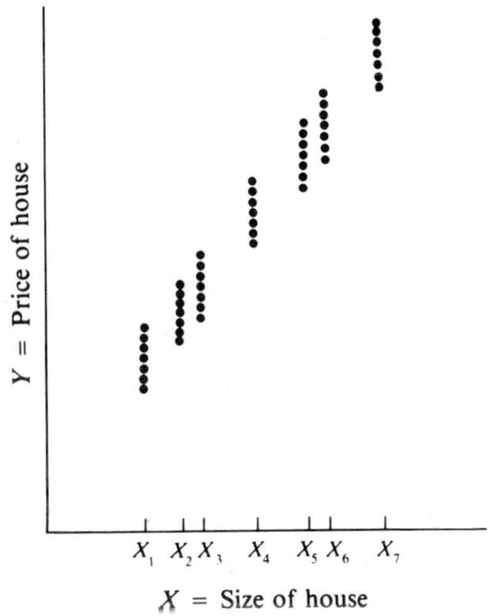

calculated. We will call this mean $\mu_{Y \cdot X}$, the Greek letter *mu* representing a population mean, the subscript Y indicating that the mean Y-value is being found, and the "dot-X" reading "for a given X-value." Imagine calculating $\mu_{Y \cdot X}$ for each X-value in the population. *If the relationship between X (size) and Y (price) is really linear, then all the $\mu_{Y \cdot X}$-values will lie on a straight line.* That is, we assume that we have a perfect linear relationship in the population between X and the mean Y-value for each X-value. The equation of the line that passes through each mean Y-value is thus

$$\mu_{Y \cdot X} = \alpha + \beta X,$$

as in Figure 2.2. If we substitute a size (X), the equation will give the *average* price ($\mu_{Y \cdot X}$) for that size. By connecting the average Y-values by a straight line we can read values for α and β from the graph, or β can be determined by finding two $(X_i, \mu_{Y \cdot X_i})$ points and using the formula

$$\beta = \frac{\mu_{Y \cdot X_2} - \mu_{Y \cdot X_1}}{X_2 - X_1} = \frac{\text{change in average value of } Y}{\text{change in } X}.$$

Thus, the linear relationship between X and the average Y for each X can be

FIGURE 2.2
Determining the Underlying Linear Relationship
$\mu_{Y \cdot X} = \alpha + \beta X$

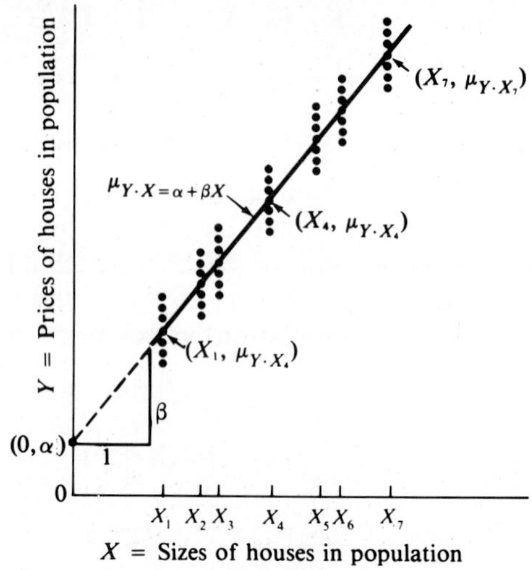

determined; this is what we have been calling the underlying linear relationship. Naturally, *all* points do not lie on the line, because for a given size house, prices will deviate from the average price because of what we have called error: Many predictors have been omitted and also random variation is present. (We do, however, have a truly linear relationship.) Since we are using only one predictor and since we cannot eliminate random variation, we shall be doing well to learn the average price for houses of a given size, even if we cannot get the exact price of any specific house. If the linear relationship is strong, the error in any specific instance will be small, and the average price will be close to the actual price.

Observed Values
Mean Values

In the scatter diagram of Figure 2.2, we notice two different kinds of Y-values corresponding to the X-values. There are **observed Y-values** and there are **mean Y-values.** The observed Y-values corresponding to any X-value are denoted Y_i and are represented by points on the scatter diagram. Then for any given X-value, if we find the average of the associated Y_i-values we obtain $\mu_{Y \cdot X_i}$, and this value lies on the line $\mu_{Y \cdot X} = \alpha + \beta X$. (Note that the mean value $\mu_{Y \cdot X_i}$ may or may not equal any observed Y_i.) Thus, for a specific house of size X_i, the difference between the actual price and the mean price,

$$Y_i - \mu_{Y \cdot X_i},$$

is the error involved in predicting the price from the size alone. Note also that the line need not actually touch any of the points in order to give a good general description.

Two observations that might seem obvious are nevertheless worth mentioning. If a straight line is to describe the underlying relationship apparent in a scatter diagram, it must pass through the mass of points and it must follow the lay of the points. Figure 2.3 shows some examples of freehand lines that do and do not indicate well the underlying linear relationships.

FIGURE 2.3
Straight Lines Providing Good and Poor Descriptions of the Underlying Linear Relationships

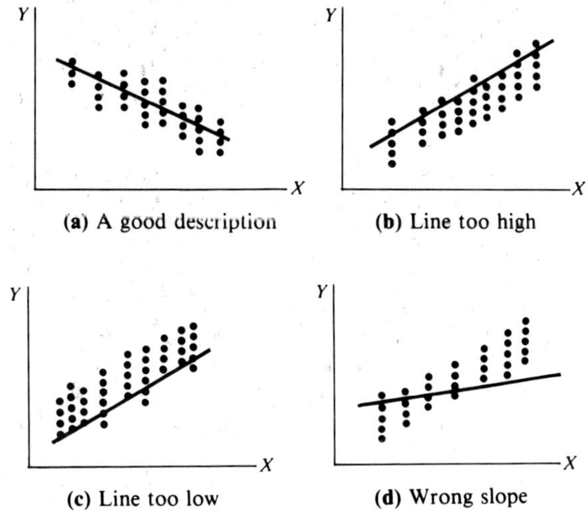

DETERMINING THE UNDERLYING RELATIONSHIP FROM A SAMPLE

We now drop the false assumption that we can consider the entire population of houses. In reality we will have only a sample of the houses within the specified range of sizes. For any size house in the population, the sample might record only one of the various prices. Furthermore, the sample may omit some sizes altogether. In other words, the sample may omit some X-values, and it certainly will omit many Y-values corresponding to each X-value sampled. The sample will consist of a randomly selected subset of the points in Figure 2.1, and its scatter diagram will be similar to that of Figure 1.2. Having only this sample of information, we will never know exactly what α and β are. But with this sample we can find an estimate a of α and an estimate b of β. With these, we can fit a straight line

$$Y = a + bX$$

through the points on the scatter diagram. As before, this line should pass through the middle of the mass of points, so that we can predict the mean Y-value for a given value of X. If we denote this *predicted,* or *estimated,* mean Y-value by $\hat{Y}$, the sample regression equation will be

$$\hat{Y} = a + bX.$$

Also as before, for any X-value in the sample there are two corresponding Y-values: Y_i, the observed value of Y corresponding to the observed X_i; and $\hat{Y}_i$, the predicted mean value of Y for X_i. (See Figure 2.4.) Then (X_i, Y_i) is a point on the scatter diagram, and $(X_i, \hat{Y}_i)$ is a point on the line $\hat{Y} = a + bX$. The difference $Y_i - \hat{Y}_i$ is an estimate of the error involved in predicting the price (Y_i) from the size (X_i) of a given house. The quantity $Y_i - \hat{Y}_i$ is called a

Residual **residual.** One may think of a residual as a measure of how much of the value of Y is left unexplained by the corresponding value of X. This "left over" amount is attributed to factors other than X that affect Y, i.e. to what we have called error.

FIGURE 2.4
Relationship Between Y_i and $\hat{Y}_i$ for a Given X_i

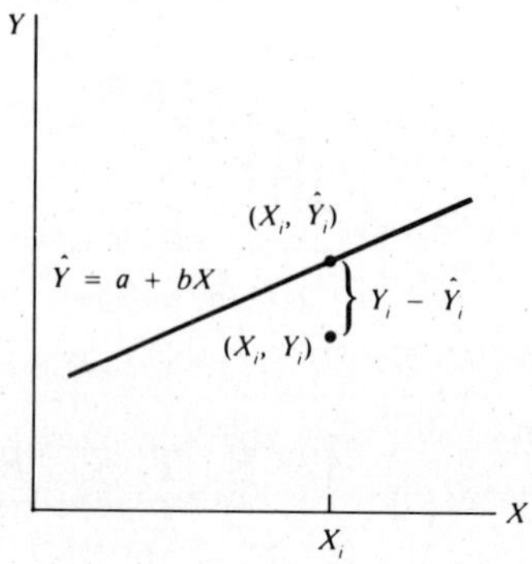

What has been said so far is actually after the fact. It assumes that values of a and b have been determined. We now turn to the question of how to determine these values for any particular problem; that is, how do we find the equation of the regression line?

SOME PROPOSED CRITERIA FOR THE "BEST" LINE

One might simply draw a freehand straight line through the mass of points on the scatter diagram and determine the slope and intercept from points on the graph. However, we can do better. What we need is a criterion for estimating the underlying linear relationship in the sample of points.

For every size (X_i) in our sample, we want the deviation, or residual, $Y_i - \hat{Y}_i$ to be as near zero as possible. if $Y_i - \hat{Y}_i = 0$, this means that the line passes through the point (X_i, Y_i). Recall, however, that the line need not pass through any point on the scatter diagram, and certainly a straight line cannot pass through all the points. Thus, all $Y_i - \hat{Y}_i$ values cannot be zero. What if we use the criterion that the lines pass through as many points as possible? Figure 2.5 shows the error in this reasoning. Lines A, B, and C all pass through four points, and line C obviously does not describe the underlying relationship.

Another criterion that might seem reasonable is to be sure that as many points fall above the line as below. One can easily imagine that this criterion

FIGURE 2.5
Poor Descriptions Arising from Lines Passing through as Many Points as Possible (A, B, C) and from Having as Many Points Above as Below the Line (D)

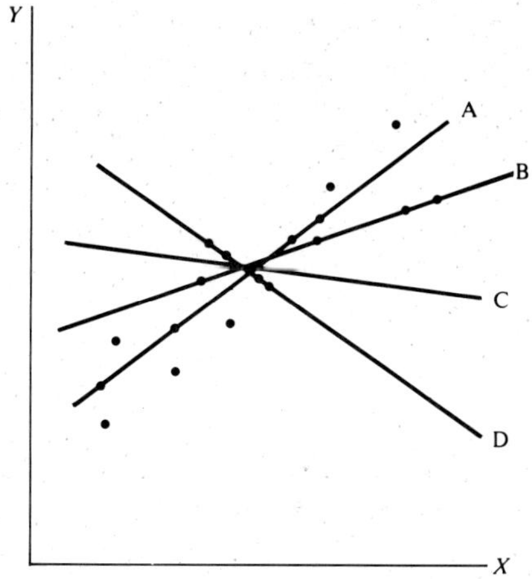

will produce several "best" lines, some like D in Figure 2.5 having the wrong slope.

Since the objective is to predict the mean Y-value for a given X-value, the line should certainly pass through $(\overline{X}, \overline{Y})$, the mean of both X and Y. While this is desirable, it is no criterion, because an infinity of lines can pass through the point $(\overline{X}, \overline{Y})$. We need something more.

Suppose we must decide which of the two lines in Figure 2.6 better describes that set of points. Both pass through $(\overline{X}, \overline{Y})$. Note that for the point (X_1, Y_1) line A is closer than line B. Thus, line A gives a more accurate prediction of Y_1 than line B. However, line B is closer to (X_2, Y_2) and thus line B will better predict Y_2. We might draw a third line, line C, which would be better in predicting both Y_1 and Y_2 than either lines A or B, but then line C would be further off for some other Y_i than either line A or line B.

FIGURE 2.6
Comparison of Two Lines

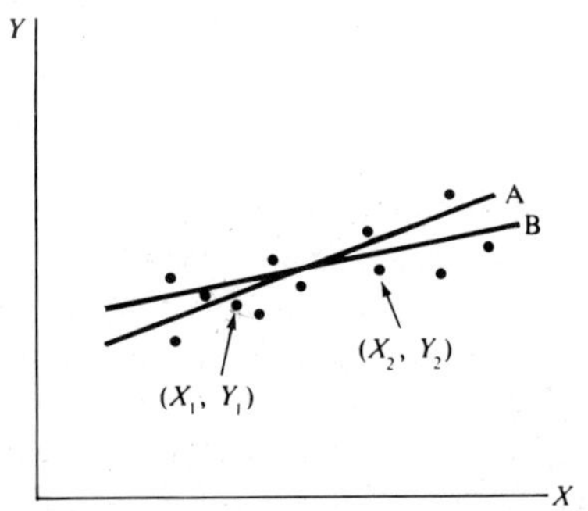

Thus, we cannot concentrate on getting the line close to any individual point, but must *consider all points simultaneously*. But how can this be done?

Suppose, as in Figure 2.7, we try to decide between lines A and B. We might calculate the deviations $Y_i - \hat{Y}_{i(A)}$ from each point to line A, and also the deviations $Y_i - \hat{Y}_{i(B)}$ from each point to line B. Now suppose we average the $Y_i - \hat{Y}_{i(A)}$ values, which can be positive, negative, or zero, depending upon

FIGURE 2.7 Deviations of Points From Two Competing Lines

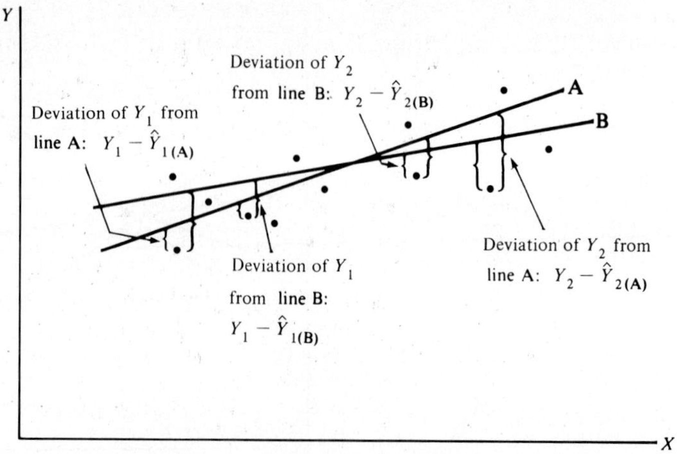

whether the point is above, below, or on the line. The average of these deviations from line A would tell *how much the points deviate from line* A, *on the average*. If we do the same thing for line B, the average of $Y_i - \hat{Y}_{i(B)}$ values will tell the average amount that the points deviate from line B. If we could repeat this procedure over and over with all possible lines, we could by elimination determine the **"best" line:** the line which is, on the average, closest to all the points. This seems to be a reasonable criterion for determining the line that gives the best description of the underlying linear relationship.

"Best" Line

Of course, it is impossible to make these comparisons for all possible lines. However, as we shall see in the next section, the methods of calculus enable us to do essentially that.

2.3 THE METHOD OF LEAST SQUARES AND NORMAL EQUATIONS

THE LEAST SQUARES CRITERION

The objective is to find the line

$$\hat{Y} = a + bX$$

in which the values *a* and *b* make the *average deviation* from the points to the line a minimum. That is, we want to find values a and b so that

$$\frac{1}{n}\sum_{i=1}^{n}(Y_i - \hat{Y}_i) = \frac{1}{n}\sum_{i=1}^{n}(Y_i - a - bX_i)$$

(since $\hat{Y}_i = a + bX_i$) is a minimum. First, we note that for a fixed sample size n, we need only consider minimizing the numerator $\sum_{i=1}^{n}(Y_i - \hat{Y}_i)$ in minimizing the left-hand side of the equation.

The problem is that as long as the line passes through $(\overline{X}, \overline{Y})$, as we said it should, this sum will always be zero, regardless of the values of a and b. This is because if $X_i = \overline{X}$, then $\hat{Y}_i = a + bX_i = a + b\overline{X}_i = \overline{Y}$, so that†

$$\begin{aligned}\sum_{i=1}^{n}(Y_i - \hat{Y}_i) = \sum_{i=1}^{n}[Y_i - (a + bX_i)] &= \sum_{i=1}^{n} Y_i - \left(na + b\sum_{i=1}^{n} X_i\right)\\ &= n\overline{Y} - [na + b(n\overline{X})]\\ &= n[\overline{Y} - (a + b\overline{X})] = n(\overline{Y} - \overline{Y}) = 0.\end{aligned}$$

Thus, if the numerator is zero, the average deviation is zero. Since the average deviation is *always* zero when the line passes through $(\overline{X}, \overline{Y})$, there seems to be no advantage in trying to find a and b so that $\sum_{i=1}^{n}(Y_i - \hat{Y}_i)$ is minimized.

What has happened is that if $\hat{Y}$ is the mean Y-value for a given X-value, then $\hat{Y}$ is so situated that deviations below the line (negative values) cancel deviations above the line (positive values). This must be so, since a mean is a center of gravity, or balancing point. (Recall the notion of variance from introductory statistics. In finding the average deviation of the values from their mean, it was noted that $\frac{1}{n}\sum_{i=1}^{n}(Y_i - \overline{Y})$ is always zero, because the mean is a center of gravity. The same thing applies here, although we are using slightly different notation.) However, since we do not care whether any particular point is above or below the line, as long as it is reasonably close to the line, we might just look at the *distances*, or absolute values of the deviations. By averaging positive numbers and zeros, we will find the average distance from the points to the line. Thus, we want to find a and b to minimize

$$\sum_{i=1}^{n}|Y_i - \hat{Y}_i| = \sum_{i=1}^{n}|Y_i - a - bX_i|.$$

(Recall that the quantity $\sum_{i=1}^{n}|Y_i - \hat{Y}_i|$ is analogous to the numerator of the *mean deviation* $\frac{1}{n}\sum_{i=1}^{n}|Y_i - \overline{Y}|$ from your introductory course.)

The methods of calculus used to achieve this minimization involve absolute values and thus present a problem that is not trivial.‡ In order to

† A review of the algebra of summation notation is given in Appendix A.

‡ It is only in the last several years that the least absolute value criterion has been formulated as a linear programming problem and a solution obtained through that approach.

simplify the minimization procedure, we will minimize, instead, the quantity

$$\sum_{i=1}^{n} (Y_i - \hat{Y}_i)^2 = \sum_{i=1}^{n} (Y_i - a - bX_i)^2.$$

By squaring the deviations we eliminate the direction of the differences, since squared real numbers are always positive or zero. (Note that the quantity $\sum_{i=1}^{n} (Y_i - \hat{Y}_i)^2$ is analogous to the numerator of the sample variance $\frac{1}{n-1}\sum_{i=1}^{n} (Y_i - \bar{Y})^2$ from your introductory course, since $\hat{Y}_i$ is also a mean.)

Squared Residuals

Our criterion has changed slightly: We now define the "best" line as the one for which **squared residuals,** or the squared deviations, $(Y_i - Y_i)^2$ are the smallest, on the average. This criterion is similar to the original one: (1) It considers all points simultaneously, and (2) closeness is defined as a function of the vertical distance between the points and the line. We choose to use the average squared deviation because the minimization procedure for this criterion is much simpler than minimization of the mean absolute deviation. In addition, you will recall from your introductory course that a calculating formula is available for a quantity such as $\sum_{i=1}^{n} (Y_i - \bar{Y})^2$, so that this quantity is much easier to calculate for large samples than is $\sum_{i=1}^{n} |Y_i - \bar{Y}|$. With this criterion, we use the **method of least squares** to find the "best" line, the one which is, on the average, closest to all the points. The derivation of the name *least squares* should be clear: least = minimize; squares = squared residuals.†

Method of Least Squares

NORMAL EQUATIONS

Normal Equations

Methods of calculus can be used to find values a and b that minimize $\sum (Y_i - \hat{Y}_i)^2 = \sum (Y_i - a - bX_i)^2$ for a given set of data. Those familiar with differential calculus can refer to Appendix B for the details of the minimization procedure. Suffice it to say here that the procedure reduces to the solution of simultaneous linear equations called the **normal equations:**

$$\sum_{i=1}^{n} Y_i = na + b\sum_{i=1}^{n} X_i$$

$$\sum_{i=1}^{n} X_iY_i = a\sum_{i=1}^{n} X_i + b\sum_{i=1}^{n} X_i^2.$$

Solving these equations simultaneously for a and b will give the line $\hat{Y} = a + bX$ that is, on the average, closest to all the points in terms of squared deviations.

† We have now advanced our mathematical skill to the early nineteenth century. The method of least squares was first published in 1806 by the Frenchman Adrien-Marie Legendre (1752 – 1833). However, the method had been independently invented, although not published, some years earlier by the great German mathematician Carl Friedrich Gauss (1777 – 1855). After Legendre's publication of the method of least squares, Gauss commented that he had invented the method himself several years previously. This statement infuriated Legendre, who felt that a man with as many credits to his name as Gauss did not need to steal his (Legendre's) thunder.

Before proceeding to the solution of the normal equations, we might make several observations. First, the equation

$$\hat{Y} = a + bX$$

is *linear,* because a and b, the unknowns, are raised only to the first power —not because X is raised to no power greater than 1. Thus, even should the scatter diagram indicate a curvilinear regression, an equation of the form

$$\hat{Y} = a + bX + cX^2$$

is still a linear equation, because a, b, and c, the constants, are raised only to the first power. Although it is the equation of a parabola, it is still referred to as *curvilinear.* In this case, we would need to solve three simultaneous equations in the three unknowns, a, b, and c. These normal equations also would be derived by the methods of calculus and would be simultaneous linear equations — linear in the unknowns. Any curvilinear equation in any number of unknowns can be fitted to the data points by the same procedure.

NOTICE THE PATTERN FOR OBTAINING NORMAL EQUATIONS

Writing normal equations without using calculus becomes quite easy once a memory trick is mastered. Consider the two normal equations for fitting the line $\hat{Y} = a + bX$:

$$\sum Y_i = na + b\sum X_i$$

$$\sum X_iY_i = a\sum X_i + b\sum X_i^2.$$

(The indices on the summation signs have been omitted for the sake of simplicity. The subscripts on X and Y are retained to help distinguish them as variables from the constants a and b.) The trick to writing down (memorizing) these two equations involves the following three steps.

STEP 1 Write down the equation you wish to fit. We want to fit

$$\hat{Y} = a + bX.$$

STEP 2 Refer to Step 1. Find the coefficient on the first unknown. Multiply each term in the equation by this coefficient. Place a summation sign in front of every term of this new equation, and use summation rules.

In this case, the first unknown is a, and its coefficient is 1. Multiplying $\hat{Y} - a + bX$ by 1 gives

$$Y = a + bX$$

(ignore the hat on Y). Placing a summation sign in front of every term gives

$$\sum Y_i = \sum a + \sum bX_i,$$

or

$$\sum Y_i = na + b \sum X_i,$$

which is the first normal equation.

STEP 3 Repeat Step 2, applying it to the second unknown, b. In this case, the coefficient is X. Multiplying each term in Step 1 by X gives

$$XY = aX + bX^2.$$

Including the summation sign and applying rules of summation notation gives us the second normal equation,

$$\sum X_i Y_i = \sum aX_i + \sum bX_i^2,$$

or

$$\sum X_i Y_i = a \sum X_i + b \sum X_i^2.$$

This trick for obtaining the two normal equations works for any linear equation. For example, if we want to fit

$$\hat{Y} = a + bX + cX^2,$$

then the first normal equation is

$$1(Y) = 1(a) + 1(bX) + 1(cX^2),$$

$$\sum Y_i = \sum a + \sum bX_i + \sum cX_i^2,$$

or

$$\sum Y_i = na + b \sum X_i + c \sum X_i^2.$$

The second is

$$XY = aX + bX^2 + cX^3,$$

or

$$\sum X_i Y_i = a \sum X_i + b \sum X_i^2 + c \sum X_i^3,$$

and the third is

$$X^2 Y = aX^2 + bX^3 + cX^4,$$

or

$$\sum X_i^2 Y = a \sum X_i^2 + b \sum X_i^3 + c \sum X_i^4,$$

so that we would solve the following three equations simultaneously for a, b, and c:

$$\sum Y_i = na + b \sum X_i + c \sum X_i^2,$$

$$\sum X_i Y_i = a \sum X_i + b \sum X_i^2 + c \sum X_i^3,$$

$$\sum X_i^2 Y_i = a \sum X_i^2 + b \sum X_i^3 + c \sum X_i^4.$$

This trick also works for multiple regression, in which we predict Y from several X-variables, as we shall see.

NORMAL EQUATIONS FOR HOUSE SIZE-PRICE EXAMPLE

We now return to the problem of predicting house prices from sizes according to the data in Table 1.1. From these data we must calculate the following quantities to substitute into the normal equations:

$$\sum Y_i, \quad n, \quad \sum X_i, \quad \sum X_iY_i, \quad \sum X_i^2.$$

The quantity $\sum X_iY_i$ indicates that for each house in the sample, we multiply the size X_i by the price Y_i, and add the products. Note that

$$\sum X_iY_i \neq (\sum X_i)(\sum Y_i),$$

since $(\sum X_i)(\sum Y_i)$ indicates that we total all the sizes $(\sum X_i)$, total all the prices $(\sum Y_i)$ and then multiply these totals. Similarly, $\sum X_i^2$ indicates that we square each size and total the squares. Note also that

$$\sum X_i^2 \neq (\sum X_i)^2,$$

since $(\sum X_i)^2$ indicates that the sizes are totaled and the total is then squared.

Necessary calculations from the data are given in Table 2.1. (Note that Table 2.1 includes Y^2 figures, which are not needed for solving the normal

TABLE 2.1
Calculations for House Size-Price Problem

House Number	X = Size (thousands of sq ft)	Y = Price (thousands of dollars)	X^2	XY	Y^2
1	1.8	32	3.24	57.6	1024
2	1.0	24	1.00	24.0	576
3	1.7	27	2.89	45.9	729
4	2.8	47	7.84	131.6	2209
5	2.2	35	4.84	77.0	1225
6	0.8	17	0.64	13.6	289
7	3.6	52	12.96	187.2	2704
8	1.1	20	1.21	22.0	400
9	2.0	38	4.00	76.0	1444
10	2.6	45	6.76	117.0	2025
11	2.3	44	5.29	101.2	1936
12	0.9	19	0.81	17.1	361
13	1.2	25	1.44	30.0	625
14	3.4	50	11.56	170.0	2500
15	1.7	30	2.89	51.0	900
16	2.5	43	6.25	107.5	1849
17	1.4	27	1.96	37.8	729
18	3.3	50	10.89	165.0	2500
19	2.2	37	4.84	81.4	1369
20	1.5	28	2.25	42.0	784
Sum	40.0	690	93.56	1554.9	26,178

equations but will be needed later.) We find there that $n = 20$ and that

$$\sum Y_i = 690, \quad \sum X_i = 40, \quad \sum X_i^2 = 93.56, \quad \sum X_i Y_i = 1554.9.$$

Substituting these values into the normal equations gives

$$\begin{aligned} 690 &= 20a + 40b \\ 1554.9 &= 40a + 93.56b. \end{aligned}$$

These two equations must now be solved simultaneously for a and b. Any one of several equivalent methods can be used. Perhaps the easiest for now is to perform operations on the equations so that one of the unknowns has the same coefficient in both equations. This is simply done if we multiply the top equation by 2, to obtain a coefficient of 40 on a in both equations:

$$1380 = 40a + 80b$$

$$1554.9 = 40a + 93.56b.$$

(Of course, multiplying the top equation by 93.56 and the bottom by 40 would give the coefficient $40(93.56) = 3742.40$ on b in both equations, should one choose to eliminate b first.) Now we simply subtract one equation from the other. Since the top equation has the smaller values, subtracting the top from the bottom will result in positive numbers, although it does not matter which is subtracted. We obtain upon subtraction

$$174.9 = 0a + 13.56b,$$

or

$$174.9 = 13.56b,$$

which is one equation in one unknown. The solution is

$$\frac{174.9}{13.56} = b = 12.8982.$$

To find a, substitute $b = 12.8982$ into one of the original equations. Since the top equation has smaller numbers,

$$690 = 20a + 40(12.8982)$$

$$690 - 515.9280 = 20a$$

$$\frac{174.072}{20} = a = 8.7036.$$

CHECKING YOUR RESULTS

Since an arithmetic error is always possible, one should substitute the values for a and b back into the original equations. Both equations must check. Checking,

$$20(8.7036) + 40(12.8982) = 690.0000$$

$$40(8.7036) + 93.56(12.8982) = 1554.8996$$

$$\approx 1554.9.$$

Noting that the second equation does not check exactly to four decimal places brings up the question of rounding error. Accuracy depends upon the number of decimal places carried in calculations; the more places carried, the more accurate the results. Clearly, then, one carries as many decimals as feasible, and this is no great burden with the wide availability of electronic calculators. When reporting results, however, one needs to be practical. If a variable is measured in, say, thousands of dollars, then a figure such as 10.98765 means \$10,987.65. However, if the units were dollars instead of thousands of dollars, the figure 10.98765 would denote \$10.98765, which would be better expressed as \$10.99.

The number of decimal places recorded in the data often indicates the form of the result. If the house prices, for example, are recorded in round thousands of dollars, such as 32 = \$32,000, it would make little sense to predict a particular house as selling for, say, \$28,935.64. Perhaps \$28,900 would be as close as you would care to estimate; as a rule of thumb, carry at least two more decimal places than shown in your data, and round results correct to one more place.

Rounding error comes about when divisions are made and when square roots are taken. Thus, error can be minimized by postponing divisions and extraction of roots as long as possible. For example, to evaluate

$$\frac{a}{b/c}$$

by first dividing b by c and then dividing a by their quotient involves two divisions and thus two sources of rounding error. It would be better done as ac/b, which involves only one division. Similarly, $\sqrt{a}\sqrt{b}$ is better calculated as $\sqrt{ab}$, since this latter form involves extracting only one square root.

A NOTE ON USING YOUR CALCULATOR

Table 2.1 presents in detail th calculations required to obtain $\sum X_i$, $\sum Y_i$, $\sum X_i^2$, $\sum X_iY_i$, and $\sum Y_i^2$. Table 2.1 is so detailed to ensure that the student understands exactly how these quantities are obtained. In practice, however, one would not write down each individual product X_iY_i or each individual value of X_i^2 in a table like Table 2.1. In the first place, producing such a table is time-consuming; secondly, one is prone to transposing digits or making other recording errors when transcribing many values. We assume that you have access to a calculator. If you are aware of its capabilities, the labor involved in finding these sums (and in some cases, in finding a and b in the regression equation) can be considerably reduced.

Of course, there are many brands of calculators on the market, and within a given brand there are many models, ranging from very simple to very sophisticated. We will illustrate here some techniques for calculating sums of squares and products that work on many calculators. Refer to your instruction booklet for key sequences specific to your calculator.

Almost any calculator will find $\sum X_iY_i$ and/or $\sum X_i^2$ and $\sum Y_i^2$ without your having to write down each separate X_iY_i or X_i^2 or Y_i^2 value, re-enter these, and add. Suppose, for example, that you wish to find $\sum X_iY_i$ for the data in the house size-price problem. For many calculators, the sequence shown in Table 2.2 will work.

TABLE 2.2 Using a Calculator to Find $\sum X_iY_i$

Enter	Key	
1.8	×	
32	+	multiplies 1.8 by 32 and adds:
1.0	×	
24	+	1.0 times 24 and adds:
1.7	×	
27	+	1.7 times 27 and adds:
		.
		.
		.
1.5	×	
28	=	1.5 times 28.
	Result: 1554.9	

To find $\sum X_i^2$, use the sequence shown in Table 2.3.

TABLE 2.3 Using a Calculator to Find $\sum X_i^2$

Enter	Key
1.8	X^2, +
1.0	X^2, +
1.7	X^2, +
.	
.	
.	
1.5	X^2, =
Result: 93.56	

Try these operations on your calculator. If they don't work, refer to your instruction booklet.

If your calculator has at least one memory, then it is probably possible for you to get two of the needed sums simultaneously. For example, you can get $\sum X_i$ and $\sum X_i^2$ in one operation if you sum the X_i-values to the memory after you enter them but before you square them. Try the sequence of operations in Table 2.4.

TABLE 2.4 Using a Calculator to Find $\sum X_i$ and $\sum X_i^2$ Simultaneously

Enter	Key	
1.8	M^+, X^2, +	
1.0	M^+, X^2, +	
⋮		
1.5	M^+, X^2, =	Result: 93.56
	RM	Result: 40.00

(Here we use the symbol M^+ to denote a key that adds values into a memory. There may be a similar key on your calculator that is labeled differently.) The symbol RM in Table 2.4 denotes a key that recalls what is in the memory. On the display after keying = will be $\sum X_i^2$; after recalling the memory, the display will show $\sum X_i$.

A similar process can be used to obtain $\sum Y_i$ and $\sum X_iY_i$ in a single operation. When a value for Y is entered, sum it to a memory before multiplying by the corresponding X-value.

More sophisticated calculators have built-in functions for obtaining the intercept and slope of the regression equation. Keys may be labeled a and b, or m and b, or INT and SLOPE, or something else. In these machines, you enter your X- and Y-values in pairs, keying them in such a way that the calculator knows which is which (see your instruction booklet). By simply pressing the appropriate buttons, the slope and intercept appear on your display. Additionally, depending on how many memories the calculator has, all or some of the quantities $\sum X_i$, $\sum X_i^2$, $\sum X_iY_i$, $\sum Y_i$, $\sum Y_i^2$, and n can be obtained by simply recalling the appropriate memory. Most calculators that are this sophisticated will also give automatically other quantities of interest in a regression problem.

While for multiple regression problems, some curvilinear regression problems, and regressions on very large amounts of data, a computer is generally needed, most simple regression problems can be handled by a calculator with a built-in linear regression function. Whether or not your calculator is sophisticated enough to perform simple linear regressions, you should still investigate its capabilities and use it so as to minimize your labor.

GRAPHING THE REGRESSION EQUATION

For the sake of simplicity, let's round our answers in the house example to

$$a = 8.7, \qquad b = 12.9,$$

which yield the regression equation

$$\hat{Y} = 8.7 + 12.9X.$$

This is the line that, on the average, is closest to all the points in our data in terms of squared deviations. Plotting the regression equation on the original scatter diagram, as in Figure 2.8, shows that the line $\hat{Y} = 8.7 + 12.9X$ both follows the lay of the points and passes through the middle of the mass of points. Plotting the regression equation on the scatter diagram is also a good way to catch any gross errors in addition or squaring or multiplying in the original data. Note in Figure 2.8 that the regression equation is labeled but that points chosen to plot it are not labeled.

FIGURE 2.8
Regression Line Superimposed on Scatter Diagram

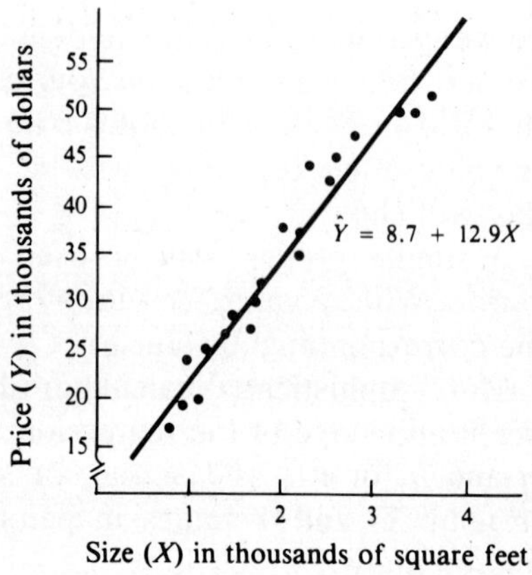

We can also verify that the least squares line passes through the point $(\bar{X}, \bar{Y})$. In our example, $\bar{X} = 40/20 = 2$ and $\bar{Y} = 690/20 = 34.5$. Substituting $\bar{X}$ for X in the regression equation produces

$$\begin{aligned}\hat{Y} &= 8.7 + 12.9(2)\\ &= 8.7 + 25.8\\ &= 34.5 = \bar{Y},\end{aligned}$$

so that the line does pass through $(\bar{X}, \bar{Y})$. This can also be demonstrated by locating the point (2, 34.5) in Figure 2.8.

The next section will show an alternative method for finding the regression equation.

EXERCISES

2.1 What is the difference between the points (X_i, Y_i) and $(X_i, \hat{Y}_i)$?

2.2 Use the three rules of summation notation in Appendix A to find equivalent forms for the following expressions. Assume all sums are from $i = 1$ to n. Example:

$\sum (X_i - 4) = \sum X_i - 4n$.

a. $\sum (X_i^2 - X_i + 4)$; b. $\sum (3X_i + Y_i)$;

c. $\sum (X_i - 4)^2$; d. $\sum (3X_i - 4Y_i)^2$;

e. $\sum (3X_i + Y_i)(4X_i - 5Y_i)$;

f. $\sum (Y_i - a - bX_i)$, for a and b constants;

g. $\sum (X_iY_i - aX_i - bX_i^2)$, for a and b constants;

h. $\sum (Y_i - a - bX_i)(X_i)$, for a and b constants;

i. $\sum (X_i - \bar{X})^2$, for $\bar{X}$ a constant;

j. $\sum (X_i - \bar{X})(Y_i - \bar{Y})$, for $\bar{X}$ and $\bar{Y}$ constants.

2.3 Using rules of summation, show that

a. $\sum (X_i - \bar{X}) = 0$;

b. $$\sum (X_i - \bar{X})^2 = \frac{n \sum X_i^2 - (\sum X_i)^2}{n} = \sum X_i^2 - \frac{(\sum X_i)^2}{n};$$

c. $$\frac{\sum (X_i - \bar{X})^2}{n - 1} = \frac{n \sum X_i^2 - (\sum X_i)^2}{n(n-1)};$$

d. $$\sum (X_i - \bar{X})(Y_i - \bar{Y}) = \frac{n \sum X_iY_i - (\sum X_i)(\sum Y_i)}{n} = \sum X_iY_i - \frac{(\sum X_i)(\sum Y_i)}{n};$$

e. $$\sum (Y_i - \bar{Y})^2 = \frac{n \sum Y_i^2 - (\sum Y_i)^2}{n} = \sum Y_i^2 - \frac{(\sum Y_i)^2}{n}.$$

2.4 Use the data X : 1, 2, 3, 4, 5.

a. Calculate $\frac{1}{n} \sum_{i=1}^{n} (X_i - \bar{X})$ and show that this quantity is zero.

b. Calculate the mean deviation

$$\text{MD} = \frac{1}{n} \sum_{i=1}^{n} |X_i - \bar{X}|.$$

c. Calculate the variance

$$s_X^2 = \frac{1}{n-1} \sum_{i=1}^{n} (X_i - \bar{X})^2.$$

2.5 Use X: 1, 2, 3, 4, 5.

a. Show that $\sum X_i^2 \neq (\sum X_i)^2$.

b. Calculate s_X^2 by using the formula

$$s_X^2 = \frac{n \sum X_i^2 - (\sum X_i)^2}{n(n-1)} = \frac{\sum X_i^2 - (\sum X_i)^2/n}{n - 1}.$$

Compare your answer to that of Exercise 2.4c. Which method of calculating s_X^2 was easier and quicker?

c. True or false: $\sum X_iY_i = (\sum X_i)(\sum Y_i)$?

2.6 Why is the regression line obtained by the method of least squares considered to be the best line that can be fitted to the data?

2.7 Write down the normal equations that one would have to solve in order to fit the following equations to a set of data:

a. $\hat{Q} = a + bP$;
b. $\hat{E} = a + bT + cT^2$;
c. $\hat{Y} = a + bX + cX^2 + dX^3$;
d. $\hat{Y} = a + bX + cZ$;
e. $\hat{Y} = b_0 + b_1X_1 + b_2X_2$.

2.8 Refer to the Concession Data in Exercise 1.5.
a. Set up and solve the normal equations needed to fit a straight line, $\hat{Y} = a + bX$, relating syrup use to ticket sales.
b. Plot the regression equation on the scatter diagram you constructed in Exercise 1.5.
c. Verify that the equation passes through the point $(\overline{X}, \overline{Y})$.

2.9 Refer to the Sales Data in Exercise 1.6.
a. Fit a simple linear regression equation for these data.
b. Plot this equation on the scatter diagram constructed in Exercise 1.6.
c. Verify that the equation passes through the point $(\overline{X}, \overline{Y})$.

2.10 Refer to the Heating Data in Exercise 1.8.
a. Fit a simple linear regression equation to these data.
b. Plot the regression equation on the scatter diagram constructed in Exercise 1.8.
c. Verify that the equation passes through the point $(\overline{X}, \overline{Y})$.

2.11 The foreman of an assembly operation has been reprimanded by his supervisor because of large costs for repairing equipment used in assembly. The foreman, however, believes that these large repair costs are incurred because management is unwilling to have routine maintenance performed on the equipment as often as necessary. He draws a simple random sample of records on ten similar machines that had been repaired during the last year and presents the evidence shown in Table 2.5.

TABLE 2.5
Repairs Data

Hours of Operation Since Last Maintenance (hundreds)	Repair Cost (hundreds of dollars)
55	2.7
40	2.0
63	3.0
50	2.6
37	1.5
57	3.0
51	2.5
60	3.1
42	2.0
45	2.3

a. Find the least squares regression equation relating repair costs to operating hours since the last routine maintenance was performed.
b. Does the foreman appear to have a case? Explain.

2.12 In forestry, one is often interested in the diameter of a tree at breast height ($4\frac{1}{2}$ ft above the ground) *inside* the bark (DBH_{ib}), but this is very difficult to measure. Diameter at breast height *outside* the bark (DBH_{ob}) is more easily measured. It is possible, but not very easy, to obtain a measurement of the thickness of the bark at a single point (single bark thickness), and it is commonplace to estimate DBH_{ib} by subtracting twice the single bark thickness from DBH_{ob}.

If a simple linear regression equation were found relating DBH_{ib} to DBH_{ob}, the inside diameters of other trees could be estimated from their outside diameters without having to measure the thickness of the bark. The data in Table 2.6, kindly provided by Professor John C. Rennie, Department of Forestry, Wildlife, and Fisheries, the University of Tennessee, are for a random sample of pines in the Cherokee Woodlot, Knoxville, Tennessee.

TABLE 2.6
Bark Data

	Tree							
	1	**2**	**3**	**4**	**5**	**6**	**7**	**8**
DBH_{ob} (cm)	29.2	23.6	38.9	29.7	25.6	20.3	36.6	24.1
Single bark thickness (cm)	1.5	1.4	2.0	1.4	1.3	1.0	1.6	1.3

	Tree						
	9	**10**	**11**	**12**	**13**	**14**	**15**
DBH_{ob}	30.5	37.6	28.7	37.6	32.3	23.9	30.7
Single bark thickness (cm)	1.5	2.1	1.4	1.9	1.6	0.9	1.5

a. Calculate values of DBH_{ib} by

$$DBH_{ib} = DBH_{ob} - 2\text{ (bark thickness)}.$$

b. Find a simple linear regression equation relating DBH_{ib} to DBH_{ob}.

2 . 4 SHORTCUT FORMULAS

The regression equation can always be found by using the normal equations, regardless of how many predictor variables are used or whether the relationship is linear or curvilinear. In simple regression, the regression line can also be found by using two shortcut formulas. These formulas eliminate the need for solving simultaneous equations. However, their greatest value lies in the fact that they involve calculating quantities that will be used extensively throughout the remainder of our analysis. Since these quantities must be calculated eventually, why not find them first and also use them to obtain values for a and b?

USEFUL NOTATION

At this point, we digress briefly to introduce some notation that will help us simplify many formulas. First recall that the rules of summation notation let us obtain a calculating formula for the quantity $\sum_{i=1}^{n} (X_i - \bar{X})^2$:

$$\sum_{i=1}^{n} (X_i - \bar{X})^2 = \sum_{i=1}^{n} (X_i^2 - 2X_i\bar{X} + \bar{X}^2) \qquad \text{(expanding the square)}$$

$$= \sum_{i=1}^{n} X_i^2 - 2\bar{X}\sum_{i=1}^{n} X_i + n\bar{X}^2$$

$$= \sum_{i=1}^{n} X_i^2 - 2\left(\frac{\sum_{i=1}^{n} X_i}{n}\right)\sum_{i=1}^{n} X_i + n\left(\frac{\sum_{i=1}^{n} X_i}{n}\right)^2 \quad \text{since} \quad \left(\bar{X} = \frac{\sum_{i=1}^{n} X_i}{n}\right)$$

$$= \sum_{i=1}^{n} X_i^2 - 2\frac{\left(\sum_{i=1}^{n} X_i\right)^2}{n} + \frac{\left(\sum_{i=1}^{n} X_i\right)^2}{n}$$

$$= \sum_{i=1}^{n} X_i^2 - \frac{\left(\sum_{i=1}^{n} X_i\right)^2}{n}$$

$$= \frac{n\sum_{i=1}^{n} X_i^2 - \left(\sum_{i=1}^{n} X_i\right)^2}{n}.$$

Let

$$S_{XX} = \sum_{i=1}^{n} (X_i - \bar{X})^2 = \sum_{i=1}^{n} X_i^2 - \frac{\left(\sum_{i=1}^{n} X_i\right)^2}{n}.$$

The S_{XX} notation is used *only* to shorten the amount of writing that must be done. Note that $\sum_{i=1}^{n} (X_i - \bar{X})^2$ is a sum of squares (or sum of squared quantities) involving X-values; it is called a **corrected sum of squares** (corrected for the mean), as opposed to the **raw sum of squares,** $\sum_{i=1}^{n} X_i^2$. The S_{XX} notation reminds us of this, if we think of the letter S as denoting sum, and the XX subscript as denoting squares of quantities involving X-values ($X \cdot X = X^2$).†

Corrected Sum of Squares

Raw Sum of Squares

Note also that

$$\sum_{i=1}^{n} (X_i - \bar{X})^2 = \sum_{i=1}^{n} X_i^2 - \frac{\left(\sum_{i=1}^{n} X_i\right)^2}{n}$$

† Some texts denote the quantity $\sum_{i=1}^{n} (X_i - \bar{X})^2$ as $\sum_{i=1}^{n} x_i^2$, where $x_i = X_i - \bar{X}$. Still other texts give the calculating formula as $\sum_{i=1}^{n} X_i^2 - n\bar{X}^2$, but the form given above will be more convenient for us in most situations. Care should be taken when reading another text to get the notation straight.

Sample Variance

is the numerator of the calculating formula for the sample variance of the variable X. Thus we could write the **sample variance** as

$$\hat{\sigma}_X^2 = \frac{S_{XX}}{n},$$

or

$$S_X^2 = \frac{S_{XX}}{n-1},$$

if an unbiased estimator† of the population variance is required.

This sum of squares notation is quite versatile. For example, replacing X_i by Y_i gives

$$S_{YY} = \sum_{i=1}^{n} (Y_i - \overline{Y})^2 = \sum_{i=1}^{n} Y_i^2 - \frac{\left(\sum_{i=1}^{n} Y_i\right)^2}{n}.$$

In addition, if we note that

$$S_{XX} = \sum_{i=1}^{n} (X_i - \overline{X})^2 = \sum_{i=1}^{n} (X_i - \overline{X})(X_i - \overline{X}),$$

or

$$S_{XX} = \sum_{i=1}^{n} X_i^2 - \frac{\left(\sum_{i=1}^{n} X_i\right)^2}{n}$$

$$= \sum_{i=1}^{n} X_i X_i - \frac{\left(\sum_{i=1}^{n} X_i\right)\left(\sum_{i=1}^{n} X_i\right)}{n},$$

then by simply replacing one of the X's in S_{XX} by a Y, we get

$$S_{XY} = \sum_{i=1}^{n} (X_i - \overline{X})(Y_i - \overline{Y})$$

$$= \sum_{i=1}^{n} X_i Y_i - \frac{\left(\sum_{i=1}^{n} X_i\right)\left(\sum_{i=1}^{n} Y_i\right)}{n}.$$

Corrected Sum of Cross-Products

S_{XY} is a **corrected sum of cross-products** (corrected for the means of X and Y) as opposed to the raw sum of cross-products, $\sum X_i Y_i$. Some readers might recognize that

$$S_{XY} = \sum_{i=1}^{n} (X_i - \overline{X})(Y_i - \overline{Y})$$

Covariance

is the numerator of the sample **covariance** between two variables X and Y and

† See Section 6.2 for a review of the meaning of *unbiased*.

is a measure of the extent to which the two variables act alike. We will have much more to say about this later.

SHORTCUT FORMULAS

With this shorthand notation, we are ready to derive the shortcut formulas for finding a and b in the equation $Y = a + bX$. Keep in mind that these formulas work only in the case of simple linear regression, as will be apparent from their derivation. Indices of summation will be omitted from now on, unless they are needed for clarity. Unless otherwise indicated, all sums are from 1 to n.

We begin by solving the two normal equations algebraically.

$$\sum Y_i = na + b\sum X_i$$

$$\sum X_iY_i = a\sum X_i + b\sum X_i^2.$$

Multiplying the first equation by $\sum X_i$ and the second by n will enable us to eliminate a in both equations:

$$(\sum X_i)(\sum Y_i) = na\sum X_i + b(\sum X_i)^2$$

$$n\sum X_iY_i = na\sum X_i + nb\sum X_i^2.$$

Subtracting the top equation from the bottom,

$$n\sum X_iY_i - (\sum X_i)(\sum Y_i) = nb\sum X_i^2 - b(\sum X_i)^2$$

$$= b[n\sum X_i^2 - (\sum X_i)^2].$$

Dividing both sides by n, we obtain

$$S_{XY} = bS_{XX},$$

or

$$b = S_{XY}/S_{XX},$$

which is the shortcut formula for b. Since we will use S_{XY} and S_{XX} many times in our analysis, we can save time by calculating S_{XY} and S_{XX} first and then using them to find b. The result will be the same as that obtained by solving the normal equations.

Once b is found, substitution into the first normal equation yields a value for a:

$$\sum Y_i = na + b\sum X_i,$$

$$\sum Y_i - b\sum X_i = na,$$

$$\overline{Y} - b\overline{X} = a,$$

where $\overline{Y} = (\sum Y_i)/n$ and $\overline{X} = (\sum X_i)/n$ are the mean values of the dependent and independent variables, respectively.

USING SHORTCUT FORMULAS IN HOUSE SIZE-PRICE EXAMPLE

In our example about sizes and prices of houses,

$$S_{XY} = 1554.9 - \frac{(40)(690)}{20} = 174.9,$$

$$S_{XX} = 93.53 - \frac{(40)^2}{20} = 13.56.$$

Then

$$b = \frac{174.9}{13.56} = 12.8982,$$

which is just what we found by using the normal equations. (Referring back to the calculation of b by using normal equations, we see that we found $b = 174.9/13.56 = S_{XY}/S_{XX}$. In some instances, the actual numbers in the ratio when solving normal equations would be nS_{XY}/nS_{XX}, or the *numerators* of S_{XY} and S_{XX}, because the n's in the denominators cancel. Thus, in our example, we could just as well have found b by $3498/271.2 = 12.8982$. Because nS_{XY}/nS_{XX} involves two fewer divisions than S_{XY}/S_{XX}, the use of formulas may introduce some additional rounding error.)

Continuing with our example,

$$\begin{aligned} a &= \overline{Y} - b\overline{X} \\ &= \frac{690}{20} - 12.8982\left(\frac{40}{20}\right) \\ &= 34.5 - 25.7964 \\ &= 8.7036, \end{aligned}$$

which is also the same value for a that we found earlier.

To summarize,

1. The same values for a and b (except for possible rounding error) will be found by solving normal equations and by using shortcut formulas;
2. Shortcut formulas can be used only in simple linear regression;
3. Normal equations can be used in *any* regression problem;
4. In simple regression, use of formulas often saves time, since S_{XY} and S_{XX} will have to be calculated eventually anyway.

A CLOSER LOOK AT THE SHORTCUT FORMULA FOR *b*

Let us pause for a moment to examine the formula for b,

$$b = \frac{S_{XY}}{S_{XX}} .$$

Recall that b, as the slope of the line $\hat{Y} = a + bX$, measures the number of units and the direction of change in Y for a one-unit change in X. Let us see how the shortcut formula for b reflects this fact.

First, note that S_{XX}, being a sum of squares, can never be negative. It can be zero, but only if all X-values are the same; that is, if we have looked at only one X-value. In any realistic regression problem, then, because S_{XX} is always positive, the sign of b must be determined by S_{XY}.

What determines the sign of S_{XY}? First suppose that X and Y are directly related. Then large X-values tend to occur with large Y-values and small X-values with small Y-values. If the X-value is large, then $X_i - \overline{X}$ will be positive. If the Y-value is large, the difference $Y_i - \overline{Y}$ will also be positive, and so will the product $(X_i - \overline{X})(Y_i - \overline{Y})$. On the other hand, if X and Y are both small, $X_i - \overline{X}$ will be negative and so will $Y_i - \overline{Y}$; their product, however, will be positive. Thus, if X and Y are directly related, $\sum (X_i - \overline{X})(Y_i - \overline{Y})$ will be the sum of mostly positive values and will itself be positive.

Suppose now that X and Y are inversely related. Then when one is large, the other tends to be small. If $X_i - \overline{X}$ is positive, the associated value of $Y_i - \overline{Y}$ will tend to be negative and the product will be negative. The sum $\sum (X_i - \overline{X})(Y_i - \overline{Y})$ will thus be negative if X and Y are inversely related.

Finally, if the two variables are not related, then regardless of the value of X, Y tends to remain constant. Most values of Y are thus equal to $\overline{Y}$ and the differences $Y_i - \overline{Y}$ are close to zero. Regardless of whether X is large or small, then, the products $(X_i - \overline{X})(Y_i - \overline{Y})$ will be near zero and so will the sum $\sum (X_i - \overline{X})(Y_i - \overline{Y})$. So if X and Y are not related, b will be zero (or close to it, when random variability is taken into account).

We have seen how the direction of the relationship determines the sign of S_{XY} and thus the sign of b.

The shortcut formula $b = S_{XY}/S_{XX}$ also reminds us that b measures change in Y for change in X. It does this because its units of measurement are Y-units over X-units. That this is so can be easily seen:

$$S_{XY} = \sum (X_i - \overline{X})(Y_i - \overline{Y})$$

in X-units times Y-units and $S_{XX} = \sum (X_i - \overline{X})^2$ is in squares of X-units. Their ratio is then in Y-units over X-units.

EXERCISES

2.13 Write out the formulas for

a. S_{TT}; **b.** S_{RR}; **c.** S_{RT}; **d.** S_{TR}.

2.14 Refer to the Concession Data in Exercise 1.5.

a. Find S_{XY}, S_{XX}, and S_{YY} for these data.

b. Use these values to find a and b for a simple linear regression equation. Check your answers against those in Exercise 2.8.

2.15 Refer to the Heating Data in Exercise 1.8.

a. Find S_{XY}, S_{XX}, and S_{YY} for these data.

b. Use these values to find the slope and intercept of a simple linear regression equation. Compare to values found in Exercise 2.10.

2.16 "Suds," a chain of beer parlors catering mostly to the college crowd, is located in several states in the Midwest. The management of "Suds" is contemplating opening up new franchises in other college towns and is trying to decide where to locate. Since it seems reasonable to assume that the larger the student body, the higher the gross sales, the management of "Suds" collected the information in Table 2.7 from a simple random sampling of its stores in order to estimate the relationship.

TABLE 2.7 "Suds" Data

	Store								
	1	**2**	**3**	**4**	**5**	**6**	**7**	**8**	**9**
Student body size (thousands)	11	10	17	4	10	4	15	2	15
Yearly gross sales (thousands of dollars)	83.7	79.2	120.0	53.7	85.0	48.2	100.7	41.6	105.3

	Store								
	10	**11**	**12**	**13**	**14**	**15**	**16**	**17**	**18**
Students	5	12	3	20	6	25	7	22	20
Sales	56.4	87.6	44.1	134.0	61.7	152.6	65.1	142.0	131.2

Use shortcut formulas to find the slope and intercept for a simple linear regression equation relating sales to the size of the student body.

2.17 A certain professor has been chided by her department chairman for receiving low ratings in teaching effectiveness. She proposes to defend herself by demonstrating that the reason her ratings are low is that she has had to teach very large sections. Her contention is that students don't like being taught in very large sections, which are impersonal and don't allow class discussion, and therefore her low ratings are simply a function of class size. To back this up, the professor asks the campus Learning Evaluation Center to select a simple random sample of classes in which a standard evaluation form has been administered and to provide her with information on class size and average rating of the instructor for each class. The center provided the information in Table 2.8.

TABLE 2.8
Class Size Data

Number of students	10	12	15	21	23	27	32	32	35	37	40
Avg. rating of instructor	3.6	3.3	3.6	3.5	3.6	4.0	3.4	3.1	2.9	3.2	3.3
Students	40	41	43	45	47	52	57	91	113	127	146
Rating	3.0	3.1	3.4	3.2	3.3	3.4	3.2	3.0	2.7	2.5	2.3

a. Use shortcut formulas to find the equation of a linear function describing the relationship between class size and rating of the instructor.
b. Does the professor appear to be correct in her contention that instructors get low ratings from large classes?
c. Plot the data and the regression equation.
d. Examine the graph in (c). Are the necessary assumptions being met? Is there anything that raises questions concerning what kinds of conclusions might be drawn from this analysis?

2.5

EFFECTS OF TRANSFORMATIONS OF THE *X*-VALUES

Transformed

Sometimes we can save ourselves a lot of calculating if the predictor variable is appropriately **transformed.** For example, an economist constructing a time series across several years might have to work with the figures

$$1969 \quad 1970 \quad 1971 \quad 1972 \quad 1973.$$

Having to add or square these values would be cumbersome. It would be easier if the economist coded the years thus:

$$1969 = -2, \quad 1970 = -1, \quad 1971 = 0, \quad 1972 = 1, \quad 1973 = 2.$$

Nothing has really been changed in the problem as long as the economist remembers that the value 0 denotes the year 1971, and so on.

An engineer testing the breaking strength of a certain metal alloy rod might subject it to pressures of 100, 150, and 200 pounds. Since the difference between the low and medium pressures is equal to the difference between the medium and high pressures—that is, the experimental levels are evenly spaced—the engineer can code the three respective pressures −1, 0, and +1, values that are much easier to work with.

CENTERING X-VALUES

In the time series example, if we denote the coded values by $U_i = X_i - \overline{X}$, then

$$\sum (U_i) = 0.$$

In effect, the coding was achieved by finding the mean of the original values and subtracting this mean from each of the original values. Thus if $X_1 = 1969$,

$$U_1 = X_1 - \overline{X} = 1969 - 1971 = -2$$
$$U_2 = X_2 - \overline{X} = 1970 - 1971 = -1$$
$$U_3 = X_3 - \overline{X} = 1971 - 1971 = 0$$
$$U_4 = X_4 - \overline{X} = 1972 - 1971 = 1$$
$$U_5 = X_5 - \overline{X} = 1973 - 1971 = 2.$$

Centering Values

Since the mean is subtracted from each value, we say that such a coding scheme **centers** the X-values, or corrects the X-values for their mean.

What effect will this centering have on our values for a and b in the regression equation? There should be none. Let us check it out: If we center the X-values, then using U-values,

$$S_{XX} - S_{UU} - \sum U_i^2 - \frac{(\sum U_i)^2}{n} = \sum U_i^2$$

and

$$S_{XY} = S_{UY} = \sum U_i Y_i - \frac{(\sum U_i)(\sum Y_i)}{n}$$
$$= \sum U_i Y_i,$$

since

$$\sum U_i = 0.$$

Then

$$b = \frac{S_{XY}}{S_{XX}} = \frac{\sum U_i Y_i}{\sum U_i^2}$$

and

$$a = \overline{Y} - b\overline{U} = \overline{Y},$$

since

$$\frac{\sum U_i}{n} = \overline{Y} = 0.$$

Note that b really has not changed, since

$$\frac{\sum U_i Y_i}{\sum U_i^2} = \frac{\sum (X_i - \overline{X}) Y_i}{\sum (X_i - \overline{X})^2}$$
$$= \frac{S_{XY}}{S_{XX}},$$

because $\sum (X_i - \bar{X})Y_i = \sum (X_i - \bar{X})(Y_i - \bar{Y})$ as shown below:

$$\begin{aligned}\sum (X_i - \bar{X})Y_i &= \sum (X_i - \bar{X})Y_i - \sum (X_i - \bar{X})\bar{Y} + \sum (X_i - \bar{X})\bar{Y} \\ &= [\sum (X_i - \bar{X})Y_i - \sum (X_i - \bar{X})\bar{Y}] + \sum (X_i - \bar{X})\bar{Y} \\ &= \sum [(X_i - \bar{X})Y_i - (X_i - \bar{X})\bar{Y}] + \bar{Y}\sum (X_i - \bar{X}) \\ &= \sum [(X_i - \bar{X})(Y_i - \bar{Y})] + \bar{Y}(0) \\ &= \sum (X_i - \bar{X})(Y_i - \bar{Y}) = S_{XY}.\end{aligned}$$

Thus, the general underlying relationship between X and Y, as indicated by b, is not affected by centering the predictor variable. The value for a, however, is changed. To understand why, (1) recall that a is the Y-intercept, the value of Y when $X = 0$; (2) note that with the centering, the Y-axis has been shifted to the mean X-value, since $U = 0$ where $X = \bar{X}$. If the vertical axis is now located at $\bar{X}$, then the line will cross the Y-axis at $\bar{Y}$, since the line goes through the point $(\bar{X}, \bar{Y})$.

When the predictor variables are centered according to the above scheme, the model is written as

$$Y = \alpha' + \beta U + \varepsilon,$$

or

$$Y = \alpha' + \beta(X - \bar{X}) + \varepsilon.$$

CENTERING X-VALUES IN HOUSE SIZE-PRICE EXAMPLE

Now let us apply the concept of centering X-values to our house example. Recall that $\bar{X} = 2$. The centered X-values and calculations for the problem are then as given in Table 2.9. Note that $\sum U_i^2 = 13.56$, which is what we got for S_{XX} when the X-values were not centered; and that $\sum U_iY_i = 174.9$, the same value we found for S_{XY}. Thus,

$$b = \frac{\sum U_iY_i}{\sum U_i^2} = \frac{174.9}{13.56} = \frac{S_{XY}}{S_{XX}} = 12.8982$$

verifies that the slope, which gives us the relationship, is unaffected by centering the predictor variable. Then

$$a' = \bar{Y} = \frac{690}{20} = 34.5,$$

which is the value of Y when $U = 0$, that is, when $X_i - \bar{X} = X_i - 2 = 0$, or $X_i = 2$. The regression equation is

$$\hat{Y} = 34.5 + 12.9U,$$

where $U = X - 2$. The intercept has changed because the vertical axis has

TABLE 2.9
Centered X-Values and Calculations for the Data in Table 1.1

House	X	$U = X - \bar{X}$	U^2	UY	Y
1	1.8	−0.2	0.04	− 6.4	32
2	1.0	−1.0	1.00	−24.0	24
3	1.7	−0.3	0.09	− 8.1	27
4	2.8	0.8	0.64	37.6	47
5	2.2	0.2	0.04	7.0	35
6	0.8	−1.2	1.44	−20.4	17
7	3.6	1.6	2.56	83.2	52
8	1.1	−0.9	0.81	−18.0	20
9	2.0	0.0	0.00	0.0	38
10	2.6	0.6	0.36	27.0	45
11	2.3	0.3	0.09	13.2	44
12	0.9	−1.1	1.21	−20.9	19
13	1.2	−0.8	0.64	−20.0	25
14	3.4	1.4	1.96	70.0	50
15	1.7	−0.3	0.09	− 9.0	30
16	2.5	0.5	0.25	21.5	43
17	1.4	−0.6	0.36	−16.2	27
18	3.3	1.3	1.69	65.0	50
19	2.2	0.2	0.04	7.4	37
20	1.5	−0.5	0.25	−14.0	28
Sum	40.0	0.0	13.56	174.9	690

Mean: $\bar{X} = \frac{40}{20} = 2$

been moved from $X = 0$ to $X = 2$. (See Figure 2.9.) The underlying linear relationship, however, is not affected, and one can easily obtain $\hat{Y} = 8.7 + 12.9X$ from the regression equation:

$$\hat{Y} = 34.5 + 12.9(X - 2) = 34.5 + 12.9X - 25.8 = 8.7 + 12.9X.$$

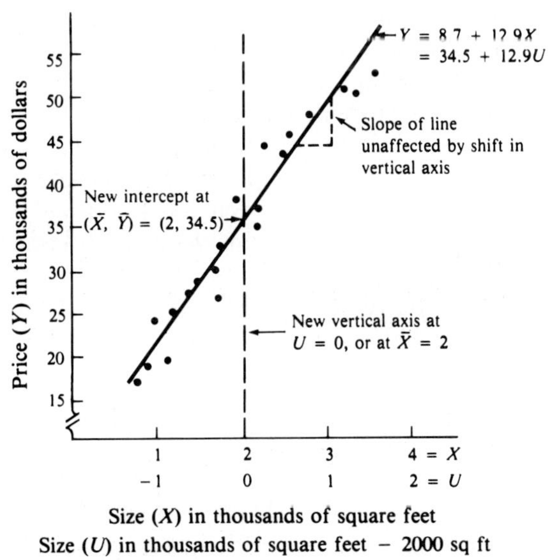

FIGURE 2.9
Effect on Slope and Intercept of Centering X

S C A L I N G X - V A L U E S

Scaling Values

A **scaling** of values is accomplished when each value is multiplied by a constant.

Let $V_i = cX_i$. What is the effect of scaling on the regression coefficients a and b? It is easy to see that

$$\sum V_i = \sum cX_i = c \sum X_i \qquad \text{so that } \overline{V} = c\overline{X}$$

and

$$\sum V_i^2 = \sum (cX_i)^2 = \sum c^2 X_i^2 = c^2 \sum X_i^2$$

and

$$\sum V_i Y_i = \sum (cX_i)(Y_i) = c \sum X_i Y_i .$$

Thus,

$$S_{VY} = \sum V_i Y_i - \frac{(\sum V_i)(\sum Y_i)}{n} = c \sum X_i Y_i - \frac{(c \sum X_i)(\sum Y_i)}{n}$$

$$= cS_{XY}$$

and

$$S_{VV} = \sum V_i^2 - \frac{(\sum V_i)^2}{n} = c^2 \sum X_i - \frac{(c \sum X_i)^2}{n}$$

$$= c^2 S_{XX}$$

so that

$$\frac{S_{VY}}{S_{VV}} = \frac{cS_{XY}}{c^2 S_{XX}} = \frac{1}{c} b .$$

Also

$$\overline{Y} - \left(\frac{1}{c} b\right) \overline{V} = \overline{Y} - \left(\frac{1}{c} b\right) (c\overline{X}) = \overline{Y} - b\overline{X} = a .$$

In this case, we see that scaling—multiplying each X-value by a constant—does not affect the intercept but does affect the slope of the regression equation.

Although you may not have noticed, we have been working with scaled values all along in the house size-price example. Note that X represents thousands of square feet, and Y, thousands of dollars. Why deal with the number 3,600 (square feet) when 3.6 (thousand square feet) will do as well? This becomes very important when one has only a hand or desk calculator (or sometimes even a computer).

What if we had left our measurements of house sizes as 1,800, 1,000, . . . , 1,500? This is equivalent to expressing values of the independent variable as

$$V_i = 1000X_i .$$

Then

$$\sum V_i = 40{,}000 = 1000 \sum X_i = 1000(40)$$

and $$\sum V_i^2 = 93{,}560{,}000 = (1000)^2 \sum X_i^2 = 1{,}000{,}000(93.56).$$

Many small calculators cannot handle such large numbers, and besides, who wants to drag all those zeros along?

Continuing, however, when fitting a regression of Y on V, we would have

$$S_{VY} = 1000 S_{XY} = 1000(174.9) = 174{,}900,$$

$$S_{VV} = (1000)^2 S_{XX} = 1{,}000{,}000(13.56) = 13{,}560{,}000,$$

and $$\frac{S_{VY}}{S_{VV}} = \frac{174{,}900}{13{,}560{,}000} = 0.0129 = \frac{1}{1000}(12.9) = \frac{1}{1000} b.$$

However,

$$\bar{Y} - \frac{b}{1000}\bar{V} = 34.5 - 0.0129\left(\frac{40{,}000}{20}\right)$$
$$= 34.5 - 0.0129(2000) = 8.7 = a.$$

That the intercept is unaffected by a change of scale makes sense: The intercept is the value of Y when the independent variable is zero. For V to be zero means that X must be zero, since $V = 1000X$. The origin has not been affected by change of scale; there has been no shift of the vertical axis as there was when X was centered.

It also makes sense that the slope *is* affected by scaling. If $B = 12.9$ means that for a 1,000 sq ft increase in size, price increases 12.9 thousand dollars, then for each 1 ft increase in size, price must increase $12.9 \div 1000 = 0.129$ thousand dollars.

Scaling is often performed for convenience. When the predictor variable is scaled by a constant c, the simple linear regression model may be written as

$$Y = \alpha + \beta' V + \varepsilon,$$

or $$Y = \alpha + \beta'(cX) + \varepsilon.$$

STANDARDIZING X-VALUES

It is possible to employ both a change of location (centering) and a change of scale on the independent variable in one operation. It should be clear to the reader that such a transformation will affect the numerical values of both the intercept and the slope of the regression equation.

While X may be transformed by adding any constant and multiplying by any constant, the most common transformation involving both change of location and scale is a **standardization**. In a standardization the variable is first centered, and then the centered values are divided by the standard deviation of the original X-values. (Recall that the standard deviation is the positive

Standardization

square root of the variance.) It is common to denote a standardized variable by the letter Z:

$$Z_i = \frac{X_i - \bar{X}}{s_X}.$$

Note that Z_i measures how many standard deviations the value X_i is above or below the mean of the set.

Standardized values are common in fields such as psychology and educational testing, as they make it possible to compare, for example, scores of a student on two tests when both the mean scores and standard deviations of scores of all students taking the tests are different on the two tests. Additionally, a convenient feature of standard scores is that they have mean zero and standard deviation 1. To see this:

$$\sum Z_i = \sum \frac{X_i - \bar{X}}{s_X} = \frac{1}{s_X}\sum (X_i - \bar{X}) = \frac{1}{s_X}(0) = 0$$

and

$$\sum Z_i^2 = \sum \left(\frac{X_i - \bar{X}}{s_X}\right)^2 = \frac{1}{s_X^2}\sum (X_i - \bar{X})^2 = \frac{S_{XX}}{s_X^2}.$$

But recall that

$$s_X^2 = \frac{S_{XX}}{n - 1},$$

so that

$$\sum Z_i^2 = \frac{S_{XX}}{S_{XX}/(n - 1)} = n - 1.$$

Thus,

$$s_Z^2 = \frac{S_{ZZ}}{n - 1} = \frac{1}{n - 1}\left[\sum Z_i^2 - \frac{(\sum Z_i)^2}{n}\right] = \frac{n - 1}{n - 1} = 1$$

and

$$s_Z = \sqrt{1} = 1.$$

What is the effect of using standardized values on the slope and intercept of the regression equation? For the slope, note that

$$S_{ZY} = \sum Z_i Y_i - \frac{(\sum Z_i)(\sum Y_i)}{n} = \sum Z_i Y_i$$

$$= \sum \left(\frac{X_i - \bar{X}}{s_X}\right) Y_i = \frac{1}{s_X}\sum (X_i - \bar{X})(Y_i) = \frac{1}{s_X} = S_{XY}$$

and

$$S_{ZY} = \sum Z_i Y_i - \frac{(\sum Z_i)(\sum Y_i)}{n} = \sum Z_i Y_i$$

Thus,

$$\frac{S_{ZY}}{S_{ZZ}} = \frac{S_{XY}/s_X}{S_{XX}/s_X^2} = \frac{b}{s_X^2}$$

and

$$\overline{Y} - \left(\frac{1}{s_X} b\right) \overline{Z} = \overline{Y} = a,$$

and we see that both the intercept and the slope are affected by standardizing X-values.

USEFULNESS OF TRANSFORMATIONS

Transformations on values of X are employed primarily for convenience. In the house size-price example we saw that it was much more convenient to express sizes of houses in thousands of square feet than in square feet. This made the arithmetic considerably less laborious.

However, the most important convenience achieved by transforming X is not necessarily computational, especially since most large regression problems are now analyzed using a computer. It is in development of the theoretical basis of regression, especially when more complicated models are considered, that centering and standardization can considerably reduce the amount of work required and the number of details to be considered. How this is achieved will be apparent later in this text.

EXERCISES

2.18 Refer to the Concession Data in Exercise 1.5.

a. Calculate values of $U_i = X_i - \overline{X}$.

b. Find the regression equation $\hat{Y} = a' + bU$ by solving normal equations.

c. Find S_{UY}, S_{UU}, S_{YY}, and compare to values S_{XY}, S_{XX}, and S_{YY} from Exercise 2.14.

d. Using shortcut formulas, find the regression equation $\hat{Y} = a' + bU$. Compare to (b).

e. Why is a' different from the intercept obtained in Exercise 2.8 or Exercise 2.14 for these same data?

f. Substituting $U = X - \overline{X}$ into $\hat{Y} = a' + bU$, show the regression equation $\hat{Y} = a + bX$.

2.19 Use the Concession Data in Exercise 1.5.

a. Calculate s_X, the standard deviation of the values of X.

b. Calculate values of a variable $V = X/s_X$.

c. Calculate S_{VY} and S_{VV}, and compare to S_{XY} and S_{XX} from Exercise 2.14.

d. Find the regression equation relating Y to V.

e. Compare the slope and intercept of this regression equation to the slope and intercept of the regression equation relating Y to X (Exercise 2.8 or Exercise 2.14).

f. Find $\hat{Y} = a + bX$ by substituting $V = X/s_X$ into $\hat{Y} = a' + b'V$.

2.20 An experiment was run in order to determine the relationship between the number of pounds of fertilizer applied per acre and the number of bushels yield for a certain crop. On three randomly selected one-acre plots, 15 lb of fertilizer was applied; each of another three plots received 20 lb; and 25 lb was applied to each of three other acres. The results are given in Table 2.10.

TABLE 2.10
Fertilizer Data

X Fertilizer (lb)	Y Yield (bu)
15	45
15	43
15	46
20	58
20	62
20	64
25	70
25	78
25	76

a. Center the X-values using the following scheme.

Let μ = average number of pounds of fertilizer applied to all nine acres,
d = spacing between levels of fertilizer application.

Then $$Z = \frac{X - \mu}{d}.$$

b. Plot the data, in terms of Z and Y, on a scatter diagram.

c. Find the regression equation

$$\hat{Y} = a' + b'\left(\frac{X - \mu}{d}\right) = a' + b'Z$$

and plot it on the scatter diagram.

d. What is the estimated average yield if 15 lb of fertilizer are applied?

e. Express the regression equation in terms of X and Y by substituting the values of μ and d into

$$\hat{Y} = a' + b'\left(\frac{X - \mu}{d}\right).$$

State the equation $\hat{Y} = a + bX$.

f. Using the equation $\hat{Y} = a + bX$, estimate the average yield if 15 lb of fertilizer are applied. Does your estimate agree with the one obtained in (d)?

g. Can the regression equations obtained in this problem be used to predict the average yield for acres to which 17 lb of fertilizer are applied? Explain.

2.21 An experiment was performed to study the relationship between the number of hours food deprivation for laboratory rats and the time required to learn a simple task for a food reward. Five different degrees of food deprivation were selected for study: slight, moderately slight, moderate, moderately great, and great. For convenience, slight deprivation is denoted by -2, moderately slight by $-1, \ldots,$ great by $+2$. Results of the study are given in Table 2.11.

TABLE 2.11
Deprivation Data

Degree of Deprivation	Time to Learn Task (minutes)
−2	73
−2	68
−1	49
−1	54
0	32
0	25
1	42
1	40
2	60
2	65

a. If the average duration of food deprivation was 72 hours, and each successive group of rats was deprived 24 hours longer than the previous group, to how many hours deprivation do the values $-2, -1, 0, 1, 2$ correspond? Use the formula

$$Z = \frac{X - \mu}{d}.$$

b. Plot the data, in terms of Z (centered X-values) and Y, on a scatter diagram. Does it appear that time to learn task is a linear function of hours deprivation for these levels of deprivation?

c. Fit a curvilinear equation of the form $\hat{Y} = a' + b'Z + c'Z^2$ to the data.

d. Plot the equation obtained in (c) on the scatter diagram.

e. Is it always true that if $\sum Z_i = 0$, then $\sum Z_i^3 = 0$

(1) If the Z-values are evenly spaced?

(2) If the Z-values are not evenly spaced?

2.22 For the house size-price example, calculate s_X, the standard deviation of the sizes of the houses. Standardize each size by $Z = (X - \overline{X})/s_X$. Using Z and Y, fit a simple linear regression equation. Show how the intercept and slope of this equation are related to $a = 8.7036$ and $b = 12.8982$. Show that $Y = a' + b'Z = a + bX$.

2 . 6

S U M M A R Y

We can now find a simple linear regression equation either by solving normal equations or by using shortcut formulas. If we are willing to solve simultaneous equations, even curvilinear regression or multiple regression models can be fitted to data, because normal equations can be written to correspond to any regression equation. When more than two equations in two unknowns are involved, however, it is not easy to solve simultaneous equations. In the next chapter we shall look at the regression problem in terms of matrices with the view to obtaining a general solution for all regression problems.

THREE

THE MATRIX APPROACH

3.1

A BRIEF INTRODUCTION TO MATRICES

Matrices do not save any great amount of time or labor in simple regression problems. At first, they may even appear to be more trouble than they are worth. But for more complicated problems, such as polynomial or multiple regression, our work will be greatly reduced if we can use matrices to formulate the regression problem. For one thing, matrix notation provides a shorthand that lets us reduce many cumbersome formulas down to just one. Furthermore, once one understands how the various elements of matrices combine and why, one can get a much better feeling for some of the fascinating things that happen in multiple or polynomial regressions. Finally, anyone who anticipates reading statistical regression literature in almost any field will not get past the first sentence without an understanding of the matrix form of the regression model.

We begin the matrix approach in conjunction with simple regression. We then show how the approach extends to *any* linear regression problem, regardless of the form of the model.

3 . 2

SOME DEFINITIONS

MATRICES, THEIR ELEMENTS AND DIMENSIONS

Matrix A **matrix** is merely a rectangular array of numbers, typically enclosed by square or round brackets. If $a, b, \ldots, f$ are numbers, then a matrix might look like

$$\begin{bmatrix} a & b & c \\ d & e & f \end{bmatrix}.$$

For example,

$$\begin{bmatrix} 1 & 2 & 3 \\ 7 & 6 & 5 \\ 9 & 1 & 4 \end{bmatrix}, \quad \begin{bmatrix} 1 & 7 & 3 & 9 \\ 1 & 4 & -2 & 7 \\ 1 & 0 & 4 & -3 \end{bmatrix}, \quad \text{and} \quad \begin{pmatrix} 1 & 0 \\ 2 & 7 \\ 3 & 9 \\ 4 & 3 \end{pmatrix}$$

are matrices. A matrix is usually denoted by a boldface capital letter, such as

$$\mathbf{A} = \begin{bmatrix} a & b & c \\ d & e & f \end{bmatrix}.$$

Elements **Dimension** The numbers in a matrix, called **elements,** are denoted by lowercase letters such as a, b, and so on. The **dimension** of a matrix tells the number of rows, r, and the number of columns, c, in a given matrix. The number of rows always comes first. Thus, the matrix **A** above is a 2×3 (two-by-three) matrix, since it has two rows and three columns of numbers. If $r = c$, as in the 3×3 matrix

$$\mathbf{B} = \begin{bmatrix} b_{11} & b_{12} & b_{13} \\ b_{21} & b_{22} & b_{23} \\ b_{31} & b_{32} & b_{33} \end{bmatrix},$$

Square Matrix then the matrix is a **square matrix.** A square matrix with n rows and n columns is often referred to as an *n-square* matrix. Note that the elements of **B** are denoted by the symbols b_{ij}, where i indicates the row and j indicates the column. Thus, the element b_{23} is found in the second row and third column of the matrix, while b_{32} is found in the third row and second column.

In the matrix

$$\mathbf{C} = \begin{bmatrix} 7 & 4 & -3 & 1 \\ 2.5 & -7 & 9.3 & 26 \\ -4 & 8 & 0 & 5 \end{bmatrix},$$

$c_{13} = -3$, $c_{23} = 9.3$, $c_{32} = 8$, $c_{34} = 5$, and so on. Similarly, we could refer to the (1, 3) element of **C** or the (2, 3) element or any other in this manner.

Vector A matrix that has only one row *or* only one column is called a **vector** and is denoted by a boldface lowercase letter. For example,

$$\mathbf{v}_1 = [a \quad b \quad c \quad d]$$

Row Vector has four columns but only one row, so $\mathbf{v}_1$ is called a 1×4 **row vector.** Similarly,

$$\mathbf{v}_2 = \begin{bmatrix} a \\ b \\ c \end{bmatrix}$$

Column Vector is a 3×1 **column vector,** since it has three rows but only one column. By convention, a vector is usually written as a column vector unless otherwise specified.

Scalar A matrix having only one row *and* only one column is called a **scalar.** An example of a scalar is

$$\underset{1\times 1}{[3]} = 3.$$

Note that sometimes the dimension of a matrix is listed below the lower right-hand corner, for emphasis.

EQUALITY OF MATRICES

Equality Two matrices are said to be **equal** (1) if they are of the same dimension, and (2) if they contain the same elements in the same positions. For example, if

$$\mathbf{A} = \begin{bmatrix} a & b & c \\ d & e & f \end{bmatrix} \quad \text{and} \quad \mathbf{B} = \begin{bmatrix} u & v & w \\ x & y & z \end{bmatrix},$$

A = **B** if and only if $a = u$, $b = v$, $c = w$, $d = x$, $e = y$, and $f = z$. The two matrices

$$\begin{bmatrix} 2 & 2 & 3 \\ 6 & 4 & 6 \end{bmatrix} \quad \text{and} \quad \begin{bmatrix} 2 & 2 & 4 \\ 6 & 4 & 6 \end{bmatrix}$$

are not equal because the (1, 3) element of the first is 3 but the (1, 3) element of the second is 4. Also,

$$\begin{bmatrix} 2 & 2 & 3 \\ 6 & 4 & 6 \end{bmatrix} \neq \begin{bmatrix} 2 & 2 \\ 6 & 4 \end{bmatrix},$$

because these two matrices are not of the same dimension: one has dimension 2×3 and the other 2×2.

TRANSPOSES

Transpose

The **transpose** of an $r \times c$ matrix is a $c \times r$ matrix obtained by writing the rows as columns and the columns as rows, while retaining the orders of the elements in each row and column. For example, the transpose of the matrix **A** above, denoted **A′** or $\mathbf{A}^T$, is

$$\mathbf{A}' = \begin{bmatrix} a & d \\ b & e \\ c & f \end{bmatrix},$$

a 3×2 matrix. Naturally, the transpose of a column vector is a row vector and vice versa, and the transpose of a scalar is just the scalar. Since vectors are conventionally written as column vectors, a symbol **v** will denote a column vector, and **v′** will denote a row vector.

SOME SPECIAL SQUARE MATRICES

Diagonal

Diagonal Elements

Off-Diagonal Elements

There are some special definitions that apply only to square matrices. The (principal) **diagonal** of an n-square matrix is the set of elements, in order, from the (1, 1) element to the (2, 2) element to the (3, 3) element, . . . , to the (n, n) element. For example, in **S,** the 4×4 matrix below, the diagonal is indicated by the dashed line. Elements on the diagonal are called **diagonal elements.** Those that do not lie on the diagonal are called **off-diagonal elements.**

$$\mathbf{S} = \begin{bmatrix} a & b & c & d \\ e & f & g & h \\ i & j & k & l \\ m & n & o & p \end{bmatrix}.$$

Diagonal Matrix

A matrix whose off-diagonal elements are zero, but whose diagonal elements are not all zero, is called a **diagonal matrix.** For example,

$$\mathbf{D} = \begin{bmatrix} d_1 & 0 & 0 & 0 \\ 0 & d_2 & 0 & 0 \\ 0 & 0 & d_3 & 0 \\ 0 & 0 & 0 & d_4 \end{bmatrix} = \text{diag}\,(d_1, d_2, d_3, d_4)$$

Zero Matrix

is a 4×4 diagonal matrix. (If the diagonal elements were all also zero, so that every element in the matrix is a zero, then we have a **zero matrix,** denoted by **0**.† A zero matrix need not be square, however.) A special diagonal matrix is one in which all the diagonal elements are 1's. The matrix

$$\mathbf{I} = \begin{bmatrix} 1 & 0 & \cdots & 0 \\ 0 & 1 & \cdots & 0 \\ \vdots & \vdots & & \vdots \\ 0 & 0 & \cdots & 1 \end{bmatrix}$$

Identity Matrix is called the **identity matrix,** and we will see presently that it is the multiplicative identity. An $n \times n$ identity matrix is sometimes denoted $\mathbf{I}_n$.

Symmetric Matrix A **symmetric matrix** is a square matrix whose (i, j) and (j, i) elements are equal. That is, if the matrix is folded in half along the principal diagonal, then the coincident elements will be equal. For example, the matrix **C** is symmetric:

$$\mathbf{C} = \begin{bmatrix} a & d & e \\ d & b & f \\ e & f & c \end{bmatrix}.$$

In **C**, the (1, 2) element is d and the (2, 1) element is also d; the (1, 3) element and the (3, 1) element are both e, and the (2, 3) and (3, 2) elements are both f. Note that if **C** is a symmetric matrix, then $\mathbf{C} = \mathbf{C}'$.

EXERCISES

3.1 Can any of the elements of a matrix be

a. Negative numbers? **b.** Imaginary numbers?
c. Zeros? **d.** Fractions?

3.2 Give the dimensions of the following matrices:

a. $\begin{bmatrix} 2 & 2 & 3 & 6 \\ 4 & 6 & 5 & 1 \\ 1 & 6 & 9 & 0 \end{bmatrix}$; **b.** $\begin{bmatrix} 6 & 3 & 6 \\ 5 & 5 & 2 \\ 3 & 0 & 5 \\ 5 & 1 & 2 \end{bmatrix}$;

c. $\begin{bmatrix} 9 & 7 & 4 \\ 2 & 5 & 5 \\ 6 & 10 & 4 \end{bmatrix}$; **d.** $\begin{bmatrix} a_{11} & a_{12} & \cdots & a_{19} \\ a_{21} & a_{22} & \cdots & a_{29} \\ \vdots & \vdots & & \vdots \\ a_{61} & a_{62} & \cdots & a_{69} \end{bmatrix}$;

e. diag $(a_1, a_2, \ldots, a_{12})$; **f.** $[a_1, a_2, \ldots, a_{12}]$;

g. $\begin{bmatrix} a \\ b \\ c \\ d \end{bmatrix}$; **h.** $[e]$.

3.3 In what row and column are the following elements found?

a. e_{47}; **b.** e_{74}; **c.** e_{77}; **d.** e_{ij}.

† Sometimes the zero matrix is denoted by $\boldsymbol{\theta}$ or by $\boldsymbol{\emptyset}$.

3.4 Which of the following pairs of matrices are equal?

a. $\begin{bmatrix} 9 & 7 & 4 & 2 \\ 5 & 5 & 6 & 1 \end{bmatrix}$ and $\begin{bmatrix} \frac{18}{2} & \frac{14}{2} & \frac{16}{4} & \frac{12}{6} \\ \frac{25}{5} & \frac{20}{4} & \frac{18}{3} & \frac{12}{12} \end{bmatrix}$;

b. $\begin{bmatrix} 9 & 7 & 4 & 2 \\ 5 & 5 & 6 & 1 \end{bmatrix}$ and $\begin{bmatrix} 9 & 5 \\ 7 & 5 \\ 4 & 6 \\ 2 & 1 \end{bmatrix}$;

c. $\begin{bmatrix} 9 & 7 & 4 & 2 \\ 5 & 5 & 6 & 1 \end{bmatrix}$ and $\begin{bmatrix} \frac{18}{2} & \frac{25}{5} \\ \frac{14}{2} & \frac{20}{4} \\ \frac{16}{4} & \frac{18}{3} \\ \frac{12}{6} & \frac{12}{12} \end{bmatrix}$;

d. $\begin{bmatrix} 1 & 0 & 0 \\ 0 & 1 & 0 \\ 0 & 0 & 1 \end{bmatrix}$ and $\begin{bmatrix} 5 & 0 & 0 \\ 0 & 5 & 0 \\ 0 & 0 & 5 \end{bmatrix}$;

e. $\begin{bmatrix} 1 & 0 & 0 \\ 0 & 1 & 0 \\ 0 & 0 & 1 \end{bmatrix}$ and $\begin{bmatrix} 1 & 0 \\ 0 & 1 \end{bmatrix}$.

3.5 Give the transposes of the following matrices.

a. $[1 \quad 2 \quad 3 \quad 4 \quad 5]$;

b. $\begin{bmatrix} 5 \\ 4 \\ 3 \\ 2 \\ 1 \end{bmatrix}$;

c. $\begin{bmatrix} 6 & 9 & 0 & 1 \\ 2 & 4 & 6 & 0 \end{bmatrix}$;

d. $\begin{bmatrix} a & b \\ c & d \\ e & f \end{bmatrix}$;

e. $\begin{bmatrix} a_{11} & a_{12} & a_{13} \\ a_{21} & a_{22} & a_{23} \\ a_{31} & a_{32} & a_{33} \end{bmatrix}$;

f. $\begin{bmatrix} 1 & 2 \\ 2 & 3 \end{bmatrix}$;

g. $\begin{bmatrix} 9 & 8 & 7 \\ 8 & 6 & 5 \\ 7 & 5 & 4 \end{bmatrix}$.

3.6 Indicate the (principal) diagonal of the following:

a. $\begin{bmatrix} a_{11} & a_{12} & a_{13} & a_{14} \\ a_{21} & a_{22} & a_{23} & a_{24} \\ a_{31} & a_{32} & a_{33} & a_{34} \\ a_{41} & a_{42} & a_{43} & a_{44} \end{bmatrix}$;

b. $\begin{bmatrix} 9 & 8 & 7 \\ 8 & 6 & 5 \\ 7 & 5 & 4 \end{bmatrix}$;

c. $\begin{bmatrix} 1 & 0 \\ 0 & 1 \end{bmatrix}$;

d. $\begin{bmatrix} 1 & 0 & 0 & 0 \\ 0 & 1 & 0 & 0 \\ 0 & 0 & 2 & 0 \\ 0 & 0 & 0 & 4 \end{bmatrix}$;

e. $\begin{bmatrix} 1 & 0 & 0 & 0 \\ 0 & 1 & 0 & 0 \\ 0 & 0 & 2 & 0 \end{bmatrix}$.

3.7 Write out all the elements of the following matrices:

a. diag(1, 1, 2, 3, 5);
b. diag(7, 0, 3, 4, 1, 0);
c. diag(1, 1, 1, 1);
d. diag(5, 5, 5);
e. diag($a_1, a_2, \ldots, a_n$).

3.8 Fill in the missing elements so as to make the matrices symmetric.

a. $\begin{bmatrix} 5 & 2 & 7 \\ & 1 & 1 \\ & & 2 \end{bmatrix}$;
b. $\begin{bmatrix} 9 & 9 & 2 & 9 \\ & 2 & 1 & 1 \\ & & 2 & 4 \\ & & & 0 \end{bmatrix}$;

c. $\begin{bmatrix} a & & \\ b & c & \\ d & e & f \end{bmatrix}$;
d. $\begin{bmatrix} a_{11} & a_{12} & a_{13} & a_{14} \\ & a_{22} & a_{23} & a_{24} \\ & & a_{33} & a_{34} \\ & & & a_{44} \end{bmatrix}$;

e. $\begin{bmatrix} 0 & & & \\ 0 & 6 & & \\ 9 & 2 & 4 & \\ 3 & 7 & 9 & 1 \end{bmatrix}$;
f. $\begin{bmatrix} 1 & 2 & 3 \\ & 4 & 5 \end{bmatrix}$.

3.3 MATRIX OPERATIONS

ADDITION

Matrix addition can be performed if two matrices have the same dimension. They are added by summing corresponding elements:

$$\begin{bmatrix} a & b & c \\ d & e & f \end{bmatrix} + \begin{bmatrix} g & h & i \\ j & k & l \end{bmatrix} = \begin{bmatrix} a+g & b+h & c+i \\ d+j & e+k & f+l \end{bmatrix}.$$

Similarly, one matrix can be subtracted from another by subtracting elements of one from the corresponding elements of the other:

$$\begin{bmatrix} 4 & -3 & 7 \\ -8 & 0 & 9 \end{bmatrix} - \begin{bmatrix} 7 & 4 & -3 \\ -1 & 7 & 0 \end{bmatrix} = \begin{bmatrix} 4-7 & -3-4 & 7-(-3) \\ -8-(-1) & 0-7 & 9-0 \end{bmatrix}$$

$$= \begin{bmatrix} -3 & -7 & 10 \\ -7 & -7 & 9 \end{bmatrix}.$$

SCALAR MULTIPLICATION

The product of a scalar and a matrix is a matrix containing the products of the scalar times each element. If s is a scalar, then

$$s\begin{bmatrix} a & d \\ b & e \\ c & f \end{bmatrix} = \begin{bmatrix} a & d \\ b & e \\ c & f \end{bmatrix} s = \begin{bmatrix} as & ds \\ bs & es \\ cs & fs \end{bmatrix} = \begin{bmatrix} sa & sd \\ sb & se \\ sc & sf \end{bmatrix}.$$

For example,

$$-2\begin{bmatrix} 4 & -3 & 7 \\ -8 & 0 & 9 \end{bmatrix} = \begin{bmatrix} -8 & 6 & -14 \\ 16 & 0 & -18 \end{bmatrix}.$$

The same rules apply, of course, to addition of vectors and multiplication of vectors by a scalar.

MATRIX MULTIPLICATION

Comformable Matrices

Multiplication of matrices is a bit more complicated. Two matrices can be multiplied together only if their dimensions are **conformable.** This means that the number of columns of the first matrix must equal the number of rows of the second. For example, if a matrix **E** is of dimension 2 × 4 and a matrix **F** is of dimension 4 × 3, then the product **EF** can be formed, because **E** has four columns and **F** has four rows. However, if **G** is 3 × 6 and **H** is 5 × 2, the product **GH** cannot be formed. If **J** is 3 × 6, the product **GJ** cannot be formed, either. We will see the reason for this shortly. A good way to check for conformable dimensions is to subscript the matrices with their dimensions and see that the two "inside" numbers are equal. For example,

$$\mathbf{E}_{2\times 4}\mathbf{F}_{4\times 3}$$
$$\uparrow\uparrow$$
"inside" numbers equal.

Notice that if **E** is 2 × 4 and **F** is 4 × 3, the product **EF** can be formed, but the product $\mathbf{F}_{4\times 3}\mathbf{E}_{2\times 4}$ cannot. That is to say, matrix multiplication is *not commutative*. Indeed, even when two square matrices (having dimensions that are conformable in either order) are multiplied, such as $\mathbf{S}_{4\times 4}\mathbf{T}_{4\times 4}$ or $\mathbf{T}_{4\times 4}\mathbf{S}_{4\times 4}$, the two products will usually be different, so that $\mathbf{S}_{4\times 4}\mathbf{T}_{4\times 4} \neq \mathbf{T}_{4\times 4}\mathbf{S}_{4\times 4}$.

When two conformable matrices are multiplied, the resulting product is a matrix with as many rows as the first matrix and as many columns as the second matrix. Thus, if a 2 × 4 matrix is multiplied on the right by a 4 × 3 matrix, the resulting matrix is a 2 × 3 matrix. For example, $\mathbf{E}_{2\times 4}\mathbf{F}_{4\times 3} = \mathbf{K}_{2\times 3}$.

Thus, the "outside" numbers of the subscripts of the matrices determine the dimension of the product matrix.

How is this matrix multiplication actually performed? First study the product below:

$$\underset{3\times 3}{\begin{bmatrix} a & b & c \\ d & e & f \\ g & h & i \end{bmatrix}} \underset{3\times 2}{\begin{bmatrix} j & m \\ k & n \\ l & o \end{bmatrix}} = \underset{3\times 2}{\begin{bmatrix} aj + bk + cl & am + bn + co \\ dj + ek + fl & dm + en + fo \\ gj + hk + il & gm + hn + io \end{bmatrix}}$$

To obtain the (1, 1) element of the product matrix, we work with the *first* row of the left matrix and the *first* column of the right matrix. Both have the same number of elements—in this case, three. We multiply the first elements aj, the second elements bk, and the third elements cl; then add up these products. Thus, the (1, 1) element of the product matrix is $aj + bk + cl$.

To obtain the (1, 2) element of the product matrix, we again use the first row of the left matrix, but we use the *second* column of the right matrix; again multiply corresponding elements, and sum. Thus the (1, 2) element of the product is $am + bn + co$.

For the (2, 1) element of the product, use the *second* row of the left matrix and the *first* column of the right, to get $dj + ek + fl$. For the (3, 2) element of the product, use the third row of the left matrix and the second column of the right, obtaining $gm + hn + io$, and so on.

A more sophisticated way to indicate the product of the previous two matrices is

$$\begin{bmatrix} a_{11} & a_{12} & a_{13} \\ a_{21} & a_{22} & a_{23} \\ a_{31} & a_{32} & a_{33} \end{bmatrix} \begin{bmatrix} b_{11} & b_{12} \\ b_{21} & b_{22} \\ b_{31} & b_{32} \end{bmatrix}$$

$$= \begin{bmatrix} a_{11}b_{11} + a_{12}b_{21} + a_{13}b_{31} & a_{11}b_{12} + a_{12}b_{22} + a_{13}b_{32} \\ a_{21}b_{11} + a_{22}b_{21} + a_{23}b_{31} & a_{21}b_{12} + a_{22}b_{22} + a_{23}b_{32} \\ a_{31}b_{11} + a_{32}b_{21} + a_{33}b_{31} & a_{31}b_{12} + a_{32}b_{22} + a_{33}b_{32} \end{bmatrix}$$

$$= \begin{bmatrix} \sum_{k=1}^{3} a_{1k}b_{k1} & \sum_{k=1}^{3} a_{1k}b_{k2} \\ \sum_{k=1}^{3} a_{2k}b_{k1} & \sum_{k=1}^{3} a_{2k}b_{k2} \\ \sum_{k=1}^{3} a_{3k}b_{k1} & \sum_{k=1}^{3} a_{3k}b_{k2} \end{bmatrix}$$

and this can easily be extended to the products of matrices with different dimensions.

As a numerical example, consider

$$\underset{3\times 3}{\begin{bmatrix} 2 & 2 & 3 \\ 6 & 4 & 6 \\ 5 & 1 & 1 \end{bmatrix}} \underset{3\times 3}{\begin{bmatrix} 6 & 9 & 0 \\ 3 & 7 & 4 \\ 6 & 1 & 1 \end{bmatrix}}$$

$$= \begin{bmatrix} 2(6)+2(3)+3(6) & 2(9)+2(7)+3(1) & 2(0)+2(4)+3(1) \\ 6(6)+4(3)+6(6) & 6(9)+4(7)+6(1) & 6(0)+4(4)+6(1) \\ 5(6)+1(3)+1(6) & 5(9)+1(7)+1(1) & 5(0)+1(4)+1(1) \end{bmatrix}_{3\times 3}$$

$$= \begin{bmatrix} 36 & 35 & 11 \\ 84 & 88 & 22 \\ 39 & 53 & 5 \end{bmatrix}_{3\times 3}.$$

To show that matrix multiplication is generally not commutative, let us form the product in reverse order.

$$\begin{bmatrix} 6 & 9 & 0 \\ 3 & 7 & 4 \\ 6 & 1 & 1 \end{bmatrix}_{3\times 3} \begin{bmatrix} 2 & 2 & 3 \\ 6 & 4 & 6 \\ 5 & 1 & 1 \end{bmatrix}_{3\times 3}$$

$$= \begin{bmatrix} 6(2)+9(6)+0(5) & 6(2)+9(4)+0(1) & 6(3)+9(6)+0(1) \\ 3(2)+7(6)+4(5) & 3(2)+7(4)+4(1) & 3(3)+7(6)+4(1) \\ 6(2)+1(6)+1(5) & 6(2)+1(4)+1(1) & 6(3)+1(6)+1(1) \end{bmatrix}_{3\times 3}$$

$$= \begin{bmatrix} 66 & 48 & 72 \\ 68 & 38 & 55 \\ 23 & 17 & 25 \end{bmatrix}_{3\times 3} \neq \begin{bmatrix} 36 & 35 & 11 \\ 84 & 88 & 22 \\ 39 & 53 & 5 \end{bmatrix}_{3\times 3}$$

The same rules apply when multiplying vectors. An $r \times 1$ column vector times a $1 \times c$ row vector will yield an $r \times c$ matrix. But a $1 \times c$ row vector times an $r \times 1$ column vector will yield a (1×1) scalar (if $r = c$). For example,

$$\begin{bmatrix} 2 & 2 & 3 \end{bmatrix}_{1\times 3} \begin{bmatrix} 6 \\ 4 \\ 5 \end{bmatrix}_{3\times 1} = [12 + 8 + 15] = [35]_{1\times 1},$$

but

$$\begin{bmatrix} 6 \\ 4 \\ 5 \end{bmatrix}_{3\times 1} \begin{bmatrix} 2 & 2 & 3 \end{bmatrix}_{1\times 3} = \begin{bmatrix} 12 & 12 & 18 \\ 8 & 8 & 12 \\ 10 & 10 & 15 \end{bmatrix}_{3\times 3},$$

and again we see that matrix multiplication is not commutative.

THE MULTIPLICATIVE IDENTITY

Multiplicative Identity

We can now verify that $\mathbf{I} = \text{diag}(1, 1, \ldots, 1)$ is the **multiplicative identity;** that is, for any matrix $\mathbf{A}_{r\times c}$,

$$\mathbf{A}_{r\times c}\mathbf{I}_{c\times c} = \mathbf{I}_{r\times r}\mathbf{A}_{r\times c} = \mathbf{A}.$$

For example, let

$$\mathbf{A} = \underset{2\times 3}{\begin{bmatrix} a & b & c \\ d & e & f \end{bmatrix}}.$$

Then

$$\mathbf{A}_{2\times 3}\mathbf{I}_{3\times 3} = \begin{bmatrix} a & b & c \\ d & e & f \end{bmatrix} \begin{bmatrix} 1 & 0 & 0 \\ 0 & 1 & 0 \\ 0 & 0 & 1 \end{bmatrix}$$

$$= \begin{bmatrix} a(1) + b(0) + c(0) & a(0) + b(1) + c(0) & a(0) + b(0) + c(1) \\ d(1) + e(0) + f(0) & d(0) + e(1) + f(0) & d(0) + e(0) + f(1) \end{bmatrix}$$

$$= \begin{bmatrix} a & b & c \\ d & e & f \end{bmatrix} = \mathbf{A}_{2\times 3}.$$

Also,

$$\mathbf{I}_{2\times 2}\mathbf{A}_{2\times 3} = \begin{bmatrix} 1 & 0 \\ 0 & 1 \end{bmatrix} \begin{bmatrix} a & b & c \\ d & e & f \end{bmatrix}$$

$$= \begin{bmatrix} 1(a) + 0(d) & 1(b) + 0(e) & 1(c) + 0(f) \\ 0(a) + 1(d) & 0(b) + 1(e) & 0(c) + 1(f) \end{bmatrix}$$

$$= \begin{bmatrix} a & b & c \\ d & e & f \end{bmatrix} = \mathbf{A}_{2\times 3}.$$

THE PRODUCTS OF A MATRIX AND ITS TRANSPOSE

Note that for any matrix $\mathbf{M}_{r\times c}$, the products $\mathbf{M}'_{c\times r}\mathbf{M}_{r\times c}$ and $\mathbf{M}_{r\times c}\mathbf{M}'_{c\times r}$ can always be formed (although they will not be equal). Also, the products $\mathbf{M}'\mathbf{M}$ and $\mathbf{M}\mathbf{M}'$ will always be square and symmetric. To illustrate, let

$$\mathbf{M} = \underset{2\times 3}{\begin{bmatrix} 9 & 7 & 4 \\ 2 & 5 & 5 \end{bmatrix}}.$$

Then

$$\mathbf{M}'\mathbf{M} = \begin{bmatrix} 9 & 2 \\ 7 & 5 \\ 4 & 5 \end{bmatrix} \begin{bmatrix} 9 & 7 & 4 \\ 2 & 5 & 5 \end{bmatrix}$$

$$= \begin{bmatrix} 81 + 4 & 63 + 10 & 36 + 10 \\ 63 + 10 & 49 + 25 & 28 + 25 \\ 36 + 10 & 28 + 25 & 16 + 25 \end{bmatrix} = \begin{bmatrix} 85 & 73 & 46 \\ 73 & 74 & 53 \\ 46 & 53 & 41 \end{bmatrix},$$

which is square and symmetric. Then

$$\mathbf{MM}' = \begin{bmatrix} 9 & 7 & 4 \\ 2 & 5 & 5 \end{bmatrix} \begin{bmatrix} 9 & 2 \\ 7 & 5 \\ 4 & 5 \end{bmatrix}$$

$$= \begin{bmatrix} 81 + 49 + 16 & 18 + 35 + 20 \\ 18 + 35 + 20 & 4 + 25 + 25 \end{bmatrix}$$

$$= \begin{bmatrix} 146 & 73 \\ 73 & 54 \end{bmatrix},$$

which is also square and symmetric.

EXERCISES

3.9 Form the following sums:

a. $\begin{bmatrix} 2 & 4 \\ 9 & 7 \end{bmatrix} + \begin{bmatrix} 5 & 7 \\ -6 & 0 \end{bmatrix}$;

b. diag(3, 3, 1, 7) + diag(6, 7, 1, 4);

c. $\begin{bmatrix} 1 & -3 & 2 & 7 \\ 1 & 0 & -4 & 1 \\ 1 & 5 & 2 & -2 \\ -5 & 6 & 6 & 4 \end{bmatrix} + \begin{bmatrix} 1 & -5 & 2 & 2 \\ 7 & 5 & 6 & -3 \\ 1 & -8 & 7 & 8 \\ 5 & 8 & 3 & 5 \end{bmatrix}$;

d. $\begin{bmatrix} 1 & -3 & 2 & 7 \\ 1 & 0 & -4 & 1 \\ 1 & 5 & 2 & -2 \\ -5 & 6 & 6 & 4 \end{bmatrix} - \begin{bmatrix} 1 & -5 & 2 & 2 \\ 7 & 5 & 6 & -3 \\ 1 & -8 & 7 & 8 \\ 5 & 8 & 3 & 5 \end{bmatrix}$;

e. $\begin{bmatrix} 1 & 0 & 5 \\ 4 & 0 & 9 \\ 2 & 1 & 1 \end{bmatrix} + \begin{bmatrix} 1 & 0 \\ 5 & 9 \\ 9 & 1 \end{bmatrix}$;

f. $\begin{bmatrix} 1 & 0 \\ 5 & 9 \\ 9 & 1 \end{bmatrix} + \begin{bmatrix} 1 & 5 & 9 \\ 0 & 9 & 1 \end{bmatrix}$;

g. $\begin{bmatrix} 2 \\ 6 \\ 1 \\ 7 \\ 2 \end{bmatrix} + \begin{bmatrix} 5 \\ 1 \\ 1 \\ 5 \\ 8 \end{bmatrix}$;

h. $\begin{bmatrix} 2 \\ 6 \\ 1 \\ 7 \\ 2 \end{bmatrix} + [5 \quad 1 \quad 1 \quad 5 \quad 8]$;

i. [2 6 1 7 2] − [2 0 7 7 4];

j. 4 + [2 6 1 7 2].

3.10 Form the following products:

a. $4\begin{bmatrix} 4 & -7 & 4 \\ 5 & 8 & 1 \\ -5 & 8 & 2 \end{bmatrix}$;

b. $\frac{1}{4}\begin{bmatrix} 4 & -7 & 4 \\ 5 & 8 & 1 \\ -5 & 8 & 2 \end{bmatrix}$;

c. $-3\begin{bmatrix} 6 & 5 \\ 7 & 4 \\ 5 & 2 \end{bmatrix}$;

d. $7[8 \quad 4 \quad 9 \quad 2 \quad 3]$;

e. $-5\begin{bmatrix} 7 \\ 2 \\ 1 \\ -2 \\ 0 \\ 7 \end{bmatrix}$.

3.11 Which of the following pairs of matrices are conformable for multiplication in the order given? For conformable pairs, state dimension of the product.

a. $\begin{bmatrix} 8 & 4 & 9 \\ 2 & 3 & 9 \end{bmatrix}\begin{bmatrix} 6 & 7 \\ 5 & 3 \\ 7 & 8 \end{bmatrix}$;

b. $\begin{bmatrix} 6 & 7 \\ 5 & 3 \\ 7 & 8 \end{bmatrix}\begin{bmatrix} 8 & 4 & 9 \\ 2 & 3 & 9 \end{bmatrix}$;

c. $\begin{bmatrix} 3 & 6 & 8 \\ 4 & 0 & 3 \\ 9 & 1 & 1 \end{bmatrix}\begin{bmatrix} 2 & 2 & 2 \\ 2 & 1 & 4 \\ 3 & 9 & 5 \end{bmatrix}$;

d. $\begin{bmatrix} 2 & 2 & 2 \\ 2 & 1 & 4 \\ 3 & 9 & 5 \end{bmatrix}\begin{bmatrix} 3 & 6 & 8 \\ 4 & 0 & 3 \\ 9 & 1 & 1 \end{bmatrix}$;

e. $[1 \quad 2 \quad 1 \quad 1]\,[0 \quad 6 \quad 2 \quad 3]$;

f. $[1 \quad 2 \quad 1 \quad 1]\begin{bmatrix} 0 \\ 6 \\ 2 \\ 3 \end{bmatrix}$;

g. $\begin{bmatrix} 1 \\ 2 \\ 1 \\ 1 \end{bmatrix}[0 \quad 6 \quad 2 \quad 3]$;

h. $\begin{bmatrix} 1 \\ 2 \\ 1 \\ 1 \end{bmatrix}\begin{bmatrix} 0 \\ 6 \\ 2 \\ 3 \end{bmatrix}$;

i. $[4]\begin{bmatrix} 1 \\ 2 \\ 1 \end{bmatrix}$.

3.12 Form the indicated products for the conformable pairs you found in Exercise 3.11.

3.13 Referring to Exercise 3.12, find examples to show that, in general, $\mathbf{AB} \neq \mathbf{BA}$ if

a. Either **A** or **B** or both are not square matrices;
b. Both **A** and **B** are square matrices.

3.14 Show the following for the matrix

$$\mathbf{S} = \begin{bmatrix} 1 & 1 & 1 & 1 \\ 1 & 6 & 8 & 3 \\ 1 & 2 & 7 & 4 \end{bmatrix}.$$

a. If $\mathbf{S}'$ is the transpose of $\mathbf{S}$, then the products $\mathbf{SS}'$ and $\mathbf{S}'\mathbf{S}$ can both be formed;
b. Both $\mathbf{SS}'$ and $\mathbf{S}'\mathbf{S}$ are symmetric;
c. $\mathbf{SS}' \neq \mathbf{S}'\mathbf{S}$.

3.15 Let $\mathbf{I}_n$ denote the n-square identity matrix. For each of the following matrices $\mathbf{M}$, show that $\mathbf{I}$ is the multiplicative identity by forming the products $\mathbf{I}_r\mathbf{M} = \mathbf{M}\mathbf{I}_c = \mathbf{M}$, where r = number of rows in $\mathbf{M}$ and c = number of columns in $\mathbf{M}$.

For example, if

$$\mathbf{M} = \begin{bmatrix} 3 & 1 \\ 6 & 2 \\ 9 & 4 \\ 0 & 6 \end{bmatrix}_{4\times 2},$$

show that

$$\mathbf{I}_4\mathbf{M}_{4\times 2} = \begin{bmatrix} 1 & 0 & 0 & 0 \\ 0 & 1 & 0 & 0 \\ 0 & 0 & 1 & 0 \\ 0 & 0 & 0 & 1 \end{bmatrix}\begin{bmatrix} 3 & 1 \\ 6 & 2 \\ 9 & 4 \\ 0 & 6 \end{bmatrix} = \begin{bmatrix} 3 & 1 \\ 6 & 2 \\ 9 & 4 \\ 0 & 6 \end{bmatrix}$$

and that

$$\mathbf{M}_{4\times 2}\mathbf{I}_2 = \begin{bmatrix} 3 & 1 \\ 6 & 2 \\ 9 & 4 \\ 0 & 6 \end{bmatrix}\begin{bmatrix} 1 & 0 \\ 0 & 1 \end{bmatrix} = \begin{bmatrix} 3 & 1 \\ 6 & 2 \\ 9 & 4 \\ 0 & 6 \end{bmatrix}.$$

a. $\begin{bmatrix} 2 & 1 \\ 1 & 9 \end{bmatrix}$;

b. $\begin{bmatrix} 1 & 2 & 7 \\ 1 & 4 & 5 \end{bmatrix}$;

c. $\begin{bmatrix} 6 & 3 & 5 \\ 7 & 2 & 9 \end{bmatrix}$;

d. $\begin{bmatrix} 5 & 1 \\ 7 & 5 \\ 2 & 6 \end{bmatrix}$;

e. $\begin{bmatrix} a_{11} & a_{12} & a_{13} & a_{14} \\ a_{21} & a_{22} & a_{23} & a_{24} \\ a_{31} & a_{32} & a_{33} & a_{34} \end{bmatrix}$;

f. $[a \quad b \quad c]$;

g. $[a \quad b \quad c \quad d]$.

3 . 4

DETERMINANTS

Determinant A useful feature of a square matrix is its **determinant.** A determinant is a number (scalar) that can represent a matrix. It is determined by all elements of the matrix. For a matrix $\mathbf{A}$, the determinant is denoted $|\mathbf{A}|$. Straight vertical bars around an array of numbers also indicate the determinant of the array. The determinant of a 1×1 matrix, or scalar, is just the scalar itself. For example,

$$|u| = u.$$

DETERMINANT OF A 2 × 2 MATRIX

For a 2 × 2 matrix, it is also quite easy to calculate the determinant. If

$$\mathbf{A} = \begin{bmatrix} a & b \\ c & d \end{bmatrix},$$

then

$$|\mathbf{A}| = \begin{vmatrix} a & b \\ c & d \end{vmatrix} = ad - bc.$$

For example, if

$$\mathbf{A} = \begin{bmatrix} 2 & 2 \\ 3 & 6 \end{bmatrix},$$

then

$$|\mathbf{A}| = \begin{vmatrix} 2 & 2 \\ 3 & 6 \end{vmatrix} = 12 - 6 = 6.$$

SINGULARITY

Nonsingular Matrix
Singular Matrix

If for any square matrix **A** the determinant is nonzero, then **A** is a **nonsingular matrix.** If $|\mathbf{A}| = 0$, then **A** is called a **singular matrix.** If, for example,

$$\mathbf{A} = \begin{bmatrix} 1 & 2 \\ 2 & 4 \end{bmatrix},$$

then

$$|\mathbf{A}| = \begin{vmatrix} 1 & 2 \\ 2 & 4 \end{vmatrix} = 4 - 4 = 0,$$

and **A** is singular. Note that in the above matrix any row or column is a multiple of some other row or column.

Linear Combination

In general, if any row(s) or column(s) in a matrix is a **linear combination** of any other row(s) or column(s), the matrix is singular. If one row is a linear combination of the other rows, then that row will be obtained by adding multiples of other rows together. Arguing on columns, suppose that $\mathbf{v}_1$, $\mathbf{v}_2$, . . . , $\mathbf{v}_c$ are the c columns of a matrix (note that they are column vectors) and that $k_1, k_2, \ldots, k_c$ is a set of c constants (scalars). Pick one of the columns, say $\mathbf{v}_j$. If

$$\mathbf{v}_j = k_1\mathbf{v}_1 + k_2\mathbf{v}_2 + \cdots + k_{j-1}\mathbf{v}_{j-1} + k_{j+1}\mathbf{v}_{j+1} + \cdots + k_c\mathbf{v}_c,$$

then $\mathbf{v}_j$ is a linear combination of the remaining $c - 1$ columns: $\mathbf{v}_j$ can be obtained by adding together multiples of the other columns.

In the matrix

$$\mathbf{A} = \begin{bmatrix} 1 & 2 \\ 2 & 4 \end{bmatrix}$$

we can see that

$$\begin{bmatrix} 1 \\ 2 \end{bmatrix} = \frac{1}{2}\begin{bmatrix} 2 \\ 4 \end{bmatrix}$$

or that $[2 \quad 4] = 2[1 \quad 2]$. Suppose we start with two columns,

$$\mathbf{c}_1 = \begin{bmatrix} 1 \\ 3 \\ 7 \end{bmatrix} \quad \text{and} \quad \mathbf{c}_2 = \begin{bmatrix} 2 \\ 4 \\ 5 \end{bmatrix},$$

of a 3×3 matrix and combine them by

$$k_1\mathbf{c}_1 + k_2\mathbf{c}_2 = 2\mathbf{c}_1 + (-1)\mathbf{c}_2 = \mathbf{c}_3$$

$$2\begin{bmatrix} 1 \\ 3 \\ 7 \end{bmatrix} + (-1)\begin{bmatrix} 2 \\ 4 \\ 5 \end{bmatrix} = \begin{bmatrix} 2 \\ 6 \\ 14 \end{bmatrix} - \begin{bmatrix} 2 \\ 4 \\ 5 \end{bmatrix} = \begin{bmatrix} 0 \\ 2 \\ 9 \end{bmatrix}.$$

Then the matrix

$$\begin{bmatrix} 1 & 2 & 0 \\ 3 & 4 & 2 \\ 7 & 5 & 9 \end{bmatrix}$$

will be singular; its determinant will be zero, since the third column is a linear combination of the first two. This will have significance when we turn to the question of solving simultaneous equations, especially in the case of multiple regression.

DETERMINANTS OF LARGER MATRICES

Finding a determinant becomes progressively more complicated as the dimension of the matrix becomes larger. We will illustrate the method of evaluating the determinant of a 3×3 matrix by **expansion by cofactors.** Let

Expansion by Cofactors

$$\mathbf{A} = \begin{bmatrix} a_{11} & a_{12} & a_{13} \\ a_{21} & a_{22} & a_{23} \\ a_{31} & a_{32} & a_{33} \end{bmatrix}.$$

Choose an element of **A,** say a_{11}. Strike out of **A** the row and column in which a_{11} appears; this results in a 2×2 matrix,

$$\begin{bmatrix} a_{22} & a_{23} \\ a_{32} & a_{33} \end{bmatrix}.$$

The determinant

$$\begin{vmatrix} a_{22} & a_{23} \\ a_{32} & a_{33} \end{vmatrix} = a_{22}a_{33} - a_{23}a_{32}$$

Minor of this 2 × 2 "submatrix" is called the **minor** of the element a_{11}. Other minors corresponding to other elements of **A** are

Element	Minor
a_{12}	$\begin{vmatrix} a_{21} & a_{23} \\ a_{31} & a_{33} \end{vmatrix} = a_{21}a_{33} - a_{23}a_{31}$
a_{23}	$\begin{vmatrix} a_{11} & a_{12} \\ a_{31} & a_{32} \end{vmatrix} = a_{11}a_{32} - a_{12}a_{31}$
a_{33}	$\begin{vmatrix} a_{11} & a_{12} \\ a_{21} & a_{22} \end{vmatrix} = a_{11}a_{22} - a_{12}a_{21}$

and so on.

Cofactor A **cofactor** is a minor that has been multiplied by either +1 or −1. In order to determine whether a given minor gets multiplied by plus or minus one, count how many horizontal or vertical steps would be taken in order to move the corresponding element to the (1, 1) position of the matrix. If zero or an even number of steps are required, the minor is multiplied by +1; if an odd number of steps must be taken, multiply the minor by −1.

For example, since a_{11} is already in the (1, 1) position, its minor,

$$\begin{vmatrix} a_{22} & a_{23} \\ a_{32} & a_{33} \end{vmatrix} = a_{22}a_{33} - a_{23}a_{32}$$

gets multiplied by +1 so that the cofactor corresponding to the element a_{11} is $+(a_{22}a_{33} - a_{23}a_{32})$. In order to move the element a_{12} to the (1, 1) position, one horizontal step to the left must be taken, so the cofactor of a_{12} is $-(a_{21}a_{33} - a_{23}a_{31})$. Moving a_{23} to the (1, 1) position takes two horizontal steps and one vertical step, so the cofactor of a_{23} is $-(a_{11}a_{32} - a_{12}a_{31})$. To move the element a_{33} to the (1, 1) position, two horizontal and two vertical steps must be taken so the minor is multiplied by +1, and so on.†

Label cofactors with subscripted capital letters corresponding to the elements with which they are associated. For example,

Element	Cofactor
a_{11}	$A_{11} = (a_{22}a_{33} - a_{23}a_{32})$
a_{12}	$A_{12} = -(a_{21}a_{33} - a_{23}a_{31})$
a_{23}	$A_{23} = -(a_{11}a_{32} - a_{12}a_{31})$
a_{33}	$A_{33} = (a_{11}a_{22} - a_{12}a_{21})$

and so on. Of course, there is a cofactor corresponding to each element of the matrix **A**.

† Equivalently, add the subscripts on the element whose minor has been found. If the sum is even, multiply the minor by +1; if the sum is odd, multiply the minor by −1. For example, the minor of a_{11} will be multiplied by +1, since 1 + 1 = 2 is even, and the minor of a_{23} will be multiplied by −1, since 2 + 3 = 5 is odd.

Once the cofactors are determined, the determinant of **A** can be found by picking any row or column of **A**, multiplying each element of that row or column by its corresponding cofactor, and then summing these products. For example, if we choose the first row, then

$$|\mathbf{A}| = a_{11}A_{11} + a_{12}A_{12} + a_{13}A_{13}.$$

If the third row is chosen, then

$$|\mathbf{A}| = a_{31}A_{31} + a_{32}A_{32} + a_{33}A_{33};$$

or if the second column is chosen,

$$|\mathbf{A}| = a_{12}A_{12} + a_{22}A_{22} + a_{32}A_{32}.$$

Regardless of which row or column is chosen, the numerical value of $|\mathbf{A}|$ will be the same.

Let us try a numerical example. Let

$$\mathbf{A} = \begin{bmatrix} 2 & 2 & 1 \\ 1 & 0 & 1 \\ 2 & 1 & 2 \end{bmatrix}.$$

Suppose we decide to determine $|\mathbf{A}|$ by expansion of cofactors across the first row:

Element	**Minor**	**Cofactor**
$a_{11} = 2$	$\begin{vmatrix} 0 & 1 \\ 1 & 2 \end{vmatrix} = 0 - 1 = -1$	$+(-1) = -1 = A_{11}$
$a_{12} = 2$	$\begin{vmatrix} 1 & 1 \\ 2 & 2 \end{vmatrix} = 2 - 2 = 0$	$-(0) = 0 = A_{12}$
$a_{13} = 1$	$\begin{vmatrix} 1 & 0 \\ 2 & 1 \end{vmatrix} = 1 - 0 = 1$	$+(1) = 1 = A_{13}$

Then

$$\begin{aligned} |\mathbf{A}| &= a_{11}A_{11} + a_{12}A_{12} + a_{13}A_{13} \\ &= 2(-1) + 2(0) + 1(1) \\ &= -2 + 0 + 1 \\ &= -1. \end{aligned}$$

Expanding down the first column gives us

Element	**Minor**	**Cofactor**
$a_{11} = 2$	$\begin{vmatrix} 0 & 1 \\ 1 & 2 \end{vmatrix} = 0 - 1 = -1$	$+(-1) = -1 = A_{11}$
$a_{21} = 1$	$\begin{vmatrix} 2 & 1 \\ 1 & 2 \end{vmatrix} = 4 - 1 = 3$	$-(3) = -3 = A_{21}$
$a_{31} = 2$	$\begin{vmatrix} 2 & 1 \\ 0 & 1 \end{vmatrix} = 2 - 0 = 2$	$+(2) = 2 = A_{31}$

Then

$$\begin{aligned}|\mathbf{A}| &= a_{11}A_{11} + a_{21}A_{21} + a_{31}A_{31} \\ &= 2(-1) + 1(-3) + 2(2) \\ &= -2 - 3 + 4 = -1,\end{aligned}$$

which is the same value for $|\mathbf{A}|$ that we obtained by expanding across the first row.

Let us now verify that the matrix

$$\begin{bmatrix} 1 & 2 & 0 \\ 3 & 4 & 2 \\ 7 & 5 & 9 \end{bmatrix},$$

in which the third column was obtained by multiplying the first column by 2 and then subtracting off the second column, has zero determinant. Expanding across the first row,

Element	Minor	Cofactor
$a_{11} = 1$	$\begin{vmatrix} 4 & 2 \\ 5 & 9 \end{vmatrix} = 36 - 10 = 26$	$+(26) = 26 = A_{11}$
$a_{12} = 2$	$\begin{vmatrix} 3 & 2 \\ 7 & 9 \end{vmatrix} = 27 - 14 = 13$	$-(13) = -13 = A_{12}$

(Note that it is not necessary to find A_{13}, since the product $a_{13}A_{13}$ will be zero.) Then

$$\begin{aligned}|\mathbf{A}| &= a_{11}A_{11} + a_{12}A_{12} + a_{13}A_{13} \\ &= 1(26) + 2(-13) + 0 = 26 - 26 = 0.\end{aligned}$$

What if we change some element in the third column of this matrix so that the third column becomes *almost,* but not exactly, a linear combination of the first two columns? What will this do to the value of the determinant? Suppose we change the (3, 3) element to 9.1, thus:

$$\begin{bmatrix} 1 & 2 & 0 \\ 3 & 4 & 2 \\ 7 & 5 & 9.1 \end{bmatrix}.$$

Then to find the determinant,

Element	Minor	Cofactor
$a_{11} = 1$	$\begin{vmatrix} 4 & 2 \\ 5 & 9.1 \end{vmatrix} = 36.4 - 10 = 26.4$	$A_{11} = +(26.4) = 26.4$
$a_{12} = 2$	$\begin{vmatrix} 3 & 2 \\ 7 & 9.1 \end{vmatrix} = 27.3 - 14 = 13.3$	$A_{12} = -(13.3) = -13.3$

and

$$\begin{aligned}|\mathbf{A}| &= a_{11}A_{11} + a_{12}A_{12} + a_{13}A_{13} \\ &= 1(26.4) + 2(-13.3) + 0 = 26.4 - 26.6 = -0.2.\end{aligned}$$

We see that the determinant is no longer zero, but that it is not very far away from zero, either. A consequence of this is discussed in the next section.

The method presented above extends directly to the problem of evaluating the determinant of a 4×4 or higher dimension matrix. The calculations required, however, become more and more involved. For example, if $\mathbf{A}$ is a 4×4 matrix, then the minor associated with any element is the determinant of a 3×3 matrix, and this minor itself must be evaluated by the cofactor expansion. Thus, to find $|\mathbf{A}_{4\times 4}|$, four 3×3 determinants must be evaluated, and each involves finding three 2×2 determinants. Thus, finding such determinants by hand is a tedious process that requires quite a bit of bookkeeping. Fortunately, there are computer programs that can evaluate determinants quite efficiently.

DETERMINANTS OF DIAGONAL MATRICES

Diagonal matrices are nice, because regardless of their dimension, their determinants are simply the products of their diagonal elements. Thus, if $\mathbf{D} = \text{diag}(d_1, d_2, \ldots, d_n)$, then $|\mathbf{D}| = d_1 d_2 \ldots d_n$. Let us use a 3×3 diagonal matrix to see how this works. Let

$$\mathbf{D} = \begin{bmatrix} 2 & 0 & 0 \\ 0 & 6 & 0 \\ 0 & 0 & 1 \end{bmatrix}.$$

Suppose we choose to expand across the first row. Then the only cofactor we need to find is A_{11}, since the products $a_{12}A_{12}$ and $a_{13}A_{13}$ will be zero.

$$A_{11} = \begin{vmatrix} 6 & 0 \\ 0 & 1 \end{vmatrix} = 6(1) - (0)(0) = 6(1).$$

Thus $|\mathbf{D}| = a_{11}A_{11} = 2(6)(1)$, the product of the diagonal elements. The exercises that follow will help you verify for yourself that this rule applies for diagonal matrices of other dimensions.

Our primary use for determinants will come in finding inverses of matrices, which we will look at in the next section.

EXERCISES

3.16 Find the determinants of the following matrices:

a. $\begin{bmatrix} 2 & 3 \\ 9 & 1 \end{bmatrix}$;

b. $\begin{bmatrix} 1 & 0 \\ -9 & 7 \end{bmatrix}$;

c. $\begin{bmatrix} a_{11} & a_{12} \\ a_{21} & a_{22} \end{bmatrix}$;

d. $\begin{bmatrix} \sum n & \sum X_i \\ \sum X_i & \sum X_i^2 \end{bmatrix}$;

e. $\begin{bmatrix} 1 & 2 \\ 3 & 4 \end{bmatrix}$; f. $\begin{bmatrix} 1 & 2 \\ 2 & 4 \end{bmatrix}$;

g. $\begin{bmatrix} 6 & 0 \\ 0 & 7 \end{bmatrix}$; h. $\begin{bmatrix} 1 & 5 & 3 \\ 0 & 0 & 8 \end{bmatrix}$;

i. [8]; j. diag(a_1, a_2).

3.17 Which of the matrices in Exercise 3.16 are
a. Singular? b. Nonsingular?

3.18 Find the determinants of the following matrices:

a. $\begin{bmatrix} 1 & 0 & 7 \\ 8 & 2 & 5 \\ 0 & 6 & 7 \end{bmatrix}$; b. $\begin{bmatrix} 1 & 0 & 0 \\ 1 & 5 & 4 \\ 9 & 2 & 4 \end{bmatrix}$;

c. $\begin{bmatrix} 5 & 7 & 0 \\ 7 & 3 & 7 \\ 0 & 7 & 3 \end{bmatrix}$; d. $\begin{bmatrix} 1 & 2 & 3 \\ 4 & 5 & 6 \\ 7 & 8 & 9 \end{bmatrix}$;

e. $\begin{bmatrix} 1 & 2 & 3 \\ 2 & 4 & 6 \\ 4 & 8 & 12 \end{bmatrix}$; f. $\begin{bmatrix} 4 & 5 \\ 7 & 2 \\ 3 & 5 \end{bmatrix}$;

g. $\begin{bmatrix} 9 & 1 & 5 \\ 6 & 2 & 1 \end{bmatrix}$; h. $\begin{bmatrix} 7 & 0 & 0 \\ 0 & 5 & 0 \\ 0 & 0 & 8 \end{bmatrix}$;

i. diag(7, 5, 3); j. diag(d_1, d_2, d_3).

3.19 Which of the matrices in Exercise 3.18 are
a. Singular? b. Nonsingular?

3.20 Find the determinants of the following matrices:

a. $\begin{bmatrix} 1 & 0 & 0 & 0 \\ 6 & 0 & 6 & 2 \\ 0 & 5 & 4 & 0 \\ 6 & 6 & 8 & 2 \end{bmatrix}$; b. $\begin{bmatrix} 5 & 6 & 2 & 4 \\ 0 & 8 & 1 & 7 \\ 0 & 4 & 1 & 5 \\ 0 & 4 & 0 & 3 \end{bmatrix}$;

c. $\begin{bmatrix} 5 & 6 & 2 & 4 \\ 0 & 8 & 1 & 7 \\ 0 & 4 & 1 & 5 \\ 0 & 12 & 3 & 15 \end{bmatrix}$; d. $\begin{bmatrix} 1 & 2 & 3 & 4 \\ 2 & 4 & 6 & 8 \\ 4 & 8 & 12 & 16 \\ 8 & 16 & 24 & 32 \end{bmatrix}$;

e. $\begin{bmatrix} 1 & 5 & 9 & 6 \\ 0 & 7 & 5 & 2 \\ 4 & 6 & 1 & 7 \\ 8 & 9 & 1 & 7 \end{bmatrix}$; f. $\begin{bmatrix} 3 & 0 & 0 & 0 \\ 0 & 6 & 0 & 0 \\ 0 & 0 & 7 & 0 \\ 0 & 0 & 0 & 4 \end{bmatrix}$;

g. diag(7, 7, 8, 5); h. diag$(a_1, a_2, \ldots, a_n)$;

i. $\begin{bmatrix} 2 & 5 & 3 & 8 \\ 7 & 9 & 6 & 3 \\ 1 & 5 & 2 & 7 \\ 2 & 9 & 0 & 1 \\ 0 & 9 & 8 & 1 \end{bmatrix}$; j. $\begin{bmatrix} 1 & 2 & 3 & 4 & 5 \\ 2 & 4 & 6 & 8 & 10 \\ 4 & 8 & 12 & 16 & 20 \\ 8 & 16 & 24 & 32 & 40 \\ 16 & 32 & 48 & 64 & 60 \end{bmatrix}$;

k. $\begin{bmatrix} 4 & 7 & 8 & 8 & 2 \\ 5 & 0 & 6 & 0 & 0 \\ 6 & 2 & 9 & 7 & 4 \\ 9 & 9 & 4 & 9 & 2 \\ 8 & 8 & 9 & 3 & 6 \end{bmatrix}$.

3.21 Which of the matrices in Exercise 3.20 are

a. Singular? **b.** Nonsingular?

3 . 5

INVERSES

Inverse Matrix

Determinants are useful in finding the (multiplicative) inverse of a matrix. Given a nonsingular matrix **A**, the **inverse** of **A** is a matrix $\mathbf{A}^{-1}$ such that

$$\mathbf{AA}^{-1} = \mathbf{A}^{-1}\mathbf{A} = \mathbf{I},$$

where **I** is the identity matrix.

GENERAL METHOD

We will illustrate the method of finding an inverse using a 3 × 3 matrix. Let

$$\mathbf{A} = \begin{vmatrix} a_{11} & a_{12} & a_{13} \\ a_{21} & a_{22} & a_{23} \\ a_{31} & a_{32} & a_{33} \end{vmatrix},$$

and let A_{ij} denote the cofactor of the element a_{ij}. Then

$$\mathbf{A}^{-1} = \frac{1}{|\mathbf{A}|}\begin{bmatrix} A_{11} & A_{21} & A_{31} \\ A_{12} & A_{22} & A_{32} \\ A_{13} & A_{23} & A_{33} \end{bmatrix}.$$

Adjoint Matrix

Note that the factor $1/|\mathbf{A}|$ requires that $|\mathbf{A}|$ be nonzero. Thus, only nonsingular matrices have inverses. Note also that the matrix given in the formula for the inverse is the *transpose* of the cofactor matrix; that is, the first subscript on A_{ij} now indicates the column and the second subscript indicates the row. The transpose of the cofactor matrix, often called the **adjoint matrix,** is denoted by adj(A).

To find $\mathbf{A}^{-1}$ where

$$\mathbf{A} = \begin{bmatrix} 2 & 2 & 1 \\ 1 & 0 & 1 \\ 2 & 1 & 2 \end{bmatrix},$$

we must first find the cofactor of every element.

Element	Minor	Sign	Cofactor
$a_{11} = 2$	$\begin{vmatrix} 0 & 1 \\ 1 & 2 \end{vmatrix} = 0 - 1 = -1$	+	$-1 = A_{11}$
$a_{12} = 2$	$\begin{vmatrix} 1 & 1 \\ 2 & 2 \end{vmatrix} = 2 - 2 = 0$	−	$0 = A_{12}$
$a_{13} = 1$	$\begin{vmatrix} 1 & 0 \\ 2 & 1 \end{vmatrix} = 1 - 0 = 1$	+	$1 = A_{13}$
$a_{21} = 1$	$\begin{vmatrix} 2 & 1 \\ 1 & 2 \end{vmatrix} = 4 - 1 = 3$	−	$-3 = A_{21}$
$a_{22} = 0$	$\begin{vmatrix} 2 & 1 \\ 2 & 2 \end{vmatrix} = 4 - 2 = 2$	+	$2 = A_{22}$
$a_{23} = 1$	$\begin{vmatrix} 2 & 2 \\ 2 & 1 \end{vmatrix} = 2 - 4 = -2$	−	$2 = A_{23}$
$a_{31} = 2$	$\begin{vmatrix} 2 & 1 \\ 0 & 1 \end{vmatrix} = 2 - 0 = 2$	+	$2 = A_{31}$
$a_{32} = 1$	$\begin{vmatrix} 2 & 1 \\ 1 & 1 \end{vmatrix} = 2 - 1 = 1$	−	$-1 = A_{32}$
$a_{33} = 2$	$\begin{vmatrix} 2 & 2 \\ 1 & 0 \end{vmatrix} = 0 - 2 = -2$	+	$-2 = A_{33}$

Then

$$|\mathbf{A}| = \sum_{i=1}^{3} a_{i1}A_{i1} = 2(-1) + 1(-3) + 2(2)$$

$$= -2 - 3 + 4 = -1,$$

by expanding down the first column. The matrix of cofactors is

$$\begin{bmatrix} A_{11} & A_{12} & A_{13} \\ A_{21} & A_{22} & A_{23} \\ A_{31} & A_{32} & A_{33} \end{bmatrix} = \begin{bmatrix} -1 & 0 & 1 \\ -3 & 2 & 2 \\ 2 & -1 & -2 \end{bmatrix};$$

and its transpose is

$$\begin{bmatrix} A_{11} & A_{21} & A_{31} \\ A_{12} & A_{22} & A_{32} \\ A_{13} & A_{23} & A_{33} \end{bmatrix} = \begin{bmatrix} -1 & -3 & 2 \\ 0 & 2 & -1 \\ 1 & 2 & -2 \end{bmatrix},$$

so that

$$\mathbf{A}^{-1} = \frac{1}{-1}\begin{bmatrix} -1 & -3 & 2 \\ 0 & 2 & -1 \\ 1 & 2 & -2 \end{bmatrix} = \begin{bmatrix} 1 & 3 & -2 \\ 0 & -2 & 1 \\ -1 & -2 & 2 \end{bmatrix}.$$

Since it is no easy job to invert a matrix, it is always a good idea to make sure that $\mathbf{AA}^{-1} = \mathbf{A}^{-1}\mathbf{A} = \mathbf{I}$. Since matrix multiplication is not generally commutative, both $\mathbf{AA}^{-1}$ and $\mathbf{A}^{-1}\mathbf{A}$ must be formed, because if a mistake has been made it is possible that $\mathbf{AA}^{-1} = \mathbf{I}$ but $\mathbf{A}^{-1}\mathbf{A} \neq \mathbf{I}$, or vice versa. For our example,

$$\mathbf{A}\mathbf{A}^{-1} = \begin{bmatrix} 2 & 2 & 1 \\ 1 & 0 & 1 \\ 2 & 1 & 2 \end{bmatrix}\begin{bmatrix} 1 & 3 & -2 \\ 0 & -2 & 1 \\ -1 & -2 & 2 \end{bmatrix}$$

$$= \begin{bmatrix} 2+0-1 & 6-4-2 & -4+2+2 \\ 1+0-1 & 3+0-2 & -2+0+2 \\ 2+0-2 & 6-2-4 & -4+1+4 \end{bmatrix} = \begin{bmatrix} 1 & 0 & 0 \\ 0 & 1 & 0 \\ 0 & 0 & 1 \end{bmatrix} = \mathbf{I};$$

and

$$\mathbf{A}^{-1}\mathbf{A} = \begin{bmatrix} 1 & 3 & -2 \\ 0 & -2 & 1 \\ -1 & -2 & 2 \end{bmatrix}\begin{bmatrix} 2 & 2 & 1 \\ 1 & 0 & 1 \\ 2 & 1 & 2 \end{bmatrix}$$

$$= \begin{bmatrix} 2+3-4 & 2+0-2 & 1+3-4 \\ 0-2+2 & 0+0+1 & 0-2+2 \\ -2-2+4 & -2+0+2 & -1-2+4 \end{bmatrix} = \begin{bmatrix} 1 & 0 & 0 \\ 0 & 1 & 0 \\ 0 & 0 & 1 \end{bmatrix} = \mathbf{I}.$$

THE INVERSE OF A 2 × 2 MATRIX

The same method applies to inversion of a 4 × 4 or higher matrix; however, the process becomes even more tedious, since the cofactors involve determinants of 3 × 3 or higher matrices. Applying this method to a 2 × 2 matrix, however, results in a neat little formula. Let

$$\mathbf{A} = \begin{bmatrix} a & b \\ c & d \end{bmatrix}.$$

Then

Element	Minor	Sign	Cofactor
a	d	$+$	d
b	c	$-$	$-c$
c	b	$-$	$-b$
d	a	$+$	a

and $|\mathbf{A}| = ad + b(-c) = ad - bc$, as we have already noted. Then

$$\mathbf{A}^{-1} = \frac{1}{|\mathbf{A}|}\begin{bmatrix} d & -c \\ -b & a \end{bmatrix}' = \frac{1}{|\mathbf{A}|}\begin{bmatrix} d & -b \\ -c & a \end{bmatrix} = \frac{1}{ad-bc}\begin{bmatrix} d & -b \\ -c & a \end{bmatrix}$$

$$= \begin{bmatrix} \dfrac{d}{ad-bc} & \dfrac{-b}{ad-bc} \\ \dfrac{-c}{ad-bc} & \dfrac{a}{ad-bc} \end{bmatrix}.$$

Note that the denominator of each term is $|\mathbf{A}| = ad - bc$; that a and d switch places but retain the same sign; and that b and c retain their positions but change sign. To check that an inverse has been found,

$$\mathbf{AA}^{-1} = \begin{bmatrix} a & b \\ c & d \end{bmatrix} \begin{bmatrix} \dfrac{d}{ad - bc} & \dfrac{-b}{ad - bc} \\ \dfrac{-c}{ad - bc} & \dfrac{a}{ad - bc} \end{bmatrix}$$

$$= \begin{bmatrix} \dfrac{ad}{ad - bc} - \dfrac{bc}{ad - bc} & \dfrac{-ab}{ad - bc} + \dfrac{ab}{ad - bc} \\ \dfrac{cd}{ad - bc} - \dfrac{cd}{ad - bc} & \dfrac{-bc}{ad - bc} + \dfrac{ad}{ad - bc} \end{bmatrix} = \begin{bmatrix} 1 & 0 \\ 0 & 1 \end{bmatrix} = \mathbf{I};$$

and $$\mathbf{A}^{-1}\mathbf{A} = \begin{bmatrix} \dfrac{d}{ad - bc} & \dfrac{-b}{ad - bc} \\ \dfrac{-c}{ad - bc} & \dfrac{a}{ad - bc} \end{bmatrix} \begin{bmatrix} a & b \\ c & d \end{bmatrix}$$

$$= \begin{bmatrix} \dfrac{ad - bc}{ad - bc} & \dfrac{bd - bd}{ad - bc} \\ \dfrac{-ac + ac}{ad - bc} & \dfrac{-bc + ad}{ad - bc} \end{bmatrix} = \begin{bmatrix} 1 & 0 \\ 0 & 1 \end{bmatrix} = \mathbf{I}.$$

THE INVERSE OF A DIAGONAL MATRIX

If the matrix to be inverted is a diagonal matrix, the procedure is simple:

$$\mathbf{D} = \text{diag}(d_1, d_2, \ldots, d_n) = \begin{bmatrix} d_1 & 0 & 0 & \cdots & 0 \\ 0 & d_2 & 0 & \cdots & 0 \\ 0 & 0 & d_3 & \cdots & 0 \\ \vdots & \vdots & \vdots & & \vdots \\ 0 & 0 & 0 & \cdots & d_n \end{bmatrix},$$

$$\mathbf{D}^{-1} = \text{diag}\left(\frac{1}{d_1}, \frac{1}{d_2}, \ldots, \frac{1}{d_n}\right) = \begin{bmatrix} 1/d_1 & 0 & 0 & \cdots & 0 \\ 0 & 1/d_2 & 0 & \cdots & 0 \\ 0 & 0 & 1/d_3 & \cdots & 0 \\ \vdots & \vdots & \vdots & & \vdots \\ 0 & 0 & 0 & \cdots & 1/d_n \end{bmatrix}.$$

For example, if

$$\mathbf{D} = \begin{bmatrix} a & 0 \\ 0 & d \end{bmatrix},$$

then $|\mathbf{D}| = ad$ and

$$\mathbf{D}^{-1} = \begin{bmatrix} \dfrac{d}{ad} & \dfrac{0}{ad} \\ \dfrac{0}{ad} & \dfrac{a}{ad} \end{bmatrix} = \begin{bmatrix} \dfrac{1}{a} & 0 \\ 0 & \dfrac{1}{d} \end{bmatrix}.$$

The reader can verify that this is always the case either by applying the inversion rule to a 3×3 or larger diagonal matrix, or by completing the following check that $\mathbf{DD}^{-1} = \mathbf{D}^{-1}\mathbf{D} = \mathbf{I}$:

$$\mathbf{DD}^{-1} = \begin{bmatrix} d_1 & 0 & \cdots & 0 \\ 0 & d_2 & \cdots & 0 \\ \vdots & \vdots & & \vdots \\ 0 & 0 & \cdots & d_n \end{bmatrix} \begin{bmatrix} 1/d_1 & 0 & \cdots & 0 \\ 0 & 1/d_2 & \cdots & 0 \\ \vdots & \vdots & & \vdots \\ 0 & 0 & \cdots & 1/d_n \end{bmatrix}$$

$$= \begin{bmatrix} 1 + 0 + \cdots + 0 & 0 + 0 + \cdots + 0 & \cdots & 0 + 0 + \cdots + 0 \\ 0 + 0 + \cdots + 0 & 0 + 1 + \cdots + 0 & \cdots & 0 + 0 + \cdots + 0 \\ \vdots & \vdots & & \vdots \\ 0 + 0 + \cdots + 0 & 0 + 0 + \cdots + 0 & \cdots & 0 + 0 + \cdots + 1 \end{bmatrix}$$

$$= \begin{bmatrix} 1 & 0 & \cdots & 0 \\ 0 & 1 & \cdots & 0 \\ \vdots & \vdots & & \vdots \\ 0 & 0 & \cdots & 1 \end{bmatrix} = \mathbf{I}.$$

In the next section, we will use what we have learned about matrices to solve systems of simultaneous equations.

EXERCISES

3.22 Referring to the exercises at the end of Section 3.4, find the inverse of each nonsingular matrix in the following exercises. Check your answers.
a. Exercise 3.16; **b.** Exercise 3.18; **c.** Exercise 3.20 a–j.

3.23 **a.** Find the inverse of the matrix

$$\begin{bmatrix} 1 & 0 \\ 0 & 6 \end{bmatrix}.$$

b. Find the inverse of the matrix

$$\begin{bmatrix} 1 & 0 & 0 \\ 0 & 2 & 3 \\ 0 & 9 & 1 \end{bmatrix}.$$

Compare to the answer of Exercise 3.16a. Note that the matrix here can be written

$$\left[\begin{array}{c|cc} 1 & 0 & 0 \\ \hline 0 & & \\ 0 & \multicolumn{2}{c}{\mathbf{M}} \\ 0 & & \end{array}\right], \quad \text{where } \mathbf{M} = \begin{bmatrix} 2 & 3 \\ 9 & 1 \end{bmatrix}.$$

Can you make a generalization?

c. Find the inverse of the matrix

$$\begin{bmatrix} 1 & 0 & 0 & 0 \\ 0 & 1 & 0 & 7 \\ 0 & 8 & 2 & 5 \\ 0 & 0 & 6 & 7 \end{bmatrix}.$$

Compare to the answer of Exercise 3.18a. Does this answer support the generalization in (b) above?

d. Complete the rule:

$$\left[\begin{array}{c|ccc} 1 & 0 & \cdots & 0 \\ \hline 0 & & & \\ \cdot & & & \\ \cdot & & \mathbf{M} & \\ \cdot & & & \\ 0 & & & \end{array}\right]^{-1} = ?$$

e. Complete the rule:

$$\left[\begin{array}{c|ccc} d & 0 & \cdots & 0 \\ \hline 0 & & & \\ \cdot & & & \\ \cdot & & \mathbf{M} & \\ \cdot & & & \\ 0 & & & \end{array}\right]^{-1} = ?$$

3.6 APPLICATION TO SOLVING SIMULTANEOUS EQUATIONS

WRITING SIMULTANEOUS EQUATIONS IN MATRIX FORM

Matrix theory evolved from the study of simultaneous equations. Let us see how this relationship works. Suppose we have the following system of equations to solve for X and Y, where a_i, b_i, and c_i are constants ($i = 1, 2$):

$$a_1X + b_1Y = c_1$$
$$a_2X + b_2Y = c_2.$$

This system can be more compactly written as

$$\mathbf{Ax} = \mathbf{c},$$

where

$$\mathbf{A}_{2\times 2} = \begin{bmatrix} a_1 & b_1 \\ a_2 & b_2 \end{bmatrix}, \quad \mathbf{x}_{2\times 1} = \begin{bmatrix} X \\ Y \end{bmatrix}, \quad \text{and} \quad \mathbf{c}_{2\times 1} = \begin{bmatrix} c_1 \\ c_2 \end{bmatrix},$$

since

$$\underset{2\times 2}{\begin{bmatrix} a_1 & b_1 \\ a_2 & b_2 \end{bmatrix}} \underset{2\times 1}{\begin{bmatrix} X \\ Y \end{bmatrix}} = \underset{2\times 1}{\begin{bmatrix} c_1 \\ c_2 \end{bmatrix}}, \quad \text{or} \quad \underset{2\times 1}{\begin{bmatrix} a_1X + b_1Y \\ a_2X + b_2Y \end{bmatrix}} = \underset{2\times 1}{\begin{bmatrix} c_1 \\ c_2 \end{bmatrix}},$$

so that

$$a_1X + b_1Y = c_1 \quad \text{and} \quad a_2X + b_2Y = c_2.$$

SOLVING SIMULTANEOUS EQUATIONS

Solving the matrix equation $\mathbf{Ax} = \mathbf{c}$ for $\mathbf{x}$ (that is, for X and Y) gives

$$\mathbf{A}^{-1}\mathbf{Ax} = \mathbf{A}^{-1}\mathbf{c}$$
$$\mathbf{Ix} = \mathbf{A}^{-1}\mathbf{c}$$
$$\mathbf{x} = \mathbf{A}^{-1}\mathbf{c}.$$

Note that we have multiplied both sides of the equation by $\mathbf{A}^{-1}$ *on the left*. In this case, we say that $\mathbf{A}^{-1}\mathbf{Ax}$ **premultiplies** the vector $\mathbf{Ax}$ by $\mathbf{A}^{-1}$. If we had an expression such as **BA,** then the matrix **B** would be **postmultiplied** by the matrix **A.**

Premultiplication

Postmultiplication

Note that in solving $\mathbf{Ax} = \mathbf{c}$, both sides of the equation are premultiplied by $\mathbf{A}^{-1}$; it would not work to form

$$\mathbf{AxA}^{-1} = \mathbf{cA}^{-1},$$

because it is not the case that $\mathbf{AxA}^{-1} = \mathbf{x}$, even if the dimensions had been conformable for multiplication in that order. Remember when solving matrix equations that if one side is premultiplied by a matrix (or vector), then the other side must also be premultiplied by that same matrix in order to preserve the equality. The same reasoning applies to postmultiplication.

LARGER SYSTEMS

Note also that while we started out with two equations in two unknowns,

$$a_1X + b_1Y = c_1$$
$$a_2X + b_2Y = c_2,$$

matrix notation reduced the problem to that of solving only one equation in one unknown: $\mathbf{Ax} = \mathbf{c}$. Furthermore, any set of n equations in n unknowns can be expressed in the same form. For example, for four equations in four unknowns,

$$a_{11}X_1 + a_{12}X_2 + a_{13}X_3 + a_{14}X_4 = c_1$$
$$a_{21}X_1 + a_{22}X_2 + a_{23}X_3 + a_{24}X_4 = c_2$$
$$a_{31}X_1 + a_{32}X_2 + a_{33}X_3 + a_{34}X_4 = c_3$$
$$a_{41}X_1 + a_{42}X_2 + a_{43}X_3 + a_{44}X_4 = c_4,$$

we get

$$\mathbf{A} = \begin{bmatrix} a_{11} & a_{12} & a_{13} & a_{14} \\ a_{21} & a_{22} & a_{23} & a_{24} \\ a_{31} & a_{32} & a_{33} & a_{34} \\ a_{41} & a_{42} & a_{43} & a_{44} \end{bmatrix}, \quad \mathbf{x} = \begin{bmatrix} X_1 \\ X_2 \\ X_3 \\ X_4 \end{bmatrix}, \quad \text{and} \quad \mathbf{c} = \begin{bmatrix} c_1 \\ c_2 \\ c_3 \\ c_4 \end{bmatrix},$$

and the system can be written simply as

$$\mathbf{Ax} = \mathbf{c}.$$

The solution to the matrix equation

$$\mathbf{x} = \mathbf{A}^{-1}\mathbf{c}$$

holds regardless of the dimension of the system. Thus, this single equation is the solution to *all* systems of simultaneous linear equations. We need not study as separate topics the solutions to two equations in two unknowns, three equations in three unknowns, or 140 equations in 140 unknowns: we have already solved them all!

NUMERICAL EXAMPLES

For example, if the system of equations is

$$2X + 5Y = -5$$
$$X - 2Y = 5,$$

then $\quad \mathbf{A} = \begin{bmatrix} 2 & 5 \\ 1 & -2 \end{bmatrix}, \quad \mathbf{x} = \begin{bmatrix} X \\ Y \end{bmatrix}, \quad \mathbf{c} = \begin{bmatrix} -5 \\ 5 \end{bmatrix},$

and $\quad |\mathbf{A}| = -4 - 5 = -9$

and

$$\mathbf{A}^{-1} = -\frac{1}{9}\begin{bmatrix} -2 & -5 \\ -1 & 2 \end{bmatrix} = \begin{bmatrix} \frac{2}{9} & \frac{5}{9} \\ \frac{1}{9} & -\frac{2}{9} \end{bmatrix},$$

so that

$$\begin{aligned} \mathbf{x} &= \mathbf{A}^{-1}\mathbf{c} \\ &= \begin{bmatrix} \frac{2}{9} & \frac{5}{9} \\ \frac{1}{9} & -\frac{2}{9} \end{bmatrix}\begin{bmatrix} -5 \\ 5 \end{bmatrix} \\ &= \begin{bmatrix} -\frac{10}{9} + \frac{25}{9} \\ -\frac{5}{9} - \frac{10}{9} \end{bmatrix} = \begin{bmatrix} \frac{15}{9} \\ -\frac{15}{9} \end{bmatrix} = \begin{bmatrix} \frac{5}{3} \\ -\frac{5}{3} \end{bmatrix}. \end{aligned}$$

Thus,

$$X = \frac{5}{3}, \qquad Y = -\frac{5}{3}.$$

Checking,

$$2\left(\frac{5}{3}\right) + 5\left(-\frac{5}{3}\right) = \frac{10}{3} - \frac{25}{3} = -\frac{15}{3} = -5$$

$$1\left(\frac{5}{3}\right) - 2\left(-\frac{5}{3}\right) = \frac{5}{3} + \frac{10}{3} = \frac{15}{3} = 5.$$

Note that the same solutions could be reached by multiplying the second equation by 2, subtracting, and thus eliminating X:

$$\begin{aligned} 2X + 5Y &= -5 \\ 2X - 4Y &= 10 \\ \hline 9Y &= -15 \\ Y &= -\frac{15}{9} = -\frac{5}{3} \end{aligned}$$

and

$$2X + 5\left(-\frac{5}{3}\right) = -5$$

$$2X = -5 + \frac{25}{3} = \frac{-15 + 25}{3} = \frac{10}{3}$$

$$X = \frac{10/3}{2} = \frac{5}{3}.$$

Matrices may not save much time if there are only two equations in two unknowns, but for three or more simultaneous equations to be solved, the matrix approach is as efficient as any other method, and often more efficient. For example, suppose we are to solve

$$\begin{aligned} 2X_1 + 2X_2 + X_3 &= 1 \\ X_1 \qquad + X_3 &= -2 \\ 2X_1 + X_2 + 2X_3 &= 0, \end{aligned}$$

where X_1, X_2, and X_3 are unknown. Then

$$\mathbf{A} = \begin{bmatrix} 2 & 2 & 1 \\ 1 & 0 & 1 \\ 2 & 1 & 2 \end{bmatrix}, \quad \mathbf{x} = \begin{bmatrix} X_1 \\ X_2 \\ X_3 \end{bmatrix}, \quad \text{and} \quad \mathbf{c} = \begin{bmatrix} 1 \\ -2 \\ 0 \end{bmatrix}.$$

Recognizing that $\mathbf{A}$ is the matrix we inverted earlier,

$$\mathbf{A}^{-1} = \begin{bmatrix} 1 & 3 & -2 \\ 0 & -2 & 1 \\ -1 & -2 & 2 \end{bmatrix},$$

we have

$$\mathbf{x} = \mathbf{A}^{-1}\mathbf{c} = \begin{bmatrix} 1 & 3 & -2 \\ 0 & -2 & 1 \\ -1 & -2 & 2 \end{bmatrix} \begin{bmatrix} 1 \\ -2 \\ 0 \end{bmatrix}$$

$$= \begin{bmatrix} 1 - 6 + 0 \\ 0 + 4 + 0 \\ -1 + 4 + 0 \end{bmatrix} = \begin{bmatrix} -5 \\ 4 \\ 3 \end{bmatrix} = \begin{bmatrix} X_1 \\ X_2 \\ X_3 \end{bmatrix}.$$

Checking,

$$2(-5) + 2(4) + 1(3) = -10 + 8 + 3 = 1$$
$$1(-5) + 0(4) + 1(3) = -5 + 0 + 3 = -2$$
$$2(-5) + 1(4) + 2(3) = -10 + 4 + 6 = 0.$$

WHEN SYSTEMS ARE AND ARE NOT SOLVABLE

If there are n unknowns, then n different equations are required to solve the system. By n different equations, we mean that each equation must carry different information about the n unknowns. In the foregoing example, all three equations carry different information:

$$2X_1 + 2X_2 + X_3 = 1$$

carries different information about X_1, X_2, and X_3 than does

$$X_1 + 0X_2 + X_3 = -2;$$

and both carry information different from

$$2X_1 + X_2 + 2X_3 = 0.$$

An example of two equations in X_1, X_2, and X_3 that do not carry different information is

$$2X_1 + 2X_2 + X_3 = 1 \quad \text{and} \quad 4X_1 + 4X_2 + 2X_3 = 2.$$

We note that the second equation is obtained by multiplying the first equation by 2, so it gives the same information as the first. (For example, we would not gain any information on the height of an object if we first measure its height in feet and then measure its height in inches, since the second measurement will just be 12 times the first.) Without different information about the three unknowns, we cannot solve for them. For example, if our set of simultaneous equations were

$$\begin{aligned} 2X_1 + 2X_2 + X_3 &= 1 \\ 4X_1 + 4X_2 + 2X_3 &= 2 \\ 2X_1 + X_2 + 2X_3 &= 0, \end{aligned}$$

then

$$\mathbf{A} = \begin{bmatrix} 2 & 2 & 1 \\ 4 & 4 & 2 \\ 2 & 1 & 2 \end{bmatrix},$$

and $|\mathbf{A}|$ will be zero, because one row is a linear combination of the others: row 2 $=$ 2(row 1) $+$ 0(row 3). Thus $\mathbf{A}^{-1}$ will not exist and the three equations cannot be solved. We essentially have only two equations in the three unknowns.

Suppose our system of equations is

$$\begin{aligned} 2X_1 + 2X_2 + X_3 &= 1 \\ 4X_1 + 4X_2 + 2X_3 &= 3 \\ 2X_1 + X_2 + 2X_3 &= 0. \end{aligned}$$

The only difference between these equations and the previous ones is that the constant on the right-hand side in the second equation has been changed from 2 to 3. In this case, two of the equations carry *contradictory,* rather than redundant, information. The first equation implies that

$$4X_1 + 4X_2 + 2X_3 = 2,$$

and this contradicts the second equation, which says that

$$4X_1 + 4X_2 + 2X_3 = 3.$$

Thus, when any row of a matrix is a linear combination of the other rows, the associated system of simultaneous equations contains either redundant or contradictory information and thus cannot be solved.

What if some column of the coefficient matrix is a linear combination of some other columns? For example, consider the system

$$\begin{aligned} 2X_1 + 2X_2 + X_3 &= 1 \\ 4X_1 + 5X_2 + 2X_3 &= 4 \\ 6X_1 + X_2 + 3X_3 &= 6. \end{aligned}$$

We can see that in the coefficient matrix

$$\mathbf{A} = \begin{bmatrix} 2 & 2 & 1 \\ 4 & 5 & 2 \\ 6 & 1 & 3 \end{bmatrix}$$

the first column is twice the third column. Essentially, this means that we have more equations than unknowns: any information given about X_1 is the same as the information given about X_3, and we will reach a contradiction when trying to solve the equations. For example, solving the first and second equations together, we have, upon multiplying the first equation by 2,

$$\begin{array}{r} 4X_1 + 4X_2 + 2X_3 = 2 \\ 4X_1 + 5X_2 + 2X_3 = 4 \\ \hline X_2 \qquad\quad = 2. \end{array}$$

But taking the first and third equations together (after multiplying the first by 3),

$$\begin{array}{rcr} 6X_1 + 6X_2 + 3X_3 & = & 3 \\ 6X_1 + X_2 + 3X_3 & = & 6 \\ \hline 5X_2 & = & -3 \\ X_2 & = & -\frac{3}{5}, \end{array}$$

and this contradicts the previous result.

In the next section, we shall apply what we have learned about systems to solving normal equations.

EXERCISES

3.24 Using matrices, solve the following systems of equations. Check your answers. (*Hint:* Refer to Exercise 3.22a.)

a. $2a + 3b = 1$
$9a + b = 5;$

b. $x = 1$
$-9x + 7y = 2;$

c. $m + 2n - 2 = 0$
$3m + 4n + 2 = 0;$

d. $x + 2y = 1$
$2x + 4y = 3;$

e. $6p = 7$
$7q = 3;$

f. $x_1 + 5x_2 + 3x_3 + 3 = 0$
$x_3 = 0;$

g. $8a = 5.$

3.25 Solve the following systems of equations using matrices. Check your answers. (*Hint:* Refer to Exercise 3.22b.)

a. $1x_1 + 0x_2 + 7x_3 = 3$
$8x_1 + 2x_2 + 5x_3 = -3$
$0x_1 + 6x_2 + 7x_3 = 9;$

b. $x = 3$
$x + 5y + 4z = 7$
$9x + 2y + 4z = 6;$

c. $$\begin{aligned} 5a + 7b - 1 &= 0 \\ 7a + 3b + 7c + 8 &= 0 \\ 7b + 3c - 4 &= 0; \end{aligned}$$

d. $$\begin{aligned} m + 2n + 3p &= 3 \\ 2m + 4n + 6p &= 3 \\ 4m + 8n + 12p &= 9; \end{aligned}$$

e. $$\begin{aligned} 4r + 5s &= 2 \\ 7r + 2s &= 5 \\ 3r + 5s &= 5; \end{aligned}$$

f. $$\begin{aligned} 7a_1 &= 5 \\ 5a_2 &= 5 \\ 8a_3 &= 5. \end{aligned}$$

3.26 Using matrices, solve the following systems of equations. Check your answers. (*Hint:* Refer to Exercise 3.22c.)

a. $$\begin{aligned} a &= 7 \\ 6a + 6c + 2d &= 0 \\ 5b + 4c &= 3 \\ 6a + 6b + 8c + 2d &= 7; \end{aligned}$$

b. $$\begin{aligned} 5x_1 + 6x_2 + 2x_3 + 4x_4 - 4 &= 0 \\ 8x_2 + x_3 + 7x_4 - 9 &= 0 \\ 4x_2 + x_3 + 5x_4 - 5 &= 0 \\ 4x_2 + 3x_4 - 2 &= 0; \end{aligned}$$

c. $$\begin{aligned} m + 2n + 3p + 4q &= 1 \\ 2m + 4n + 6p + 8q &= 5 \\ 4m + 8n + 12p + 16q &= 2 \\ 8m + 16n + 24p + 32q &= 4. \end{aligned}$$

d. $$\begin{aligned} x_1 + 5x_2 + 9x_3 + 6x_4 &= -6 \\ 7x_2 + 5x_3 + 2x_4 &= 7 \\ 4x_1 + 6x_2 + x_3 + 7x_4 &= 7 \\ 8x_1 + 9x_2 + x_3 + 7x_4 &= -4; \end{aligned}$$

e. $$\begin{aligned} 2x_1 + 5x_2 + 3x_3 + 8x_4 &= 2 \\ 7x_1 + 9x_2 + 6x_3 + 3x_4 &= 7 \\ x_1 + 5x_2 + 2x_3 + 7x_4 &= 3 \\ 2x_1 + 9x_2 + x_4 &= 6 \\ 9x_2 + 8x_3 + x_4 &= 7. \end{aligned}$$

3.7 APPLICATION TO SIMPLE LINEAR REGRESSION

Since finding the values a and b to substitute into the regression equation $Y = a + bX$ requires solving the normal equations, which are simultaneous equations, we can reformulate the regression problem in terms of matrices.

WRITING THE MODEL IN MATRIX FORM

Recall that the model for a simple linear regression,

$$Y = \alpha + \beta X + \varepsilon,$$

says that each Y-value is linearly related to an X-value, but imperfectly. Thus, for every one of the N items in the population,

$$
\begin{aligned}
Y_1 &= \alpha + \beta X_1 + \varepsilon_1 \\
Y_2 &= \alpha + \beta X_2 + \varepsilon_2 \\
&\vdots \\
Y_N &= \alpha + \beta X_N + \varepsilon_N,
\end{aligned}
$$

and we have a system of equations which we can write in matrix form as

$$\mathbf{y} = \mathbf{X}\boldsymbol{\beta} + \boldsymbol{\varepsilon},$$

where

$$\mathbf{y}_{N\times 1} = \begin{bmatrix} Y_1 \\ Y_2 \\ \vdots \\ Y_N \end{bmatrix}, \quad \boldsymbol{\beta}_{2\times 1} = \begin{bmatrix} \alpha \\ \beta \end{bmatrix}, \quad \boldsymbol{\varepsilon}_{N\times 1} = \begin{bmatrix} \varepsilon_1 \\ \varepsilon_2 \\ \vdots \\ \varepsilon_N \end{bmatrix}$$

and

$$\mathbf{X}_{N\times 2} = \begin{bmatrix} 1 & X_1 \\ 1 & X_2 \\ \vdots & \vdots \\ 1 & X_N \end{bmatrix}.$$

Note that in order to be conformable for multiplication with the vector $\boldsymbol{\beta}$, the matrix **X** must have two columns, and the column of 1's in X is necessary to pick up the α in the vector $\boldsymbol{\beta}$:

$$\mathbf{X}\boldsymbol{\beta} = \underset{N\times 2}{\begin{bmatrix} 1 & X_1 \\ 1 & X_2 \\ \vdots & \vdots \\ 1 & X_N \end{bmatrix}} \underset{2\times 1}{\begin{bmatrix} \alpha \\ \beta \end{bmatrix}} = \underset{N\times 1}{\begin{bmatrix} \alpha + \beta X_1 \\ \alpha + \beta X_2 \\ \vdots \\ \alpha + \beta X_N \end{bmatrix}}$$

Then we can see that

$$\underset{N\times 1}{\begin{bmatrix} Y_1 \\ Y_2 \\ \vdots \\ Y_N \end{bmatrix}} = \underset{N\times 1}{\begin{bmatrix} \alpha + \beta X_1 \\ \alpha + \beta X_2 \\ \vdots \\ \alpha + \beta X_N \end{bmatrix}} + \underset{N\times 1}{\begin{bmatrix} \varepsilon_1 \\ \varepsilon_2 \\ \vdots \\ \varepsilon_N \end{bmatrix}}$$

gives the set of N equations we started with.

NORMAL EQUATIONS IN MATRIX FORM

Since a sample of n observations (X_i, Y_i) drawn from the population of N items produces the estimating equation

$$\hat{Y} = a + bX,$$

the sample information is written in the following form:

$$\mathbf{y} = \underset{n\times 1}{\begin{bmatrix} Y_1 \\ Y_2 \\ \vdots \\ Y_n \end{bmatrix}}, \qquad \mathbf{X} = \underset{n\times 2}{\begin{bmatrix} 1 & X_1 \\ 1 & X_2 \\ \vdots & \vdots \\ 1 & X_n \end{bmatrix}}, \qquad \hat{\boldsymbol{\beta}} = \underset{2\times 1}{\begin{bmatrix} a \\ b \end{bmatrix}}.$$

Recall that the "hat" on $\boldsymbol{\beta}$ indicates that *estimates* are being made and in order to find the value of $\hat{\boldsymbol{\beta}}$ (the values of a and b), we must come up with a system of normal equations. Recall that the normal equations are

$$\sum Y_i = na + b\sum X_i$$

$$\sum X_iY_i = a\sum X_i + b\sum X_i^2.$$

How do we write these equations in matrix form? (Remember, the *unknowns* here are a and b.) In matrices we get

$$\underset{2\times 2}{\begin{bmatrix} n & \sum X_i \\ \sum X_i & \sum X_i^2 \end{bmatrix}} \underset{2\times 1}{\begin{bmatrix} a \\ b \end{bmatrix}} = \underset{2\times 1}{\begin{bmatrix} \sum Y_i \\ \sum X_iY_i \end{bmatrix}}.$$

In terms of our data, $\mathbf{y}$, $\mathbf{X}$, and $\hat{\boldsymbol{\beta}}$, clearly $\begin{bmatrix} a \\ b \end{bmatrix} = \hat{\boldsymbol{\beta}}$. Also, the vector on the right-hand side involves both X's and Y's. Since $\mathbf{X}$ is $n \times 2$ and $\mathbf{y}$ is $n \times 1$, the two matrices are not conformable for multiplication. However, $\mathbf{X}'$ is $2 \times n$ and the product $\mathbf{X}'\mathbf{y}$ can be formed, and this product is 2×1, the same dimension as $\hat{\boldsymbol{\beta}}$. Forming the product $\mathbf{X}'\mathbf{y}$,

$$\mathbf{X}'\mathbf{y} = \begin{bmatrix} 1 & 1 & \cdots & 1 \\ X_1 & X_2 & \cdots & X_n \end{bmatrix} \begin{bmatrix} Y_1 \\ Y_2 \\ \vdots \\ Y_n \end{bmatrix} = \begin{bmatrix} Y_1 + Y_2 + \cdots + Y_n \\ X_1Y_1 + X_2Y_2 + \cdots + X_nY_n \end{bmatrix}$$

$$= \begin{bmatrix} \sum Y_i \\ \sum X_iY_i \end{bmatrix},$$

and this is the vector on the right-hand side of the equation. The matrix

$$\begin{bmatrix} n & \sum X_i \\ \sum X_i & \sum X_i^2 \end{bmatrix}$$

involves the square of the X-variable. It is also symmetric. Note that since $\mathbf{X}$ is $n \times 2$, $\mathbf{X}'$ is $2 \times n$ and $\mathbf{X}'\mathbf{X}$ is 2×2, which is conformable for multiplication on the right by $\hat{\boldsymbol{\beta}}$. Then

$$\mathbf{X}'\mathbf{X} = \begin{bmatrix} 1 & 1 & \cdots & 1 \\ X_1 & X_2 & \cdots & X_n \end{bmatrix} \begin{bmatrix} 1 & X_1 \\ 1 & X_2 \\ \vdots & \vdots \\ 1 & X_n \end{bmatrix}$$

$$= \begin{bmatrix} 1 + 1 + \cdots + 1 & X_1 + X_2 + \cdots + X_n \\ X_1 + X_2 + \cdots + X_n & X_1^2 + X_2^2 + \cdots + X_n^2 \end{bmatrix}$$

$$= \begin{bmatrix} n & \sum X_i \\ \sum X_i & \sum X_i^2 \end{bmatrix}.$$

Thus, the normal equations can be written in matrix notation as

$$(\mathbf{X}'\mathbf{X})\hat{\boldsymbol{\beta}} = \mathbf{X}'\mathbf{y}.$$

This formulation of normal equations is the same for every regression problem, regardless of the number of predictors and regardless of whether we are fitting a straight line or a curve. The matrix $\mathbf{X}$ for multiple regression or curvilinear regression will be defined differently, as will the vector $\hat{\boldsymbol{\beta}}$; but as long as $\mathbf{X}$ and $\hat{\boldsymbol{\beta}}$ are appropriately defined, the normal equations can always be written as

$$(\mathbf{X}'\mathbf{X})\hat{\boldsymbol{\beta}} = \mathbf{X}'\mathbf{y}.$$

For example, if we want to fit an equation of the form

$$\hat{Y} = a + bX + cX^2,$$

then we need to solve the normal equations

$$\sum Y_i = na + b\sum X_i + c\sum X_i^2$$

$$\sum X_iY_i = a\sum X_i + b\sum X_i^2 + c\sum X_i^3$$

$$\sum X_i^2Y_i = a\sum X_i^2 + b\sum X_i^3 + c\sum X_i^4.$$

We first define

$$\mathbf{y} = \begin{bmatrix} Y_1 \\ Y_2 \\ \vdots \\ Y_n \end{bmatrix}, \quad \hat{\boldsymbol{\beta}} = \begin{bmatrix} a \\ b \\ c \end{bmatrix}, \quad \text{and} \quad \mathbf{X} = \begin{bmatrix} 1 & X_1 & X_1^2 \\ 1 & X_2 & X_2^2 \\ \vdots & \vdots & \vdots \\ 1 & X_n & X_n^2 \end{bmatrix}.$$

Note that there is a column in $\mathbf{X}$ corresponding to the coefficient of each unknown in the equation to be fitted. Then

$$\mathbf{X}'\mathbf{y} = \begin{bmatrix} 1 & 1 & \cdots & 1 \\ X_1 & X_2 & \cdots & X_n \\ X_1^2 & X_2^2 & \cdots & X_n^2 \end{bmatrix} \begin{bmatrix} Y_1 \\ Y_2 \\ \vdots \\ Y_n \end{bmatrix} = \begin{bmatrix} \sum Y_i \\ \sum X_iY_i \\ \sum X_i^2Y_i \end{bmatrix}$$

and

$$\mathbf{X}'\mathbf{X} = \begin{bmatrix} 1 & 1 & \cdots & 1 \\ X_1 & X_2 & \cdots & X_n \\ X_1^2 & X_2^2 & \cdots & X_n^2 \end{bmatrix} \begin{bmatrix} 1 & X_1 & X_1^2 \\ 1 & X_2 & X_2^2 \\ \vdots & \vdots & \vdots \\ 1 & X_n & X_n^2 \end{bmatrix}$$

$$= \begin{bmatrix} n & \sum X_i & \sum X_i^2 \\ \sum X_i & \sum X_i^2 & \sum X_i^3 \\ \sum X_i^2 & \sum X_i^3 & \sum X_i^4 \end{bmatrix};$$

and we can see that

$$(\mathbf{X}'\mathbf{X})\hat{\boldsymbol{\beta}} = \mathbf{X}'\mathbf{y}$$

gives the normal equations.

In general, $\mathbf{y}$ will simply be a vector containing the n Y-values. $\hat{\boldsymbol{\beta}}$ will be a vector containing the unknowns. For p unknowns in the regression model $\hat{\boldsymbol{\beta}}$ will be a $p \times 1$ vector and $\mathbf{X}$ will be an $n \times p$ matrix containing columns corresponding to the coefficients of the unknowns in the regression model. Once $\mathbf{y}$, $\hat{\boldsymbol{\beta}}$, and $\mathbf{X}$ are appropriately defined for any problem, the normal equations are always given by

$$(\mathbf{X}'\mathbf{X})\hat{\boldsymbol{\beta}} = \mathbf{X}'\mathbf{y}.$$

Their solutions are given by

$$(\mathbf{X}'\mathbf{X})^{-1}(\mathbf{X}'\mathbf{X})\hat{\boldsymbol{\beta}} = (\mathbf{X}'\mathbf{X})^{-1}\mathbf{X}'\mathbf{y}$$

$$\mathbf{I}\hat{\boldsymbol{\beta}} = \hat{\boldsymbol{\beta}} = (\mathbf{X}'\mathbf{X})^{-1}\mathbf{X}'\mathbf{y},$$

as long as $(\mathbf{X}'\mathbf{X})$ is nonsingular so that its inverse, $(\mathbf{X}'\mathbf{X})^{-1}$, exists.

THE SOLUTION TO NORMAL EQUATIONS FOR SIMPLE REGRESSION

In the simple regression problem, we have

$$\mathbf{X}'\mathbf{X} = \begin{bmatrix} n & \sum X_i \\ \sum X_i & \sum X_i^2 \end{bmatrix}, \quad \text{and} \quad |\mathbf{X}'\mathbf{X}| = n \sum X_i^2 - (\sum X_i)^2 = nS_{XX},$$

so that

$$(\mathbf{X}'\mathbf{X})^{-1} = \begin{bmatrix} \dfrac{\sum X_i^2}{nS_{XX}} & \dfrac{-\sum X_i}{nS_{XX}} \\ \dfrac{-\sum X_i}{nS_{XX}} & \dfrac{n}{nS_{XX}} \end{bmatrix}.$$

(Note that both $\mathbf{X'X}$ and $(\mathbf{X'X})^{-1}$ are symmetric. It will always be true that the inverse of a symmetric matrix is also symmetric.) Then

$$(\mathbf{X'X})^{-1}\mathbf{X'y} = \begin{bmatrix} \dfrac{\sum X_i^2}{nS_{XX}} & \dfrac{-\sum X_i}{nS_{XX}} \\ \dfrac{-\sum X_i}{nS_{XX}} & \dfrac{n}{nS_{XX}} \end{bmatrix} \begin{bmatrix} \sum Y_i \\ \sum X_i Y_i \end{bmatrix}$$

$$= \begin{bmatrix} \dfrac{\sum X_i^2 \sum Y_i - \sum X_i \sum X_i Y_i}{nS_{XX}} \\ \dfrac{-\sum X_i \sum Y_i + n \sum X_i Y_i}{nS_{XX}} \end{bmatrix}.$$

The numerator of the bottom element is easily recognizable as nS_{XY}. Thus, the bottom element is $nS_{XY}/nS_{XX} = S_{XY}/S_{XX} = b$, which is what we obtained before. It is a little more difficult to see that the top value is $a = \bar{Y} - b\bar{X}$:

$$\frac{\sum X_i^2 \sum Y_i - \sum X_i \sum X_i Y_i}{nS_{XX}}$$

$$= \frac{\sum X_i^2 \sum Y_i - \sum X_i \sum X_i Y_i - (\sum X_i)^2 \bar{Y} + (\sum X_i)^2 \bar{Y}}{nS_{XX}}$$

$$= \frac{[\sum X_i^2 \sum Y_i - (\sum X_i)^2(\sum Y_i/n)] - [\sum X_i \sum X_i Y_i - (\sum X_i)^2(\sum Y_i/n)]}{nS_{XX}}$$

$$= \frac{S_{XX} \sum Y_i - S_{XY} \sum X_i}{nS_{XX}}$$

$$= \bar{Y} - b\bar{X} = a.$$

Thus, the matrix approach gives the same values for a and b as we obtained before.

THE HOUSE SIZE-PRICE PROBLEM USING MATRICES

Let us return to the house size-price example, using matrices for the problem we worked out in Chapter 2. We find

$$\mathbf{y} = \underset{20\times 1}{\begin{bmatrix} 32 \\ 24 \\ 27 \\ 47 \\ \vdots \\ 37 \\ 28 \end{bmatrix}}, \qquad \hat{\boldsymbol{\beta}} = \underset{2\times 1}{\begin{bmatrix} a \\ b \end{bmatrix}}, \qquad \mathbf{X} = \underset{20\times 2}{\begin{bmatrix} 1 & 1.8 \\ 1 & 1.0 \\ 1 & 1.7 \\ 1 & 2.8 \\ \vdots & \vdots \\ 1 & 2.2 \\ 1 & 1.5 \end{bmatrix}},$$

$$\mathbf{X}'\mathbf{X} = \underset{2\times 2}{\begin{bmatrix} 20 & 40 \\ 40 & 93.56 \end{bmatrix}}, \qquad \mathbf{X}'\mathbf{y} = \underset{2\times 1}{\begin{bmatrix} 690 \\ 1554.9 \end{bmatrix}}.$$

The normal equations are $\mathbf{X}'\mathbf{X}\boldsymbol{\beta} = \mathbf{X}'\mathbf{y}$, or

$$\begin{bmatrix} 20 & 40 \\ 40 & 93.56 \end{bmatrix}\begin{bmatrix} a \\ b \end{bmatrix} = \begin{bmatrix} 690 \\ 1554.9 \end{bmatrix},$$

which yields

$$\begin{aligned} 20a + 40b &= 690 \\ 40a + 93.56b &= 1554.9, \end{aligned}$$

as before. To solve the normal equations,

$$|\mathbf{X}'\mathbf{X}| = 20(93.56) - (40)^2 = 271.2 = nS_{XX},$$

so that

$$(\mathbf{X}'\mathbf{X})^{-1} = \begin{bmatrix} \dfrac{93.56}{271.2} & \dfrac{-40}{271.2} \\ \dfrac{-40}{271.2} & \dfrac{20}{271.2} \end{bmatrix} = \begin{bmatrix} 0.344985 & -0.147493 \\ -0.147493 & 0.073746 \end{bmatrix}$$

and

$$\begin{aligned} \hat{\boldsymbol{\beta}} = (\mathbf{X}'\mathbf{X})^{-1}\mathbf{X}'\mathbf{y} &= \begin{bmatrix} 0.344985 & -0.147493 \\ -0.147493 & 0.073746 \end{bmatrix}\begin{bmatrix} 690 \\ 1554.9 \end{bmatrix} \\ &= \begin{bmatrix} 0.344985(690) - 0.147493(1554.9) \\ -0.147493(690) + 0.073746(1554.9) \end{bmatrix} \\ &= \begin{bmatrix} 8.7035 \\ 12.8982 \end{bmatrix} = \begin{bmatrix} a \\ b \end{bmatrix}. \end{aligned}$$

Note that these results are almost identical to what we found in Chapter 2. The slight difference in the value for a (8.7035 here vs. 8.7036 before) is due to rounding error.

CORRECTING X FOR ITS MEAN

Suppose X is corrected for its mean, so that our model is

$$\hat{Y} = \alpha' + \beta(X - \bar{X}) + \varepsilon.$$

Let us denote the **X**-matrix for centered X-values by $\mathfrak{X}$. Then

$$\mathfrak{X} = \begin{bmatrix} 1 & X_1 - \bar{X} \\ 1 & X_2 - \bar{X} \\ \vdots & \vdots \\ 1 & X_n - \bar{X} \end{bmatrix};$$

the vector $\hat{\boldsymbol{\beta}}$ is

$$\hat{\boldsymbol{\beta}} = \begin{bmatrix} a' \\ b \end{bmatrix},$$

and **y** remains unchanged. Then

$$\mathfrak{X}'\mathbf{y} = \begin{bmatrix} 1 & 1 & \cdots & 1 \\ X_1 - \bar{X} & X_2 - \bar{X} & \cdots & X_n - \bar{X} \end{bmatrix} \begin{bmatrix} Y_1 \\ Y_2 \\ \vdots \\ Y_n \end{bmatrix}$$

$$= \begin{bmatrix} Y_1 + Y_2 + \cdots + Y_n \\ Y_1(X_1 - \bar{X}) + Y_2(X_2 - \bar{X}) + \cdots + Y_n(X_n - \bar{X}) \end{bmatrix}$$

$$= \begin{bmatrix} \sum Y_i \\ \sum Y_i(X_i - \bar{X}) \end{bmatrix}.$$

Recall from Section 2.4 that $\sum Y_i(X_i - \bar{X}) = \sum (Y_i - \bar{Y})(X_i - \bar{X}) = S_{XY}$, so that if X is corrected for its mean,

$$\mathfrak{X}'\mathbf{y} = \begin{bmatrix} \sum Y_i \\ S_{XY} \end{bmatrix}.$$

Now

$$\mathfrak{X}'\mathfrak{X} = \begin{bmatrix} 1 & 1 & \cdots & 1 \\ X_1 - \bar{X} & X_2 - \bar{X} & \cdots & X_n - \bar{X} \end{bmatrix} \begin{bmatrix} 1 & X_1 - \bar{X} \\ 1 & X_2 - \bar{X} \\ \vdots & \vdots \\ 1 & X_n - \bar{X} \end{bmatrix}$$

$$= \begin{bmatrix} 1 + \cdots + 1 & (X_1 - \bar{X}) + \cdots + (X_n - \bar{X}) \\ (X_1 - \bar{X}) + \cdots + (X_n - \bar{X}) & (X_1 - \bar{X})^2 + \cdots + (X_n - \bar{X})^2 \end{bmatrix}$$

$$= \begin{bmatrix} n & \sum (X_i - \bar{X}) \\ \sum (X_i - \bar{X}) & \sum (X_i - \bar{X})^2 \end{bmatrix} = \begin{bmatrix} n & 0 \\ 0 & S_{XX} \end{bmatrix}.$$

Note that in this case, $\mathfrak{X}'\mathfrak{X}$ is diagonal. The normal equations are still $\mathfrak{X}'\mathfrak{X}\hat{\boldsymbol{\beta}} = \mathfrak{X}'\mathbf{y}$, so that

$$\begin{bmatrix} n & 0 \\ 0 & S_{XX} \end{bmatrix} \begin{bmatrix} a' \\ b \end{bmatrix} = \begin{bmatrix} \sum Y_i \\ S_{XY} \end{bmatrix},$$

or

$$na' = \sum Y_i$$

$$S_{XX}b = S_{XY}.$$

These are two separate equations in two separate unknowns, so

$$a' = \frac{\sum Y_i}{n} = \bar{Y}$$

$$b = \frac{S_{XY}}{S_{XX}},$$

as before. Should one desire to solve the normal equations as

$$\hat{\boldsymbol{\beta}} = (\mathfrak{X}'\mathfrak{X})^{-1}\mathfrak{X}'\mathbf{y},$$

inversion of $\mathfrak{X}'\mathfrak{X}$ is simplified quite a bit because $\mathfrak{X}'\mathfrak{X}$ is diagonal:

$$(\mathfrak{X}'\mathfrak{X})^{-1} = \begin{bmatrix} \frac{1}{n} & 0 \\ 0 & \frac{1}{S_{XX}} \end{bmatrix}$$

and

$$\hat{\boldsymbol{\beta}} = \begin{bmatrix} \frac{1}{n} & 0 \\ 0 & \frac{1}{S_{XX}} \end{bmatrix} \begin{bmatrix} \sum Y_i \\ S_{XY} \end{bmatrix} = \begin{bmatrix} \frac{\sum Y_i}{n} \\ \frac{S_{XY}}{S_{XX}} \end{bmatrix}$$

$$= \begin{bmatrix} \bar{Y} \\ \frac{S_{XY}}{S_{XX}} \end{bmatrix} = \begin{bmatrix} a' \\ b \end{bmatrix}$$

Note the great reduction in labor achieved if X is corrected for its mean, or centered, so that $\mathfrak{X}'\mathfrak{X}$ is diagonal. A similar reduction is achieved if X is scaled to $Z = (X - \bar{X})/d$. The reduction in labor is even greater in some multiple regression problems, where $\mathfrak{X}'\mathfrak{X}$ is 3×3 or greater.

EXERCISES

3.27 Use the Concession Data in Exercise 1.5.

a. Display the matrix $\mathbf{X}$ and the vectors $\mathbf{y}$ and $\hat{\boldsymbol{\beta}}$ for fitting a simple linear regression equation.

b. Find $\mathbf{X}'\mathbf{X}$ and $\mathbf{X}'\mathbf{y}$.

c. Find $(\mathbf{X}'\mathbf{X})^{-1}$.

d. Solve for $\hat{\boldsymbol{\beta}}$. State the regression equation, and compare to the one found in Exercise 2.8 or 2.14.

3.28 Refer to Exercise 2.18 (Concession Data), in which X was corrected for its mean and a regression equation $\hat{Y} = a' + bU$ was found. Using values of U and Y,

a. Display the matrix $\mathbf{U}$ and the vectors $\mathbf{y}$ and $\hat{\boldsymbol{\beta}}$ for fitting a simple linear regression equation.

b. Find $\mathbf{U}'\mathbf{U}$ and $\mathbf{U}'\mathbf{y}$.

c. Find $(\mathbf{U}'\mathbf{U})^{-1}$.

d. Solve for $\hat{\boldsymbol{\beta}}$ and compare to regression coefficients in Exercise 2.18.

e. Substitute $U = X - \bar{X}$ and transform the equation $\hat{Y} = a' + bU$ to the equation $\hat{Y} = a + bX$.

3.29 Use the Heating Data in Exercise 1.8.

a. Show the matrix $\mathbf{X}$ and the vectors $\mathbf{y}$ and $\hat{\boldsymbol{\beta}}$ for fitting a simple linear regression equation.

b. Calculate $\mathbf{X}'\mathbf{X}$ and $\mathbf{X}'\mathbf{y}$.

c. Calculate $(\mathbf{X}'\mathbf{X})^{-1}$.

d. Solve for $\hat{\boldsymbol{\beta}}$. State the regression equation, and compare to the one obtained in Exercise 2.10 or 2.15.

3.30 Use the Repairs Data in Exercise 2.11.

a. Show the matrix $\mathbf{X}$ and the vectors $\mathbf{y}$ and $\hat{\boldsymbol{\beta}}$ for fitting a simple linear regression equation.

b. Calculate $\mathbf{X}'\mathbf{X}$ and $\mathbf{X}'\mathbf{y}$.

c. Calculate $(\mathbf{X}'\mathbf{X})^{-1}$.

d. Solve for $\hat{\boldsymbol{\beta}}$ and state the regression equation. Compare to the one found in Exercise 2.11.

3.31 A mail order sporting equipment company is looking for a quick way to estimate the number of orders received on any given day so that a decision can be made early in the day whether to pull some personnel off other jobs to help with processing orders. Someone suggested that it would be simple to weigh the sacks

TABLE 3.1
Mail Data

	Day 1	2	3	4	5	6	7	8
Mail (pounds)	211	310	207	266	417	327	289	186
Orders (hundreds)	10.3	13.0	10.2	12.0	16.7	14.0	12.8	9.7

	Day 9	10	11	12	13	14	15
Mail (pounds)	238	418	205	424	325	264	403
Orders (hundreds)	11.5	16.2	10.7	17.0	13.5	12.1	16.2

of mail and use that information to estimate the number of orders received. For three weeks, the weights of mail bags and the numbers of orders are recorded. (Note $n = 15$, because orders are not processed on weekends.)

Using the matrix formulation, find the simple linear regression equation relating number of orders to weight of the mail.

3.32 Refer to the Police Data in Exercise 1.7.

a. Display the matrix **X** and the vectors **y** and $\hat{\boldsymbol{\beta}}$ necessary to fit a curvilinear equation $\hat{Y} = a + bX + cX^2$ to the data.

b. Find $\mathbf{X}'\mathbf{X}$ and $\mathbf{X}'\mathbf{y}$.

c. Find $(\mathbf{X}'\mathbf{X})^{-1}$.

d. Find $\hat{\boldsymbol{\beta}}$. State the regression equation.

3.33 Refer to the Fertilizer Data in Exercise 2.20. Using given X-values,

a. Display the matrix **X** and the vectors **y** and $\hat{\boldsymbol{\beta}}$ necessary to fit a linear equation $\hat{Y} = a + bX$ to the data.

b. Find $\mathbf{X}'\mathbf{X}$ and $\mathbf{X}'\mathbf{y}$.

c. Find $(\mathbf{X}'\mathbf{X})^{-1}$.

d. Find $\hat{\boldsymbol{\beta}}$. State the regression equation. Compare the answer to that of Exercise 2.20e.

3.34 Repeat Exercise 3.33, but scale the X-values. Compare the regression equation to the one found above.

3.35 Refer to the Deprivation Data in Exercise 2.21.

a. Display the matrix **Z** and the vectors **y** and $\hat{\boldsymbol{\beta}}$ necessary to fit a curvilinear equation $\hat{Y} = a' + b'Z + c'Z^2$ to the data.

b. Find $\mathbf{Z}'\mathbf{Z}$ and $\mathbf{Z}'\mathbf{y}$.

c. Find $(\mathbf{Z}'\mathbf{Z})^{-1}$.

d. Find $\hat{\boldsymbol{\beta}}$. State the regression equation. Compare to answer of Exercise 2.21c.

3 . 8

SUMMARY

We have now explored three equivalent methods for finding a least squares regression equation. Normal equations have been solved as simultaneous equations, by using algebraic formulas, and finally by using matrices. All three methods produce the same results, although at first they may appear to be different.

Since the methods presented in this chapter have universal applicability, they lay the foundation for the study of multiple regression. Understanding some of the problems involved in solving simultaneous equations will also provide insight into the interrelationships among variables in multiple regression.

While we have in theory solved all solvable sets of simultaneous equations, there remains the tedious job of actually obtaining a solution for a real data problem. We shall next see how to make a computer do this work.

FOUR

THE USE OF COMPUTER PROGRAMS IN SIMPLE REGRESSION

4.1 INTRODUCTION

So far, our simple regression problems have been small enough to do with the aid of only a calculator. The larger n becomes, however, the more time-consuming and error-prone the calculations become. You probably are not very eager to calculate $\sum X_i Y_i$ even if $n = 20$: imagine your feelings if n were 175! When we get to multiple regression, the computations become enormous. Thus, it is important to be able to use a computer for the calculations, in the interest of both time and accuracy. While it may seem unnecessary to run a simple regression problem through the computer, one must begin with relatively simple problems.

You may appreciate that it is not easy to program a computer to perform a regression analysis. Fortunately, programming experts have written programs to perform a wide variety of complicated statistical analyses, including regression, and have made these programs available. Four widely available sets of programs in large research or university computing centers are the Biomedical Computer Programs (BMD), Minitab, the Statistical Analysis System (SAS), and the Statistical Package for the Social Sciences (SPSS). In this chapter we introduce each of these packages and show how to use them to perform simple linear regression analyses. It is not expected that the reader will learn all four packages, although all are easy enough to use. In some instances, your choice of package will be determined by what is available in your computing center; otherwise you may tend to use a particular package because of a preference for the way the instructions are written or for the output available.

If you are using a personal computer or a package other than the ones discussed here, you will probably be able to obtain results similar to those we display. While the concern of this chapter is to show quickly how to use BMD, Minitab, SAS, and SPSS, the emphasis of this book is to teach the reader how to interpret the computer results.

We assume that the reader is familiar enough with computing to be able to punch cards or to compose programs on a console. For historical reasons,

Card
Printout

we refer to a line in a program as a **card.** Similarly, we often refer to **printout,** although this term does not refer only to hard copy. On our written worksheets we are careful to distinguish between the letter I and the number 1, and also between the letter "oh," denoted by O, and the number zero, written Ø. (When proofreading cards, notice that the letter O will typically be more squarish in computer type than the number 0). Most of the sample programs we show will accept data entered in a free format. For some problems, however, more exact formatting is required. A brief review of format statements is given in Appendix C.

4 . 2

BIOMEDICAL (BMD) P-SERIES PACKAGES

The Department of Biomathematics at UCLA has developed many computer programs to perform statistical analyses and has made these programs available to computing centers across the country. Using these "canned" programs, we can perform very complicated statistical analyses by merely giving the computer some simple instructions and the data to be analyzed. Complete instructions for using all the BMDP† programs are found in W. J. Dixon, ed., *BMDP Statistical Software, 1981* (Berkeley: University of California Press, 1981). In this text we will give some general instructions for running BMDP programs and will show how to perform various regression analyses.

Figure 4.1 shows a worksheet for using the BMDP1R program to perform a simple linear regression analysis of the house size-price example. (The 1R appended to the BMDP indicates that this is the first (1) regression (R) program listed in the manual. There are several other regression programs, for use in more complicated analyses.) We will go through this worksheet step-by-step and explain how to submit data for a simple linear regression analysis. Each line on the worksheet indicates a separate card to be punched.

† The name BMDP is a combination of an abbreviation of the word *biomedical* and an abbreviation for *parameter control language*.

FIGURE 4.1 Worksheet for Performing a Simple Linear Regression Analysis on House Size-Price Data Using BMDP1R

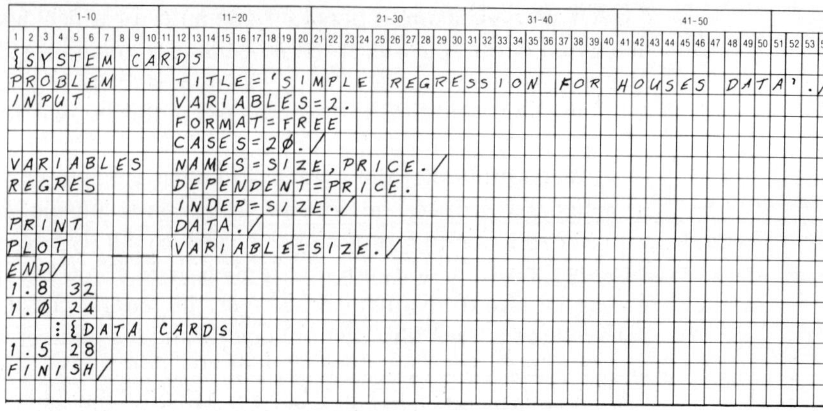

GENERAL SETUP OF A BMDP PROGRAM

System/JCL Cards First, several **system cards,** or **JCL** (job control language) **cards,** precede any program submitted to a computing center. These cards contain information concerning the user's account number, the work space required for the analysis, and the language or canned program to be used. Since the specifics of how these cards are to be punched vary from one computing center to the next, you should ascertain this information for your center. One of these cards will contain the statement

```
...EXEC BIMED,PROG=BMDP1R...
```

which instructs the computer to use (execute) the 1R program to perform this analysis.

The cards `PROBLEM` through `END/` define a BMDP problem to be analyzed. These cards instruct the computer as to what kinds of analyses to perform on your data and how to read your data cards. The 20 cards following `END/` contain the data to be analyzed. Finally, the `FINISH/` card tells the computer that you wish to perform no further analyses. The difference between `END/` and `FINISH/` is that if you wanted to analyze another problem, then after the data for the first problem you could enter another set of cards, `PROBLEM` through `END/`, and another data set. Only when you have done all the analyses you want to, and are ready to sign off, would you insert the `FINISH/` card.

Let us first discuss the data cards. There are several ways that data can be entered for BMDP analysis. Perhaps the easiest way in most cases is to enter each pair of observations on a separate card. It does not matter whether you enter X-values first or Y-values first, as long as you are consistent from card to card. Data may be in a fixed format (see Appendix C) or a free format (see the next section) as long as the values are separated by at least one space and no spaces occur within a value. On the worksheet we see that X-values (sizes) are entered in columns 1–3, with the decimals punched, and that

Y-values (prices) are entered in columns 5 and 6. For other methods of entering data (for example, what to do if it takes more than one card to enter values for a single observation), refer to the BMDP manual.

THE BMDP LANGUAGE

The heart of the worksheet is the statement of the problem. BMDP problems are written in sentences and paragraphs. Each sentence within a paragraph ends with a period, and each paragraph ends with a slant (/). We see in Figure 4.1 that this problem contains seven paragraphs: PROBLEM, INPUT, VARIABLES, REGRES, PRINT, PLOT, and END. On the worksheet, paragraph names begin in column 1, and sentences within paragraphs are indented for ease of reading. This format is not mandatory, however. For example, the entire REGRES paragraph could have been written on one card as

```
REGRES DEPENDENT=PRICE.INDEP=SIZE.
```

Words must be separated by blanks or by punctuation or a symbol, but the number of blanks between words or around symbols is arbitrary. Blanks may or may not be left around the equals sign but *not* around other punctuation. In addition, the equals sign may be replaced by either ARE or IS. If any label or title contains punctuation or spaces—any symbol other than a letter or a number—the title or label must be enclosed in single quotation marks. Most words that define sentences or paragraphs may be abbreviated to the first three or four letters. For example, PROBLEM may be abbreviated as PROB, and VARIABLES as VAR. Refer to the BMDP manual for acceptable abbreviations.

The first three paragraphs in Figure 4.1 define the problem and instruct the computer how to read your data. These first three paragraphs must be included in any BMDP program. The first paragraph tells the computer that a problem is to be defined and gives the problem a title. Notice that the title, SIMPLE REGRESSION FOR HOUSES DATA, is enclosed in single quotation marks, since there are spaces between the words in the title. The TITLE sentence ends with a period, and it is the only sentence in the PROBLEM paragraph. The paragraph ends with a slant. The INPUT paragraph tells the computer that your data consist of the values of two variables, that data are entered in a free format, and that there are 20 observations. By a **free format** is meant that each pair of values is entered on a separate card or line, the two values are separated by at least one space, no spaces are left within a value, and the decimal point (if any) is punched. In Figure 4.1 we see that the values of *X* are entered in columns 1–3 with the decimal point in column 2; column 4 is left blank; and values of *Y* are entered in columns 5 and 6. (It is not necessary to always enter *X*-values first in every program, but if *X*-values are entered first on one card, they must be entered first on all cards.) Since the data are entered consistently in this manner, the format sentence could have been written to indicate data in a **fixed format** as

Free Format

Fixed Format

```
FORMAT='(F3.1,F3.0)'.
```

(Refer to Appendix C if you are not familiar with how to write fixed-format statements.) As in the TITLE sentence, since the FORMAT definition contains symbols other than numbers and letters, it must be enclosed in single quotation marks.

Finally, the VARIABLES paragraph tells the computer that the first number entered on any data card is a value of the variable SIZE and that the second number is a value of the variable PRICE. Variable names must begin with a character but may contain numbers. They may not exceed eight characters in length, and if they contain spaces or any other symbols, they must be enclosed within single quotation marks.

Now that the computer knows how to read your data, you may instruct it to perform some analyses. The REGRESSION paragraph (abbreviated to REGRES on the worksheet) instructs the computer to perform a regression analysis. The sentences in this paragraph tell the computer which variable is the dependent variable and which is the independent. If the sentence INDEP=SIZE. were omitted, the computer would assume that SIZE was to be the independent variable, since it is the only other variable left after PRICE was specified as the dependent variable. (In multiple regression problems, as we shall see, if no INDEP sentence is specified, all other variables entered will be used as independent variables; otherwise, using an INDEP sentence will allow you to select a subset of the possible independent variables to be used as predictors.) In all cases, the DEPENDENT sentence must be used to specify the response variable.

OUTPUT FROM THE REGRESSION PARAGRAPH

Let us pause now and see what the results from a BMDP1R regression analysis of the house size-price data would look like. Your printout will contain a page or two that reproduces the instructions you entered on cards. If you have made an error somewhere, notice of this and a clue as to what kind of error was made will appear on these first pages.

Figure 4.2 shows the printout resulting from the REGRES paragraph. In Figure 2(a), the first information printed is a table of simple descriptive statistics for both variables. The table should be self-explanatory, as the mean, standard deviation, and minimum and maximum values should be familiar to the reader. The coefficient of variation is the ratio of the standard deviation to the mean.

The regression analysis is shown in Figure 4.2(b). Note that there is a message:

```
DEPENDENT VARIABLE...2 PRICE
```

The 2 indicates that PRICE was the second variable entered on your data cards. MULTIPLE R, MULTIPLE R-SQUARE, STD ERROR OF EST, and ANALYSIS OF VARIANCE will be explained later. What is of interest to us now is that under COEFFICIENT we see the value 8.70355 across from INTERCEPT and the value 12.89823 across from SIZE 1. (The 1 is

FIGURE 4.2 BMDP1R Regression Analysis

(a) Simple statistics from the BMDP1R analysis

NUMBER OF CASES READ.............. 20

VARIABLE	MEAN	STANDARD DEVIATION	COEFFICIENT OF VARIATION	MINIMUM	MAXIMUM
1 SIZE	2.00000	0.84480	0.42240	0.80000	3.60000
2 PRICE	34.49994	11.17563	0.32393	17.00000	52.00000

(b) BMDP1R regression analysis of house size-price data

REGRESSION TITLE IS
SIMPLE REGRESSION FOR HOUSES DATA

DEPENDENT VARIABLE...............2 PRICE
TOLERANCE.........................0.0100
ALL DATA CONSIDERED AS A SINGLE GROUP

MULTIPLE R	0.9750	STD ERROR OF EST	2.5506
MULTIPLE R-SQUARE	0.9507		

ANALYSIS OF VARIANCE

	SUM OF SQUARES	DF	MEAN SQUARE	F RATIO	P(TAIL)
REGRESSION	2255.8992	1	2255.8992	346.765	0.0000
RESIDUAL	117.1001	18	6.5056		

VARIABLE	COEFFICIENT	STD ERROR	STD REG COEFF	T	P(2 TAIL)	TOLERANCE
INTERCEPT	8.70355					
SIZE 1	12.89823	0.69265	0.975	18.622	0.0	1.00000

for the first variable entered on data cards.) Thus, we have that the regression equation is

$$\hat{Y} = 8.70355 + 12.89823X.$$

The rest of the information for the regression analysis will be explained later.

THE PRINT PARAGRAPH AND ITS OUTPUT

The PRINT paragraph is included (see Figure 4.1) so that you can check the accuracy of your keypunching and obtain some other information important to understanding what is going on in your data.

The PRINT DATA command results in the printout shown in Figure 4.3. At the right we see values of X (SIZE) and Y (PRICE). You should carefully check these values to make sure you made no errors in entering your data. Asterisks next to values indicate values that are more than 1, 2, or 3 standard deviations from their mean. These are useful in helping you to notice unusual cases that might have resulted from errors in recording data, or they might indicate an observation so unusual that it could be considered a special case or to have come from a different population from the one you had thought

FIGURE 4.3 Printout From the PRINT DATA Command in BMDP1R

```
LIST OF PREDICTED VALUES, RESIDUALS, AND VARIABLES
   NOTE—NEGATIVE CASE NUMBER DENOTES A CASE WITH MISSING VALUES.
         THE NUMBER OF STANDARD DEVIATIONS FROM THE MEAN IS DENOTED BY UP TO 3
         ASTERISKS TO THE RIGHT OF EACH RESIDUAL OR VARIABLE.
MISSING VALUES AND VALUES OUT OF RANGE ARE DENOTED BY VALUES
GREATER THAN OR EQUAL TO 0.2127E 38 IN ABSOLUTE VALUE.
```

CASE LABEL	NO.	RESIDUAL	PREDICTED VALUE	VARIABLES 1 SIZE	2 PRICE
	1	0.7965E-01	31.92	1.800	32.00
	2	2.398	21.60	1.000*	24.00
	3	-3.631*	30.63	1.700	27.00
	4	2.181	44.82	2.800	47.00*
	5	-2.080	37.08	2.200	35.00
	6	-2.022	19.02	0.8000*	17.00*
	7	-3.137*	55.14	3.600*	52.00*
	8	-2.892*	22.89	1.100*	20.00*
	9	3.500*	34.50	2.000	38.00
	10	2.761*	42.24	2.600	45.00
	11	5.631**	38.37	2.300	44.00
	12	-1.312	20.31	0.9000*	19.00*
	13	0.8186	24.18	1.200	25.00
	14	-2.558*	52.56	3.400*	50.00*
	15	-0.6305	30.63	1.700	30.00
	16	2.051	40.95	2.500	43.00
	17	0.2389	26.76	1.400	27.00
	18	-1.268	51.27	3.300*	50.00*
	19	-0.7964E-01	37.08	2.200	37.00
	20	-0.5089E-01	28.05	1.500	28.00

```
SERIAL CORRELATION OF RESIDUALS = 0.0324
```

you sampled. (See the section on Outliers in Chapter 9.) In addition to the data being reproduced, the PRINT DATA command will also give you values of $\hat{Y}$ corresponding to each X-value in the sample (PREDICTED VALUE) and values of $Y - \hat{Y}$ (RESIDUAL). These values are for determining whether or not a simple linear model is best for these data.

Finally, SERIAL CORRELATION OF RESIDUALS is a very useful statistic when your data were collected in order over time. Since this is not the case with the house size-price data, this number is meaningless in this problem. (See the section on Serial Correlation in Chapter 9 for an explanation of the meaning of a serial correlation.)

THE PLOT PARAGRAPH AND ITS OUTPUT

A picture is worth a thousand words, so the saying goes, and thus we would like the computer to draw some graphs of the data and the results of the regression analysis. The VARIABLE=SIZE sentence in the PLOT paragraph (Figure 4.1) will produce Figure 4.4. Figure 4.4(a) is a scatter diagram of the

FIGURE 4.4 BMDP1R Graphs

(a) BMDP1R plot of data and regression line

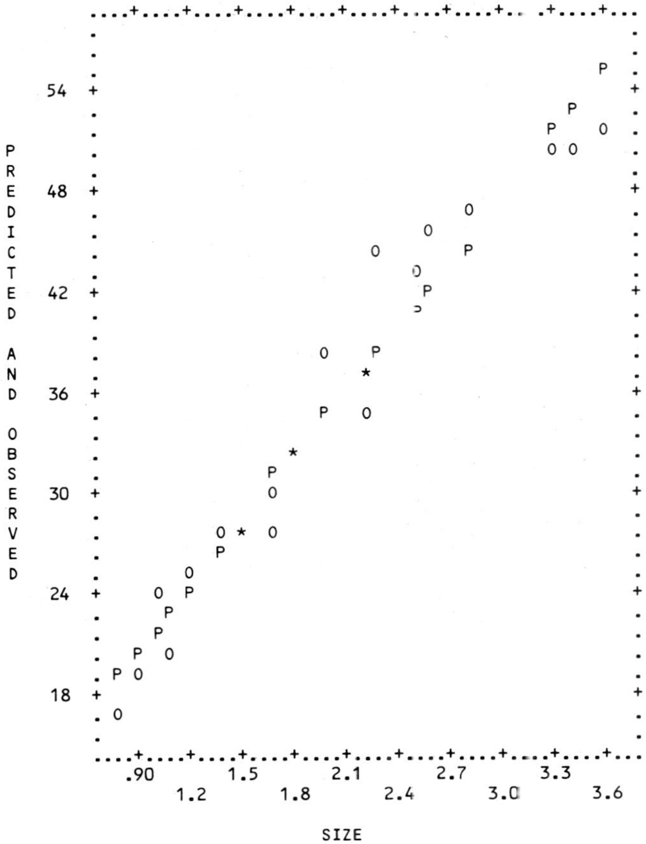

(b) Plot of residuals versus size using BMDP1R

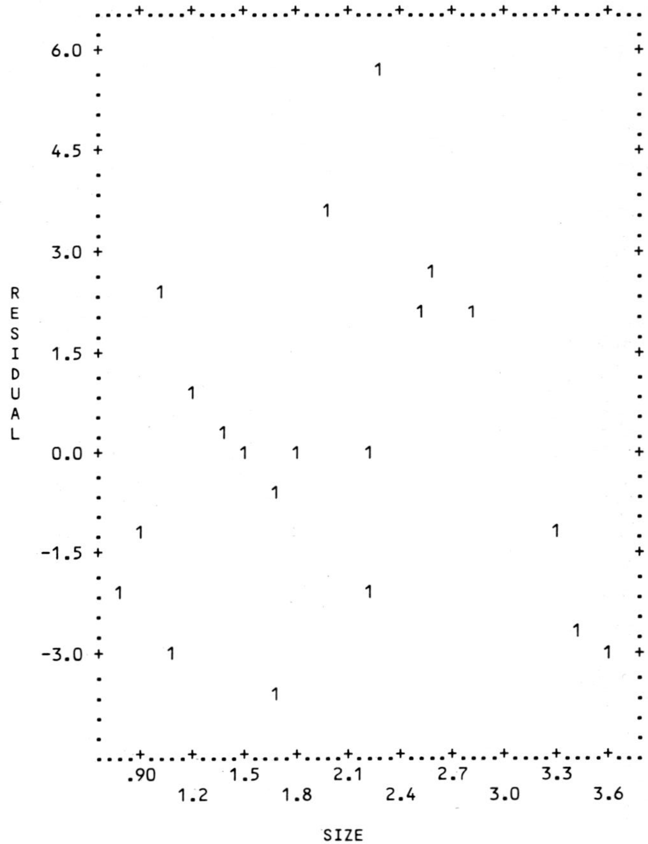

data points (indicated by the letter O for observed value) with the regression line (indicated by P for predicted value) superimposed. Where observed and predicted values coincide, so that the printer cannot distinguish them, the symbol * is printed. The P's do not all lie exactly on a straight line, because the graph is drawn with a typewriter-like machine, so that points cannot be located completely accurately. Compare this graph to Figure 2.8.

Figure 4.4(b) is a graph of the residuals, $Y - \hat{Y}$, versus size, X. The number of residuals at a given location is indicated by a number. Since there are no coincident points in Figure 4.4(b), all points are plotted with the symbol 1. Had there been two coincident points, they would have been plotted with the symbol 2, and so on. This second graph is essentially the same one as Figure 4.4(a), as if the regression line had been rotated so that it had a zero intercept and horizontal slope, but with the points maintaining their positions relative to the line. We will see in Chapter 9 that it is important to look at graphs of residuals to evaluate the appropriateness of a simple straight-line model for these data.

The analysis produced by the program in Figure 4.1 is a very simple one, given the capabilities of the BMDP1R program. In later chapters we will see how this program can be used in multiple regression analysis and also how other BMDP regression programs can be used to fit more complicated models and perform more sophisticated analyses.

EXERCISES

4.1 Refer to the Concession Data in Exercise 1.5. Use the BMDP1R computer program to perform a simple linear regression analysis. Obtain a scatter diagram and a graph of the regression equation. Compare results to those in Exercises 1.5 and 2.8 or 2.14. Also plot residuals against size, X.

4.2 Use BMDP1R to fit a simple linear regression equation to the Smokers Data in Exercise 1.9. Obtain a scatter diagram and a graph of the regression equation. Also plot residuals against X.

4.3 Use BMDP1R to perform a simple linear regression analysis of the Class Size Data in Exercise 2.17. Obtain a scatter diagram and a graph of the regression equation. Compare to Exercise 2.17. Also plot the residuals against X.

4.4 Appraised values and sale prices for a simple random sample of single family residences sold during the last year in a certain city are shown in Table 4.1. Use BMDP1R to obtain a simple linear regression analysis of the relationship between sale price (Y) and appraised value (X). Graph the data and the regression equation, and graph the residuals.

TABLE 4.1
Appraisal Data

House No.	Appraised Value (dollars)	Sale Price (thousands of dollars)	House No.	Appraised Value (dollars)	Sale Price (thousands of dollars)
1	27,070	60.0	13	23,540	42.5
2	26,740	52.5	14	31,140	55.9
3	28,160	53.0	15	26,270	46.0
4	23,750	45.5	16	23,400	41.5
5	34,310	63.8	17	31,760	53.5
6	34,240	60.9	18	42,160	68.2
7	34,880	61.9	19	30,640	51.3
8	28,690	51.3	20	35,850	58.5
9	28,010	49.9	21	32,940	54.0
10	41,990	71.0	22	25,530	44.0
11	25,020	44.9	23	28,540	48.0
12	23,790	42.9	24	28,040	47.0

4 . 3
MINITAB

GETTING STARTED

Batch Mode

Interactively

Minitab is "a general purpose statistical computing system, designed especially for students and researchers who have no previous experience with computers."† While it can be used in **batch mode,** which means that you write an entire program, submit it to be executed, and receive all your results at once, Minitab is really designed to be used **interactively.** This means that a command is immediately carried out and results are shown at your terminal. Most installations allow Minitab to be used in either mode; you should check with your own computing center to ascertain if this is the case. We will illustrate the use of Minitab in the interactive mode. If only batch mode is available to you, don't worry—there is no difference in the way data are entered or commands are written. What you will have to determine are the procedures for submitting your program for execution. Even if you are using Minitab interactively, you will have to ascertain the procedures in effect at your installation for connecting to the system, as these vary from one computing center to another.

Assuming that you are using Minitab interactively, once you have made connection with the Minitab system, your terminal will prompt you with a dash (—). This indicates that Minitab is awaiting your instructions. To use Minitab, instruct the computer to read your data, and then type commands

† Thomas A. Ryan, Jr., Brian L. Joiner, and Barbara F. Ryan: *Minitab Reference Manual* (1981), p. 1. Minitab Project, Statistics Department, 215 Pond Laboratory, The Pennsylvania State University, University Park, PA 16802.

for analyses to be performed on the data. Data are entered into numbered "columns" and these columns are referred to in the commands designating analyses.

E N T E R I N G   D A T A

Suppose we want to use Minitab to analyze the house size-price data. (Refer to Figure 4.5.) When prompted by the dash, type

```
READ DATA INTO COLUMNS C1 AND C2
```

and press the carriage return. Then begin entering pairs of values of the variables size and price, with decimal points typed and with at least one space between the two values. Commas may be used to separate values, with or without spaces around them. Thus, commas should not be used within a number. For example, the number 22,364 should be entered as 22364 or else it will be read as two numbers, 22 and 364. After each pair of values is entered, press the carriage return so that each pair of values is on a different line. When all data have been entered, type on a separate line the END OF DATA command.

FIGURE 4.5
Entering Data With Minitab (CR = Carriage Return)

```
—READ DATA INTO COLUMNS C1 AND C2  (CR)
—1.8 32  (CR)
—1.0 24  (CR)
—1.7 27  (CR)
—2.8 47  (CR)
—2.2 35  (CR)
—0.8 17  (CR)
—3.6 52  (CR)
—1.1 20  (CR)
—2.0 38  (CR)
—2.6 45  (CR)
—2.3 44  (CR)
—0.9 19  (CR)
—1.2 25  (CR)
—3.4 50  (CR)
—1.7 30  (CR)
—2.5 43  (CR)
—1.4 27  (CR)
—3.3 50  (CR)
—2.2 37  (CR)
—1.5 28  (CR)
—END OF DATA  (CR)
```

Once you get used to Minitab, you can abbreviate a lot. Minitab has a dictionary of commands that it recognizes by the first four letters of the commands only. Thus, if a command like MULTIPLY has more than four letters, you need only type the first four: MULT. If you misspell command names (as long as you get the first four letters right), it won't make any difference.

In addition, you need only indicate the numbers of the columns to be used. For example, instead of writing

```
READ DATA INTO COLUMNS C1 and C2,
```

we could have written simply

```
READ C1 C2
```

The rest of the text is solely for your own ease of reading or writing and is ignored by the computer.

If Minitab does not understand what you are trying to tell it to do, it will let you know with an error message, and you can try again. You need not worry about starting commands or data lines in any particular column on your terminal; and if you forget to leave spaces or make other similar kinds of errors, Minitab will often be able to figure out what you intended.

Data may also be entered by using the SET command. The SET command reads all data into a single column. For example,

SET THE FOLLOWING DATA INTO C1 (or SET C1)

2 2 3 6 5 6 1 1

will form a column, C1, containing those numbers in that order.

Columns may be given names, which makes it easier to remember which values belong to which variable and produces output that is easier to read, too. Columns may be named as the data set is being defined or after the data are already entered. To name columns as the data set is being entered, amend READ C1 C2 to

```
READ 'SIZE' C1 'PRICE' C2
```

where the variable names are enclosed in single quotation marks. After the data are entered, the columns can be named or renamed by using the NAME command. For example, if data were entered using READ C1 C2, then you could write

```
NAME C1 'SIZE' C2 'PRICE'
```

In subsequent commands, data in column 1 could then be used by typing either C1 or 'SIZE', and data in column 2 could be called up by typing either C2 or 'PRICE'.

If several variables are to be used in an analysis, requiring many columns for storage, a list of consecutive column numbers may be indicated by using a dash. For example, you could type

READ C1, C2, C3, C4 or READ C1-(C4

CHECKING THE DATA

To check the data to make sure you have entered numbers correctly, you may want to use the PRINT command. Simply type PRINT and the column number. Figure 4.6(a) shows what happens if you type just PRINT C1; Figure 4.6(b) shows the result of PRINT C1 C2.

FIGURE 4.6 Printing With Minitab

(a) Printing one column

```
—PRINT C1
COLUMN          C1
COUNT           20
          1.80000     1.00000   1.70000   2.80000   2.20000   0.80000
          3.60000     1.10000   2.00000   2.60000   2.30000   0.90000
          1.20000     3.40000   1.70000   2.50000   1.40000   3.30000
          2.20000     1.50000
```

(b) Printing more than one column

```
—PRINT C1 C2
COLUMN          C1        C2
COUNT           20        20
ROW
 1         1.80000       32.
 2         1.00000       24.
 3         1.70000       27.
 4         2.80000       47.
 5         2.20000       35.
 6         0.80000       17.
 7         3.60000       52.
 8         1.10000       20.
 9         2.00000       38.
10         2.60000       45.
11         2.30000       44.
12         0.90000       19.
13         1.20000       25.
14         3.40000       50.
15         1.70000       30.
16         2.50000       43.
17         1.40000       27.
18         3.30000       50.
19         2.20000       37.
20         1.50000       28.
```

If you have made an error in entering data, it may be corrected by using the SUBSTITUTE command. For example,

SUBSTITUTE 1.9 INTO ROW 1 OF C1 or SUBS 1.9 ROW 1 C1

will change the 1.8 value of the variable "size" for house 1 to 1.9.

SIMPLE STATISTICS

Minitab will count the number of observations in each column of data and will calculate simple statistics such as the sum, the mean, the sum of squares, standard deviation, maximum, and minimum. Figure 4.7 shows these statistics for the house size-price data.

FIGURE 4.7
Simple Statistics From Minitab

```
—COUNT C1
 COUNT   =        20.000
—SUM C1
 SUM     =        40.000
—AVERAGE C1
 AVERAGE =         2.000
—STANDARD C1
 ST DEV  =         0.8448
—MAXI C1
 MAXIMUM =         3.600
—MINI C1
 MINIMUM =         0.800
—SSQ C1
 SSQ     =        93.560
—COUNT C2
 COUNT   =        20.000
—SUM C2
 SUM     =       690.000
—AVER C2
 AVERAGE =        34.500
—STANDARD C2
 ST DEV  =        11.176
—MAXI C2
 MAXIMUM =        52.000
—MINI C2
 MINIMUM =        17.000
—SSQ C2
 SSQ     =     26178.000
```

In Figure 4.7 values of the variable "size" were stored in column 1; those for "price" were put in column 2. When Minitab is used interactively, the simple statistics are printed following the command and are not automatically stored for further use. To store these values, simply include in the command statement the *constant number* (K) of the value you want stored. For example, to save $\overline{X}$ as a constant, K3, type

`AVERAGE THE VALUES IN COLUMN C1, STORE IN CONSTANT K3` , or

`AVER C1 K3`

Then $\overline{X}$ can be used in further calculations by simply referring to K3 in a command.

USING MINITAB TO PERFORM A SIMPLE LINEAR REGRESSION

Once data are entered, commands for analysis can be written. To regress Y = price on X = size, you must type the REGRES command, indicate the name or column number of the dependent variable, indicate how many predictors are to be used, and tell the computer the names or column numbers of the predictors. Suppose the data have been entered simply by the command READ C1 C2, where the values of the variable "size" are in column 1 and those for "price" are in column 2. Then the regression command is

REGRES Y IN C2 ON 1 X IN C1 or REGR C2 1 C1

Had column 1 been named 'SIZE' and column 2 'PRICE', then the command could be written as above or as

REGR 'PRICE' 1 'SIZE'

Unless otherwise specified, Minitab will print out a "brief" summary of the resulting analysis. To obtain *all* pertinent information, precede the REGRES command by a NOBRIEF command. The resulting printout is shown in Figure 4.8. The regression equation is easy to spot in Figure 4.8: It is given as Y = 8.70 + 12.9 X1. Additionally, values of a and b with more decimal places are given under the heading COEFFICIENT as 8.704 and 12.8982, respectively. Most of the rest of the information on the printout will be explained later. Toward the middle of the page, however, we see row numbers, or observation numbers, and values of X and Y printed out. (X is called X1 on the printout. Had the columns been named "size" and "price," respectively, then instead of C1, SIZE would have appeared under X1, and instead of C2 would be PRICE.) Figures in the column labeled PRED Y VALUE are values of $\hat{Y}_i$, obtained by substituting the value X_i into the regression equation. For example, for house 1, 8.704 + 12.8982 (1.80) = 31.920. In the column RESIDUAL the value for house 1 is $Y_1 - \hat{Y}_1 = 32.000 - 31.920 = 0.080$. The standardized residuals (ST RES) are the residuals divided by their standard deviation. You can verify that the standard deviation of the residuals is 2.5507, so that for house 1 the standardized residual is 0.080/2.5507 = 0.03.

You can also verify that the mean of the residuals is zero (except possibly for rounding error). Thus, standardized residuals express how many standard deviations each residual is from its mean, and in what direction. We shall say more about residuals later. In addition, the values in the column labeled ST DEV PRED Y will be discussed in Chapter 6.

Before leaving the inspection of residuals, note that two of the residuals are flagged—one with an X and one with an R. These flags are explained below the table: R denotes an observation with a large standardized residual and will

FIGURE 4.8 Simple Linear Regression Analysis With Minitab

```
—NOBRIEF

—REGRES Y IN C2 ON 1 PREDICTOR IN C1

THE REGRESSION EQUATION IS
Y = 8.70 + 12.9 X1

                                      ST DEV        T-RATIO =
       COLUMN        COEFFICIENT      OF COEF       COEF/S.D.
                       8.704          1.498          5.81
X1     C1             12.8982         0.6926        18.62

THE ST DEV OF Y ABOUT REGRESSION LINE IS
S = 2.551
WITH (20 -2) = 18 DEGREES OF FREEDOM

R-SQUARED = 95.1 PERCENT
R-SQUARED = 94.8 PERCENT, ADJUSTED FOR DF

ANALYSIS OF VARIANCE

DUE TO              DF        SS                  MS=SS/DF
REGRESSION           1       2255.900             2255.900
RESIDUAL            18        117.100                6.506
TOTAL               19       2373.000

       X1      Y          PRED Y    ST DEV
ROW    C1      C2         VALUE     PRED Y     RESIDUAL     ST RES

  1    1.80    32.000     31.920    0.587        0.080        0.03
  2    1.00    24.000     21.602    0.897        2.398        1.00
  3    1.70    27.000     30.631    0.607       -3.631       -1.47
  4    2.80    47.000     44.819    0.795        2.181        0.90
  5    2.20    35.000     37.080    0.587       -2.080       -0.84
  6    0.80    17.000     19.022    1.008       -2.022       -0.86
  7    3.60    52.000     55.137    1.246       -3.137       -1.41X
  8    1.10    20.000     22.892    0.845       -2.892       -1.20
  9    2.00    38.000     34.500    0.570        3.500        1.41
 10    2.60    45.000     42.239    0.706        2.761        1.13
 11    2.30    44.000     38.369    0.607        5.631        2.27R
 12    0.90    19.000     20.312    0.952       -1.312       -0.55
 13    1.20    25.000     24.181    0.795        0.819        0.34
 14    3.40    50.000     52.558    1.125       -2.558       -1.12
 15    1.70    30.000     30.631    0.607       -0.631       -0.25
 16    2.50    43.000     40.949    0.667        2.051        0.83
 17    1.40    27.000     26.761    0.706        0.239        0.10
 18    3.30    50.000     51.268    1.066       -1.268       -0.55
 19    2.20    37.000     37.080    0.587       -0.080       -0.03
 20    1.50    28.000     28.051    0.667       -0.051       -0.02

R DENOTES AN OBS WITH A LARGE ST RES.
X DENOTES AN OBS WHOSE X VALUE GIVES IT LARGE INFLUENCE.

DURBIN-WATSON STATISTIC = 1.94

(X-PRIME X)INVERSE

          0              1

  0     0.344985
  1    -0.147493       0.073746
```

appear whenever a residual is greater than 2 or less than -2, i.e., whenever a residual is more than 2 standard deviations away from its mean. Your attention is called to such values because they may indicate the presence of outliers—unusually large or small values—in the data. Outliers are discussed in Chapter 9. The X flag denotes an observation whose X-value gives it a large influence. This is done because the least squares line can be pulled upward or downward quite a bit by points at either end of the range of X-values. Such instances are also discussed in Chapter 9. Intelligent use of computer programs requires that output be examined with an eye toward the appropriateness of the analysis performed.

Finally, note that the $(\mathbf{X}'\mathbf{X})^{-1}$ matrix is shown as part of the Minitab output if the NOBRIEF command is used. Since $(\mathbf{X}'\mathbf{X})^{-1}$ is symmetric, only the off-diagonal element below the diagonal is printed.

PLOTS USING MINITAB

To obtain a scatter diagram of the data, simply write

```
PLOT Y IN C2 VS X IN C1   or   PLOT C2 C1
```

The first variable mentioned will be plotted on the vertical axis.

In order to obtain a graph of the regression equation, we need to plot values of $\hat{Y}$ versus X. However, we have only values of X and Y in our data set—values of $\hat{Y}$ are obtained as a result of the regression analysis. We can, however, add $\hat{Y}$-values to the data set by making an adjustment to the REGRES command. In a similar fashion, the residuals and the regression coefficients a and b can be added to the data set. Consider the command

```
REGRES Y IN C2 on 1 X in C1\STORE ST RESID IN C3\
   STORE PRED IN C4\STORE COEFFS IN C5
```

or simply

```
REGR C2 1 C1\C3\C4\C5
```

In general, standardized residuals will be stored in whichever column appears first after a backwards slant, the predicted values will be stored in the next column mentioned, and the regression coefficients in the third column indicated. These columns may be named, as well. For example, one could write

```
NAME C3 'R' C4 'YHAT' C5 'COEFFS'
```

Then, provided that columns C1 and C2 have also been given the names 'SIZE' and 'PRICE', respectively, a graph of the residuals against X is obtained by writing

```
PLOT 'R' VS 'SIZE'
```

This graph is shown in Figure 4.9.

FIGURE 4.9
Minitab Plot of Residuals Versus X

```
-- PLOT 'R' VS 'SIZE'

      R
   3.0+
      -
      -
      -
      -                    *
   2.0+
      -
      -
      -                *
      -                        *
   1.0+    *                     *
      -                      *
      -
      -      *
      -
   0.0+         **   *    *
      -            *
      -
      -  *                             *
      - *                 *
  -1.0+
      -     *                            *
      -            *                       *
      -
      -
  -2.0+
        +---------+---------+---------+---------+---------+SIZE
      0.80      1.60      2.40      3.20      4.00      4.80
```

To obtain a scatter diagram of the data with the regression line superimposed, use the multiple plot command, MPLOT. Figure 4.10 shows output resulting from the command

```
MPLOT 'PRICE' VS 'SIZE' AND 'YHAT' VS 'SIZE'
```

In Figure 4.10 (X, Y) points are indicated by A, since the first plot requested was for these points. The B symbols denote the $(X, \hat{Y})$ points, which are points on the straight line $\hat{Y} = 8.70 + 12.9X$. The B's do not all lie in a straight line in Figure 4.10, because the graph is drawn by a typewriter-like device, which can only print at certain places; it is getting the symbol as close as it can to the point it is supposed to represent. The symbols 2 and 3 in Figure 4.10 represent two or three points, respectively, that fall close enough together so that the typewriter would have to print them on the same space.

ENDING A SESSION

When you have obtained the analysis you want, sign off from Minitab by typing the command STOP. Once you do this, however, all the data and the results will be erased. It is possible to save your data and to have your results printed out as hard copy (on paper).

To save your data, assign them a *file name*. The house size-price data could be called "house." Then, before the STOP command, type

SAVE \'HOUSE.DAT'\ (DAT means "data")

F I G U R E 4 . 1 0
Minitab Plot of Scatter Diagram With Regression Equation

```
-- MPLOT 'PRICE' VS 'SIZE' AND 'YHAT' VS 'SIZE'

PRICE
   60.+
      -
      -                                   B
      -
      -                               B B A
   50.+                               A A
      -                         A
      -                       A
      -                   A A   B
      -                       B
   40.+                     B
      -               A  3B
      -                  A
      -               B
      -             2
   30.+           3
      -         A2 A
      -      A  B
      -    A B
      -    BB
   20.+ B2  A
      - A
      -
      -
      -
   10.+
       +---------+---------+---------+---------+---------+SIZE
     0.80      1.60      2.40      3.20      4.00      4.80
```

Then if you want to work with these data in a later session, you will not have to re-enter all 20 pairs of sizes and prices. At the beginning of the next session type RETRIEVE \'HOUSE.DAT'\. Minitab will prompt you with the dash when it has retrieved these data, and you may then proceed to type commands as usual.

In order to have the results of your analysis printed as hard copy, consult with your particular computing center for specific instructions.

E X E R C I S E S

4.5 Use Minitab to obtain a simple linear regression analysis of the Concession Data in Exercise 1.5. In addition to the regression analysis, produce a plot of the data with the regression line superimposed and a plot of the residuals versus X. Compare to results from Exercises 1.5 and 2.8 or 2.14.

4.6 Fit a simple linear regression equation to the Smokers Data in Exercise 1.9 using Minitab. Graph the data and the regression equation. Plot the residuals.

4.7 Using Minitab, perform a simple linear regression analysis on the Class Size Data in Exercise 2.17. Graph the data, the regression equation, and the residuals. Compare to results from Exercise 2.17.

4.8 Refer to the Appraisal Data in Exercise 4.4. Using Minitab, fit a simple linear regression equation relating sale price (Y) to appraised value (X). Obtain plots of the data with the regression equation, and of the residuals.

4 . 4 STATISTICAL ANALYSIS SYSTEM (SAS)

The Statistical Analysis System (SAS) provides several programs for performing regression analyses.† In this section we will discuss how to write SAS programs in general, and how to use the regression procedure, PROC REG, to perform simple linear regression analyses. In later chapters we will look at some of the other regression programs provided by SAS.

Figure 4.11 shows a worksheet for performing a simple linear regression analysis for the house size-price data and for obtaining a scatter diagram of the data with a superimposed plot of the regression equation.

FIGURE 4.11 Worksheet for Performing a Simple Linear Regression Analysis on House Size-Price Data Using SAS

```
    {SYSTEM CARDS
DATA;
INPUT    SIZE PRICE;
CARDS;
1.8  32
1.Ø  24
    :{DATA CARDS
1.5  28
PROC  PRINT;
        TITLE  HOUSES  DATA;
PROC  MEANS;
PROC  REG;
        TITLE  SIMPLE  REGRESSION  OF  PRICE  ON  SIZE;
        MODEL  PRICE=SIZE/P  XPX  I;
        OUTPUT  OUT=NEW  P=YHAT  R=RESID;
PROC  PLOT  DATA=NEW;
        TITLE  PLOT  OF  RESIDUALS  VS  SIZE;
        PLOT  RESID*SIZE/VREF=Ø;
PROC  PLOT;
        PLOT  PRICE*SIZE  YHAT*SIZE='*'/OVERLAY;
        TITLE  SCATTER  DIAGRAM  WITH  REGR  EQUN;
```

† For instructions on the use of SAS programs, refer to *SAS User's Guide: Basics* (1982), *SAS User's Guide: Statistics* (1982), and *SAS Introductory Guide* (1983). SAS Institute, Inc., Box 8000, Cary, NC 27511.

System/JCL Cards

The first several cards are **system cards,** or **JCL cards.** They contain information such as your account number and instructions to the computer that an SAS program is to be run. The rules for JCL cards vary somewhat from one installation to another, so you should confer with your own computing center to find out exactly how to write system cards. One of these cards will contain the command

```
...EXEC SAS...
```

which tells the computer to perform (execute) an SAS analysis.

THE LOGIC AND LANGUAGE OF SAS

First note that *with the exception of data cards,* every SAS statement or command ends with a semicolon. It does not matter in which column of a card you begin a statement, as the computer does not read cards by lines but rather from one semicolon to another. In the worksheet in Figure 4.11 we have indented some statements for ease in reading the program, but all statements could have begun in column 1, or they could have been continued on the previous card. For example, what we wrote as

```
PROC PRINT;
     TITLE HOUSES DATA;
```

could just as well have been written

```
PROC PRINT;
TITLE HOUSES DATA;
```

or

```
PROC PRINT; TITLE HOUSES DATA;
```

Similarly, SAS is not too picky about whether or not you leave spaces around symbols such as semicolons, equals signs, or slants. For example, the command

```
PROC PLOT DATA=NEW;
```

could have been written

```
PROC PLOT DATA = NEW ;
```

As for data cards, each observation must be on a separate card. In our example, the first data card gives the size of house 1, followed by its price. Then comes another card giving the size and price, respectively, of house 2, and so on. In our example, the size and price are separated by a space, and the decimal points are punched on the cards. Just as you can look at a card and recognize that there are two values punched when data are entered this way, so can SAS. (If data must be punched without spaces or decimal points, or if it takes more than one card to enter a single observation, special instructions are needed. See *SAS User's Guide: Basics* (1982) for these instructions.) Finally, it is not necessary to end an SAS program with any particular statement such as END or FINISH. Just stop writing statements, and SAS will realize that your program is complete.

With these general instructions in mind, begin an SAS program by defining a data set. That is, simply give the computer a set of values and let it know which set of values go with which variable. The statement DATA alerts the computer that a data set is to be created. INPUT SIZE PRICE says that the first numbers on the cards will be sizes and the second numbers will be prices. Be careful to specify variable names in their proper order. Variable names can be anything you like as long as they begin with a character (not a number), have no more than eight characters and/or numbers, and contain no spaces. CARDS then prepares the computer to read the data which follow. After reading these first three statements and the data cards, the computer awaits your instructions as to what to do with these data.

It is possible, and sometimes desirable, to give your data set a name. For example, DATA HOUSES would have told the computer that the data set it will work with is named "houses." If you don't give your data set a name, the computer will name it WORK.DATA1. While the significance of this will not be apparent until later, remember that unless otherwise instructed, the computer will always use the *last* data set created.

SAS PRINT AND MEANS PROCEDURES AND TITLE STATEMENTS

Now that we have defined a data set, we may request that the computer perform some analyses on it. The various analyses are called *procedures*, abbreviated PROC. For instance, PROC PRINT directs the computer to print out the data set. It is always a good idea to obtain such a listing in order to check for keypunching errors, to make sure the right names got associated with the variables, and so on. Should you want descriptive titles, include a TITLE statement after the PROC statement. Any TITLE statement will remain in effect until another TITLE statement is encountered, and use of TITLE statements is optional. Simply write TITLE, followed by whatever you want the title to be, and end with a semicolon. We have not shown the printout from the PROC PRINT procedure here. It is simply a page with the title "Houses Data," the values of the variables "size" and "price," and the observations numbered 1 through 20.

PROC MEANS directs the computer to print some summary statistics for the variables in the data set. The table shown in Figure 4.12 contains some useful information, such as the mean value, standard deviation, and maximum and minimum value of each variable.

FIGURE 4.12
Printout From the SAS Procedure PROC MEANS

SUMMARY STATISTICS

VARIABLE	N	MEAN	STANDARD DEVIATION	MINIMUM VALUE	MAXIMUM VALUE
SIZE	20	2.00000000	0.84479833	0.80000000	3.60000000
PRICE	20	34.50000000	11.17563138	17.00000000	52.00000000

STD ERROR OF MEAN	SUM	VARIANCE	C.V.
0.18890265	40.00000000	0.71368421	42.240
2.49894715	690.00000000	124.89473684	32.393

THE SAS REGRESSION PROCEDURE

The regression procedure, PROC REG, instructs the computer to perform a regression analysis on the data. (A title card has been included after the PROC REG statement, but this is optional.) The computer must know which variable is to be the dependent variable and which the independent, so a MODEL statement must be included after PROC REG. A statement MODEL PRICE=SIZE would be sufficient to produce most of the analyses shown in Figure 4.13. Note (in Figure 4.11) that the MODEL statement is of the form: $Y = f(X)$, or Y (price) is a simple linear function of X (size).

While we are not yet in position to understand all the information in Figure 4.13, we can pick out values for a and b. Under a heading VARIABLE we see INTERCEP and SIZE. Under a heading PARAMETER ESTIMATE we see the familiar values 8.703540 and 12.898230 across from INTERCEP and SIZE, respectively. (The word *intercept* has nine letters; SAS abbreviates it to eight letters.)

OPTIONS IN THE MODEL STATEMENT

Note that the MODEL statement in Figure 4.11 is not as simple as MODEL PRICE=SIZE. In order to look at the problem more closely, and to be able to get some plots of the data and the regression equation, we have specified some options in the MODEL statement. It is often of interest to be able to examine the matrices $\mathbf{X'X}$ and $(\mathbf{X'X})^{-1}$ in a regression problem. The options XPX and I (or INVERSE) following the slant in the MODEL statement cause these two matrices to be printed, along with $\mathbf{X'y}$ and some other quantities. The printout from specifying these options is shown in Figure 4.14 and is worth looking at because it is not easy to sort out all the information at one

FIGURE 4.13
Printout From the SAS Procedure PROC REG

SIMPLE REGRESSION OF PRICE ON SIZE

DEP VARIABLE: PRICE

SOURCE	DF	SUM OF SQUARES	MEAN SQUARE	F VALUE	PROB > F
MODEL	1	2255.900	2255.900000	346.767	0.0001
ERROR	18	117.100	6.5055310		
C TOTAL	19	2373.000			

ROOT MSE	2.550594	R-SQUARE	0.9507
DEP MEAN	34.500000	ADJ R-SQ	0.9479
C.V.	7.393027		

VARIABLE	DF	PARAMETER ESTIMATE	STANDARD ERROR	T FOR H0: PARAMETER=0	PROB > |T|
INTERCEP	1	8.703540	1.498103	5.810	0.0001
SIZE	1	12.898230	0.692646	18.622	0.0001

OBS	ACTUAL	PREDICT VALUE	RESIDUAL
1	32.000	31.920	0.079646
2	24.000	21.602	2.398000
3	27.000	30.631	-3.631000
4	47.000	44.819	2.181000
5	35.000	37.080	-2.080000
6	17.000	19.022	-2.022000
7	52.000	55.137	-3.137000
8	20.000	22.892	-2.892000
9	38.000	34.500	3.500000
10	45.000	42.239	2.761000
11	44.000	38.369	5.631000
12	19.000	20.312	-1.312000
13	25.000	24.181	0.818584
14	50.000	52.558	-2.558000
15	30.000	30.631	-0.630531
16	43.000	40.949	2.051000
17	27.000	26.761	0.238938
18	50.000	51.268	-1.268000
19	37.000	37.080	-0.079646
20	28.000	28.051	-0.050885

SUM OF RESIDUALS	-1.88294E-13
SUM OF SQUARED RESIDUALS	117.0996

FIGURE 4.14
$X'X$ and $(X'X)^{-1}$ Matrices as Options in SAS Procedure PROC REG

MODEL CROSSPRODUCTS X'X X'Y Y'Y

X'X	INTERCEP	SIZE	PRICE
INTERCEP	20	40	690
SIZE	40	93.56	1554.9
PRICE	690	1554.90	26178.0

X'X INVERSE, B, SSE

INVERSE	INTERCEP	SIZE	PRICE
INTERCEP	0.3449853	-0.14749300	8.70354
SIZE	-0.1474930	0.07374631	12.89823
PRICE	8.7035400	12.89823000	117.0996

glance. In the top part of Figure 4.14 we have a 3 × 3 matrix, which is of the form

$$\left[\begin{array}{cc|c} n & \sum X & \sum Y \\ \sum X & \sum X^2 & \sum XY \\ \hline \sum Y & \sum XY & \sum Y^2 \end{array}\right].$$

We see that this matrix contains **X′X**, **X′y**, the transpose of **X′y**, and the product $\mathbf{y}'\mathbf{y} = \sum Y^2$:

$$\left[\begin{array}{c|c} \mathbf{X}'\mathbf{X} & \mathbf{X}'\mathbf{y} \\ \hline \mathbf{y}'\mathbf{X} & \mathbf{y}'\mathbf{y} \end{array}\right].$$

In the lower portion of Figure 4.14 we see that this 3 × 3 matrix contains $(\mathbf{X}'\mathbf{X})^{-1}$, the vector $\hat{\boldsymbol{\beta}}$ (which the printout calls B) and its transpose, and a quantity SS(E):

$$\left[\begin{array}{c|c} (\mathbf{X}'\mathbf{X})^{-1} & \hat{\boldsymbol{\beta}} \\ \hline \hat{\boldsymbol{\beta}}' & \text{SS(E)} \end{array}\right].$$

We will see in Chapter 6 that $\text{SS(E)} = \sum (Y - \hat{Y})^2$, the sum of squared errors, or residuals.

The option P (in the MODEL statement in Figure 4.11) causes the values of Y_i, $\hat{Y}_i$, and $Y - \hat{Y}_i$ to be calculated and printed, as can be seen in the lower portion of Figure 4.13. Under ACTUAL are given observed values of Y; PREDICT VALUE shows a list of values of $\hat{Y}$ obtained by substituting corresponding values of X into the regression equation; and in the column labeled RESIDUAL are $Y_i - \hat{Y}_i$. The sum of the residuals is shown to be −1.88294E-13, or -1.88294×10^{-13}, which is, except for the rounding error, zero, as it should be. (The sum of squared residuals is there for a check on the accuracy of the computing algorithm employed by this program. It should equal a quantity "sum of squares for error"—if not, there is a serious error in your data. This will be explained later.)

USING SAS TO OBTAIN GRAPHS

The foregoing steps are sufficient to give a very complete analysis of a simple regression problem. However, there is no substitute for actually looking at the data, and we would also like to have the computer draw a scatter diagram and plot the regression equation on it. In addition, it is hard to comprehend a list of 20 residual values, so we might also want to see them graphed. The rest of the steps shown in the worksheet in Figure 4.11 are concerned with obtaining such plots.

First, recall that SAS is working with a data set containing two variables, "size" and "price." While this data set contains all the information necessary to construct a scatter diagram, it does not contain enough information to plot the regression equation or the residuals. Thus, we need to create a new data set that does contain this information.

Recall that the print option, P, in the MODEL statement caused values of $\hat{Y}$ and of $Y - \hat{Y}$ to be calculated. In order to plot residuals, we need to add values of $Y - \hat{Y}$ to our data set. Since plotting values of $\hat{Y}$ against X will graph the regression equation, we also want to add these values to our data set. The OUTPUT statement allows us to add to our present data set quantities that were calculated in the regression analysis. (Note that unless you specify the option P in the MODEL statement, you cannot use an OUTPUT statement.) OUT=NEW says that this new data set will be named NEW. The variables to be added are predicted values, which we will call YHAT (P=YHAT), and residuals, which we will call RESID (R=RESID). That is, we have created a new data set, named NEW, which contains values of SIZE, PRICE, YHAT, and RESID. By referring to this data set, we can obtain the desired graphs.

We will plot the residuals against X = SIZE first, because the instructions are simpler. The command PROC PLOT DATA=NEW instructs the computer to draw a graph. Note that we have specified in the PROC PLOT command that we are referring to the data set NEW, which contains values of X, Y, $\hat{Y}$, and $Y - \hat{Y}$. This was done for emphasis and clarity here, but since SAS will always use the most recently created data set, a simple PROC PLOT would have sufficed here. In the worksheet in Figure 4.11 we have asked for this graph to be titled "Plot of Residuals Versus Size." PLOT RESID*SIZE tells the computer that the variable RESID is to be plotted on the vertical axis, and that SIZE is to be shown on the horizontal axis. Although it is not necessary, it makes the graph easier to read if we draw in a horizontal line at $Y - \hat{Y} = 0$, since the mean of the residuals is zero. This is accomplished by including the option VREF=0 (vertical reference at zero) after a slant on the PLOT statement in Figure 4.11. The printout from this series of instructions is shown in Figure 4.15. Plots of residuals will be of great interest to us later, when we want to know whether or not we have correctly specified the form of the relationship.

Note in Figure 4.15 that points are plotted with the letter A. Had there been two coincident points, they would have been plotted with the letter B, and so on.

To obtain a scatter diagram with a plot of the regression equation, we again call the PLOT procedure. Note in Figure 4.11 that in the PLOT statement we ask for a plot of PRICE on the vertical axis for SIZE on the horizontal axis and also for a plot of the predicted values, YHAT, by SIZE, on the same graph. The OVERLAY option after a slant in the PLOT statement says that

FIGURE 4.15 SAS Plot of Residuals Versus X

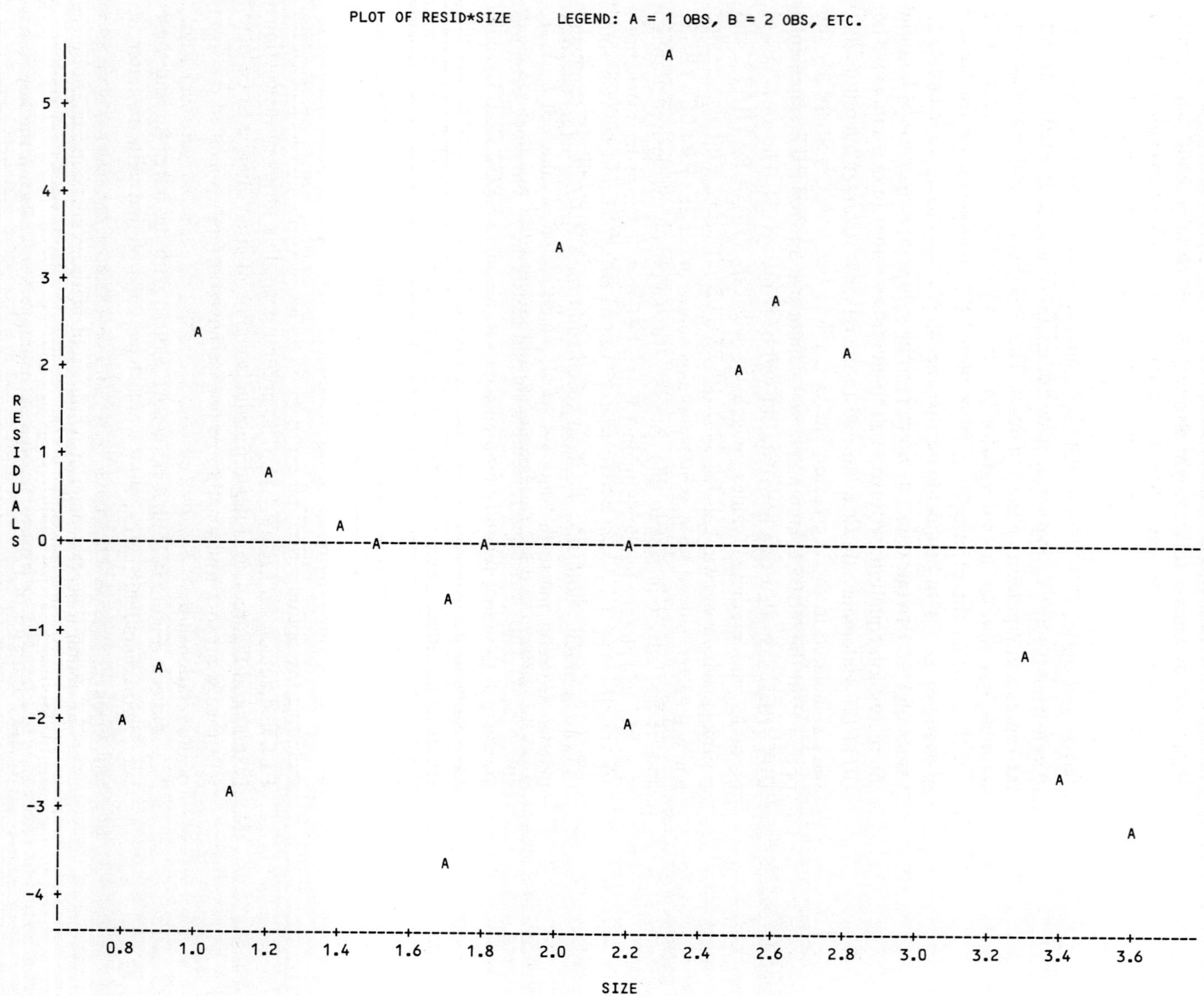

both graphs are to be made on the same page. Now, unless otherwise specified, the computer would plot all points with the symbol A—both those points representing (X, Y)-values and those representing (X, $\hat{Y}$)-values. Thus, it would be difficult, if not impossible, to tell the points on the scatter diagram from those which lie on the regression line. Therefore, we want to use a different symbol to indicate points on one of these plots. The instruction `YHAT*SIZE='*'` indicates that values of (X, $\hat{Y}$) are to be printed with the symbol *. Figure 4.16 shows a scatter diagram with the regression equation. The asterisks do not lie in a straight line, as points on the regression equation should, but remember that the picture was drawn with a device similar to a typewriter, so that it is not possible to plot points completely accurately.

The message `NOTE: 2 OBS HIDDEN` in Figure 4.16 indicates that there were two instances in which predicted values are identical, because the X-values are the same. (Note that you can count only 18 asterisks, not 20, in Figure 4.16.)

ADDITIONAL INFORMATION

With a little practice, it is very easy to learn how to use SAS procedures. You should pay special attention to the preliminary information given on the printout. There, the computer reads your program and notes any errors. It also describes data sets as they are created and tells the amounts of time required to execute each procedure.

EXERCISES

4.9 Use SAS to perform a regression analysis, with graphs of the data and regression equation, and a plot of the residuals, using the Concession Data in Exercise 1.5. Compare results to those of Exercises 1.5 and 2.8 or 2.14.

4.10 With the Smokers Data in Exercise 1.9, use SAS to obtain a regression analysis. Plot Y against X and, on the same graph, $\hat{Y}$ against X. On a different graph, plot residuals against X.

4.11 Obtain a regression analysis, a plot of the data with the regression equation, and a plot of residuals using SAS and the Class Size Data from Exercise 2.17. Compare results to those in Exercise 2.17.

4.12 Refer to the Appraisal Data in Exercise 4.4. Let Y = sale price, and X = appraised value, and use SAS to fit a regression equation $\hat{Y} = a + bX$. Plot the data with the regression line. Plot residuals.

FIGURE 4.16 SAS Scatter Diagram With Regression Equation

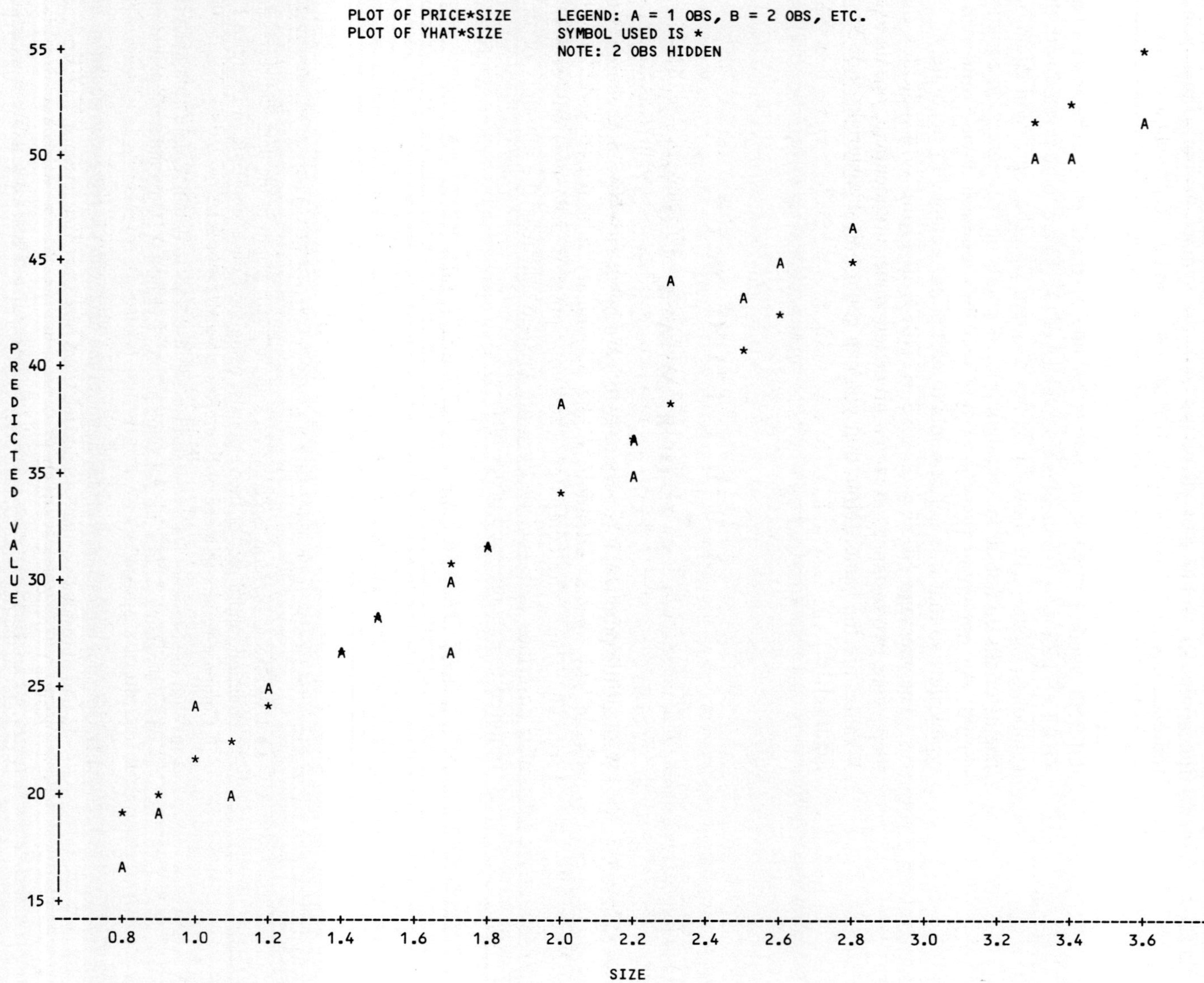

4 . 5

STATISTICAL PACKAGE FOR THE SOCIAL SCIENCES (SPSS^X)

The Statistical Package for the Social Sciences, or SPSS, developed by the University of Chicago, provides many programs for statistical analyses.† We will outline here how to submit data for analysis and how to perform a simple linear regression analysis using the 1983 version, SPSS^X.

ACCESS TO SPSS^X

System/JCL Cards

The first several cards in any SPSS^X computer program contain information concerning the user's name and account number, the job name, and instructions to the computer to use SPSS^X programs. These **JCL cards,** or **system cards,** differ in form from one computing installation to another, so you should check with your computing facility about how to write them to enable you to use SPSS^X. One of the system cards will contain a phrase,

```
EXEC SPSSX...
```

which instructs the computer to refer to these programs.

SUBMITTING DATA

Figure 4.17 shows a worksheet for entering the house size-price data and for performing a regression analysis using SPSS^X. We will first discuss the language of SPSS^X and how to submit data. In the next section we will explain how to perform a regression analysis.

Generally, the language of SPSS^X consists of commands and subcommands. A command instructs SPSS^X to perform a procedure or an analysis. Subcommands state the particulars of which variables are to be analyzed, and how. Commands and subcommands are denoted by keywords, which tell the computer which analysis to perform. A command must begin in column 1 of a card or line, and *only* commands may begin in column 1; subcommands or continuation lines must be indented to at least column 2. Several subcommands within a command are separated from each other by slants. Thus, since the

† Refer to the manual *User's Guide, SPSS^X* (1983), published by the McGraw-Hill Book Company for SPSS, Inc., Chicago.

FIGURE 4.17 Worksheet for Performing a Simple Linear Regression Analysis on House Size-Price Data Using SPSS^X

```
   {SYSTEM CARDS
UNNUMBERED
TITLE SPSS-X SIMPLE LINEAR REGRESSION FOR HOUSES DATA
DATA LIST LIST/
           SIZE PRICE
BEGIN DATA
1.8 32
1.Ø 24
 :  {DATA CARDS
1.5 28
END DATA
LIST
SCATTERGRAM PRICE WITH SIZE
STATISTICS ALL
REGRESSION DESCRIPTIVES=MEAN STDDEV COV XPROD/
           VARIABLES=SIZE PRICE/
           DEPENDENT=PRICE/
           ENTER/
           CASEWISE=ALL/
           SCATTERPLOT (PRICE,SIZE) (*PRED,SIZE) (*RESID,SIZE)
```

computer will look for commands beginning in column 1 and slants separating subcommands, SPSS^X is very lenient concerning spacing between or around words and symbols. The only exception to this is that sometimes columns 73–80 are reserved for numbering the statements in a program.

Referring to Figure 4.17, we see that the first command is

```
UNNUMBERED
```

This command tells SPSS^X that columns 73–80 are *not* being reserved to number the program statements, and thus any symbols punched in those columns are to be read as part of the program.

If you would like to have each page of printout titled with a description of the analysis being performed and/or the data being analyzed, include a `TITLE` command. In Figure 4.17 the title is `SPSS-X SIMPLE LINEAR REGRESSION FOR HOUSES DATA`. It is not mandatory to request a title.

The computer must know how to read the data you submit to it for analysis. The `DATA LIST` command contains instructions for this. In Figure 4.17 there are two subcommands instructing the computer how the data are entered on cards. The `LIST` subcommand says that each case (in this instance, a case is a pair of observations—size and price—on each of the houses in the sample) is entered on a separate data card, with at least one space separating the values, and with decimal points punched. The next subcommand, `SIZE PRICE`, says that the first number entered on each card is the variable "size" and the second number on a card is a value of "price." Names of variables may be eight or fewer characters in length, should start with a letter (not a number), and should contain no spaces or symbols such as =, +, /. Note that a slant separates the LIST subcommand from the subcommand naming the variables.

The BEGIN DATA command indicates that data cards are to follow, and END DATA denotes the end of the deck of data cards. Note that the data cards, as shown in Figure 4.17, conform to the subcommands given in the DATA LIST command. That is, there are 20 separate data cards, one for each house in the sample. On each card the size is given first, followed by the price. One space separates the two values on cards on our worksheet, but more spaces could have been left between values. Furthermore, if you were hurriedly punching data cards and left two spaces between values on some cards but only one between others, the computer would still be able to read your data. Note also that decimal points are punched in for values of "size."

It is always a good idea to request a printout of your data, so that you can check that you made no keypunching errors and that your instructions as to which values go with which variables were understood. The LIST command accomplishes this. While we have not shown the printout resulting from the LIST command here, if you run your own program you will see a list of X- and Y-values, headed by the names of the variables (SIZE and PRICE) and the note NUMBER OF CASES READ = 20. NUMBER OF CASES LISTED = 20.

SCATTER DIAGRAMS

Perhaps the easiest way to perform a simple linear regression analysis using SPSS^X is to use the SCATTERGRAM command. On the worksheet in Figure 4.17

```
SCATTERGRAM PRICE WITH SIZE
```

instructs the computer to construct a scatter diagram of the data. The first variable listed will be plotted on the vertical axis; the variable following WITH will be plotted horizontally. The graph is shown in Figure 4.18. Note that the graph is divided into thirds vertically and horizontally. This is done only for ease of reading and locating individual points and has no other significance.

Use of the command STATISTICS ALL in conjunction with the SCATTERGRAM command produces the information at the bottom of the graph in Figure 4.18. We can easily find

INTERCEPT (A)-8.70354 and SLOPE (B)-12.89823

so that the regression equation is $\hat{Y} = 8.70354 + 12.89823X$, where X = size, and Y = price. Also given are several other statistics, whose meaning we shall learn later. For a simple linear regression, as we shall see, the SCATTERGRAM procedure gives us just about all the information we need to analyze the relationship completely. However, SCATTERGRAM becomes inadequate if a polynomial or multiple regression model is being investigated. Thus, we want to analyze this problem again using the more powerful REGRESSION command.

FIGURE 4.18 Printout from SPSS^{X} Scattergram Procedure Applied to House Size-Price Data

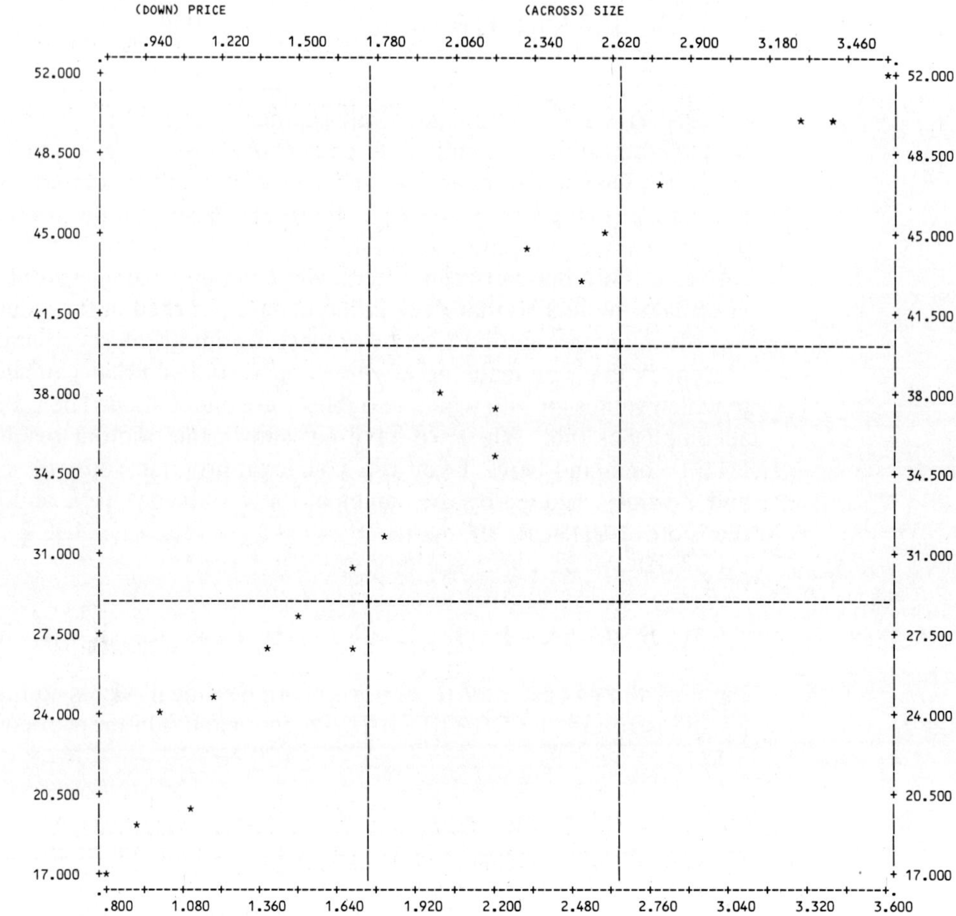

STATISTICS..

CORRELATION (R)-	.97501	R SQUARED -	.95065	SIGNIFICANCE -	.00000
STD ERR OF EST -	2.55059	INTERCEPT (A) -	8.70354	SLOPE (B) -	12.89823
PLOTTED VALUES -	20	EXCLUDED VALUES-	0	MISSING VALUES -	0

'********' IS PRINTED IF A COEFFICIENT CANNOT BE COMPUTED.

REGRESSION ANALYSIS

The REGRESSION command instructs the computer to perform a regression analysis. Several subcommands specify which variables are to be the independent and dependent variables, which descriptive statistics you would like to see, and so on. The REGRESSION command can perform a great variety of regressions—such as multiple and polynomial regressions, as well as simple linear regression—depending upon which subcommands are used. There are three subcommands which *must* be used to perform a regression analysis: VARIABLES, DEPENDENT, and a "method" subcommand: in this case, ENTER.

The VARIABLES subcommand tells the computer which variables are to be used in the analysis. In this case, we have entered only two variables, SIZE and PRICE, and we want to regress one on the other, but it is possible to select a subset of variables from your DATA LIST to analyze. That PRICE is to be the dependent variable is specified by the subcommand

```
DEPENDENT = PRICE/
```

Finally, ENTER instructs the computer to enter all the variables in the VARIABLES subcommand, as independent variables in the regression, except the one denoted as the dependent variable. These three subcommands are sufficient to produce the regression analysis shown in Figure 4.19.

FIGURE 4.19 SPSS^X Regression Analysis of House Size-Price Data

```
EQUATION NUMBER 1     DEPENDENT VARIABLE..  PRICE
DESCRIPTIVE STATISTICS ARE PRINTED ON PAGE 8
BEGINNING BLOCK NUMBER 1.  METHOD:  ENTER
VARIABLE(S) ENTERED ON STEP NUMBER  1..    SIZE

MULTIPLE R          0.97501   ANALYSIS OF VARIANCE
R SQUARE            0.95065                      DF  SUM OF SQUARES  MEAN SQUARE
ADJUSTED R SQUARE   0.94791   REGRESSION          1  2255.90044      2255.90044
STANDARD ERROR      2.55059   RESIDUAL           18   117.09956         6.50553

                              F = 346.76654          SIGNIF F = 0.0000

VARIABLES IN THE EQUATION
VARIABLE          B          SE B       BETA        T      SIG T
SIZE         12.89823     0.69265    0.97501    18.622    0.0000
(CONSTANT)    8.70354     1.49810                5.810    0.0000

FOR BLOCK NUMBER 1 ALL REQUESTED VARIABLES ENTERED.
```

Inspection of Figure 4.19 shows a lot of information that we have not yet learned how to interpret (but soon will). Under VARIABLES IN THE EQUATION in the lower part of Figure 4.19, however, we can pick out the regression coefficients *a* and *b*. Under VARIABLE are listed SIZE and (CONSTANT), that is, *b* and *a*, respectively. Across from these, under the heading B are the values 12.89823 and 8.70354, so that we can read off the regression equation $\hat{Y} = 8.70354 + 12.89823X$, or Price = 8.70354 + 12.89823 (size).

Let us note before returning to the consideration of other subcommands in Figure 4.17 that any subcommand keyword may be abbreviated to its first three letters, that is, VAR instead of VARIABLES, DEP for DEPENDENT, and ENT for ENTER. (This is not so, however, for command keywords. For example, you *cannot* abbreviate REGRESSION as REG.) Also, the equals signs after the keywords could have been omitted as long as at least one space were left between the keyword and subsequent specifications. Spaces could have been left on either side of the equals sign, as well. Finally, in the list of variable names in the VARIABLES subcommand, commas with or without spaces could have been used to separate variable names. However, subcommands must be separated by slants.

DESCRIPTIVE STATISTICS

To obtain a bit more information from the regression analysis, much of which will be useful to us later, it is helpful to supplement the three basic regression subcommands with the DESCRIPTIVES subcommand. Note that DESCRIPTIVES must precede the VARIABLES subcommand.

Consider the DESCRIPTIVES subcommand. In the worksheet in Figure 4.17 we have asked to see the means (MEAN) and standard deviation (STDDEV) of both variables, their variance-covariance matrix (COV), and a matrix showing S_{XY}, S_{XX}, and S_{YY} (XPROD). Figure 4.20 shows printout resulting from this subcommand. (On the actual printout the items seen in Figure 4.20 would precede those in Figure 4.19, since the DESCRIPTIVES subcommand preceded the VAR, DEP, and ENTER subcommands.) You can verify that entries in Figure 4.20 correspond to the entries shown in Table 4.2.

FIGURE 4.20 SPSS^X Descriptive Statistics for House Size-Price Data

```
VARIABLE LIST NUMBER 1  LISTWISE DELETION OF MISSING DATA
                        MEAN            STD DEV          LABEL
      SIZE              2.000            0.845
      PRICE            34.500           11.176

N OF CASES =  20

COVARIANCE, CROSS-PRODUCT:

                        SIZE            PRICE
      SIZE               0.714            9.205
                        13.560          174.900
      PRICE              9.205          124.895
                       174.900         2373.000
```

The specifications MEAN, STDDEV, and COV are the *defaults* for the DESCRIPTIVES subcommand. That is, had we written just DESCRIPTIVES or DESCRIPTIVES=DEFAULTS, we would have obtained $\overline{X}$, $\overline{Y}$, s_X, s_Y, and a matrix containing elements $S_{ij}/(n - 1)$. However, specifying DESCRIPTIVES=XPROD would have shown only a matrix with elements S_{ij}. Other descriptive statistics are available; the reader is referred to the SPSS^X manual. Note that when the DESCRIPTIVES subcommand is used, SPSS^X will tell you before the regression analysis on which page of printout the descriptive statistics can be found. In Figure 4.19: DESCRIPTIVE STATISTICS ARE PRINTED ON PAGE 8.

TABLE 4.2 Explanation of SPSS^X Descriptive Statistics

	MEAN	STD DEV
SIZE	$\overline{X}$	$\left(\frac{S_{XX}}{n-1}\right)^{1/2}$
PRICE	$\overline{Y}$	$\left(\frac{S_{YY}}{n-1}\right)^{1/2}$

N OF CASES = n

COVARIANCE, CROSS-PRODUCT:

	SIZE	PRICE
SIZE	$\frac{S_{XX}}{n-1}$ S_{XX}	$\frac{S_{XY}}{n-1}$ S_{XY}
PRICE	$\frac{S_{XY}}{n-1}$ S_{XY}	$\frac{S_{YY}}{n-1}$ S_{YY}

PREDICTED AND RESIDUAL VALUES

While we have entered values of two variables (X and Y) on data cards, as a result of the regression analysis the computer has created several more variables. Figure 4.21 shows summary statistics for the 12 variables created by the analysis of the house size-price data. While most of these variables have not yet been defined in this text, *PRED and *RESID are values of $\hat{Y}$ (predicted value) and $Y - \hat{Y}$ (residual), respectively. The asterisk denotes that these are *temporary* variables. To see more than just maximum, minimum, and mean values of these, we insert the CASEWISE=ALL subcommand (see Figure 4.17). This produced Figure 4.22.

FIGURE 4.21 Summary Statistics for "Temporary Variables" Created by SPSS^X Regression Analysis

```
EQUATION NUMBER 1   DEPENDENT VARIABLE..  PRICE
RESIDUALS STATISTICS:
               MIN        MAX        MEAN      STD DEV     N
*PRED        19.0221    55.1372    34.5000    10.8964     20
*ZPRED       -1.4205     1.8939     0.0000     1.0000     20
*SEPRED       0.5703     1.2464     0.7814     0.2052     20
*ADJPRED     19.3964    56.1213    34.5803    11.0037     20
*MAHAL        0.0000     3.5870     0.9500     1.0408     20
*COOK D       0.0000     0.3117     0.0585     0.0752     20
*RESID       -3.6305     5.6305     0.0000     2.4826     20
*ZRESID      -1.4234     2.2075     0.0000     0.9733     20
*SRESID      -1.4655     2.2728    -0.0146     1.0259     20
*DRESID      -4.1213     5.9686    -0.0803     2.7645     20
*SDRESID     -1.5176     2.6158     0.0002     1.0780     20
*LEVER        0.0000     0.1888     0.0500     0.0548     20
TOTAL CASES = 20
```

FIGURE 4.22 SPSS^X Casewise Plot of Observed Values of Y, Predicted Values, and Residuals

```
EQUATION NUMBER 1    DEPENDENT VARIABLE..  PRICE

*: SELECTED   M: MISSING

        -3.0           0.0            3.0
CASE #  0:..............:..............:0      PRICE       *PRED      *RESID
     1   .              *              .       32.00     31.9204       .0796
     2   .              .    *         .       24.00     21.6018      2.3982
     3   .       *      .              .       27.00     30.6305     -3.6305
     4   .              .   *          .       47.00     44.8186      2.1814
     5   .          *   .              .       35.00     37.0796     -2.0796
     6   .          *   .              .       17.00     19.0221     -2.0221
     7   .        *     .              .       52.00     55.1372     -3.1372
     8   .        *     .              .       20.00     22.8916     -2.8916
     9   .              .      *       .       38.00     34.5000      3.5000
    10   .              .    *         .       45.00     42.2389      2.7611
    11   .              .          *   .       44.00     38.3695      5.6305
    12   .           *  .              .       19.00     20.3119     -1.3119
    13   .              . *            .       25.00     24.1814       .8186
    14   .         *    .              .       50.00     52.5575     -2.5575
    15   .             *.              .       30.00     30.6305     -.6305
    16   .              .   *          .       43.00     40.9491      2.0509
    17   .              *              .       27.00     26.7611       .2389
    18   .            * .              .       50.00     51.2677     -1.2677
    19   .              *              .       37.00     37.0796     -.0796
    20   .              *              .       28.00     28.0509     -.0509
CASE #  0:..............:..............:0      PRICE       *PRED      *RESID
        -3.0           0.0            3.0
```

We see the actual price (Y), predicted price ($\hat{Y}$), and residual ($Y - \hat{Y}$) for each of the 20 houses. In addition, there is a casewise plot of the *standardized* residual. Recall that a standardized value is one that measures how many standard deviations above or below the mean a given value lies. In this case, the mean of the residuals is zero, so standardized residuals are computed by dividing each of the residuals by the standard deviation of the set of residuals. Since most values in any set lie within 3 standard deviations of their mean, most standardized values should be between -3 and $+3$; if not, such a value is called an outlier and is cause for further study (see Chapter 9).

GRAPHS

The SCATTERPLOT subcommand of the REGRESSION command can graph both variables submitted on data cards and temporary variables. Enclose the pairs of variables to be graphed in parentheses, separated from each other by commas. The first variable of each pair will be graphed vertically; the second, horizontally. Precede the name of any temporary variable by an asterisk. On our worksheet in Figure 4.17, we have requested three graphs: Y by X, $\hat{Y}$ by X, and $Y - \hat{Y}$ by X. These three graphs are shown in Figure 4.23. All these graphs use *standardized* values. Nevertheless, comparing the graph of Y by X (standardized) in Figure 4.23(a) to the graph produced by SCATTERGRAM (Figure 4.18) using actual data values, we see that the pictures are the same, except for scale.

FIGURE 4.23 SPSS^X Scatterplots

(a) Scatterplot of price versus size

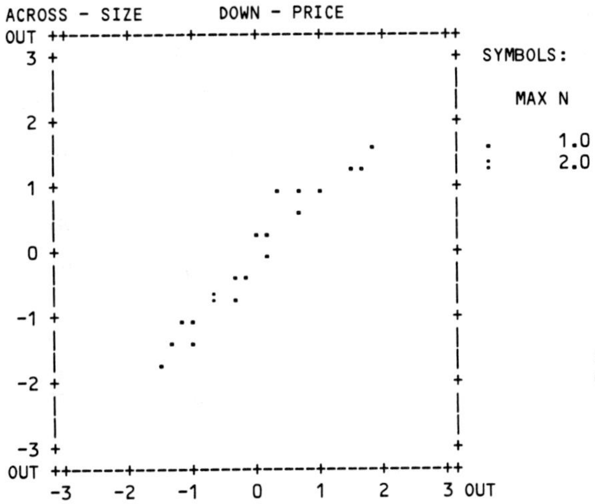

(b) Scatterplot of predicted *Y* versus *X*

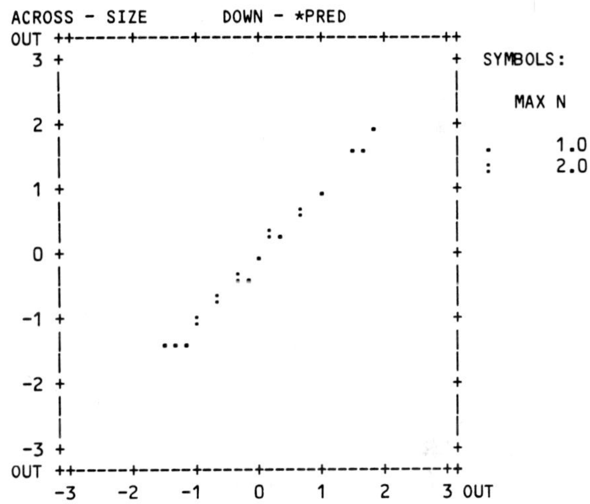

(c) Scatterplot of residual values versus *X*

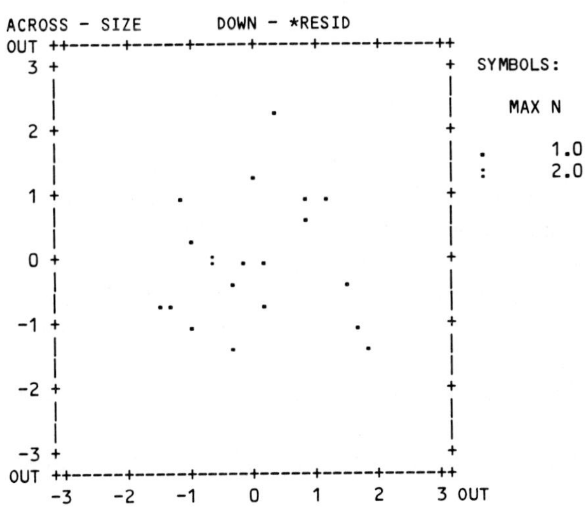

It is not possible to obtain a graph of the regression equation plotted on the scatter diagram of the data. The best we can do is to get graphs of Y by X and $\hat{Y}$ by X (Figure 4.23(b)) side by side to see how well the straight line describes the lay of the data points. However, the plot of residual values by X in Figure 4.23(c) shows the same pattern of points around the line $Y - \hat{Y} = 0$ as would be seen with actual points around the regression equation. Any unusual (nonrandom) pattern in the residuals would indicate that a straight line, $\hat{Y} = a + bX$, was not a good description of these data.

EXERCISES

4.13 Using SPSS^X, analyze the Concession Data of Exercise 1.5. Obtain a scatter diagram, graph the residuals, and compare results to those in Exercises 1.5 and 2.8 or 2.14.

4.14 Use SPSS^X to fit a simple linear regression equation to the Smokers Data of Exercise 1.9. Produce graphs of the data and the residuals.

4.15 Refer to the Class Size Data of Exercise 2.17. Use SPSS^X to perform a simple linear regression analysis. Construct a scatter diagram and a graph of residuals plotted against X. Compare results to those in Exercise 2.17.

4.16 Using SPSS^X find a simple linear regression equation relating Y = sale price to X = appraised value for the Appraisal Data in Exercise 4.4. Plot the data and the residuals.

4.6 INVERTING A MATRIX

As we have seen, inverting a matrix larger than 2×2 is surely a job for a computer. Actually, the regression packages we have discussed all invert matrices to find the regression equation, but they do not all display this inversion as an essential part of the output. Seeing the inverted matrix, however, helps one understand the workings of the regression problem.

IMSL PACKAGES

Almost every center using an IBM/360-370 or comparable computer has access to the packages of the International Mathematical and Statistical Libraries, Inc. (IMSL). These contain a great variety of standard mathematical and statistical calculations.† If these are not available at your computing center, then another program for matrix inversion probably is.

Recall that in solving the normal equations to fit the straight line $\hat{Y} = a + bX$ to the data of the house size-price problem, we had to invert the matrix

$$\mathbf{X'X} = \begin{bmatrix} 20 & 40 \\ 40 & 93.56 \end{bmatrix}.$$

Figure 4.24 shows a worksheet using IMSL for inverting this matrix. To discuss this program in detail would constitute a course in FORTRAN programming. But we can point out a few of its limitations and other pertinent features.

FIGURE 4.24 Worksheet to Invert ($\mathbf{X'X}$) for the House Size-Price Problem, Calling IMSL INVERT Routine

```
     }SYSTEM CARDS
      REAL*4 MATRIX(8,8)
      LOGICAL    SING
      INTEGER    ORDER,DIM
      READ(5,1ØØ) N
      ORDER = N
      DIM = 8
      DO 2Ø I=1,N
      READ(5,1Ø)(MATRIX(I,J),J=1,N)
 2Ø   CONTINUE
      WRITE(6,4Ø)
      DO 3Ø I=1,N
      WRITE(6,5Ø)(MATRIX(I,J),J=1,N)
 3Ø   CONTINUE
      CALL INVERT(MATRIX,ORDER,DIM,SING)
      IF(SING) GO TO 88
      WRITE(6,8Ø)
      DO 6Ø I=1,N
      WRITE(6,5Ø)(MATRIX(I,J)J=1,N)
 6Ø   CONTINUE
      GO TO 99
 88   WRITE(6,7Ø)
 1Ø   FORMAT(8F1Ø.4)
 4Ø   FORMAT(1H1,T2Ø,'MATRIX GIVEN')
 5Ø   FORMAT(1HØ,8(5X,F15.7))
 7Ø   FORMAT(1HØ,T1Ø,'MATRIX GIVEN IS SINGULAR')
 8Ø   FORMAT(1HØ,////,T2Ø,'INVERSE COMPUTED')
 1ØØ  FORMAT(I3)
 99   STOP
      END
     }SYSTEM CARD(S)
ØØ2
   2Ø.Ø         4Ø.Ø
   4Ø.Ø         93.56
```

† For a detailed description of what is available, refer to the *Library 1 Reference Manual*, available from IMSL, Suite 510, 6200 Hillcraft, Houston, TX 77036.

The statement `REAL*4 MATRIX(8,8)` says that this program can be used to invert a matrix of dimension 8 × 8 or smaller. Those familiar with FORTRAN can easily modify the program to handle larger matrices, but the 8 × 8 should be sufficient for most of our purposes. The statement numbered 10,

```
10...FORMAT(8F10.4)
```

says that any data card is divided into eight groups of 10 columns each, the last four columns of each group containing decimal places. Thus, the first value in any row of the matrix is punched in columns 1–10, the second value in columns 11–20, and so on. One data card is entered for each *row* of the matrix to be inverted.

The order, or dimension, of the matrix to be inverted, is punched on the first data card, in the first three columns, right-justified. With the maximum dimension of 8 × 8, the largest number that can be punched on the first data card is 008.

The computer will read in the data and then reproduce the matrix given, under the title `MATRIX GIVEN`. This lets you check on keypunching. Should the matrix be singular, the computer will also print out this message. If it is nonsingular, the computer will find the inverse and will display it under the heading `INVERSE COMPUTED`.

The results of inverting

$$\mathbf{X'X} = \begin{bmatrix} 20 & 40 \\ 40 & 93.56 \end{bmatrix}$$

are shown in Figure 4.25. Note that even though the value 93.56 was punched as the (2, 2) element of the matrix, the value of the (2, 2) position of `MATRIX GIVEN` is 93.5599976. This sometimes occurs because the computer converts decimal system values to the binary number system.† In such a conversion and reconversion to the decimal system, some rounding error occasionally creeps in, but it is nothing to worry about. The IMSL matrix inversion program called for in Figure 4.25 is a single-precision program. This means that any number entered occupies one storage location in the computer and has eight significant digits. Had we used a double-precision matrix inversion program, each number would occupy two adjacent storage locations and would have 16 significant digits. Obviously, we could obtain greater accuracy with a double-

FIGURE 4.25
Output of Program Inverting a 2 × 2 Matrix

```
                MATRIX GIVEN
20.0000000                      40.0000000
40.0000000                      93.5599976
              INVERSE COMPUTED
 0.3449844                      -0.1474922
-0.1474922                       0.0737461
```

† Those interested in the binary number system should refer to Appendix D.

precision program. We use the single-precision program here because a few regression programs are only single-precision, and you should realize that some rounding error can creep into them. However, any unusual or unexpected value should always be investigated and not attributed automatically to rounding error.

Comparing the results of the printout to what we found in Section 3.7,

$$(\mathbf{X'X})^{-1} = \begin{bmatrix} \dfrac{93.56}{271.2} & \dfrac{-40}{271.2} \\ \dfrac{-40}{271.2} & \dfrac{20}{271.2} \end{bmatrix} = \begin{bmatrix} 0.344985 & -0.147493 \\ -0.147493 & 0.073746 \end{bmatrix}.$$

This is identical to the values on the printout up to the sixth decimal place, where some rounding error occurs.

The particular program given will work for any matrix up to dimension 8 × 8. Since **X′X** is always symmetric in regression problems, it might be better to use one of the IMSL programs designed expressly for inverting symmetric matrices. However, this is a more general procedure.

To invert a matrix of dimension greater than 2 × 2, the only necessary changes are in the data cards. Enter the dimension of the matrix on the first card; then enter the appropriate number (same as the dimension) of data cards, each containing one row of the matrix, in order. If it is desired to invert a matrix larger than 8 × 8, readers who are familiar with FORTRAN should have no difficulty amending the given program to suit their needs.

M I N I T A B

To read a matrix into Minitab, type the `READ` command, followed by the number of rows and columns of the matrix, followed by the letter `M` (for matrix) and a number. For example, to enter

$$\mathbf{X'X} = \begin{bmatrix} 20 & 40 \\ 40 & 93.56 \end{bmatrix}$$

and to call it "matrix number 1," type

`READ THE 2 BY 2 MATRIX M1` or `READ 2 2 M1`

Then type in each row of the matrix, one at a time (pressing the carriage return after each row). As shown in Figure 4.26(a), the command `PRINT M1` will then reproduce the matrix you have entered.

To invert a matrix using Minitab, simply type the command `INVERT`, followed by the name of the matrix to be inverted and the name by which you wish to call the inverse. In Figure 4.26(b) the command `INVERT M1 M2` says to invert the matrix M1 and call the inverse M2. Then `PRINT M2` prints out the inverse. Should you try to invert a singular matrix, Minitab will print the message `*ERROR * INVERSE DOES NOT EXIST.`

FIGURE 4.26
Matrices in Minitab

(a) Reading in a matrix

```
—READ  2 2 M1
—20 40
—40 93.56
—PRINT M1

  MATRIX M1     2 ROWS BY 2 COLUMNS
    20.0000     40.0000
    40.0000     93.5600
```

(b) Inverting a matrix

```
—INVERT M1 M2
—PRINT M2

  MATRIX M2     2 ROWS BY 2 COLUMNS
    0.344985     -0.147493
   -0.147493      0.073746
```

If all that is desired in the way of matrix inversion is to see the $(\mathbf{X}'\mathbf{X})^{-1}$ matrix for a regression problem, this inverse will be printed automatically as part of the regression output if the `NOBRIEF` command is used. See Figure 4.8. Note that the rows and columns of $(\mathbf{X}'\mathbf{X})^{-1}$ are numbered with zero for the first row and column and 1 for the second row and column. We shall see later that it is convenient to denote $(\mathbf{X}'\mathbf{X})^{-1}$ as

$$(\mathbf{X}'\mathbf{X})^{-1} = \begin{bmatrix} c_{00} & c_{01} \\ c_{10} & c_{11} \end{bmatrix}.$$

SAS MATRIX PROCEDURE

Matrix inversion with the SAS system is very easy. In order to invert the matrix for the house size-price problem, all one needs are the appropriate system cards and the following three statements:

```
PROC MATRIX PRINT;
A = 20 40 / 40 93.56;
B = INV(A);
```

The printout in Figure 4.27 is the result.

SAS allows you to perform many manipulations of matrices. Adding the word `PRINT` will ensure that all steps are printed out. After calling the `MATRIX` procedure, simply name the matrix (**A** in this case) and list the elements in the matrix, separating the rows by a slant. If you take up all the columns on the card before entering all the elements in the matrix, simply continue on to the next card. The semicolon tells the computer when you are through listing elements. To ask for the inverse of the matrix **A** to be printed,

FIGURE 4.27
SAS Matrix Inversion

A	COL1	COL2
ROW1	20	40
ROW2	40	93.56

B	COL1	COL2
ROW1	0.344985	-0.147493
ROW2	-0.147493	0.0737463

simply write INV(A). We have named this matrix **B**. Note that the SAS inversion procedure is a double-precision procedure. If the matrix is singular, SAS will print a message stating so.

INVERSION USING THE SAS REGRESSION PROCEDURE

One of the nicest features of the SAS regression procedure is that you can request that the **X′X** and $(\mathbf{X'X})^{-1}$ matrices for the model be printed right along with the rest of the regression analysis. The procedure is quite simple. Refer to Figure 4.11, and recall that in order to run the regression procedure, one first writes PROC REG and follows this by a MODEL statement. Options follow the slant in the MODEL statement. For example, P asks for the optional print of residuals. The MODEL statement

```
MODEL Y = X/P XPX I
```

not only asks for the print option but also for **X′X** to be printed (XPX) and for $(\mathbf{X'X})^{-1}$ to be shown (the I—for inverse—option).

INVESTIGATING MATRICES

Now that matrix inversion has been simplified through the availability of computer packages, let us investigate some properties of matrices a bit more fully.

Recall that a singular matrix is one whose determinant is zero. Also, to invert a matrix, every element of the adjoint matrix must be divided by the determinant of the matrix to be inverted. Since division by zero is not defined, a singular matrix has no inverse. Let us consider for a moment the matrix

$$\begin{bmatrix} 1 & 2 & 3 \\ 2 & 4 & 6 \\ 4 & 8 & 12 \end{bmatrix}$$

from Exercise 3.18e. It is clearly singular, since the second row is twice the first row and the third row is twice the second row. In Chapter 3 we found that this determinant is zero and that the inverse does not exist. Suppose this matrix is inverted using the IMSL program in Figure 4.24. The results are shown in Figure 4.28.

FIGURE 4.28
IMSL Inversion of a 3 × 3 Singular Matrix

```
                              MATRIX GIVEN
 1.0000000               2.0000000                     3.0000000
 2.0000000               4.0000000                     6.0000000
 4.0000000               8.0000000                    12.0000000
                            INVERSE COMPUTED
 **************          1118481.0000000           -279620.1875000
 1118481.0000000          -69905.0000000           -244667.6875000
 -372826.9375000         -326223.6875000            256318.5625000
```

The computer should have informed us that the matrix is singular. However, because of rounding error in a single-precision program, the computer found the determinant to be not exactly zero. Note how large all the entries in the matrix are. The asterisks in the (1, 1) position of the inverse indicate that the number that belongs there is so large that it cannot be fitted into the print format. This is a clue that the given matrix is singular.

Now let us consider the matrix

$$\begin{bmatrix} 1 & 2 & 3 & 4 \\ 2 & 4 & 6 & 8 \\ 4 & 8 & 12 & 16 \\ 8 & 16 & 24 & 32 \end{bmatrix}$$

from Exercise 3.20d. We know that this matrix is singular, and so does the IMSL program, as seen in Figure 4.29. Let us change it enough to make the rows/columns *not quite* linear combinations of each other. We want to construct a matrix that is *almost*, but not exactly, singular. For example, let

$$\mathbf{A} = \begin{bmatrix} 1 & 2 & 3 & 4 \\ 2.001 & 4 & 6 & 7.999 \\ 4 & 7.999 & 12 & 16 \\ 8 & 16 & 24.001 & 32 \end{bmatrix}.$$

FIGURE 4.29
IMSL Recognizes a Singular Matrix

```
                              MATRIX GIVEN
 1.0000000         2.0000000          3.0000000          4.0000000
 2.0000000         4.0000000          6.0000000          8.0000000
 4.0000000         8.0000000         12.0000000         16.0000000
 8.0000000        16.0000000         24.0000000         32.0000000

 MATRIX GIVEN IS SINGULAR
```

The result of running this matrix through the SAS inversion program, where rounding error is minimized, is shown in Figure 4.30. Note the very large and very small values in the inverse matrix. We can conclude that if a matrix is nearly singular, which means that a row or column gives essentially the same information as some other row or column, the elements of its inverse will be very large or very small in absolute value in relation to the elements in the original matrix.

FIGURE 4.30 SAS Inversion of a Nearly Singular Matrix

A	COL1	COL2	COL3	COL4
ROW1	1	2	3	4
ROW2	2.001	4	6	7.999
ROW3	4	7.999	12	16
ROW4	8	16	24.001	32

B	COL1	COL2	COL3	COL4
ROW1	1600.2	800	400	-600
ROW2	4000	-889039E-16	-1000	-265920E-15
ROW3	-8000	3552621E-16	-710395E-15	1000
ROW4	3600.2	-200	400	-600

Ill-Conditioned

Linear Dependencies

If the matrix of coefficients from a system of simultaneous equations turns out to be nearly singular, then we say that the system of equations, or the matrix of coefficients, is **ill-conditioned.** Then we should look for **linear dependencies** in the system—that is, for rows or columns that are linear combinations of other rows or columns, or are very nearly so. If, for example, one equation in a system is redundant, we should try to replace it by another equation that gives unique information; otherwise the system of equations will have no solution. In multiple regression, the problem of ill-conditioned matrices is very important.

EXERCISES

4.17 Use an available matrix inversion program to invert these square matrices. Compare results to those found in Exercise 3.22b.

a. $\begin{bmatrix} 1 & 0 & 7 \\ 8 & 2 & 5 \\ 0 & 6 & 7 \end{bmatrix}$; **b.** $\begin{bmatrix} 1 & 0 & 0 \\ 1 & 5 & 4 \\ 9 & 2 & 4 \end{bmatrix}$; **c.** $\begin{bmatrix} 5 & 7 & 0 \\ 7 & 3 & 7 \\ 0 & 7 & 3 \end{bmatrix}$;

d. $\begin{bmatrix} 1 & 2 & 3 \\ 4 & 5 & 6 \\ 7 & 8 & 9 \end{bmatrix}$; **e.** $\begin{bmatrix} 7 & 0 & 0 \\ 0 & 5 & 0 \\ 0 & 0 & 8 \end{bmatrix}$; **f.** diag(7, 5, 3).

4.18 Use an available matrix inversion program to invert these square matrices. Compare results with those of Exercise 3.22c.

a. $\begin{bmatrix} 1 & 0 & 0 & 0 \\ 6 & 0 & 6 & 2 \\ 0 & 5 & 4 & 0 \\ 6 & 6 & 8 & 2 \end{bmatrix}$; **b.** $\begin{bmatrix} 5 & 6 & 2 & 4 \\ 0 & 8 & 1 & 7 \\ 0 & 4 & 1 & 5 \\ 0 & 4 & 0 & 3 \end{bmatrix}$;

c. $\begin{bmatrix} 5 & 6 & 2 & 4 \\ 0 & 8 & 1 & 7 \\ 0 & 4 & 1 & 5 \\ 0 & 12 & 3 & 15 \end{bmatrix}$; **d.** $\begin{bmatrix} 1 & 5 & 9 & 6 \\ 0 & 7 & 5 & 2 \\ 4 & 6 & 1 & 7 \\ 8 & 9 & 1 & 7 \end{bmatrix}$;

e. $\begin{bmatrix} 3 & 0 & 0 & 0 \\ 0 & 6 & 0 & 0 \\ 0 & 0 & 7 & 0 \\ 0 & 0 & 0 & 4 \end{bmatrix}$; **f.** diag(7, 7, 8, 5);

g. $\begin{bmatrix} 1 & 2 & 3 & 4 & 5 \\ 2 & 4 & 6 & 8 & 10 \\ 4 & 8 & 12 & 16 & 20 \\ 8 & 16 & 24 & 32 & 40 \\ 16 & 32 & 48 & 64 & 60 \end{bmatrix}$; **h.** $\begin{bmatrix} 4 & 7 & 8 & 8 & 2 \\ 5 & 0 & 6 & 0 & 0 \\ 6 & 2 & 9 & 7 & 4 \\ 9 & 9 & 4 & 9 & 2 \\ 8 & 8 & 9 & 3 & 6 \end{bmatrix}$.

4.19 Invert the following matrix, using an available program.

$$\begin{bmatrix} 20 & 40 & 430 \\ 40 & 93.56 & 844.7 \\ 430 & 844.7 & 10{,}990 \end{bmatrix}$$

4.7

CHOICE OF A PACKAGE

Often the choice of which computer package to use for a regression analysis or to invert a matrix is restricted by what is available. Most large computing centers, however, have all the packages we have discussed. Then the choice becomes largely a mattter of personal preference.

As we go along, you may want to refer to the four packages in this chapter to see which gives the assortment of statistics you require. The reader who has none of these available should run the house size-price data through whatever package is available and compare results to those presented here. Since all regression packages give essentially the same output, you should be able to figure out what yours gives by comparing it to one or more of the programs discussed here.

4 . 8

SUMMARY

The choice of which computer package to use boils down to three considerations:

1. What is available through your computing center;
2. What graphs, statistics, and tests you need;
3. Personal preference as to layout of results, ease of writing program, and so on.

We have now solved simultaneous equations and fitted regression equations to data in a large variety of ways. Very little of this is new to the reader who has had reasonably good training in mathematics or some exposure to computers. It is time to get back to statistics and see what the regression equation tells us about our variables of interest.

FIVE

THE USE AND INTERPRETATION OF THE REGRESSION EQUATION

5.1 INTRODUCTION

We now consider what the regression equation we have found actually means with reference to a specific set of data. We return to our house size-price problem, this time to illustrate how to use the regression equation for making estimates and, perhaps more important, how *not* to use it.

5.2 PREDICTING AN AVERAGE

Until now our primary emphasis has been on the use of mathematical techniques to fit a straight line $\hat{Y} = a + bX$ through a sample of points. Now we turn to the practical interpretation of our mathematical results.

ESTIMATING PRICE FROM SIZE

For describing the relationship between sizes and prices of houses, we found the equation

$$\hat{Y} = 8.7 + 12.9X.$$

It is simple enough to estimate the price of a house given its size; just substitute the size for X and solve the equation. For example, we estimate the price of a house 2,000 sq ft in size to be

$$\hat{Y}_2 = 8.7 + 12.9(2) = 34.5 \quad (=\$34{,}500).$$

Notice that we substitute $X = 2$, not $X = 2000$, into the equation, since the units on X are given as thousands of square feet (Table 1.1). Similarly, we estimate that a house 2,200 sq ft in size will sell for

$$\hat{Y}_{2.2} = 8.7 + 12.9(2.2) = 37.08 \quad (=\$37{,}080).$$

Recall that the regression equation predicts or estimates the *average* price for houses of a given size. Thus, $\hat{Y}_2 = \$34{,}500$ means that the average price of houses 2,000 sq ft in size is $34,500; or, equivalently, the typical house 2,000 sq ft in size will sell for that amount. The prediction says nothing about any specific house. In Table 1.1, house 9 is 2,000 sq ft in area, and its price is $38,000. Thus, house 9 is more expensive than the typical house that size, because of (1) other factors that we have not taken into account and (2) random variation.

Similarly, houses 5 and 19, both having 2,200 sq ft, cost $35,000 and $37,000, respectively. And both cost less than the average house of their size, which we found to cost $37,080. One might ask why the estimated average value obtained by using $\hat{Y} = 8.7 + 12.9(2.2)$ is not somewhere between the actual sample values, $35,000 and $37,000. After all, doesn't the regression equation give an average? The scatter diagram of Figure 2.8, reproduced in Figure 5.1, helps answer the question. Locate the points (2.2, 35) and (2.2, 37), corresponding to houses 5 and 19, respectively. Note that the line passes above both points. Slightly to the left of these points lies the point (2, 38), which corresponds to house 9. These three houses are nearly the same size. The fact that house 9 is relatively more expensive than the other two is what raises the line above the points for houses 5 and 19; in other words, the line tends to average all three houses, not just houses 5 and 19.†

† The term *regression* actually came about because only averages can be predicted. Studying Darwin's theory of evolution, Sir Francis Galton (1822–1911) figured that if tall parents tend to produce tall children and short parents tend to produce short children, then humans should evolve into one race of giants and one race of midgets. Actual observations showed, however, that his conclusion would not hold; both unusually tall and unusually short sets of parents tend to produce children of roughly average height. Thus, the children's heights tend to converge, or regress, to the average height for that generation.

FIGURE 5.1
Regression Line Superimposed on Scatter Diagram, From Figure 2.8

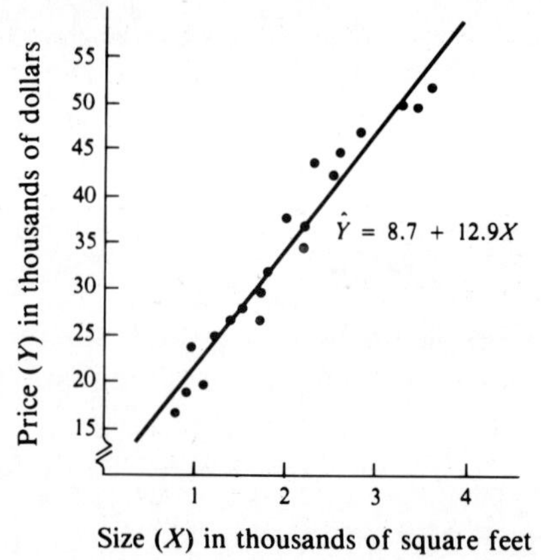

THE RELEVANT RANGE

Since a random sample of houses was taken and the size and price were recorded for each sampled item, the independent variable, size, is a random variable. Thus, we consider that we have a representation of all sizes within the relevant range (800 to 3,600 sq ft). Therefore, any size within this range can be substituted into the regression equation to obtain a predicted price. The size need not be one of those sampled.

One must be extremely cautious, however, in using this sample to predict the price of a house that has less than 800 sq ft or more than 3,600 sq ft. Why? Does not the straight line $\hat{Y} = 8.7 + 12.9X$ extend infinitely in both directions?

Although mathematically speaking, a straight line can be extended as far as one likes, we must remember that we are using mathematics only as a tool to describe a real-world problem and that the interpretations are limited by what makes sense in the problem. Extending the line indiscriminately can lead to some absurdities. Suppose we extend the line to $X = 0$. Then $\hat{Y}_0 = 8.7 + 12.9(0) = 8.7$, and the interpretation would be that a house zero square feet in size typically sells for \$8,700. Or if $X = -0.5$,

$$\hat{Y}_{-0.5} = 8.7 + 12.9(-0.5) = 2.25 \quad (=\$2{,}250),$$

so that a house of minus 500 sq ft size will cost on the average \$2,250. Obviously these interpretations are nonsense.

But what about a house 500 sq ft, or one 4,000 sq ft? Why could one not predict their average price? The reason is that *we have information about prices of houses only for houses between 800 and 3,600 sq ft in size. We do not know anything about prices of larger or smaller houses.* How do you know that the same relationship that exists for houses between 800 and 3,600 sq ft also exists for larger or smaller houses? On the other hand, how do you know that this same relationship does not exist? The point is, we have no information, so we just don't know if this same relationship holds or not. Our equation was determined as a good general description of the relationship *in this range.* It is not intended to be some law about the relationship between all sizes and prices. The situation might be as shown in Figure 5.2, in which very small houses cost very little but great mansions are fantastically expensive. To describe the relationship between size and price for *all* houses, an S-shaped curve might be appropriate. However, in the range from 800 to 3,600 sq ft, a straight line is adequate. Using the regression equation to predict outside of the sampled range can result in some very gross errors, as is illustrated in Figure 5.2.

FIGURE 5.2
Straight Line Provides a Good Description Only Over a Limited Portion of the Entire Range

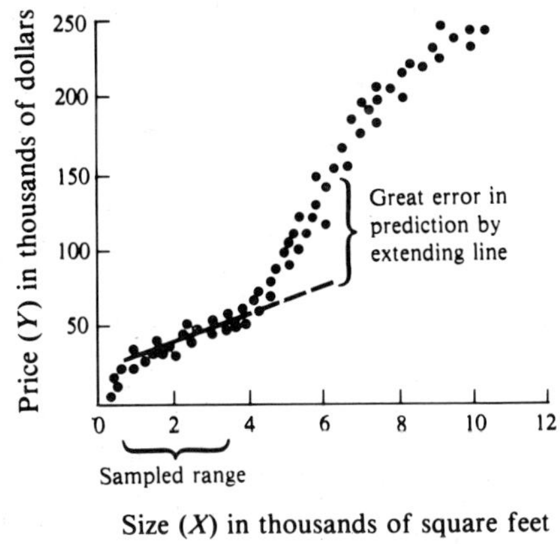

What about going just a little bit outside the range? For example, suppose the newcomer to town were particularly interested in a house of 4,000 sq ft. Since this is only a little bit larger than the largest house in the sample, could one not use the regression equation to estimate the price of this house? Maybe. Usually one would not expect the relationship to change very drastically very fast, at least not without some indication of a change beginning at the end of the scatter diagram, so maybe the newcomer could get a ball-park estimate of the price. But then again, maybe not. We just don't know anything about houses other than those in the sampling range. Anyone interested in houses larger than 3,600 sq ft should include them in the sample.

Sometimes there can be a theoretical structure to the problem that dictates that the same relationship found in the sample must continue past the sampled range. If this is the case, then it is safe (provided the theory is correct) to make predictions outside the range using the regression equation. In the absence of a theoretical framework, however, it is best to stick to situations that you know something about. Note that the regression line in Figure 5.1 is not extended past the points, showing that the relationship has been found only for those data.

One of the most common misuses of a regression is extrapolating beyond the data. Certainly there is no reason to believe that a relationship between IQ and some other variable found for mentally retarded children with IQ's between, say, 60 and 80, should also hold for children with IQ's between 90 and 120. The two groups of children can be thought of as two different populations. Laws of science, whether they are laws of physics, chemistry, or economics, are established on the basis of much more than the evidence from a single sample from a restricted range.

5.3

PREDICTING THE AVERAGE IN A DESIGNED EXPERIMENT

REGULATED VERSUS RANDOM VARIABLES

Recall from Chapter 1 the distinction between the regression study in which X is a random variable and one in which X is a preselected, or regulated, variable. In our example about house size and price, a range of house sizes of interest was defined and a random sample of houses in this size range was taken. Thus, X, as well as Y, was a random variable, since we had no way of foretelling which X-values we would end up looking at. Our sample of houses from throughout the relevant range is considered to give information about *all* sizes within the range. Therefore, we can use our regression equation to predict the price of any house in the range.

The situation is slightly changed, however, if the predictor values are regulated, or preselected, as often happens in a controlled experiment. Laboratory scientists, agricultural experimenters, experimental psychologists, and others have control over the values of the independent variable (the "experimental level," the "treatment level," the "stimulus intensity"); they can run their experiments under whatever conditions they choose. Often the levels of the independent variable are chosen at evenly spaced intervals, as was illustrated in Chapter 2 when we discussed scaling the X-values. This might be done for the sake of reducing computational complexity but is more often done according to the dictates of **experimental design.** Briefly, an experimental design is a method for running an experiment in order to obtain the most, and the least ambiguous, information possible at the least cost.

Experimental Design

This gives laboratory scientists a distinct advantage over social scientists, because experimental designs often let laboratory scientists know the separate effects of each of several variables on a response. Social scientists rarely enjoy such a situation. The advantage held by laboratory scientists, however, is not universal. Laboratory scientists who preselect values of the independent variable choose specific X-values from a range of interest to study. One might say, then, that laboratory scientists are actually interested in only those levels of the predictor variable that they choose to use in the experiment—not in any X-value between any two experimental levels.

ESTIMATES IN DESIGNED EXPERIMENTS

Since laboratory scientists work with only a few X-values, the resulting predictions of the average response levels (Y-values) can be made *only* for those X-values studied. For example, suppose an agronomist applies 20, 25, and 30 lb of fertilizer to various plots of soil, measures the yield of a certain crop, and finds a regression equation relating intensity of fertilization to yield. He can predict the average yield for the respective plots to which 20, 25, and 30 lb of fertilizer are applied. He cannot make a valid prediction of yield for any amount of fertilizer between 20 and 25 lb or between 25 and 30 lb unless he is absolutely certain that his model is correct. Thus, by preselecting experimental levels, laboratory scientists lose a degree of generality in the results. Social scientists have the advantage in this regard through their ability to generalize for any value in the range.†

In practice laboratory scientists do generalize for any value in the range of X-values; this is usually the purpose of the study. But they run the risk of making bad estimates and missing important features of the relationship. Suppose, for example, that the actual relationship is as shown in Figure 5.3.

FIGURE 5.3 Picking Points at Evenly Spaced Intervals Might Miss Interesting Features of the Relationship

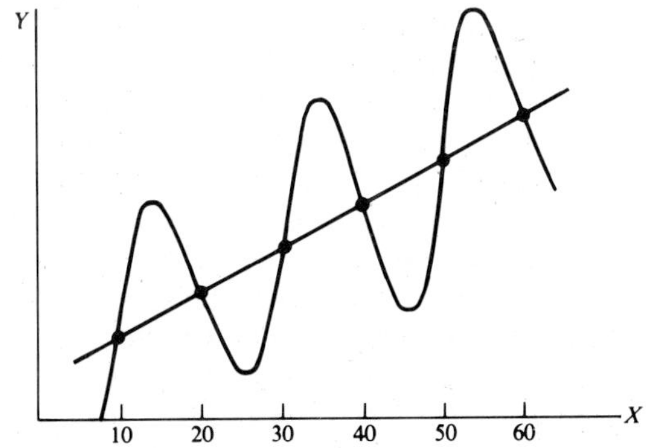

† The reader who is familiar with fixed and random effects models in the analysis of variance can recognize the analogy between our discussion and the generalizations that can be made in these models.

A researcher who conducts an experiment as in Figure 5.3 at $X = 10, 20, \ldots, 60$ would certainly feel justified in concluding that Y is linearly related to X. An estimate of Y for $X = 35$, however, would not be a very accurate prediction. While an underlying relationship such as this one might be relatively unlikely, and while it might be still more unlikely that the experimenter would be so unlucky as to choose the X-levels shown, such an instance is not impossible. Whenever one generalizes between preselected levels, there is a risk that the predictions will be false.

5.4 THE MEANING OF THE INTERCEPT

THE INTERCEPT MIGHT NOT BE INTERPRETABLE

Regression Constant

We noted earlier that the intercept, or **regression constant,** in our house size-price example has no physical meaning: It makes no sense to say that the average price of houses of zero square feet is $\hat{Y}_0 = 8.7 = \$8{,}700$.

How, then, do we interpret the meaning of $a = 8.7$ in the equation? Obviously, the value is needed to get a reasonably accurate prediction of price from size. It is a reference, or starting point, but where does it come from? We don't have enough information to answer this question exactly. Perhaps it is some kind of base, or minimum, price, or a realtor's fee, or some such. It is a real value affecting price determined by *something outside the relationship between size and price*.

WHEN THE INTERCEPT CAN BE INTERPRETED

Can the intercept ever be meaningful to the relationship? Suppose X = winter average daily temperature and Y = amount of electricity used for heating. Then the value a in the regression equation will give the average amount of electricity used on days when the average temperature is zero. The intercept will have physical meaning to the problem and will be the average value of Y when X is zero provided that *both* of two conditions hold: (1) It must be physically possible for X to equal zero; and (2) Data must be collected around $X = 0$.

When X = winter average daily temperature and Y = amount of electricity used for heating, X satisfies both conditions, since (1) it is possible for the average daily temperature in winter to be 0° F and (2) we suppose that data have been collected for such days. If, however, the data were collected around Miami, Florida, during the summer, then condition 2 would fail, and a would not be the average amount of electricity used for heating on zero-degree days; instead it would be a constant determined by something besides temperature. For the house size and price example, condition 1 fails because it is impossible to have a house of no size. If condition 1 fails, then, of course, condition 2 also fails.

FITTING THE REGRESSION EQUATION THROUGH THE ORIGIN

Suppose that one recognizes in a given problem that it is not physically possible for X to be zero, so that it seems logical that if $X = 0$, Y should also be zero. For example, if the size of a house is zero, should not the price also be zero? Or, if X = number of items purchased = 0 and Y = total amount paid, then Y should also be zero. In such cases, it seems reasonable that the intercept should be zero and that the regression equation should be of the form

$$\hat{Y} = 0 + b^*X, \quad \text{or} \quad \hat{Y} = b^*X.$$

Fitting Through the Origin

We denote the slope by b^* to remind ourselves that it will not be the same as if the line has a nonzero intercept. A regression line with zero intercept is said to be **fitted through the origin.** How would this be accomplished, in our house size-price example?

To find the line passing through (0, 0), which is on the average closest to all the points (in terms of squared deviations), we need to find the value of b^* that minimizes

$$Q_1 = \sum (Y_i - \hat{Y}_i)^2 = \sum (Y_i - b^*X_i)^2.$$

With calculus it can be shown that

$$b^* = \frac{\sum X_i Y_i}{\sum X_i^2}.$$

For the interested reader, Appendix F gives the details of obtaining this result using methods of calculus.

Equivalently, in matrix form, to fit the equation

$$\hat{Y} = 0 + b^*X,$$

we have

$$\mathbf{y} = \begin{bmatrix} Y_1 \\ Y_2 \\ \vdots \\ Y_n \end{bmatrix}, \qquad \mathbf{X} = \begin{bmatrix} X_1 \\ X_2 \\ \vdots \\ X_n \end{bmatrix}, \quad \text{and} \quad \hat{\boldsymbol{\beta}} = b^*.$$

Note that the matrix **X** contains no column of ones and that the vector $\hat{\boldsymbol{\beta}}$ consists of the single scalar b^*. Then

$$\mathbf{X}'\mathbf{X} = [X_1, X_2, \cdots, X_n] \begin{bmatrix} X_1 \\ X_2 \\ \vdots \\ X_n \end{bmatrix} = \sum X_i^2,$$

$$(\mathbf{X}'\mathbf{X})^{-1} = \frac{1}{\sum X_i^2},$$

$$\mathbf{X}'\mathbf{y} = [X_1, X_2, \cdots, X_n] \begin{bmatrix} Y_1 \\ Y_2 \\ \vdots \\ Y_n \end{bmatrix} = \sum X_i Y_i,$$

and $$b^* = \hat{\boldsymbol{\beta}} = (\mathbf{X}'\mathbf{X})^{-1}\mathbf{X}'\mathbf{y} = \left(\frac{1}{\sum X_i^2}\right)\left(\sum X_i Y_i\right) = \frac{\sum X_i Y_i}{\sum X_i^2}.$$

Notice that this value for b^* is of the same form as $b = S_{XY}/S_{XX}$, although the two quantities are not equal unless X has been corrected for its mean. Inspection of the scatter diagram in Figure 5.1 reveals that if the line is forced to pass through the origin, then the slope must change in order for the line to pass through the mass of dots.

FITTING THROUGH THE ORIGIN IN THE HOUSE SIZE-PRICE EXAMPLE

In our example, $\sum X_i Y_i = 1554.9$ and $\sum X_i^2 = 93.56$, so that the regression equation fitted through the origin is

$$\hat{Y} = 16.62X.$$

How well does this equation describe the underlying relationship? In Figure 5.4 we see that this line does not follow the lay of the dots as well as our original

FIGURE 5.4
Regression Through the Origin for House Size-Price Example

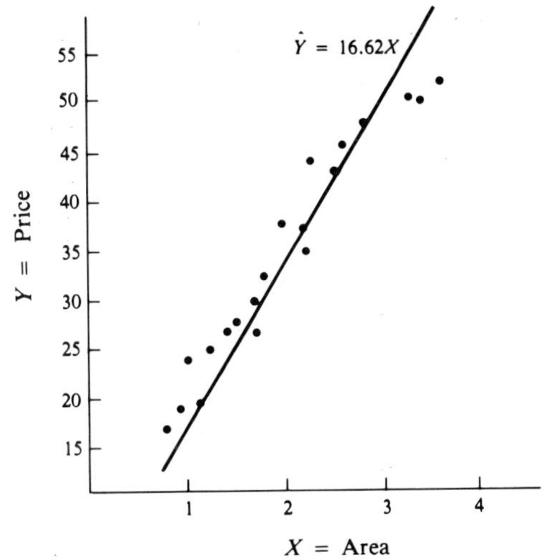

regression line did. Thus, even though it might seem reasonable for Y to be 0 if $X = 0$, such a strong assumption should be made only after careful consideration. Otherwise the equation obtained might not provide a very good description of the relationship. One problem in this example is that zero square feet is probably not the lower limit on house size; perhaps a dwelling must be at least, say, 300 sq ft in area before it can be classified as a house. In addition, by fitting the line through the origin, we are assuming that the same relationship that is found for houses between 800 and 3,600 sq ft in size also holds for houses between 0 and 800 sq ft in size. As we saw in Section 5.1, this assumption is probably not justified.

FITTING THROUGH THE ORIGIN USING THE COMPUTER

All the regression programs we have studied can be easily altered to force the regression equation to pass through the origin:

1. BMDP1R. Add a sentence in the `REGRESSION` paragraph: `TYPE=ZERO.`
2. Minitab. Insert the command `NOCONSTANT` prior to the `REGRESSION` command. This `NOCONSTANT` command will say in effect for all regressions you perform until overridden by a command `CONSTANT`.
3. SAS. Add an option `NOINT` (no intercept) after the slash in the `MODEL` statement.

4. SPSS^X. Add a subcommand `ORIGIN` under the `REGRESSION` command. This subcommand must come after the `VARIABLES` subcommand and before the `DEPENDENT` subcommand.

5.5 MEANING OF THE SLOPE

THE SLOPE DESCRIBES THE RELATIONSHIP

The intercept of the regression equation gives us one specific piece of information: It is either the value of Y when $X = 0$, or it is a starting point determined outside the relationship. As such, the intercept really reveals no more about the relationship than any other $(X, \hat{Y})$-value would. What really informs us about the overall relationship is the slope of the line; it tells how much and in what direction Y changes with X, i.e., what the relationship is.

Regression Coefficient

The value b is usually referred to as the **regression coefficient,** rather than slope, since we are trying to leave the realm of pure mathematics and return to a real-world problem. We will have to change some of our terminology and be careful of our language when interpreting the value of the regression coefficient in a specific problem.

CROSS-SECTIONAL VERSUS LONGITUDINAL DATA

Cross-Sectional Study

Longitudinal Study

Note first that the sample was taken at one point in time, and thus our data are *not time-dependent*. That is, one day we looked at the sizes and prices of 20 houses; we did not look at one house on 20 different days. In comparison to psychological development and similar studies, we have a **cross-sectional study** rather than a **longitudinal** one. Thus, *we have not looked at how either the price or the size of any house has changed over time*. Rather, we have looked at 20 houses that are at one time different from each other in size and also different in price. This is in contrast to a time series study, in which an economic variable, GNP for example, is recorded for several successive years and one is interested in the growth or decline in GNP over a period of time. Both the assumptions and interpretations in regression studies differ from those in time series studies.

A COMMON MISINTERPRETATION

It is important to keep in mind that in a regression study we are observing existing differences and not changes over time. Otherwise, our results might be misinterpreted. To illustrate why this is so, suppose a statistician has been hired by a group of local merchants to analyze sales. As part of her analysis, the statistician has obtained a regression equation $\hat{Y} = 79 + 8X$, relating Y = weekly sales in hundreds of dollars to X = number of employees in retail stores. When explaining to the merchants what the regression equation means, the statistician might say that as the number of employees in a store increases by 1, sales increase $800 per week.

Now, what the statistician is really saying is that if she has observed one store with one more employee than another store, she has also observed a difference in sales of $800, the larger store having the greater sales. Note carefully that the relationship between sales and number of employees is probably a function of the *size* of the store, the larger store having both more employees and greater sales.

This is what the statistician is saying, but it is probably not what the untrained observer is hearing. The merchant could easily reason, "Well, if sales increase by $800 for every additional employee, I'll hire 25 or 30 more employees and my sales will really increase!" He could easily fill his store so full of employees that there would be no room for customers.

You might think that this is an extreme instance, but in fact it is not all that unusual for people to try to use a regression equation in this way. The problem is in the language. The words *increase*, *decrease*, *change*, and even *predict* carry a connotation of a time-dependent activity. Thus, to be strictly proper, and to avoid misleading an untrained listener, we should learn how to interpret the regression coefficient without using terms that have a time connotation.

INTERPRETING THE REGRESSION COEFFICIENT IN THE HOUSE SIZE-PRICE EXAMPLE

It would be easy to interpret $b = 12.9$ in our regression equation by saying that as the size increases by 1,000 sq ft, the price increases by 12.9 thousand dollars. As long as we recognize that we really do not expect any house to grow and increase its size by 1,000 sq ft, there is nothing wrong with such an interpretation. However, the word *increase* implies some kind of a change over time, and this is what we often need to avoid. How can we interpret the regression coefficient without using the word *increase*?

Rather than talk about a change, we need to talk about two things that are, at a given point in time, different from each other. Thus, *if two houses differ in size by 1,000 sq ft* (1 X-unit), *then on the average the larger house will sell for $12,900* (12.9 Y-units) *more* (since $b = 12.9$ is a positive number) *than the smaller house*. Note that, rather than saying that an increase in size causes an increase in price, we can only report that houses of different sizes sell for different amounts of money, larger houses tending to cost more than smaller houses.

Note also that we can talk about an *average* difference in price for a 1,000 sq ft difference in size, because not every pair of houses 1,000 sq ft different in size will have prices differing by exactly $12,900. For example, house 13 has 1,200 sq ft and sells for $25,000, while house 19 has 2,200 sq ft and costs $37,000. Although the size difference is 1,000 sq ft, the price difference is $12,000 instead of $12,900, because we do not have a perfect linear relationship between the two variables.

STANDARDIZED REGRESSION COEFFICIENTS

Recall that a standardized variable is of the form

$$Z = \frac{X - \bar{X}}{s_X}$$

and measures how many standard deviations the value X_i is from the mean of the set. Related to a standardized variable is a **standardized regression coefficient.** Recall that the regression coefficient, b, measures how many *units* Y changes for each 1-unit increase in X. A standardized regression coefficient measures how many *standard deviations* Y changes for an increase of 1 standard deviation in X. Standardized regression coefficients are called by many terms, most common among which are *beta coefficient* or *beta weight*. (The term *beta* is not to be confused with the regression parameter β in the model $Y = \alpha + \beta X + \varepsilon$, however.) Let us denote the standardized regression coefficient by the symbol BW. Then

Standardized Regression Coefficient

$$BW = b\frac{s_X}{s_Y}.$$

In our house size-price example, the sample regression coefficient $b = 12.8982$ means (loosely) that price increases 12.8982 thousand dollars for each 1,000 sq ft increase in size. The standardized regression coefficient is

$$BW = 12.8982\left(\frac{0.845}{11.176}\right) = 0.975.$$

Note that BW has no units: The units on b are Y/X, those on s_X are X, and those on s_Y are Y, so this standardization has canceled out units. The interpretation of $BW = 0.975$ is that price increases 0.975 standard deviations for each increase of 1 standard deviation in size. That is, if size increases 0.845 thousand sq ft (1 standard deviation in X), then price on the average increases $0.975\,(11.176) = 10.8966$ thousand dollars (0.975 standard deviations in Y).

This method for interpreting the regression coefficient is very appealing to many people, as it explicitly takes into account how much variability there is in both variables. In addition, we shall see that in simple regression the standardized regression coefficient is equivalent to a *correlation coefficient* (to be defined in Chapter 8), which can take on values only between -1 and $+1$ and is used as a measure of the degree of association or strength of the relationship between X and Y. Standardized regression coefficients are most useful in multiple regressions (in which case they are generally *not* the same thing as correlation coefficients) for the purpose of comparing the relative strengths of two predictors whose units of measurement are different. For example, suppose in a two-predictor multiple regression we find

$$\hat{Y} = 20 + 2.7X_1 + 3.8X_2,$$

where Y = sales of convenience food stores in thousands of dollars, X_1 = number of hours of daily operation, and X_2 = number of customers per week. Which is the stronger predictor of sales: hours or number of customers? You cannot compare $b_1 = 2.7$ to $b_2 = 3.8$, because the first measures change in sales for a 1-hour increase in operations, but the latter measures change in sales for a 1-customer increase. That would be like comparing a measurement made in miles to one made in kilometers. If the two regression coefficients were converted to standardized regression coefficients, however, they would become comparable in assessing the effect on Y of an increase of 1 standard deviation in the predictors.

To find standardized regression coefficients on computer printouts, look for `STD REGR COEF` in BMD and `BETA` in SPSS^X. SAS and Minitab do not routinely print values of standardized regression coefficients.

5 . 6

RELATIONSHIPS AND CAUSAL RELATIONSHIPS

As illustrated in the example about the relationship between the number of employees in a store and the weekly sales, careless interpretation of a relationship often leads to misuse of a regression study: namely, the assumption that the existence of a relationship implies a *cause-effect* relationship.

As we have seen, the fact that stores with large numbers of employees also have large sales volumes does not necessarily mean that a change in the number of employees will cause a change in sales. Instead, the size of the store probably determines both the size of the sales force and the sales volume.

Is the size of a house the cause of its price? Does cigarette smoking cause lung cancer? Does depriving a rat of food cause it to learn to press a bar? Does fertilizer cause corn to grow? What is meant by *cause,* anyway?

HOW CAUSE-EFFECT RELATIONSHIPS ARE ESTABLISHED

"Everyone knows" that cigarette smoking causes lung cancer in humans. But if you ask how everyone knows this, you are likely to get one of two answers. The first is, "The Surgeon-General says so." This is all well and good, except that appeal to authority has not been an accepted method of proof since the Dark Ages. The other typical answer is that it has been established that the more cigarettes people smoke, the greater their incidence of lung cancer. Here, an appeal is made to the relationship between the number of cigarettes smoked and the incidence of lung cancer at a given age. But is this proof that smoking *causes* cancer? How would you go about setting up a scientific experiment, following the rules of the scientific method, in order to prove that smoking causes cancer? One method might be to form four groups of children, evenly matched as to a variety of physiological attributes so that one group is as much like another as possible. The children of one group are prevented from smoking entirely, or from even breathing someone else's smoke, during their entire lifetimes. The second group can be exposed to a little smoke, from someone else's cigarette or a very occasional smoke of their own. The third group is forced to smoke between one-half and $1\frac{1}{2}$ packs per day, and the fourth group is forced to smoke between $1\frac{1}{2}$ and 4 packs a day as long as they live.

Periodically, individuals from all four groups are "sacrificed" and studies made of changes in the lung tissue. For those individuals not sacrificed, their ages at time of death and causes of death are recorded. All deceased are autopsied and studies made of lung tissue.

From a controlled experiment such as that described, scientists should be able to determine whether or not smoking causes lung cancer. Obviously, however, such a study is not likely to be made.

Almost everybody does believe that smoking causes lung cancer, but for more reasons than just that a statistical relationship has been found to exist. Physicians can observe changes in the cilia lining of the throat and lungs resulting from exposure to cigarette smoke, smokers themselves can feel the effects of the smoke, and so on.† Our objective is not to dispute the unhealthy effects of cigarette smoking but to point out that *a causal relationship cannot be inferred from only knowing that two variables are related.* The existence of a statistical relationship may—and often does—provide an indication that a cause and an effect are present. However, *experimentation, and not the existence of the statistical relationship, is necessary to establish a cause.*

SITUATIONS IN WHICH *X* AND *Y* WILL BE RELATED

Under what conditions might we observe that two variables are related?

1. *X* and *Y* will be related if *X* causes *Y*.

   Suppose X = the number of checkout counters open in a supermarket, and Y = the number of customers per lane. Then we will see that *X* and *Y* are inversely related, because should the store manager open up an additional checkout lane, some customers will leave their lanes to go to the newly opened one, thus causing the number of customers per lane to decrease.

2. *X* and *Y* will be related if *Y* causes *X*.

   Using the same variables as before, how do you know that the situation is not just the opposite? If the number of customers per lane becomes large, this will cause the store manager to open up additional lanes, and we will see that *X* and *Y* are directly related.

3. *X* and *Y* will be related if *X* and *Y* affect each other.

   If we carry the two examples a bit further, we see that an increase in the number of customers per lane will cause more lanes to be opened up. But opening additional lanes will cause the number of customers per lane to decrease, and this decrease in the sizes of the lines will result in some counters being closed down. Then the fewer lanes being open, the more customers per lane, and the cycle begins anew. Thus, because *X* and *Y* affect each other; we observe that they are related.

4. *X* and *Y* will be related if both are caused by a third variable, *Z*.

   Perhaps both the number of lanes open and the number of customers per lane are determined by the number of customers in the store. Thus, when the store is busy, there will be many lanes open and a large number of

† For the story of how it has been determined that "cigarette smoking is dangerous to your health" the reader is referred to the essay, "Statistics, Scientific Method, and Smoking," by B.W. Brown, Jr., in *Statistics: A Guide to the Unknown,* edited by Judith M. Tanur et al., Holden-Day, Inc., 1972.

customers in each lane—just the opposite when there are few shoppers. Still, we will see a relationship between X and Y.

Spurious Relationship

In such a case, we say that the relationship between X and Y is **spurious.** Spurious means "wrongly attributed": We observe that two variables are related, but we have missed something. For example, suppose that for cities in a certain state, X is the number of school teachers, and Y is the number of liquor stores. It will be the case that those cities in which there are a large number of school teachers also have a large number of liquor stores, but this is not to say that school teachers are necessarily heavy drinkers. Obviously, Z = the population size of the city has something to do with both variables. Thus, X and Y *are* related—but only through this third variable, Z.

5. X and Y will be related if they act alike just by chance.

Suppose the growth rates of baby chicks and piglets happen to follow the same pattern. This does not mean that pig growth influences chick growth or vice versa. In fact, all growth curves look very similar, regardless of whether an organism or an investment is growing. Suppose we have two variables for which the values in Table 5.1 have been observed. Clearly, X and Y are related; they are even perfectly related, since $Y = 100X$. What if you are now told that X = the number of pencils sold by a street vendor in New York City, and Y = the number of butterflies born in a laboratory in Calcutta? It is hard to see how any causal relationship exists between the two variables; it is just chance that the two patterns are alike. This is what we mean by a relationship—merely that two variables act alike, to some extent at least. The two variables need not have anything to do with each other; still, we can often find a statistical relationship between them.

TABLE 5.1
Two Perfectly Related Variables

	X	Y
Sunday	2	200
Monday	5	500
Tuesday	6	600
Wednesday	3	300
Thursday	1	100
Friday	7	700
Saturday	2	200

6. X and Y can appear to be related even when they are not because of a nonrepresentative (biased) sample.

Sometimes, even though two variables are not related in the population, the sample drawn makes it appear that a relationship exists. Suppose, as in Figure 5.5, there is no relationship between X and Y; the average Y-value $\mu_{Y \cdot X}$ for each X-value lies on the horizontal line $\mu_{Y \cdot X} = \alpha$ for which $\beta = 0$. It is possible that a simple random sample could choose

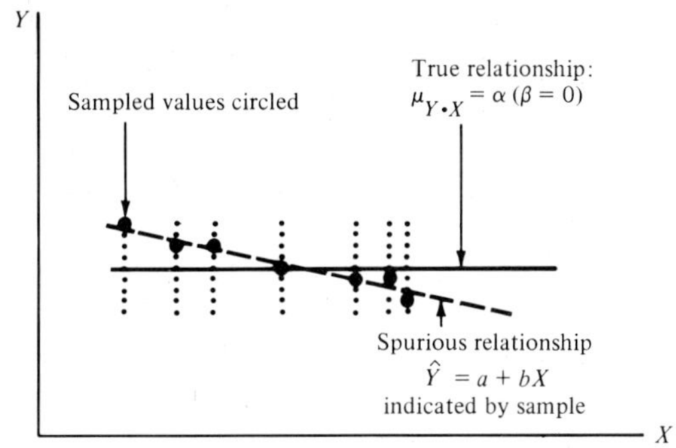

FIGURE 5.5 How a Simple Random Sample That Is Biased Can Indicate a Relationship That Really Does Not Exist

those items circled in Figure 5.5. Since the only available information comes from the sample, one could conclude that an inverse relationship exists, when in fact there is no relationship.

We will see in Chapter 7 that one can test for the *significance* of the relationship, so that we can be reasonably confident that the relationship seen in the sample is not due to an aberration of our sample. In each of the other five cases just listed, further exploration is necessary, beyond statistical analysis on the given data, in order to determine whether or not a causal relationship exists and, if so, the kind of relationship. Note that we are not claiming that cause and effect do not exist. We are merely saying that causes and effects are established by investigation and experimentation and not by statistics calculated from sample data. A relationship between two variables may lead one to suspect a causal link but does not establish that such a link exists.

EXERCISES

5.1 The regression equation relating amount of soft drink syrup to advance ticket sales (thousands) for the Concession Data in Exercise 1.5 was

$$\hat{Y} = 0.989 + 1.45X.$$

For the data analyzed, ticket sales ranged from 54,600 to 74,900.

a. Estimate how much syrup the concession stand manager should have on hand if advance ticket sales were 60,000.

b. If, for the game against the arch-rival team, advance sales are 80,000 tickets, should this regression equation be used to estimate how much syrup will be needed?

c. What does $b = 1.45$ tell you?

d. What does $a = 0.989$ tell you?

e. Does it seem reasonable that the intercept should be zero? Explain.

f. Fit the regression equation through the origin. Plot this line on the scatter diagram, and examine how well it describes the relationship. Comment.

5.2 We found the regression equation $Y = -0.3256 + 0.3842X$ to describe the relationship between X = number of employees, and Y = average weekly retail sales (thousands of dollars) in the Sales Data in Exercise 1.6. The number of employees ranged from 10 to 48 in the 15 stores sampled.

a. Estimate the average weekly sales of a store that employs 17 people.
b. Interpret the meaning of the estimate in (a).
c. Compare the estimate in (a) to the actual sales of the two stores in the sample that had 17 employees, and explain any differences.
d. Suppose that another store employed 3 people. Could the regression equation be used to estimate sales for that store? Why, or why not?
e. If another store employed 40 people, could the regression equation be used to estimate its sales? Explain.
f. If another store employed 75 people, could the regression equation be used to estimate the sales for that store? Explain.
g. Interpret the meaning of the regression constant in this problem.
h. Does it seem reasonable in this problem to use a regression line fitted through the origin? Explain.
i. Interpret the meaning of the regression coefficient.
j. The manager of one of the sampled stores has been unhappy that his sales have not been larger. From looking at the results of the analysis of these data, he concludes that he can increase his sales by hiring more employees. Comment.

5.3 For the Police Data of Exercise 1.7, we found the regression equation $Y = 2010.2296 - 35.3156X + 0.2096X^2$. Here X = number of police officers (between 50 and 90), and Y = number of robberies in cities of comparable size.

a. Estimate the number of robberies in cities employing 67 police officers.
b. Interpret the meaning of the estimate in (a).
c. Compare the estimate in (a) to the actual number of robberies, 560 and 630, respectively, in cities 3 and 7, which both employed 67 policemen. Explain any differences.
d. Suppose that another city of comparable size employed 107 policemen. Could the regression equation be used to estimate the number of robberies in that city? Why, or why not?
e. Suppose that some other city of comparable size employed 45 police officers. Could the regression equation be used to estimate the number of robberies in that city? Why, or why not?
f. If another city employed 60 police officers, could the regression equation be used to predict the number of robberies there? Explain.
g. Can it be said that if a city wants to reduce the number of robberies, then it should employ more policemen? Why, or why not?
h. Interpret the meaning of the regression constant in this problem.

5.4 We found an equation $Y = 0.7348 - 0.008X$, to describe the relationship for the Heating Data of Exercise 1.8. Here, X = average daily temperature, which ranged from -2 to 50 degrees, and Y = daily heating oil bill, in dollars.

a. Estimate the daily heating oil bill for days on which the average temperature was 33 degrees.
b. Interpret the meaning of the estimate in (a).
c. Compare the estimate in (a) to the actual bill, 50¢, on day 5, on which the temperature averaged 33 degrees. Explain any differences.
d. Suppose that one day in January the temperature dropped to -15 degrees. Could the regression equation be used to estimate the bill on that day? Explain.

e. In April, the temperature was 76 degrees on several days. Could the regression equation be used to estimate the oil bill on those days? Why, or why not?
f. Although the sample taken did not include any days for which the average temperature was 20 degrees, could the regression equation be used to estimate the bill on days in which the temperature was 20 degrees? Why, or why not?
g. Interpret the meaning of the regression constant.
h. Would it be appropriate to use a regression line fitted through the origin in this problem? Explain.
i. Interpret the meaning of the regression coefficient.

5.5 Refer to the Fertilizer Data in Exercise 2.20. In this experiment, different plots of ground received 15, 20, or 25 lb of fertilizer in order to determine the effect on crop yield. Using X-values as given, the regression equation $\hat{Y} = 0.2222 + 3X$ was obtained; if X was scaled, the regression equation was $\hat{Y} = 60.2222 + 15Z$.
a. Estimate the yield per acre of the plots to which 25 lb of fertilizer was applied.
b. Interpret the meaning of the estimate in (a).
c. Compare the estimate in (a) to the actual yield of the three plots that received 25 lb of fertilizer. Explain any differences.
d. Can the regression equation $\hat{Y} = a' + b'Z$, or $\hat{Y} = a + bX$, found in Exercise 2.20 be used to estimate the yield if no fertilizer is applied? Why, or why not?
e. Can the regression equation be used to estimate the yield from plots receiving 17 lb of fertilizer? Explain.
f. Interpret the meaning of the regression constant in $\hat{Y} = a' + b'Z$.
g. Interpret the meaning of the regression constant in $\hat{Y} = a + bX$.
h. Would it seem reasonable in this problem to fit a line $\hat{Y} = b^*X$ through the origin? Comment.
i. Interpret the meaning of the regression coefficient b' in $\hat{Y} = a' + b'Z$.
j. Interpret the meaning of the regression constant b in $\hat{Y} = a + bX$.

5.6 Refer to the Repairs Data in Exercise 2.11.
a. Interpret the meaning of the slope and the intercept in the regression equation $\hat{Y} = -0.3790 + 0.056980X$ relating repair costs (hundreds of dollars) to the number of hours of operation (hundreds) since the last maintenance was performed.
b. Can the foreman infer that the cause of high repair costs is that management allows the assembly machines to go too long without routine maintenance?

5.7 Refer to the "Suds" Data in Exercise 2.16, in which yearly gross sales were related to student body size by $\hat{Y} = 30.698 + 4.99778X$. ($X$ is thousands of dollars, Y is thousands of students.)
a. What do the numbers 30.698 and 4.99778 tell you?
b. If "Suds" is considering opening a franchise near a campus with 18,000 students, what are that store's anticipated yearly gross sales?
c. If "Suds" will only open a new franchise if anticipated sales are at least $75,000, what is the minimum campus size they should consider?

5.8 Refer to the Class Size Data in Exercise 2.17. A professor had set out to demonstrate that instructors of large sections tend to be rated low in teaching effectiveness by the students. Her data gave the following regression equation relating X = class size to Y = average rating for a random sample of 22 sections taught: $\hat{Y} = 3.637411 - 0.008677X$. Does this support the teacher's contention that the reason she received low ratings was that she taught large sections?

5.9 With regard to the Mail Data in Exercise 3.31,

a. Interpret the meaning of the intercept, 4.2846, and the slope, 0.0293164, of the regression equation relating the number of orders (hundreds) to the weight of mail received (pounds).

b. Could you argue for a cause-effect relationship in this example?

5.10 For the Smokers Data in Exercise 1.9, the regression equation relating number of days sick leave (Y) to average number of cigarettes smoked per day (X) was found to be $\hat{Y} = 6.285 + 0.05552X$.

a. What do the slope and the intercept of this equation tell you?

b. Do you think that the number of days sick leave is related to how much an employee smokes?

5.11 The diameter of a tree at breast height inside the bark (DBH_{ib}) can be estimated from the diameter outside the bark (DBH_{ob}) by the following equation, according to the Bark Data in Exercise 2.12:

$$\widehat{DBH_{ib}} = 0.0908 + 0.89726(DBH_{ob}).$$

Recall that data for DBH_{ib} were found by subtracting twice the single bark thickness from DBH_{ob} (measurements in centimeters).

a. What do the slope and the intercept of the regression equation tell you?

b. If a tree has outside diameter of 30 cm, what do you estimate its diameter inside the bark to be?

5.12 A savings and loan company looked at records for a simple random sample of ten weeks during the last two years and found the information in Table 5.2 concerning the mortgage interest rates and the number of mortgage loan applications it received.

TABLE 5.2
Loans Data

	Week									
	1	2	3	4	5	6	7	8	9	10
Interest rate (%)	12.5	13.0	10.5	12.0	13.5	10.0	10.0	11.0	13.0	12.0
No. of loan applications	50	43	68	47	39	74	63	61	48	52

a. Find a simple linear regression equation to estimate the number of loan applications from the interest rate.

b. If in a given week the interest rate is 12.75%, how many loan applications might be expected?

c. Suppose that the interest rate soars to 15%. What would this regression equation predict the number of applications to be? Do you think this is reasonable?

d. If the interest rate were to drop to only 8%, would you use this equation? Explain.

e. Can you argue that an increase in interest rates causes a decrease in the number of mortgage loans applied for? Explain.

5.13 In the Appraisal Data in Exercise 4.4, the sale prices (thousands of dollars) for single-family residences were related to the appraised values (dollars) as follows:

$$\hat{Y} = 10.433 + 0.0014203X.$$

What does this equation tell you about *how* sale price is related to appraised value?

5.14 Sales of a product on trial for six weeks were used to estimate the first year's sales. The regression equation was

$$\hat{Y} = 2.167 + 10.0662X,$$

where both variables are measured in hundreds of dollars. If the company will not market the new product in a given area unless the trial-period sales indicate that expected first year sales are at least $15,000, how well must the product perform during the trial period?

5.15 In a study to determine the extent to which a certain preservative in food is related to the degree of hyperactivity in young children, a random sample of children was selected from among children whose parents agreed to their taking part in the study. The children were given food containing the preservative, and the amount of preservative ingested (in grams) was calculated from measuring how much remained in the uneaten portion. The degree of hyperactivity was measured by a function involving the amount of fidgeting, pestering other children, and so on, one hour after the food was consumed. Results are shown in Table 5.3.

TABLE 5.3
Hyperactivity Data

Amount of preservative (gm)	39.7	90.6	43.0	52.2	21.5	26.0	66.1	63.1	42.0	33.4	83.1
Hyperactivity level	4.4	5.4	3.6	2.9	0.9	1.7	4.6	5.0	2.2	4.2	5.8

a. Find a simple linear regression equation relating hyperactivity level to amount of preservative ingested.
b. Does this prove that this food additive causes increased hyperactivity? For all children? For children in this sample?
c. What does the intercept, 0.7410, mean?

5.16 Can Maison Française estimate how many dinner guests will arrive on a given evening by counting the number of dinner reservations made by 5 P.M. that day? Table 5.4 gives some information from the last several months.

TABLE 5.4
Reservations Data

Day	No. of Reservations	No. of Guests	Day	No. of Reservations	No. of Guests
1	132	148	13	127	145
2	212	235	14	226	250
3	107	126	15	229	250
4	240	270	16	137	154
5	196	213	17	174	190
6	179	200	18	120	130
7	251	281	19	117	138
8	168	186	20	209	234
9	183	200	21	152	172
10	142	161	22	206	224
11	157	170	23	146	167
12	146	163	24	206	224

a. Find the simple linear regression equation relating the number of guests to the number of reservations.

b. What does the equation tell you about how many guests the restaurant can expect for dinner?

5.17 Stock market analysts perform what they call a beta analysis to measure the performance of a particular stock relative to the market as a whole. A defensive stock is one that is not very sensitive to market changes: Its increase in yield is not as great as the average market increase in times of a "bull" market, nor does its return decline as much as the market average in "bearish" times. An aggressive stock, on the other hand, reacts more noticeably to moods of optimism and pessimism than does the market as a whole. Thus, if an investor anticipates that stock prices on the average will rise, he or she will purchase aggressive stocks but will sell them in favor of defensive stocks if a downturn is felt to be imminent.

In order to classify a stock as aggressive or defensive, a simple linear regression equation is found relating the return of this stock to the average market return over some time period. (The average market return may be measured by a stock market index such as *Standard and Poor's 500 Stock Index.*) The slope of the regression equation, which financial analysts refer to as Beta, measures whether the stock is aggressive or defensive. If the slope of the regression equation is 1, this means that the stock's return moves just like the market average: For each 1 percent increase (or decrease) in the market return, there is a 1 percent increase (or decrease) in the stock's return. If Beta is greater than 1, then the stock is aggressive, since for a 1 percent change in the market there is a greater than 1 percent change in the stock. If Beta is less than 1, the stock is classified as defensive. (It is uncommon to see Betas of zero, which would indicate complete insensitivity to market movements; a few stocks, such as gold mining stocks, move in the opposite direction from the market and thus have negative Betas.)

While the slope of the regression equation is taken as a measure of the effect of general economic conditions on this stock, the intercept may be interpreted as a measure of the stock's general level of return, which is not related to the market level but rather to factors specific to the firm.

Suppose that the following data represent 24 consecutive months' figures on the return of XZQ Stock and a common market index.

TABLE 5.5
Stock Data

Month	Market Return(%)	XZQ Stock Return (%)	Month	Market Return (%)	XZQ Stock Return (%)
1	6.0	6.5	13	−1.5	−5.4
2	0.8	2.8	14	3.9	10.4
3	−4.2	−2.6	15	−1.6	−8.6
4	2.1	7.0	16	5.2	2.7
5	2.1	9.4	17	1.6	9.5
6	−0.4	−0.1	18	8.1	4.6
7	1.6	−8.0	19	2.7	5.4
8	−6.9	−14.8	20	−0.3	−3.9
9	−8.6	−13.6	21	1.5	−0.5
10	2.8	2.7	22	−0.9	−4.3
11	2.2	−2.0	23	1.7	3.9
12	−0.9	−2.0	24	−0.7	−0.6

a. Find the simple linear regression equation relating XZQ Stock's return to the average market return.
b. Interpret the meaning of the regression coefficient, in terms of whether XZQ Stock is aggressive or defensive.
c. Interpret the meaning of the regression constant.

5.18 A new pressure gauge has been purchased, and it is suspected from past experience with such instruments that this new gauge may need some adjustments. To see if adjustments are necessary, and if so, what kinds; several pressure readings are made using both the new instrument and a gauge that is already in service. Results are shown in Table 5.6.

TABLE 5.6
Gauge Data

Reading	New Gauge (lb)	Old Gauge (lb)	Reading	New Gauge (lb)	Old Gauge (lb)
1	387	386	11	230	229
2	279	278	12	277	276
3	168	168	13	397	393
4	187	183	14	140	137
5	148	142	15	195	193
6	155	157	16	290	287
7	150	146	17	178	174
8	130	127	18	123	120
9	355	353	19	350	344
10	230	226	20	119	117

a. Which set of readings should be the dependent variable (Y) and which should be the independent variable (X)? Why?
b. If the new gauge is in proper adjustment, what would you expect the slope and the intercept of the regression equation to be?
c. Fit a simple linear regression equation, and interpret what the slope and the intercept tell you about how (if) the new instrument needs adjustment.

5.19 For each of the relationships described in (a) to (e), tell which condition (i to vi) best describes the relationship. The conditions are

(i) X causes Y,
(ii) Y causes X,
(iii) X and Y affect each other,
(iv) X and Y are both caused by Z,
(v) X and Y are related by chance,
(vi) the relationship between X and Y is only apparent, because of a nonrepresentative sample.

a. X = number of television commercials run per week,
Y = weekly sales of advertised product,
X and Y are directly related.
b. X = dosage level of pain-killing drug,
Y = number of hours relief obtained by patients,
X and Y are directly related.
c. X inches growth per month of pine saplings,
Y − inches growth per month of cedar saplings,
X and Y are directly related.

d. X = number of employees in a store,
Y = sales volume of store,
X and Y are directly related.

e. X = number of flat tires by motorists on a certain turnpike,
Y = number of shoes thrown by race horses at a certain track,
X and Y are directly related.

5.7 SUMMARY

In this chapter we began to examine the interpretation of the regression equation found for a given set of data. We noticed that the meaning of the regression equation depends to a great extent upon the data from which it is obtained. Only a careful examination of the data can tell the researcher what information the equation can accurately convey. Perhaps the most important concept for any student to grasp is that just because two variables are found to be related, one cannot infer that a cause-effect relationship exists without more information than just a list of X- and Y-values.

If the regression equation is to be used to estimate values of one variable from values of another, a question immediately arises: How accurately can such estimates be made? The next chapter addresses this question.

SIX

MEASURING ERROR IN ESTIMATION

6.1 INTRODUCTION

Now that we know how to predict the average Y-value for any given value of X (within the range of X-values sampled, of course), the question arises as to how much variability about this mean we can expect. That is, while we can estimate the average price for houses of a specified size, how close to this average price can we expect most of the actual prices to be?

What we need is a measure of the scatter, or error. In other words, we need to determine how much the actual Y-values differ from the mean Y value because of omitted predictors, random variation, and inaccuracy of the form of the model.

6.2 ASSUMPTIONS

Let us return to the conceptualization of the population as illustrated in Figures 2.1 and 2.2, reproduced here in Figures 6.1 and 6.2. We note that for any X-value in the population, there is a subpopulation of Y-values (Figure 6.1) and the underlying linear relationship,

$$\mu_{Y \cdot X} = \alpha + \beta X,$$

is a straight line passing through the mean Y-value for each X-value (Figure 6.2). Note also in both figures that the scatter of the subpopulation is homogeneous, or uniform.

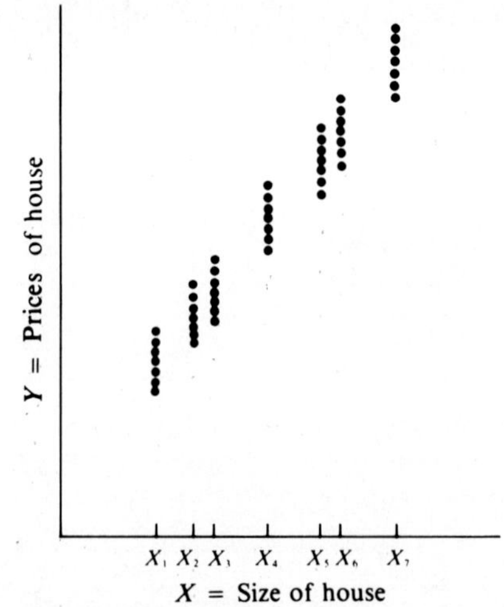

FIGURE 6.1
Scatter Diagram for the Population of Houses, From Figure 2.1

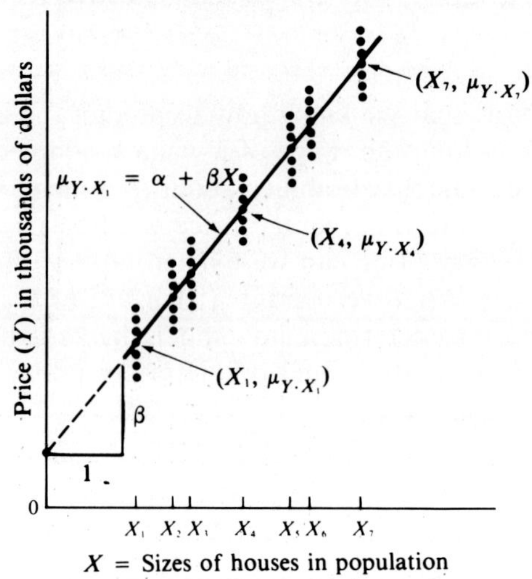

FIGURE 6.2
Determining the Underlying Linear Relationship $\mu_{Y \cdot X} = \alpha + \beta X$, From Figure 2.2

THE VARIANCE OF THE *Y*-VALUES FOR ANY GIVEN *X*-VALUE

If we had the entire population at our disposal, then for any *X*-value we could measure the variability in *Y*-values by simply finding the variance of the subpopulation of *Y*-values associated with each *X*-value. For example, consider houses of size X_i. There are several different prices, $Y_{i1}, Y_{i2}, \ldots, Y_{iN_i}$, for which houses of size X_i sell (N_i is the number of houses of size X_i). Consider these *Y*-values as a small population of interest. The mean of this population is $\mu_{Y \cdot X_i}$, as we have previously determined. We denote the variance of this subpopulation by

$$\sigma^2_{Y \cdot X_i}.$$

To measure this variance, we proceed as usual and find the average squared deviation of the values from their mean:

$$\sigma^2_{Y \cdot X_i} = \frac{\sum_{j=1}^{N_i} (Y_{ij} - \mu_{Y \cdot X_i})^2}{N_i}.$$

That is, there are N_i Y-values associated with the X-value X_i, and we denote these as Y_{ij}, $j = 1, 2, \ldots, N_i$, the first subscript indicating that all Y-values are associated with the ith X-value.

If the entire population were available, we could measure $\sigma^2_{Y \cdot X_i}$ for each X-value and thus could know the variability of prices for every different house. This could be done regardless of whether or not the scatter of the dots is homogeneous for all X-values. If the scatter is not uniform, then each $\sigma^2_{Y \cdot X_i}$-value will be different, but if scatter is homogeneous, all $\sigma^2_{Y \cdot X_i}$-values will be the same. In either case, we would have a measure of how much each price differs from the average for houses of a specified size.

ESTIMATING THE VARIANCE FROM A SAMPLE

All of the above assumes that the entire population is available for our inspection, but in reality we will have only a sample. For any X-value in our sample, we will have at best only a few observed Y-values, and in most cases only one Y-value. In addition, we will have an *estimate* of the mean Y-value for that X-value, given by

$$\hat{Y}_i = a + bX_i.$$

Recall from the scatter diagram of Figure 1.1 that for any X-value we have at best only two observed Y-values, and in all but one instance we have only one observed price for each size house. Pick any X-value for which there is only one Y-value. How can we measure the variability in observed Y-values for that X-value if there is only one Y-value? This can be done only if we make a very important assumption—that the scatter of the points around the line is the same (homogeneous) for every X-value. This assumption is often referred to as the assumption of **homoscedasticity.** (This is a term coined by economists, whose data in many cases do not display uniform scatter, and who must work in situations in which **heteroscedasticity,** or nonhomogeneous scatter, is present.)

Homoscedasticity

Heteroscedasticity

How will the assumption of homoscedasticity allow an estimate of the variance to be made from the sample? Recall that if scatter is uniform in the population, then all $\sigma^2_{Y \cdot X_i}$-values are equal, regardless of which X_i is being considered. Thus, we may regard our sample as a collection of subsamples from populations having equal variances and may *pool* these subsamples to get the best possible estimate of the common variance.

The pooling of sample variances in a regression problem is a relatively straightforward procedure. For each subsample, find how far the observed value, Y_i, differs from the estimated mean value, $\hat{Y}_i$, and average these (squared) deviations across all X-values. Referring to the scatter diagram and regression line in Figure 2.8, reproduced in Figure 6.3, we find the vertical distance of each point from the line, square these distances, and average. Then $s^2_{Y \cdot X}$ is the sample variance of the point around the line (their mean). Note that $s^2_{Y \cdot X}$ does not measure total variability in the response variable Y, but measures variability in Y *after it has been taken into account* (by the line) *that Y varies with X*. Simply put, it measures variability around the line. Note also that this pooling could not be accomplished unless the assumption of homoscedasticity is met. For example, if $\sigma^2_{Y \cdot X_1}$ were not equal to $\sigma^2_{Y \cdot X_2}$, then $(Y_i - \hat{Y}_i)^2$ at $X = X_1$ would give information about $\sigma^2_{Y \cdot X_1}$, and $(Y_i - \hat{Y}_i)^2$ at $X = X_2$ would give information about a *different* quantity, namely $\sigma^2_{Y \cdot X_2}$. It would then make no sense to put these two pieces of information together, since they do not give information about the same thing.

FIGURE 6.3
Regression Line Superimposed on Scatter Diagram, From Figure 2.8

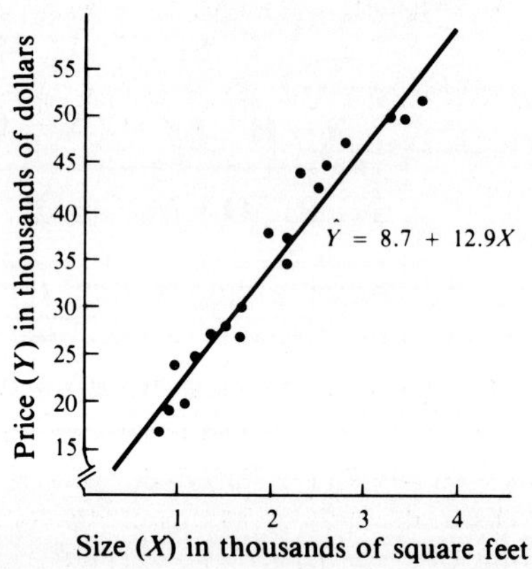

THE VARIANCE OF THE POINTS AROUND THE LINE

The mathematical definition of the sample variance of Y, given X, is

$$s^2_{Y \cdot X} = \frac{\sum_{i=1}^{n} (Y_i - \hat{Y}_i)^2}{n - 2}.$$

Note the analogy of $s^2_{Y \cdot X}$ to $\sigma^2_{Y \cdot X}$. The reason for $n - 2$ in the denominator, instead of n, is that we want $s^2_{Y \cdot X}$ to be an **unbiased estimator** of $\sigma^2_{Y \cdot X}$. This means that $s^2_{Y \cdot X}$ tends neither to overestimate nor to underestimate $\sigma^2_{Y \cdot X}$. It is accurate on the average. For any particular sample, $s^2_{Y \cdot X}$ may be larger or smaller than $\sigma^2_{Y \cdot X}$ because of chance, but not because of anything inherent in $s^2_{Y \cdot X}$ itself. Were the denominator just n, then we would tend to underestimate $\sigma^2_{Y \cdot X}$, and this underestimation would not be a function of chance alone.

Unbiased Estimator

DEGREES OF FREEDOM

Degrees of Freedom

The quantity $n - 2$ in the denominator $s^2_{Y \cdot X}$ is called **degrees of freedom.** For an intuitive understanding of what is meant by degrees of freedom, consider first the population variance of a random variable Y:

$$\sigma^2_Y = \frac{\sum_{i=1}^{N} (Y_i - \mu_Y)^2}{N}.$$

In the population, the true mean, μ_Y, is known; and a variance is measured by finding the average squared deviation from the mean, which is considered a typical value. In a sample, however, the true mean can only be *estimated* using the sample itself. We would calculate a sample variance using $\sum_{i=1}^{n} (Y_i - \bar{Y})^2$ as the numerator. From this estimation procedure, we *lose a degree of freedom*. One might think of the situation as having to pick one sample value to stand in for the typical value, leaving only $n - 1$ other sample values to compare to this typical value. Thus, the unbiased estimator of the sample variance is calculated as

$$s^2_Y = \frac{\sum_{i=1}^{n} (Y_i - \bar{Y})^2}{n - 1}.$$

It can be proven mathematically that the $(n - 1)$ denominator in s^2_Y is necessary to make s^2_Y an *unbiased* estimator of σ^2_Y; that is, one that is accurate on the average, in the long run. While in any given sample, s^2_Y will almost surely either underestimate or overestimate σ^2_Y, it will *tend* to do neither. (This is what unbiasedness means.) Conversely, if n were used as the denominator of s^2_Y, then s^2_Y would tend to underestimate σ^2_Y more often than it would overestimate. As an intuitive justification of why this would be so, imagine the population of heights of all male undergraduate students on a university campus. It would not be too unreasonable to assume that the heights range from, say, 4′6″ to 7′6″, although there are not very many heights at either of these extremes. Suppose a sample of 25 to 50 male students is randomly selected. It is possible that this sample could contain students with heights 4′6″ and 7′6″, but we feel that this would be an unlikely occurrence. Most probably, the smallest value in the sample will be larger than the smallest value in the population, and the largest sample value will be smaller than the largest value in the population. Thus, a sample variance calculated as $s^2_Y = \sum_{i=1}^{n} (Y_i - \bar{Y})^2/n$ is very likely to underestimate the population variance. This tendency can be corrected simply by using the denominator $(n - 1)$, which results in a larger ratio than if n were used as the denominator.

In a regression problem we are also trying to measure how much the various values differ from a typical value. Only in this case, the typical value is a point on the line. In order to find a line, we need two points, thus leaving only $n - 2$ other values to compare to the typical one. We have used up *two* points in order to find a typical value. Or, in order to find the line, two parameters, α and β, must be estimated by a and b from the sample. Since two estimates had to be made, 2 degrees of freedom were lost.

The general rule is: You lose 1 degree of freedom every time you have to estimate a parameter. If μ_Y must be estimated by $\overline{Y}$ in order to find s_Y^2, then 1 degree of freedom is lost; if $\mu_{Y \cdot X}$ must be estimated by $Y = a + bX$, where a estimates α and b estimates β, then 2 degrees of freedom are lost. This rule continues to work in multiple regression, where we predict Y using some number, say k, of X-variables. The denominator of $s_{Y \cdot X}^2$ will be $n - (k + 1)$, since there will be a β associated with each of the k X's and also an α. Generally, the denominator is $n - p$, where p is the number of parameters, $p = k + 1$.

THE STANDARD ERROR OF ESTIMATE

Returning to the problem at hand, any variance is a rather artificial measure in that it averages *squared* deviations. Thus, the units of a variance are the squares of the units of the variable; as such, they often have no physical meaning or are not relevant to the problem under study. For example, if Y is price of house in thousands of dollars, then $\hat{Y}$ is also in thousands of dollars, the difference $Y - \hat{Y}$ is in thousands of dollars, and $(Y - \hat{Y})^2$ has units "square thousands of square dollars." The variance is then the average of square thousands of square dollars. To get this measure of variability back into meaningful terms, the square root is taken to form a standard deviation, whose units are the same as the units on the original variable. The standard deviation of the Y-values around the line is called the **standard error of estimate** and is denoted $s_{Y \cdot X}$:

Standard Error of Estimate

$$s_{Y \cdot X} = \sqrt{s_{Y \cdot X}^2}.$$

Note that the terminology here is poor, because the standard error of estimate is *not* the standard error of any estimate. It is rather an estimate of the standard deviation of the points around the regression line and is a measure of error variation.

EXERCISE

6.1 Referring to the Textbook Data in Exercise 1.10, explain why the methods presented in the text for estimating error cannot be used.

6 . 3

CALCULATING A FORMULA FOR THE STANDARD ERROR OF ESTIMATE

DERIVATION OF THE FORMULA

It is too laborious to calculate $s_{Y \cdot X}$ by taking the square root of

$$s^2_{Y \cdot X} = \frac{\sum_{i=1}^{n} (Y_i - \hat{Y}_i)^2}{n - 2},$$

since this formula requires that we first find $\hat{Y}_i$ for every X-value in the sample, subtract $\hat{Y}_i$ from Y_i for all n observed Y-values, square, add, and divide. Since $s^2_{Y \cdot X}$ is a variance, there is a shortcut calculating formula for it, as there are for all variances. Let us derive a calculating formula. Consider the numerator:

$$\sum_{i=1}^{n} (Y_i - \hat{Y}_i)^2 = \sum_{i=1}^{n} (Y_i - a - bX_i)^2.$$

Recall that $a = \overline{Y} - b\overline{X}$, so we can write

$$\begin{aligned}
\sum_{i=1}^{n} [Y_i - (\overline{Y} - b\overline{X}) - bX_i]^2 &= \sum_{i=1}^{n} [(Y_i - \overline{Y}) - b(X_i - \overline{X})]^2 \\
&= \sum_{i=1}^{n} [(Y_i - \overline{Y})^2 - 2b(Y_i - \overline{Y})(X_i - \overline{X}) \\
&\qquad + b^2(X_i - \overline{X})^2] \\
&= \sum_{i=1}^{n} (Y_i - \overline{Y})^2 - 2b \sum_{i=1}^{n} (Y_i - \overline{Y})(X_i - \overline{X}) \\
&\qquad + b^2 \sum_{i=1}^{n} (X_i - \overline{X})^2 \\
&= S_{YY} - 2bS_{XY} + b^2 S_{XX}.
\end{aligned}$$

Now recall that $b = S_{XY}/S_{XX}$, so that

$$b^2 S_{XX} = \left(\frac{S_{XY}}{S_{XX}}\right)^2 S_{XX} = \frac{(S_{XY})^2}{(S_{XX})^2} S_{XX} = \frac{(S_{XY})^2}{S_{XX}}$$

$$= \frac{S_{XY}}{S_{XX}} S_{XY} = bS_{XY}.$$

Then we have that

$$\sum_{i=1}^{n} (Y_i - \hat{Y}_i)^2 = S_{YY} - 2bS_{XY} + bS_{XY} = S_{YY} - bS_{XY}.$$

Calculating Formula for $s^2_{Y \cdot X}$

The **calculating formula for $s^2_{Y \cdot X}$** is

$$s^2_{Y \cdot X} = \frac{S_{YY} - bS_{XY}}{n - 2}.$$

STANDARD ERROR OF ESTIMATE FOR HOUSE SIZE-PRICE PROBLEM

Let us illustrate the use of this formula by calculating the standard error of estimate for the problem concerning the relationship between house size and price. One of the beauties of this formula is that most of the quantities involved have already been calculated. We found in Chapter 2 that

$$S_{XY} = 174.9, \qquad b = 12.8982, \quad \text{and} \quad n = 20.$$

We must now find

$$S_{YY} = \sum_{i=1}^{n} Y_i^2 - \frac{\left(\sum_{i=1}^{n} Y_i\right)^2}{n}.$$

In Table 2.1 we calculated $\sum_{i=1}^{n} Y_i$ and $\sum_{i=1}^{n} Y_i^2$, although we have not yet used $\sum_{i=1}^{n} Y_i^2$. Thus,

$$S_{YY} = 26{,}178 - \frac{(690)^2}{20} = 2373.$$

Then

$$s^2_{Y \cdot X} = \frac{2373 - 12.8982(174.9)}{20 - 2} = \frac{117.1048}{18} = 6.5058,$$

and the standard error of estimate is

$$s_{Y \cdot X} = \sqrt{6.5058} = 2.5506,$$

or approximately \$2,550.

EXAMINATION OF THE CALCULATING FORMULA

Clearly, the calculating formula requires much less effort than the formula

$$s_{Y \cdot X}^2 = \frac{\sum_{i=1}^{n} (Y_i - \hat{Y}_i)^2}{n - 2},$$

since this definitional formula requires the calculation of 20 $\hat{Y}_i$ values, 20 subtractions $Y_i - \hat{Y}_i$, 20 squares $(Y_i - \hat{Y}_i)^2$, and then finally an averaging of these values. Additionally, the calculating formula helps explain exactly what is being calculated when a standard error of estimate is found. Note that S_{YY} is a measure of the total amount of variation in the Y-values, since $s_Y^2 = S_{YY}/(n - 1) = \sum_{i=1}^{n} (Y_i - \overline{Y})^2/(n - 1)$ is the variance of Y and thus tells the average squared amount that the Y-values differ from their mean. In the standard error of estimate, we do not want to measure total variability in Y, but we want to measure how much the Y-values vary *around the line*, or how much Y-values vary *after we have taken into account that Y varies partly because it is related to X, and X also varies*. Naturally, then, we expect $s_{Y \cdot X}^2$ to be smaller than s_Y^2. So from S_{YY}, the total variability in Y, we subtract bS_{XY}, a measure of how much Y varies because of X. This latter quantity involves the slope of the line, b, and the numerator of the covariance of X and Y, S_{XY}. The units of the product bS_{XY} are

$$\frac{Y\text{-units}}{X\text{-units}} (X \text{ units} \times Y \text{ units}),$$

or Y-units squared, and thus measure variance of Y because of relationship with X. The difference $S_{YY} - bS_{XY}$ thus measures variance of Y, having adjusted for, or taken out, variation in X. Much more will be made of this concept when we study correlation in Chapter 8.

THE MATRIX-VECTOR FORM

Recall that the simple linear model can be written in matrix notation as

$$\mathbf{y} = \mathbf{X}\boldsymbol{\beta} + \boldsymbol{\varepsilon}$$

where

$$\mathbf{y} = [Y_1 \quad Y_2 \cdots Y_n]', \qquad \boldsymbol{\varepsilon} = [\varepsilon_1 \quad \varepsilon_2 \cdots \varepsilon_n]',$$

$$\boldsymbol{\beta} = [\alpha \quad \beta]', \quad \text{and} \quad \mathbf{X} = \begin{bmatrix} 1 & X_1 \\ 1 & X_2 \\ \vdots & \vdots \\ 1 & X_n \end{bmatrix}.$$

The normal equations for finding estimates a and b of α and β, respectively, can be written as

$$\mathbf{X}'\mathbf{X}\hat{\boldsymbol{\beta}} = \mathbf{X}'\mathbf{y}$$

where

$$\mathbf{y} = [Y_1 \quad Y_2 \cdots Y_n]', \qquad \hat{\boldsymbol{\beta}} = [a \quad b]', \quad \text{and} \quad \mathbf{X} = \begin{bmatrix} 1 & X_1 \\ 1 & X_2 \\ \vdots & \vdots \\ 1 & X_n \end{bmatrix}.$$

The solution to the normal equations is

$$\hat{\boldsymbol{\beta}} = (\mathbf{X}'\mathbf{X})^{-1}\mathbf{X}'\mathbf{y},$$

giving numerical values of a and b in the regression equation $\hat{Y} = a + bX$.

Consider the product

$$\mathbf{X}\hat{\boldsymbol{\beta}} = \begin{bmatrix} 1 & X_1 \\ 1 & X_2 \\ \vdots & \vdots \\ 1 & X_n \end{bmatrix} \begin{bmatrix} a \\ b \end{bmatrix} = \begin{bmatrix} a + bX_1 \\ a + bX_2 \\ \vdots \\ a + bX_n \end{bmatrix} = \begin{bmatrix} \hat{Y}_1 \\ \hat{Y}_2 \\ \vdots \\ \hat{Y}_n \end{bmatrix}.$$

We see that this product gives the estimated values of Y corresponding to the values of $X_1, X_2, \ldots, X_n$. Thus, the vector

$$\mathbf{y} - \mathbf{X}\hat{\boldsymbol{\beta}} = \begin{bmatrix} Y_1 \\ Y_2 \\ \vdots \\ Y_n \end{bmatrix} - \begin{bmatrix} \hat{Y}_1 \\ \hat{Y}_2 \\ \vdots \\ \hat{Y}_n \end{bmatrix} = \begin{bmatrix} Y_1 - \hat{Y}_1 \\ Y_2 - Y_2 \\ \vdots \\ Y_n - \hat{Y}_n \end{bmatrix} = \begin{bmatrix} e_1 \\ e_2 \\ \vdots \\ e_n \end{bmatrix}$$

is the vector of residuals. Finally, the product $(\mathbf{y} - \mathbf{X}\hat{\boldsymbol{\beta}})'(\mathbf{y} - \mathbf{X}\hat{\boldsymbol{\beta}})$ will give the sum of squared residuals:

$$(\mathbf{y} - \mathbf{X}\hat{\boldsymbol{\beta}})'(\mathbf{y} - \mathbf{X}\hat{\boldsymbol{\beta}}) = [Y_1 - \hat{Y}_1 \quad Y_2 - \hat{Y}_2 \cdots Y_n - \hat{Y}_n]\begin{bmatrix} Y_1 - \hat{Y}_1 \\ Y_2 - \hat{Y}_2 \\ \vdots \\ Y_n - \hat{Y}_n \end{bmatrix}$$

$$= (Y_1 - \hat{Y}_1)^2 + (Y_2 - \hat{Y}_2)^2 + \cdots + (Y_n - \hat{Y}_n)^2$$

$$= \sum (Y_i - \hat{Y}_i)^2 = \sum e_i^2.$$

We see that this is the numerator of $s^2_{Y \cdot X}$. Thus, $s^2_{Y \cdot X}$ can be written in matrix notation as

$$s^2_{Y \cdot X} = \frac{(\mathbf{y} - \mathbf{X}\hat{\boldsymbol{\beta}})'(\mathbf{y} - \mathbf{X}\hat{\boldsymbol{\beta}})}{n - 2}.$$

We will see how easily the formula in matrix notation handles the variance of the points around the regression line for more complicated models.

Before leaving this discussion, consider again the vector of estimates $\mathbf{X}\hat{\boldsymbol{\beta}}$:

$$\mathbf{X}\hat{\boldsymbol{\beta}} = [\hat{Y}_1 \quad \hat{Y}_2 \cdots \hat{Y}_n]'.$$

Recall that $\hat{\boldsymbol{\beta}} = (\mathbf{X}'\mathbf{X})^{-1}\mathbf{X}'\mathbf{y}$, so that we may write

$$\mathbf{X}\hat{\boldsymbol{\beta}} = \mathbf{X}[(\mathbf{X}'\mathbf{X})^{-1}\mathbf{X}'\mathbf{y}] = [\mathbf{X}(\mathbf{X}'\mathbf{X})^{-1}\mathbf{X}']\mathbf{y}$$

or

$$\mathbf{X} = \mathbf{H}\mathbf{y}$$

Hat Matrix

where $\mathbf{H} = \mathbf{X}(\mathbf{X}'\mathbf{X})^{-1}\mathbf{X}'$ is often referred to as the **hat matrix.** The hat matrix is very important to the study of the theory of linear regression and to more advanced techniques of the analysis of residuals. We mention it here for the future reference of those readers who will go on to more advanced courses.

Finally, in Chapter 7 we will see the sum of squared residuals written as $\mathbf{y}'\mathbf{y} - \hat{\boldsymbol{\beta}}'\mathbf{X}'\mathbf{y}$. Let us show that this is the same as $(\mathbf{y} - \mathbf{X}\hat{\boldsymbol{\beta}})'(\mathbf{y} - \mathbf{X}\hat{\boldsymbol{\beta}})$. First, note that $(\mathbf{y} - \mathbf{X}\hat{\boldsymbol{\beta}})$ can be written as $\mathbf{y}' - (\mathbf{X}\boldsymbol{\beta})'$. Now, there is a rule in matrix algebra that the transpose of a product of two matrices is the product of the transposes *in reverse order* (to keep dimensions conformable), so that

$$(\mathbf{X}\hat{\boldsymbol{\beta}})' = \hat{\boldsymbol{\beta}}'\mathbf{X}'.$$

Thus, we have that

$$(\mathbf{y} - \mathbf{X}\hat{\boldsymbol{\beta}})' = \mathbf{y}' - \hat{\boldsymbol{\beta}}'\mathbf{X}'.$$

Then

$$(\mathbf{y} - \mathbf{X}\hat{\boldsymbol{\beta}})'(\mathbf{y} - \mathbf{X}\hat{\boldsymbol{\beta}}) = (\mathbf{y}' - \hat{\boldsymbol{\beta}}'\mathbf{X}')(\mathbf{y} - \mathbf{X}\hat{\boldsymbol{\beta}})$$

$$= \mathbf{y}'\mathbf{y} - \mathbf{y}'\mathbf{X}\hat{\boldsymbol{\beta}} - \hat{\boldsymbol{\beta}}'\mathbf{X}'\mathbf{y} + \hat{\boldsymbol{\beta}}'\mathbf{X}'\mathbf{X}\hat{\boldsymbol{\beta}}.$$

Consider $\mathbf{y}'\mathbf{X}\hat{\boldsymbol{\beta}}$. Since $\mathbf{y}$ is of dimension $1 \times n$, $\mathbf{X}$ is $n \times p$, and $\hat{\boldsymbol{\beta}}$ is $p \times 1$, $\mathbf{y}' \times \boldsymbol{\beta}$ is 1×1—a scalar. The transpose of a scalar is just the scalar, so that

$$(\mathbf{y}'\mathbf{X}\hat{\boldsymbol{\beta}}) = (\mathbf{y}'\mathbf{X}\hat{\boldsymbol{\beta}})' = \hat{\boldsymbol{\beta}}'\mathbf{X}'\mathbf{y}.$$

We now have that

$$(\mathbf{y} - \mathbf{X}\hat{\boldsymbol{\beta}})'(\mathbf{y} - \mathbf{X}\hat{\boldsymbol{\beta}}) = \mathbf{y}'\mathbf{y} - 2\hat{\boldsymbol{\beta}}'\mathbf{X}'\mathbf{y} + \hat{\boldsymbol{\beta}}'\mathbf{X}'\mathbf{X}\hat{\boldsymbol{\beta}}.$$

Now consider the last term above. Recall that the normal equations are

$$\mathbf{X}'\mathbf{X}\hat{\boldsymbol{\beta}} = \mathbf{X}'\mathbf{y},$$

so that we can write

$$\hat{\boldsymbol{\beta}}'\mathbf{X}'\mathbf{X}\boldsymbol{\beta} = \hat{\boldsymbol{\beta}}'(\mathbf{X}'\mathbf{X}\hat{\boldsymbol{\beta}}) = \hat{\boldsymbol{\beta}}'\mathbf{X}'\mathbf{y}.$$

Finally, then

$$(\mathbf{y} - \mathbf{X}\hat{\boldsymbol{\beta}})'(\mathbf{y} - \mathbf{X}\hat{\boldsymbol{\beta}}) = \mathbf{y}'\mathbf{y} - 2\hat{\boldsymbol{\beta}}'\mathbf{X}'\mathbf{y} + \hat{\boldsymbol{\beta}}'\mathbf{X}'\mathbf{y}$$

$$= \mathbf{y}\mathbf{y} - \hat{\boldsymbol{\beta}}'\mathbf{X}'\mathbf{y},$$

which was what we set out to show.

Compare the expression $\mathbf{y}'\mathbf{y} - \hat{\boldsymbol{\beta}}'\mathbf{X}'\mathbf{y}$ to the calculating formula $S_{YY} - bS_{XY}$ and note the similarity in form. Also compare the form of $(\mathbf{y} - \mathbf{X}\hat{\boldsymbol{\beta}})'(\mathbf{y} - \mathbf{X}\hat{\boldsymbol{\beta}})$ to $\sum (Y_i - \hat{Y}_i)^2 = \sum [Y_i - (a + bX_i)]^2$ and note the similarity.

6.4 INTERPRETATION OF THE STANDARD ERROR OF ESTIMATE

We have seen that $s_{Y \cdot X}$ measures the variation in Y after we have accounted for variation because of the relationship with X. What does this mean in terms of the problem of relating prices to sizes of houses?

The relationship of Y with X is described by the regression equation $\hat{Y} = a + bX$, which gives the estimated average price of houses of a given size. Naturally one does not expect all prices to be equal to the mean price, because of random variation and such other characteristics as location and age, which have not been considered. The \$2,550 figure found for $s_{Y \cdot X}$ is a measure of how much the prices in the sample, on the average, differ from the mean price because of the factors we call error. Given our assumption that scatter is homogeneous around the line, then regardless of the size of house being considered, the easiest way to interpret $s_{Y \cdot X} = 2.55$ is that *the average amount that the actual prices differ from the estimated mean price is \$2,550.* Some houses will differ more than \$2,550 from the predicted mean price, some houses will be closer to the estimated mean price than \$2,550; some houses may even cost exactly as much as the estimated mean price. But if we should calculate how much each house in the sample differs from the predicted mean price and then average these differences, this average difference would be about \$2,550. (Actually, the average amount that the actual prices differ from the estimated mean price is measured by $\sum |Y_i - \hat{Y}_i|/n$, and not by $s_{Y \cdot X} = \left[\sum (Y_i - \hat{Y}_i)^2/(n - 2)\right]^{1/2}$. The correct interpretation of $s^2_{Y \cdot X}$ is the unbiased estimate of the average squared deviation (variance) about the line, but this technically correct interpretation requires reasonably extensive training in statistics to understand.)

SIZE OF THE ERROR

Is this amount of variation excessive? Is it so large that the estimates are essentially useless? Is \$2,550 a large amount of error, or is it small enough that we can have some measure of confidence that our estimates are close enough to be useful? Clearly, too large an error in estimation would indicate a very weak relationship between the sizes and the prices of the houses under consideration, and we would not gain much information concerning price from knowing the size of a house.

An error of \$2,550 can be relatively small or quite large. For example, an error of \$2,550 in estimating the cost of putting an astronaut on the moon would be almost negligible, since the total cost would run to millions or billions of dollars. But an error of \$2,550 in estimating the cost of a hamburger is so large as to be ridiculous. Thus, the error is large or small depending upon the magnitude of the values involved. If the houses in our sample cost between \$17,000 and \$52,000, is \$2,550 a large or a small amount of error?

We might compare the size of the error to the average, or typical, price from our sample. We found

$$\overline{Y} = \frac{690}{20} = 34.5 \quad (= \$34{,}500)$$

Thus, as compared to the typical price, the error in estimation is

$$\frac{2.55}{34.5} = 0.0739$$

or $7\frac{1}{3}$ percent. You may notice that this measure of the percentage error, $s_{Y \cdot X}/\overline{Y}$, is analogous to the **coefficient of variation,** $s_Y/\overline{Y}$, from your introductory course, where it was computed to determine the amount of variability relative to the sizes of the values.

Coefficient of Variation

Now, whether $7\frac{1}{3}$ percent error in estimation is excessive or not depends upon the nature and intended use of the data at hand. If for a given problem the percentage error in estimation is considered too large, can we find another straight line equation for which the standard error of estimate is smaller? No, because the method of least squares gives the line that is, on the average, closest to all the data points in terms of squared deviations; it gives the smallest average squared errors and the most accurate predictions possible. Recall that in Chapter 2 we found the regression line by minimizing the quantity

$$Q = \sum (Y_i - \hat{Y}_i)^2 = \sum (Y_i - a - bX_i)^2.$$

We recognize this now as the line that minimizes error in predictions. It is the line for which $s_{Y \cdot X}$ is smallest.

EXERCISES

6.2 For the Concession Data in Exercise 1.5,

a. Find the standard error of estimate and state its units;

b. Interpret the meaning of the standard error of estimate;

c. Calculate the coefficient of variation. Does the error seem large?

6.3 In Exercise 3.27 the Concession Data were analyzed using the matrix algebra approach. Referring to $\mathbf{X}$, $\mathbf{y}$, and $\hat{\boldsymbol{\beta}}$ from Exercise 3.27, calculate the standard error of estimate. Compare to the value in Exercise 6.2.

6.4 Refer to the Heating Data in Exercise 1.8.

a. Find the standard error of estimate. State its units.

b. What does this number mean?

c. Calculate the coefficient of variation, and interpret its meaning.

6.5 Refer to the Smokers Data in Exercise 1.9.

a. Find the standard error of estimate. State its units.

b. Calculate the coefficient of variation. State its units.

c. How well can the number of days sick leave taken by an employee be estimated from knowing how many cigarettes that employee smokes each day?

6.6 In the Mail Data of Exercise 3.31, a company wanted to estimate how many orders needed to be processed on a given day by weighing the sacks of mail received. How well do you think this will work?

6.7 A company estimates the success of its products in a given marketing area by first putting a product on sale on a trial basis for six weeks in that area. Information on sales during the trial period and sales after one year from a simple random sampling of past cases is given in Table 6.1.

TABLE 6.1
Marketing Data

Trial period sales (hundreds of dollars)	13.2	25.1	21.2	10.7	24.0	19.6	11.9
First-year sales (hundreds of dollars)	134.7	258.7	220.1	109.6	251.7	198.4	125.2
Trial period sales	25.5	17.9	25.9	25.2	11.9	11.6	
First-year sales	248.7	175.6	259.9	258.7	125.2	115.0	

How well do trial period sales predict the success of the product in terms of first-year sales?

6.8 In Exercise 3.32 a quadratic regression equation of the form $\hat{Y} = a + bX + cX^2$ was fitted to the police data.

a. Find the standard error of estimate.

b. How many degrees of freedom does $s^2_{Y \cdot X}$ have? Why?

6.9 Suppose a model $Y = \alpha + \beta_1 X_1 + \beta_2 X_2 + \beta_3 X_2^2 + \varepsilon$ were fitted to some data.

a. How would $s^2_{Y \cdot X}$ be calculated?

b. How many degrees of freedom would $s^2_{Y \cdot X}$ have?

6 . 5

INTERVAL ESTIMATES: A REVIEW

In this section we shall review some ideas concerning the distribution of sample means and the formation of confidence intervals for the mean of a population based on the mean of a sample. Here, we will consider a univariate problem in which we have a simple random sample of values Y_i from a population whose mean we wish to estimate. In the next section we shall see how to extend these techniques to the regression problem, in which we want to use information about the independent variable X to aid in estimating the mean value of Y. Those students with good backgrounds in statistical inference should be able to skim quickly through this section.

THE NORMALITY ASSUMPTION

For many variables Y, it is not unreasonable to assume that the distribution of values is at least approximately **normal.** A normally distributed random variable Y is one for which the probability density function is

Normal Distribution

$$f(y) = \frac{1}{\sqrt{2\pi}\sigma} \exp\left[-\frac{(y-\mu)^2}{2\sigma^2}\right] \quad \text{for} \quad -\infty < y < \infty,$$

where exp[ ] indicates the number e raised to the power within the brackets. We note that a normal variable Y is **continuous**—its values can be any real numbers—with mean value μ and standard deviation σ. The function above plots a smooth, symmetric, bell-shaped curve, centered at μ and whose inflection points occur at distance 1σ either side of μ. (See Figure 6.4.)

Continuous

FIGURE 6.4
Graph of a Normal Distribution With Mean μ and Standard Deviation σ

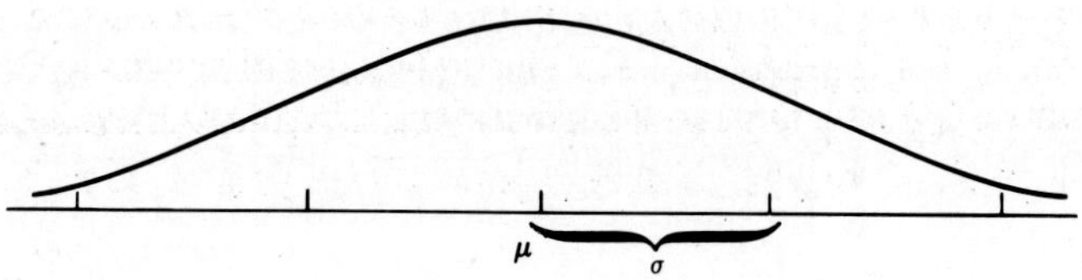

Recall that if Y is normally distributed with mean μ and standard deviation σ, written in shorthand as

$$Y \sim N(\mu, \sigma),$$

1. Approximately 68 percent of the values lie within a 1-standard deviation distance from the mean,
2. Approximately 95 percent of the values lie within 2 standard deviations of the mean;
3. More than 99 percent of the values lie within 3 standard deviations of the mean.

Symbolically, the foregoing relationships may be expressed in several ways:

1. Approximately 68 percent of all normal values lie in the interval $\mu - \sigma$ to $\mu + \sigma$;
2. Approximately 95 percent of all normal values lie in the interval $\mu \pm 2\sigma$;
3. If P represents probability, $P(\mu - 3\sigma < Y < \mu + 3\sigma) > 0.99$.

Exact probabilities can be found in tables of the standard normal distribution, found in all elementary texts. While standard normal tables can be set up in several different ways, they all give essentially the probabilities of obtaining values within some number of standard deviations, Z, from the mean. For example, *exactly* 95 percent of all normally distributed values lie within 1.96 (approximately 2) standard deviations of their mean.

The normal distribution can serve as a population model for a large variety of variables. For the assumption of normality to be reasonable, the variable Y should first be continuous. That is, values of Y are measurements, instead of categories or counts, that could theoretically be made to any desired degree of accuracy. Secondly, the distribution of values should be bell-shaped. This means that most values tend to be near average, and the farther from average a value is, the less likely it is to occur. Finally, values below the mean behave just as values above the mean (symmetry).

The normal distribution got its name from the "normal" or "typical" behavior of many kinds of measurements and was developed through study of how measurement errors in astronomy occurred. It is true that many kinds of measurements, not only in astronomy, physics, and other physical sciences but also in sociology, psychology, anthropology, and other social sciences, display normal (or at least approximately normal) distributions. However, *not all* measurements are normally distributed. Some continuous variables display distributions that are **skewed.** For example, relatively small values for some variables may be quite common, but the larger the value, the less likely it is to occur. This does not make the variable "abnormal," however. The statistical definition of *normal* is not synonymous with the general use of this word.

Skewed

THE CENTRAL LIMIT THEOREM

Consider two people, each taking independent simple random samples of size n from $N(\mu, \sigma)$, a normal population with mean μ and standard deviation σ. Each of two samples will contain different items, and thus if $\bar{Y}$, the sample mean, is calculated for each sample, it would be surprising indeed if both $\bar{Y}$'s were equal to each other. It would also be surprising to find that either $\bar{Y}$ was exactly equal to μ. What we would expect is that both values of $\bar{Y}$ would be reasonably close to μ and thus not too different from each other.

Imagine extending this example to the situation in which a very large number of people each drew a simple random sample of size n from $N(\mu, \sigma)$, so that we would have a very large collection of (potentially) different values for $\bar{Y}$. What would the **sampling distribution** of these values **for $\bar{Y}$** look like?

Sampling Distribution for $\bar{Y}$

If a sample is a simple random sample from a normal population, then since we would expect it to be representative of the population, we would expect a majority of sample values to be near μ, with a few somewhat larger and a few somewhat smaller than the rest. The mean of these values should be near μ. It is possible, but not as likely, that a simple random sample would contain a large preponderance of relatively large values from the population, so that the mean of this sample would be larger than μ. A similar statement can be made concerning samples that happened by chance to include a disproportionately high percentage of the smallest values in the population.

It is not too hard to imagine, then, that the collection of $\overline{Y}$-values would contain very many values very close to μ, while the farther $\overline{Y}$ were from μ in either direction, the fewer such values there would be. The distribution of $\overline{Y}$-values, then, should be bell-shaped and symmetric, centered at μ, and with a small standard deviation, relative to the standard deviation of the population. That the standard deviation will be small is apparent if you consider that the mean of any sample can neither be as large as the largest value nor as small as the smallest value in the sample. Thus, although the actual values may vary quite a bit, the means will not vary as much as the actual values, since the mean finds the *middle* of the set.

It can be shown that the standard deviation of the sampling distribution of the sample mean is $\sigma/\sqrt{n}$. Note that the larger the sample, the smaller the standard deviation of $\overline{Y}$, since $\sqrt{n}$ is in the denominator of the standard deviation. This makes sense, as the larger the size of the sample, the more accurate $\overline{Y}$ should be in estimating μ.

Thus, the collection of means from possible samples drawn from a normal population will themselves follow a normal distribution. In fact, it has been proved mathematically that this is so. Further, probabilities concerning sample means for samples from normal populations can be calculated by standardizing the sample mean,

$$Z = \frac{\overline{Y} - \mu}{\sigma/\sqrt{n}},$$

and referring to tables of the standard normal distribution.

Central Limit Theorem

A most useful theorem, the **Central Limit Theorem,** states that even if the population from which the sample is drawn is not normal, the random variable

$$Z = \frac{\overline{Y} - \mu}{\sigma/\sqrt{n}}$$

will still be approximately distributed as a normal random variable with mean zero and standard deviation 1. The approximation is better, the closer the population is to normal and the larger the size of the sample. Thus, whenever one is dealing with means of samples, their behavior is well modeled by the normal distribution. For most of our work in regression, however, we will need to make the assumption that the population itself is normal.

CONFIDENCE INTERVALS FOR POPULATION MEANS

The Central Limit Theorem describes the theoretical behavior of means of samples drawn from normal populations. In practice, we will have only *one* mean from only *one* sample of size n from an *assumed* normal population.

The reason the sample was drawn was to obtain information about the population: for example, what is the mean value in the population? Suppose for the moment that the population mean is unknown, but that (somehow) the population standard deviation is known, and that we want to use the information in a sample of size n to estimate the unknown population mean, μ.

Point Estimate

A **point estimate** of μ would be the sample mean $\bar{Y}$, since the mean of all possible $\bar{Y}$'s is μ. (This says that $\bar{Y}$ is unbiased for μ: in any sample, $\bar{Y}$ could overestimate or underestimate μ, but it tends to do neither. One could say $\bar{Y}$ is "trying to find" μ.) But in any sample, $\bar{Y}$ will probably miss the true value of μ by some amount. By how much? The standard deviation of $\bar{Y}$, $\sigma/\sqrt{n}$, tells the expected amount by which $\bar{Y}$ will miss μ, since a standard deviation is (loosely) the average amount that values in a set differ from the mean of the set. Since $\bar{Y}$ is normal (Figure 6.5(a)), we can calculate the **probable size of the error** in using $\bar{Y}$ to estimate μ. For example, since 95 percent of all normal values lie within a 1.96-standard deviation distance of their mean (Figure 6.5(b)), the probability that $\bar{Y}$ is within a distance of $1.96\sigma/\sqrt{n}$ either side of μ is 0.95. That is,

Probable Size of Error

$$P\left(\mu - \frac{1.96\sigma}{\sqrt{n}} < \bar{Y} < \mu + \frac{1.96\sigma}{\sqrt{n}}\right) = 0.95.$$

Confidence Interval

This statement can be used to obtain a 95% **confidence interval** for μ. Suppose that $\bar{Y}$ falls within $1.96\sigma/\sqrt{n}$ units of μ. Then the interval

$$\bar{Y} \pm \frac{1.96\sigma}{\sqrt{n}}$$

will cover μ (see Figure 6.5(c)). Since the probability that $\bar{Y}$ is within a $1.96\sigma/\sqrt{n}$ distance of μ is 0.95, then the probability that the interval $\bar{Y} \pm 1.96\sigma/\sqrt{n}$ covers μ is 0.95.

On the other hand, suppose the sample is by chance a strange one, so that the value for $\bar{Y}$ is more than 1.96 standard deviations from μ. Then the interval $\bar{Y} \pm 1.96\sigma/\sqrt{n}$ will not cover μ (see Figure 6.5(d)). But this will happen only in 5 percent of all possible samples.

In any given situation, you will have one value for $\bar{Y}$ and one interval $\bar{Y} \pm 1.96\sigma/\sqrt{n}$. You will never know whether or not the interval covers μ. But the procedure you are using provides that, on the average, 95 out of 100 intervals so calculated will cover the true mean; so your bounds can legitimately be called 95% confidence bounds.

Confidence intervals with confidence coefficients other than 0.95 can be formed by replacing 1.96 by the appropriate value from the normal tables.

The confidence intervals formed here are rather artificial, because it is not often the case that the standard deviation in the population is known when the objective is to estimate the population mean. Typically, the population standard deviation is unknown as well and is estimated by the sample variance

$$s = \left[\frac{\sum (Y_i - \bar{Y})^2}{n - 1}\right]^{1/2} = \left[\frac{S_{YY}}{n - 1}\right]^{1/2}.$$

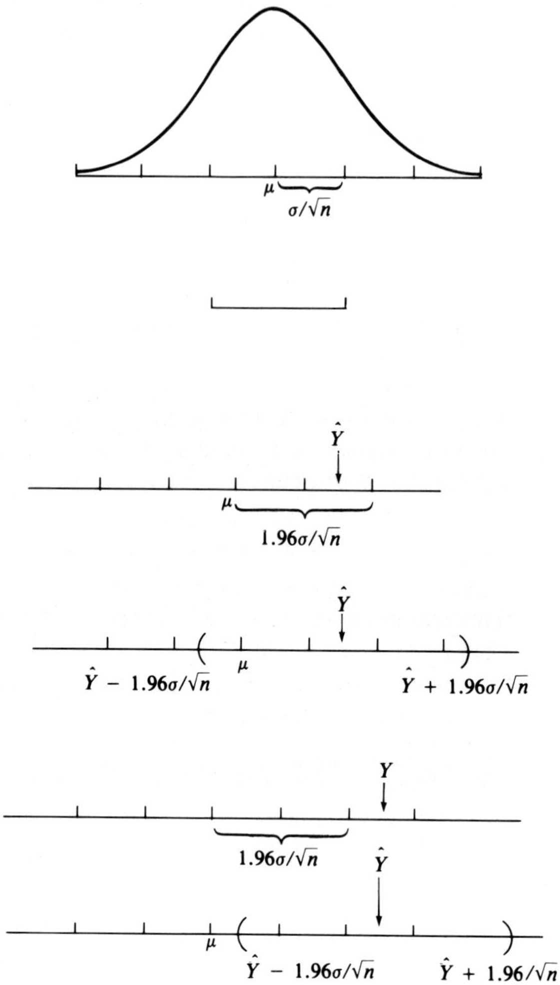

FIGURE 6.5
Confidence Intervals on a Population Mean

(a) The sampling distribution of $\overline{Y}$ is normal, with mean μ and standard deviation $\sigma/\sqrt{n}$

(b) Length of a $1.96\sigma/\sqrt{n}$ unit distance

(c) Suppose $\overline{Y}$ falls within $1.96\ \sigma/\sqrt{n}$ units of μ. Then the interval $\overline{Y} \pm 1.96\sigma/\sqrt{n}$ will cover μ

(d) Suppose $\overline{Y}$ falls farther than $1.96\sigma/\sqrt{n}$ units from μ. Then the interval $\overline{Y} \pm 1.96\sigma/\sqrt{n}$ will not cover μ

To use s in place of σ in the confidence interval will introduce more variability into our estimate of μ because we must consider not only how much $\overline{Y}$ varies from sample to sample, but also how much s varies from sample to sample. The interval estimate of μ, using s to estimate σ, must thus be wider than $\overline{Y} \pm 1.96\sigma/\sqrt{n}$ to take into account that s is also subject to sampling variability and in any one sample may be either underestimating or overestimating σ. The coefficient 1.96 must be replaced by a somewhat larger value if we wish to obtain an interval which has 0.95 probability of being correct. This coefficient is obtained from tables of Student's t distribution.

6 . 6

THE t DISTRIBUTION

COMPARISON OF THE t TO THE STANDARD NORMAL

Student's t Distribution

If σ is not known but must be estimated by s, we can no longer refer to the normal distribution but must instead refer to the **Student's t distribution.**† **The** t distribution looks very much like the standard normal distribution. It is also bell-shaped, symmetrical, and centered at zero. However, a t distribution is shorter and fatter, and its variance is larger than that of the standard normal (which is 1). Further, while there is only one standard normal distribution, there are many t distributions, one for each different positive integer. In this case, the integers are called *degrees of freedom*. The smaller the number of the degrees of freedom, the shorter and fatter (more variable) the distribution. The greater the number of degrees of freedom, the taller and skinnier the distribution, until for an infinite number of degrees of freedom, the t and the normal are the same. Figure 6.6 compares some selected t distributions with the standard normal distribution.

FIGURE 6.6
Comparison of Two t Distributions With the Standard Normal Distribution

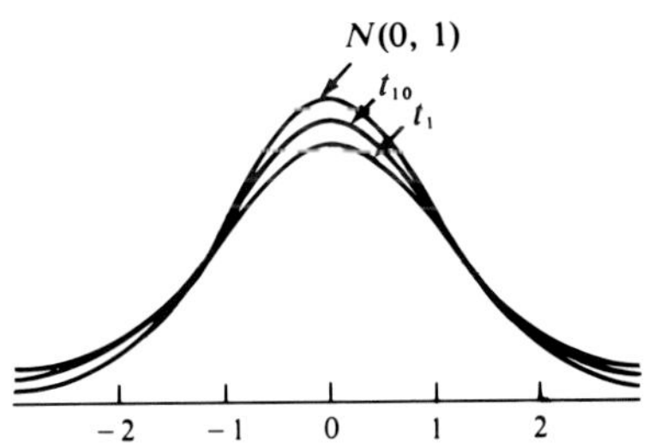

We can see in the figure that the distribution of a t with 1 degree of freedom (denoted t_1) has more area in its "tails" than does the standard normal. Thus, to find the points that cut off $2\frac{1}{2}$ percent of the area under the upper and lower tails we have to go farther out than ± 1.96. But we do not have to go as far out on t_{10} as on t_1. Let $t_{\nu,\alpha}$ denote that value of t with ν

† The t distribution was invented by W. S. Gossett, who was forced to use the pseudonym Student because his employer, the Guinness Brewery of Dublin, Ireland, would not permit him to publish his research under his own name.

degrees of freedom, which cuts off 100α percent of the area under the upper tail of the distribution. That is, by definition,

$$P(t_\nu > t_{\nu,\alpha}) = \alpha.$$

(See Figure 6.7.) By symmetry around zero,

$$P(t_\nu < -t_{\nu,\alpha}) = \alpha,$$

FIGURE 6.7
Definition of $t_{\nu,\alpha}$

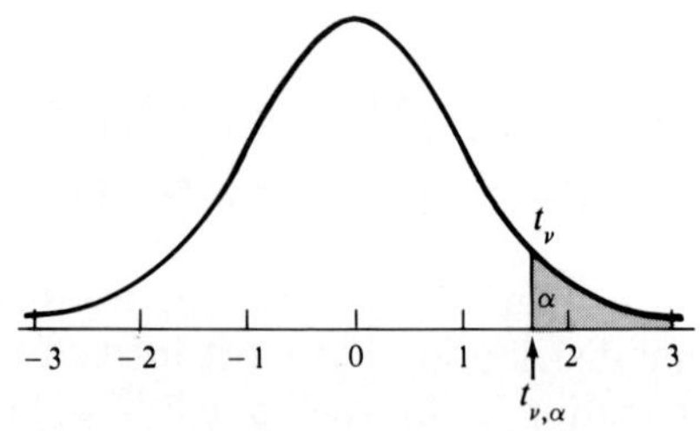

or if $t_{\nu,\alpha}$ cuts off the 100α percent of the area under the upper tail, then $-t_{\nu,\alpha}$ cuts off 100α percent of the area under the lower tail. (Do not confuse this α with the intercept in the model. Here, α represents a small probability, or a small percentage. We believe the context will make clear which α is being referred to.)

TABLES OF THE *t* DISTRIBUTION

Appendix H gives values of the *t* distribution. Across the top row are values of α, and down the outside column are listed degrees of freedom. From the example given on the tables, we see that

$$P(t_{10} > 2.228) = 0.025.$$

By symmetry around zero,

$$P(t_{10} < -2.228) = 0.025.$$

(The reader should be careful when using other tables—not all of them are set up like this one. It is necessary to look closely at the table headings in order to determine exactly what areas are being given.) As another example, the *t*-values which cut off the upper and lower 5 percent of the area under a distribution with 20 degrees of freedom are

$$t_{20,0.05} = 1.725 \quad \text{and} \quad -t_{20,0.05} = -1.725,$$

respectively. For degrees of freedom not given in the tables, one can usually use the *t*-value corresponding to the next lowest number. Note that for degrees of freedom equal to infinity, the values ± 1.96 cut off the upper and lower $2\frac{1}{2}$ percent of the area. Thus, the numbers corresponding to the standard normal numbers are given in the bottom row of the t-tables. Note also that for any of the percentages indicated in the table, all t-values are larger than the standard normal values. Using t-values in a confidence interval will thus give a wider interval than will using standard normal values. However, since for $\nu \geq 30$ the t-values and normal values are quite close in size, it is common to use normal values in samples of size greater than 30.

CONFIDENCE INTERVALS USING THE *t* DISTRIBUTION

The number of degrees of freedom, ν, for the t is determined by the degrees of freedom in the denominator of the estimator of the population variance. In the case of estimating σ^2 by s^2, the denominator of s^2 is $n - 1$, so that we enter the tables in the row $\nu = n - 1$ and find the appropriate coefficient in the column labeled $\alpha/2$, where $\alpha = 1 -$ confidence coefficient. For example, if we have a sample of size $n = 25$ and wish to form a 95% confidence interval on μ, we look up

$$t_{24,0.025} = 2.064$$

in the tables. The 95% confidence interval for μ would then be

$$\overline{Y} \pm \frac{t_{24,0.025}\, s}{\sqrt{n}} = \overline{Y} \pm \frac{2.064s}{\sqrt{25}}.$$

Note that the validity of the values given in the t-tables depends on the assumption that the population from which the sample was drawn is a normal population. However, if the sample is large enough—say, so that $\nu \geq 30$ and we are using normal distribution values—then because of the Central Limit Theorem the confidence intervals so constructed will be accurate in terms of their nominal confidence.

EXERCISE

6.10 Using Appendix H, find the indicated t-values.

a. $t_{20,0.05}$

b. $-t_{27,0.01}$

c. The value of t with 12 degrees of freedom that cuts off the smallest 10 percent of the values;

d. The value of t with 25 degrees of freedom that is exceeded by only 0.5 percent of the values;

e. Those values of t with 6 degrees of freedom that cut off the smallest $2\frac{1}{2}$ percent and the largest $2\frac{1}{2}$ percent of the values.

6.7 CONFIDENCE INTERVALS ON A MEAN RESPONSE

In order to form a confidence interval for the mean value of Y for a given value of X in a regression problem, we need to first make an additional assumption concerning the distribution of Y-values at any fixed value of X.

Recall that we have already assumed uniform scatter of points around the regression line. This enabled us to say that the variance of the Y-values was constant, regardless of the associated value of X, so that

$$s_{Y \cdot X}^2 = \frac{\sum (Y_i - \hat{Y}_i)^2}{n - 2}$$

could be calculated as an estimate of the constant variance $\sigma_{Y \cdot X}^2$. Also, we have already assumed that the mean Y-value at any given X-value is a linear function of X: $\mu_{Y \cdot X_i} = \alpha + \beta X_i$. We now need to assume that at each value of X, Y-values are *normally* distributed around the regression line.

Figure 6.8 shows the model for our house size-price problem, illustrating all these assumptions. It is not unreasonable to assume that for any fixed size house, the distribution of prices would be bell-shaped and symmetric, the prices tending to concentrate around a mean price but some houses having or lacking features that make them relatively more or less expensive than other houses the same size.

With this final assumption, we are now ready to form confidence intervals on the mean price of houses of size X_i. Recall that the general form of a confidence interval on a population mean based solely on the information provided in a sample is

Point estimate $\pm$ (t-value) (estimate of standard deviation of point estimate).

The point estimate of $\mu_{Y \cdot X_i}$ is $\hat{Y}_i = a + bX_i$. The t-value will have $\nu = n - 2$ degrees of freedom, since $n - 2$ is the degrees of freedom for $s_{Y \cdot X}^2$, which estimates $\sigma_{Y \cdot X}^2$. We now need only an estimate of the standard deviation of $\hat{Y}_i$. It can be shown that the variance of $\hat{Y}_i$ is

$$\sigma_{Y \cdot X}^2 \left[\frac{1}{n} + \frac{(X_i - \overline{X})^2}{S_{XX}} \right]$$

FIGURE 6.8
Values of Y Are Normally Distributed About the Line at Each X-Value

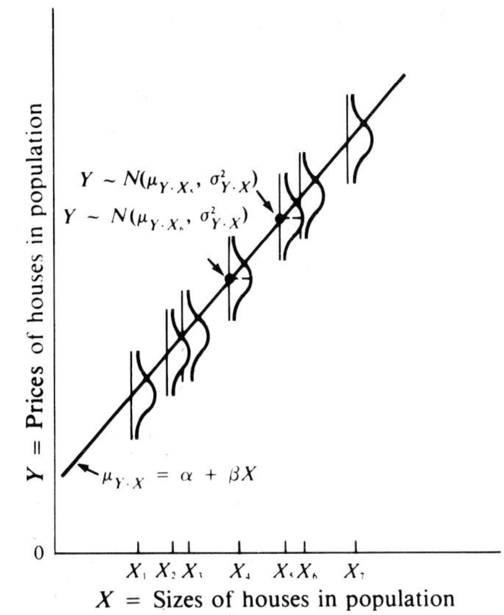

which is estimated by replacing $\sigma^2_{Y \cdot X}$ by $s^2_{Y \cdot X}$. The estimated standard deviation of $\hat{Y}_i$ is the square root of

$$s^2_{Y \cdot X}\left[\frac{1}{n} + \frac{(X_i - \overline{X})^2}{S_{XX}}\right]$$

and measures (roughly) how much our estimates of the mean price of houses of size X_i would vary from sample to sample. Thus, a $100(1 - \alpha)\%$ confidence interval on $\mu_{Y \cdot X_i}$ is given by

$$\hat{Y}_i - t_{n-2,\alpha/2}\sqrt{\frac{1}{n} + \frac{(X_i - \overline{X})^2}{S_{XX}}} < \mu_{Y \cdot X_i} < \hat{Y}_i + t_{n-2,\alpha/2}\sqrt{\frac{1}{n} + \frac{(X_i - \overline{X})^2}{S_{XX}}}.$$

CONFIDENCE INTERVAL FOR HOUSE SIZE-PRICE EXAMPLE

Let us work an example, using our house size-price data. Suppose we want a 95% confidence interval on the mean price of houses 2,200 sq ft in size. We have found

$$\hat{Y}_i = 8.7036 + 12.8982(2.2) = 37.0796;$$

$$s_{Y \cdot X} = 2.5506, \qquad n = 20, \qquad S_{XX} = 13.56;$$

$$t_{18,0.025} = 2.101, \qquad \overline{X} = 2.$$

Then

$$\hat{Y}_i \pm t_{18,0.025}s_{Y\cdot X}\sqrt{\frac{1}{n}+\frac{(X_i-\overline{X})^2}{S_{XX}}} = 37.0796 \pm 2.101(2.5506)\sqrt{\frac{1}{20}+\frac{(2.2-2)^2}{13.56}}$$
$$= 37.0796 \pm 1.2331.$$

We are 95 percent sure that the mean price of houses 2,200 sq ft in size is between

$$37.0796 - 1.2331 = 35.8465 \quad (\approx \$35{,}846)$$

and

$$37.0796 + 1.2331 = 38.3127 \quad (\approx \$38{,}313).$$

EXAMINING THE CONFIDENCE LIMITS

Note that the farther X_i is from $\overline{X}$, the larger the interval will become because of the term $(X_i - \overline{X})^2$ in the numerator of the variance of $\hat{Y}_i$. For example, if $X_i = 2.2$, the interval is of width $2(1.2331) = 2.4662$; but if $X_i = 3.5$, then the interval is of width

$$2(2.101)(2.5506)\sqrt{\frac{1}{20}+\frac{(3.5-2)^2}{13.56}} = 2(2.101)(2.5506)\sqrt{0.05+0.16}$$
$$= 2(2.4557) = 4.9114.$$

Confidence Bands

Table 6.2 shows confidence limits for several selected values of X_i. Thus, if we plotted $100(1-\alpha)\%$ confidence limits for each X_i in the sample, these limits would not lie on straight lines parallel to the regression line; rather, they would curve, as illustrated in Figure 6.9. We call these lines **confidence bands.** Note that the vertical distance between the upper band and the lower band is smallest at $X_i = \overline{X}$, and this distance becomes greater the farther X_i is from $\overline{X}$. What this says, essentially, is that we have less trustworthy information

TABLE 6.2 Confidence Limits for $\mu_{Y\cdot X_i}$ for Selected Values of X_i

X_i	$(X_i-\overline{X})^2$	$\frac{(X_i-\overline{X})^2}{S_{XX}}$	$\left[\frac{1}{n}+\frac{(X_i-\overline{X})^2}{S_{XX}}\right]^{1/2}$	$ts_{Y\cdot X}\left[\frac{1}{n}+\frac{(X_i-\overline{X})^2}{S_{XX}}\right]^{1/2}$	$\hat{Y}_i$	Lower Limit	Upper Limit
0.8	1.44	0.1062	0.3952	2.1178	19.0222	16.9044	21.1400
1.0	1.00	0.0737	0.3517	1.8847	21.6018	19.7171	23.4865
1.5	0.25	0.0184	0.2615	1.4013	28.0509	26.6496	29.4522
2.0	0.00	0.0000	0.2236	1.1982	34.5000	33.3018	35.6982
2.2	0.04	0.0030	0.2302	1.2336	37.0796	35.8460	38.2030
2.5	0.25	0.0184	0.2615	1.4013	40.9491	39.5478	42.3504
3.0	1.00	0.0737	0.3517	1.8847	47.3982	45.5135	49.2829
3.5	2.25	0.1659	0.4646	2.4897	53.8473	51.3576	56.3370
3.6	2.56	0.1888	0.4887	2.6188	55.1371	52.5183	57.7559

FIGURE 6.9
Confidence Bands for $\mu_{Y \cdot X_i}$

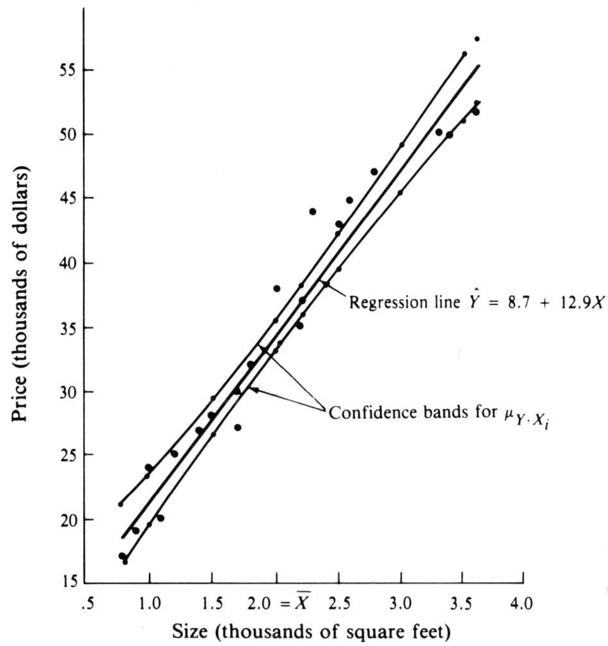

about the relationship between size and price for houses at the lower and upper ranges of sizes in our sample; we trust the regression line more for those houses in the middle range. Imagine rotating the regression line shown in Figure 6.9 clockwise, pivoting at the point $(\overline{X}, \overline{Y})$, until the line touches the confidence bands. Then rotate the line counterclockwise until it touches the bands again. We feel that the true line can lie anywhere between these two extremes. Thus we can pinpoint $\mu_{Y \cdot X_i}$ fairly accurately as long as X_i is near $\overline{X}$ but not so accurately if X_i is rather far away from $\overline{X}$.

EXERCISES

6.11 Refer to the Concession Data in Exercise 1.5. Give a 95% confidence interval on the average number of gallons of syrup required by the football stadium soft drink concession on weeks in which 65,000 advance tickets are sold.

6.12 Refer to the Heating Data of Exercise 1.8. Give a 99% confidence interval on the average oil bill for days on which the average temperature is 20° F. Explain what this interval means.

6.13 Refer to the Smokers Data in Exercise 1.9.

a. Give a 90% confidence interval on the mean number of sick days taken by nonsmokers.

b. Give a 90% confidence interval on the mean number of sick days taken by employees who smoke one pack (20 cigarettes) per day.
c. What do these intervals tell you?

6.14 For the Reservations Data in Exercise 5.16, calculate 95% confidence bands for the mean number of guests. Draw a scatter diagram of the data, graph the regression equation, and graph the confidence bands.

6.15 Refer to the Loans Data in Exercise 5.12.
a. Calculate 99% confidence bands for the average number of loans applied for.
b. Graph the data, the regression equation, and the confidence bands.

6.16 Confidence bands on $\mu_{Y \cdot X_i}$ are the shortest distance apart at $X_i = \overline{X}$, and the farther X_i is from $\overline{X}$, the wider the bands become, for a specified confidence coefficient. Discuss the implication of this to the use of the regression equation to estimate Y-values for X-values outside the range of data sampled.

6.8

CONFIDENCE INTERVAL FOR AN INDIVIDUAL RESPONSE

We can use the regression equation $\hat{Y} = a + bX$ to estimate the price of a single house of any size within our sampled range. While we realize that our estimated price, $\hat{Y}_i$, is an estimate of the mean price of houses of size X_i, we would like to obtain an interval in which we expect some large percentage of *all* prices to lie for houses that size, not just the mean price. Let Y_i^* denote the actual price of a house of size X_i. We want to find an interval of the form

$$\text{number} < Y_i^* < \text{number}$$

in which we can be, say, 95 percent sure that the actual price will lie. Some people call this a **prediction interval.**

Prediction Interval

This interval will be wider than the one for $\mu_{Y \cdot X_i}$, because sample means tend to cluster much closer around the true mean value than do individual observations. It can be shown that if we use $\hat{Y}_i$ to estimate the true value of Y for a given $X = X_i$, then an estimate of the variance of $\hat{Y}_i$ is

$$s_{Y \cdot X}^2 \left[1 + \frac{1}{n} + \frac{(X_i - \overline{X})^2}{S_{XX}} \right].$$

Again, the closer X_i is to $\overline{X}$, the smaller this interval will be. If we choose $X_i = 2.2$, then a 95% confidence interval on an individual Y_i-value, Y_i^*, is given by

$$\hat{Y}_i \pm t_{n-2,0.025} s_{Y \cdot X} \sqrt{1 + \frac{1}{n} + \frac{(X_i - \overline{X})^2}{S_{XX}}}$$

$$= 37.0796 \pm 2.101(2.5506) \sqrt{1 + \frac{1}{20} + \frac{(2.2 - 2)^2}{13.56}}$$

$$= 37.0796 \pm 5.4990.$$

Thus we claim that 95 percent of all houses 2,200 sq ft in size will sell for between

$$37.0796 - 5.4990 = 31.5806 \quad (\approx \$31{,}581)$$

and

$$37.0796 + 5.4990 = 42.5786 \quad (\approx \$42{,}579).$$

Referring to our sample, we note that houses 5 and 19 are both 2,200 sq ft in size and that they sold for \$35,000 and \$37,000, respectively. Since both of these prices lie in the interval from \$31,581 to \$42,579, we see that neither house is priced unusually high or low for its size (at least, they are not among the $2\frac{1}{2}$ percent lowest or $2\frac{1}{2}$ percent highest priced houses).

EXERCISES

6.17 Suppose that on a given week, 65,000 advance tickets were sold for a football game. Using the Concession Data of Exercise 1.5, give a 95% confidence interval on the number of gallons of syrup required by the soft drink concession that week. Compare to the interval constructed in Exercise 6.11 and explain the difference.

6.18 Refer to the Mail Data in Exercise 3.31. If on a given day, 250 lb of mail was received by the mail order house, give a 99% confidence interval on the number of orders to be processed. Interpret this interval.

6.19 In Exercise 6.14, 95% confidence bands for the average number of guests were graphed, along with the points and regression line for the Reservations Data. Calculate 95% confidence bands for the individual number of guests, graph these on the same graph, and compare.

6.20 For the Marketing Data in Exercise 6.7, calculate 99% confidence bands for the first-year sales of individual products. Graph the data, the regression equation, and the confidence bands.

6.9 FINDING THE STANDARD ERROR OF ESTIMATE ON COMPUTER PRINTOUT

All the simple regression computer programs we discussed in Chapter 4 give the standard error of estimate, and it is usually not very difficult to locate.

BMD PROGRAM

For the BMDP1R program for our house size-price problem, refer to Figure 4.2(b). It is hard to miss that STD ERROR OF EST 2.5506, since this is printed at the top of the analysis.

MINITAB PROGRAM

Minitab prints out a message in Figure 4.8: THE ST DEV OF Y ABOUT REGRESSION LINE IS S = 2.551 WITH (20 - 2) = 18 DEGREES OF FREEDOM. Additionally, the variance of the points about the regression line can be found in the column headed MS = SS/DF and in the row labeled RESIDUAL in the table headed ANALYSIS OF VARIANCE. There we find the value 6.506 for $s^2_{Y \cdot X}$.

SAS PROGRAM

For the SAS printout, refer to Figure 4.13. Under ERROR MEAN SQUARE we find 6.505531 for $s^2_{Y \cdot X}$, and below and to the left of this figure we find ROOT MSE 2.550594 for the standard error of estimate. Note that under a column labeled STANDARD ERROR we find two figures, but these have to do with how much a and b vary from sample to sample from the same population and are not equal to $s_{Y \cdot X}$. We will see in the next chapter that these are important measures. SAS calculates the coefficient of variation, C.V. = 7.393027, as well.

SPSS^X PROGRAM

To find the standard error of estimate when using the SPSS^X package, refer to the list of simple statistics given at the bottom of the scatter diagram in Figure 4.18. The second statistic given is `STD ERR OF EST - 2.55059`, so there is no trouble finding it here. In addition, the standard error can be found in Figure 4.14 in the regression routine, printed out under the label `STANDARD ERROR`.

USING THE COMPUTER TO CALCULATE CONFIDENCE LIMITS

BMD There is no simple way to instruct the BMDP1R program to calculate confidence limits for either the mean Y-value or an individual Y-value for a given X-value.

MINITAB Minitab does not automatically calculate confidence limits for either a mean response or an individual response, but it does most of the hard work for you. If the `NOBRIEF` command is used, a table containing values of X, Y, $\hat{Y}$, standard deviations of the predicted Y-values, residuals, and standardized residuals is printed. (See Figure 4.8.) Of interest to us here are the standard deviations of the predicted Y-values. These are values of the square root of

$$s_{Y\cdot X}^2\left[\frac{1}{n}+\frac{(X_i-\overline{X})^2}{S_{XX}}\right].$$

Thus, confidence limits on the mean value of Y at a given value of X, X_i, can be obtained by multiplying these values by the appropriate value of $t_{n-2,\alpha/2}$, adding, and subtracting these products from $\hat{Y}$.

To form confidence intervals on an individual response, square the value in the column labeled `STD DEV PRED Y` and divide by $s_{Y\cdot X}^2$ to obtain

$$\frac{1}{n}+\frac{(X_i-\overline{X})^2}{S_{XX}}.$$

Then add 1, multiply by $s_{Y\cdot X}^2$, take the square root, and proceed as before.

The problem is that there is no way to get the column of standard deviations of $\hat{Y}$ stored for use in further computation. The following Minitab program shows how to calculate confidence limits on individual responses.

Suppose that the house size-price data have been analyzed as in Figure 4.8. Data were entered so that values of "size" are in column 1 and values of "price" are in column 2. When the regression was performed, standardized residuals were stored in column 3 and predicted values in column 4. Let us now enter some values that remain constant in this series of calculations. These are $\overline{X} = 2$, $S_{XX} = 13.56$, $n = 20$, $t = 2.101$, and $s_{Y \cdot X} = 2.551$. To enter a constant, use the K command, with an accompanying number. For example,

```
LET K1 = 2
```

defines a constant K1 with the value 2. Similarly, we can write

```
LET K2 = 13.56, K3 = 20, K4 = 2.101, K5 = 2.551
```

(Note that constants may not be given names, as columns could.)

We want to subtract $\overline{X}$ from each value of X. This is accomplished by writing

```
SUBTRACT K1 FROM C1, PUT IN C5
```

Note that instead of defining K1 = 2, we could have written

```
SUBTRACT 2 FROM C1, PUT INTO C5
```

but it is often handy to know how to store constants for future use. Then we want to square the value of $X_i - \overline{X}$. Write

```
RAISE C5 TO THE POWER 2, PUT IN C6
```

Then to divide $(X_i - \overline{X})^2$-values by S_{XX}, write

```
DIVIDE C6 BY K2, PUT IN C7
```

Now calculate $1/n$ and store that as a constant:

```
DIVIDE 1 BY K3, PUT IN K6
```

(The computer will do so and tell you that K6 = 0.0500.) Now write

```
ADD K6 TO C7, PUT IN C8
```

Column 8 now contains values $1/n + (X_i - \overline{X})^2/S_{XX}$. To these values we add 1 by

```
ADD 1 TO C8, PUT IN C9
```

We now take the square root by

```
RAISE C9 TO THE POWER 0.50, PUT IN C10
```

These values are multiplied by t and s by

```
MULTIPLY C10 BY K4, PUT IN C11
```

and

```
MULTIPLY C11 BY K5, PUT IN C12
```

Then the lower limits are given by

```
SUBTRACT C12 FROM C4, PUT IN C13
```

and the upper limits by

```
ADD C12 TO C4, PUT IN C14
```

If we then

`NAME C13 'LOWER_ I' C14 'UPPER_ I'` and `PRINT C13 C14`

the confidence limits on an individual Y for the associated value of X are given as in Figure 6.10. The coefficient of variation can be easily obtained by instructing Minitab to divide $s_{Y \cdot X}$ by $\overline{Y}$.

FIGURE 6.10 Confidence Limits for an Individual Response, Using Minitab

COLUMN COUNT ROW	LOWER_I 20	UPPER_I 20
1	26.4206	37.4201
2	15.9202	27.2834
3	25.1212	36.1399
4	39.2045	50.4327
5	31.5799	42.5794
6	13.2591	24.7852
7	49.1718	61.1025
8	17.2455	28.5377
9	29.0080	39.9920
10	36.6779	47.7999
11	32.8601	43.8788
12	14.5913	26.0326
13	18.5673	29.7955
14	46.6997	58.4154
15	25.1212	36.1399
16	35.4091	46.4891
17	21.2001	32.3221
18	45.4589	57.0765
19	31.5799	42.5794
20	22.5109	33.5909

S A S The SAS procedure `PROC REG` can be instructed to give confidence limits either on the mean response or on the individual response for each X-value in the sample. Simply insert as an option after the slant in the `MODEL` statement either `CLI` (confidence limits for individual response) or `CLM` (confidence limits for the mean response). These will be 95% confidence limits.

S P S S ^{X} There is no simple way to instruct the SPSS^X regression program to calculate confidence limits either on the mean response or on an individual response.

6.10

SUMMARY

We now know how far we can expect our estimates of Y based on X to differ from the actual observed Y-values, and we can use this information to obtain interval estimates on both the typical Y-value and an individual Y-value at a given value of X.

The strength of a relationship is determined not only by the slope of the regression line but also by the amount of scatter (error) around the line. In order to see how strong the relationship really is, and especially to see if it is strong enough to rule out the possibility that it is only a chance function of the sample we happened to draw, we turn to the study of tests of significance in Chapter 7.

SEVEN

INFERENCES CONCERNING THE REGRESSION COEFFICIENTS

7.1 INTRODUCTION

Central to the study of regression are hypothesis tests to determine whether or not we can be reasonably confident that two variables are, in fact, related. We recognize the possibility that we could have drawn a sample indicating a relationship that does not exist in the population. We will review statistical tests of hypotheses, illustrating each step with reference to the house size-price example. The confidence intervals we consider this time have to do with the parameters of the model rather than with predicted values.

7.2 CAN WE BE CONFIDENT THAT A RELATIONSHIP EXISTS?

STRENGTH OF THE LINEAR RELATIONSHIP

In obtaining the regression equation relating prices to sizes of houses, we found $b = 12.9$. This was interpreted to mean that if two houses differed in size by 1,000 sq ft, then the larger house would cost on the average \$12,900 more than the smaller house. Our standard error of estimate, $s_{Y \cdot X}$ = \$2550, gave the average amount of error we could expect in estimating price from size alone.

Now, a difference of 1,000 sq ft in the floor space of two houses is considerable. While a price difference of \$12,900 also seems large, is it *really* a large difference in price for two houses that are 1,000 sq ft different in size? Is a difference of \$12.90 per square foot really large when we consider that the average variation around the mean price is \$2,550, regardless of the size of the house? In other words, *taking the size of the error into account, is $b = 12.9$ large enough for us to be reasonably sure that a linear relationship between size and price really exists?*

We know that it is possible to draw a sample indicating a relationship when, in fact, there is no relationship in the population. With a simple random sample, there is always a small chance that the observed relationship is only a function of the values that happened to be included in the sample. How can we tell in a particular problem if the value obtained for b is large enough (in absolute value) for us to be reasonably sure that our sample did not come from a population in which $\beta = 0$? That is, in a given problem, is b **significantly different** from zero? Are the chances of observing $b = 12.9$ when there is no relationship so small that we can be quite confident that the relationship is real? In order to answer this question, we turn to the problem of **hypothesis tests,** or **significance tests,** on β.

Significant Difference

Hypothesis Tests/ Significance Tests

Actually, this question should be the first one answered in any regression study. If it is found that a relationship actually does exist, then it makes sense to use the regression equation to make estimates and to measure the size of the error involved in estimation. However, without reasonable confidence that the two variables are related, it hardly makes sense to predict one from the other or to measure error. We have delayed this central question until now only because it was necessary to calculate b, $s_{Y \cdot X}$, and several other quantities in order to answer the question.

ADEQUACY OF THE MODEL

No less central a question is whether or not the data fit the proposed model. For our house size-price data, we assumed a simple linear regression model $Y = \alpha + \beta X + \varepsilon$ and found the regression equation and standard error of estimate corresponding to this model. We need to make reasonably certain that this model is appropriate to describe the relationship. From our inspection of the scatter diagram of the data, the relationship appeared to be reasonably linear; but we noted that price appears to increase with size at a decreasing rate—a leveling-off of prices can be seen from the largest houses in the sample. We opted for the simple linear model, but we should make sure our choice was appropriate.

We first address the question of whether β is significantly different from zero; later we use similar techniques to test the adequacy of the model.

7 . 3 HYPOTHESIS TESTS ON THE REGRESSION COEFFICIENT

Some review of the procedure for testing hypotheses may be in order. Generally, there are five steps to any hypothesis test, as outlined below.

OUTLINE OF THE PROCEDURE

STEP 1 *Set Up Null and Alternative Hypotheses.*

Null Hypothesis The **null hypothesis** is the hypothesis of no effect, often formulated for the express purpose of being rejected, or disproved. In simple regression problems the null hypothesis will typically be

$$H_0 : \beta = 0.$$

This states that there is no relationship between X and Y in the population. Typically, regression studies are undertaken because it is suspected that the variables *are* related in the population, so that the objective usually is to demonstrate that $H_0 : \beta = 0$ is a false statement.

Alternative Hypothesis The **alternative hypothesis** is a statement of the experimental hypothesis, or what we suspect is true and want to test. We hope our sample will provide sufficient evidence so that we can *accept* the alternative hypothesis. For simple regression studies, the alternative may be any of the following, depending on the particular problem:

$$H_1 : \beta > 0,$$

$$\text{or} \quad H_1 : \beta < 0,$$

$$\text{or} \quad H_1 : \beta \neq 0.$$

If we hope to establish a *direct* relationship, then our alternative hypothesis is $H_1:\beta > 0$; but if we seek an inverse relationship, the alternative $H_1:\beta < 0$ is appropriate. In some instances we have no prior suspicion, or are indifferent to whether the relationship is direct or inverse. We simply want to know if any relationship exists. Then we can use the alternative $H_1:\beta \neq 0$.

By setting up the hypotheses, we state what the problem is. In the house size-price example, we probably want to know if $b = 12.9$ indicates a true direct relationship. Thus, the null and alternative hypotheses are

$$H_0:\beta = 0 \quad \text{(no relationship)},$$

$$H_1:\beta > 0 \quad \text{(direct relationship)}.$$

Most computer programs use the alternative $H_1:\beta \neq 0$ so that the same program can be used whether the problem involves an inverse or a direct relationship.

STEP 2 *Choose a Significance Level*

We can never be 100 percent certain of our conclusion unless we have the entire population at hand, so we must specify what level of certainty is required in a particular problem in order to use the regression equation to make predictions. A certainty of 95 percent or 99 percent is often required, although a minimum certainty of 90 percent is sometimes acceptable. It is not the problem of the statistician to specify the required degree of certainty; rather, it is up to the researcher. Any one researcher can employ a variety of degrees of certainty for various problems, and any two researchers can require different degrees of confidence for the same problem. The researcher's personality and the nature of the particular problem and of its consequences determine how certain one must be of the results: life-or-death consequences require higher degrees of confidence than do preliminary pilot studies, and researchers with cautious personalities tend to require the higher degrees of certainty than do their less conservative colleagues.

Significance Level

Once the degree of confidence has been specified for a particular problem, the **significance level** is the difference between this percentage and 100 percent: for example, if 95 percent certainty is required, then the significance level, denoted by α (not to be confused with the α in our model $Y = \alpha + \beta X + \varepsilon$), is 0.05, or 5 percent. This is the probability of rejecting H_0 (and thus accepting H_1) when, in fact H_0 is true. That is, α is the probability of concluding that a relationship exists when there is no relationship in the population. If we *can* reject H_0, then we can be at least $100(1 - \alpha)$ percent (e.g., 95 percent) sure that a relationship does exist. However, if we cannot reject H_0, this does not mean that we are, say, 95 percent sure that there is no relationship. *If we cannot reject H_0, then we cannot be at least 95 percent sure that a relationship does exist.* We might be only 93 percent sure of a relationship, but if α is specified to be 0.05, then being 93 percent sure is not enough. The significance level is specified prior to collection and analysis of data and, as stated before, is the *largest* probability of falsely rejecting a null hypothesis deemed acceptable in the problem at hand. If 95 percent certainty is specified, the researcher cannot hedge and say, "Oh well, I guess 93 percent is okay." If 93 percent certainty is acceptable, this should have been specified at the outset and *not* after analysis of data. Otherwise one commits the scientific sin of forcing the data to show what is wanted, and objectivity is lost.

STEP 3 *Determine the Critical Region*

Critical Region
Rejection Region

The **critical region,** or **rejection region,** consists of a set of values that constitute sufficient evidence for the researcher to be $100(1 - \alpha)$ percent sure that H_1 is true. Usually, and always for our purposes, the critical region must be determined by using statistical tables. The statement in the alternative hypothesis and the level of significance must both be taken into account in determining the rejection region. For regression problems, we shall use tables of the Student's t distribution and Fisher's ***F* distribution**† (see Appendices H and I).

***F* Distribution**

CRITICAL REGIONS USING THE *t* DISTRIBUTION

In Chapter 6 we used the t distribution to form confidence intervals. It can also be used to test hypotheses concerning the values of α and β in the model $Y = \alpha + \beta X + \varepsilon$. As we shall see, the t-statistic essentially compares the sample regression coefficient b to a hypothesized value of β, usually zero. A difference $b - \beta$ close to zero is evidence that H_0 is true. Values far away from zero, either positive or negative, give evidence that H_0 is false and thus H_1 is true. If $H_1{:}\beta > 0$, we need to see that t is much *larger* than zero; if $H_1{:}\beta < 0$, we need to see that t is considerably *less* than zero: and $H_1{:}\beta \neq 0$ can be established as true if t is *either* a positive number sufficiently larger than zero *or* a negative number much less than zero. To determine how large (in absolute value) t must be in order to be sufficiently different from zero, we refer to the significance level, α.

We defined $t_{\nu,\alpha}$ to be that value of t with ν degrees of freedom that cuts off the largest 100α percent of all t-values. (See diagram in Appendix H.) That is, only 100α percent of all t_ν-values are greater than $t_{\nu,\alpha}$. Then if we observe a value greater than or equal to $t_{\nu,\alpha}$, chances are at least $100(1 - \alpha)$ percent that β really is greater than zero; i.e., that $H_1{:}\beta > 0$ is true. The critical region corresponding to the hypotheses

$$H_0{:}\beta = 0$$

$$H_1{:}\beta > 0$$

† The F distribution was named for its inventor, Sir Ronald A. Fisher, one of the founders of statistics as a tool for science.

is thus

$$\text{Reject } H_0 \quad \text{if} \quad t > t_{\nu,\alpha}.$$

Recall that regardless of the number of degrees of freedom, the t distribution is symmetric around the point $t = 0$. Thus, if $t_{\nu,\alpha}$ cuts off the largest 100α percent of all t_ν-values, $-t_{\nu,\alpha}$ cuts off the smallest 100α percent. Since chances are small that we would observe one of the 100α percent smallest possible t-values if $H_0{:}\beta = 0$ is true, we take such an observation as evidence that H_0 is false and that $H_1{:}\beta < 0$ is true. That is, if

$$H_0{:}\beta = 0$$
$$H_1{:}\beta < 0,$$

then the critical region is

$$\text{Reject } H_0 \quad \text{if} \quad t < -t_{\nu,\alpha}.$$

For the two-sided test, either very large *or* very small t-values give evidence that $H_0{:}\beta = 0$ is false. Looking for the most extreme 100α percent of all t-values, we look at the same time for the $100(\alpha/2)$ percent largest and the $100(\alpha/2)$ percent smallest. Thus, if

$$H_0{:}\beta = 0$$
$$H_1{:}\beta \neq 0,$$

then

$$\text{Reject } H_0 \quad \text{if} \quad t > t_{\nu,\alpha/2} \quad \text{or if} \quad t < -t_{\nu,\alpha/2}.$$

For example, if $\nu = 15$, $\alpha = 0.05$, and

$$H_0{:}\beta = 0$$
$$H_1{:}\beta > 0,$$

then the critical region is

$$\text{Reject } H_0 \quad \text{if} \quad t > t_{15,0.05} = 1.753.$$

If

$$H_0{:}\beta = 0$$
$$H_1{:}\beta < 0,$$

then

$$\text{Reject } H_0 \quad \text{if} \quad t < -t_{15,0.05} = -1.753.$$

If

$$H_0{:}\beta = 0$$
$$H_1{:}\beta \neq 0,$$

then, since $0.05/2 = 0.025$,

$$\text{Reject } H_0 \quad \text{if} \quad t > t_{15,0.025} = 2.131 \quad \text{or if} \quad t < -t_{15,0.025} = -2.131.$$

Note carefully that the two-sided t test has a two-sided critical region.

CRITICAL REGIONS USING THE *F* DISTRIBUTION

Sometimes the hypotheses $H_0{:}\beta = 0$, $H_1{:}\beta \neq 0$ are tested using an F-statistic instead of a t-statistic. In simple regression an F-statistic is the square of a t-statistic. Thus, F's are always positive, and large positive F-values are evidence that H_1 is true. Since F is never negative, we cannot test for the direction of the relationship but can use F only when the alternative is $H_1{:}\beta \neq 0$. Standard computer programs usually test using the two-sided alternative and the F-test.

An F-statistic has two different degrees of freedom, denoted ν_1 and ν_2, and usually called degrees of freedom for the numerator and degrees of freedom for the denominator, respectively. (F is a ratio of variances, as will be seen in step 4.) In simple regression ν_1 is always 1 and ν_2 is always $n - 2$. We recognize $n - 2$ as the degrees of freedom for $s^2_{Y \cdot X}$, and when we study multiple regression, ν_1 will be the number of independent variables, k, and ν_2 will be $n - p = n - (k + 1)$.

Appendix I gives critical values of the F distribution. Note that there is one table for $\alpha = 0.01$ and another for $\alpha = 0.05$. More extensive F-tables are available elsewhere. Both tables list ν_1-values across the top and ν_2 values down the left column. If $F_{\nu_1,\nu_2,\alpha}$ is that value of F with ν_1 and ν_2 degrees of freedom which cuts off the largest 100α percent of all F_{ν_1,ν_2}-values, then to test

$$H_0{:}\beta = 0$$
$$H_1{:}\beta \neq 0,$$
$$\text{Reject } H_0 \quad \text{if} \quad F > F_{\nu_1,\nu_2,\alpha}.$$

For example, if $\alpha = 0.05$ and $\nu_1 = 1$, $\nu_2 = 15$, then reject H_0 if $F > F_{1,15,0.05} = 4.54$. Note that $t_{15,0.025}$ was found to be 2.131, and that $(2.131)^2 = 4.54$; that is, $t^2_{\nu,\alpha/2} = F_{1,\nu,\alpha}$ and the two-sided t- and F-tests are equivalent. Note carefully that the two-sided F-test has a *one-sided* critical region.

STEP 4 *Take a Sample and Calculate the Value of a Test Statistic*

The hypotheses, significance level, and critical region have all been determined prior to the actual collection of data. That is, steps 1–3 define the problem and the evidence necessary to determine if a relationship exists. In step 4 we actually gather evidence and boil the entire sample down into one value, a t- or an F-statistic.

THE t-STATISTIC

If $H_1{:}\beta > 0$ or if $H_1{:}\beta < 0$, we will test with a t-statistic. The alternative

t-statistic

$H_1{:}\beta \neq 0$ can be tested with either statistic, as noted in step 3. The ***t*-statistic** is

$$t = \frac{b - \beta}{\sqrt{s^2_{Y \cdot X}/S_{XX}}}$$

The numerator, $b - \beta$, compares the sample evidence b to a hypothesized value for β, the *value stated in the null hypothesis*. If we are testing for the existence of a relationship, then

$$H_0{:}\beta = 0$$

and β will be zero in the t-statistic.

(It is also possible to test $H_0{:}\beta = \beta_0$, where β_0 is some value other than zero, using the t-statistic; in such cases one substitutes the value β_0 for β in the t-statistic. For example, suppose that for a given city an equation $\hat{Y} = 1.25 + 0.75X$ is found relating income (X) in hundreds of dollars and weekly amount spent on beer (Y) in dollars. Then $b = 0.75$ is a *marginal propensity to consume* (MPC), and the equation is called a *consumption function*. Suppose further that in a nationwide study, the MPC for beer is found to be 0.50, and the question arises as to whether the MPC for this city is significantly greater than that for the nation. Then

$$H_0{:}\beta = 0.50$$

$$H_1{:}\beta > 0.50$$

and the numerator of t would be $b - \beta = 0.75 - 0.50$. While such tests are easily done using the t-statistic, our primary concern in regression is to show that $H_0{:}\beta = 0$ is false. Note that F can test H_0 only for $\beta = 0$.)

The numerator of t is in units Y/X, since b and β both measure differences in Y for differences in X. Thus, the size of the difference $b - \beta$ depends upon the units in which X and Y are measured. It is thus necessary to take units of measurement into account when deciding if the difference between b and β is large enough to be attributable to something other than chance. The denominator of t is also in units Y/X: $s^2_{Y \cdot X}$ is in Y^2-units, S_{XX} is in X^2-units, and then the square root is taken. Thus, the denominator of t takes units of measurements into account and corrects for them, making the comparison of b to $\beta = 0$ independent of the units of measurement of either X or Y.

For example, if X is an income variable measured in dollars, then b might be 0.02. But if the same data are used with X measured in thousands of dollars, so that an income of \$16,000 is called just 16, then b would be 20. Both values of b measure the same relationship although they seem different because of the difference in units. The t-statistic eliminates reflection of units in b and enables a unitless comparison of b to zero to be made.

Standard Error of the Regression Coefficient

The denominator of t, $[s^2_{Y \cdot X}/S_{XX}]^{1/2}$, called the **standard error of the regression coefficient,** estimates how much b will vary from sample to sample of the same size from the same population. Imagine several different people all sampling 20 randomly chosen houses from this same city. Because each sample is different, each person will obtain a slightly different value for b. If the standard deviation of all these b-values were calculated, then we would have an estimate of how much b varies from sample to sample, and this would be the standard error of b. Using this standard error in the denominator of the t-statistic gives us a statistic that estimates how many standard deviations b is away from $\beta = 0$. The tabulated values of t tell how many (estimated) standard deviations difference is considered too extreme to be attributed solely to chance.

Further study of the denominator of t shows where the degrees of freedom, ν, for the t-statistic come from. In the denominator of t is $s^2_{Y \cdot X}$, whose own denominator (which we called degrees of freedom in Chapter 6) is $n - p$; thus t has $\nu = n - 2$ degrees of freedom in simple regression problems.

In our example concerning prices of houses of varying sizes, we found

$$b = 12.8982,$$

$$S_{XX} = 13.56,$$

$$s^2_{Y \cdot X} = 6.5058.$$

The value $b = 12.9$ appears to be larger than zero, but we must remove units to make the comparison. Calculating the t-statistic,

$$t = \frac{12.8982 - 0}{\sqrt{\dfrac{6.5058}{13.56}}} = 12.8982\sqrt{\frac{13.56}{6.5058}}$$

$$= 12.8982\sqrt{2.0843} = 12.8982(1.4437)$$

$$= 18.6212.$$

This value represents a condensation of all the sample data into one figure, which we may compare in size to zero. Note that we estimate that b is 18.6212 standard deviations above $\beta = 0$.

THE F-STATISTIC

F-statistic The ***F*-statistic** can only be used to test $H_0{:}\beta = 0$, and it accomplishes the same "deunitizing" as does the t-statistic. Since an F is the square of a t in simple regression, consider

$$t^2 = \left[\frac{b - 0}{\sqrt{s^2_{Y \cdot X}/S_{XX}}}\right]^2 = \frac{b^2 S_{XX}}{s^2_{Y \cdot X}} = \frac{(S^2_{XY}/S^2_{XX})S_{XX}}{s^2_{Y \cdot X}}$$

$$= \frac{S^2_{XY}/S_{XX}}{s^2_{Y \cdot X}} = \frac{bS_{XY}}{s^2_{Y \cdot X}} = F.$$

The numerator of F measures the relationship between X and Y, while the denominator measures the variation in Y due to error. Thus, F compares variation in Y due to its relationship with X to variation in Y due to error. If the relationship is strong enough to overcome error variation, then F will be large enough for us to reject $H_0: \beta = 0$.

In the denominator of F is $s_{Y \cdot X}^2$, with degrees of freedom $n - p$, or $n - 2$ in simple regression. This accounts for the value of ν_2, the denominator degrees of freedom for the F-ratio. The term in the numerator represents the relationship of Y to *one* predictor variable X. Thus $\nu_1 = 1$. We shall see in multiple regression that ν_1 will always equal the number of predictor variables, and that the relationship of Y with all k predictors will be measured in the numerator of F.

SUMS OF SQUARES AND MEAN SQUARES

Regression Mean Square
Error Mean Square

Often, the quantity bS_{XY} is referred to as the mean square due to regression, or **regression mean square.** The quantity $s_{Y \cdot X}^2$ in the denominator is called the mean square due to error, or **error mean square.** F then is the ratio of the regression mean square to the error mean square, or

$$F = \frac{MS(R)}{MS(E)}.$$

The reasons for this terminology are fairly obvious. Consider

$$s_{Y \cdot X}^2 = \frac{\sum (Y_i - \hat{Y}_i)^2}{n - 2}.$$

The numerator, $\sum (Y_i - \hat{Y}_i)^2$, is a sum of squares. When a sum is divided by the number of terms, a mean is obtained; thus $s_{Y \cdot X}^2$ is a mean of squares, or mean square. Since the squared quantities are errors, $s_{Y \cdot X}^2$ is an error mean square. Similarly, $bS_{XY} = b^2 S_{XX}$, as was shown in the derivation of F from the square of a t, and $S_{XX} = \sum (X_i - \overline{X})^2$ is also a sum of squares. Since S_{XX} gets multiplied by b^2, the numerator measures relationship of Y with X(regression). The degrees of freedom are 1, so that the sum of squares gets divided by 1 to become a mean square. Another name for mean square is simply variance. Thus, the F-statistic compares variation due to regression to variation due to error.

ANALYSIS OF VARIANCE

Analysis of Variance (ANOVA) Table

All these pieces of information can be shown nicely in their respective roles if one calculates the F-statistic by means of an **analysis of variance (ANOVA) table.** As the name implies, we are analyzing the variation in Y into its component parts: one part due to relationship with X and one part due to error. The general form of the analysis of variance table for a simple regression is shown in Table 7.1. Note that total variation in Y is the variance of Y, or $s_Y^2 = S_{YY}/(n - 1) = SS(T)/(n - 1)$.

TABLE 7.1 Analysis of Variance (ANOVA)

Source of Variation	Sum of Squares	Degrees of Freedom	Mean Square	F
Regression	$SS(R) = bS_{XY}$	1	$MS(R) = \frac{bS_{XY}}{1} = bS_{XY}$	$\frac{MS(R)}{MS(E)} = \frac{bS_{XY}}{s_{Y \cdot X}^2}$
Error	$SS(E) = S_{YY} - bS_{XY}$	$n - 2$	$MS(E) = \frac{S_{YY} - bS_{XY}}{n - 2} = s_{Y \cdot X}^2$	
Total	$SS(T) = S_{YY}$	$n - 1$		

Our example about house sizes and prices produces Table 7.2. Here, $F = 346.75$ means that the variation due to regression is an estimated 346.75 times as great as the variation due to error. Note that we found $t = 18.6212$, and that $t^2 = 346.75 = F$.

TABLE 7.2 ANOVA for House-Size Price Data

Source	SS	df	MS	F
Regression	12.8982(174.9) = 2255.8952	1	2255.8952	346.75
Error	117.1048 (by subtraction)	18	6.5058	
Total	2373	19		

STEP 5 *Compare Test Statistic to the Critical Region and Make Appropriate Conclusion*

TESTING FOR DIRECT RELATIONSHIPS

We might want to test whether or not a significant *direct* relationship exists between prices and sizes of houses. The hypotheses are

$$H_0: \beta = 0$$

$$H_1: \beta > 0.$$

If α is chosen to be 0.05, then we will reject H_0 if $t > t_{18,0.05} = 1.734$. In step 4, we found $t = 18.6212$; since $t > 1.734$, we reject H_0 and conclude we are more than 95 percent sure that there is a direct relationship between the size of a house and its price: The larger the house, generally the greater its price. Thus, we may use the regression equation to estimate price from size and can know the average error involved in the estimates.

Suppose, on the other hand, that t had been found to be some small number, say $t = 1.234$. Then since $t \not> 1.734$, we *could not* reject H_0 and we *could not* be 95 percent sure that house prices are related to size. In this case, we would not be sufficiently certain that a relationship exists to be willing to use the regression equation to estimate price from size alone.

Note that if t is not greater than the critical value, we *do not* conclude that we are 95 percent sure that no relationship exists. This is not at all the case. There is *some* evidence of a relationship but not enough to be 95 percent confident that it is real.

TESTING FOR INVERSE RELATIONSHIPS

In the house size-price problem one would probably not hypothesize the existence of an inverse relationship, but should one wish to test

$$H_0: \beta = 0$$
$$H_1: \beta < 0,$$

then H_0 will be rejected if $t < -t_{18,0.05} = -1.734$. We found $t = 18.6212$, and t is *not* less than -1.734. Thus, we cannot be 95 percent sure that an inverse relationship exists.

Let us consider an instance in which it would be reasonable to test for an inverse relationship. Suppose X = price charged for an item in various locations, and Y = number of items sold per week. Suppose further that the regression equation, based on 32 observations, was found to be

$$\hat{Y} = 178.4 - 4.96X,$$

with $s^2_{Y \cdot X} = 4$ and $S_{XX} = 16$. We want to see if we can be 95 percent sure that fewer items are sold in locations charging higher prices. Then

$$H_0: \beta = 0$$
$$H_1: \beta < 0,$$

and if $\alpha = 0.05$, we will reject H_0 if $t < -t_{30,0.05} = -1.697$. The test statistic is

$$t = \frac{-4.96 - 0}{\sqrt{4/16}} = \frac{-4.96}{\sqrt{1/4}} = \frac{-4.96}{1/2}$$
$$= 2(-4.96)$$
$$= -9.92.$$

Since $-9.92 < -1.697$, we reject H_0 and conclude we are more than 95 percent sure that locations charging higher prices sell fewer items than those charging less.

TESTING TWO-SIDED ALTERNATIVES

If the objective is to determine whether or not *any* relationship is present, either direct or inverse, then the hypotheses are

$$H_0{:}\beta = 0$$

$$H_1{:}\beta \neq 0;$$

and either a t- or an F-test can be used. For the t-test in the house size-price example, H_0 is rejected if $t > t_{18,0.025} = 2.101$ or if $t < -t_{18,0.025} = -2.101$. Then since $t = 18.6212 > 2.101$, we reject H_0 and conclude we are more than 95 percent sure that the price of a house is related to its size. (Note that in the conclusion we do not specify the direction of the relationship, although we can tell the direction by noting the sign of t.)

If the F-test is used, then we must be testing the two-sided alternative to $H_0{:}\beta = 0$ and we reject H_0 if $F > F_{1,18,0.05} = 4.41$. Since $F = 346.75 > 4.41$, reject H_0 and conclude we are more than 95 percent sure that the sizes and the prices of houses are related. (Recall that we cannot tell, just by looking at the value of F, whether the relationship is direct or inverse.)

The first question to be answered in a regression problem is whether or not a relationship exists. Unless one can be reasonably confident that the two variables are related, it certainly does not make sense to estimate one from the other, measure the error involved, and interpret the regression constant and the regression coefficient. The reader can see, however, that the question of significance of the relationship had to be postponed until after the study of error, because so many intermediate quantities had to be calculated in order to find the t- and F-statistics.

EXERCISES

7.1 Refer to the Concession Data in Exercise 1.5. The manager of the soft drink concession wanted to be able to estimate how many soft drinks would be sold at a football game from knowing how many tickets were sold by Wednesday.

a. What are the null and alternative hypotheses? Why?

b. Use a t-test to test the hypotheses in (a) with significance level $\alpha = 0.05$.

c. Why cannot an F-test be used?

7.2 For the Heating Data in Exercise 1.8,

- **a.** What are H_0 and H_1? Why?
- **b.** Use a t-test to test these hypotheses with significance level $\alpha = 0.05$. State your conclusion.
- **c.** Could you have tested these hypotheses with an F-test? Why, or why not?

7.3 Refer to the Smokers Data in Exercise 1.9.

- **a.** What do you think were the null and alternative hypotheses that prompted them to be collected?
- **b.** What conclusion can be reached concerning how the number of sick days taken is related to the number of cigarettes smoked?

7.4 Regarding the Repairs Data in Exercise 2.11, the foreman claimed that one reason repair costs were high was that routine maintenance was not performed often enough.

- **a.** Can he be 99 percent sure that the greater the number of hours since a machine was worked on, the higher its repair costs?
- **b.** Does the conclusion in (a) prove that the foreman was correct?

7.5 Regarding the Class Size Data in Exercise 2.17, the professor claimed that the reason she received low ratings was because she taught very large classes.

- **a.** Can you be 99 percent sure that teachers of large sections tend to receive lower ratings than teachers of small sections?
- **b.** Has the professor proved her point?

7.6 For the Appraisal Data of Exercise 4.4, test to see if there is *any* relationship between appraised value and sale price, with $\alpha = 0.05$.

- **a.** Use a t-test.
- **b.** Use an F-test.
- **c.** State the conclusion in nontechnical terms.

7.7 For the Hyperactivity Data in Exercise 5.15 use significance level $\alpha = 0.01$ to test if there is *any* relationship between the amount of food preservative ingested and the hyperactivity level of children.

- **a.** Use an F-test.
- **b.** Use a t-test.
- **c.** State your conclusion in nontechnical terms.

7.8 Refer to the Stock Data in Exercise 5.17. Assume that the 24 months for which returns were recorded can be considered as a random sample from all possible 24-month periods of potential interest. It was found that the slope of the regression equation was greater than 1, indicating that XZQ should be classified as an aggressive stock. Can you conclude that this is true, in general, for XZQ Stock, or was this conclusion only a function of the time period (sample) that was chosen for analysis? Test $H_0:\beta = 1$ and $H_1:\beta > 1$, and state your conclusion.

7.9 Refer to the Gauge Data in Exercise 5.18. If the new pressure gauge gives the same readings as the standard gauge, then the regression equation relating readings made on the same pressure with both instruments should have intercept zero and slope 1, except for sampling error and errors made in using or reading the gauges. It was found that $\hat{Y} = 2.5108 + 1.000402X$, where Y = readings made by the new gauge, and X = readings made by the old gauge. Using $\alpha = 0.01$, test $H_0:\beta = 1$ against $H_1:\beta \neq 1$.

7.10 In forestry, it is often important to measure the diameters of trees. An easily made measurement is diameter at breast height, outside bark (DBH_{ob}). Breast height is defined as $4\frac{1}{2}$ ft above ground level. While students are first learning how to measure tree diameters, it is important to evaluate the accuracy with which they make their measurements. The data in Table 7.3 are a sample of measurements for selected pines in the Cherokee Woodlot made by a student and by the instructor. All measurements are in centimeters.

TABLE 7.3
Diameters Data

	Tree							
	1	2	3	4	5	6	7	8
Student	29.8	24.0	38.1	29.7	25.1	22.2	36.4	25.2
Instructor	29.2	23.6	38.9	29.7	25.6	20.3	36.6	24.1
	9	10	11	12	13	14	15	
Student	29.3	37.7	28.7	35.8	32.3	24.8	30.6	
Instructor	30.5	37.6	28.7	37.6	32.3	23.9	30.7	

The accuracy of the measurements made by the student can be evaluated by regressing the student's measurements on the instructor's measurements and testing to see if the intercept is zero and the slope is 1. (Data and problem kindly provided by Professor John C. Rennie, Department of Forestry, Wildlife, and Fisheries, The University of Tennessee.)

Find the regression equation relating the student's measurements to the instructor's measurements and test to see if the slope is significantly different from 1. What does this tell you about the accuracy with which the student is making measurements?

7.4 TESTS ON THE REGRESSION CONSTANT

It is also possible to test hypotheses concerning α, the regression constant. However, at best, α gives very specific, as opposed to general, information about the relationship—it gives the average Y-value at a specified X-value, $X = 0$; but often α doesn't even have any physical meaning in the problem. There is, however, an instance in which one is interested in testing a hypothesis concerning the regression constant. For some pairs of variables, if $X = 0$, it seems reasonable that Y should also be zero. For example, if X = number of items sold and Y = revenue from sales of the item, then it stands to reason that if no items are sold, then no revenue will be accrued. However, because of sampling variability, the fitted regression line could show an intercept different from zero. If the nature of the problem is such that it seems reasonable that the line should pass through the point (0, 0), then we might want to test

$$H_0{:}\alpha = 0$$

against some alternative ($H_1{:}\alpha < 0$ or $H_1{:}\alpha > 0$ or $H_1{:}\alpha \neq 0$). If H_0 cannot be rejected, then we have no evidence that the line should not be fitted through the origin, and thus failure to reject H_0 can provide some justification for adopting the model

$$Y = \beta^* X + \varepsilon$$

and fitting an equation of the form

$$\hat{Y} = b^* X,$$

where $b^* = \sum X_i Y_i / \sum X_i^2$ (refer to Chapter 5 and Appendix F).

The statistic for testing a hypothesis concerning α is

$$t = \frac{a - \alpha}{\sqrt{s_{Y \cdot X}^2(1/n + \bar{X}^2/S_{XX})}}$$

with $n - 2$ degrees of freedom. The mechanics of the test are exactly the same as for the t-test on β. It is not always necessary to test $H_0{:}\alpha = 0$. The test statistic allows for an hypothesized value other than zero—but it is seldom of interest to test any other null hypothesis. The denominator of the t-statistic above is the **standard error of the regression constant.** It estimates how much the value of a varies from sample to sample.

Standard Error of the Regression Constant

In our house size-price example, it might seem reasonable to suppose that a house of no size should sell for zero dollars. Let us test to see whether we might want to fit the regression line through the origin:

$$H_0{:}\alpha = 0,$$

$$H_1{:}\alpha > 0.$$

If $\alpha = 0.05$, reject H_0 if $t > t_{18,0.05} = 1.734$.

$$t = \frac{8.7036 - 0}{\sqrt{6.5058[1/20 + (2)^2/13.56]}} = \frac{8.7036}{1.4981} = 5.8098.$$

Since $5.8098 > 1.734$, we can be more than 95 percent sure that the intercept is above zero. Thus we will not adopt the model $Y = \beta^* X + \varepsilon$. Recall that in Chapter 5 we fitted a regression line through the origin for these data and saw that this line did not follow the data points very well at all. Perhaps zero thousand square feet is not the actual lower bound on size.

EXERCISES

7.11 Refer to the Concession Data in Exercise 1.5.

a. Test to see if the intercept is different from zero, at the $\alpha = 0.05$ significance level. Would you use a regression line fitted through the origin? Why, or why not?

b. If the regression line is fitted through the origin, test $H_0:\beta^* = 0$ versus $H_1:\beta^* > 0$, using $\alpha = 0.05$ significance level.

7.12 Refer to the Smokers Data in Exercise 1.9.

a. Use significance level $\alpha = 0.01$ to test $H_0:\alpha = 0$ versus $H_1:\alpha \neq 0$.

b. State the null and alternative hypotheses in nontechnical terms.

c. State your conclusion in terms specific to the problem.

7.13 Refer to the Gauge Data in Exercise 5.18. Test $H_0:\alpha = 0$ against $H_1:\alpha > 0$. What does this tell you about how the new gauge performs?

7.14 For the Diameters Data in Exercise 7.10, test to see if the intercept is significantly different from zero. What does this tell you about the accuracy of the student's measurements?

7.15 Test to see if the intercept is significantly greater than zero for the Reservations Data in Exercise 5.16. What are the implications of the results of this test?

7.16 For the Hyperactivity Data in Exercise 5.15, is the intercept significantly greater than zero? What does this tell you concerning hyperactivity and the amount of food preservative ingested?

7 . 5 THE ATTAINED SIGNIFICANCE LEVEL (p-VALUE)

When a significance level, α, is chosen for a test of hypotheses, it is a criterion applied by the experimenter to the maximum risk he or she is willing to take of rejecting a null hypothesis that is in fact true. When a critical value corresponding to an α-level is found and a rule such as "Reject H_0 if $F > F_{\alpha,1,n-2}$" is formulated, this rule assures the researcher that if H_0 is rejected, the probability that a wrong conclusion has been reached is less than α, the specified maximum acceptable risk. One might refer to α as the **specified (maximum) significance level.**

Specified Significance Level

When H_0 can be rejected, then, we know that the risk of having made an error is less than α. Often the question arises as to *how much less* than α the actual risk is: What is the probability of an error in this particular case? This risk is often referred to as the **attained significance level,** or ***p*-value** of the test.

Attained Significance Level

***p*-Value**

For example, suppose that to test $H_0: \beta = 0$ against $H_1: \beta > 0$, a sample of size 22 is taken. The rejection region for this test, if α is specified to be 0.10, is

Reject H_0

if

$$t > t_{20,0.10} = 1.325.$$

Suppose that when the data are analyzed, we find that $t = 2.10$. Then we could reject H_0 and conclude that we are more than 90 percent sure that $\beta > 0$. How much more than 90 percent sure? Because of limitations of the t-tables in Appendix H, giving t-values corresponding only to selected upper-tail probabilities, we often cannot answer questions such as this very specifically. However, it is possible to be more specific than just "more than 90 percent sure." Look in the row of the t-tables for 20 degrees of freedom and find two values that bracket the observed $t = 2.10$. We find

$$t_{20,0.025} = 2.086, \quad \text{and} \quad t_{20,0.01} = 2.528.$$

Since our observed t-value lies between 2.068 and 2.528, the probability of observing a value this large or larger by chance must be between 0.01 and 0.025. The attained, or actual, significance level of this test is thus between 0.01 and 0.025, usually written as

$$0.01 < p < 0.025.$$

Since p, the attained significance, is less than α, the maximum acceptable risk, our decision is to reject H_0. Thus, we are between 97.5 and 99 percent sure that $\beta > 0$.

As another example, suppose that in a test of $H_0: \beta = 0$ against $H_1: \beta \neq 0$ based on a sample of size 30, the calculated F-value is $F = 9.75$. If $\alpha = 0.05$, what is the conclusion? Using the specified significance level approach, since $9.75 > F_{1,28,0.05} = 4.20$, H_0 is rejected. Using the attained significance level approach, we look at critical values of F with 1 and 28 degrees of freedom for varying significance levels. We see $F_{1,28,0.05} = 4.20$ and $F_{1,28,0.01} = 7.64$, and our tables don't go any further. Since the calculated F is greater than $F_{0.01}$, we conclude that $p < 0.01$. If the researcher has specified that $\alpha = 0.05$, then since $p < \alpha$, H_0 can be rejected.

Often in published research p-values of tests are reported. This is so that readers of the report can evaluate for themselves the results of the study. Since a significance level is selected by an individual on the basis of his or her own personality and the seriousness of the consequences of erroneously concluding that a relationship exists, different people would interpret results differently. The researcher cannot impose a personal significance level upon the reader.

As we shall see in Section 7.8, computer printouts typically report attained significance levels. These are much more specific than those we can typically find using printed tables, because the computer can easily calculate a p-value to several decimal places.

EXERCISES

7.17 Find the attained significance levels (p-values) corresponding to each of these t-values:

a. $t = 2.179$ with 12 *df* for testing a one-sided alternative;
b. $t = 2.179$ with 12 *df* for testing a two-sided alternative;
c. $t = 2.513$ with 25 *df* for testing a one-sided alternative;
d. $t = 2.513$ with 25 *df* for testing a two-sided alternative;
e. $t = -2.007$ with 18 *df* for testing a one-sided alternative;
f. $t = -2.007$ with 18 *df* for testing a two-sided alternative;
g. $t = 5.167$ with 5 *df* for testing a one-sided alternative;
h. $t = 5.167$ with 5 *df* for testing a two-sided alternative;
i. $t = 1.217$ with 21 *df* for testing a one-sided alternative;
j. $t = 1.217$ with 21 *df* for testing a two-sided alternative.

7.18 Find the attained significance levels (p-values) corresponding to each of these F-values:
(*Hint:* Compare answers to those in 7.17b, d, f, h, and j.)

a. $F = 4.75$ with 1 and 12 *df;*
b. $F = 6.32$ with 1 and 25 *df;*
c. $F = 4.03$ with 1 and 18 *df;*
d. $F = 26.70$ with 1 and 5 *df;*
e. $F = 1.48$ with 1 and 21 *df.*

7.19 If $\alpha = 0.05$, for which of the t-values in Exercise 7.17 would H_0 be rejected?

7.20 If $\alpha = 0.05$, for which of the F-values in Exercise 7.18 would H_0 be rejected?

7.21 State the conclusions to the following exercises in terms of the attained significance levels:

a. Exercise 7.1;
b. Exercise 7.2;
c. Exercise 7.3;
d. Exercise 7.6;
e. Exercise 7.11;
f. Exercise 7.12.

7.6 ANALYSIS OF VARIANCE USING MATRICES

THE ANALYSIS OF VARIANCE TABLE

When the regression problem is formulated in terms of matrices and vectors, we should be able to do the hypothesis tests in terms of matrices and vectors, as well. This will be helpful to us when the need arises to test hypotheses concerning regression parameters in more complicated models.

Recall that we could write and solve normal equations using

$$\mathbf{y} = \begin{bmatrix} Y_1 \\ Y_2 \\ \vdots \\ Y_n \end{bmatrix}, \quad \hat{\boldsymbol{\beta}} = \begin{bmatrix} a \\ b \end{bmatrix}, \quad \text{and} \quad \mathbf{X} = \begin{bmatrix} 1 & X_1 \\ 1 & X_2 \\ \vdots & \vdots \\ 1 & X_n \end{bmatrix}.$$

Consider the following quantities:

$$\mathbf{y}'\mathbf{y} = [Y_1 \quad Y_2 \quad \cdots \quad Y_n] \begin{bmatrix} Y_1 \\ Y_2 \\ \vdots \\ Y_n \end{bmatrix} = Y_1^2 + Y_2^2 + \cdots + Y_n^2 = \sum Y_i^2$$

and

$$\hat{\boldsymbol{\beta}}'\mathbf{X}'\mathbf{y} = [a \quad b] \begin{bmatrix} \sum Y_i \\ \sum X_i Y_i \end{bmatrix} = a \sum Y_i + b \sum X_i Y_i.$$

We will show that $\mathbf{y}'\mathbf{y}$ can be used as a total sum of squares and $\hat{\boldsymbol{\beta}}'\mathbf{X}'\mathbf{y}$ can be used as a regression sum of squares. Then the analysis of variance table in terms of matrices and vectors can be written as in Table 7.4. This is not quite

TABLE 7.4 ANOVA in Matrix Form

Source	*SS*	*df*
Regression	$\hat{\boldsymbol{\beta}}'\mathbf{X}'\mathbf{y}$	2
Error	$\mathbf{y}'\mathbf{y} - \hat{\boldsymbol{\beta}}'\mathbf{X}'\mathbf{y}$	$n - 2$
Total	$\mathbf{y}'\mathbf{y}$	n

the same as our original analysis of variance table: for one thing, the regression sum of squares has 2 degrees of freedom instead of 1, and the total sum of squares has n degrees of freedom instead of $n - 1$. In addition, the total sum of squares is $\sum Y_i^2$ instead of S_{YY}. We can make a few alterations in this table to show how it is not really very different from the one we previously used.

Consider again the quantity

$$\hat{\boldsymbol{\beta}}'\mathbf{X}'\mathbf{y} = a \sum Y_i + b \sum X_i Y_i.$$

Recall that $a = \overline{Y} - b\overline{X}$, so that

$$\begin{aligned}\hat{\boldsymbol{\beta}}'\mathbf{X}'\mathbf{y} &= (\overline{Y} - b\overline{X})\left(\sum Y_i\right) + b \sum X_i Y_i \\ &= \frac{\left(\sum Y_i\right)^2}{n} - \frac{b\left(\sum X_i\right)\left(\sum Y_i\right)}{n} + b \sum X_i Y_i \\ &= \frac{\left(\sum Y_i\right)^2}{n} + bS_{XY}.\end{aligned}$$

In our previous ANOVA table, the regression sum of squares involved only bS_{XY}, containing only the regression coefficient, and not the regression constant. That regression sum of squares had 1 degree of freedom associated with the slope; $\hat{\boldsymbol{\beta}}'\mathbf{X}'\mathbf{y}$ contains an intercept term as well as the slope and thus has 2 degrees of freedom. If we formed an F ratio as

$$F = \frac{MS(R)}{MS(E)} = \frac{\hat{\boldsymbol{\beta}}'\mathbf{X}'\mathbf{y}/2}{(\mathbf{y}'\mathbf{y} - \hat{\boldsymbol{\beta}}'\mathbf{X}'\mathbf{y})/(n-2)},$$

this would test

$$H_0: \boldsymbol{\beta} = \begin{bmatrix} \alpha \\ \beta \end{bmatrix} = \begin{bmatrix} 0 \\ 0 \end{bmatrix},$$

i.e., that *both* the intercept and slope are zero. This is not generally our interest, however: we just want to test $H_0: \beta = 0$.

To do this, we can break down $\hat{\boldsymbol{\beta}}'\mathbf{X}'\mathbf{y}$ as in Table 7.5. To obtain an analysis of variance table that tests only $H_0: \beta = 0$, if we rewrite the regression sum of squares as

$$SS(R) = \hat{\boldsymbol{\beta}}'\mathbf{X}'\mathbf{y} - \frac{\left(\sum Y_i\right)^2}{n},$$

TABLE 7.5
Decomposition of Regression Sum of Squares

Source	*SS*	*df*
Regression	$\hat{\boldsymbol{\beta}}'\mathbf{X}'\mathbf{y}$	2
{ Intercept	$\left(\sum Y_i\right)^2/n$	1
{ Slope	bS_{XY}	1

then it will have 1 degree of freedom. But then

$$SS(R) + SS(E) = SS(T),$$

so

$$\left[\hat{\boldsymbol{\beta}}'\mathbf{X}'\mathbf{y} - \frac{(\sum Y_i)^2}{n}\right] + [\mathbf{y}'\mathbf{y} - \hat{\boldsymbol{\beta}}'\mathbf{X}'\mathbf{y}] = \mathbf{y}'\mathbf{y} - \frac{(\sum Y_i)^2}{n} = \sum Y_i^2 - \frac{(\sum Y_i)^2}{n} = S_{YY}$$

and $SS(T) = S_{YY}$ will have $n - 1$ degrees of freedom. In summary, we can write the analysis of variance table using vectors and matrices as shown in Table 7.6 and have a table equivalent to the one originally presented.

TABLE 7.6 The General Form of an ANOVA Table

Source	*SS*	*df*
Regression	$\hat{\boldsymbol{\beta}}'\mathbf{X}'\mathbf{y} - \frac{(\sum Y_i)^2}{n}$	1
Error	$SS(T) - SS(R)$	$n - 2$
Total	$\mathbf{y}'\mathbf{y} - \frac{(\sum y_i)^2}{n}$	$n - 1$

A NUMERICAL EXAMPLE

For our house size-price example,

$$\mathbf{y} = \begin{bmatrix} 32 \\ 24 \\ \vdots \\ 28 \end{bmatrix}, \quad \mathbf{X} = \begin{bmatrix} 1 & 1.8 \\ 1 & 1.0 \\ \vdots & \vdots \\ 1 & 1.5 \end{bmatrix}, \quad \mathbf{X}'\mathbf{y} = \begin{bmatrix} 690 \\ 1554.9 \end{bmatrix}$$

and $\hat{\boldsymbol{\beta}}$ was found to be

$$\hat{\boldsymbol{\beta}} = \begin{bmatrix} a \\ b \end{bmatrix} = \begin{bmatrix} 8.7036 \\ 12.8982 \end{bmatrix}.$$

Now,

$$\mathbf{y}'\mathbf{y} = [32 \quad 24 \quad \cdots \quad 28]\begin{bmatrix} 32 \\ 24 \\ \vdots \\ 28 \end{bmatrix} = 26{,}178,$$

and

$$\hat{\boldsymbol{\beta}}'\mathbf{X}'\mathbf{y} = [8.7036 \quad 12.8982]\begin{bmatrix} 690 \\ 1554.9 \end{bmatrix} = 26{,}060.8952.$$

The analysis of variance table (Table 7.7) is then the same as that obtained in Section 7.3.

TABLE 7.7
ANOVA for House Size-Price Example

Source	*SS*	*df*
Regression	$26{,}060.8952 - \frac{(690)^2}{20} = 2255.8952$	1
Error	(by subtraction) $= 117.1048$	18
Total	$26{,}178 - \frac{(690)^2}{20} = 2373$	19

When we turn to the consideration of more complicated models, such as polynomial or multiple regressions, the most convenient way to set up an analysis of variance table will be by using matrices and vectors. The algebraic expressions for the regression sum of squares in these more complicated models will be so involved that we will be glad for a simple expression like $\hat{\boldsymbol{\beta}}'\mathbf{X}'\mathbf{y} - (\sum Y_i)^2/n$.

The next section illustrates this when we fit an even simpler model—one with no intercept.

ANALYSIS OF VARIANCE FOR MODELS WITH ZERO INTERCEPT

If a model has been adopted in which the regression equation must pass through the origin, recall that

$$\mathbf{y} = \begin{bmatrix} Y_1 \\ Y_2 \\ \vdots \\ Y_n \end{bmatrix}, \quad \mathbf{X} = \begin{bmatrix} X_1 \\ X_2 \\ \vdots \\ X_n \end{bmatrix}, \quad \text{and} \quad \hat{\boldsymbol{\beta}} = b^*,$$

so that $\mathbf{X}'\mathbf{X} = \sum X_i^2$, $(\mathbf{X}'\mathbf{X})^{-1} = 1/(\sum X_i^2)$, $\mathbf{X}'\mathbf{y} = \sum X_iY_i$, and the solution to the normal equation is $\hat{\boldsymbol{\beta}} = \sum X_iY_i/\sum X_i^2$.

Then for the analysis of variance table (Table 7.8),

$$SS(R) = \hat{\boldsymbol{\beta}}'\mathbf{X}'\mathbf{y} = \frac{\sum X_iY_i}{\sum X_i^2}\sum X_iY_i$$

$$= \frac{(\sum X_iY_i)^2}{\sum X_i^2}$$

TABLE 7.8
ANOVA for Regression Through the Origin

Source	SS	df
Regression	$\hat{\boldsymbol{\beta}}'\mathbf{X}'\mathbf{y} = \dfrac{(\sum X_iY_i)^2}{\sum X_i^2}$	1
Error	(by subtraction)	$n - 1$
Total	$\mathbf{y}'\mathbf{y} = \sum Y_i^2$	n

and the F-statistic for testing $H_0\colon \beta^* = 0$ will have 1 and $n - 1$ degrees of freedom.

Note that since $\hat{\boldsymbol{\beta}}'\mathbf{X}'\mathbf{y}$ contains only one term, it measures the estimated contribution of the regression coefficient β^* only and contains no terms involving the intercept. Thus $\hat{\boldsymbol{\beta}}'\mathbf{X}'\mathbf{y}$ has only 1 degree of freedom, and since there is no intercept term to correct for, $H_0\colon \beta^* = 0$ is tested using $SS(R) = \hat{\boldsymbol{\beta}}'\mathbf{X}'\mathbf{y}$.

Suppose we adopted the model

$$Y = \beta^* X + \varepsilon$$

for our house size-price example (although, as we have seen, this would not be a very appropriate model to use). Then

$$\mathbf{y} = \begin{bmatrix} 32 \\ 24 \\ \vdots \\ 28 \end{bmatrix}, \quad \mathbf{X} = \begin{bmatrix} 1.8 \\ 1.0 \\ \vdots \\ 1.5 \end{bmatrix}, \quad \text{and} \quad \hat{\boldsymbol{\beta}} = b^*.$$

We find $\mathbf{y}'\mathbf{y} = 26.178$, $\mathbf{X}'\mathbf{y} = 1554.9$, and $\mathbf{X}'\mathbf{X} = 93.56$. Then

$$\hat{\boldsymbol{\beta}} = b^* = 1554.9\left(\frac{1}{93.56}\right) = 16.6193$$

and $\hat{\boldsymbol{\beta}}'\mathbf{X}'\mathbf{y} = 16.6193(1554.9) = 25{,}841.3496$. The analysis of variance is shown in Table 7.9. Since $F = 1458.44 > F_{1,19,0.01} = 8.18$, we are more than 99 percent sure that $H_0\colon \beta^* = 0$ is false. That is, the model

$$Y = \beta^* X + \varepsilon$$

is to be preferred to the model

$$Y = \varepsilon.$$

TABLE 7.9
Numerical Example of Fitting Through the Origin

Source	SS	df	MS	F
Regression	25,841.3496	1	25,841.3496	1458.44
Error	336.6504	19	17.7184	
Total	26,178	20		

CORRECTING X FOR ITS MEAN

Of course, this approach is not affected if the X-values are corrected for their mean. If X is corrected for its mean, then

$$\hat{\boldsymbol{\beta}} = \begin{bmatrix} a' \\ b \end{bmatrix}, \qquad \mathbf{X} = \begin{bmatrix} 1 & X_1 - \overline{X} \\ 1 & X_2 - \overline{X} \\ \vdots & \vdots \\ 1 & X_n - \overline{X} \end{bmatrix}, \qquad \mathbf{X}'\mathbf{y} = \begin{bmatrix} \sum Y_i \\ S_{XY} \end{bmatrix},$$

so that

$$\hat{\boldsymbol{\beta}}'\mathbf{X}'\mathbf{y} = [a' \quad b]\begin{bmatrix} \sum Y_i \\ S_{XY} \end{bmatrix} = a' \sum Y_i + bS_{XY}.$$

Since $a' = \overline{Y}$,

$$\hat{\boldsymbol{\beta}}'\mathbf{X}'\mathbf{y} = \frac{(\sum Y_i)^2}{n} + bS_{XY},$$

as before; thus the analysis of variance table and the tests of significance are unaffected if X is corrected for its mean.

FINDING STANDARD ERRORS OF THE REGRESSION COEFFICIENTS FROM $(\mathbf{X}'\mathbf{X})^{-1}$

Before we leave the consideration of the regression problem in matrix form, let us look again at $(\mathbf{X}'\mathbf{X})^{-1}$. It turns out that the standard errors of a and b can be found by looking at the diagonal elements of this matrix.

Recall that for the simple linear regression model,

$$\mathbf{X}'\mathbf{X} = \begin{bmatrix} n & \sum X_i \\ \sum X_i & \sum X_i^2 \end{bmatrix},$$

so that

$$(\mathbf{X}'\mathbf{X})^{-1} = \frac{1}{n\sum X_i^2 - (\sum X_i)^2}\begin{bmatrix} \sum X_i^2 & -\sum X_i \\ -\sum X_i & n \end{bmatrix}$$

$$= \frac{1}{nS_{XX}}\begin{bmatrix} \sum X_i^2 & -\sum X_i \\ -\sum X_i & n \end{bmatrix}.$$

Consider that the first row and column of $(\mathbf{X}'\mathbf{X})^{-1}$ correspond to the intercept, a, and that the second row and column correspond to b. Thus, the element in the (1, 1) position,

$$\frac{\sum X_i^2}{nS_{XX}},$$

corresponds to a, while the element in the (2, 2) position,

$$\frac{n}{nS_{XX}} = \frac{1}{S_{XX}},$$

corresponds to b.

Now recall that the standard error of b is

$$\sqrt{\frac{s_{Y\cdot X}^2}{S_{XX}}} = \sqrt{s_{Y\cdot X}^2\left(\frac{1}{S_{XX}}\right)},$$

and that the factor $1/S_{XX}$ comes from the (2, 2) position of $(\mathbf{X}'\mathbf{X})^{-1}$. It is not so obvious, but it can be shown that the standard error of a is

$$\sqrt{s_{Y\cdot X}^2 c_{11}}$$

where $c_{11} = \sum X_i^2/nS_{XX}$, the element in the (1, 1) position of $(\mathbf{X}'\mathbf{X})^{-1}$. To show that this is so, consider

$$\frac{\sum X_i^2}{nS_{XX}} = \frac{1}{nS_{XX}}\left[\sum X_i^2 - \frac{(\sum X_i)^2}{n} + \frac{(\sum X_i)^2}{n}\right]$$

$$= \frac{1}{nS_{XX}}\left[S_{XX} + \frac{(\sum X_i)^2}{n}\right] = \frac{1}{n} + \frac{(\sum X_i)^2}{n^2 S_{XX}}$$

$$= \frac{1}{n} + \frac{\overline{X}^2}{S_{XX}},$$

and we used

$$\sqrt{s_{Y\cdot X}^2\left(\frac{1}{n} + \frac{\overline{X}^2}{S_{XX}}\right)}$$

as the standard error for the intercept in Section 7.4.

It will always be the case that the variance of the regression coefficients in *any* linear regression model can be found using the diagonal elements of $(\mathbf{X}'\mathbf{X})^{-1}$. This will be very helpful when it comes to performing hypothesis tests in more complicated situations, such as polynomial or multiple regressions.

Covariance Between Regression Coefficients

What about the off-diagonal elements of $(\mathbf{X}'\mathbf{X})^{-1}$? Just as the diagonal elements tell about the variance of a and b, the off-diagonal elements (when multiplied by $s_{Y\cdot X}^2$) give the **covariance between the regression coefficients** a and b. That is, a and b cannot be independently estimated in a given sample. Change the value for a, and you change the value for b. We saw that this was so when we fitted a regression line through the origin in the house size-price example: forcing the intercept to be zero resulted in a different value for b than we originally obtained. The off-diagonal element of $(\mathbf{X}'\mathbf{X})^{-1}$, lying in a column corresponding to a and a row corresponding to b (or vice versa), measures the extent to which values of a and b vary together.

One reason that we might want to formulate a regression model as

$$Y = \alpha' + \beta(X - \overline{X}) + \varepsilon,$$

that is, using X corrected for its mean, or centered, is that by so doing we can (1) separate the intercept term, which we are typically not interested in, from the slope, and (2) estimate the intercept and the slope independently. Recall that if X is corrected for its mean, then

$$\mathbf{X} = \begin{bmatrix} 1 & X_1 - \overline{X} \\ 1 & X_2 - \overline{X} \\ \vdots & \vdots \\ 1 & X_n - \overline{X} \end{bmatrix},$$

so that

$$\mathbf{X}'\mathbf{X} = \begin{bmatrix} n & \sum (X_i - \overline{X}) \\ \sum (X_i - \overline{X}) & \sum (X_i - \overline{X})^2 \end{bmatrix} = \begin{bmatrix} n & 0 \\ 0 & S_{XX} \end{bmatrix}$$

and

$$(\mathbf{X}'\mathbf{X})^{-1} = \begin{bmatrix} \frac{1}{n} & 0 \\ 0 & \frac{1}{S_{XX}} \end{bmatrix}.$$

From this form of $(\mathbf{X}'\mathbf{X})^{-1}$ we see that the (2, 2) element, corresponding to b, is still $1/S_{XX}$. When X is corrected for its mean, the pseudo-intercept, α', is estimated by $\overline{Y}$, and the variance of $\overline{Y}$ is obtained using the (1, 1) element, $1/n$. The off-diagonal elements are zero, meaning that $\overline{Y}$ and b are independent. Formulating a more complicated model in this manner helps us to get to the heart of the matter more easily, since we don't have to worry about the intercept term when our primary interest is in the regression coefficients. This will become more apparent in Chapter 10 and those following.

EXERCISES

7.22 Consider the two predictor model $Y = \alpha + \beta_1 X_1 + \beta_2 X_2 + \varepsilon$.

a. Show $\mathbf{y}$, $\boldsymbol{\beta}$, and $\mathbf{X}$ to fit the regression equation.

b. Show $\mathbf{X}'\hat{\mathbf{X}}$ and $\mathbf{X}'\mathbf{y}$.

c. The solution to the normal equations is, as always,

$$\hat{\boldsymbol{\beta}} = (\mathbf{X}'\mathbf{X})^{-1}\mathbf{X}'\mathbf{y} = \begin{bmatrix} a \\ b_1 \\ b_2 \end{bmatrix}.$$

Show $\hat{\boldsymbol{\beta}}'\mathbf{X}'\mathbf{y}$ in algebraic form.

d. How many degrees of freedom will be associated with $\hat{\boldsymbol{\beta}}'\mathbf{X}'\mathbf{y}$?

e. If $SS(R)$ is defined as $SS(R) = \hat{\boldsymbol{\beta}}'\mathbf{X}'\mathbf{y}$, what null hypothesis would be tested by the F-test?

f. Show the analysis of variance table for testing $H_0{:}\beta_1 = \beta_2 = 0$.

7.23 Consider the two-predictor model with no intercept, $Y = \beta_1 X_1 + \beta_2 X_2 + \varepsilon$.

a. Show $\mathbf{y}$, $\hat{\boldsymbol{\beta}}$, and $\mathbf{X}$ to fit the regression equation.

b. Show $\mathbf{X}'\mathbf{X}$ and $\mathbf{X}'\mathbf{y}$.

c. The solution to the normal equations is

$$\hat{\boldsymbol{\beta}} = (\mathbf{X}'\mathbf{X})^{-1}\mathbf{X}'\mathbf{y} = \begin{bmatrix} b_1 \\ b_2 \end{bmatrix}.$$

Show $\hat{\boldsymbol{\beta}}'\mathbf{X}'\mathbf{y}$ in algebraic form.

d. How many degrees of freedom will be associated with $\hat{\boldsymbol{\beta}}'\mathbf{X}'\mathbf{y}$?

e. If $SS(R)$ is defined as $SS(R) = \hat{\boldsymbol{\beta}}'\mathbf{X}'\mathbf{y}$, what null hypothesis would be tested by the F-test?

f. Show the analysis of variance table for testing $H_0{:}\beta_1 = \beta_2 = 0$.

7.24 Consider the two-predictor model $Y = \alpha' + \beta_1(X_1 - \overline{X}_1) + \beta_2(X_2 - \overline{X}_2) + \varepsilon$.

a. Show $\mathbf{y}$, $\hat{\boldsymbol{\beta}}$, and $\mathbf{X}$ to fit the regression equation.

b. Show $\mathbf{X}'\mathbf{X}$ and $\mathbf{X}'\mathbf{y}$.

c. The solution to the normal equations is

$$\hat{\boldsymbol{\beta}} = (\mathbf{X}'\mathbf{X})^{-1}\mathbf{X}'\mathbf{y} = \begin{bmatrix} a' \\ b_1 \\ b_2 \end{bmatrix}.$$

Show $\hat{\boldsymbol{\beta}}'\mathbf{X}'\mathbf{y}$ in algebraic form.

d. How many degrees of freedom will be associated with $\hat{\boldsymbol{\beta}}'\mathbf{X}'\mathbf{y}$?

e. If $SS(R)$ is defined as $SS(R) = \hat{\boldsymbol{\beta}}'\mathbf{X}'\mathbf{y}$, what null hypothesis will be tested by the F-test?

f. Show the analysis of variance table for testing $H_0{:}\beta_1 = \beta_2 = 0$.

7.25 Suppose that a regression equation, $\hat{Y} = a + b_1 X_1 + b_2 X_2$ is fitted to a set of data, for which

$$(\mathbf{X}'\mathbf{X})^{-1} = \begin{bmatrix} c_{00} & c_{01} & c_{02} \\ c_{10} & c_{11} & c_{12} \\ c_{20} & c_{21} & c_{22} \end{bmatrix}.$$

The error mean square for the model is $MS(E) = s^2_{Y \cdot X}$.

a. Find the standard error of the regression constant, a.

b. Find the standard error of the regression coefficient b_1.

c. Find the standard error of the regression coefficient b_2.

7 . 7 CONFIDENCE INTERVALS ON THE PARAMETERS IN THE MODEL

CONFIDENCE INTERVALS FOR β

Internal Estimate of β

Often it is desired to have an **interval estimate of β** rather than just a point estimate, since we know that b, being based only on sample information, gives only an estimate of the true relationship between X and Y.

Recall that

$$t = \frac{b - \beta}{\sqrt{s_{Y \cdot X}^2 / S_{XX}}}$$

follows the Student's t distribution with $\nu = n - 2$ degrees of freedom. Then since $100(1 - \alpha)\%$ of all t_ν-values lie between $-t_{\nu,\alpha/2}$ and $+t_{\nu,\alpha/2}$, we can be $100(1 - \alpha)$ percent confident that

$$-t_{\nu,\alpha/2} < \frac{b - \beta}{\sqrt{s_{Y \cdot X}^2 / S_{XX}}} < t_{\nu,\alpha/2}.$$

Playing around with the above inequality to obtain β by itself in the middle,

$$-t_{\nu,\alpha/2}\sqrt{\frac{s_{Y \cdot X}^2}{S_{XX}}} < b - \beta < t_{\nu,\alpha/2}\sqrt{\frac{s_{Y \cdot X}^2}{S_{XX}}},$$

$$-b - t_{\nu,\alpha/2}\sqrt{\frac{s_{Y \cdot X}^2}{S_{XX}}} < -\beta < -b + t_{\nu,\alpha/2}\sqrt{\frac{s_{Y \cdot X}^2}{S_{XX}}},$$

$$b + t_{\nu,\alpha/2}\sqrt{\frac{s_{Y \cdot X}^2}{S_{XX}}} > \beta > b - t_{\nu,\alpha/2}\sqrt{\frac{s_{Y \cdot X}^2}{S_{XX}}},$$

$$b - t_{\nu,\alpha/2}\sqrt{\frac{s_{Y \cdot X}^2}{S_{XX}}} < \beta < b + t_{\nu,\alpha/2}\sqrt{\frac{s_{Y \cdot X}^2}{S_{XX}}}.$$

If we require a 95% confidence interval on β, in the house size-price example,

$$t_{\nu,\alpha/2} = t_{18,0.025} = 2.101$$

and

$$12.8982 - 2.101\sqrt{\frac{6.5058}{13.56}} < \beta < 12.8982 + 2.101\sqrt{\frac{6.5058}{13.56}},$$

$$12.8982 - 2.101(0.6927) < \beta < 12.8982 + 2.101(0.6927),$$

$$12.8982 - 1.4553 < \beta < 12.8982 + 1.4553,$$

$$11.4429 < \beta < 14.3535, \quad \text{or}$$

$$11.44 < \beta < 14.35.$$

Thus, we can say with 95 percent confidence that if one house is 1,000 sq ft larger than another, it will sell for between \$11,440 and \$14,350 more than the smaller house. This is a very strong statement of the relationship between the sizes and prices of houses.

The reader might note that if no relationship exists—if $H_0{:}\beta = 0$ cannot be rejected—then the confidence interval on β will straddle zero. For example, suppose we had found $b = 1.23$ in our example, with $s^2_{Y\cdot X}$ and S_{XX} remaining the same. Then the 95% confidence interval would be 1.23 ± 1.45, or $-0.22 < \beta < 2.68$, which would say with 95 percent confidence that a house 1,000 square feet larger in size than another will cost between \$220 *less* and \$2,680 *more* than the smaller house. This kind of statement is equivalent to saying that no relationship exists.

CONFIDENCE INTERVALS FOR α

Confidence Interval on α

Recognizing that the regression constant can vary from sample to sample, one might want to form a **confidence interval on α** as well, especially when the intercept has meaning. As with the confidence interval on β, we begin with

$$P\left(-t_{n-2,\alpha/2} < \frac{a - \alpha}{\sqrt{s^2_{Y\cdot X}(1/n + \overline{X}^2/S_{XX})}} < t_{n-2,\alpha/2}\right) = 1 - \alpha,$$

and then manipulate the inequality to obtain α by itself in the middle of the interval. The $100(1 - \alpha)\%$ confidence interval on α is found to be

$$a - t_{n-2,\alpha/2}\sqrt{s^2_{Y\cdot X}\left(\frac{1}{n} + \frac{\overline{X}^2}{S_{XX}}\right)} < \alpha < a + t_{n-2,\alpha/2}\sqrt{s^2_{Y\cdot X}\left(\frac{1}{n} + \frac{\overline{X}^2}{S_{XX}}\right)}.$$

Although it is artificial to find a confidence interval on α for the house size-price problem, since it has no physical meaning, we will construct one for illustration. For a 95% confidence interval on α, $t_{18,0.025} = 2.101$; so we have

$$8.7036 \pm 2.101\sqrt{6.5058\left[\frac{1}{20} + \frac{(2)^2}{13.56}\right]},$$

$$8.7036 \pm 3.1476,$$

$$5.5560 < \alpha < 11.8512.$$

We are 95 percent sure that the intercept of the regression equation lies between 5.5560 and 11.8512. (There is no interpretation in terms of sizes and prices of houses, since it makes no sense to talk about houses zero thousand square feet in size.)

Note that the standard error of the regression constant,

$$s_a = s_{Y \cdot X}\sqrt{\frac{1}{n} + \frac{\overline{X}^2}{S_{XX}}}$$

is a special case of the standard error of $\hat{Y}$, when $\hat{Y}$ is used to estimate a mean value of Y at $X = X_0$:

$$s_{Y \cdot X}\sqrt{\frac{1}{n} + \frac{(X_0 - \overline{X})^2}{S_{XX}}}.$$

(Refer to Section 6.7, where we constructed confidence intervals on $\mu_{Y \cdot X}$.) Recall that the intercept, a, is mathematically the estimated mean value of Y when $X = X_0 = 0$. Thus, substituting $X_0 = 0$ into the formula for the confidence interval on $\mu_{Y \cdot X}$ gives the confidence interval for α.

Even though it does not make sense to say that the mean price of houses zero thousand square feet in size is between $5,556 and $11,851, the confidence interval can be interpreted as saying that an intercept is needed in the model —it takes into account the combined effect of other predictors not explicitly considered by the model, and without the intercept term, we would consistently underestimate the mean price as a function of size. Of course, this conclusion is completely equivalent to the conclusion drawn by failure to reject $H_0{:}\alpha = 0$.

EXERCISES

7.26 Refer to the Concession Data in Exercise 1.5.

a. Give a 95% confidence interval on β. Interpret its meaning in terms specific to the problem.

b. Give a 99% confidence interval on α for the Concession Data. What does this interval tell you?

7.27 For the Heating Data in Exercise 1.8,

a. Give a 99% confidence interval on β. What does this confidence interval tell you?

b. Give a 95% confidence interval on α for the Heating Data. What does this interval tell you?

7.28 For the Smokers Data in Exercise 1.9,
- **a.** Give a 90% confidence interval on the slope.
- **b.** Give a 90% confidence interval on the intercept.
- **c.** What do these intervals tell you about the relationship between number of cigarettes smoked and number of sick days taken?

7.29 For the Gauge Data in Exercise 5.18,
- **a.** Give a 99% confidence interval on the slope.
- **b.** Give a 99% confidence interval on the intercept.
- **c.** What do these two intervals tell you about the relative performances of the old and new gauges?

7.30 For the Diameters Data in Exercise 7.10,
- **a.** Give a 95% confidence interval on α.
- **b.** Give a 95% confidence interval on β.
- **c.** What do these two intervals tell you about the accuracy with which the student measures diameters?

7.31 Give a 90% confidence interval on the regression coefficient for the Stock Data in Exercise 5.17. What does this tell you about the aggressiveness of XZQ stock?

7.8 TESTING FOR SIGNIFICANCE USING THE COMPUTER

BMD

The BMDP1R program performs both the F- and t-tests of $H_0:\beta = 0$. Refer to Figure 4.2(b) for the BMDP1R analysis of the house size-price problem. The analysis of variance table is easy to locate. Entries in the table should look familiar, although the sum of squares and degrees of freedom for total variation in Y are not given, and BMD uses `RESIDUAL` instead of our term *error*. The value of the F-statistic is given under `F RATIO` as 346.765. To the right of the F-value is shown `P(TAIL)` = 0.0000. This denotes the area in the upper tail of the F distribution with 1 and 18 degrees of freedom beyond the value $F = 346.765$. To four decimal places, this area is zero. Thus, a value of F as large as 346.765 or larger would occur with probability zero (to four decimal places) if $H_0:\beta = 0$ were true. This is very strong evidence that $\beta \neq 0$.

The t-statistic is also given in Figure 4.2(b). For the variable `SIZE`, the regression coefficient is shown to be 12.89823, and its standard error is 0.69265.

Then the t-statistic, which is formed as 12.89823/0.69265, is shown to be t = 18.622. The computer calculates the attained significance level for testing H_0:β = 0 against the two-sided alternative H_1:β ≠ 0 (note the heading P(2 TAIL)) and finds this probability to be essentially zero.

BMDP1R does not provide a t-test of H_0:α = 0. Furthermore, since the standard error of the regression constant is not given, we cannot obtain from the printout sufficient information to calculate the value of the t-statistic ourselves.

Since the standard error for b is shown, confidence intervals on β can be calculated by hand by finding the appropriate t-coefficients from the tables. We cannot do this to obtain a confidence interval on α, however.

M I N I T A B

The standard regression program in Minitab (see Figure 4.8) gives both t- and F-values for testing H_0:β = 0 and also the t-value for testing H_0:α = 0. Just after the regression equation, Minitab presents a table showing the regression coefficients, their standard errors (labeled ST DEV OF COEF), and the t-values. Minitab also notes that the t-ratio is obtained by dividing the coefficient by its standard error. In Figure 4.8 we find a t-value of 5.81 corresponding to the regression constant. The t-value for testing H_0:β = 0 is found to be t_b = 18.62. While most computer programs will print an attained significance level for these t-values, Minitab does not. You must refer to t-tables to find the critical value corresponding to the appropriate alternative hypothesis.

To perform an F-test, the value of the F-statistic must be calculated from information given from the REGRESSION command. Minitab shows mean squares for regression and for residual (error) only. From Figure 4.8 we can find

$$F = \frac{2255.900}{6.506} = 346.74.$$

Compare this value to the critical value looked up in the F-tables with numerator df = 1 and denominator df = 18 at the appropriate significance level.

Confidence intervals on α and β can be constructed by

COEFFICIENT ± (t-value) (ST DEV OF COEF)

and can be computed using Minitab interactively.

S A S

The SAS regression procedure performs both t- and F-tests of H_0:β = 0 and the t-test of H_0:α = 0. Figure 4.13 shows the SAS regression analysis of the house size-price problem. The analysis of variance table is easy to find, although MODEL is used for the source of variation we have called *regres-*

sion, and C TOTAL reminds us that the total sum of squares is $S_{YY} = \sum (Y_i - \bar{Y})^2$—that is, the sum of squares of Y-values corrected for their mean—instead of $\mathbf{y}'\mathbf{y} = \sum Y^2$. Also, the DF column in the SAS ANOVA table precedes the SUM OF SQUARES column, but that is merely a matter of preference. The F-value is given as $F = 346.767$, and the computer looks up the attained significance level. PROB > F = 0.0001 means that the probability of observing an F-value greater than, or equal to, 346.767 with 1 and 18 degrees of freedom if $H_0{:}\beta = 0$ is true is less than 0.0001. (This probability is actually much smaller than 0.0001, but this is the smallest significance probability that SAS will print.)

The details of the t-tests of $H_0{:}\alpha = 0$ and $H_0{:}\beta = 0$ are shown below the analysis of variance table. To test $H_0{:}\alpha = 0$, we see that $a = 8.703540$, the standard error of a is 1.498103, and thus the t-value is $t = 8.703540/1.498103 = 5.810$. Similarly, the t-statistic for testing $H_0{:}\beta = 0$ has value b/(standard error of b) $= 12.898230/0.692646 = 18.622$. The computer also calculates significance probabilities for these t-tests. Note that the heading PROB > |T| indicates that these probabilities correspond to tests of the two-sided alternatives $H_1{:}\alpha \neq 0$ and $H_1{:}\beta \neq 0$.

Confidence intervals on α and β can be formed by hand, using the information on the computer printout and looking up the appropriate t-value in the t-tables.

SPSS^X

An analysis of variance table and F-test for $H_0{:}\beta = 0$ are provided by SPSS^X, as are also t-tests of $H_0{:}\alpha = 0$ and $H_0{:}\beta = 0$. Refer to Figure 4.19 for the SPSS^X regression analysis of the house size-price data.

The analysis of variance table is easy to locate in Figure 4.19, although the sum of squares and degrees of freedom for total variation have been omitted, and SPSS^X uses the heading RESIDUAL instead of the term *error.* In the SPSS^X ANOVA table, the DF column precedes the sum of squares column, and the F-value is shown below the rest of the table. The F-value is 346.76654, and next to the F-value is shown SIGNIF F = 0.0000. The computer calculates the significance probability to four decimal places, so it is not necessary to refer to F-tables. We see that $H_0{:}\beta = 0$ can be rejected with almost 100 percent certainty.

Under VARIABLES IN THE EQUATION in Figure 4.19 is all the information concerning t-tests of $H_0{:}\alpha = 0$ and $H_0{:}\beta = 0$. Across from SIZE we find $b = 12.89823$, the standard error of the regression coefficient, SE B $= 0.69265$, and $t = 12.89823/0.69265 = 18.622$. The probability that b would be 18.622 standard deviations away from $\beta = 0$ by chance is essentially zero, so we would reject $H_0{:}\beta = 0$. Similarly, the t-value for testing $H_0{:}\alpha = 0$ is given by $t = a$/(standard error of a) $= 8.70354/1.49810 = 5.810$, and SIG T $= 0.0000$ indicates that α is almost certainly not equal to zero. Note that the significance probabilities for the t-statistics are for testing two-sided alternatives so as to make the t- and F-tests of $H_0{:}\beta = 0$ equivalent.

If the appropriate t-value from the t-tables is obtained, $100(1 - \alpha)\%$ confidence intervals for α and β can be constructed from the information provided on the printout.

EXERCISES

7.32 Refer to the computer analysis of the Concession Data from Chapter 4.

a. From information on the printout, use an F-test to test $H_0:\beta = 0$ versus $H_1:\beta \neq 0$. State your conclusion if you are using significance level $\alpha = 0.05$. State the attained significance level.

b. From information on the printout, use a t-test to test $H_0:\beta = 0$ versus $H_1:\beta > 0$, using significance level $\alpha = 0.05$. State the attained significance level.

c. Use a t-test to test $H_0:\alpha = 0$ versus $H_1:\alpha \neq 0$ at significance level $\alpha = 0.05$. State the attained significance level.

d. Give a 95% confidence interval on β. Interpret.

e. Give a 95% confidence interval on α. Interpret.

7.33 Refer to the computer analysis of the Smokers Data from Chapter 4, and using information on the printout,

a. State the conclusion from the F-test of $H_0:\beta = 0$.

b. At significance level $\alpha = 0.10$, test $H_0:\beta = 0$ against $H_1:\beta > 0$. What is the attained significance level?

c. Test $H_0:\alpha = 0$ against $H_1:\alpha \neq 0$, and state your conclusion. What is the attained significance level?

7.34 Refer to the computer analysis of the Class Size Data from Chapter 4, and using information on the printout,

a. What null and alternative hypotheses are being tested by the F-test? What is the conclusion reached? State attained significance level.

b. Test $H_0:\beta = 0$ against $H_1:\beta < 0$, and state your conclusion. What is the attained significance level?

c. Test $H_0:\alpha = 0$ against $H_1:\alpha > 0$, and state your conclusion. What is the attained significance level?

d. Construct and interpret a 99% confidence interval on β.

e. Construct and interpret a 99% confidence interval on α.

7 . 9

SUMMARY

We test an hypothesis concerning the regression coefficient in a simple regression problem to decide whether or not we can be reasonably sure that two variables are related. If a relationship is found, then it makes sense to use one variable to predict the other, to measure the expected error in prediction, and to explain the data by interpreting the regression equation. By testing hypotheses or forming confidence intervals on the regression constant, we can decide whether or not the regression equation should be fitted through the origin. These concerns are usually central to the study of the simple linear model.

Often, however, our objective is not really to estimate the value of one variable from the value of another. It might be sufficient just to know the extent to which two variables are related. In the next chapter we shall see that relationships can be studied without even finding a regression equation.

The house size-price problem that we have been using for illustration is so well-behaved that so far it has afforded us small opportunity to see what can go wrong when we try to analyze actual data. Our study of the simple linear model will not be complete until we tackle this problem in Chapter 9.

EIGHT

CORRELATION

8.1 INTRODUCTION

The tests of hypotheses considered in Chapter 7 answer the central question in regression analysis: Can we infer with reasonable confidence that the relationship apparent in the sample exists in the population from which the sample was drawn? Unless the answer to this question is yes, there is no reason to proceed with the analysis, as we would be studying only a phenomenon peculiar to this particular sample and not any true relationship between the variables. Keep in mind that the question of statistical significance has to do only with whether or not a relationship exists in the population and not with how strong or useful the relationship may be. A relationship is of **practical significance** if, in the judgment of the researcher, it is strong enough to be useful or informative.

Practical Significance

For a relationship to have practical significance implies that it is also a statistically significant (real) relationship, but not vice versa. It is often the case that very weak relationships can be shown to be statistically significant if the sample is large enough. If, for example, a sample regression coefficient is calculated from a very large sample, you trust that this sample coefficient will be very close to the true (population) value of β, whether this value is near zero or far from zero. Suppose that from a very large sample, we find $b = 0.01$. Then we are very confident that β is *not* zero, even though it is not very much different from zero. That is, we conclude that the variables *are* related in the population, although perhaps not very strongly. The question of practical significance is: If a one-unit increase in X results in a 0.01-unit increase in Y, does X give me enough information about Y for my purposes? Can I estimate the value of Y well enough from knowing the value of X?

To address the question of practical significance, we want to introduce a measure of the strength of the relationship. Such a measure is typically referred to as a **measure of correlation.** It tells us the extent to which two variables behave alike or vary together in a linear fashion. While reference to b and $s_{Y \cdot X}$ or the coefficient of variation can tell us the degree of linear association between two variables, their values are unit-dependent and thus somewhat difficult to interpret. On the other hand, a measure of correlation is

Measures of Correlation

unitless. We will first define the coefficient of determination and then the Pearson product-moment correlation coefficient.

8.2 THE COEFFICIENT OF DETERMINATION

THE PROPORTION OF EXPLAINED VARIATION

When developing the calculating formula for $s^2_{Y \cdot X}$ and when discussing the F-statistic used to test $H_0{:}\beta = 0$, we noted that the total variation in Y could be broken down into the sum of two parts: variation in Y due to its relationship with X and variation in Y due to error. If

σ^2_Y = population variance of Y,

$\sigma^2_{Y \cdot X}$ = population variance of Y, given X (i.e., variance of points around the regression line),

then

$$\sigma^2_Y = \sigma^2_{Y \text{ due to } X} + \sigma^2_{Y \cdot X}$$

is the mathematical statement of the component parts of the variance of Y. If the objective is to measure the strength of the relationship, then we want to measure the *proportion of variation in Y that can be explained by its relationship with X*,

$$\frac{\sigma^2_{Y \text{ due to } X}}{\sigma^2_Y} = \frac{\text{amount of variation in } Y \text{ explained by } X}{\text{total amount of variation in } Y}$$

Population Coefficient of Determination

We call this measure the **population coefficient of determination** and denote it by ρ^2 (ρ is the Greek letter rho).

Then

$$\rho^2 = \frac{\sigma^2_{Y \text{ due to } X}}{\sigma^2_Y} = \frac{\sigma^2_Y - \sigma^2_{Y \cdot X}}{\sigma^2_Y}$$

$$= 1 - \frac{\sigma^2_{Y \cdot X}}{\sigma^2_Y}.$$

This last form for ρ^2 indicates that the percentage of variation in Y due to X is 100 percent of variation in Y minus the percentage due to error.

Suppose that X and Y are perfectly linearly related. Then all points lie on the straight line $\mu_{Y \cdot X} = \alpha + \beta X$, and $\sigma^2_{Y \cdot X} = 0$. Thus,

$$\rho^2 = 1 - \frac{0}{\sigma^2_Y} = 1;$$

in other words, variation in X explains 100 percent of the variation in Y, which varies only because X varies. At the other extreme, if X and Y are not related, then the only reason Y varies is because of error, so that

$$\sigma^2_{Y \cdot X} = \sigma^2_Y, \qquad \frac{\sigma^2_{Y \cdot X}}{\sigma^2_Y} = 1, \quad \text{and} \quad \rho^2 = 1 - 1 = 0.$$

In this case, variation in X explains 0 percent of the variation in Y. Thus,

$$0 \le \rho^2 \le 1,$$

regardless of the units on X or Y, and the closer ρ^2 is to 1, the stronger the relationship between the two variables.

SAMPLE COEFFICIENT OF DETERMINATION

Sample Coefficient of Determination

In order to estimate ρ^2 from a sample, we define the **sample coefficient of determination**

$$r^2 = 1 - \frac{SS(E)}{SS(T)} = \frac{SS(T) - SS(E)}{SS(T)} = \frac{SS(R)}{SS(T)},$$

where $SS(E)$ is the error sum of squares, $SS(R)$ is the regression sum of squares, and $SS(T)$ is the total sum of squares. (Sums of squares are used instead of mean squares because we are not interested in making estimates of the population variances at this point; thus there is no need to adjust for degrees of freedom. That is, we may understand the denominator of n in both $MS(E)$ and $MS(T)$, and these n's will cancel out in the ratio. A statistic usually called **adjusted r^2**, which takes degrees of freedom into account, is formed as

Adjusted r^2

$$1 - \frac{s^2_{Y \cdot X}}{s^2_Y}.$$

This is shown on some computer printouts.)

The sample coefficient of determination is calculated as $SS(R)$ divided by $SS(T)$ for any regression model—simple, linear, quadratic, multiple linear, and so on—and can be found easily once the analysis of variance table has been constructed. It measures the effectiveness of the model in explaining variation in Y.

In the case of a simple linear model, $Y = \alpha + \beta X + \varepsilon$, there is a simple form for $SS(R)$:

$$SS(R) = bS_{XY}.$$

Thus, for a linear bivariate relationship, we have

$$r^2 = \frac{SS(R)}{SS(T)} = \frac{bS_{XY}}{S_{YY}} = \frac{(S_{XY}/S_{XX})S_{XY}}{S_{YY}}$$

$$= \frac{(S_{XY})^2}{S_{XX}S_{YY}},$$

which gives a quick calculating formula for r^2. Note that this formula does not require a knowledge of b. In a correlation study the value of b may never be calculated if the researcher is not interested in making estimates of Y from X. Note also that since S_{XY} is the only quantity in r^2 that could be negative, and since S_{XY} is squared, r^2 must always be positive. It cannot detect the direction of the relationship.

A NUMERICAL EXAMPLE

In our example about house sizes and prices, the following quantities were found:

$$S_{XY} = 174.9, \qquad S_{XX} = 13.56, \qquad S_{YY} = 2373.$$

Then

$$r^2 = \frac{(174.9)^2}{(13.56)(2373)} = 0.9507 \approx 95\%.$$

Equivalently, this can be found by reading from the ANOVA table as

$$r^2 = \frac{SS(R)}{SS(T)} = \frac{2255.8952}{2373} = 0.9507.$$

Thus, differences in the sizes of houses account for about 95 percent of the variation in prices. Obviously this indicates a very strong relationship. All other variables together, including chance, can account for at most an additional 5 percent of the variation in price.

HOW LARGE MUST r^2 BE?

What if we had found $r^2 = 0.33$? Is this interpreted as strong, weak, moderate, or what? Since $0 \le r^2 \le 1$, and since $0.33 < 0.50$, one might say than an r^2 of 0.33 indicates a moderately weak relationship. However, it means that X by itself explains 33 percent of the total variability in Y, leaving only two-thirds of the variability to be explained by *all other* possible predictors taken together. While the interpretation of how strong is strong is rather subjective, it can be argued that $r^2 = 0.33$ indicates a moderately strong relationship.

It takes a bit of exposure for one to be able to appreciate how strong relationships are. Let us look at some graphs. In the following series of figures, all X-values are between 1 and 5, and Y-values range from 4 to 17. Figure 8.1 shows three situations in which the coefficients of determination are zero. Data points on each graph let one verify this. In Figure 8.1(a) we see that for a given X-value, Y is sometimes relatively large and sometimes relatively small, but *on the average* Y is constant regardless of the value of X. In this instance, X does not really explain anything about Y, so $r^2 = 0$.

FIGURE 8.1 Examples of Coefficients of Determination Equal to Zero

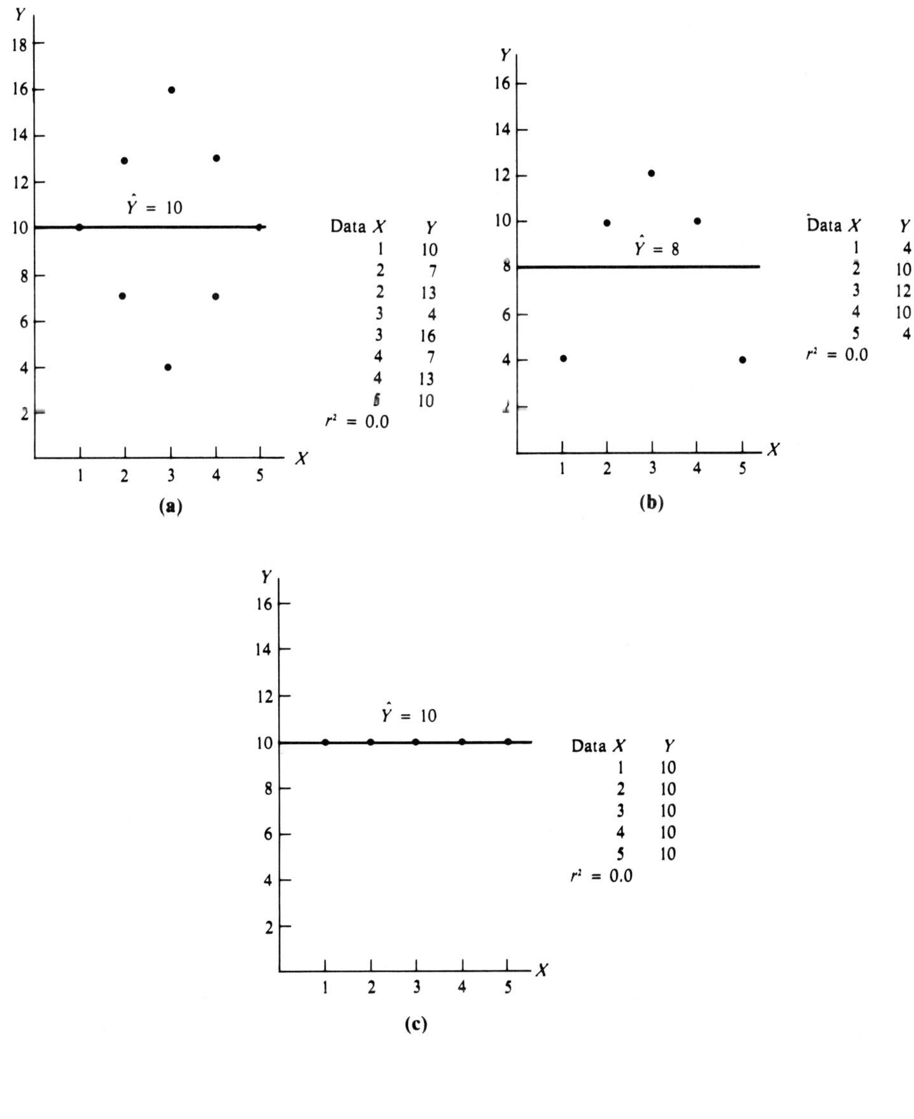

In Figure 8.1(b) we see that Y very definitely is related to X: As X increases, Y first increases and then decreases. It would be possible to fit a parabola exactly through these points. However, there is *no linear relationship* between the two variables: regardless of the value of X, the average value of Y is constant. It is important to keep in mind that

$$r^2 = \frac{(S_{XY})^2}{S_{XX}S_{YY}}$$

measures the strength of a *linear* association, since it is only for the model $Y = \alpha + \beta X + \varepsilon$ that $SS(R) = bS_{XY}$ and the above formula for r^2 was derived. This is not to say, however, that r^2 cannot measure the strength of a curvilinear relationship. If a model $Y = \alpha + \beta_1 X + \beta_2 X^2 + \varepsilon$ were adopted, then r^2, calculated as $r^2 = SS(R)/SS(T)$ measures the proportion of variation in Y explained by X and X^2. In this case, $SS(R)$ is not calculated as bS_{XY}, however.

In Figure 8.1(c) all points lie on the line. So we might be tempted to say that we have a perfect linear relationship and the coefficient of determination should be 1 instead of 0. However, since $Y = 10$ everywhere, X carries no information about Y, and $r^2 = 0$.

At the other end of the scale are situations in which $r^2 = 1$. Two of these are shown in Figure 8.2. In Figure 8.2(a) the variables are perfectly directly related, and in Figure 8.2(b) the variables are perfectly inversely related. Whether Y increases or decreases as X increases is irrelevant. What matters is that all points lie on a straight nonhorizontal line so that if X is known, Y is known without error. Thus, X tells 100 percent of what there is to know about Y.

FIGURE 8.2
Examples of Data for Which $r^2 = 1$

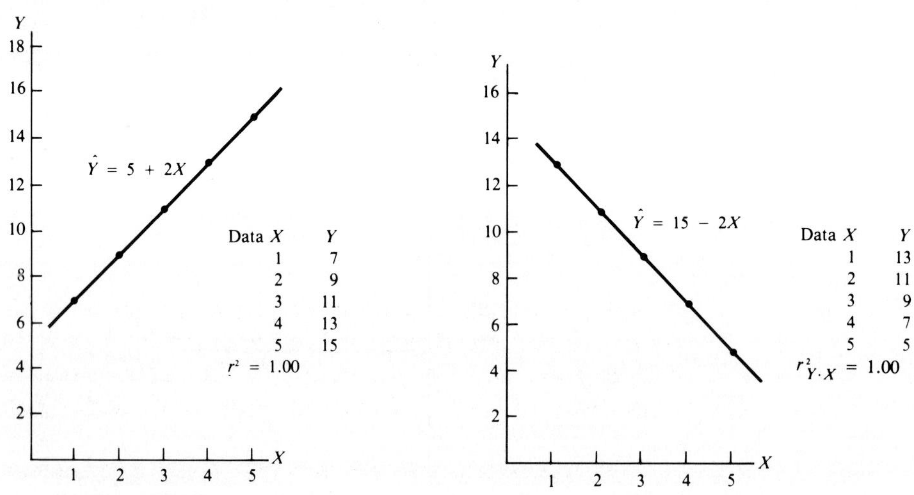

Intermediate values of r^2 are shown in Figure 8.3. Since all three graphs are drawn to the same scale, we can compare the scatter of the points around the lines. While all three graphs show direct relationships, the direction makes no difference to r^2.

FIGURE 8.3 Data for Which r^2 Is Approximately 0.25, 0.50, and 0.75

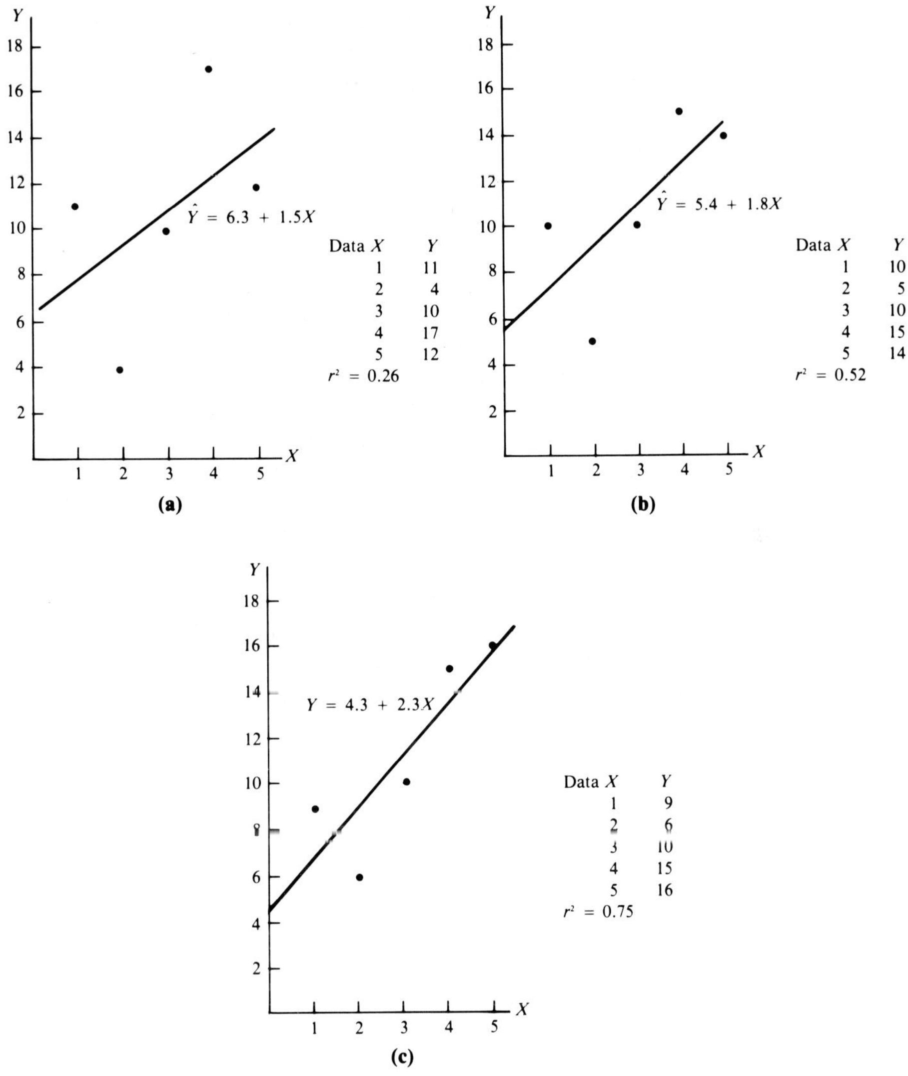

In Figure 8.3(a) some points are about six units away from the line. While there is a definite tendency for Y to increase with X, one could get only a very rough approximation to the value of Y by knowing the value of X. You can verify that $r^2 \approx 0.25$, so 26 percent of the variation in Y can be explained by X.

The coefficient of determination doubles in Figure 8.3(b). Here the farthest point from the line is about 4 units away. Predictions of Y from X can be made much more accurately in this case, since variation in X explains about 52 percent of the variation in Y.

Finally, in Figure 8.3(c) $r^2 = 75\%$. No point is more than about 3 units from the line, and we are a lot closer to knowing Y if we know X than we were in either of the two previous cases.

While the relationship in Figure 8.3(c) is certainly stronger than the one in Figure 8.3(a), it is arguable whether the one in (a) can really be considered weak. It should, of course, be kept in mind that only five data points were used in the example for the sake of simplicity. Still, in (a), one variable alone can explain 26 percent of the variation in another, leaving over 74 percent to be explained by all other predictors, how numerous or few they may be.

It should be clear, however, that one cannot simply place r^2 on a scale from 0 to 1 and interpret the strength of the relationship by referring to how close it is to the endpoints of the scale. It must be kept in mind what this statistic is measuring. In the next section we will present further support of the contention that $r^2 = 0.26$ is not necessarily indicative of a weak relationship.

PROPORTIONAL REDUCTION IN ERROR

We saw that the coefficient of determination, r^2, can be interpreted as the proportion of variation in Y explained by variation in X. As such, it is an intuitively interpretable measure of the strength of the relationship between X and Y. There is another way in which r^2 can be interpreted, as well, which is very appealing to many people.

Suppose you were asked to estimate the price of a house for sale, but the only information you had to help you make this estimate was the prices of other houses that had recently come on the market. You had no information about the sizes, ages, or locations of any of these houses. The only reasonable estimate you could make would be μ_Y, the average price of all houses. Then, on the average, your estimate would be off by

$$\frac{\sum |Y_i - \mu_Y|}{N},$$

or using a variance to measure the average size of error by σ_Y^2.

Now suppose that you are allowed to use knowledge of the relationship, $Y = \alpha + \beta X + \varepsilon$, between price and size to predict price, and that you are given information on the sizes as well as the prices of the houses. Using this information, the error in your estimates of price would be measured by $\sigma_{Y \cdot X}^2$. Assuming $\beta \neq 0$, you would expect to be much closer to actual prices when using knowledge of X than when simply predicting the average price in each case. That is, $\sigma_{Y \cdot X}^2$ should be less than σ_Y^2. The stronger the relationship between X and Y, the smaller $\sigma_{Y \cdot X}^2$ will be in comparison to σ_Y^2. The difference $\sigma_Y^2 - \sigma_{Y \cdot X}^2$ measures the amount of reduction in error when using X to predict Y. Thus,

$$\rho^2 = \frac{\sigma_Y^2 - \sigma_{Y \cdot X}^2}{\sigma_Y^2}$$

Proportional Reduction in Error

measures the **proportional reduction in error (PRE)** when using knowledge of X to predict Y over using knowledge of Y alone.

In our house size-price example, we found $r^2 = 0.95$. This can be interpreted as a PRE measure: The sizes of the errors in estimating prices of houses can be reduced 95 percent when using knowledge of their sizes, as opposed to making such estimates without knowledge of sizes. Clearly, this indicates a very strong relationship between price and size.

What if $r^2 = 0.33$? A 33 percent reduction in error using knowledge of X to help estimate Y does not sound like the relationship between X and Y is an especially weak one, does it? It appears that r^2 does not have to be extremely close to 1 in order for it to indicate that the relationship is at least moderately strong. The correlation coefficient, discussed in the next section, will give us yet another way to interpret the strength of the relationship.

EXERCISES

8.1 Calculate and interpret the coefficient of determination for the Concession Data of Exercise 1.5.

8.2 For the Heating Data of Exercise 1.8, calculate and interpret the coefficient of determination.

8.3 Calculate and interpret the coefficient of determination for the Smokers Data of Exercise 1.9.

8.4 Calculate and interpret the coefficient of determination for the Mail Data of Exercise 3.31.

8.5 For the Police Data of Exercise 1.7,

a. Calculate and interpret the coefficient of determination, assuming that a quadratic regression equation has been fitted to the data.

b. Calculate adjusted r^2.

8.6 **a.** Suppose that a simple linear regression equation is fitted to $n = 15$ pairs of values and $SS(R) = 500$, $SS(T) = 830$. Find the values of r^2 and adjusted r^2, and compare.

b. Suppose that a four-predictor multiple regression equation is fitted to $n = 15$ pairs of values, and $SS(R) = 500$, $SS(T) = 830$. Find the values of r^2 and adjusted r^2, and compare.

c. Do parts (a) and (b) give you a clue as to the usefulness of adjusted r^2?

8 . 3

THE CORRELATION COEFFICIENT

Since the sample coefficient of determination, r^2, is always nonnegative, it cannot indicate whether the relationship is direct or inverse. When we need to know the direction of the relationship as well as its strength, another measure of correlation is used.

Population Correlation Coefficient

The **population correlation coefficient,** ρ, (formally known as the Pearson product-moment correlation coefficient), is defined as for linear bivariate models

$$\rho = \pm\sqrt{\rho^2},$$

so that

$$-1 \leq \rho \leq 1.$$

When $\rho = -1$ a perfect inverse relationship exists; $\rho = 0$ indicates no relationship; $\rho = 1$ indicates a perfect direct relationship; and values of r close to $+1$ or -1 indicate strong relationships. Note that while ρ^2 is a proportion, or percentage, ρ is *not*. It is a ''pure'' number whose sole function is to be compared to ± 1.

Sample Correlation Coefficient

In a simple linear regression, the **sample correlation coefficient,** r, is calculated:

$$r = \frac{S_{XY}}{\sqrt{S_{XX}S_{YY}}},$$

and the sign on S_{XY} determines the sign on r.

In our example,

$$r = \frac{174.9}{\sqrt{(13.56)(2373)}} = 0.975,$$

or

$$r = \pm\sqrt{r^2} = \pm\sqrt{0.9507} = \pm 0.975$$

and the plus sign is retained upon inspection of the sign of S_{XY}. Since r is very close to $+1$, a very strong direct relationship is indicated. Note that only for the model $Y = \alpha + \beta X + \varepsilon$ is $SS(R)$ calculated as bS_{XY}, so that r^2 is calculated as $S_{XY}^2/S_{XX}S_{YY}$. Thus, $r = S_{XY}/\sqrt{S_{XX}S_{YY}}$ measures the strength of a *linear bivariate* relationship only.

Equivalently, in a simple linear regression r may be calculated as b adjusted for the standard deviations of X and Y. That is, in a simple linear regression, a correlation coefficient is the same as a standardized regression coefficient or beta coefficient:

$$r = \frac{S_{XY}}{\sqrt{S_{XX}S_{YY}}} = \frac{S_{XY}}{\sqrt{S_{XX}S_{YY}}} \cdot \frac{S_{XX}}{S_{XX}} = \frac{S_{XY}}{S_{XX}} \cdot \frac{S_{XX}}{\sqrt{S_{XX}S_{YY}}}$$

$$= b\sqrt{\frac{S_{XX}}{S_{YY}}} = b\sqrt{\frac{S_{XX}/(n-1)}{S_{YY}/(n-1)}}$$

$$= b\frac{S_X}{S_Y}.$$

In our example,

$$r = 12.8982\left(\frac{0.8448}{11.1756}\right) = 0.975.$$

To interpret the strength of the relationship indicated by the value of r, one commonly thinks of that segment of the real-number line from -1 to 1 (see Figure 8.4). Those values between -1 and 0 indicate inverse relationships,

FIGURE 8.4
A Scale from -1 to $+1$

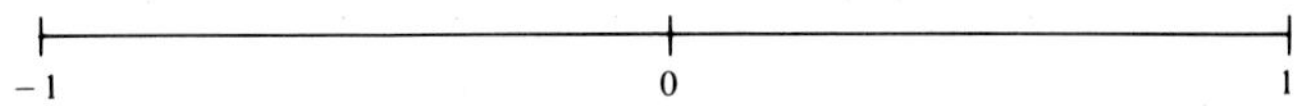

and those between 0 and 1 indicate direct relationships. At the ends of the segment, ± 1 indicate perfect relationships, while in the middle, at zero, there is no relationship. If we define "moderate" to be halfway between none and perfect, then moderate would be located at $\pm \frac{1}{2}$. Then perhaps $\pm \frac{3}{4}$ would stand for moderately strong and $\pm \frac{1}{4}$ would denote moderately weak (see Figure 8.5).

FIGURE 8.5
Interpreting r

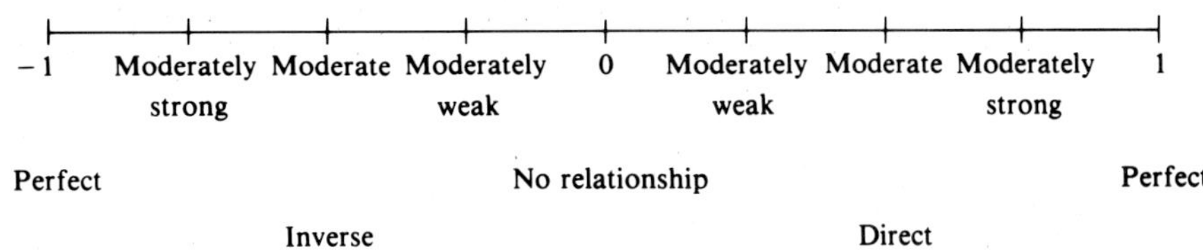

Suppose $r^2 = 0.33$ and $r = -0.5744$. The negative sign indicates an inverse relationship, and since r is over halfway from 0 to -1, we would interpret $r = -0.5744$ as a moderately strong inverse relationship. This backs up our interpretation of $r^2 = 0.33$ indicating a moderately strong relationship, since r and r^2 are two different ways of measuring the strength of the *same* relationship.

In Figure 8.3(a) we found $r^2 = 0.26$. Since the relationship is direct, $r = +\sqrt{0.26} = 0.51$, which we see as a moderate relationship, rather than a weak one. Similarly, in Figure 8.3(b) $r = +\sqrt{0.52} = 0.72$ indicates a moderately strong direct relationship. Finally, the $r^2 = 0.75$ from Figure 8.3(c) yields $r = 0.87$, which probably translates as strong.

Note that r and r^2 are two different methods for estimating the strength of the relationship between two variables, and different standards are used to interpret them. Moreover, while r^2 can be interpreted in terms of the problem, r has no physical meaning.

8 . 4 TEST OF SIGNIFICANCE

The Pearson correlation coefficient is intended as a measure of the strength of linear relationship; there is also a test of the significance of the relationship based on r. Since sampling error can cause one to think that a relationship exists even if it does not, how large does r have to be (in absolute value) for us to be reasonably certain that a relationship exists? While this question has already been answered by the hypothesis tests on β, we can run an equivalent test using r instead of b.

In order to test for a significant correlation, we test $H_0:\rho = 0$ against the appropriate alternative: $H_1:\rho > 0$ if we seek a direct relationship, $H_1:\rho < 0$ for an inverse relationship, or $H_1:\rho \neq 0$ for either a positive or negative correlation. The test statistic is a t:

$$t = \frac{r}{\sqrt{(1 - r^2)/(n - 2)}} = r\sqrt{\frac{n - 2}{1 - r^2}},$$

where t has $n - 2$ degrees of freedom. The critical region corresponding to the significance level and the alternative hypothesis is determined in the same way as for the t-test on β.

In our example,

$$H_0:\rho = 0$$

$$H_1:\rho > 0$$

seem to be the reasonable hypothesis to test, so if $\alpha = 0.05$, we will reject H_0 if $t > t_{18,0.05} = 1.734$. Then

$$t = 0.975\sqrt{\frac{20 - 2}{1 - 0.9507}} = 18.6301.$$

Since $18.63 > 1.734$, reject H_0 and conclude you are more than 95 percent sure that there is a positive correlation between house size and price—generally the larger the house, the higher the price.

COMPARISON TO TEST ON β

Notice that this test on ρ is almost exactly the same as the test for $\beta > 0$. The critical region is the same, the conclusion is the same, and the value of the t-statistic is almost the same. Only the approaches are different. A researcher who is interested only in the strength of the relationship will utilize the correlation approach. Working just with r and r^2, and never even calculating a, b, or $S_{Y \cdot X}$, one can tell if the relationship is significant, determine its direction, and measure its strength.

LIMITATIONS OF THE TEST STATISTIC

The foregoing test statistic can only be used to test a null hypothesis that the correlation is *zero*. If we want to test an hypothesis such as $H_0:\rho = 0.5$, we must use something other than the t-statistic given here. There are other statistics available for testing such hypotheses. Since these can be found in introductory texts, they will be omitted here, because our primary interest is detecting the existence of a relationship, and thus we will typically want to test $H_0:\rho = 0$.

ASSUMPTIONS

The hypothesis test given here depends for its validity on a very important assumption. Recall that when we studied confidence intervals on $\mu_{Y \cdot X}$ and Y_i^* we made the assumption that for each X-value in the population, the subpopulation of associated Y-values was a *normal* subpopulation. When conducting a correlation analysis, it is assumed that X, as well as Y, must be a random variable. In order for the correlation coefficient and its test of significance to be valid, we must make the very important assumption that *the X-values, as well as the Y-values, come from a normal population.*

In many studies of relationships this assumption of normality is not too difficult to justify (although it should be checked out by methods presented in the next chapter) if the data collected represent *measurements* (or sometimes *counts*) of the values of a random variable. In other instances, the data may consist of rankings or classifications instead of measurements. An example of data consisting of rankings might be students' ranks in graduating class; examples of data consisting of classifications might be 5 = strongly agree; 4 = agree; . . . , 1 = strongly disagree; or 1 = male, 0 = female. For ranked data or categorical data special measures of correlation have been developed, and the reader is referred to texts dealing with the study of nonparametric statistics.

PRACTICAL VERSUS STATISTICAL SIGNIFICANCE

We found a very large positive correlation for the house size-price problem, but it must be remembered that these were artificial data. Seldom in practice does one find such a high correlation (positive or negative) between two variables. For large samples, r can be relatively small and still be significantly different from zero. For example, if $n = 200$ and we are testing the alternative $H_1:\rho > 0$ with $\alpha = 0.05$, then $H_0:\rho = 0$ will be rejected if

$$t = r\sqrt{\frac{200 - 2}{1 - r^2}} > t_{198,0.05} \approx 1.658.$$

How large, then, must r or r^2 be? Solving,

$$r^2\left(\frac{198}{1 - r^2}\right) > (1.658)^2 = 2.749$$

$$198r^2 > 2.749(1 - r^2) = 2.749 - 2.749r^2$$

$$(198 + 2.749)r^2 > 2.749$$

$$r^2 > \frac{2.749}{200.749} = 0.0137$$

$$|r| > 0.117.$$

Thus, even though $r = 0.12$ is not very large, if $n = 200$, we can be 95 percent sure that ρ is really greater than zero, and that this nonzero value for r did not occur just by chance. Another sample of size 200 would be expected to produce a correlation coefficient approximately this same size. If, however, a sample of only ten pairs of observations had resulted in a correlation of 0.12, it would be much more likely that a nonzero correlation was observed just by chance and, in fact, the true correlation is zero. Another sample of size 10 might not reproduce a correlation coefficient as large as this one. The point is that a low correlation found to be statistically significant in a large sample only means that we are reasonably sure that a weak relationship is most likely present, one not due just to chance. It is another matter to determine whether or not X in this case will be practically useful in predicting Y.

EXERCISES

8.7 For the Concession Data of Exercise 1.5,
  a. Calculate and interpret the correlation coefficient.
  b. Test $H_0:\rho = 0$ against the appropriate alternative, and state your conclusion. Compare conclusion to the test of $H_0:\beta = 0$.
  c. Square r and compare to Exercise 8.1.

8.8 For the Heating Data of Exercise 1.8,
  a. Calculate and interpret the correlation coefficient.
  b. Test $H_0:\rho = 0$ against the appropriate alternative, and state your conclusion. Compare conclusion to the test of $H_0:\beta = 0$.
  c. Square r and compare to Exercise 8.2.

8.9 Referring to the Smokers Data of Exercise 1.9,
  a. Calculate and interpret the correlation coefficient.
  b. Test $H_0:\rho = 0$ against $H_1:\rho \neq 0$, and state your conclusion.
  c. Square r and compare to Exercise 8.3.

8.10 For the Mail Data of Exercise 3.31,
  a. Calculate and interpret the correlation coefficient.
  b. Test $H_0:\rho = 0$ against $H_1:\rho > 0$.
  c. Square r and compare to Exercise 8.4.

8.11 For the Police Data of Exercise 1.7,

a. Calculate and interpret the correlation coefficient.

b. Square the correlation coefficient and compare to r^2 in Exercise 8.5. Why are they different?

8.12 For samples of sizes 5, 10, 15, 20, 25, 30, 50, 75, and 100, determine how large r must be in order to reject $H_0: \rho = 0$ in favor of $H_1: \rho > 0$ at $\alpha = 0.05$.

8.13 Suppose that X and Y are both expressed in standardized form:

$$Z_X = \frac{X - \overline{X}}{s_X}, \qquad Z_Y = \frac{Y - \overline{Y}}{s_Y}.$$

Show that

$$\frac{1}{n - 1} S_{Z_1 Z_2} = \frac{[\sum Z_1 Z_2 - (\sum Z_1)(\sum Z_2)/n]}{(n - 1)} = r_{XY}.$$

8 . 5 CORRELATION USING THE COMPUTER

It is easy to locate the coefficient of determination and the correlation coefficient in the three programs we studied in Chapter 4.

BMD PROGRAMS

For the BMDP1R program for the house size-price example, refer to Figure 4.2(b). The correlation coefficient is given by `MULTIPLE R` = 0.9750, although the adjective *multiple* is not applicable when we have only one predictor. Note that r is calculated as the *positive* square root of r^2 by this program, regardless of the direction of the relationship. One must look at the sign of b to determine the sign of r. The "multiple" coefficient of determination is also prominently displayed in Figure 4.12.

MINITAB

Minitab shows both the coefficient of determination and the adjusted r^2 just above the analysis of variance table in Figure 4.8. Minitab does not calculate a correlation coefficient as part of the regression program. However, to find the simple linear bivariate correlation coefficient between size and price, you can use the CORRELATION command. Suppose that values of size are stored in column 1 and prices are in column 2. Then

```
CORR C1 C2
```

will cause the correlation coefficient to be printed.

S A S P R O G R A M S

Figure 4.13 shows the simple regression analysis for the house size-price problem using the SAS. One can easily find

R-SQUARE 0.9507 and ADJ R-SQ 0.9479

under the analysis of variance table. In order to find the correlation coefficient, the square root of r^2 must be taken, and one must look at the regression coefficient to determine the sign of the root.

S P S S^X P R O G R A M S

The SPSS^X SCATTERGRAM printout for this problem is shown in Figure 4.18. At the bottom of the diagram both r and r^2 are given in the table of statistics. We find

CORRELATION (R) 0.97501 and R SQUARED 0.95065.

The SCATTERGRAM procedure also shows the attained significance level for the t-test of $H_0{:}\rho = 0$ against $H_1{:}\rho \neq 0$. SIGNIFICANCE 0.00000 says that, to five decimal places, the probability that $H_0 : \rho = 0$ is true is zero.

The SPSS^X regression analysis shown in Figure 4.19 also gives the coefficient of determination, correlation coefficient, and adjusted r^2. The correlation coefficient in Figure 4.19 is labeled MULTIPLE R: This program is designed to do more than just a simple linear regression, so the labeling is more general than is appropriate to our simple linear regression analysis. Also note that MULTIPLE R is calculated as the positive square root of r^2 here (we shall see in Chapter 12 why this is so) and thus will always be a positive number. You must refer to the sign of the regression coefficient, b, in a simple regression to know the sign of the correlation coefficient.

E X E R C I S E S

8.14 Using the computer analysis of the Concession Data from Chapter 4, find r and r^2.

8.15 Using the computer analysis of the Smokers Data from Chapter 4, find r and r^2.

8.16 Find r and r^2 from the computer analysis of the Class Size Data from Chapter 4.

8.17 Find r and r^2 from the computer printout of the analysis of the Appraisal Data in Chapter 4.

8.18 From a computer analysis fitting a quadratic regression equation to the Police Data (Chapter 4), find

a. r^2;

b. Adjusted r^2;

c. The simple linear bivariate correlation coefficient.

8 . 6

SUMMARY

Methods of correlation enable us to study the strength of the relationship between variables and test for the significance of the relationship without calculating a regression equation.

Correlation analysis is more than an alternative to regression. It is essential to the understanding of the interrelationships among the variables in a multiple regression.

We shall turn to the study of multiple regression after examining some problems that can crop up when using simple regression to analyze actual data.

NINE
APPROPRIATENESS OF THE MODEL

9.1
INTRODUCTION

Our house size-price problem as well as the problems in the exercises are artificially constructed. Seldom would one encounter real data that are so well behaved. Even with the house size-price problem, however, we noted that some of our assumptions really should be checked out. For example, should a simple linear model have been used, or is there a leveling-off of prices for higher-priced houses? Is the scatter of the points around the line truly uniform? Can we justify the assumption that for each size house, the prices are normally distributed around the line?

It is now time to check out our assumptions for the house size-price problem and see what corrective measures can be taken if they are not met. Of course, these assumptions normally would be verified before fitting the regression equation. Some new examples will be introduced to illustrate the kinds of problems that often arise with real data.

9.2
THE ASSUMPTION OF UNIFORM SCATTER

In Chapter 6 we assumed that regardless of the value of X, the variability in the associated Y-values remained constant. Otherwise we could not have estimated the error in estimation. Without uniform scatter, the accuracy of our estimates would depend on the value of X for which we were trying to estimate Y.

LOOKING AT THE SCATTER DIAGRAM

In order to determine whether or not the scatter of the points is uniform around the regression line, we can look at the scatter diagram with the regression equation graphed, as in Figure 2.8. The data points fit snugly within two lines parallel to and equidistant from the regression line, so it appears that the scatter is uniform for our house size-price example. Had the scatter been greater for some size houses than for others, nonhomogeneous scatter would be indicated.

PLOTTING THE RESIDUALS

If a curvilinear equation had been fitted to the data, we could still inspect the scatter diagram for uniform scatter by seeing if the dots lay snugly within curved lines following the regression equation. However, this inspection might not be as easy as it sounds. Indeed, the fact that the regression line in Figure 2.8 rises might somewhat obscure the pattern of variation we are looking for. Inspection of a graph for uniform scatter can be made easier if we adopt a method whereby we need not take into account any rise, fall, or curve in the regression equation. Toward this purpose, we plot the residuals, $Y_i - \hat{Y}_i$, against the X-values. The residuals can be read from one of the computer printouts obtained in Chapter 4.

For our house size-price data, refer to the list of residuals in the SAS analysis in Figure 4.13. The plot of the residuals was obtained from SAS in Figure 4.15, which is reproduced in Figure 9.1. Note that the residuals are deviations about the horizontal line $Y - \hat{Y} = 0$. By plotting residuals against X, then, we have rotated the regression line in Figure 2.8 until it is horizontal, carrying the dots along in their same relative positions. This way, we can look just at the vertical scatter of the dots.

OUTLIERS

From the plot of residuals, we see something that was not readily apparent from the scatter diagram: all but one of the residuals lie between -4 and $+4$. That is, there is one residual that seems to be unusually large in absolute value

FIGURE 9.1 SAS Plot of Residuals from Figure 4.15

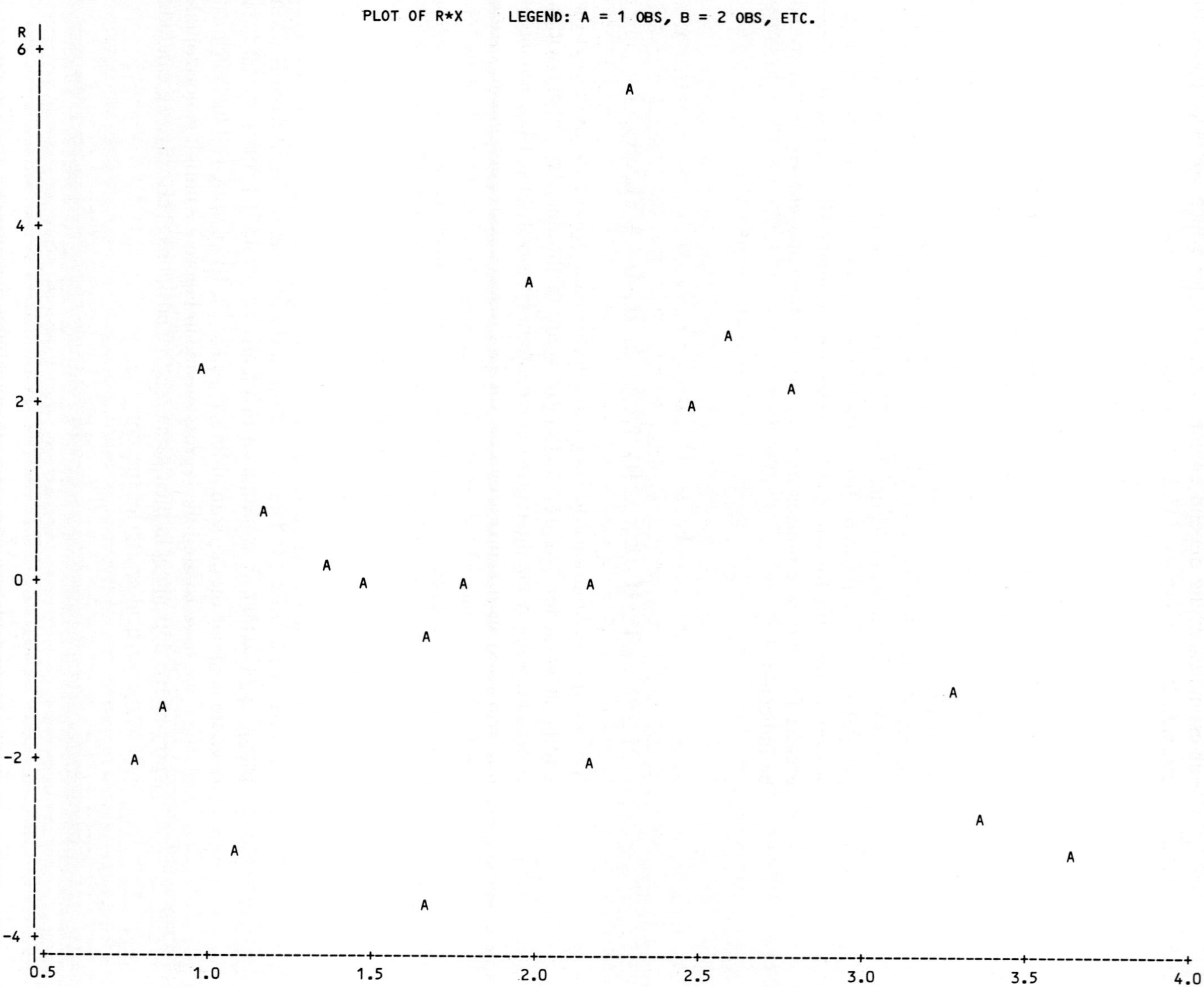

when compared to the others. Such an unusual case is referred to as an

Outlier

outlier—it is not typical of the rest of the data.

An outlier in the data warns us to stop and investigate. How could this have come about? One would first make sure that an outlier did not result from an error in the data. Data are sometimes misrecorded, digits can be transposed, the measuring instrument can malfunction, the observer may be distracted. If we find that the outlier resulted from a mistake, then our only course of action is to throw the observation away. This is a drastic action, because each piece of information bears a cost. But misinformation distorts a relationship.

If one value is extremely different from all the others, its effect can be considerable. As an example, consider

X	Y
0	3
1	4
2	5
3	6
4	2

A quick inspection shows that the first four values all lie on the line $\hat{Y} = 3 + X$, but that the point (4, 2) lies below this line. (The reader might want to graph these points.) Suppose that X and Y are, in fact, perfectly related as in the above equation, and that the observation (4, 2) really should have been recorded as (4, 7). If we fit a straight line to these data, our calculations are as shown in Table 9.1, and

$$S_{XY} = 40 - \frac{10(20)}{5} = 0,$$

$$S_{XX} = 30 - \frac{(10)^2}{5} = 10,$$

$$\overline{Y} = \frac{20}{5} = 4, \ \overline{X} = \frac{10}{5} = 2.$$

Then

$$b = 0/10 = 0 \quad \text{and} \quad a = 4 - (0)(2) = 4.$$

TABLE 9.1
Calculations for Outlier Example

	X	Y	X^2	XY
	0	3	0	0
	1	4	1	4
	2	5	4	10
	3	6	9	18
	4	2	16	8
Sum	10	20	30	40

The regression equation is found to be $\hat{Y} = 4$, which is quite a different relationship from the one that actually exists!

This is not to say that we automatically discard all outliers. On the contrary, the more information we have, the more accurate our description of the relationship. If it cannot be established that an outlier is the result of an error in data collection, then we have to live with this extreme observation and try to discover what it is telling us.

Why would this particular house be atypically high-priced for its size? There are many possible reasons: perhaps it has a terrific view, or a pool, tennis courts, and stable, or all new carpet and paint. An outlier (or several outliers) indicates that we should take other variables into account. For accurate predictions, we should choose a multiple regression model in preference to the simple linear model in order to take these other factors into account.

On the other hand, this one particular house might not really be an outlier. Usually a true outlier has a residual 3 standard deviations or more from zero. Recall our assumption that at each X-value, the Y-values are normally distributed around the regression line. In a normal distribution, more than 99 percent of all values lie within 3 standard deviations of their mean. Thus, it is unusual to see an observation 3 standard deviations greater or less than the mean, and it is quite rare to find a value 4 or more standard deviations different from the mean. In our example, house 11 has a residual of about 5.63 (see Figure 9.1, for example). Since the standard error of estimate is about 2.5, this residual is only a little over 2 standard deviations from zero. How can we tell whether or not to consider it an outlier? One method is to fit a regression equation to the other $n - 1 = 19$ observations, holding out the questionable one. Using this new regression equation, find a $100(1 - \alpha)\%$ confidence interval for the prices of houses this size. If the actual price lies outside the confidence limits, then we can conclude that this observation did not come from the same population as the rest of our data and it is an outlier.

IS HOUSE NO. 11 AN OUTLIER?

In our example, if we hold out house 11 and redo the analysis, we find, for $n = 19$

$$\sum X_i = 37.7, \qquad \sum Y_i = 646, \qquad \sum X_iY_i = 1{,}453.7,$$

$$\sum X_i^2 = 88.27, \qquad \sum Y_i^2 = 24{,}242.$$

Then

$$S_{XX} = 88.27 - \frac{(37.7)^2}{19} = 13.4653,$$

$$S_{YY} = 24{,}242 - \frac{(646)^2}{19} = 2278,$$

$$S_{XY} = 1{,}453.7 - \frac{(37.7)(646)}{19} = 171.9,$$

and

$$b = 12.7661,$$

$$a = 34 - 12.7661(1.9842) = 8.6694.$$

Our new regression equation is

$$\hat{Y} = 8.67 + 12.77X.$$

The new standard error will be the square root of

$$s^2_{Y \cdot X} = \frac{2278 - 12.7661(171.9)}{17} = 4.9122,$$

$$s_{Y \cdot X} = 2.2163.$$

For a 99% confidence interval on the price of a house 2,300 sq ft in area,

$$\hat{Y}_{2 \cdot 3} = 8.6694 + 12.7761(2.3) = 38.0544,$$

$$t_{17,0.005} = 2.898,$$

$$38.0544 \pm 2.898(2.2163)\sqrt{1 + \frac{1}{19} + \frac{(2.3 - 1.9842)^2}{13.4653}},$$

$$38.0544 \pm 6.6128,$$

$$31.3316 < \hat{Y}_{2 \cdot 3} < 44.6672.$$

The actual price of house 11 is \$44,000. Since this is within the range of \$31,442 to \$44,667, we conclude that house 11 is not an outlier. (At least, we cannot be 99 percent sure that it is; we used a 99% confidence interval instead of a 95% confidence interval because we want to be quite certain that we really have an outlier before we throw an observation away.) Thus, we will retain house 11 in our data, using the original analysis based on all 20 observations. We are aware, however, that to improve the accuracy of our estimates of prices, we need to include variables other than size in our model. Chapter 11 and all those following will address the problem of multiple linear regression.

There are other methods for detecting outliers, among them a quick small-sample test known as the Dixon criterion.

THE DIXON CRITERION

Dixon Criteria

The **Dixon criterion** is a small-sample test ($n \leq 25$) for outliers that are either extremely large or extremely small. The procedure is as follows. First, choose

α as the largest risk you are willing to take of throwing out an observation that really does belong in the set. That is,

$$H_0\text{: observation is not an outlier,}$$

$$H_1\text{: observation is an outlier.}$$

Arrange the set of values V_i in order from smallest to largest. Call the smallest $V_{(1)}$ and the largest $V_{(n)}$, where n is the number of values in the set.

Next, compute a comparison of ranges, r_{ij}, depending upon the size of n: if

$$3 \le n \le 7, \quad \text{compute } r_{10},$$

$$8 \le n \le 10, \quad \text{compute } r_{11},$$

$$11 \le n \le 13, \quad \text{compute } r_{21},$$

$$14 \le n \le 25, \quad \text{compute } r_{22},$$

where r_{ij} values are computed as shown in Table 9.2.

TABLE 9.2
Test Statistics for Dixon Test

r_{ij}	If Smallest Value Is Suspect	If Largest Value Is Suspect
r_{10}	$\frac{V_{(2)} - V_{(1)}}{V_{(n)} - V_{(1)}}$	$\frac{V_{(n)} - V_{(n-1)}}{V_{(n)} - V_{(1)}}$
r_{11}	$\frac{V_{(2)} - V_{(1)}}{V_{(n-1)} - V_{(1)}}$	$\frac{V_{(n)} - V_{(n-1)}}{V_{(n)} - V_{(2)}}$
r_{21}	$\frac{V_{(3)} - V_{(1)}}{V_{(n-1)} - V_{(1)}}$	$\frac{V_{(n)} - V_{(n-2)}}{V_{(n)} - V_{(2)}}$
r_{22}	$\frac{V_{(3)} - V_{(1)}}{V_{(n-2)} - V_{(1)}}$	$\frac{V_{(n)} - V_{(n-2)}}{V_{(n)} - V_{(3)}}$

Critical values for selected values of α are shown in Appendix J. If the computed value for r_{ij} exceeds the tabulated value, you can be $100(1 - \alpha)$ percent sure that the observation is an outlier.

Let us use the Dixon test to determine if the residual for house 11 is an outlier. Table 9.3 shows residuals from fitting a simple linear regression equation arranged in order from smallest, $V_{(1)}$, to largest, $V_{(20)}$. Since $n = 20$, and the largest value is suspect, we will compute

$$r_{22} = \frac{V_{(n)} - V_{(n-2)}}{V_{(n)} - V_{(3)}} = \frac{V_{(20)} - V_{(18)}}{V_{(20)} - V_{(3)}}$$

$$= \frac{5.6305 - 2.7611}{5.6305 - (-2.8916)} = \frac{2.8694}{8.5221} = 0.3367.$$

Since $0.3367 < 0.450$, we are less than 90 percent sure that the residual for house 11 is an outlier.

TABLE 9.3
Residuals Arranged in Ascending Order of Magnitude

$-3.6305 = V_{(1)}$	-0.0509
$-3.1372 = V_{(2)}$	0.0796
$-2.8916 = V_{(3)}$	0.2389
-2.5575	0.8186
-2.0796	2.0509
-2.0221	2.1814
-1.3119	2.3982
-1.2677	$2.7611 = V_{(18)}$
-0.6305	$3.5000 = V_{(19)}$
-0.0796	$5.6305 = V_{(20)}$

A TEST FOR HOMOGENEOUS SCATTER

Having decided that house 11 is not an outlier, we return to the problem of whether or not the scatter can be considered homogeneous. Besides a visual interpretation of the plot of residuals, there are statistical tests to help us decide whether or not the variances of the Y-values remain constant regardless of the value of X. One of these tests involves a **split-sample analysis.**† Split the sample in half, calculate $s^2_{Y \cdot X}$ for each of the halves, and compare the two values for $s^2_{Y \cdot X}$ using an F-test. How is the sample split in half? This depends upon where you suspect the differences in scatter might be.

Split-Sample Analysis

Suppose the plot of residuals showed a pattern as in Figure 9.2. The scatter appears to be larger for large X-values than for smaller X-values. Then as one split of the sample, we would choose the $n/2$ observations with the smallest X-values, and put the $n/2$ observations with the largest X-values in the other split. (If n is an odd number, then one split will have one more observation than the other, but it doesn't matter which split is the larger.) A similar split would be used if the plot of residuals showed larger scatter at small X-values than at large X-values.

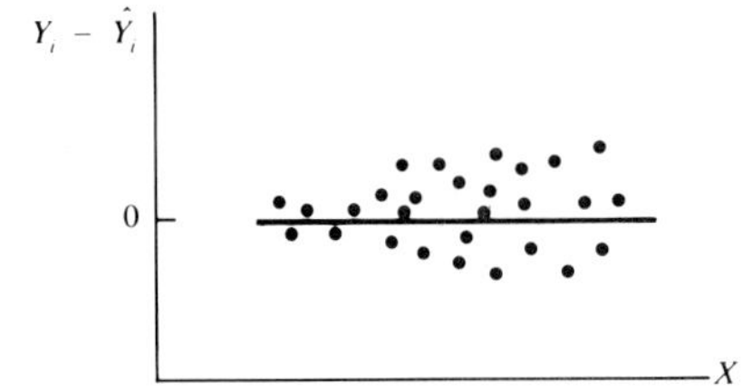

FIGURE 9.2
Plot of Residuals Showing Scatter Increasing With X

Referring to Figure 9.1, the plot of residuals for the house size-price data, the greatest scatter seems to occur at middle-sized values of X. Thus, as one split of the sample we could choose the middle ten observations, and

† The term *split sample* here has a different connotation than in discriminant analysis, in which part of the sample is used to verify results found for the other part.

put the five largest and five smallest houses together in the other split. Let us try it as shown in Table 9.4. Are the two variances in the table, 5.9581 and 6.4642, different enough to indicate a real difference in scatter, or is the difference just due to sampling error? To answer this question, let s_1^2 and s_2^2 denote the two variances. Form an F-test statistic,

$$F = \frac{s_1^2}{s_2^2},$$

in order to test

$$H_0: \sigma_1^2 = \sigma_2^2$$
$$H_1: \sigma_1^2 \neq \sigma_2^2,$$

where σ_1^2 and σ_2^2 denote the true variances of the points around the regression lines in the two subsamples. Let n_1 denote the number of observations in subsample 1, and n_2 the observations in the other subsample. Then s_1^2 has $n_1 - 2$ degrees of freedom and s_2^2 has $n_2 - 2$ degrees of freedom, and the F-statistic will have $\nu_1 = n_1 - 2$ degrees of freedom in the numerator and $\nu_2 = n_2 - 2$ degrees of freedom in the denominator.

TABLE 9.4 Splitting the Sample to Test for Uniform Scatter

	Split 1			Split 2		
	House No.	**Size (X)**	**Price (Y)**	**House No.**	**Size (X)**	**Price (Y)**
Small	6	0.8	17	17	1.4	27
	12	0.9	19	20	1.5	28
	2	1.0	24	3	1.7	27
	8	1.1	20	15	1.7	30
	13	1.2	25	1	1.8	32
Large	7	3.6	52	9	2.0	38
	14	3.4	50	5	2.2	35
	18	3.3	50	19	2.2	37
	4	2.8	47	11	2.3	44
	10	2.6	45	16	2.5	43
Sum		20.7	349		19.3	341
Sum of squares		55.11	14,189		38.45	11,989
Sum of products		877.5			677.4	
S_{XX}		12.261			1.201	
S_{YY}		2008.9			360.9	
S_{XY}		155.07			19.27	
$s_{Y \cdot X}^2$		5.9581			6.4642	

We have chosen a two-sided alternative hypothesis so that we can detect if $\sigma_1^2 > \sigma_2^2$ as well as if $\sigma_2^2 > \sigma_1^2$. We thus need a two-sided critical region, for values of F either considerably less than 1 *or* considerably greater than 1 can be taken as evidence that $\sigma_1^2 \neq \sigma_2^2$. Thus, we need to find F-values that cut off the upper $100(\alpha/2)$ percent of the area under the F-curve and the lower $100(\alpha/2)$ percent, as illustrated in Figure 9.3.

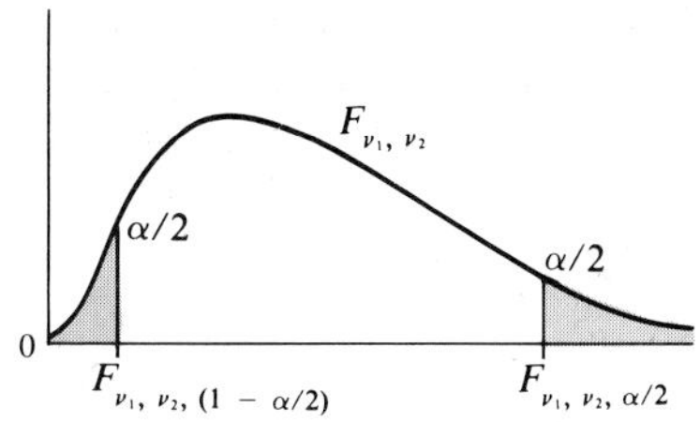

FIGURE 9.3 Finding the Two-Sided Critical Region for an *F*-Test

Our tables give only *upper-tail* percentage points, but we can still use them to find the lower-tail critical values. Suppose $\alpha = 0.10$. Then we want to cut off $\alpha/2 = 0.05$ of the area under the upper tail and 5 percent of the area under the lower tail. From the *F*-tables, we can easily find $F_{\nu1,\nu2,0.05}$. The problem is to cut off 5 percent of the area under the lower tail, which is equivalent to cutting off 95 percent of the area under the upper tail. Thus we can denote this percentage point as $F_{\nu1,\nu2,0.95}$. The upper-tail tables do not include $\alpha = 0.95$. However, we can find $F_{\nu1,\nu2,0.95}$ by means of the relation

$$F_{\nu1,\nu2,0.95} = \frac{1}{F_{\nu2,\nu1,0.05}}.$$

Note that numerator and denominator degrees of freedom are switched. It is very important to remember that the *F*-test for equality of variances is extremely sensitive to departures from the normality assumption. This means that unless the populations from which the samples were drawn are normal, the probabilities for the calculated *F*-values are not accurate. The further from normal the populations, the less accurate the probabilities, and these probabilities become inaccurate quickly with small departures from normality. The researcher should thus use this test only when he or she is quite confident that the distributions of the residuals around the regression line are normal, or else take the computed probabilities as rough estimates rather than as accurate measures of probabilities of occurrence of *F*-values this large or larger by chance alone.

TEST FOR HOMOGENEOUS SCATTER IN HOUSE SIZE-PRICE EXAMPLE

In our example, $\nu_1 = \nu_2 = 8$, so

$$F_{8,8,0.95} = \frac{1}{F_{8,8,0.05}} = \frac{1}{3.44} = 0.29.$$

Thus, if α is chosen to be 0.10, we will reject H_0 if

$$F > F_{8,8,0.05} = 3.44 \quad \text{or if} \quad F < F_{8,8,0.95} = 0.29.$$

The test statistic can be formed either as

$$F = \frac{5.9581}{6.4642} = 0.9217 \quad \text{or as} \quad F = \frac{6.4642}{5.9581} = 1.0849,$$

depending upon which split sample was denoted as sample 1. Whichever way F is formed, we cannot reject $H_0{:}\sigma_1^2 = \sigma_2^2$.

Thus, we cannot be 90 percent sure that the scatter is nonuniform. In the absence of sufficient evidence to conclude nonuniform scatter, we will assume that the scatter is close enough to being homogeneous.

TRANSFORMATIONS TO OBTAIN HOMOGENEOUS SCATTER

What if we had a problem in which $H_0{:}\sigma_1^2 = \sigma_2^2$ was rejected, and we were forced to conclude that the scatter is not homogeneous? In this case, $s_{Y \cdot X}$ is meaningless, because the error in estimation depends on which X-value you are using to make the estimate. Then everything that depends on $s_{Y \cdot X}$, such as significance tests, is also invalid. This can make us believe that the variables are not related when in fact they are, for example, because the variance of the regression coefficient may be inflated. Unfortunately, heteroscedastic data are common in economics, especially when X is an income variable; it often happens that $\sigma_{Y \cdot X}$ is proportional to X, that is,

$$\sigma_{Y \cdot X_i} = cX_i,$$

for some constant c. In this case, we can use a transformation on the data to make the scatter homogeneous. Such a transformation is called a **variance-stabilizing transformation.**

Variance-Stabilizing Transformation

The first problem is to determine if the standard error is proportional to the X-value, and if so, to estimate the constant c. Again a split-sample type analysis can be employed. Divide the data into several subsamples, depending on the sizes of the X-values, and calculate $s_{Y \cdot X}$ for each subsample. By comparing the $s_{Y \cdot X}$-values, one can determine whether or not the standard error is proportional to X and obtain an estimate of c. For example, suppose we divide a set of data into four subsamples and find the values of $s_{Y \cdot X}$ shown in Table 9.5. We see that as X increases by 10, $s_{Y \cdot X}$ increases by 2. (Of course, real data would seldom show such a clear-cut relationship.) That is, $s_{Y \cdot X} \approx 0.2X_i$.

Once we have determined that the scatter around the regression line is proportional to the value of X, we can transform the model so that in the transformed model the scatter is uniform. Begin with the original model

$$Y = \alpha + \beta X + \varepsilon,$$

TABLE 9.5 Relating the Standard Error to the Size of X

Subsample	X-Values	$s_{Y \cdot X}$
1	0–9	1
2	10–19	3
3	20–29	5
4	30–39	7

in which the variance of the errors, or residuals, is $\sigma^2_{Y \cdot X} = c^2X^2$. Instead of regressing Y on X, regress the ratio of Y to X on the reciprocal of X (regress Y/X on $1/X$). Of course, this will yield different values for the slope and intercept and for the errors. Denoting these by β', α', and ε', respectively, we can write the model as

$$\frac{Y}{X} = \alpha' + \beta' \frac{1}{X} + \varepsilon',$$

or

$$Y' = \alpha' + \beta' X' + \varepsilon'.$$

How does this model relate to the original one? Multiplying through by X, in the transformed model,

$$Y = \alpha' X + \beta' + \varepsilon' X,$$

so that α' corresponds to the slope and β' corresponds to the intercept of a simple linear model. That is,

$$Y' = \frac{Y}{X}, \quad X' = \frac{1}{X}, \quad \alpha' = \beta, \quad \beta' = \frac{\alpha}{X}, \quad \text{and} \quad \varepsilon' = \frac{\varepsilon}{X}.$$

In this transformed model the variance of the new errors is $1/X^2$ times the variance of the original errors, or

$$\frac{1}{X^2}(c^2X^2) = c^2.$$

Thus, the variance of the errors in the transformed model is constant—it does not depend on the value of X—and the scatter will be uniform in the transformed model.†

† We used the theorem that if each value in a set of data is multiplied by a constant, then the variance of the new set will be the square of the constant times the variance of the original set of data. For example, suppose that X_i, $i = 1, 2, \ldots, n$ are all multiplied by a constant k, yielding a new set of values $kX_1, kX_2, \ldots, kX_n$. The mean of this new set will be

$$\frac{1}{n}\sum_{i=1}^{n} kX_i = k\left(\frac{1}{n}\sum_{i=1}^{n} X_i\right) = k\overline{X},$$

and the variance will be

$$\frac{1}{n-1}\sum_{i=1}^{n}(kX_i - k\overline{X})^2 = \frac{1}{n-1}\sum_{i=1}^{n}[k(X_i - \overline{X})]^2$$

$$= \frac{1}{n-1}\sum_{i=1}^{n}[k^2(X_i - \overline{X})^2]$$

$$= k^2\frac{1}{n-1}\sum_{i=1}^{n}(X_i - \overline{X})^2 = k^2s^2.$$

AN EXAMPLE OF TRANSFORMING TO OBTAIN HOMOGENEITY

Suppose we have the following data relating Y = weekly beer expenditure in dollars, to X = weekly income in dollars. In Table 9.6 the symbol $\sum(\cdot)$ indicates that a sum is taken, either of X-values or of Y-values. $\sum(\cdot)^2$, of course, indicates a sum of squares.

TABLE 9.6 Hypothetical Data Relating Beer Expenditure to Income

Family	Average Weekly Beer Expenditure (dollars)		Weekly Income (hundreds of dollars)
1	2.40		0.9
2	3.00		1.0
3	2.50		1.1
4	3.10		1.2
5	3.20		1.3
6	2.40		1.3
7	3.20		1.5
8	2.40		1.8
9	3.60		2.0
10	2.60		2.1
11	2.40		2.6
12	4.10		2.8
13	1.80		3.0
14	3.00		3.2
15	5.00		3.2
$\sum(\cdot)$	44.7		29.0
$\sum(\cdot)^2$	142.15		65.82
$\sum X_iY_i$		89.41	

$S_{YY} = 8.944 \quad S_{XX} = 9.7533 \quad S_{XY} = 2.99$

From the table, we find

$$b = \frac{2.99}{9.7533} = 0.3066,$$

$$a = 2.98 - (0.3066)(1.9333) = 2.3973,$$

so that the regression equation is

$$\hat{Y} = 2.3873 + 0.3066X.$$

The data and regression line are plotted in Figure 9.4. There we can see that the scatter of the points around the line increases with X. Thus, the error

FIGURE 9.4
Scatter Diagram and Regression Equation for Beer Consumption Example

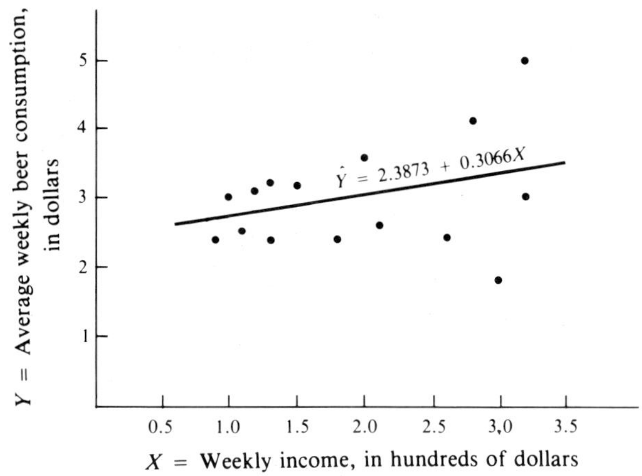

involved in estimating beer consumption depends upon what income we are talking about.

We also see that the regression line does not have a very steep slope. To test $H_0{:}\beta = 0$, we find $s^2_{Y \cdot X} = 0.61749$ and $F = 1.4844$, which would not allow us to conclude that beer consumption expenditures are related to weekly income. The standard error of the regression coefficient is $(s^2_{Y \cdot X}/S_{XX})^{1/2} = 0.2516$, which is nearly as large as $b = 0.3066$ itself. This large standard error and corresponding small F-value might be due in part to the apparent nonuniform scatter of the points around the regression line. To see if the error is proportional to the value of X, let us divide the data into five groups of three families each, according to ascending values of income, and calculate $s^2_{Y \cdot X}$ for each group. Calculations are shown in Table 9.7, where we see that the ratio of the standard error to the mean X-value is about 0.44 for each group. Thus, $\sigma_{Y \cdot X} \approx 0.44X$, or $\sigma^2_{Y \cdot X} \approx 0.2X^2$. The transformation

$$Y' = \frac{Y}{X}, \qquad X' = \frac{1}{X}$$

is thus appropriate for these data. For the transformed values in Table 9.8, we find the regression equation

$$b' = \frac{2.3944}{0.9966} = 2.4026$$

$$a' = 1.7882 - (2.4026)(0.6194) = 0.3$$

$$\hat{Y}' = 0.3 + 2.4026X'.$$

Figure 9.5 shows the transformed data and their corresponding regression equation plotted. We note that the scatter of the transformed data around their regression line is reasonably uniform.

Further, if we calculate $s^2_{Y \cdot X} = 0.11380$, then we find that $F = 50.5534$, which allows us to reject $H_0{:}\beta' = 0$ and conclude that there *is* a relationship

TABLE 9.7
Beer Consumption Data Divided into Five Groups

Families	Calculations		$s^2_{Y \cdot X}$	$\overline{X}$	$s_{Y \cdot X}/\overline{X}$
1–3	$\sum Y_i = 7.90$	$\sum X_iY_i = 7.91$	0.2017	1.0000	0.4491
	$\sum Y_i^2 = 21.01$	$S_{YY} = 0.2067$			
	$\sum X_i = 3.00$	$S_{XX} = 0.0200$			
	$\sum X_i^2 = 3.02$	$S_{XY} = 0.0100$			
4–6	$\sum Y_i = 8.7$	$\sum X_iY_i = 11.00$	0.3203	1.2667	0.4468
	$\sum Y_i^2 = 25.61$	$S_{YY} = 0.3800$			
	$\sum X_i = 3.80$	$S_{XX} = 0.0067$			
	$\sum X_i^2 = 4.82$	$S_{XY} = -0.0200$			
7–9	$\sum Y_i = 9.20$	$\sum X_iY_i = 16.32$	0.7120	1.7667	0.4776
	$\sum Y_i^2 = 28.96$	$S_{YY} = 0.7467$			
	$\sum X_i = 5.30$	$S_{XX} = 0.1267$			
	$\sum X_i^2 = 9.49$	$S_{XY} = 0.0667$			
10–12	$\sum Y_i = 9.1$	$\sum X_iY_i = 23.18$	1.0155	2.5000	0.4031
	$\sum Y_i^2 = 29.33$	$S_{YY} = 1.7267$			
	$\sum X_i = 7.50$	$S_{XX} = 0.2600$			
	$\sum X_i^2 = 19.01$	$S_{XY} = 0.4300$			
13–15	$\sum Y_i = 9.80$	$\sum X_iY_i = 31.00$	2.0057	3.2000	0.4426
	$\sum Y_i^2 = 37.24$	$S_{YY} = 5.2267$			
	$\sum X_i = 9.40$	$S_{XX} = 0.0267$			
	$\sum X_i^2 = 29.48$	$S_{XY} = 0.2933$			

TABLE 9.8
Beer Consumption Data Transformed to Give Uniform Scatter

Family	$Y' = \dfrac{Y}{X}$	$X' = \dfrac{1}{X}$
1	2.6667	1.1111
2	3.0000	1.0000
3	2.2727	0.9091
4	2.5833	0.8333
5	2.4615	0.7692
6	1.8462	0.7692
7	2.1333	0.6667
8	1.3333	0.5556
9	1.8000	0.5000
10	1.2381	0.4762
11	0.9231	0.3846
12	1.4643	0.3571
13	0.6000	0.3333
14	0.9375	0.3125
15	1.5625	0.3125
$\sum (\cdot)$	26.8225	9.2904
$\sum (\cdot)^2$	55.1956	6.7507
$\sum X'Y'$	19.0072	

$S_{Y'Y'} = 7.2325$ $S_{X'X'} = 0.9966$ $S_{X'Y'} = 2.3944$

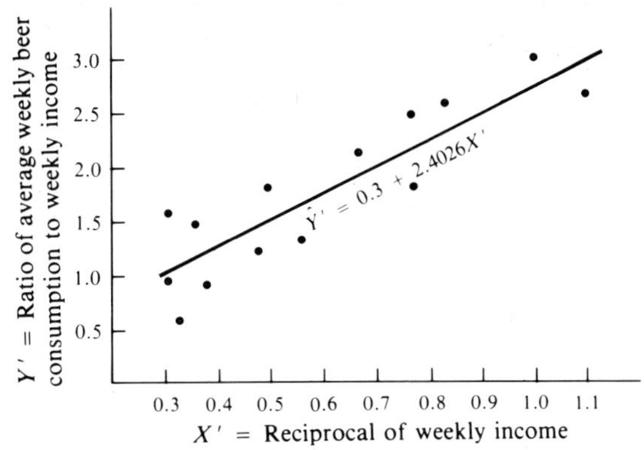

FIGURE 9.5 Scatter Diagram and Regression Line for Transformed Beer Consumption Data

between Y' and X'. The standard error of b' can be shown to be 0.33795. While this standard error is larger than the one found for the untransformed data (there, $s_b^2 = 0.25162$), it is *smaller* in relation to the size of the regression coefficient ($b' = 2.4$ for transformed data, as compared to $b = 0.31$).

We now have an equation relating the ratio of consumption to income to the reciprocal of income. How can this equation be used to estimate consumption from income? Suppose a family has weekly income of \$200. Then, by using the original equation, we estimate

$$\hat{Y} = 2.3873 + 0.3066(2) = 3.0005,$$

although this would not be a valid operation, since we were unable to demonstrate that Y was related to X. Equivalently, since there is a significant relationship between Y' and X', we could transform $X = 2$ to $X' = 1/2 = 0.5$ and use the regression equation for the transformed data:

$$\hat{Y}' = 0.3 + 2.4028(0.5) = 1.5014.$$

Then from $Y' = Y/X$, we have $Y'X = Y$; and $\hat{Y} = 2(1.5014) = 3.0028$. An easier way, however, is to interchange a' and b' in the equation for the transformed data:

$$\hat{Y} = 2.4028 + 0.3X,$$

so that for $X = 2$,

$$\hat{Y} = 2.4028 + 0.3(2) = 3.0028.$$

This latter method requires much less calculation.

EXERCISES

9.1 In order to see if its secretaries were being paid according to their abilities, a large company sampled 20 of its secretaries, obtaining an efficiency rating and monthly salary for each. The ratings, on a scale from 1.0 to 5.0, were obtained from input from supervisors on various attributes. The results are shown in Table 9.9.

TABLE 9.9
Secretaries Data

Employee	Rating	Salary ($)	Employee	Rating	Salary ($)
Mrs. Adams	3.9	790	Mrs. Irving	4.4	770
Ms. Bright	3.6	740	Mr. Jacobs	4.8	780
Mrs. Collins	3.1	750	Miss Knight	2.3	700
Miss Dale	4.9	820	Ms. Lawley	2.5	730
Ms. Evans	3.1	720	Ms. Moore	3.6	850
Mr. Fisher	2.4	700	Mrs. Owen	2.5	690
Mrs. Foster	4.9	790	Miss Paul	3.0	720
Miss Greene	3.6	770	Mrs. Priestly	2.3	720
Miss Hill	4.4	800	Ms. Williams	3.2	730
Mrs. Hustings	2.2	720	Mrs. Wilson	4.0	750

a. Fit a simple linear regression equation to these data, using rating to predict salary.
b. Plot the residuals from the equation obtained in (a).
c. Office gossip has it that Ms. Moore enjoys preferential treatment because she is a relative of a senior executive of the company. From examination of the plot of residuals, does Ms. Moore's salary appear to be out of line?
d. Form a 99% confidence interval to see if Ms. Moore's salary represents an outlier observation.
e. Use the Dixon test to determine if Ms. Moore's salary is out of line.

9.2 Refer to the Textbook Data of Exercise 1.10.
a. Fit a simple linear regression equation and plot the residuals.
b. Divide the data into two groups: in one group, use the seven textbooks with the smallest number of pages; in the other, use the eight longest books. Calculate $s^2_{Y \cdot X}$ for each group separately, and test for homogeneous scatter.
c. Transform the data by $Y' = Y/X$ and $X' = 1/X$, and fit a straight line to the transformed data. Plot the residuals and examine for uniform scatter.
d. Using the same groups, test for uniform scatter in the transformed data.
e. Test for a significant linear relationship, using transformed data.

9.3 Plot the residuals from fitting a straight line to the Appraisal Data of Exercise 4.4. Examine for uniform scatter.

9.4 Plot the residuals from fitting a straight line to the Reservations Data of Exercise 5.16. Examine for uniform scatter.

9.3 THE ASSUMPTION OF THE SIMPLE LINEAR MODEL

We fit a simple linear regression equation to a set of data because we have hypothesized the model $Y = \alpha + \beta X + \varepsilon$, meaning that we think that Y

should be linearly related to X, except for error. Just because we hypothesize this model and can fit a simple linear regression line to any set of data does not mean that we have done the correct thing. In the last section we noted that outliers can suggest that a multiple-predictor model might be more appropriate. Plots of residuals can reveal omitted predictors in more ways than one; and even if we use only one predictor, we should check to see that the relationship is truly linear, not curvilinear. In this section we shall investigate ways to detect when the simple linear model is appropriate.

SHOULD A MULTIPLE REGRESSION MODEL BE USED?

We have seen that outliers in the data can indicate that an important predictor, or several important predictors, have not been considered. While Chapter 11 and those following will deal with multiple regression, we will briefly consider here some ways to spot the omission of predictors.

What would you think if you saw a plot like the one in Figure 9.6? Perhaps one ought to take into account that most of the state is rural, with a few large metropolitan areas.

FIGURE 9.6 Plot of Hypothetical Data and Regression Equation Relating Newspaper Circulation to Number of Registered Voters by County in a State

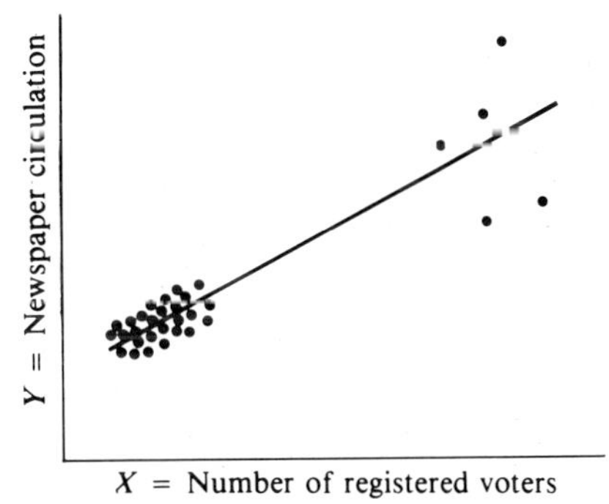

Sometimes one can identify possible sources of variation that have not been taken into account. For example, suppose Figure 9.7 shows the relationship between age and earnings for a group of people. At first glance, all seems to be in order. But what if the dots represent males and the circles females? In that case, the sex variable should be taken into account if one's objective is to make accurate estimates of earnings.

TIME DEPENDENCIES

Sometimes data cannot be collected all at once but must be obtained in a time sequence that might affect the response. For example, suppose a gymnastics

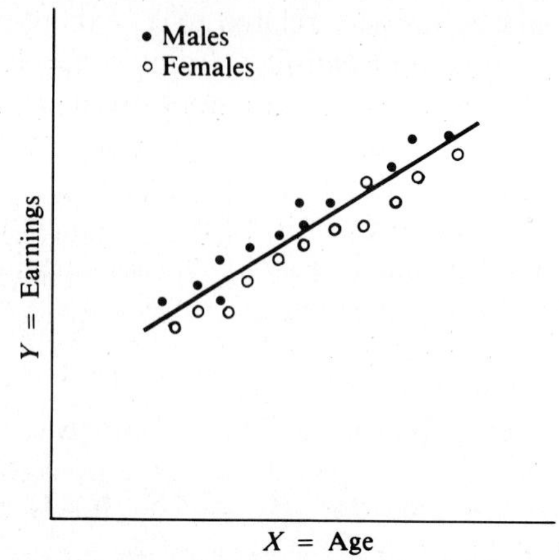

FIGURE 9.7
Scatter Diagram and Regression Line Relating Earnings to Age

coach is interested in relating the performance of one of his gymnasts to the number of hours sleep on the previous night. Suppose data are collected over several months, and a plot of performance measure by hours sleep shows a well-behaved linear regression. We are aware, however, that performance is related to total number of hours practice, or learning, so that it is quite likely that performance will improve over time, regardless of the number of hours sleep. If this is the case, then if the residuals are plotted in time order, a picture such as Figure 9.8 might well appear. In this case, time should be included as a predictor. How this can be done will be discussed in Section 9.4 and Chapters 11 through 16.

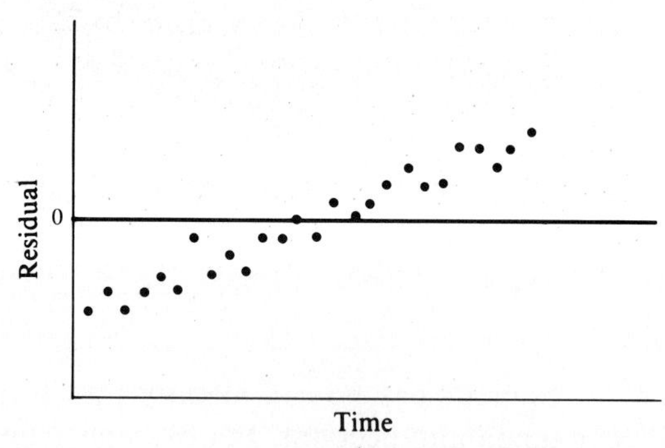

FIGURE 9.8
Residuals Plotted in Time Order, Showing a Practice Effect

SHOULD A CURVILINEAR MODEL BE USED?

Plotting the data in a scatter diagram is a first and invaluable step in determining whether a linear or a curvilinear model should be adopted. A pronounced curvature often shows up in the plot of data points. In other cases, the curvature may be more subtle or may be partially obscured by the choice of scales on the horizontal and vertical axes. One can always fit any set of data more closely with a more complicated curvilinear question, but this does not mean that a curved regression line is to be preferred to a simple linear one. We are generally interested in more than an accurate fit to the data at hand. We also want to describe the underlying relationship and interpret the meaning of the regression equation. Thus, the simpler description is usually the preferred one.

We chose a simple linear model for the house size-price example because our original plot of the data indicated that the increase in price with size was relatively constant. It could be argued, however, that there appears to be a leveling-off of the prices for the larger houses in the sample; price increases with size, but by decreasing amounts. If this is indeed the case, then we should have adopted a curvilinear model. We should examine the data more closely to see if there is significant curvature. We realize that this apparent leveling-off of prices could be just a function of the particular sample we happened to draw.

Examining Figure 2.8, in which the regression line was plotted on the scatter diagram for the house size-price data, we see that for houses between about 800 and 2,200 sq ft in size, the associated prices appear to vary reasonably randomly around the line. However, for houses larger than 2,200 sq ft, we see first one group of prices above the line and then another group of prices below the line. Such a pattern of variation could indicate systematic variation rather than random error. It might be a clue that the wrong model has been used. For example, suppose we have a definitely curvilinear relationship, as illustrated in Figure 9.9, but fit a straight-line regression equation. Here we see a very definite pattern to the scatter of the points around the line: at first there is a group of points all above the line, then a group all below the line, and then another group all above the line.

A systematic pattern often shows up better in a plot of residuals than in the scatter diagram. The plot of residuals from Figure 9.9 would first show a group of positive residuals, then a group of negative residuals, and then another

FIGURE 9.9
A Straight Line Fitted to Curvilinear Data

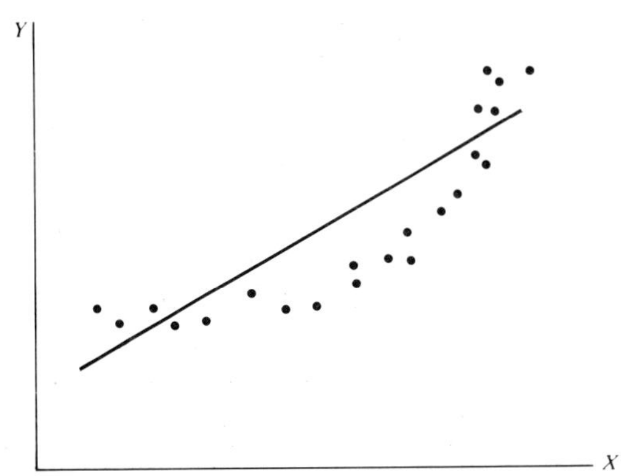

group of positive residuals. Looking at the plot of residuals from the house size-price data, we see for the larger houses first a group of positive residuals and then a group of negative residuals. Perhaps this is evidence that a curvilinear equation should have been fitted. We now need to test for significant curvature in the data. If we find it, we then need to decide on the kind of curvilinear model to adopt; or perhaps we can make a transformation on the data to achieve linearity.

LACK-OF-FIT TESTS

When we have two or more Y-values corresponding to at least one X-value, we can test to see if the linear model is appropriate. With a designed experiment, it is relatively easy to take enough observations to have *replications* in the data. In a survey, you may or may not observe two items with the same X-value but differing Y-values. The objective is to break the error sum of squares down into two pieces, one due to **pure error**, and one due to **lack of fit.** Without at least one replication, we cannot get a measure of pure error, and thus cannot separate it from lack of fit. In our example, both houses 5 and 19 have 2,200 sq ft, and both 3 and 15 have 1,700 sq ft. The differences in prices of houses 5 and 19 and of houses 3 and 15 are due to (1) omitted predictors, and (2) random variation. They are *not* due to failure of the simple linear model. Thus, we will be able to break out from the sum of squares for error a piece measuring just pure error.

Pure Error

Lack of Fit

To do this, we want to measure the variance in prices of houses the same size. Consider houses 5 and 19, whose prices are 35 and 37 thousand dollars, respectively. We can calculate the variance of these two prices by finding first their mean and then the average squared deviation of the prices from their mean. Call this mean $\overline{Y}_{2.2}$. Then

$$\overline{Y}_{2.2} = \frac{35 + 37}{2} = 36;$$

and their variance will be

$$s_{2.2}^2 = \frac{(35 - 36)^2 + (37 - 36)^2}{2 - 1} = 2.$$

We can do a similar thing for the prices of houses 3 and 15, which are 27 and 30 thousand dollars, respectively:

$$\overline{Y}_{1.7} = \frac{27 + 30}{2} = 28.5,$$

$$s_{1.7}^2 = \frac{(27 - 28.5)^2 + (30 - 28.5)^2}{2 - 1} = 4.5.$$

The assumption of homogeneous variances becomes quite important now, because we are going to *pool* these two variances into a single estimate of the pure-error variance. Unless we can justify the assumption of homoscedasticity, we cannot put these two sample variances together; they would not be estimates of the same quantity.

It is actually easier in larger examples to pool sums of squares instead of variances. (In our case, it makes no difference, because the denominators are both equal to 1.) To pool the sums of squares, simply add up the numerators of $s^2_{2.2}$ and $s^2_{1.7}$, and call this sum *SS(pure error)*:

$$SS(pure\ error) = 2 + 4.5 = 6.5.$$

Of course, had we replications at other X-values, we could have simply calculated s^2 for those cases and added them in, too. The degrees of freedom associated with this sum of squares is n minus the number of *different* X- values in the sample. In our example, there are 18 different values of X, so our degrees of freedom are $20 - 18 = 2$ for pure error.

The sum of squares for lack of fit, *SS(lack of fit)*, is the difference between the error sum of squares and the sum of squares for pure error:

$$SS(lack\ of\ fit) = SS(E) - SS(pure\ error).$$

Its degrees of freedom will be the number of different X-values minus 2, since the degrees of freedom for lack of fit and the degrees of freedom for pure error must sum to $n - 2$. In our example, degrees of freedom for lack of fit are $18 - 2 = 16$.

Our analysis of variance table for the house size-price problem now looks like Table 9.10. We can now use an F-test to see whether or not the linear model is appropriate; that is, we test

$$H_0: Y = \alpha + \beta X + \varepsilon$$

against

$$H_1: Y \neq \alpha + \beta X + \varepsilon.$$

TABLE 9.10
ANOVA Showing *SS(E)* Broken into Lack of Fit and Pure Error

Source	*SS*	*df*	*MS*
Regression	2255.8952	1	
Error	117.1048	18	
Lack of Fit	110.6048	16	6.9128
Pure	6.5	2	3.250
Total	2373	19	

If we form an F-test statistic

$$F = \frac{MS(lack\ of\ fit)}{MS(pure\ error)},$$

with 16 and 2 degrees of freedom, then large values of F will indicate that more error is coming from lack of fit than from random variability and that the simple linear model is not appropriate. In our example,

$$F = \frac{6.9128}{3.250} = 2.127.$$

Comparing to the critical value $F_{16,2,0.05} \approx 19.13$, we are unable to reject

$$H_0: Y = \alpha + \beta X + \varepsilon.$$

We thus conclude that the slight curvature apparent from the scatter diagram is most probably due only to sampling variability and not to any real leveling-off effect in the population.

A caution is in order: note that we had repeated observations at only two values of X. This is not much information on which to base an estimate of the pure-error variance. It is quite possible that we have overestimated the variance due to pure error. This lack-of-fit test is more trustworthy when many observations are repeated. Even though the foregoing test does not indicate a significant lack of fit of the data to a simple linear model, we shall see in Chapter 10 a surprising result that may seem to contradict this conclusion.

This same test can be used to decide if a multiple regression model or a curvilinear model is appropriate. Note also that a rejection of $H_0: Y = \alpha + \beta X + \varepsilon$ only tells you that the simple linear model is not appropriate. It does not tell you which model you should use. If H_0 is rejected, then, you must go back to the scatter diagram or plot of residuals in order to get an idea of what model should be used. The new model can be tested out just as this simple linear model was. On the other hand, failure to conclude that the model $Y = \alpha + \beta X + \varepsilon$ does not fit the data does not necessarily imply that this is THE model. In our example, we found that the simple linear model fits the data reasonably well, and this is all we found.

This test for lack of fit can, we have said, be used only when there are repeated observations for at least one value of X. Without these, one must rely on examination of the scatter diagram or of the plot of residuals in determining the appropriateness of the proposed model.

EXERCISES

9.5 Refer to the Deprivation Data of Exercise 2.21.

a. Fit a simple linear regression equation to the data. Test for the significance of the regression.

b. Plot the residuals from the simple linear equation, and notice any pattern that indicates that the simple linear model is not appropriate.

c. Test for lack of fit to the simple linear model.

9.6 Suppose that a forester needs a quick way to estimate the volume of a stand of trees, so that when they are harvested she can know approximately how many board-feet the stand will yield. The forester decides to measure the circumferences at 24 in. above the ground of a sample of 25 trees and then record the volume for each tree when it is harvested. She hopes to be able to use circumference as an indicator of volume. Her data are shown in Table 9.11.

a. Fit a straight line to the data, and plot the residuals. Is there any pattern in the residuals that indicates a simple linear model is not appropriate?

b. Test for lack of fit. Comment on the effectiveness of the lack-of-fit test for detecting an inappropriate model.

TABLE 9.11
Trees Data

Tree	Circumference (ft)	Volume (cu ft)	Tree	Circumference (ft)	Volume (cu ft)
1	4.7	114	14	5.8	134
2	2.7	96	15	3.9	106
3	3.5	100	16	2.8	102
4	2.5	110	17	5.8	138
5	4.0	112	18	3.5	102
6	3.6	108	19	3.6	102
7	2.7	100	20	2.5	100
8	5.5	132	21	3.5	106
9	5.0	118	22	4.7	118
10	2.6	100	23	3.6	106
11	2.5	94	24	2.5	98
12	6.0	144	25	5.8	132
13	3.8	110			

9.7 In an evaluation of a typist, a recent manuscript was examined for the number of errors per page. The number of words per page were to be used as the indicator of the number of errors. Table 9.12 shows the numbers of errors recorded in the 23-page manuscript.

a. Fit a simple linear regression equation to these data. Plot the data and the regression equation.

b. The pages of the manuscript were typed in order and at one sitting. Those evaluating the study were also interested in any fatigue factor that might be

TABLE 9.12
Typist Data

Page	Number of Words	Number of Errors	Page	Number of Words	Number of Errors
1	228	10	13	185	4
2	230	11	14	209	9
3	221	9	15	193	6
4	208	5	16	209	8
5	222	8	17	207	8
6	197	4	18	208	9
7	180	1	19	229	13
8	229	10	20	206	8
9	212	7	21	208	9
10	190	3	22	207	7
11	209	6	23	185	5
12	207	6			

present. Calculate the residuals, and plot them in time order. Is there any effect of fatigue on the number of errors? Should it be taken into account in order to improve predictions of number of errors from number of words per page?

9.8 In a study to relate IQ to performance on a standardized test, 26 grade-school students were randomly selected from two classes at the same school. Each student was given the same IQ test and was taught from the same textbook. All students took the standardized test at the same time. The results are given in Table 9.13.

a. Plot the data, indicating whether each student was taught by teacher A or teacher B.

b. Fit and plot a simple linear regression equation.

c. Do you think that the teacher should be taken into account when trying to predict score on the standardized test? Explain.

TABLE 9.13
IQ Data

Student	IQ (X)	Score (Y)	Teacher	Student	IQ (X)	Score (Y)	Teacher
1	114	90	A	14	109	86	B
2	118	90	B	15	96	80	A
3	110	84	B	16	130	98	A
4	109	88	A	17	123	96	A
5	94	78	A	18	108	82	B
6	113	90	B	19	106	86	B
7	114	88	B	20	113	92	A
8	110	88	B	21	125	94	A
9	107	80	B	22	124	96	A
10	106	82	B	23	111	86	B
11	106	84	B	24	120	94	A
12	109	84	B	25	104	82	A
13	109	90	A	26	117	92	A

9.9 In the Class Size Data of Exercise 2.17 it was apparent that the larger the class size, the lower the rating of the instructor by the students. Notice, however, that 18 of the 22 classes in the study contained between 10 and 57 students; then the next largest class sizes were 91, 113, 127, and 146. There is a considerable gap between 57 and 91. Does this make any difference to the analysis? Explain. Perform additional analyses, if you think they are appropriate.

9.4 INDEPENDENCE OF ERRORS

Every item in our sample is a unique individual, and thus each of our measurements contains variation attributable to just that individual. Provided our model $Y = \alpha + \beta X + \varepsilon$ is correct in that the relationship is truly linear and no important predictors have been omitted, then ε represents the effect of only these individual differences. Recall that we could detail the model as

$$Y_1 = \alpha + \beta X_1 + \varepsilon_1$$
$$Y_2 = \alpha + \beta X_2 + \varepsilon_2$$
$$\vdots$$
$$Y_N = \alpha + \beta X_N + \varepsilon_N.$$

Then ε_1 is attributable to the individuality of the first item in the population, ε_2 is attributable to the uniqueness of the second item, and so on. Since each error term is attributable only to the individual, the size of any error ε_i should have no effect on the size of any other error ε_j. That is, the error terms are independent, provided that the model is correct. All our analyses of the regression problem are based on the assumption that the errors are independent; obviously, we must check to see if this assumption holds. If not, we must find out why, and take remedial measures if possible. If the errors are not independent, the model is usually incorrect; perhaps some predictors are omitted, or perhaps a curvilinear model should have been used.

We estimate the ε_i's by $e_i = Y_i - \hat{Y}_i$, $i = 1, 2, \ldots, n$ found by the regression equation. Now, these estimates will *not* be independent, because all n observations were used to find the regression equation and thus to find $\hat{Y}_i$; so the value of e_i is dependent upon all observations in the sample and not just the ith one. Thus, even if the model is correct, there will be some correlation among the residuals, but this will be small if the sample size is reasonably large.

In most of our examples so far we have no reason to doubt that the error terms are independent. For example, in our house size-price data, why would the individual variability in one house affect that in another? The 20 houses form a random sample chosen from a large population of houses, and the fact that the sample was random says that the choosing of one house has no effect on whether or not any other house will be chosen.

SERIAL CORRELATION

Serial Correlation
Autocorrelation

When the data are collected over time, however, it is important to see if any time effect is incorporated in the residuals. Such time effects could be growth, decay, practice, or boredom effects, change in conditions over time, and so on. Correlation over time, called **serial correlation,** or **autocorrelation,** means that the level of response in one time period affects the level of response in the next period. Thus, the residual in one time period is related to the residual in the next. In order to see if any serial correlation among the errors is present, take the residuals in pairs and find their correlation coefficient. For example, suppose we have the residuals shown in Table 9.14. We see that a group of negative residuals is followed by a group of positive ones. When residuals

TABLE 9.14 Residuals per Time Period

Time Period (t)	ε_t
1	−2.6
2	−2.1
3	−2.6
4	−1.3
5	0.6
6	−0.1
7	0.8
8	2.4
9	2.1
10	2.8

Positive Serial Correlation

Negative Serial Correlation

with the same sign tend to occur in clusters, this indicates **positive serial correlation.** This often occurs in economic time series data in which the value of the dependent variable is affected by periods of expansion and contraction in the economy. Values of the variable, and thus of the residuals, may all tend to be large during times of economic expansion but small during periods of contraction. **Negative serial correlation** is less often seen. It is indicated by residuals alternating in sign pretty much one after the other, so that, for example, a positive residual at one time period would be followed by a negative one in the next time period, and so on. Negative serial correlation is most often seen in time series data in which a seasonal pattern is present. Many economic variables, such as retail sales and unemployment rates, are affected by climatic conditions or social customs (such as Christmas always occurring on December 25) that cause the value of the variable to be higher than average during one month or one quarter but lower than average in the next. Consider residential consumption of electric power, for example. In most of the United States consumption is high in the winter because electricity is used for heating. It falls off sharply during the spring but then peaks in the summer because of the use of air conditioning for cooling. Consumption falls again in the fall, only to rise again the following winter. Since weather conditions and social customs are relatively stable from year to year, several years' worth of seasonally affected data might show a regular, saw-toothed pattern. This alternating of above-average and below-average values will result in residuals that alternate in sign.

Returning to the example in Table 9.14, we suspect that positive serial correlation is present. In order to measure this serial correlation, pair up the residuals as shown in Table 9.15. Then calculate the serial correlation using

$$r_{t,t-1} = \frac{\sum_{t=2}^{n} e_t e_{t-1}}{\sum_{t=2}^{n} e_{t-1}^2} = \frac{26.22}{30.8} = 0.8513,$$

where $r_{t,t-1}$ denotes the correlation between the residual at time t and the residual at time $t-1$.

TABLE 9.15 Calculation of Serial Correlation

Time Period (t)	Residual (e_t)	Residual for Preceding Period (e_{t-1})	$e_t e_{t-1}$
1	−2.6		
2	−2.1	−2.6	5.46
3	−2.6	−2.1	5.46
4	−1.3	−2.6	3.38
5	0.6	−1.3	−0.78
6	−0.1	0.6	−0.06
7	0.8	−0.1	−0.08
8	2.4	0.8	1.92
9	2.1	2.4	5.04
10	2.8	2.1	5.88
Sum			26.22
Sum of Squares		30.8	

A TEST FOR SERIAL CORRELATION

As we noted, there will be some correlation among residuals whenever the data are used to fit a regression equation. For a given set of data, if we suspect that an effect of time is present, we can calculate the serial correlation coefficient and then test it to see whether it is large enough (in absolute value) to be attributable to an effect of time. For this test, we use a special statistic called the **Durbin-Watson *d*-statistic.** Its form depends upon whether one wishes to test for a significant positive serial correlation or a negative one. Let $\rho_{t,t-1}$ denote the true serial correlation. To test

Durbin-Watson *d*-Statistic

$$H_0: \rho_{t,t-1} = 0$$
$$H_1: \rho_{t,t-1} > 0,$$

use

$$d = \frac{\sum_{t=2}^{n} (e_t - e_{t-1})^2}{\sum_{t=1}^{n} e_t^2}$$

To test

$$H_0: \rho_{t,t-1} = 0$$
$$H_1: \rho_{t,t-1} < 0,$$

use the test statistic

$$d^- = 4 - d.$$

Refer to the tables of the Durbin-Watson statistic in Appendix K. There are separate tables for $\alpha = 0.05$ and $\alpha = 0.01$. Also, the tables are subdivided into parts for different numbers of parameters in the model, p. The sample size begins at $n = 15$; otherwise the serial correlation would not be based on enough information to be trustworthy. Under each value of p are listed two values, d_L and d_U, denoting lower and upper limits, respectively, on the value of d. To test the alternative $H_1:\rho_{t,t-1} > 0$, the procedure is

$$\text{Reject } H_0 \quad \text{if } d < d_L,$$

$$\text{Accept } H_0 \quad \text{if } d > d_U;$$

$$\text{The test is inconclusive} \quad \text{if } d_L \le d \le d_U.$$

The procedure for testing the alternative $H_1 : \rho_{t,t-1} < 0$ is

$$\text{Reject } H_0 \quad \text{if } d^- < d_L,$$

$$\text{Accept } H_0 \quad \text{if } d^- > d_U;$$

$$\text{The test is inconclusive} \quad \text{if } d_L \le d^- \le d_U.$$

Some of the rationale behind this test is based on the fact that d has a range from 0 to 4. Values of d around 2 usually indicate that no positive serial correlation is present. (Note that the tabulated values of d_U are generally less than 2.) To test for a negative serial correlation, $d^- = 4 - d$ will convert the d-value based on data with negative serial correlation to a frame of reference comparable to the value of d for positively serially correlated data.

TESTING FOR SERIAL CORRELATION: AN EXAMPLE

Suppose that a certain department store tried out a television advertising campaign in order to see what effect, if any, television advertising had on its sales. Data on the number of minutes' advertising per week and the sales in the store were recorded for 20 successive weeks, beginning a week ahead of the advertising. Hypothetical data are given in Table 9.16. Since there could very easily be a cumulative effect of advertising over time, it is necessary to examine the residuals for serial correlation. Calculation of the serial correlation coefficient and of the Durbin-Watson statistic are shown in Table 9.17. Since we fitted a simple linear regression equation, $p = 2$. Recall that $n = 20$. If $\alpha = 0.05$, we will reject

$$H_0:\rho_{t,t-1} = 0$$

in favor of

$$H_1:\rho_{t,t-1} > 0$$

if $d < d_L = 1.20$.

TABLE 9.16 Data and Calculations for Television Advertising and Sales Example

Week (t)	Minutes of TV Advertising (X)	Weekly Sales (Y) (thousands of dollars)	Residual
1	0	8.8	−0.5363
2	5	9.0	−0.9553
3	9	9.2	−1.2505
4	4	10.0	0.1685
5	6	9.6	−0.4791
6	13	10.4	−0.5457
7	18	10.8	−0.7647
8	5	9.8	−0.1553
9	4	9.4	−0.4315
10	9	10.2	−0.2505
11	17	12.0	0.5591
12	13	11.0	0.0543
13	7	10.8	0.5971
14	9	10.8	0.3495
15	6	10.6	0.5209
16	6	11.0	0.9209
17	9	11.2	0.7495
18	7	11.0	0.7971
19	5	10.4	0.4447
20	15	11.4	0.2067
$\sum (\cdot)$	167	207.4	−0.0006
$\sum (\cdot)^2$	1813	2164.68	
$\sum XY$	1783.6		

$S_{XX} = 418.55 \quad S_{XY} = 51.81 \quad S_{YY} = 13.942$

$b = 0.1238 \quad a = 9.3363 \quad \hat{Y} = 9.3363 + 0.1238X$

Since $d = 0.6504 < 1.20$, we conclude that we are more than 95 percent sure that there is positive serial correlation in the data. This means that sales have increased over time, as well as with the amount spent on television advertising.

ADJUSTING FOR SERIAL CORRELATION

Regression on First Differences

If we want to use only television advertising to estimate sales, we must adjust the data so as to eliminate the time effect. This way we will not violate the assumption of independent error terms. One way is to perform a **regression on first differences.** A first difference is the difference between a value in one time period and the value in the preceding time period. That is, if we let X' and Y' denote the first differences for the X- and Y-values, respectively, then

$$X'_t = X_t - X_{t-1},$$

$$Y'_t = Y_t - Y_{t-1}.$$

TABLE 9.17 Calculation of Serial Correlation Coefficient and Durbin-Watson Statistic

Week (t)	e_t	e_{t-1}	$e_t e_{t-1}$	$e_t - \varepsilon_{t-1}$
1	−0.5363	—	—	—
2	−0.9553	−0.5363	0.5123	−0.4190
3	−1.2505	−0.9553	1.1946	−0.2952
4	0.1685	−1.2505	−0.2107	1.4190
5	−0.4791	0.1685	−0.0807	−0.6476
6	−0.5457	−0.4791	0.2614	−0.0666
7	−0.7647	−0.5457	0.4173	−0.2190
8	−0.1553	−0.7647	0.1188	0.6094
9	−0.4315	−0.1553	0.0670	−0.2762
10	−0.2505	−0.4315	0.1081	0.1810
11	0.5591	−0.2505	−0.1401	0.8096
12	0.0543	0.5591	0.0304	−0.5048
13	0.5971	0.0543	0.0324	0.5428
14	0.3495	0.5971	0.2087	−0.2476
15	0.5209	0.3495	0.1821	0.1714
16	0.9209	0.5209	0.4797	0.4000
17	0.7495	0.9209	0.6902	−0.1714
18	0.7971	0.7495	0.5974	0.0476
19	0.4447	0.7971	0.3545	−0.3524
20	0.2067	0.4447	0.0919	−0.2380
$\sum (\cdot)$			4.9153	
$\sum (\cdot)^2$	7.5289	7.4762		4.8967

$$r_{t,t-1} = \frac{4.9153}{7.4762} = 0.6575 \qquad d = \frac{4.8967}{7.5289} = 0.6504$$

These first differences should take the time effect into account if the value in one time period is affected by the value in the preceding time period.

First-Order Autoregressive Model

To see that this is so, consider the **first-order autoregressive model**

$$Y_t = \alpha + \beta X_t + \varepsilon_t,$$

where

$$\varepsilon_t = \varepsilon_{t-1} + \mu_t.$$

Here, we see that the error term at time t is a function of the error at the previous time period, plus some random fluctuation, μ_t. Consider now the first differences, $Y_t - Y_{t-1}$:

$$Y_t - Y_{t-1} = \alpha + \beta X_t + \varepsilon_t - (\alpha + \beta X_{t-1} + \varepsilon_{t-1})$$
$$= \beta(X_t - X_{t-1}) + (\varepsilon_t - \varepsilon_{t-1}).$$

But since $\varepsilon_t = \varepsilon_{t-1} + \mu_t$,

$$\varepsilon_t - \varepsilon_{t-1} = \varepsilon_{t-1} + \mu_t - \varepsilon_{t-1} = \mu_t,$$

so that

$$Y_t - Y_{t-1} = \beta(X_t - X_{t-1}) + \mu_t,$$

or

$$Y_t' = \beta X_t' + \mu_t,$$

and the residuals in the first-difference model are purely random.

Table 9.18 shows a regression on first differences for the television advertising and sales example. (The symbol $\sum (\cdot)(\cdot)$ denotes a cross-product in Table 9.18.) Note that the first-differences model contains no intercept term, so that the slope of our fitted regression equation is

$$b' = \frac{\sum X_t' Y_t'}{\sum (X_t')^2} = \frac{54.4}{585} = 0.093.$$

Thus, if in week 21, the store plans to purchase $X_t = 20$ minutes of advertising on television, how would we estimate sales in week 21? In week 20, $X_{t-1} = 15$ minutes. Thus, $X_t' = X_t - X_{t-1} = 20 - 15 = 5$ more minutes will be purchased in week 21 than in week 20. We estimate the *growth* in sales from week 20 to week 21 to be $\hat{Y}_t' = 0.093(5) = 0.465$ (= \$465).

TABLE 9.18 Regression on First Differences for Television Advertising and Sales Example

Week (t)	X_t	X_{t-1}	Y_t	Y_{t-1}		$X_t' = X_t - X_{t-1}$	$Y_t' = Y_t - Y_{t-1}$
1	0	—	8.8	—		—	—
2	5	0	9.0	8.8		5	0.2
3	9	5	9.2	9.0		4	0.2
4	4	9	10.0	9.2		−5	0.8
5	6	4	9.6	10.0		2	−0.4
6	13	6	10.4	9.6		7	0.8
7	18	13	10.8	10.4		5	0.4
8	5	18	9.8	10.8		−13	−1.0
9	4	5	9.4	9.8		−1	−0.4
10	9	4	10.2	9.4		5	0.8
11	17	9	12.0	10.2		8	1.8
12	13	17	11.0	12.0		−4	−1.0
13	7	13	10.8	11.0		−6	−0.2
14	9	7	10.8	10.8		2	0.0
15	6	9	10.6	10.8		−3	−0.2
16	6	6	11.0	10.6		0	0.4
17	9	6	11.2	11.0		3	0.2
18	7	9	11.0	11.2		−2	−0.2
19	5	7	10.4	11.0		−2	−0.6
20	15	5	11.4	10.4		10	1.0
					$\sum (\cdot)$	15	2.6
					$\sum (\cdot)^2$	585	9.4
					$\sum (\cdot)(\cdot)$	54.4	

$\hat{Y}_t' = 0.093X_t'$

In week 20, sales were \$11,400, so we estimate actual sales in week 21 to be \$11,400 + \$465 = \$11,865, since

$$\hat{Y}_t' = 0.465 = \hat{Y}_t - Y_{t-1} = \hat{Y}_t - 11.4$$

so that $0.465 + 11.4 = \hat{Y}_t$.

Residuals from fitting this model are shown in Table 9.19. Fitting the first-difference model assumes that residuals are correlated from one time period to the next. If the interdependence extends back beyond the previous time period, however, then first differences may not be effective in removing the time dependency. Thus, when transforming the data into first differences, one should make sure that the residuals resulting from the transformed variables are independent. If so, the transformation has effectively eliminated the time dependency; but if not, some other transformation should be made on the data (a second-difference transformation, perhaps). It is also possible for first differences to overcorrect. Then, if the original data show positive serial correlation, the transformed data show negative serial correlation, and vice versa. This is what happened in our example, as can be seen from the calculations in Table 9.19. Since $d^- = 1.13 < 1.20 = d$, we see a significant

TABLE 9.19 Residual Analysis for Television Advertising and Sales Example

	ε_t	ε_{t-1}	$\varepsilon_t - \varepsilon_{t-1}$
	—	—	—
	−0.265	—	—
	−0.172	−0.265	0.093
	1.265	−0.172	1.437
	−0.586	1.265	−1.851
	0.149	−0.586	0.735
	−0.065	0.149	−0.214
	0.209	−0.065	0.274
	−0.307	0.209	−0.516
	0.335	−0.307	0.642
	1.056	0.335	0.721
	−0.628	1.056	−1.684
	0.358	−0.628	0.986
	−0.186	0.358	−0.544
	0.079	−0.186	0.265
	0.400	0.079	0.321
	−0.079	0.400	−0.479
	−0.014	−0.079	0.065
	−0.414	−0.014	−0.400
	0.070	−0.414	0.484
$\sum(\cdot)$	4.27104	4.336365	12.264353
$\sum(\cdot)(\cdot)$	−1.828474		

$$d = \frac{12.2644}{4.2710} = 2.8716$$

$$r_{t,t-1} = \frac{-1.8285}{4.3364} = -0.4217$$

$$d^- = 1.1284$$

negative serial correlation in the transformed data. Thus, the first-difference transformation was ineffective in this case, and the first-difference equation should not be used.

Another method for removing serial correlation uses the transformation

$$Y'_t = Y_t - \hat{\rho} Y_{t-1},$$
$$X'_t = X_t - \hat{\rho} X_{t-1},$$

where $\hat{\rho}$ is an estimate of the serial correlation $\rho_{t,t-1}$. Note that the first-difference transformation can be considered a special case of this transformation, with $\rho_{t,t-1}$ estimated to be 1. We can obtain an estimate of $\rho_{t,t-1}$ from the serial correlation coefficient calculated in Table 9.17 for the original (untransformed) data. There, $r_{t,t-1} = 0.6575$. Table 9.20 shows the residuals, serial correlation, and Durbin-Watson test for the data transformed by

$$Y'_t = Y_t - 0.6575 Y_{t-1}$$
$$X'_t = X_t - 0.6575 X_{t-1}.$$

TABLE 9.20
Transformed Data and Durbin-Watson Test for Television Advertising and Sales Data

Week (t)	$Y'_t = Y_t - 0.6575Y_{t-1}$	$X'_t = X_t - 0.6567X_{t-1}$	e_t	e_{t-1}	$e_t - e_{t-1}$
1	—	—	—	—	—
2	3.2140	5.0000	−0.5897	—	—
3	3.2825	5.7125	−0.5860	−0.5897	0.0037
4	3.9510	−1.9175	0.7761	−0.5860	1.3621
5	3.0250	3.3700	−0.6305	0.7761	−1.4066
6	4.0880	9.0550	−0.0843	−0.6305	0.5462
7	3.9620	9.4525	−0.2464	−0.0843	−0.1621
8	2.6990	−6.8350	−0.0289	−0.2464	0.2175
9	2.9565	0.7125	−0.4575	−0.0289	−0.4286
10	4.0195	6.3700	0.0913	−0.4575	0.5488
11	5.2935	11.0825	0.9369	0.0913	0.8456
12	3.1100	1.8225	−0.4049	0.9369	−1.3418
13	3.5675	−1.5475	0.3590	−0.4049	0.7639
14	3.6990	4.3975	−0.0499	0.3590	−0.4089
15	3.4990	0.0825	0.1423	−0.0499	0.1922
16	4.0305	2.0550	0.4945	0.1423	0.3522
17	3.9675	5.0550	0.1588	0.4945	−0.3357
18	4.6360	1.0825	0.1884	0.1588	0.0296
19	3.1675	0.3975	−0.2178	0.1884	−0.4062
20	4.5620	11.7125	0.1481	−0.2178	0.3659
$\sum(\cdot)$	69.7300	67.0600	−0.0005		
$\sum(\cdot)^2$	262.8539	647.9824	3.5453	3.5234	8.5303
$\sum(\cdot)(\cdot)$	283.5037		−0.9047		

$$\hat{Y}' = 3.3492 + 0.0909X'$$

$$r_{t,t-1} = \frac{-0.9047}{3.5234} = -0.2568$$

$$d = \frac{8.5303}{3.5453} = 2.4061$$

$$d^- = 4 - 2.4061 = 1.5939$$

From Table 9.20 we see that $d^- = 1.5939 > 1.4 = d_U$, so we can conclude that there is no negative serial correlation in these transformed data. If this transformation had also failed to remove serial correlation, then it would have been necessary to revise our estimate of $\rho_{t,t-1}$ and employ the transformation $Y_t' = Y_t - \hat{\rho}Y_{t-1}$, $X_t' = X_t - \hat{\rho}X_{t-1}$ using this new value of $\hat{\rho}$. Usually this new value should be slightly larger than $r_{t,t-1}$ in absolute value, because $r_{t,t-1}$ tends to underestimate ρ. The idea is to keep repeating the process until a value for $\hat{\rho}$ is found so that the transformed data do not exhibit any serial correlation. This is called an *iterative process*. It is iterated, or repeated, increasing $|\hat{\rho}|$ by small amounts at each step, until a satisfactory result is obtained.

To use the equation

$$\hat{Y} = 3.3492 + 0.0909X'$$

to estimate sales in week 21, if the company plans to purchase 20 minutes of television advertising, recall that at week 20, 15 minutes were purchased. Then

$$X_t' = X_t - 0.6575X_{t-1} = 20 - 0.6575(15)$$
$$= 10.1375,$$

and

$$\hat{Y}_t' = 3.3492 + 0.0909(10.1375)$$
$$= 4.2707.$$

Then since

$$Y_t' = Y_t - 0.6575Y_{t-1},$$
$$4.2707 = Y_t - 0.6575(11.4)$$

because sales in week 20 were \$11,400. Thus, we estimate week 21 sales to be

$$\hat{Y}_t = 4.2707 + 0.7575(11.4)$$
$$= 11.766 \quad (= \$11,766).$$

USING THE COMPUTER TO CALCULATE THE DURBIN-WATSON STATISTIC AND THE SERIAL CORRELATION COEFFICIENT

Calculation of a serial correlation coefficient and a Durbin-Watson statistic by hand is tedious, even for small data sets. Fortunately, it is quite easy to amend BMD, Minitab, SAS, and SPSS^X programs to request these statistics.

B M D The serial correlation coefficient will be calculated and printed if the DATA sentence is used in the PRINT paragraph in BMDP1R. BMD regression programs do not provide values of the Durbin-Watson statistic, however.

M I N I T A B As standard output from its REGRESS command, Minitab prints the value of the Durbin-Watson statistic. If the BRIEF command is used, the Durbin-Watson statistic will be the last information given. (See Figure 4.8.)

Minitab can also be used to calculate the serial correlation coefficient, $r_{t,t-1}$ by using the LAG command. Suppose that when a regression is performed, values of X and Y are stored in columns 1 and 2 and that the standardized residuals e_t are stored in column 3. Then the command

```
LAG 1 IN C3, PUT IN C4
```

will enter values of e_{t-1} (standardized) in column 4. In column 4 there will be no value in row 1; row 2 will contain the value that is in row 1 of column 3, and so on. Then we can multiply e_t by e_{t-1} by writing

```
MULTIPLY C3 BY C4, PUT IN C5.
```

To obtain $\sum e_t e_{t-1}$,

```
SUM C5, PUT IN K1.
```

Then $\sum e_{t-1}^2$ can be obtained by writing

```
RAISE C4 TO THE POWER 2, PUT IN C6
SUM C6, PUT IN K2.
```

Finally, the serial correlation coefficient is obtained by

```
DIVIDE K1 BY K2.
```

Note that the value of $r_{t,t-1}$ is unaffected if standardized residuals instead of the actual residuals are used, since their proportional magnitudes are preserved.

S A S To obtain a value of the Durbin-Watson statistic, include the option DW after the slant in the MODEL statement. In addition to the value of the Durbin-Watson statistic, the first-order serial correlation coefficient will appear.

S P S S X Use a subcommand RESIDUALS = DURBIN in the REGRESSION command. The value of the Durbin-Watson statistic will be shown, but the serial correlation coefficient will not.

The computer programs we have studied all give the serial correlation or the value of the Durbin-Watson statistic, either automatically or as an option. However, this statistic does not always have relevance to the problem. For example, in the house size-price example, the data were *not* collected over time. The numbering of the houses in the sample was completely arbitrary.

Nevertheless, the computer could calculate Durbin-Watson statistics using the residuals in the order given. A different numbering of the houses would have resulted in a different value for the Durbin-Watson statistic, although no other results from the printout (other than the order of values in the list of residuals) would have been affected. It is only when data are entered in a time order that the test for serial correlation has any meaning.

Note that the SPSS^X computer program plots (standardized) residuals in the order given. This is a very handy feature when the data are collected in a time order.

EXERCISES

9.10 For the Typist Data of Exercise 9.7,

a. Calculate the serial correlation coefficient, and test for significant serial correlation.

b. Try to remove the serial correlation by using first differences. Test to see if this method has been effective.

9.11 Refer to the Heating Data of Exercise 1.8. Suppose that the temperature and oil bills were recorded for 15 consecutive days. Test for serial correlation.

9.12 Test for serial correlation in the Stock Data of Exercise 5.17.

9.13 The University of Tennessee, Knoxville, operates on the quarter system, with reduced class offerings during the summer. Data in Table 9.21 are numbers of students enrolled by quarter from 1971 to 1979. These data are graphed in Figure 9.10. There is a pronounced seasonal pattern in these data: Every year, enrollments are highest in the fall quarter, drop in the winter, drop again in the spring, and are lowest in the summer.

TABLE 9.21
Enrollment Data
Source: Office of Enrollment Statistics, UTK.

Year	Winter	Spring	Summer	Fall
1971	18,511	17,788	10,839	24,588
1972	22,480	21,227	11,291	23,956
1973	22,700	21,389	11,570	23,986
1974	22,827	21,865	11,833	25,371
1975	24,488	23,575	13,734	27,374
1976	26,553	25,155	12,761	28,176
1977	26,467	25,047	12,426	28,858
1978	26,343	24,791	11,765	27,721
1979	25,427	25,347	12,453	28,229

Note: Figures include day and night students but exclude professional students (those enrolled in College of Architecture and Law). *Student* is defined as a person registered in and participating in one or more classes.

FIGURE 9.10 Quarterly Enrollment of Students at the University of Tennessee, Knoxville, 1971–79.

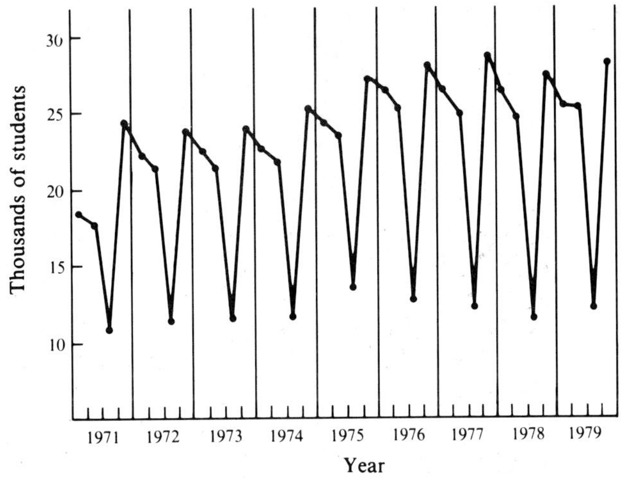

a. Would you expect that these data show serial correlation? If so, would you expect it to be positive or negative?

b. Use an available computer program, if possible, to regress enrollment versus time (let X = quarter number, where $X = 1$ in winter 1971, . . . , $X = 36$ in fall 1979), and calculate the value of the Durbin-Watson statistic.

c. Test for significant serial correlation, using the Durbin-Watson statistic.

9.14 Refer to the Enrollment Data of Exercise 9.13. Since the data display a strong quarterly seasonal pattern, it is decided to calculate *fourth* differences in enrollments, $Y_t - Y_{t-4}$. These fourth differences measure growth in enrollments from a given quarter in one year to the same quarter in the following year. Data are shown in Table 9.22 and graphed in Figure 9.11.

a. Do the lagged enrollments appear to be serially correlated? Does the serial correlation appear to be positive or negative?

b. Regress lagged enrollments on quarter number, and test for significant serial correlation.

c. Explain why lagged enrollments demonstrate positive serial correlation.

TABLE 9.22 Lagged Enrollments

Quarter	$Y_t - Y_{t-4}$	Quarter	$Y_t - Y_{t-4}$	Quarter	$Y_t - Y_{t-4}$	Quarter	$Y_t - Y_{t-4}$
1	—	10	162	19	1901	28	682
2	—	11	279	20	2003	29	−124
3	—	12	30	21	2065	30	−256
4	—	13	127	22	1580	31	−661
5	3969	14	476	23	−973	32	−1137
6	3439	15	263	24	802	33	−916
7	452	16	1385	25	−86	34	556
8	−632	17	1661	26	−108	35	688
9	220	18	1710	27	−335	36	508

FIGURE 9.11
Lagged Enrollment Figures

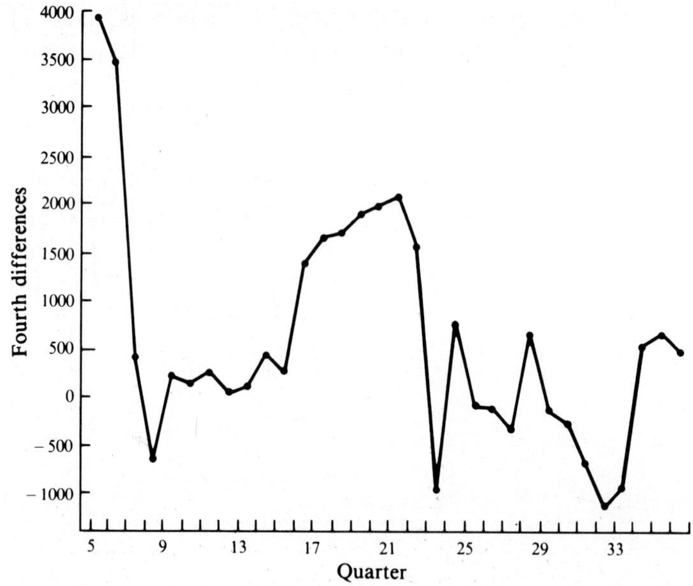

9.5 THE ASSUMPTION OF NORMALITY

THE IMPORTANCE OF THE ASSUMPTION

In order for *F*- or *t*-tests and confidence intervals to be valid, the values of the test statistics must be calculated from data that are normally distributed. This is why we made the assumption in Chapter 6 that in the population the possible *Y*-values for each *X*-value are normally distributed around the line $\mu_{Y \cdot X} = \alpha + \beta X$. Usually this is not an unreasonable assumption to make. For example, for a given size house we would expect a greater concentration of prices near the mean price than far away from it, and we would expect about as many prices above the mean as below. This kind of structure of prices for a given size gives at least a reasonably symmetric, mound-shaped distribution. While the distribution of prices for a given size might not be exactly normal, as long as the distribution is reasonably symmetric, the *F*- and *t*-statistics can be trusted. This is because *F* and *t* are what is called **robust** to the assumption of normality. This means that as long as the data are not too far from normal, the *F*-statistic will still follow the *F*-distribution reasonably well, and the

Robust

probabilities found from the t-tables will be reasonably applicable to the calculated value of the t-statistic. If we cannot assume that the data come from a normal population, then we can still look for "large" versus "small" t- or F-values in order to see if a variable is a strong predictor. But we cannot answer the question, "How large is *large*?" The probability values from the t- and F-tables will no longer be applicable, so we cannot use the tables to find critical values beyond which "large" (in absolute value) t- or F-values lie.

A TEST FOR NORMALITY

To check out the normality assumption, one quick and easy test is based on the properties of the normal distribution: about 68 percent of all normal values lie within 1 standard deviation of their mean, and so on. In a standard normal distribution, $N(0, 1)$, this means that about 68 percent of all standard normal values lie between -1 and $+1$, about 95 percent lie between -2 and $+2$, and more than 99 percent lie between -3 and $+3$.

Now if the Y-values are normally distributed at each X-value, then the errors $\varepsilon_i = Y_i - \mu_{Y \cdot X_i}$ will also be normal. Since $\mu_{Y \cdot X_i}$ is the true mean of the Y-values at $X = X_i$, the errors will have mean zero. The variance of the Y-values, and thus the variance of the errors, is $\sigma^2_{Y \cdot X_i}$. That is, $\varepsilon_i \sim N(0, \sigma^2_{Y \cdot X_i})$. If we standardize the errors by dividing them by their standard deviation, then $\varepsilon_i / \sigma_{Y \cdot X_i} \sim N(0, 1)$. In the sample, we can estimate these standardized errors by **standardized residuals,** $e_i / s_{Y \cdot X}$, and these residuals can be used to give a quick check on the normality assumption. Approximately 68 percent of the standardized residuals should lie between $+1$ and -1, and so on. Equivalently, to avoid having to calculate standardized residuals, we can check that approximately 68 percent of the residuals lie between $-1s_{Y \cdot X}$ and $+1s_{Y \cdot X}$, and so on.

Standardized Residuals

For the house size-price data, the standardized residuals are shown in Table 9.23. We see that 13 of the standardized residuals, or $13/20 = 0.65 =$ 65% are between -1 and $+1$. There are 17 standardized residuals, 95%,

TABLE 9.23 Standardized Residuals for House Size-Price Example (Residuals Taken from Figure 4.13)

House	e_i	$e_i/2.5506$	House	e_i	$e_i/2.5506$
1	0.0796	0.0312	11	5.6305	2.2077
2	2.3982	0.9403	12	−1.3119	−0.5144
3	−3.6305	−1.4235	13	0.8186	0.3210
4	2.1814	0.8553	14	−2.5575	−1.0028
5	−2.0796	−0.8154	15	−0.6305	−0.2472
6	−2.0221	−0.7929	16	2.0509	0.8042
7	−3.1372	−1.2301	17	0.2389	0.0937
8	−2.8916	−1.1338	18	−1.2677	−0.4971
9	3.5000	1.3724	19	−0.0796	−0.0312
10	2.7611	1.0826	20	−0.0509	−0.0200

between −2 and +2. One hundred percent of the standardized residuals lie between −3 and +3. From this close association between our percentages and those expected for normal data, we have little reason to doubt that our data are reasonably close to normal. Had we found poor agreement with the normal percentages, we would have needed to transform the data so that the normality assumption would hold. Fortunately, the violation of the normality assumption often occurs along with violation of the homoscedasticity assumption. Thus, the reciprocal transformation described in Section 9.2 for achieving equal variances will almost always give transformed data for which the normality assumption holds.

Another quick method for verifying that the residuals are at least approximately normally distributed is to construct a histogram of the residuals. This method provides a visual check that the distribution is mound-shaped and reasonably symmetric. It works best, of course, if the sample size is large, so that there are many residuals to be graphed. Figure 9.12 shows a frequency distribution of residuals for the house size-price example. Even though there are only 20 residuals to be tabulated, the distribution appears to be reasonably mound-shaped and symmetric; at least there is no clear evidence that the normality assumption is untenable. These methods for checking the normality assumption are among the quickest and easiest available. They are also the roughest. Other methods, such as plots on normal probability paper or goodness-of-fit tests, are also often used to check this assumption.

FIGURE 9.12 Histogram of Residuals From Fitting a Simple Linear Model to House Size-Price Data

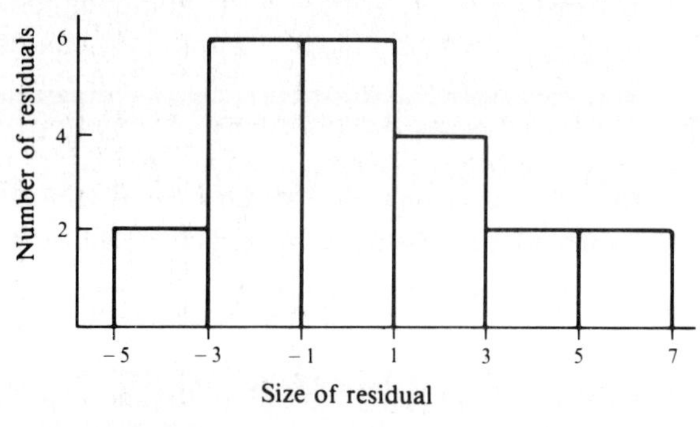

EXERCISES

9.15 Form standardized residuals and check for normality in the Appraisal Data of Exercise 4.4.

9.16 Form standardized residuals and check for normality in the Reservations Data of Exercise 5.16.

9.17 Form standardized residuals and check for normality in the Stock Data of Exercise 5.17.

9 . 6

SUMMARY

Whether a simple linear model or a more complicated one is proposed for a given set of data, it is important to check to see that the model is appropriate—or at least not too inappropriate—and that the assumptions underlying the model have been met. In this chapter we discussed several of the assumptions, ways in which they could be checked, and measures to be taken if they were not met.

Most computer packages provide extensive capabilities for *regression diagnostics*—techniques for checking the model assumptions and spotting where and how they might be violated. The explanation of these techniques could very easily form a textbook by itself (and indeed, such have been written). Thus, we shall content ourselves here with this brief introduction to methods for checking the appropriateness of the model.

In subsequent chapters we will look at curvilinear and multiple regression models, which are used if the relationship is not linear or if one predictor is found to be insufficient. It should be kept in mind that if a multiple regression or curvilinear model is used, the same kind of examination of it is necessary: Are the variances constant? Are the data normal? Does the model fit the data? Is there a time effect? Mathematically, almost any model can be used with almost any set of data—with regression equations, correlation coefficients, F- and t-values, and so on, being calculated. Just because such calculations can be made, however, does not mean they are appropriate. Whether a simple or multiple regression model, linear or curvilinear, is adopted, one should be careful to investigate whether or not the theory (model) is supported by the evidence (data). Only then can the inquiry being conducted have meaningful scientific value.

TEN

EXPONENTIAL AND POLYNOMIAL MODELS

10.1

INTRODUCTION

In the first nine chapters we have dealt almost exclusively with strictly linear models, in which Y increases or decreases with X in a linear fashion. That is, for each one-unit increase in X, Y increases or decreases by a constant amount. While a linear function may serve well as a model for many relationships among variables, there are other instances in which a more appropriate model would be one that describes a curving relationship. For example, Y may increase rather quickly with X for small values of X but more slowly for larger values of X. The researcher should consider at the outset whether a linear or a curvilinear model would be theoretically appropriate as a description of the relationship. After collection of data a scatter diagram helps verify whether the proposed model is appropriate.

Suppose that theoretical considerations and a preliminary inspection of the data indicate that a curved model is appropriate. In this chapter we will look at two types of curved models. First, exponential models will be examined: These are appropriate in cases in which Y changes with X at a constant *rate* (by a constant *percentage*) rather than by a constant amount. Then polynomial models will be considered: These are appropriate when the rate of change of Y with X is itself increasing or decreasing.

1 0 . 2

ALTERNATIVES TO THE STRAIGHT-LINE MODEL: EXPONENTIAL MODELS

EXPONENTIAL FUNCTIONS

Recall that if Y is a linear function of X, then Y increases or decreases by a constant amount with each one-unit increase in X. For example, suppose you deposit \$100 in a savings account that pays 5% *simple* interest quarterly. That is, every quarter the bank will pay you 5% of \$100. Assuming that you make no additional deposits or withdrawals, then Y (balance in the account) is a linear function of X (number of quarters the account is maintained):

$$Y = 100 + 5X, \qquad \text{since} \qquad 5\% \times 100 = 5.$$

No bank pays simple interest, however—all savings accounts pay compound interest. Thus, \$100 deposited at 5% interest compounded quarterly will pay 0.05(\$100) = \$5 the first quarter, giving a balance of \$105. At the end of the second quarter, the bank will pay 5% of \$105, or \$5.25, leaving a balance of \$105 + \$5.25 = \$110.25. Table 10.1 continues this for several quarters. From this table we see that Y (balance) is *not* a linear function of X (number of months), since as X increases by 1, Y does not increase by a constant amount. Rather, Y increases by a constant *percentage*. Note that the *amount* of increase will get larger and larger, faster and faster. Growth (or decline) by a constant percentage is called **exponential growth** (or decline) and is modeled by

Exponential Growth

$$Y = \alpha\beta^X,$$

TABLE 10.1 Quarterly Balance for \$100 at 5 Percent Interest

End of Quarter	Balance (dollars)	Amt. of Increase (dollars)	Percent Increase
0	100.00	—	—
1	105.00	5.00	5
2	110.25	5.25	5
3	115.76	5.51	5
4	121.55	5.79	5
5	127.63	6.08	5

where, for a deterministic relationship such as in the compound interest example,

$$\alpha = \text{the value of } Y \text{ when } X = 0, \qquad \text{and} \qquad \beta - 1 = \text{constant growth rate.}$$

That is, at time 0, the initial amount deposited was \$100. After each quarter, the balance was 1.05 times the balance at the previous quarter, so

$$Y = 100(1.05)^X.$$

Exponential Model

If we have data showing a pronounced curvature, as illustrated in Figure 10.1, we may choose to adopt the **exponential model**

$$Y = \alpha\beta^X\varepsilon$$

if it seems reasonable that Y should increase or decrease at a constant *rate* with X. The problem is to obtain numerical estimates of α and β and a regression equation of the form

$$\hat{Y} = ab^X.$$

FIGURE 10.1 Scatter Diagrams Suggesting Exponential Models

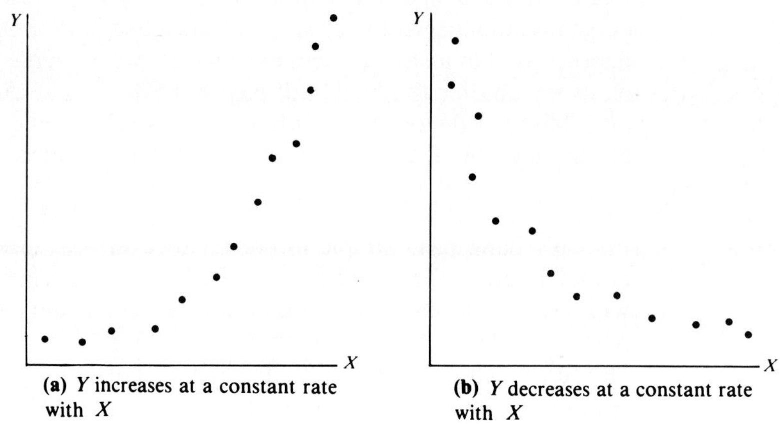

LINEARIZING THE MODEL

Intrinsically Linear

This appears to be a very complicated problem until one notices that the model is **intrinsically linear;** that is, it can be transformed into a linear model by taking logarithms:

$$Y = \alpha\beta^X\varepsilon$$

$$\log Y = \log\alpha + X\log\beta + \log\varepsilon$$

$$Y' = \alpha' + \beta'X + \varepsilon',$$

where $Y' = \log Y$, $\alpha' = \log\alpha$, $\beta' = \log\beta$, $\varepsilon' = \log\varepsilon$.

Thus, we can transform Y values into $Y' = \log Y$ and fit a straight line

$$Y' = a' + b'X$$

to the logarithms by the methods of ordinary least squares. Once this linear equation has been found, the regression coefficients of the exponential regression equation can be found by taking antilogarithms of a' and b'.

Let us illustrate with a numerical example. Consider the hypothetical data in Table 10.2 which are graphed in Figure 10.2. In Table 10.3 are natural logarithms of the values of Y. It does not matter whether natural (base e) or common (base 10) logarithms are used. While intermediate steps will differ, when antilogarithms are taken, the results will be the same. The graph of the natural logarithms (Figure 10.3) shows a *linear* tendency. Thus, we can fit an ordinary simple linear regression equation to these logarithms. The calculations required are shown in Table 10.3. The linear equation $\hat{Y}' = -0.61908 + 0.07588X$ is plotted on the graph of the logarithms in Figure 10.3.

To obtain the curved regression equation of the form $\hat{Y} = ab^X$ appropriate to the actual death rates, we find the antilogarithms of the intercept and slope

FIGURE 10.2 Death Rate as a Function of Age

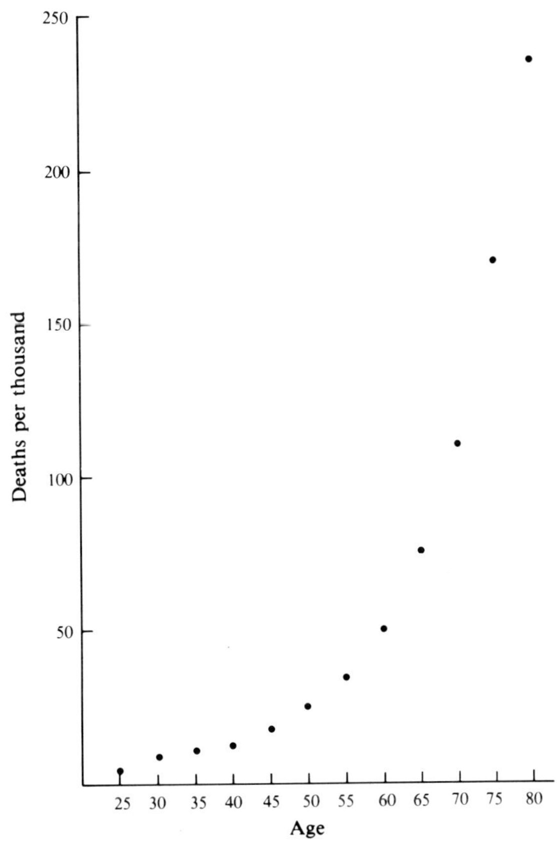

TABLE 10.2 Death Rate as a Function of Age for a Hypothetical Population

Age (X)	Deaths Per Thousand (Y)
25	3
30	7
35	8
40	10
45	15
50	25
55	35
60	50
65	75
70	110
75	160
80	235

TABLE 10.3 Calculations for Fitting a Linear Regression Equation to Logarithms of Death Rates

	X	$Y' = \ln Y$	Y	$\hat{Y}$
	25	1.09861	3	3.6
	30	1.94591	7	5.3
	35	2.07944	8	7.7
	40	2.30259	10	11.3
	45	2.70805	15	16.5
	50	3.21888	25	24.1
	55	3.55535	35	35.2
	60	3.91202	50	51.5
	65	4.31749	75	75.4
	70	4.70048	110	110.2
	75	5.07517	160	161.2
	80	5.45959	235	235.8
$\sum (\cdot)$	630	40.37357		
$\sum (\cdot)^2$	36,650	156.55829		
$\sum XY'$	2390.87327			
$S_{\cdot\cdot}$	3575	20.72284		
S_{XY}	271.26064			

$$\hat{Y}' = -0.61908 + 0.07588X$$

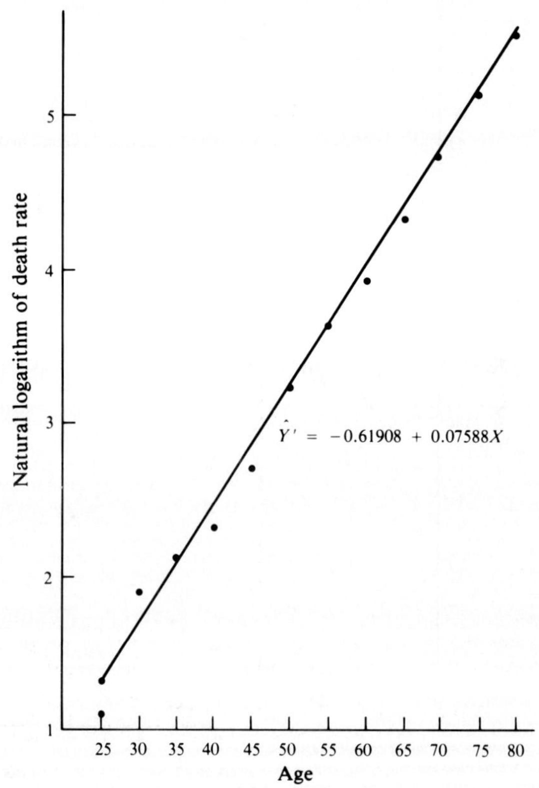

FIGURE 10.3 Logarithms of Death Rates With Linear Regression Equation

of the linear equation:

$$a = \text{antiln}\, a' = \text{antiln}\,(-0.61908) = e^{-0.61908} = 0.53844,$$

$$b = \text{antiln}\, b' = \text{antiln}\,(0.07588) = e^{0.07588} = 1.07883.$$

The exponential regression equation is then

$$\hat{Y} = (0.538)(1.079)^X.$$

The estimated death rate at age zero (whatever that means) is then 0.538 per thousand, and the death rate increases by $1.079 - 1 = 0.079$, or 7.9 percent, per year.

To graph this equation, find $\hat{Y}$ for several selected values of X, and connect these points with a smooth curve. Values of $\hat{Y}$ are shown in Table 10.3 and are graphed, along with the given data, in Figure 10.4. We see that this exponential function gives a very good description of the relationship between age and death rate. (Compare values of Y and $\hat{Y}$ in Table 10.3, and observe how closely the curve follows the points in Figure 10.4.)

FIGURE 10.4 Exponential Function Graphed With Death Rates

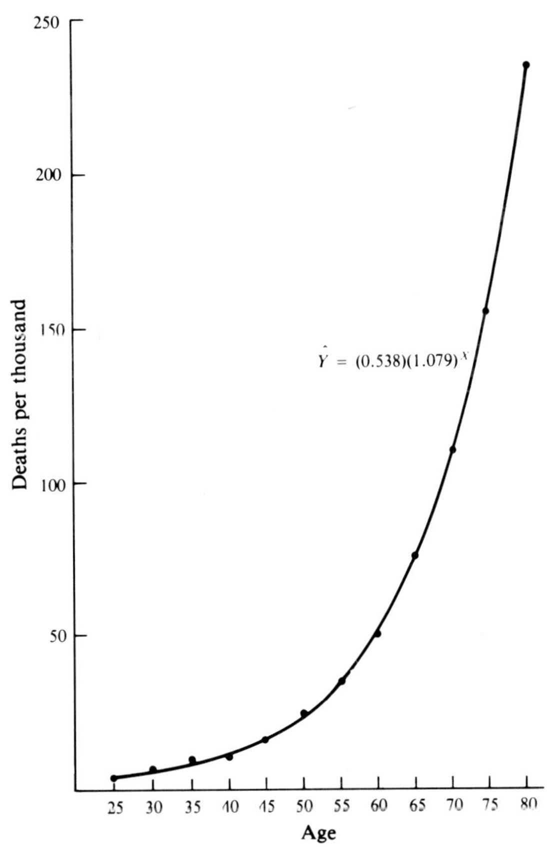

Note that the "slope," $b = 1.079$, tells us that the percentage increase in death rate per year of age is 7.9 percent. Since values of X in this example are given in five-year intervals, we might ask what the rate of increase in death rate is for each five-year increase in age. Note that it is *not* 5(7.9%) = 39.5%. Let r denote the five-year proportional increase in death rate. Then

$$(1.079)^5 = 1 + r$$

$$1.4625 = 1 + r$$

$$0.4625 = r = 46.25\%,$$

so that the death rate increases by 46.25 percent with every five additional years of age. That this is so can be verified by forming the ratio of a value of $\hat{Y}$ in Table 10.3 to the value preceding it. For example,

$$\frac{\hat{Y}_{80}}{\hat{Y}_{75}} = \frac{235.8}{161.2} = 1.4628$$

and

$$\frac{\hat{Y}_{55}}{\hat{Y}_{50}} = \frac{35.2}{24.1} = 1.4606,$$

both of which ratios are (except for rounding error) essentially 1.4625.

HINTS ON USING YOUR CALCULATOR TO FIT EXPONENTIAL FUNCTIONS

Most hand-held calculators will find either natural or common logarithms, or both, thus obviating the necessity of looking up logarithms in tables. Typically, a function key labeled "log" will find a common (base 10) logarithm, and the key labeled "ln" will find a natural (base e) logarithm. Calculators differ as to the key or sequence of keys necessary to find antilogarithms. Sometimes it is necessary to punch the inverse function ("inv") key, followed by the function key. Other calculators will have keys labeled "10^X," which will find antilogs base 10, or "e^X," which will find antilogs of natural logarithms.

If your calculator will fit a straight line automatically, there is no need for you to write down all the values of log Y as was done in Table 10.3. Simply enter values of X as if you were going to fit a simple straight line, but before entering corresponding Y-values, have your calculator convert them to logarithms. The function keys giving the intercept and slope will find a' and b', respectively. When one of these quantities appears on your display, simply apply whatever function key is necessary to find the antilog. Thus, the equation $Y = ab^X$ can be found without ever having to write down a single number.

It is easy to find values of $\hat{Y}$ if your calculator has a universal exponent key, typically labeled Y^X. Enter the values of the numbers to be raised to the power, key Y^X, then enter X, and then press the equals key. This is especially handy when X is not an integer, since the values of Y^X cannot be found in those cases by successive multiplications. Lacking this function, values of $\hat{Y}$ must be found by taking antilogarithms of values $\hat{Y}' = a' + b'X$.

HYPOTHESIS TESTS IN EXPONENTIAL REGRESSION

Note that in the exponential model

$$Y = \alpha\beta^X\varepsilon,$$

β must be greater than zero, and if $\beta > 1$, the value of Y increases with X, but if $\beta < 1$ the value of Y decreases with X. Also, Y is a function of X only if $\beta \neq 1$. Thus we may want to test the hypothesis

$$H_0\colon\beta = 1.$$

When the model is linearized to the form

$$Y' = \alpha' + \beta'X + \varepsilon',$$

testing $H_0\colon\beta = 1$ becomes equivalent to testing

$$H_0\colon\beta' = 0,$$

since $\log 1 = 0$ or $\ln 1 = 0$. Thus, all the statistics pertinent to testing $H_0\colon\beta' = 0$ for a simple linear regression of $\log Y$ on X are applicable to testing $H_0\colon\beta = 1$ in the exponential model.

Also note that if in the straight line fitted through the logarithms, b' is negative, indicating that $\log Y$ decreases with X, then $b = \text{antilog } b'$ will be less than 1. Conversely, if b' is positive, then b will be greater than 1.

Assumptions such as uniform scatter and normally distributed values of Y at each X-value should hold in the logarithms in order for the F- and t-tests to be valid. If these assumptions hold in the logarithms, it is likely that they will not hold for the given Y-values, and vice versa.

10.3 USING THE COMPUTER TO PERFORM EXPONENTIAL REGRESSION

Typically, only one alteration is needed in a computer program that will do simple linear regression to adapt it for exponential regression. We must somehow get the computer to convert *Y*-values to logarithms. Then the simple linear regression program can fit a straight line to the logarithms of the *Y*-values. Let us use the data in Table 10.2 to illustrate how to use the computer to fit an exponential function.

EXPONENTIAL REGRESSION WITH BMDP1R

The worksheet reproduced in Figure 10.5 shows instructions for transforming death rates to logarithms and fitting a straight line through the logarithms using BMDP1R. There are very few changes in this set of instructions from those given for fitting a simple straight line. In fact, all that must be done is to define a new variable, log *Y*, and instruct the computer to use this as the dependent variable.

To transform death rates, which we have named `RATE`, to logarithms, the sentence `ADD=1` must be added to the `VARIABLES` paragraph. This is to tell the computer that in addition to the values of the two variables entered on the data cards, values of another variable are to be created. Then insert a `TRANSFORM` paragraph. The sentence `LRATE=LOG(RATE)` instructs the computer to create a new variable, `LRATE`, by taking common logarithms of the values of the variable `RATE`. Had we wanted to use natural logarithms, the sentence would have read `LRATE=LN(RATE)`.

To fit a straight line through the logarithms of the death rates, we simply identify `LRATE` as the dependent variable and `AGE` as the independent variable in the `REGRESSION` paragraph. Note that since the computer now has a data set consisting of three variables—`AGE`, `RATE`, and `LRATE`—it is not permissible to omit the `INDEP=AGE` sentence.

FIGURE 10.5 Worksheet for Using BMDP1R to Fit an Exponential Regression

```
  {SYSTEM   CARDS
PROBLEM     TITLE='EXPONENTIAL  REGRESSION'./
INPUT       VARIABLES=2.
            FORMAT=FREE.
            CASES=12./
VARIABLES   NAMES=AGE,RATE,LRATE.
            ADD=1./
TRANSFORM   LRATE=LOG(RATE)./
PRINT       DATA./
REGRES      DEPENDENT=LRATE.
            INDEP=AGE./
PLOT        VARIABLE=AGE./
END/
25 3
3Ø 7
   :{DATA   CARDS
8Ø 235
FINISH/
```

Output from this program is exhibited in Figure 10.6. Note that simple statistics for all three variables are given before the analysis of variance for the regression problem. From the analysis, we find the equation of a straight line fitted through the common logarithms of the death rates to be

$$\hat{Y}' = -0.26887 + 0.032953X.$$

It is necessary to take antilogarithms of these values using either tables of logarithms or a calculator to obtain the exponential regression equation for the given data, $\hat{Y} = 0.5384(1.0788)^X$. Unfortunately, there is no way to get the computer to do this for you using this program.

FIGURE 10.6 BMDP1R Exponential Regression

```
NUMBER OF CASES READ. . . . . . . . . . . . . . . .12

VARIABLE    MEAN       STANDARD     COEFFICIENT      MINIMUM      MAXIMUM
                       DEVIATION    OF VARIATION

1  AGE      52.50000   18.02776     0.34339          25.00000      80.00000
2  RATE     61.08330   72.97128     1.19462           3.00000     235.00000
3  LRATE     1.46117    0.59609     0.40796           0.47712       2.37107

REGRESSION TITLE IS
EXPONENTIAL REGRESSION

DEPENDENT VARIABLE. . . . . . . . . . . . . . . . . 3  LRATE
TOLERANCE . . . . . . . . . . . . . . . . . . . . . 0.0100

ALL DATA CONSIDERED AS A SINGLE GROUP

MULTIPLE R          0.9966    STD ERROR OF EST    0.0515
MULTIPLE R-SQUARE   0.9932

ANALYSIS OF VARIANCE
                      SUM OF SQUARES   DF   MEAN SQUARE    F RATIO    P(TAIL)

          REGRESSION  3.8821            1   3.8821         1466.062   0.0000
          RESIDUAL    0.0265           10   0.0026

                                          STD REG
VARIABLE     COEFFICIENT    STD ERROR     COEFF        T        P(2 TAIL)   TOLERANCE

INTERCEPT   -0.26887
AGE   1       0.32953E-01   0.86064E-03   0.997     38.289      0.0         1.00000
```

If a PRINT paragraph is included, values of all three variables, with predicted and residual values, will be printed, as shown in Figure 10.7. Figure 10.8 shows graphs resulting from the PLOT paragraph. The left-hand graph shows a scatter diagram of the logarithms of *Y* plotted against *X*, with the straight line through the logarithms superimposed. The right-hand graph is a plot of residuals against *X* and shows the residuals with a definite wedge shape. Recall that if scatter in *Y* is uniform, scatter in log *Y* will not be (and vice versa), so that if inferences are to be made concerning the fit of the straight line through log *Y*, the probabilities calculated are all suspect and should not be taken at face value. It is possible, using the BMDP6D program, to obtain a plot of *Y* versus *X* to investigate homogeneity of scatter in *Y*. The interested reader is referred to the BMD manual.

EXPONENTIAL REGRESSION WITH MINITAB

In order to perform an exponential regression, we need to convert values of *Y* to logarithms and then fit a simple linear regression equation relating log *Y* to *X*. This operation is easily performed in Minitab by using either the LOGE (natural logarithm) or the LOGTEN (base 10 logarithm) command.

Suppose, for instance, that values of *X* have been entered in column 1 and values of *Y* in column 2. Then

```
LOGTEN C2, PUT INTO C3
```

FIGURE 10.7 Residuals and Predicted Values From BMDP1R Exponential Regression

```
LIST OF PREDICTED VALUES, RESIDUALS, AND VARIABLES
NOTE - NEGATIVE CASE NUMBER DENOTES A CASE WITH MISSING VALUES.
       THE NUMBER OF STANDARD DEVIATIONS FROM THE MEAN IS DENOTED BY UP TO
       3 ASTERISKS TO THE RIGHT OF EACH RESIDUAL OR VARIABLE.
MISSING VALUES AND VALUES OUT OF RANGE ARE DENOTED BY VALUES
GREATER THAN OR EQUAL TO 0.2127E 38 IN ABSOLUTE VALUE.
CASE                          PREDICTED  VARIABLES
LABEL  NO.   RESIDUAL          VALUE      1  AGE       2  RATE    3  LRATE
        1   -0.7784E-01*      0.5550     25.00*       3.000      0.4771*
        2    0.1254     **    0.7197     30.00*       7.000      0.8451*
        3    0.1860E-01       0.8845     35.00        8.000      0.9031
        4   -0.4925E-01       1.049      40.00       10.00       1.000
        5   -0.3793E-01       1.214      45.00       15.00       1.176
        6    0.1916E-01       1.379      50.00       25.00       1.398
        7    0.5198E-03       1.544      55.00       35.00       1.544
        8   -0.9343E-02       1.708      60.00       50.00       1.699
        9    0.1983E-02       1.873      65.00       75.00       1.875
       10    0.3549E-02       2.038      70.00      110.0        2.041
       11    0.1511E-02       2.203      75.00      160.0*       2.204*
       12    0.3694E-02       2.367      80.00      235.0**      2.371*
SERIAL CORRELATION OF RESIDUALS = -0.3096
```

FIGURE 10.8 Graphs of ln Y Versus X and Residuals versus X from BMDP1R Exponential Regression

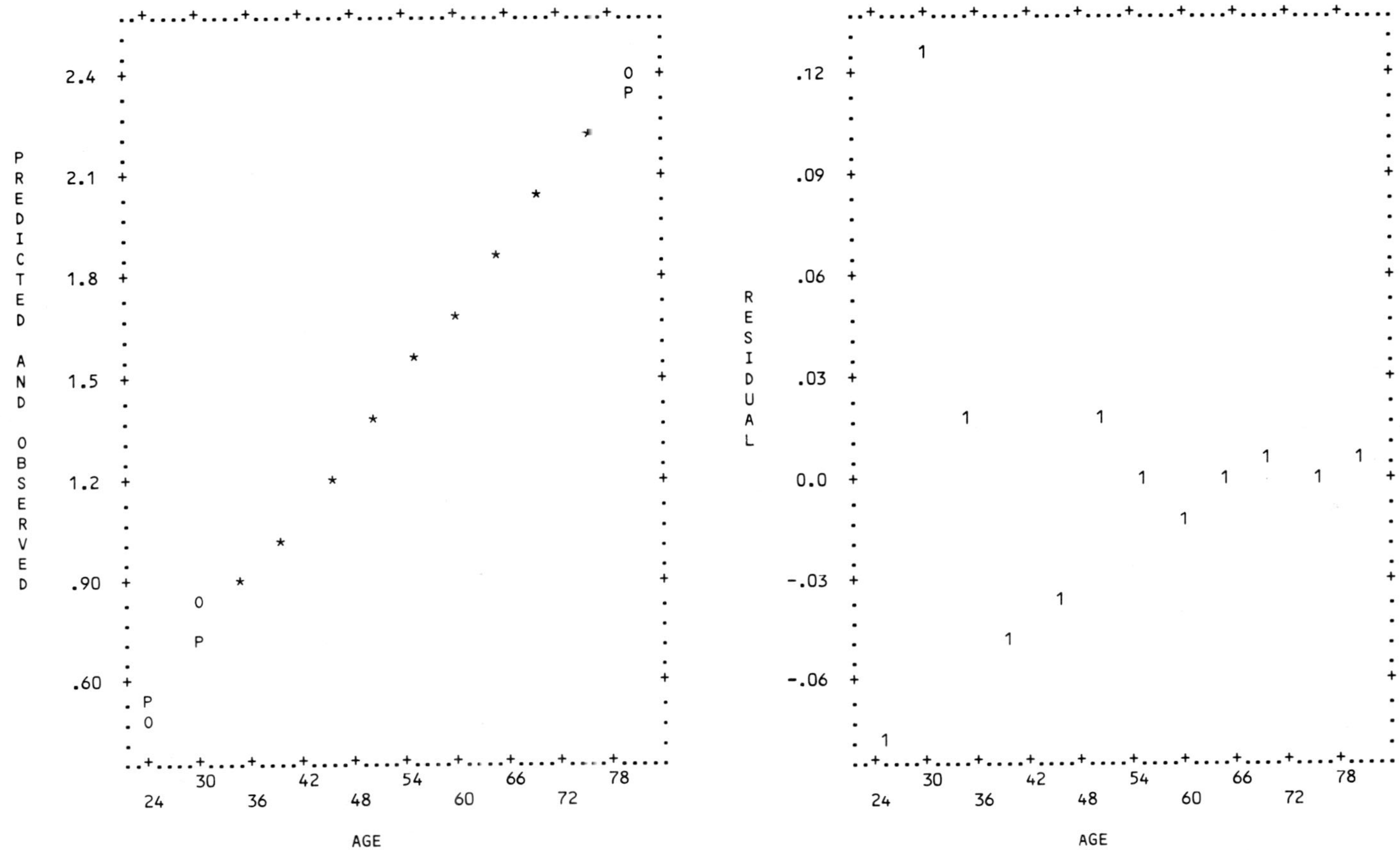

will create a column of $\log_{10}$ Y-values, C3. Then we can

```
REGRESS C3 ON 1 VAR IN C1
```

and the computer will show the analysis of the straight-line equation fitted through the logarithms. In order to obtain graphs and to be able to convert back to Y from log Y, however, residuals, predicted values, and the regression coefficients from the regression analysis must be saved. Thus, we will write

```
REGRESS C3 on 1 VAR IN C1\C4\C5\C6
```

The brief printout from this procedure is shown in Figure 10.9.

FIGURE 10.9 Minitab Exponential Regression

```
-REGRES C3 ON 1 VAR IN C1\C4\C5\C6

THE REGRESSION EQUATION IS
Y = - 0.269 + 0.0330 X1

                              ST DEV        T-RATIO =
     COLUMN    COEFFICIENT    OF COEF       COEF/S.D.

               -0.26886       0.04756       -5.65
X1   C1         0.0329530     0.0008606     38.29

THE ST DEV OF Y ABOUT REGRESSION LINE IS
S = 0.05145
WITH (12 - 2) = 10 DEGREES OF FREEDOM

R-SQUARED = 99.3 PERCENT
R-SQUARED = 99.3 PERCENT, ADJUSTED FOR DF

ANALYSIS OF VARIANCE

DUE TO        DF    SS          MS=SS/DF

REGRESSION     1    3.882095    3.882095
RESIDUAL      10    0.026475    0.002647
TOTAL         11    3.908570

      X1      Y          PRED Y     ST DEV
ROW   C1      C3         VALUE      PRED Y     RESIDUAL     ST RES

2     30.0    0.8451     0.7197     0.0244     0.1254       2.77R

R DENOTES AN OBS WITH A LARGE ST RES

DURBIN-WATSON STATISTIC = 2.31
```

At this point, suppose we name the columns as follows:

```
NAME C1 'AGE', C2 'RATE', C3 'LOGRATE',
C4 'LOGR', C5 'LOGYHAT', C6 'LOGB'
```

Figure 10.10 shows log $\hat{Y}$ plotted against X ("age"). Note that this regression equation is a straight line. To show the curved exponential regression equation, we need to take antilogs of the values in column 5. Write

```
ANTILOG C5, PUT IN C7
NAME C7 'YHAT'
```

(If base e logarithms had been used, the antilogarithms would have been

FIGURE 10.10
Minitab Plot of Log $\hat{Y}$ Versus X

```
-- NAME C1 'AGE', C2 'RATE', C3 'LOGRATE', C4 'LOGR', C5 'LOGYHAT', C6 'LOGB'

-- PLOT 'LOGYHAT' VS 'AGE'

   LOGYHAT
     2.50+
         -                                      *
         -
         -                                  *
         -
     2.00+                               *
         -                            *
         -
         -                        *
         -
     1.50+                     *
         -                  *
         -
         -              *
         -
     1.00+           *
         -        *
         -
         -    *
         - *
     0.50+
           +---------+---------+---------+---------+---------+AGE
          25.       40.       55.       70.       85.      100.
```

calculated using the command EXPONENTIAL instead of ANTILOG.) Figure 10.11 shows $\hat{Y}$ plotted against "age."

In column 6 are values of a' and b' from fitting the straight-line regression equation to $\log_{10} Y$. We can find the parameter estimates a and b for the

FIGURE 10.11
Minitab Plot of the Exponential Function

```
-- ANTILOG C5 PUT IN C7

-- NAME C7 'YHAT'

-- PLOT 'YHAT' VS 'AGE'

     YHAT
    240.+
        -
        -                                      *
        -
        -
    180.+
        -
        -                                  *
        -
        -
    120.+
        -                               *
        -
        -
        -                            *
     60.+
        -                        *
        -                     *
        -                  *
        -        *  *  *
      0.+ *  *
          +---------+---------+---------+---------+---------+AGE
         25.       40.       55.       70.       85.      100
```

exponential regression equation $\hat{Y} = ab^X$ by taking antilogs of the values in column 6.

```
ANTILOG C6, PUT IN C8
NAME C8 'BETAS'
```

Figure 10.12 shows both columns C6 and C8. We find the exponential equation to be

$$Y = 0.53844(1.07883)^X.$$

FIGURE 10.12 Regression Coefficients for the Exponential Equation, From Minitab

```
-ANTILOG C6 PUT IN C8
-NAME C8 'BETAS'
-PRINT C6 C8
COLUMN   LOGB         BETAS
COUNT    2            2
ROW
1        -0.268865    0.53844
2         0.032953    1.07883
-STOP
```

EXPONENTIAL REGRESSION WITH SAS

SAS can output results from an analysis to a data set and thus make these results available for further manipulation. Thus, we can perform a very complete analysis of an exponential regression using these programs. The reader may want to review the section entitled "The Logic and Language of SAS" in Chapter 4 before looking at the program outlined in Figure 10.13.

The objective of this program is to convert values of death rate (`RATE`) to logarithms and to fit a simple linear regression equation to these logarithms. In addition, we would like the data plotted in a scatter diagram with the regression equation superimposed. Also, it is important to examine the residuals of the logarithms of Y around the regression line and of the actual values of Y around their regression curve. Finally, we would like to be able to find the coefficients in the equation $\hat{Y} = ab^X$ as well as those in the linearized form, $\hat{Y}' = a' + b'X$.

LINEARIZING THE REGRESSION EQUATION It is fairly simple to use the SAS `REGRESSION` procedure to fit a linear regression equation to either log Y or ln Y. The program in Figure 10.13 converts values of Y (death rate) to common logarithms by adding the statement `LRATE = LOG 10 (RATE)` after the `INPUT` statement and before the `CARDS` statement. The data set created, which the computer will name `WORK.DATA1`

FIGURE 10.13 Worksheet for Using SAS to Perform an Exponential Regression Analysis

```
  {SYSTEM    CARDS
DATA;
    INPUT  AGE  RATE;
    LRATE=LOG1Ø(RATE);
    CARDS;
25 3
3Ø 7
   :  {DATA  CARDS
8Ø 235
PROC  REG  OUTEST=EST;
    TITLE  EXPONENTIAL  REGRESSION;
    MODEL  LRATE=AGE/P;
    OUTPUT  OUT=NEW  P=YHAT  R=RESID;
PROC  PRINT  DATA=EST;
DATA  CURVE;
    SET  EST;
    A=1Ø**INTERCEP;
    B=1Ø**AGE;
    PROC  PRINT  DATA=CURVE;
PROC  PLOT  DATA=NEW;
    PLOT  LRATE*AGE  YHAT*AGE='*'/OVERLAY;
    PLOT  RESID*AGE/VREF=Ø;
DATA  ANTLG;
    SET  NEW;
    ANTYHT=1Ø**YHAT;
    ANTRSD=1Ø**RESID;
PROC  PLOT,
    PLOT  RATE*AGE  ANTYHT*AGE='*'/OVERLAY;
    PLOT  ANTRSD*AGE;
```

if no other name is given to it, will contain values of AGE, RATE, and LRATE = $\log_{10}$ RATE. Writing PROC REG; MODEL LRATE=AGE/P will perform a simple linear regression relating to log Y to X and will compute predicted values and residuals. The OUTPUT statement creates a new data set, called NEW, which contains values of X, Y, $Y' = \log Y$, $\hat{Y}'$, and $Y' - \hat{Y}'$. All this is very much like a simple linear regression analysis, and the output from this analysis is shown in Figure 10.14. It is easy to find the regression coefficients for the straight line fitted to the logarithms:

$$\hat{Y}' = -0.268864 + 0.032953X,$$

and other quantities relevant to the fitting of this straight line, including residuals $Y' - \hat{Y}'$, are also shown.

FINDING REGRESSION COEFFICIENTS FOR THE EXPONENTIAL EQUATION

Note, however, that in the PROC REG statement we have added the instructions OUTEST=EST. This command takes the parameter estimates

FIGURE 10.14 SAS Regression Procedure for an Exponential Regression

```
DEP VARIABLE: LRATE

                     SUM OF         MEAN
SOURCE       DF      SQUARES        SQUARE          F VALUE       PROB > F
MODEL         1      3.882095       3.882095        1466.335      0.0001
ERROR        10      0.026475       0.002647481
C TOTAL      11      3.908570

       ROOT MSE      0.051454       R-SQUARE        0.9932
       DEP MEAN      1.461168       ADJ R-SQ        0.9925
       C.V.          3.521407

                     PARAMETER      STANDARD        T FOR H0:
VARIABLE     DF      ESTIMATE       ERROR           PARAMETER=0   PROB > |T|
INTERCEP      1      -0.268864      0.047558        -5.653        0.0002
AGE           1       0.032953      0.0008605546    38.293        0.0001

                     PREDICT
OBS     ACTUAL       VALUE          RESIDUAL
 1      0.477121     0.554961       -0.077839
 2      0.845098     0.719726        0.125372
 3      0.903090     0.884491        0.018599
 4      1.000        1.049          -0.049256
 5      1.176        1.214          -0.037930
 6      1.398        1.379           0.019154
 7      1.544        1.544           5.2E-04
 8      1.699        1.708          -0.009346
 9      1.875        1.873           0.001980
10      2.041        2.038           0.003547
11      2.204        2.203           0.001509
12      2.371        2.367           0.003692

SUM OF RESIDUALS              3.05311E-15
SUM OF SQUARED RESIDUALS      0.02647481
```

from the REG procedure and creates a data set called EST containing estimates of α' and β' in the linearized model $Y' = \alpha' + \beta' X + \varepsilon'$ as well as some other values. PROC PRINT DATA=EST produces Figure 10.15. We can find under AGE and INTERCEP the regression coefficients of the straight line fitted to the logarithms. The -1 under LRATE simply denotes that LRATE is the dependent variable, and the value labeled SIGMA is the square root of the error mean square (or standard error of estimate) from fitting the linear equation. Other information in Figure 10.15 says that there is one observation in the data set, that OLS—ordinary least squares—was used to obtain the parameter estimates, and since we did not give any label to the model fitted, MODEL is left blank.

FIGURE 10.15 Printout of the Data Set EST

```
OBS   TYPE   MODEL   SIGMA        LRATE   AGE        INTERCEP
1     OLS            0.0514537    -1      0.032953   -0.26886
```

Of interest to us in the data set EST are $a' = -0.26886$ and $b' = 0.032953$. We want to take the antilogarithms of these two values to obtain a and b in the exponential equation $Y = ab^X$. To do this, we use the data set EST to define a new data set CURVE. Using values in the set EST, we define A = antilog a', and B = antilog b' and define this new data set to be CURVE. That is, the data set CURVE contains the same values as the set EST, plus the values A and B, as shown in Figure 10.16. (Figure 10.16 was produced by the command PROC PRINT DATA=CURVE.) From Figure 10.16 we can find that the exponential equation fitted to the actual death rate values is

$$Y = (0.538438)(1.07883)^X.$$

FIGURE 10.16 Printout of the Data Set CURVE

OBS	TYPE	MODEL	SIGMA	LRATE	AGE	INTERCEP	A	B
1	OLS		0.0514537	-1	0.032953	-0.26886	0.538438	1.07883

PLOTTING THE STRAIGHT LINE THROUGH LOG Y AND ITS RESIDUALS Recall that the data set NEW contains values of X = AGE, Y = RATE, Y' = LRATE, $\hat{Y}'$ = YHAT, and RESID = $Y' - \hat{Y}'$. To plot the scatter diagram with regression equation imposed (using asterisks to plot the line) specify

```
PROC PLOT DATA=NEW;
PLOT LRATE*AGE YHAT*AGE='*'/OVERLAY;
```

as in simple linear regression. The graph is shown in Figure 10.17. Also as in simple regression, PLOT RESID*AGE/VREF=0 plots the residuals against X ("age") with a horizontal line drawn in at $Y' - \hat{Y}' = 0$. Note that it is not necessary to write another PROC PLOT command; that is, you may use multiple PLOT statements under one PROC PLOT command. These residuals should be examined for uniform scatter, and as shown in Figure 10.18, these residuals display a pronounced funnel shape.

PLOTTING THE DATA WITH THE EXPONENTIAL CURVE AND ITS RESIDUALS In order to plot the (X, Y)-values, we need to refer only to the original data set, or to the set NEW, both of which contain values of AGE and RATE. However, in order to plot the curved regression equation, we need to find the antilogarithms of the predicted values from the linearized model, and to plot the residuals from the exponential model we need the antilogarithms of the residuals from the linearized model. To get these antilogarithms, we refer to the data set NEW and define a new data set, ANTLG, by defining ANTYHT as antilog YHAT and ANTRSD as antilog RESID. Then using the data set ANTLG, we request the desired graphs, shown in Figures 10.19 and 10.20.

While this program is complicated, requiring that several new data sets be created and several new variables be defined, it does provide a very complete analysis. With a bit of practice, writing SAS programs such as this can become

FIGURE 10.17 Straight Line Fitted Through Logarithms Using SAS

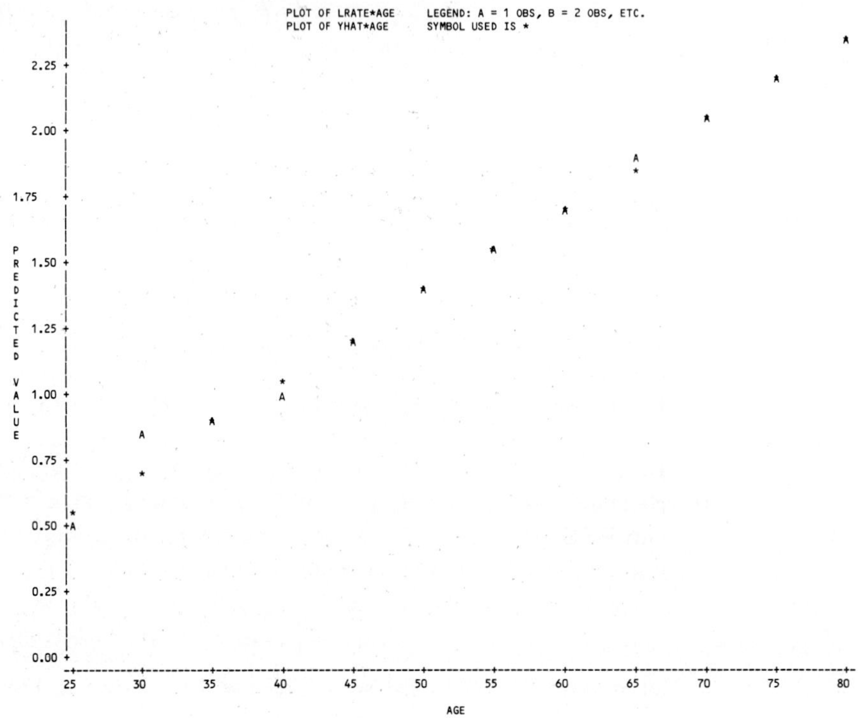

FIGURE 10.18 Residuals From Fitting a Straight Line to Logarithms Using SAS

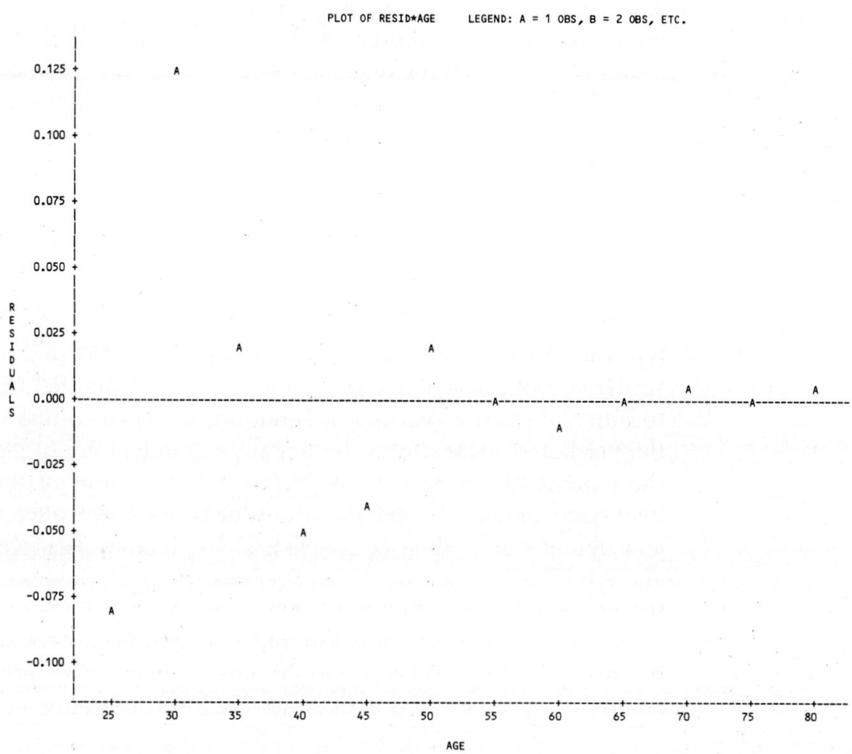

FIGURE 10.19 Exponential Function Fitted to Data Using SAS

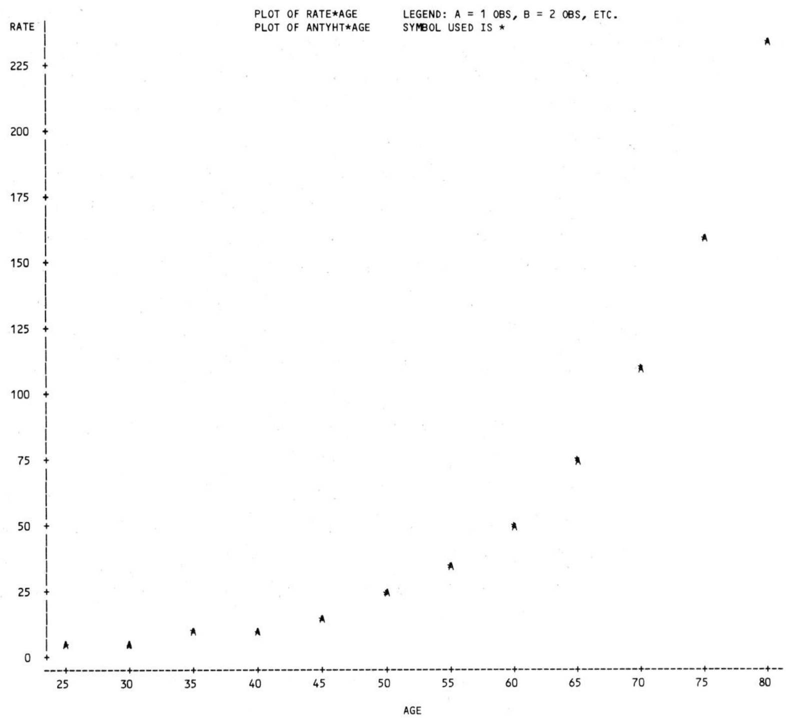

FIGURE 10.20 Residuals of Actual Values From the Exponential Regression Function

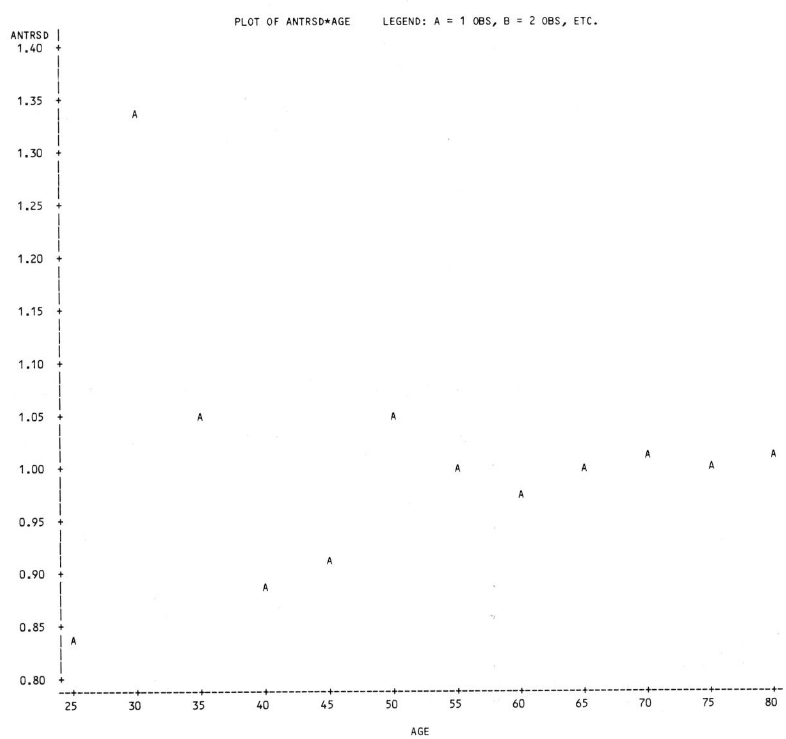

quite easy. Most regression analyses using SAS are more straightforward than this one, being written for more common analyses than fitting an exponential function.

EXPONENTIAL REGRESSION WITH SPSS^{X}

The worksheet in Figure 10.21 shows how to regress ln Y on X using SPSS^{X}. This program is very similar to the one for a simple linear regression. Values of variables AGE and RATE are indicated in the DATA LIST command.

FIGURE 10.21 Worksheet for Exponential Regression Using SPSS^{X}

```
  {SYSTEM CARDS
UNNUMBERED
TITLE SPSS-X EXPONENTIAL REGRESSION
DATA LIST LIST/
          AGE RATE
COMPUTE LRATE=LN(RATE)
BEGIN DATA
25 3
30 7
  : {DATA CARDS
80 235
END DATA
REGRESSION VARIABLES=AGE LRATE/
           DEPENDENT=LRATE/
           ENTER/
```

Before data cards are entered, the command

```
COMPUTE LRATE=LN(RATE)
```

is given. (*Note:* If common logarithms are to be used, define LRATE=L10(RATE). In our example, we have used the REGRESSION command, designating that the variables AGE and LRATE are to be used (but not RATE), with LRATE the dependent variable. Figure 10.22 shows printout from the REGRESSION command. Of course, values for the slope and intercept must be converted back from logarithms to be relevant to the original variables, AGE and RATE. Graphs and residual analyses could have been requested but are not shown here. The SCATTERGRAM procedure could have been used to fit a simple straight line through the logarithms as well.

FIGURE 10.22 SPSSX Exponential Regression

```
EQUATION NUMBER 1    DEPENDENT VARIABLE..LRATE
BEGINNING BLOCK NUMBER 1. METHOD: ENTER
VARIABLE(S) ENTERED ON STEP NUMBER 1..AGE

                                    ANALYSIS
MULTIPLE R            0.99661     OF VARIANCE
R SQUARE              0.99323                      DF   SUM OF SQUARES   MEAN SQUARE
ADJUSTED R SQUARE     0.99255     REGRESSION        1    20.58247         20.58247
STANDARD ERROR        0.11848     RESIDUAL         10     0.14037          0.01404
                                  F =  1466.33509   SIGNIF F = 0.0000

                            VARIABLES IN THE EQUATION

VARIABLE        B            SE B          BETA        T         SIG T
AGE            0.07588     1.9815E-03     0.99661    38.293     0.0000
(CONSTANT)    -0.61908     0.10951                   -5.653     0.0002

FOR BLOCK NUMBER 1   ALL REQUESTED VARIABLES ENTERED.
```

EXERCISES

10.1 Suppose that certain cancer cells are exposed to radiation in the laboratory and that the figures in Table 10.4 represent X = number of hours exposure, and Y = percent of cells killed.

TABLE 10.4 Cancer Data

X	Y
0.25	1.0
0.25	1.2
0.50	1.5
0.50	1.7
0.75	2.5
0.75	3.0
1.00	5.2
1.00	5.7
1.50	12.6
1.50	15.0
2.00	50.0
2.00	42.6

a. Plot the data. Does a linear model seem appropriate? Why, or why not?
b. Fit an exponential regression equation to these data.
c. Interpret the meaning of the values a and b in the regression equation.

10.2 In order to study the durability of a certain kind of abrasive, grinding wheels were coated with a 5 mm thickness of the abrasive and then were put into operation. Periodically, randomly chosen wheels were stopped and the thickness of remaining abrasive was measured. In Table 10.5 are data relating Y = thickness of remaining abrasive (mm) to X = number of hours of operation of grinding wheel.

TABLE 10.5
Abrasive Data

X	Y
40	3.30
40	2.60
40	2.45
50	1.65
50	0.98
50	0.78
60	0.24
60	0.56
60	0.80
70	0.10
70	0.30
70	0.40

a. Graph the data, and determine whether an exponential model seems appropriate to describe the relationship.
b. Fit an exponential model to the data.
c. Interpret the meaning of the parameter estimates.

10.3 In Table 10.6 are U.S. population figures from 1890 through 1970.

TABLE 10.6
Census Data: I
Source: U.S. Bureau of the Census

Year	Population (millions)
1890	62.95
1900	75.99
1910	91.97
1920	105.71
1930	122.78
1940	131.67
1950	150.70
1960	179.32
1970	200.25

a. Let X = number of decades since 1890, so that $X = 0$ in 1890, $X = 1$ in 1900, . . . , $X = 8$ in 1970. Fit an exponential function relating population size to X.
b. Interpret the meaning of the parameter estimates in the exponential function.
c. Project the U.S. population size in 1980. Refer to an almanac or some other source for the actual 1980 census figure to see how well the exponential function forecasted population size.

10.4 Refer to Exercise 10.3. Suppose that you are given additional information on U.S. population (Table 10.7).
a. Let X = number of decades since 1850, so that $X = 0$ in 1850, $X = 1$ in 1870, and so on. Fit an exponential function relating U.S. population to X from 1850 to 1970.
b. Plot the data and the exponential equation on the same graph. Does the exponential model appear to provide an adequate description of population growth?

TABLE 10.7 Census Data: II *Source:* U.S. Bureau of the Census

Year	Population (millions)
1850	23.19
1860	31.44
1870	39.82
1880	50.16

c. What does the exponential model forecast population to be in 1970? Compare to the actual 1970 population figure. Is the exponential model appropriate for forecasting U.S. population from 1970 on?

10.5 Refer to the Police Data in Exercise 1.7.

a. Fit an exponential regression equation to these data.

b. Interpret the values of a and b.

c. Plot the exponential function on a scatter diagram of the data and comment on the fit.

10.6 A consumer magazine obtained the information in Table 10.8 regarding miles per gallon, horsepower (ft-lb), weight (thousands of pounds), and transmission type (0 = automatic, 1 = manual) for a simple random sample of makes and models of automobiles sold in the United States in a given year.

a. Plot mpg versus horsepower and mpg versus weight. Differentiate between automobiles with manual and automatic transmissions. Are the relationships linear?

b. Fit an exponential regression equation relating mpg to horsepower. Interpret the regression coefficients. Plot the exponential function, and comment on how well it fits the data.

c. Fit an exponential regression equation relating mpg to weight. Interpret the regression coefficients. Plot the exponential equation, and comment on how well it describes the relationship.

d. Do you think that transmission type should be taken into account when analyzing these data?

TABLE 10.8 MPG Data

Automobile	mpg	HP	Weight	Transmission	Automobile	mpg	HP	Weight	Transmission
1	22	95	3.4	1	14	15	215	5.2	0
2	35	70	1.9	1	15	17	185	3.8	0
3	20	110	3.4	0	16	32	75	2.3	1
4	18	130	3.9	0	17	14	180	4.2	0
5	14	190	5.3	0	18	36	80	2.0	1
6	24	110	3.0	0	19	22	110	3.0	0
7	29	85	2.2	1	20	17	170	3.9	0
8	18	155	3.9	0	21	18	140	3.9	0
9	30	75	2.3	1	22	21	110	3.2	0
10	12	215	4.2	0	23	22	110	2.7	1
11	19	165	3.9	0	24	25	80	2.7	1
12	20	105	3.5	0	25	16	145	3.7	0
13	16	155	3.9	0					

10.4

ALTERNATIVES TO THE STRAIGHT-LINE MODEL: POLYNOMIAL MODELS

POLYNOMIAL EQUATIONS

Recall that the assumption for fitting a simple linear model is that Y changes by a *constant amount* with X and for an exponential model that Y changes at a *constant rate* (percentage) with X. It is sometimes the case, however, when curvature is observed in a scatter diagram, that Y is changing at a nonconstant rate with X. For example, the relationships suggested by Figures 10.23(a) and 10.23(b) would be well modeled neither by a linear function nor by an exponential function. Polynomial models may be adopted to describe the relationships in these cases, as well as in situations like those shown in Figures 10.23(c) and 10.23(d), in which at first glance exponential models may appear to be appropriate but in which the rate of change of Y with X turns out to be nonconstant.

Polynomial Model

A **polynomial model** of degree k is of the form

$$Y = \alpha + \beta_1 X + \beta_2 X^2 + \beta_3 X^3 + \cdots + \beta_k X^k + \varepsilon.$$

We see that the simple linear model is a polynomial model of degree 1. A polynomial equation of degree 1 will have no curve to it, a polynomial equation of degree 2 (a *quadratic equation*) will curve once, a cubic equation will curve in two directions, and so on. (See Figure 10.24.) Thus, it is possible to describe the relationship between two variables perfectly by fitting a polynomial of degree $n - 1$ to the n data points. This is hardly our aim, however, because it becomes increasingly difficult to interpret the meaning of the model as the degree of the polynomial becomes higher. We will thus concentrate on polynomial models of degree 2:

$$Y = \alpha + \beta_1 X + \beta_2 X^2 + \varepsilon.$$

In such a model the parameters are easily interpretable. The intercept, α, is the average value of Y when $X = 0$. As in the case of a simple linear regression,

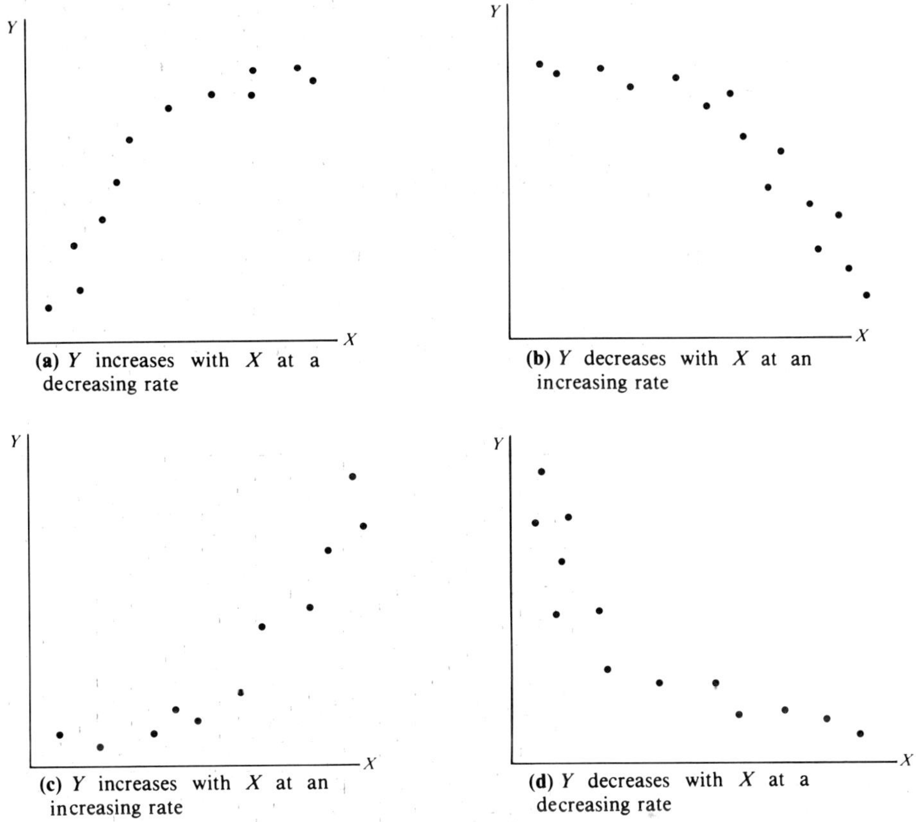

FIGURE 10.23 Some Candidates for Polynomial Models

(a) Y increases with X at a decreasing rate

(b) Y decreases with X at an increasing rate

(c) Y increases with X at an increasing rate

(d) Y decreases with X at a decreasing rate

whether or not the intercept is interpretable depends upon whether it makes sense for X to be zero and whether the data include values of X near or at zero. The parameter β_1, which would be the slope of a straight line, is in this case the slope of the tangent to the curve at the point $X = 0$. Since the quadratic equation describes a curve, the slope of the tangent to the curve changes as X changes. The parameter β_2 is one-half the number of units that the slope of the tangent to the curve changes as X increases by one unit.†

† Those readers familiar with calculus can verify these statements by recalling that the first derivative gives the slope and the second derivative gives acceleration. The first derivative with respect to X of $Y = \alpha + \beta_1 X + \beta_2 X^2$ is

$$\frac{dY}{dX} = \beta_1 + 2\beta_2 X.$$

If $X = 0$, then $dY/dX = \beta_1$, so that β_1 is the slope of the tangent to the curve at $X = 0$. The second derivative is

$$\frac{d^2Y}{dX^2} = 2\beta_2$$

so that acceleration is $2\beta_2$, or β_2 is one-half the acceleration.

FIGURE 10.24
Examples of Quadratic and Cubic Equations

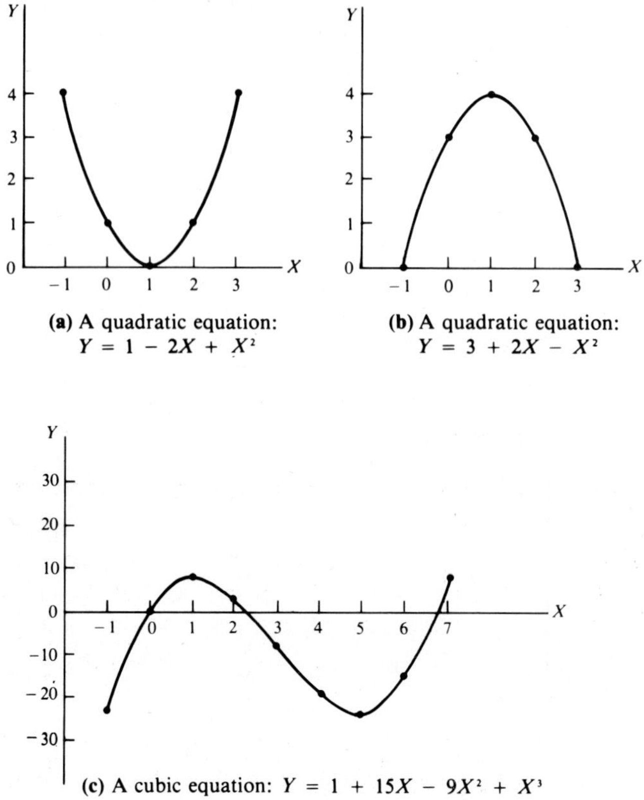

(a) A quadratic equation: $Y = 1 - 2X + X^2$

(b) A quadratic equation: $Y = 3 + 2X - X^2$

(c) A cubic equation: $Y = 1 + 15X - 9X^2 + X^3$

To illustrate this, consider the quadratic equation graphed in Figure 10.24(b)—$Y = 3 + 2X - X^2$. We see that if $X = 0$, $Y = 3$. The slope of the tangent to the curve over $X = 0$ is 2, as is illustrated in Figure 10.25. Then as X increases by 1, this slope increases by $2(-1)$; i.e., it decreases by 2. We see that the slope of the tangent is zero over $X = 0$, so that the curve has reached a maximum. As X increases from 0 to 1, the slope of the tangent decreases by another 2 and becomes -2, and so on. It requires some ingenuity to interpret these parameters with specific reference to the variables X and Y in the regression study, but it can be done. If the interpretation becomes too laborious or does not seem reasonable, this could be an indication that a polynomial model is not appropriate. We will delay interpretation of these parameters until we have a specific problem to refer to.

Note that a quadratic equation of the form $Y = \alpha + \beta_1 X + \beta_2 X^2$ plots a parabola. What we do by fitting a quadratic model to data is to fit a section of the parabola to our observations, as is illustrated in Figure 10.26. Also shown in Figure 10.26, there is some danger involved in fitting a quadratic regression equation to data if you try to predict values of Y for values of X outside the data. Extrapolation beyond the data can give wildly inaccurate estimates if the quadratic regression equation turns either upward or downward.

FIGURE 10.25 Geometric Interpretation of the Parameters in a Quadratic Equation

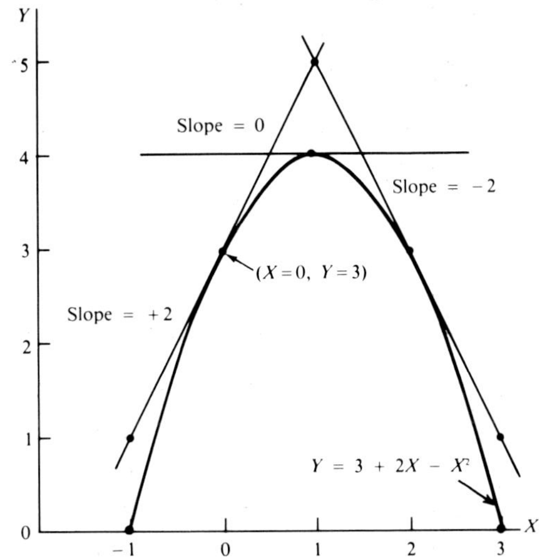

FIGURE 10.26 Fitting a Section of a Parabola to n Observations

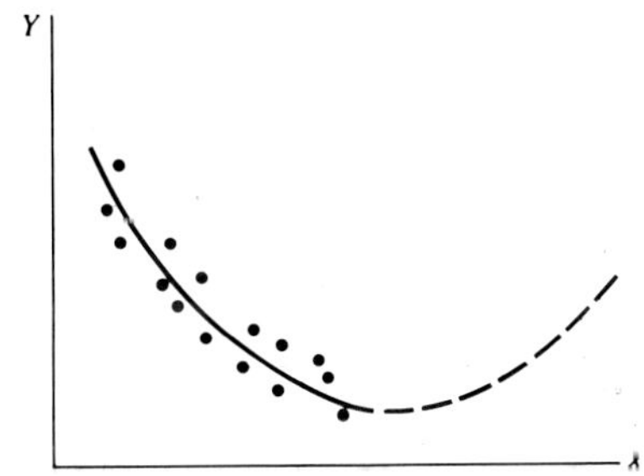

As always, apply the regression equation only to the range for which you have data, and don't be too free in extrapolating.

FITTING A POLYNOMIAL EQUATION TO DATA

We now turn to the question of how to find the regression coefficients for a quadratic regression equation. To find numerical values for a, b_1, and b_2 in

$$\hat{Y} = a + b_1X + b_2X^2$$

we will need to set up three normal equations. (Refer to the "trick" for writing normal equations explained in Section 2.3.) The normal equations are

$$\sum Y_i = na + b_1 \sum X_i + b_2 \sum X_i^2$$

$$\sum X_iY_i = a \sum X_i + b_1 \sum X_i^2 + b_2 \sum X_i^3$$

$$\sum X_i^2Y_i = a \sum X_i^2 + b_1 \sum X_i^3 + b_2 \sum X_i^4.$$

Solution of these three equations in three unknowns is a job for the computer, although it is, of course, possible to obtain the solution by hand.

The matrix formulation of the quadratic regression problem will be instructive, especially when we turn to the question of testing hypotheses concerning the regression coefficients. The quadratic regression model can be written as

$$\mathbf{y} = \mathbf{X}\boldsymbol{\beta} + \boldsymbol{\varepsilon},$$

where

$$\mathbf{y} = \begin{bmatrix} Y_1 \\ Y_2 \\ \vdots \\ Y_n \end{bmatrix} \quad \text{and} \quad \boldsymbol{\varepsilon} = \begin{bmatrix} \varepsilon_1 \\ \varepsilon_2 \\ \vdots \\ \varepsilon_n \end{bmatrix},$$

just as in simple regression, but where

$$\boldsymbol{\beta} = \begin{bmatrix} \alpha \\ \beta_1 \\ \beta_2 \end{bmatrix} \quad \text{and} \quad \mathbf{X} = \begin{bmatrix} 1 & X_1 & X_1^2 \\ 1 & X_2 & X_2^2 \\ \vdots & \vdots & \vdots \\ 1 & X_n & X_n^2 \end{bmatrix}.$$

As before, the normal equations can be written as

$$(\mathbf{X}'\mathbf{X})\hat{\boldsymbol{\beta}} = \mathbf{X}'\mathbf{y},$$

and their solution is

$$\hat{\boldsymbol{\beta}} = (\mathbf{X}'\mathbf{X})^{-1}(\mathbf{X}'\mathbf{y}),$$

where

$$\mathbf{X}'\mathbf{y} = \begin{bmatrix} 1 & 1 & \cdots & 1 \\ X_1 & X_2 & \cdots & X_n \\ X_1^2 & X_2^2 & \cdots & X_n^2 \end{bmatrix} \begin{bmatrix} Y_1 \\ Y_2 \\ \vdots \\ Y_n \end{bmatrix} = \begin{bmatrix} \sum Y_i \\ \sum X_i Y_i \\ \sum X_i^2 Y_i \end{bmatrix}$$

$$\mathbf{X}'\mathbf{X} = \begin{bmatrix} 1 & 1 & \cdots & 1 \\ X_1 & X_2 & \cdots & X_n \\ X_1^2 & X_2^2 & \cdots & X_n^2 \end{bmatrix} \begin{bmatrix} 1 & X_1 & X_1^2 \\ 1 & X_2 & X_2^2 \\ \vdots & \vdots & \vdots \\ 1 & X_n & X_n^2 \end{bmatrix} = \begin{bmatrix} n & \sum X_i & \sum X_i^2 \\ \sum X_i & \sum X_i^2 & \sum X_i^3 \\ \sum X_i^2 & \sum X_i^3 & \sum X_i^4 \end{bmatrix},$$

and obtaining $(\mathbf{X}'\mathbf{X})^{-1}$ requires inverting a 3 × 3 matrix.

Before we proceed with a numerical example of fitting a quadratic regression, let us point out that it should be clear that any higher-degree polynomial model can be formulated essentially as an extension of the methods used for the first- and second-degree models. Definition of **X**, **y**, and $\hat{\boldsymbol{\beta}}$ and calculation of $\mathbf{X}'\mathbf{y}$ and $\mathbf{X}'\mathbf{X}$ and its inverse are left as exercises for the reader.

FITTING A QUADRATIC EQUATION TO THE HOUSE SIZE-PRICE DATA

Suppose that upon inspection of the scatter diagram for the prices and sizes of 20 houses, we suspect that the increase of price with size seems to level off as size increases. That is, we want to investigate whether price increases at decreasing rate with size, and thus we want to examine the appropriateness of a quadratic model,

$$Y = \alpha + \beta_1 X + \beta_2 X^2 + \varepsilon.$$

We define

$$\mathbf{y} = \begin{bmatrix} 32 \\ 24 \\ \vdots \\ 28 \end{bmatrix}, \qquad \mathbf{X} = \begin{bmatrix} 1 & 1.8 & (1.8)^2 \\ 1 & 1.0 & (1.0)^2 \\ \vdots & \vdots & \vdots \\ 1 & 1.5 & (1.5)^2 \end{bmatrix} = \begin{bmatrix} 1 & 1.8 & 3.24 \\ 1 & 1.0 & 1.00 \\ \vdots & \vdots & \vdots \\ 1 & 1.5 & 2.25 \end{bmatrix}$$

$$\hat{\boldsymbol{\beta}} = \begin{bmatrix} a \\ b_1 \\ b_2 \end{bmatrix}.$$

Then

$$\mathbf{X}'\mathbf{y} = \begin{bmatrix} 690.0 \\ 1554.9 \\ 3965.89 \end{bmatrix}, \qquad \mathbf{X}'\mathbf{X} = \begin{bmatrix} 20 & 40 & 93.56 \\ 40 & 93.56 & 245.59 \\ 93.56 & 245.59 & 698.9576 \end{bmatrix},$$

and $(\mathbf{X}'\mathbf{X})^{-1}$ can be found to be

$$(\mathbf{X}'\mathbf{X})^{-1} = \begin{bmatrix} 2.04396 & -2.00408 & 0.43057 \\ -2.00408 & 2.10257 & -0.47051 \\ 0.43057 & -0.47051 & 0.10912 \end{bmatrix}.$$

Finally,

$$\hat{\boldsymbol{\beta}} = \begin{bmatrix} a \\ b_1 \\ b_2 \end{bmatrix} = \begin{bmatrix} 2.04396 & -2.00408 & 0.43057 \\ -2.00408 & 2.10257 & -0.47051 \\ 0.43057 & -0.47051 & 0.10912 \end{bmatrix} \begin{bmatrix} 690.0 \\ 1554.9 \\ 3965.89 \end{bmatrix} = \begin{bmatrix} 1.77590 \\ 20.46853 \\ -1.75566 \end{bmatrix}$$

so that the regression equation is

$$\hat{Y} = 1.77590 + 20.46853X - 1.75566X^2.$$

We can interpret the meaning of the regression coefficients as follows. The intercept, $a = 1.77590$, would be the average price of houses zero thousand square feet in size, if there were any such thing. All we can say is that $1,775.90 is a constant affecting price determined by something other than size. At sizes of houses around zero thousand square feet, the rate of increase of price with size is \$20,468.53 per 1,000 sq ft, but this rate of increase slows down as size increases. For each additional 1,000 sq ft increase in size, the rate of increase in price is 2(1.75566) or \$3,511.32 less.

TESTING FOR THE SIGNIFICANCE OF A QUADRATIC REGRESSION

There are several questions to be answered when fitting a quadratic regression. The *overall* significance test tests the hypothesis

$$H_0: \boldsymbol{\beta} = \mathbf{0} \quad \text{versus} \quad H_1: \boldsymbol{\beta} \neq \mathbf{0},$$

that is,

$$H_0: \boldsymbol{\beta} = \begin{bmatrix} \alpha \\ \beta_1 \\ \beta_2 \end{bmatrix} = \begin{bmatrix} 0 \\ 0 \\ 0 \end{bmatrix}.$$

We are performing a simultaneous inference: Is there *any* term in the model that can explain differences in values of Y? Note that H_0 can be rejected if any or all of the parameters of the model can be demonstrated to be significantly different from zero.

To test this hypothesis, we refer to the analysis of variance table in matrix form (Table 10.9). Note that the degrees of freedom for the total sum of squares in the matrix formulation is always n, since n is the number of terms in the sum $\mathbf{y}'\mathbf{y} = \sum_{i=1}^{n} Y_i^2$. For a quadratic regression, degrees of freedom

TABLE 10.9 ANOVA For Testing All Three Regression Parameters

Source	*SS*	*df*
Regression	$\hat{\boldsymbol{\beta}}'\mathbf{X}'\mathbf{y}$	3
Error	(by subtraction)	$n - 3$
Total	$\mathbf{y}'\mathbf{y}$	n

for regression are 3, because there are three coefficients in $\hat{\boldsymbol{\beta}}' = [a \quad b_1 \quad b_2]$, or because there are three terms in the sum $\hat{\boldsymbol{\beta}}'(\mathbf{X}'\mathbf{y}) = a\sum Y + b_1 \sum XY + b_2 \sum X^2Y$. Error degrees of freedom must then be $n - 3$. Typically, however, since the intercept does not tell us whether Y is related to X, we remove the intercept term from the regression sum of squares and obtain an analysis of variance table like Table 10.10. Thus, we are testing a reduced form of H_0, namely

$$H_0: \begin{bmatrix} \beta_1 \\ \beta_2 \end{bmatrix} = \begin{bmatrix} 0 \\ 0 \end{bmatrix}.$$

TABLE 10.10 ANOVA For Quadratic Regression

Source	SS	df
Regression	$\hat{\boldsymbol\beta}'\mathbf{X}'\mathbf{y} - \dfrac{(\sum Y_i)^2}{n}$	2
Error	(by subtraction)	$n - 3$
Total	$\mathbf{y}'\mathbf{y} - \dfrac{(\sum Y_i)^2}{n}$	$n - 1$

Filling in numerical values from our house size-price example, we have

$$\mathbf{y}'\mathbf{y} = \sum Y_i^2 = 26,178$$

$$\sum Y_i = 690$$

$$\hat{\boldsymbol\beta}'\mathbf{X}'\mathbf{y} = [1.77590 \quad 20.46853 \quad -1.75566] \begin{bmatrix} 690.0 \\ 1554.9 \\ 3965.89 \end{bmatrix} = 26,089.1338,$$

so that we get an analysis of variance table like Table 10.11. Comparing this calculated F to $F_{2,17,0.01} = 6.11$, we conclude that we are more than 99 percent sure that price is related either to size or to the square of size, or to both; That is, $\beta_1 \neq 0$, or $\beta_2 \neq 0$, or neither β_1 nor β_2 is equal to zero. Note that in Table 10.11 the regression sum of squares for the quadratic model is greater than for the simple linear model ($SS(R) = 2284.1338$ here as compared to 2255.8952 for the simple linear model) and that the error sum of squares is smaller (88.8662 here versus 117.1048 before), and that the total sum of squares $SS(T) = 2373$ remains the same regardless of the model adopted.

Since H_0 has been rejected, it is now of interest to investigate which one(s) of the regression coefficients is (are) different from zero. Had this overall test failed to reject H_0, it is extremely unlikely that individual tests on β_1 and β_2 would reveal either of them to be significantly different from zero (although such strange things can happen occasionally).

TABLE 10.11 ANOVA For Quadratic Model Fitted to House Size-Price Data

Source	SS	df	MS	F
Regression	$26,089.1338 - \dfrac{(690)^2}{20} = 2284.1338$	2	1142.0669	218.477
Error	(by subtraction) 88.8662	17	5.2274	
Total	$26,178 - \dfrac{(690)^2}{20} = 2373.0000$	19		

SEQUENTIAL TESTS ON INDIVIDUAL TERMS IN A QUADRATIC REGRESSION

Now that we have determined that either Y is a function of X, or that Y is a function of X^2, or that Y is a function of both X and X^2, the question is: Which of these three cases holds? That is, we now want to test two hypotheses

$$H_{01}:\beta_1 = 0 \quad \text{and} \quad H_{02}:\beta_2 = 0.$$

(For the sake of generality, we will assume that the alternative hypotheses are two-sided.) These tests on individual terms can be made in one of two ways, either by *sequential tests* or *partial tests*. Sequential tests will be discussed first, as they are typically more appropriate for polynomial regressions.

Sequential Test

In a **sequential test** of the hypothesis $H_0:\boldsymbol{\beta}' = [\beta_1 \;\; \beta_2] = 0$, we first specify which term in the equation should logically be the more important. This is typically the linear term. This term is allowed to explain as much of the variation in Y as it can; that is, a simple regression significance test of the regression of Y on X is performed. Then the other term (typically the quadratic) is assigned the portion of the overall regression sum of squares that was not used by the first term, and an F-test is used to test whether this additional sum of squares is large enough to attribute to anything other than chance.

Let the overall regression sum of squares, $SS(R) = 2284.1338$, be denoted as

$$SS(R)(X, X^2) = 2284.1338.$$

Then the regression sum of squares for the simple linear model is

$$SS(R)(X) = 2255.8952,$$

and the *additional sum* of squares for regression resulting from the inclusion of the quadratic term is

$$\begin{aligned} SS(R)(X^2|X) &= SS(R)(X, X^2) - SS(R)(X) \\ &= 2284.1338 - 2255.8952 = 28.2476. \end{aligned}$$

Note that this sum of squares involving X^2 is a *conditional* sum of squares. We are asking, in order, the following questions:

1. Can X explain a significant amount of variation in Y? We are testing

$$H_{01}\text{:the appropriate model is } Y = \alpha + \varepsilon$$

against

$$H_{11}\text{:the appropriate model is } Y = \alpha + \beta_1 X + \varepsilon.$$

2. Given that the model includes a linear (X) term, can the quadratic term

(X^2) add a significant amount to the explained variation in Y? We are testing

$$H_{02}\text{: the model is } Y = \alpha + \beta_1 X + \varepsilon$$

versus

$$H_{12}\text{: the model is } Y = \alpha + \beta_1 X + \beta_2 X^2 + \varepsilon.$$

The tests to answer these questions are summarized in Table 10.12. Note that each of the sequential sums of squares has 1 degree of freedom, and that the error mean square, with $n - 3$ degrees of freedom is used as the denominator of the F-statistic in both cases. The F-value for testing $H_{01}: Y = \alpha + \varepsilon$ is 431.552, and that for testing $H_{02}: Y = \alpha + \beta_1 X + \varepsilon$ is 5.404. Comparing these values to $F_{1,17,0.01} = 8.40$, we can be more than 99 percent sure that a linear term in X is effective in explaining variation in Y, but given that a linear term is to be used, we cannot be 99 percent sure that addition of a quadratic term will improve the explanatory power of the model. We thus conclude that any curvature apparent in the scatter diagram is not due to the existence of a quadratic relationship, but rather is just an aberration in this particular sample. A simple linear model should be used to describe the relationship between prices and sizes of houses.

TABLE 10.12
Sequential Tests Using House Size-Price Data

Source	*SS*	*df*	*MS*	*F*
Regression on X and X^2	2284.1338	2		
on X	2255.8952	1	2255.8952	431.552
on X^2, given X	28.2476	1	28.2476	5.404
Error	88.8662	17	5.2274	
Total	2373.0000	19		

Had we chosen to use $\alpha = 0.05$ to test the quadratic term, then we would have concluded that a quadratic model should be adopted. This would have contradicted the results of our lack-of-fit test, which indicated no significant lack of fit using the simple linear model. Recall, however, that the lack-of-fit test was not too trustworthy, since it had only 2 degrees of freedom for pure error. Also, not finding significant lack of fit to a linear model did not imply that the linear model was necessarily the correct one but only that it was *not inconsistent* with the data.

Using sequential tests requires that an a priori decision be made as to the order in which the linear and quadratic terms are to be tested. Since polynomial models become harder and harder to justify and explain the higher their degree, it is generally agreed that simpler models are to be preferred over more complicated ones. In fitting polynomial models, then, it is customary first to give the linear term a chance to "do its thing," then the quadratic, and then the cubic, fourth-degree, and so on. The results would have been quite different had we first entered the quadratic term in this example, as Table

10.13 shows. In this case, $H_{01}: Y = \alpha + \varepsilon$ would be rejected in favor of $H_{11}: Y = \alpha + \beta_2 X^2 + \varepsilon$, saying that X^2 by itself can explain variation in Y. But also $H_{02}: Y = \alpha + \beta_2 X^2 + \varepsilon$ is rejected in favor of $H_{12}: Y = \alpha + \beta_1 X + \beta_2 X^2 + \varepsilon$, which implies that even if X^2 is going to be used to explain variation in Y, addition of a linear term can improve what a quadratic term could do alone.

TABLE 10.13 Sequential Tests with a Quadratic Term Entered First

Source	*SS*	*df*	*MS*	*F*
Regression	2294.1338	2		
on X^2	2084.8722	1	2084.8722	398.835
on X, given X^2	199.2616	1	199.2616	38.119
Error	88.8662	17	5.2274	
Total	2373.0000	19		

This is not to say that lower-degree terms should always be considered first. There may indeed be situations in which it seems reasonable that the model should be

$$Y = \alpha + \beta X^2 + \varepsilon,$$

so that no linear term is even included. In other instances, the quadratic term may logically be the more important one, but it is of interest to see if a linear term should also be included.

Some difficulties can be encountered with sequential F-tests on individual terms. Suppose it is decided to test

$$H_{01}: Y = \alpha + \varepsilon$$

$$H_{11}: Y = \alpha + \beta_1 X + \varepsilon$$

and then

$$H_{02}: Y = \alpha + \beta_1 X + \varepsilon$$

$$H_{12}: Y = \alpha + \beta_1 X + \beta_2 X^2 + \varepsilon$$

and it turns out that you cannot reject H_{01}. Then what meaning does the second hypothesis have? In such a case, the order of testing the terms should probably be reversed.

PARTIAL TESTS ON INDIVIDUAL TERMS IN A QUADRATIC REGRESSION

Partial F-Tests

Partial F-tests of the individual terms in a quadratic regression do not require that the order of the terms be specified, but neither do they show the individual effect of any single term. Partial F-tests are all conditional, assuming that the other variable or variables have already been entered in the model. That is, using partial tests of a quadratic model, one tests

$$H_{01}: Y = \alpha + \beta_2 X^2 + \varepsilon$$

$$H_{11}: Y = \alpha + \beta_1 X + \beta_2 X^2 + \varepsilon$$

and

$$H_{02}: Y = \alpha + \beta_1 X + \varepsilon$$

$$H_{12}: Y = \alpha + \beta_1 X + \beta_2 X^2 + \varepsilon.$$

The F-tests for both of these hypotheses, as shown above, use the 1 degree of freedom sums of squares.

$$SS(R)(X|X^2) = SS(R)(X, X^2) - SS(R)(X^2)$$

and

$$SS(R)(X^2|X) = SS(R)(X, X^2) - SS(R)(X),$$

respectively, as their numerators, and the $n - 3$ degrees of freedom error mean square from fitting the "full model." Thus,

$$F_1 = \frac{199.2616/1}{5.2274} = 38.119 \quad \text{and} \quad F_2 = \frac{28.2476/1}{5.2274} = 5.404.$$

Note that these partial sums of squares do not total to the sum of squares for regression for the full model.

From these F-statistics we could conclude that (1) addition of a linear term to a quadratic term will result in a better model than using the quadratic term alone, but (2) addition of a quadratic term will not (with 99 percent certainty) improve over a linear model.

It is also possible to conduct partial tests of the terms using t-tests. As always, a t-statistic is calculated as

$$t_i = \frac{b_i}{\text{standard error of } b_i}.$$

We have found values for b_1 and b_2 already. To estimate the standard errors of b_1 and b_2, we must refer to the $(\mathbf{X}'\mathbf{X})^{-1}$ matrix. Recall that in a simple linear regression, $(\mathbf{X}'\mathbf{X})^{-1}$ was found to be

$$(\mathbf{X}'\mathbf{X})^{-1} = \begin{bmatrix} \dfrac{\sum X_i^2}{nS_{XX}} & \dfrac{\sum X_i}{nS_{XX}} \\ \dfrac{-\sum X_i^2}{nS_{XX}} & \dfrac{n}{nS_{XX}} \end{bmatrix}$$

and that the standard error of b was

$$\sqrt{\frac{s_{Y \cdot X}^2}{S_{XX}}} = \sqrt{s_{Y \cdot X}^2 \left(\frac{1}{S_{XX}}\right)}.$$

Note that the multiplier of $s^2_{Y \cdot X}$, $1/S_{XX}$, is found in the (2, 2) position of $(\mathbf{X}'\mathbf{X})^{-1}$. Think of the first row and column of $(\mathbf{X}'\mathbf{X})^{-1}$ corresponding to the estimate a and the second row and column as corresponding to b. Suppose we number the rows and columns starting with zero instead of with 1. Then $1/S_{XX}$ becomes the (1, 1) element. Now, in this quadratic regression, we found

$$(\mathbf{X}'\mathbf{X})^{-1} = \begin{bmatrix} 2.04396 & -2.00408 & 0.43057 \\ -2.00408 & 2.10257 & -0.47051 \\ 0.43057 & -0.47051 & 0.10912 \end{bmatrix}.$$

Think of the first row and column as corresponding to a, the second to b_1, and the third to b_2, and number rows and columns beginning with zero. Then the (1, 1) element, 2.10257, corresponds to b_1 and the (2, 2) element, 0.10912, corresponds to b_2. Denote these elements by c_{11} and c_{22}, respectively. Then the standard error of b_1 is given by

$$\sqrt{s^2_{Y \cdot X}\, c_{11}},$$

where $s^2_{Y \cdot X}$ is the error mean square for the full model. The standard error of b_1 is then

$$\sqrt{5.2274(2.10257)} = 3.3153,$$

and the standard error of b_2 is

$$\sqrt{5.2274(0.10912)} = 0.7553.$$

Thus, the t-statistic for testing

$$H_{01}: Y = \alpha + \beta_2 X^2 + \varepsilon$$

versus

$$H_{11}: Y = \alpha + \beta_1 X_1 + \beta_2 X^2 + \varepsilon$$

is

$$t_1 = \frac{20.46853}{3.3153} = 6.174,$$

and the t-statistic for testing

$$H_{02}: Y = \alpha + \beta_1 X + \varepsilon$$

versus

$$H_{12}: Y = \alpha + \beta_1 X + \beta_2 X^2 + \varepsilon$$

is

$$t_2 = \frac{-1.75566}{0.7553} = 2.324.$$

Assuming two-sided alternatives and an $\alpha = 0.01$ significance level, the critical value of t is $\pm\, t_{17,0.005} = \pm 2.898$. Thus, we may be more than 99 percent sure that the linear term should be included if the quadratic term is used, but cannot be 99 percent sure that we need a quadratic term in addition to a linear term.

A quick check of these t-values shows that they are equivalent to the partial F-values: $t_1^2 = (6.174)^2 = 38.118 = F_1$, and $t_2^2 = (-2.324)^2 = 5.401 = F_2$.

Almost all computer programs will give partial F- or t-values. Some of the programs we have studied will give sequential F- and t-values. We now look at the BMD, Minitab, SAS, and SPSS^X programs to see how to obtain the analyses we want.

10.5 USING THE COMPUTER TO FIT POLYNOMIAL MODELS

POLYNOMIAL REGRESSION WITH BMD

USING P1R TO PERFORM POLYNOMIAL REGRESSION

By inserting a `TRANSFORM` paragraph, it is possible to obtain a polynomial regression analysis using the BMDP1R program, although BMD provides a special program for polynomial regressions that most people prefer to use. The worksheet in Figure 10.27 shows how to instruct the computer to create a new variable, X^2, from the data entered on the data cards and to use this variable in the analysis. (The reader should compare this worksheet to that in Figure 10.5, where an exponential function was fitted.)

The `PROBLEM` and `INPUT` paragraphs are the same as for fitting a simple linear equation. Values of two variables, `SIZE` and `PRICE`, are entered on data cards. In the `VARIABLES` paragraph, however, the computer is alerted that it is to create an additional variable by the inclusion of the sentence `ADD=1` and that this variable will be named `SZSQ` (size squared). The `TRANSFORM` paragraph tells the computer how to create the variable `SZSQ` by writing `SZSQ=SIZE**2.` The computer now has available three variables, Y, X, and X^2, with which to work. If cubic or higher-order models

FIGURE 10.27 Worksheet for Using BMDP1R to Perform a Quadratic Regression

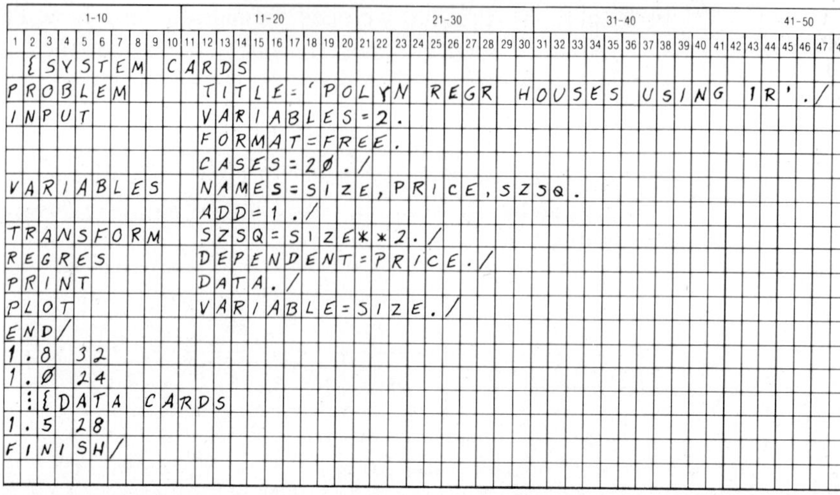

are to be investigated, it should be apparent how to instruct the computer to calculate and use values of X^3, and so on. Note that the ADD sentence in the VARIABLES paragraph must specify the number of variables the computer is to create and that all variables must be named in the NAMES sentence.

In the REGRESSION paragraph the dependent variable is specified to be PRICE, and since no INDEPENDENT sentence is used, the computer will use all other variables except PRICE—that is, SIZE and SZSQ—as independent variables. Thus, an equation of the form $\hat{Y} = a + b_1X + b_2X^2$ will be fitted to the data.

Output from the PRINT paragraph will contain values of Y, X, X^2, $\hat{Y}$, and $Y - \hat{Y}$, and will not be reproduced here. The PLOT paragraph as written in Figure 10.27 will plot Y against X, and $\hat{Y}$ against X, on a single graph, and $Y - \hat{Y}$ against X on a separate graph. This plot will not be reproduced here, as it is the same plot shown in Figure 10.30, in which the BMDP5R program is used for polynomial regression.

Figure 10.28 shows the output from the REGRESSION paragraph. (The table of simple statistics is omitted.) In addition to the analysis of variance table and measures of correlation from fitting a second-degree polynomial, we see estimates of the regression coefficients and associated statistics. Note that the t-statistics for the regression coefficients are calculated as partial t-statistics, so that they test the significance of that term in the model, given the others. The BMDP1R program does not calculate sequential sums of squares or F-statistics, so using this program we cannot see the effect of the linear term first, followed by the effect of the quadratic term. The worksheet in Figure 10.27 could be altered to show a sequential analysis, however. If the REGRESSION paragraph included a sentence INDEP=SIZE, then a simple linear regression of price on size would be performed. Adding another REGRESSION paragraph before the END paragraph, in which both X and X^2 are used as

FIGURE 10.28 BMDP1R Quadratic Regression for House Size-Price Data

```
REGRESSION TITLE IS
POLYN REGR HOUSES USING 1B
DEPENDENT VARIABLE. . . . . . . . . . . . . . . . . 2  PRICE
TOLERANCE . . . . . . . . . . . . . . . . . . . . . 0.0100
ALL DATA CONSIDERED AS A SINGLE GROUP

MULTIPLE R          0.9811    STD ERROR OF EST    2.2862
MULTIPLE R-SQUARE   0.9626

ANALYSIS OF VARIANCE
                      SUM OF SQUARES    DF    MEAN SQUARE    F RATIO    P(TAIL)
         REGRESSION   2284.1472          2    1142.0735      218.513    0.0000
         RESIDUAL       88.8518         17       5.2266

                                           STD REG
VARIABLE      COEFFICIENT    STD ERROR     COEFF        T         P(2 TAIL)    TOLERANCE
INTERCEPT       1.77599
SIZE   1       20.46852       3.31497      1.547       6.175     0.0000       0.03508
SZSQ   3       -1.75566       0.75519     -0.583      -2.325     0.0327       0.03508
```

predictors would produce another analysis, as in Figure 10.28, using the quadratic model. We would thus have first an analysis for a first-degree (linear) polynomial, followed by the analysis for a second-degree model. Comparison of the partial *t*-statistics for the regression coefficients in the two analyses would give us some indication of the worth of including the quadratic term in addition to the linear term.

While it is simple to use BMDP1R for polynomial regression, there is a program designed specifically for polynomial regression, the BMDP5R.

USING P5R TO PERFORM POLYNOMIAL REGRESSION

The BMDP5R program was designed to investigate polynomial models of various degrees by first fitting a simple linear model, then a quadratic, then a cubic, and so on. The highest-degree polynomial that can be analyzed is 15, but (unless you specify otherwise) the default conditions are that polynomials of degrees 1, 2, and 3 will be considered. Figure 10.29 shows a worksheet for using BMDP5R. Very little differs here from the worksheet shown in Figure 10.27 for performing a simple linear regression. What determines the difference in the analysis performed is that in the system cards will be the statement

```
. . . EXEC BIMED,PROG=BMDP5R . . . .
```

According to the instruction manual, the BMDP5R program *requires* that the `INDEP` sentence be included in the `REGRESSION` paragraph. (It was not always required in the P1R program.) Unless otherwise specified, P5R will fit polynomials through degree 3. To change this, insert a sentence `DEGREE = #` in the `REGRESSION` paragraph, where the "#" is number representing the highest degree you are interested in (an integer up to and including 15).

The `PRINT` and `PLOT` paragraphs contain the sentence `DEGREE=1,2,3`. This sentence says that you want to see data printed and plots made for each of the three analyses.

FIGURE 10.29 Worksheet for a BMDP5R Polynomial Regression Analysis

```
  {SYSTEM CARDS
PROBLEM    TITLE='POLYNOMIAL REGRESSION FOR HOUSES DATA'./
INPUT      VARIABLES=2.
           FORMAT=FREE.
           CASES=2Ø./
VARIABLES  NAMES=SIZE,PRICE./
REGRES     DEPENDENT=PRICE.
           INDEP=SIZE./
PRINT      DEGREE=1,2,3./
PLOT       DEGREE=1,2,3./
END/
1.8 32
1.Ø 24
  :  {DATA CARDS
1.5 28
FINISH/
```

PRINTOUT FROM P5R The table of simple statistics for "size" and "price" is the same as it was when these data were analyzed in a simple linear model; it is not repeated here. Similarly, printouts of values of X, Y, $\hat{Y}$, and $Y - \hat{Y}$ are omitted. Figure 10.30 shows the printout and plots from the BMDP5R fitting of a first-degree polynomial. First look at the right-hand side of Figure 10.30(a), labeled `POLYNOMIAL IN X`. Here we see regression coefficients, standard errors, and partial t-values. The value `0.87035395D 01` is listed across from `DEGREE 0`. `DEGREE 0` corresponds to the intercept, and the notation `D 01` indicates that the decimal point should be moved 1 space to the right. That is, a is found to be 8.7035395, as when fitting a straight line. Similarly, b is listed as `0.12898232D 02`, which is 12.898232. The associated t-values indicate that a and b are both significantly different from zero (although the computer does not print attained significance levels in P5R as it did in P1R. The requested plots in Figure 10.30(b) show exactly the same graphs as in Figure 4.4, in which P1R fitted a simple linear equation to these data.

FIGURE 10.30(a) Printout From First-Degree Polynomial Regression Using BMDP5R

```
RESULTS FOR POLYNOMIAL OF DEGREE 1

POLYNOMIAL
COEFFICIENTS        ORTHOGONAL POLYNOMIAL                        POLYNOMIAL IN X
                 REGRESSION     STANDARD   T          REGRESSION       STANDARD        T
DEGREE           COEFFICIENT    ERROR      VALUE      COEFFICIENT      ERROR           VALUE
0                154.28869045   2.55059    60.49      0.87035395D 01   0.14981D 01      5.81
1                 47.49632001   2.55059    18.62      0.12898232D 02   0.69265D 00     18.62

RESIDUAL MEAN SQUARE = 6.50553 (DF=18)
MULTIPLE R-SQUARE =    0.95065
```

FIGURE 10.30(b) Plots From First-Degree Polynomial Regression Using BMDP5R

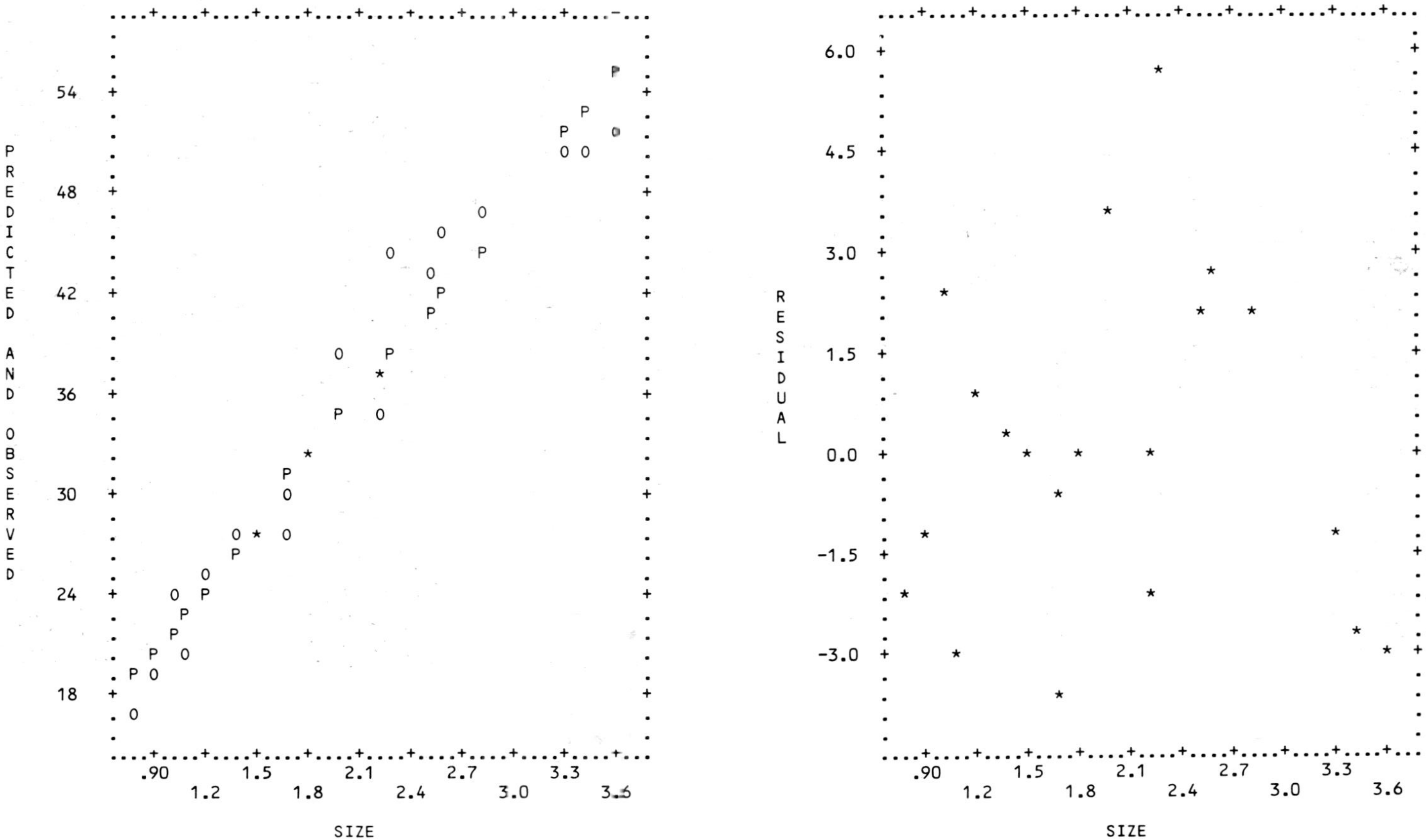

What is new to us in Figure 10.30(a) is the information under the heading ORTHOGONAL POLYNOMIAL. A brief explanation of orthogonal polynomials is given in Appendix E, and the interested reader is referred there. In summary, what orthogonal polynomials do is transform the X-values so that the matrix $\mathbf{X'X}$ can be made diagonal and thus easily inverted. Regression coefficients and related statistics are given relating Y to transformed values of X under the heading ORTHOGONAL POLYNOMIAL. Then the equation is transformed back into terms of X and relevant information is given under POLYNOMIAL IN X.

Figure 10.31(a) shows the printout from the analysis of a quadratic regression. We see that the regression equation is of the form

$$\hat{Y} = 1.7759021 + 20.468535X - 1.7556579X^2,$$

FIGURE 10.31(a) Printout From Second-Degree Polynomial Regression Using BMDP5R

RESULTS FOR POLYNOMIAL OF DEGREE 2

POLYNOMIAL COEFFICIENTS	ORTHOGONAL POLYNOMIAL			POLYNOMIAL IN X		
DEGREE	REGRESSION COEFFICIENT	STANDARD ERROR	T VALUE	REGRESSION COEFFICIENT	STANDARD ERROR	T VALUE
0	154.28869045	2.28617	67.49	0.17759021D 01	0.32685D 01	0.54
1	47.49632001	2.28617	20.78	0.20468535D 02	0.33150D 01	6.17
2	-5.31485468	2.28617	-2.32	-0.17556579D 01	0.75519D 00	-2.32

RESIDUAL MEAN SQUARE = 5.22658 (DF=17)
MULTIPLE R-SQUARE = 0.96256

with associated t-values of 0.54, 6.17, and -2.32, respectively. This indicates that the addition of the quadratic term to the model possibly improves the fit of the data to the model, that the linear term should be retained if the quadratic term is to be used, and that the intercept might possibly be zero (if that made sense). Again, these t-statistics are partial t's, conditional on inclusion of the other terms in the model. The graphs associated with this step in the BMDP5R analysis are shown in Figure 10.31(b). Note the slight curvature in the graph of the regression equation.

The printout resulting from the third step of the analysis is shown in Figure 10.32(a). Here, it appears that the intercept and cubic terms are conditionally significant, but that the linear and quadratic terms are not. Since these partial t-statistics are conditional, given all other terms in the model, they become increasingly difficult to interpret as the degree of the polynomial increases. For example, it appears in Figure 10.32(a) that we might want to adopt the model

$$\hat{Y} = \alpha + \beta_2 X^2 + \beta_3 X^3 + \varepsilon,$$

but this only brings up the question of how good the model

$$\hat{Y} = \alpha + \beta_2 X^2 + \varepsilon$$

would be, and we have no analysis of a quadratic model without a linear term. Inspection of the left-hand graph in Figure 10.32(b) shows that the estimated

FIGURE 10.31(b) Plots From Second-Degree Polynomial Regression Using BMDP5R

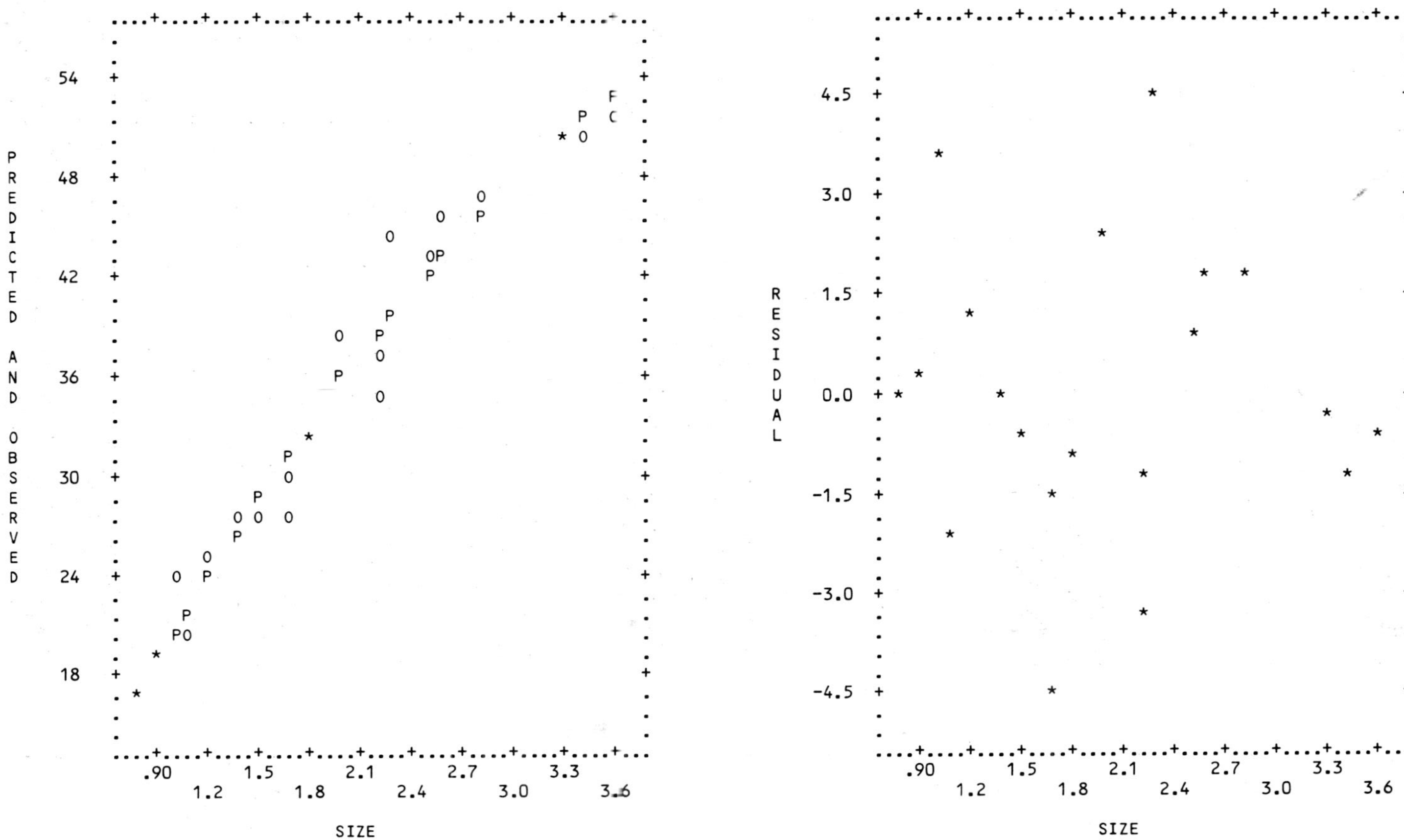

FIGURE 10.32(a) Printout From Third-Degree Polynomial Regression Using BMDP5R

```
RESULTS FOR POLYNOMIAL OF DEGREE 3

POLYNOMIAL
COEFFICIENTS          ORTHOGONAL POLYNOMIAL                          POLYNOMIAL  IN X
                  REGRESSION     STANDARD   T        REGRESSION           STANDARD       T
DEGREE            COEFFICIENT    ERROR      VALUE    COEFFICIENT          ERROR          VALUE
0                 154.28869045   2.16964    71.11     0.15361981D 02      0.85918D 01     1.79
1                  47.49632001   2.16964    21.89    -0.30816269D 01      0.14240D 02    -0.22
2                  -5.31485468   2.16964    -2.45     0.10221336D 02      0.70996D 01     1.44
3                  -3.67895701   2.16964    -1.70    -0.18334733D 01      0.10813D 01    -1.70

RESIDUAL MEAN SQUARE = 4.70732 (DF=16)
MULTIPLE R-SQUARE =    0.96826
```

regression equation is *S*-shaped, and this type of relationship may be difficult to interpret on theoretical or practical grounds. Thus, while the fit can always be improved by adopting a higher-degree model, it is unusual for cubic or higher-degree models to be proposed.

Finally, the BMDP5R program gives us some summary information, as shown in Figure 10.33. Table 10.14 explains the message above the summary table in Figure 10.33. Thus, the 3 degrees of freedom sum of squares 2297.68282 measures the improvement in using $\hat{Y} = a + bX + cX^2 + dX^2$ to predict Y over using just $\overline{Y}$ to predict Y. The 2 degrees of freedom sum of squares measures how much addition of quadratic and cubic terms improve over using just a linear term, and the 1 degree of freedom sum of squares measures how much inclusion of X^3 improves over a model using X and X^2. Thus, these sums of squares and associated F-tests are sequential but not in the way we have previously seen a sequence of sequential tests.

TABLE 10.14 Explanation of Information Given by BMDP5R Summary Table

Degree	SS	df	Explanation
0	2297.68282	3	$SS(R)(X, X^2, X^3)$
1	41.78240	2	$SS(R)(X, X^2, X^3) - SS(R)(X) = SS(R)(X^2, X^3 \mid X)$
2	13.53472	1	$SS(R)(X, X^2, X^3) - SS(R)(X, X^2) = SS(R)(X^3 \mid X, X^2)$
Resid	75.31718	16	$SS(E)(X, X^2, X^3)$

The way to interpret the F-statistics is as follows: Since $F = 162.70$ is significant ($p = 0.0000$), you should use a polynomial degree higher than zero. Since $F = 4.44$ is significant ($p = 0.0293$), you should consider using a polynomial of degree higher than 1. Since $F = 2.88$ is not significant ($p = 0.1093$), a polynomial of degree higher than 2 should not be used. Thus, we conclude that a quadratic equation probably gives the best fit to the data, and we would refer to Figure 10.31, where the analysis for the second-degree polynomial is given.

Since the P5R program provides this summary information, it is to be preferred over performing multiple analyses of polynomials of varying degrees using the P1R. Both are equally simple to run.

FIGURE 10.32(b) Plots From Third-Degree Polynomial Regression Using BMDP5R

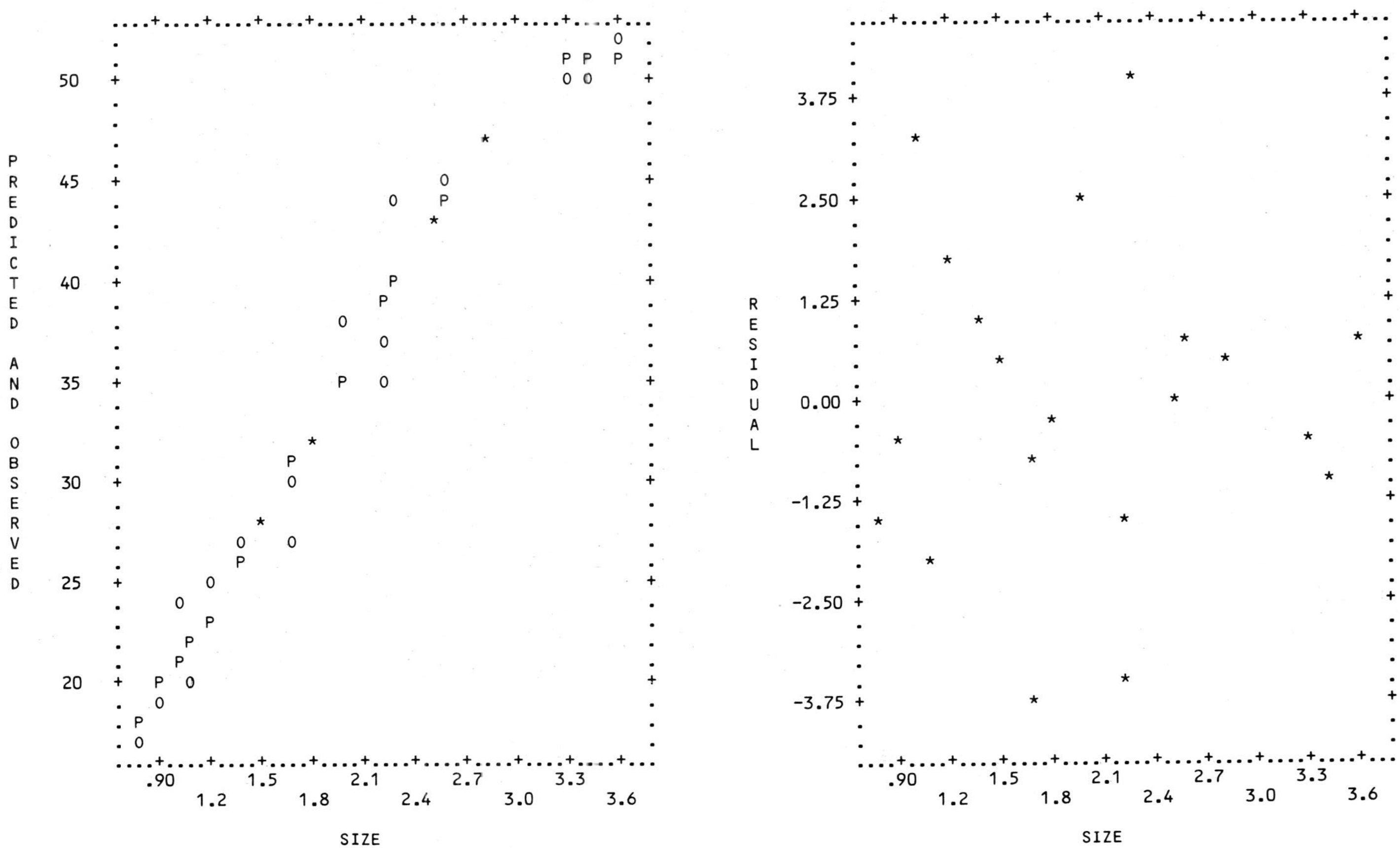

FIGURE 10.33 Summary of BMDP5R Polynomial Regression Analysis of House Size-Price Data

```
GOODNESS OF FIT TEST

FOR THE POLYNOMIAL OF EACH DEGREE, A TEST IS MADE FOR
ADDITIONAL INFORMATION IN THE ORTHOGONAL POLYNOMIALS OF
HIGHER DEGREE. THE NUMERATOR SUM OF SQUARES FOR EACH OF
THESE TESTS IS THE SUM OF SQUARES ATTRIBUTED TO ALL
ORTHOGONAL POLYNOMIALS OF HIGHER DEGREE AND THE DENOMINATOR
SUM OF SQUARES IS THE RESIDUAL SUM OF SQUARES FROM THE FIT
TO THE HIGHEST DEGREE POLYNOMIAL (FIT TO ALL ORTHOGONAL
POLYNOMIALS). A SIGNIFICANT F-STATISTIC THUS INDICATES THAT
A HIGHER DEGREE POLYNOMIAL SHOULD BE CONSIDERED.

DEGREE   SUM OF SQUARES   DF   MEAN SQUARE    F        TAIL PROBABILITY
0        2297.68282        3   765.89427     162.70    0.0000
1          41.78240        2    20.89120       4.44    0.0293
2          13.53472        1    13.53472       2.88    0.1093
RESID      75.31718       16     4.70732
```

POLYNOMIAL REGRESSION WITH MINITAB

Suppose that we want to fit a second-degree polynomial regression equation to the house size-price data using Minitab. In order to do this, we ask Minitab to fit a multiple regression using two predictors, where the second predictor is just the square of the first. That is, our model is

$$Y = \alpha + \beta_1 X_1 + \beta_2 X_2 + \varepsilon,$$

where $X_2 = X_1^2$.

Enter values of "size" and "price" into, say, columns 1 and 2, respectively. Then enter values of "size squared" into column 3:

```
RAISE C2 TO THE POWER 2, PUT IN C3
```

You may want to

```
NAME C1 'SIZE' C2 'PRICE' C3 'SIZESQ'
```

Then ask that a regression be performed using two predictors:

```
REGRES C2 ON 2 VAR IN C1 AND C3\C4\C5\C6
```

where we are requesting that the standardized residuals, predicted values, and parameter estimates be stored in columns C4, C5, and C6, respectively. Figure 10.34 shows the printout, if the NOBRIEF command is used. (Most of the table listing residuals, predicted values, and so on, has been left out here.)

The regression equation is found to be

$$\text{Price} = 1.776 + 20.469(\text{size}) - 1.7557(\text{size squared})$$

and the partial t-values for testing the parameters in the model are also given. The t-value for testing

$$H_0: Y = \beta_1 X + \beta_2 X^2 + \varepsilon$$

$$H_1: Y = \alpha + \beta_1 X + \beta_2 X^2 + \varepsilon$$

is seen to be $t = 0.54$, indicating that it might be reasonable to fit the regression

FIGURE 10.34 Second-Degree Polynomial Regression with Minitab

```
-NOBRIEF
-REGRES C2 ON 2 VAR IN C1 AND C3\C4\C5\C6

THE REGRESSION EQUATION IS
Y = 1.78 + 20.5 X1 - 1.76 X2
                                    ST DEV     T-RATIO =
       COLUMN     COEFFICIENT      OF COEF    COEF/S.D.
                     1.776          3.268        0.54
X1     SIZE         20.469          3.315        6.17
X2     SIZESQ       -1.7557         0.7552      -2.32

THE ST DEV OF Y ABOUT REGRESSION LINE IS
S = 2.286
WITH (20 - 3) = 17 DEGREES OF FREEDOM

R-SQUARED = 96.3 PERCENT
R-SQUARED = 95.8 PERCENT, ADJUSTED FOR DF

ANALYSIS OF VARIANCE

DUE TO           DF     SS          MS=SS/DF

REGRESSION        2    2284.148     1142.074
RESIDUAL         17      88.852        5.227
TOTAL            19    2373.000

FURTHER ANALYSIS OF VARIANCE
SS EXPLAINED BY EACH VARIABLE WHEN ENTERED IN THE ORDER GIVEN

DUE TO           DF     SS

REGRESSION       2     2284.148
SIZE             1     2255.900
SIZESQ           1       28.248

      X1      Y         PRED Y    ST DEV
ROW  SIZE    PRICE      VALUE     PRED Y     RESIDUAL    ST RES
 1   1.80    32.000     32.931    0.682      -0.931      -0.43
 .    .        .          .         .           .          .
 .    .        .          .         .           .          .
 .    .        .          .         .           .          .
20   1.50    28.000     28.528    0.632      -0.528      -0.24

R DENOTES AN OBS WITH A LARGE ST RES.
X DENOTES AN OBS WHOSE X VALUE GIVES IT LARGE INFLUENCE.

DURBIN-WATSON STATISTIC = 2.48

(X-PRIME X)INVERSE

        0            1           2

0    2.04396
1   -2.00408     2.10257
2    0.43057    -0.47051     0.10912
```

equation through the origin. To test

$$H_0: Y = \alpha + \beta X^2 + \varepsilon$$

$$H_1: Y = \alpha + \beta_1 X + \beta_2 X^2 + \varepsilon$$

refer to the value $t = 6.17$. The linear term clearly contributes to the model. The t-value, $t = -2.32$, might be compared to $-t_{17,0.025} = -2.110$ and to $-t_{17,0.01} = -2.567$, indicating that we can be more than 95 percent sure, but less than 98 percent sure, that the model

$$Y = \alpha + \beta_1 X + \beta_2 X^2 + \varepsilon$$

is to be preferred to the model

$$Y = \alpha + \beta X + \varepsilon.$$

The 2 degrees of freedom F-value from the analysis of variance table, $F = 1142.074/5.227 = 218.495$ says that

$$H_0: Y = \alpha + \varepsilon$$

is to be rejected in favor of

$$H_1: Y = \alpha + \beta_1 X + \beta_2 X^2 + \varepsilon.$$

If the NOBRIEF command is used, as in Figure 10.34, Minitab will produce a table labeled FURTHER ANALYSIS OF VARIANCE SS EXPLAINED BY EACH VARIABLE WHEN ENTERED IN THE ORDER GIVEN. These are, of course, sequential sums of squares. Note that the 2 degrees of freedom regression sum of squares from the analysis of variance table, $SS(R)(X, X^2) = 2284.148$, is given, and under it is a partition into two 1 degree of freedom regression sums of squares. Since the Minitab REGRESSION command specified the predictor in column C1 before the predictor in column C3 (that is, SIZE was mentioned before SIZESQ), the first 1 degree of freedom regression sum of squares is

$$SS(R)(X) = 2255.900.$$

Then the sum of squares for regression on X^2, given X, is

$$SS(R)(X^2|X) = SS(R)(X, X^2) - SS(R)(X) = 2284.148 - 2255.900 = 28.248.$$

If these sums of squares (which are also sequential regression mean squares, since they each have only 1 degree of freedom) are divided by $MS(E) = 5.227$, then we have first an F-test of

$$H_{01}: Y = \alpha + \varepsilon$$

$$H_{11}: Y = \alpha + \beta X + \varepsilon$$

and then an F-test of

$$H_{02}: Y = \alpha + \beta X + \varepsilon$$

$$H_{12}: Y = \alpha + \beta_1 X + \beta_2 X^2 + \varepsilon.$$

We would reject H_{01} in favor of H_{11}, since $F_1 = 2255.9/5.227 = 431.586$. Then $F_2 = 28.248/5.227 = 5.404$ is found to be greater than $F_{1,17,0.05} = 4.45$ but less than $F_{1,17,0.01} = 8.40$, so H_{02} would or would not be rejected, depending upon the significance level you require. Note that the t-value for testing the quadratic term, given the linear term, when it is squared gives the F-value for testing the same hypothesis:

$$t^2 = (-2.32)^2 = 5.38$$

$$F = 5.04.$$

(The difference in t^2 and F is due to rounding error.)

The commands to produce a graph of the data with the quadratic regression equation, and the printout from this command, are shown in Figure 10.35.

FIGURE 10.35 Minitab Plot of Data With Quadratic Regression Equation

```
-- NAME C5 'YHAT'

-- MPLOT 'PRICE' BY 'SIZE' AND 'YHAT' BY 'SIZE'

     PRICE
       60.+
          -
          -
          -
          -                                  B 2
       50.+                                2 A
          -                          A
          -                        A B
          -                    A A B
          -                      B
       40.+                    B
          -                A  3
          -                B  A
          -
          -            2 2
       30.+            A
          -         A2 A
          -      A  B
          -    A B
          -      B
       20.+  A BA
          - 2B
          -
          -
          -
       10.+
           +---------+---------+---------+---------+---------+---------+SIZE
         0.80      1.60      2.40      3.20      4.00      4.80
```

Suppose that a cubic model is to be investigated:

$$Y = \alpha + \beta_1 X + \beta_2 X^2 + \beta_3 X^3 + \varepsilon.$$

Then

```
RAISE C1 TO POWER 3, PUT IN C7
NAME C7 'SIZECU'
REGRES C2 on 3 VAR C1 C3 C7
```

produces Figure 10.36. The partial t-values for testing each of the parameters in the presence of all others are easily found in Figure 10.36.

The sequential sums of squares are $SS(R)(X)$, $SS(R)(X^2|X)$, and $SS(R)(X^3|X, X^2)$, respectively. Note also that Minitab, before even stating the regression equation, prints

```
NOTE-THE FOLLOWING VARIABLES ARE HIGHLY CORRELATED WITH
OTHER PREDICTOR VARIABLES, SIZEQ
```

The implications of having a model in which one predictor variable is highly correlated with another predictor will be discussed fully in Chapter 11.

FIGURE 10.36 Minitab Third-Degree Polynomial Regression

```
-NOBRIEF
-REGRES 'PRICE' ON 3 VAR 'SIZE' 'SIZESQ' 'SIZEQ'
NOTE - THE FOLLOWING VARIABLES ARE HIGHLY
CORRELATED WITH OTHER PREDICTOR VARIABLES
SIZEQ

THE REGRESSION EQUATION IS
Y = 15.4 - 3.08 X1 + 10.2 X2 - 1.83 X3

                                ST DEV     T-RATIO =
      COLUMN    COEFFICIENT     OF COEF    COEF/S.D.
                15.362           8.592      1.79
X1    SIZE      -3.08           14.24      -0.22
X2    SIZESQ    10.221           7.100      1.44
X3    SIZEQ     -1.833           1.081     -1.70

THE ST DEV OF Y ABOUT REGRESSION LINE IS
S = 2.170
WITH (20 - 4) = 16 DEGREES OF FREEDOM

R-SQUARED = 96.8 PERCENT
R-SQUARED = 96.2 PERCENT, ADJUSTED FOR DF

ANALYSIS OF VARIANCE
DUE TO         DF    SS          MS=SS/DF
REGRESSION      3    2297.683    765.894
RESIDUAL       16      75.317      4.707
TOTAL          19    2373.000

FURTHER ANALYSIS OF VARIANCE
SS EXPLAINED BY EACH VARIABLE WHEN ENTERED IN THE ORDER GIVEN
DUE TO         DF    SS
REGRESSION     3     2297.683
SIZE           1     2255.900
SIZESQ         1       28.248
SIZEQ          1       13.535

      X1      Y        PRED Y    ST DEV
ROW   SIZE    PRICE    VALUE     PRED Y    RESIDUAL    ST RES
 1    1.80    32.000   32.239    0.765     -0.239      -0.12
 .     .        .        .         .          .           .
 .     .        .        .         .          .           .
 .     .        .        .         .          .           .
20    1.50    28.000   27.550    0.833      0.450       0.22
```

FIGURE 10.36 (Continued)

```
X DENOTES AN OBS WHOSE X VALUE GIVES IT LARGE INFLUENCE.
DURBIN-WATSON STATISTIC = 2.46
(X-PRIME X)INVERSE
        0            1            2          3
0    15.6816
1   -25.6436      43.0794
2    12.4530     -21.3102     10.7077
3    -1.8404       3.1902     -1.6225    0.2484
```

SAS POLYNOMIAL REGRESSION

The worksheet in Figure 4.11 for fitting a simple linear regression equation can easily be modified to fit a polynomial equation. (See Figure 10.37.) If it is desired to fit a quadratic equation $Y = a + b_1X + b_2X^2$, then a variable X^2 must be defined. Add the statement

```
SIZESQ=SIZE*SIZE
```

FIGURE 10.37 Worksheet for an SAS Quadratic Regression

```
  {SYSTEM CARDS
DATA;
    INPUT SIZE PRICE;
    SIZESQ=SIZE*SIZE;
    CARDS;
1.8 32
1.Ø 24
   : {DATA CARDS
1.5 28
PROC REG;
    MODEL PRICE=SIZE SIZESQ/P XPX I SS1 SS2;
    OUTPUT OUT=NEW P=YHAT R=RESID;
PROC PLOT;
    PLOT PRICE*SIZE YHAT*SIZE='*'/OVERLAY;
    PLOT RESID*SIZE/VREF=Ø;
```

after the INPUT statement and before CARDS. This will add the variable SIZESQ to the data set with which the computer works. Then the MODEL statement under the PROC REG command specifies that Y is a function of X and X^2:

```
MODEL PRICE=SIZE SIZESQ
```

The matrices $\mathbf{X'X}$ and $(\mathbf{X'X})^{-1}$ may be printed as options, and predicted values and residuals may be requested as well. Predicted values and residuals will be calculated using the quadratic equation. If they are output to a new data set, predicted values can be used to plot the quadratic regression equation and the residuals can be examined. Figure 10.38 shows the SAS analysis for a quadratic regression, and Figure 10.39 shows a scatter diagram with the regression equation plotted. Higher-order polynomials can be fitted in a similar fashion.

FIGURE 10.38 SAS Quadratic Regression Using REG Procedure

DEP VARIABLE: PRICE

SOURCE	DF	SUM OF SQUARES	MEAN SQUARE	F VALUE	PROB > F
MODEL	2	2284.148	1142.074	218.513	0.0001
ERROR	17	88.851896	5.226582		
C TOTAL	19	2373.000			

ROOT MSE	2.286172	R-SQUARE	0.9626
DEP MEAN	34.500000	ADJ R-SQ	0.9582
C.V.	6.626585		

VARIABLE	DF	PARAMETER ESTIMATE	STANDARD ERROR	T FOR H0: PARAMETER=0	PROB > |T|	TYPE I SS	TYPE II SS
INTERCEP	1	1.775905	3.268476	0.543	0.5939	23805.000	1.543002
SIZE	1	20.468529	3.315002	6.175	0.0001	2255.900	199.262
SIZESQ	1	-1.755657	0.755192	-2.325	0.0327	28.247662	28.247662

OBS	ACTUAL	PREDICT VALUE	RESIDUAL
1	32.000	32.931	-0.930929
2	24.000	20.489	3.511
3	27.000	31.499	-4.499
4	47.000	45.323	1.677
5	35.000	38.309	-3.309
6	17.000	17.027	-0.027107
7	52.000	52.709	-0.709297
8	20.000	22.167	-2.167
9	38.000	35.690	2.310
10	45.000	43.126	1.874
11	44.000	39.566	4.434
12	19.000	18.775	0.224501
13	25.000	23.810	1.190
14	50.000	51.074	-1.074
15	30.000	31.499	-1.499
16	43.000	41.974	1.026
17	27.000	26.991	0.009242
18	50.000	50.203	-0.202948
19	37.000	38.309	-1.309
20	28.000	28.528	-0.528470

SUM OF RESIDUALS -1.38556E-13
SUM OF SQUARED RESIDUALS 88.8519

Note in Figure 10.38 that the *t*-statistics associated with the intercept and the variables "size" and "size squared" are *partial* *t*-statistics. That is, they test to see if the associated term adds anything to the model, given that the

FIGURE 10.39 Data and Quadratic Regression Function Plotted Using SAS

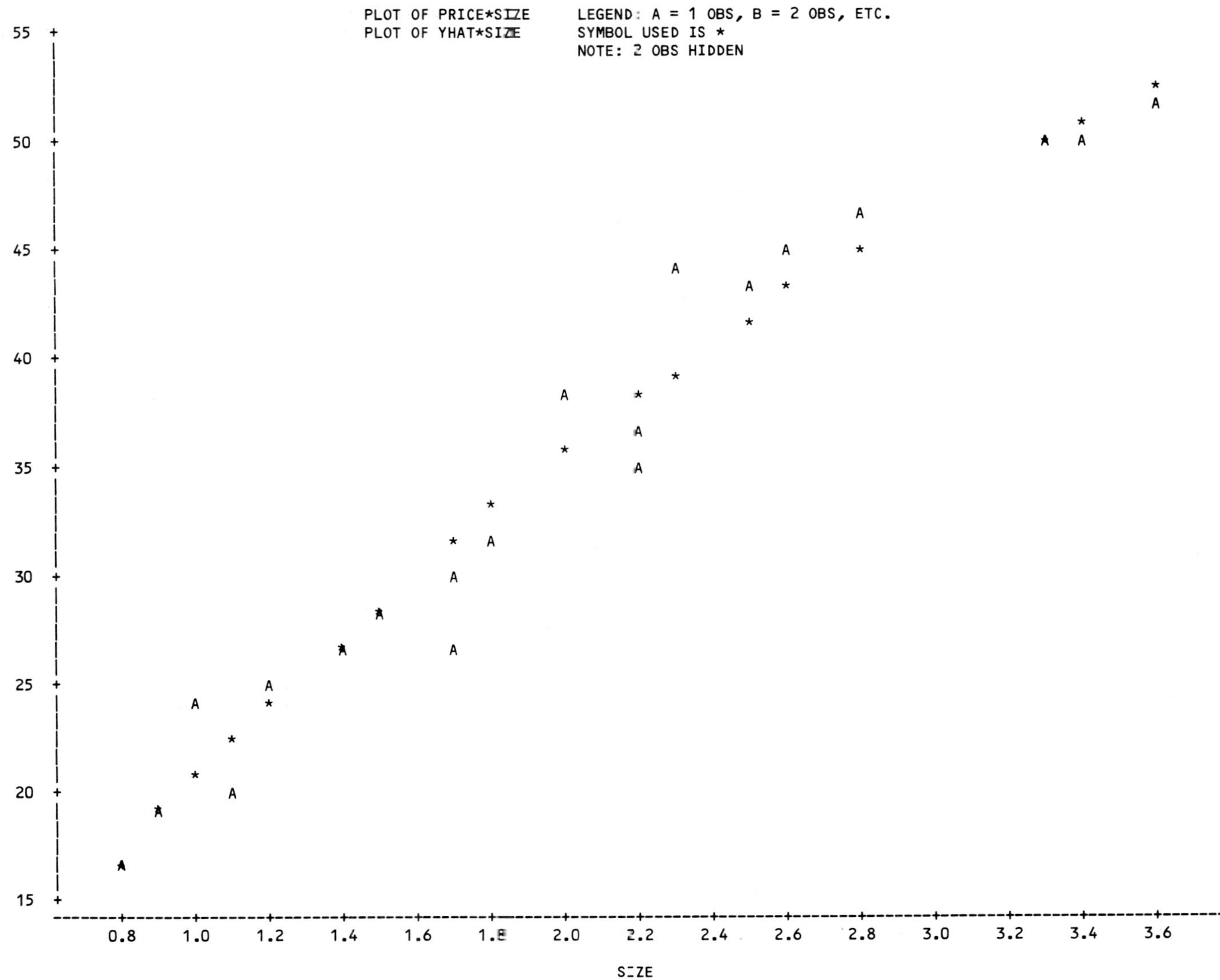

other terms will be used. Specifically, the three t-statistics test, respectively, the hypotheses

$$H_{00}: Y = \beta_1 X + \beta_2 X^2 + \varepsilon$$
$$H_{10}: Y = \alpha + \beta_1 X + \beta_2 X^2 + \varepsilon,$$

$$H_{01}: Y = \alpha + \beta_2 X^2 + \varepsilon$$
$$H_{11}: Y = \alpha + \beta_1 X + \beta_2 X^2 + \varepsilon,$$

$$H_{02}: Y = \alpha + \beta_1 X + \varepsilon$$
$$H_{12}: Y = \alpha + \beta_1 X + \beta_2 X^2 + \varepsilon.$$

SEQUENTIAL SUMS OF SQUARES USING SAS In order to be able to perform sequential test of the predictors, another option must be specified in the `MODEL` statement: `SS1` specifies that sequential sums of squares are to be printed. (The option `SS2` will cause partial sums of squares to be printed, if they are desired in addition to the partial t-statistics, which are shown automatically.) SAS will calculate sequential sums of squares according to the order in which the independent variables were listed in the `MODEL` statement. Since this example used `MODELPRICE=SIZE SIZESQ`, the sequential sums of squares are $SS(R)$(size) and $SS(R)$(size squared|size). Had we specified `MODEL PRICE=SIZESQSIZE`, the sequential sums of squares would have been $SS(R)$ (size squared) and $SS(R)$ (size|size squared) and would have been used to test H_{01}:quadratic term not needed and H_{02}:linear term not needed, given quadratic term used. The sequential F-statistics must be calculated by hand, by dividing the sequential sums of squares by the error mean square for the full model. Thus, to test $H_{01}: Y = \alpha + \varepsilon$ versus $H_{11}: Y = \alpha + \beta X + \varepsilon$,

$$F_1 = \frac{2255.9}{5.226582} = 431.62,$$

and to test $H_{02}: Y = \alpha + \beta X + \varepsilon$ versus $H_{12}: Y = \alpha + \beta_1 X + \beta_2 X^2 + \varepsilon$,

$$F_2 = \frac{28.247662}{5.226582} = 5.40,$$

and both F-statistics have 1 and 17 degrees of freedom.

The sequential sum of squares for the intercept term is

$$23{,}805 = \frac{(690)^2}{20} = \frac{(\sum Y)^2}{n}.$$

THE SAS GLM PROCEDURE There is another procedure available in SAS that will perform regression analyses and show both partial and sequential sums of squares, F-statistics, and their associated probabilities automatically. This procedure, called the General Linear Model, or GLM, procedure, can be used to perform simple or polynomial regression analyses in the same way that the REG procedure does, but it is a very powerful procedure designed for analyzing a great variety of linear models (not all of which are regression models). The only alteration to the programs thus far presented, in order to change from the REG procedure to GLM is simply to write `PROC GLM` instead of `PROC REG` (but it is not necessary to specify the options `SS1` and `SS2` in the `MODEL` statement when using GLM). Figure 10.40 shows the printout resulting from

FIGURE 10.40 SAS Quadratic Regression Using GLM Procedure

```
DEP VARIABLE: Y
                 SUM OF          MEAN
SOURCE    DF     SQUARES         SQUARE          F VALUE   PR > F       R-SQUARE   C.V.
MODEL      2     2284.14810434   1142.07405217   218.51    0.0001       0.962557   6.6266
ERROR     17       88.85189566      5.22658210             STD DEV                 Y MEAN
TOTAL     19     2372.00000000                             2.28617193              34.50000000

SOURCE    DF     TYPE I SS       F VALUE   PR > F    DF   TYPE IV SS      F VALUE   PR > F
X          1     2255.90044248   431.62    0.0001    1    199.26157592    38.12     0.0001
X*X        1       28.24766186     5.40    0.0327    1     28.24766186     5.40     0.0327

                                T FOR H0:                STD ERROR OF
PARAMETER       ESTIMATE        PARAMETER=0    PR > |T|  ESTIMATE
INTERCEPT       1.77590491      0.54           0.5939    3.26847611
X              20.46852858      6.17           0.0001    3.31500152
X*X            -1.75564671     -2.32           0.0327    0.75519175

                              PREDICT
OBS    ACTUAL                 VALUE           RESIDUAL
 1     32.00000000            32.93092863     -0.93092863
 2     24.00000000            20.48877678      3.51122322
 3     27.00000000            31.49855561     -4.49855561
 4     47.00000000            45.32343636      1.67656364
 5     35.00000000            38.30928933     -3.30928933
 6     17.00000000            17.02710748     -0.02710748
 7     52.00000000            52.70929690     -0.70929690
 8     20.00000000            22.16694173     -2.16694173
 9     38.00000000            35.69033525      2.30966475
10     45.00000000            43.12583989      1.87416011
11     44.00000000            39.56609667      4.43390333
12     19.00000000            18.77549870      0.22450130
13     25.00000000            23.80999355      1.19000645
14     50.00000000            51.07351057     -1.07351057
15     30.00000000            31.49855561     -1.49855561
16     43.00000000            41.97437195      1.02562805
17     27.00000000            26.99075778      0.00924222
18     50.00000000            50.20294770     -0.20294770
19     37.00000000            38.30928933     -1.30928933
20     28.00000000            28.52847019     -0.52847019

SUM OF RESIDUALS                               -0.00000000
SUM OF SQUARED RESIDUALS                       88.85189566
SUM OF SQUARED RESIDUALS - ERROR SS            -0.00000000
FIRST ORDER AUTOCORRELATION                    -0.24754441
DURBIN-WATSON D                                 2.47899672
```

```
PROC GLM;
MODEL PRICE=SIZE SIZESQ;
```

for comparison of the output from the GLM procedure to that from the REG procedure.

QUADRATIC REGRESSION WITH SPSS^X

In order to perform a quadratic regression using SPSS^X, we must instruct the computer to create the variable X^2 from our input data and to use it as a predictor in the regression equation. Figure 10.41 shows a worksheet for a quadratic regression using SPSS^X. Note that after values of SIZE and PRICE are indicated on the DATA LIST command, the command

```
COMPUTE SIZESQ = SIZE**2
```

is given, forming the variable SIZESQ (or X^2).

FIGURE 10.41 Worksheet for Quadratic Regression Using Both the ENTER and FORWARD Methods in SPSS^X

```
1-10      11-20     21-30     31-40
   {SYSTEM  CARDS
UNNUMBERED
TITLE SPSS-X QUADRATIC REGR FOR HOUSES DATA
DATA LIST LIST/
          SIZE PRICE
COMPUTE SIZESQ=SIZE**2
BEGIN DATA
1.8 32
1.Ø 24
  :  {DATA  CARDS
1.5 28
END DATA
REGRESSION VARIABLES=SIZE PRICE SIZESQ/
           DEPENDENT=PRICE/
           ENTER/
           DEPENDENT=PRICE/
           FORWARD/
```

ENTER METHOD It is not appropriate to use the SCATTERGRAM procedure in this case, as the statistics given there are for fitting a straight line only. Thus, we must use the REGRESSION command. Note that all three variables, SIZE, PRICE, and SIZESQ, are to be used in the equation, with PRICE being the dependent variable. If the method subcommand ENTER is then used, the computer fits the equation

$$\hat{Y} = a + b_1X + b_2X^2,$$

using both X and X^2 to predict Y, and we see *partial* t-statistics, testing $H_{01}: Y = \alpha + \beta X^2 + \varepsilon$ and $H_{02}: Y = \alpha + \beta X + \varepsilon$, respectively, as shown in Figure 10.42.

FIGURE 10.42 SPSS^X Quadratic Regression Using the ENTER Method

```
EQUATION NUMBER 1 DEPENDENT VARIABLE..PRICE
BEGINNING BLOCK NUMBER 1. METHOD: ENTER
VARIABLE(S) ENTERED ON STEP NUMBER 1..SIZESQ
                                   2..SIZE

MULTIPLE R           0.98110   ANALYSIS OF VARIANCE    DF   SUM OF SQUARES   MEAN SQUARE
R SQUARE             0.96256
ADJUSTED R SQUARE    0.95815   REGRESSION               2   2284.14810        1142.07405
STANDARD ERROR       2.28617   RESIDUAL                17     88.85190           5.22658
                               F = 218.51260          SIGNIF F = 0.0000

VAR-COVAR MATRIX OF REGRESSION COEFFICIENTS (B)
BELOW DIAGONAL:  COVARIANCE    ABOVE:  CORRELATION

           SIZESQ      SIZE

SIZESQ     0.57031    -0.98231
SIZE      -2.45917    10.98924

                  VARIABLES IN THE EQUATION

VARIABLE        B            SE B        BETA           T       SIG T

SIZESQ       -1.75566     0.75519     -0.58257      -2.325     0.0327
SIZE         20.10853     3.31600      1.54728       6.175     0.0000
(CONSTANT)    1.77590     3.26848                    0.543     0.5939

FOR BLOCK NUMBER 1 ALL REQUESTED VARIABLES ENTERED.
```

FORWARD METHOD To obtain a sequential analysis it is necessary to use a method subcommand other than ENTER. Note in the worksheet in Figure 10.41 that we have included a second DEPENDENT=PRICE subcommand, followed by the method subcommand FORWARD. The FORWARD subcommand instructs the computer to determine first whether the linear or the quadratic term by itself is the more powerful in predicting Y. This is done by performing two simple regressions: one of Y on X, and one of Y on X^2, and selecting the predictor that gives the larger F-value. While the computer does not show us all the work it has done, the linear term was the stronger of the two, as is seen by VARIABLE(S) ENTERED ON STEP NUMBER 1 . . SIZE in Figure 10.43(a). The simple linear regression analyses of "price" on "size" is then shown. The t-statistic is a sequential t, testing $H_0: Y = \alpha + \varepsilon$, or $H_0: \beta_1 = 0$.

Figure 10.43(a) looks just like the printout for the simple linear regression analysis in Figure 4.19, except that there is a new section entitled VARIABLES NOT IN THE EQUATION. Here we see information concerning the

FIGURE 10.43(a) Step 1 of Quadratic Regression Using the FORWARD Method of SPSS^X

```
EQUATION NUMBER 2 DEPENDENT VARIABLE..PRICE
BEGINNING BLOCK NUMBER 1. METHOD: FORWARD
VARIABLE(S) ENTERED ON STEP NUMBER 1..SIZE
MULTIPLE R            0.97501   ANALYSIS OF VARIANCE
R SQUARE              0.95065                          DF   SUM OF SQUARES     MEAN SQUARE
ADJUSTED R SQUARE     0.94791   REGRESSION              1         2255.90044 2255.90044
STANDARD ERROR        2.55059   RESIDUAL               18          117.09956    6.50553
                                F = 346.76654               SIGNIF F = 0.0000
VAR-COVAR MATRIX OF REGRESSION COEFFICIENTS (B)
BELOW DIAGONAL:  COVARIANCE    ABOVE:  CORRELATION
       SIZE
SIZE   0.47976
                  VARIABLES IN THE EQUATION                                            VARIABLES NOT IN THE EQUATION
VARIABLE       B          SE B       BETA        T          SIG T    VARIABLE    BETA IN     PARTIAL     MIN TOLER    T         SIG T
SIZE        12.89823   0.69265    0.97501     18.622     0.0000   SIZESQ      -0.58257    -0.49115    0.03507      -2.325    0.0327
(CONSTANT)   8.70354   1.49810                 5.810     0.0000
```

variable `SIZESQ`. `BETA IN` is what the standardized regression coefficient for the quadratic term would be if `SIZESQ` were added to `SIZE` as a predictor. `PARTIAL` is the partial correlation of `SIZESQ` with `PRICE`, controlling for `SIZE`. `MIN TOLER` (minimum tolerance) is a measure designed to make sure that the $(\mathbf{X}'\mathbf{X})^{-1}$ matrix for the model containing both linear and quadratic terms is not too close to being singular (we will say more about this in Chapter 15), and the t-statistic is the partial t for the quadratic term, given that the linear term is already in the model.

In Figure 10.43(b) we see both the linear and quadratic terms entered into the model. The F-statistic tests $H_0: \beta_1 = \beta_2 = 0$, and we recognize the t-statistics as *partial* t's, testing $H_{01}: \beta_1 = 0 | \beta_2$, and $H_{02}: \beta_2 = 0 | \beta_1$. Note that when we used the method `ENTER` instead of `FORWARD`, the printout looked like `EQUATION NUMBER 2` in Figure 10.43(b).

FIGURE 10.43(b) Step 2 of Quadratic Regression Using the `FORWARD` Method of SPSS^X

```
EQUATION NUMBER 2 DEPENDENT VARIABLE..PRICE
VARIABLE(S) ENTERED ON STEP NUMBER 2..SIZESQ

MULTIPLE R          0.98110   ANALYSIS OF VARIANCE
                                                  DF   SUM OF SQUARES     MEAN SQUARE
R SQUARE            0.96256
ADJUSTED R SQUARE   0.95815   REGRESSION               2    2284.14810   1142.07405
STANDARD ERROR      2.28617   RESIDUAL            17             88.85190      5.22658

                              F = 218.51260             SIGNIF F = 0.0000

VAR-COVAR MATRIX OF REGRESSION COEFFICIENTS (B)
BELOW DIAGONAL:  COVARIANCE    ABOVE:  CORRELATION

            SIZE         SIZESQ

SIZE       10.98924     -0.98231
SIZESQ     -2.45917      0.57031

                     VARIABLES IN THE EQUATION

VARIABLE         B             SE B          BETA          T        SIG T

SIZE          20.46853      3.31500       1.54728      6.175     0.0000
SIZESQ        -1.75566      0.75519      -0.58257     -2.325     0.0327
(CONSTANT)     1.77590      3.26848                    0.543     0.5939

FOR BLOCK NUMBER 1 ALL REQUESTED VARIABLES ENTERED.
```

In summary, using the `FORWARD` method subcommand, SPSS^X allows us to see a sequential analysis and a partial analysis of the quadratic model. In the first step the stronger predictor is used and a complete unconditional analysis is shown for that model. In the second step the other term is added, and the terms are tested for significance, conditional on the inclusion of the other. It is not the researcher who specifies in which order the variables will enter, but rather the variables themselves determine that, based on their own strengths. In this example the linear term by itself was stronger than the quadratic term by itself, so the first analysis was of the model $Y = \alpha + \beta_1 X + \varepsilon$. In another example the model $Y = \alpha + \beta_2 X^2 + \varepsilon$ may be fitted first, followed by the full model, $Y = \alpha + \beta_1 X + \beta_2 X^2 + \varepsilon$, if the quadratic term contributes more by itself than the linear term does. Thus, in some instances it may not be possible to add linear terms first, then quadratic, then cubic, and so on. For example, Figure 10.44 shows what happens when a cubic model

$$Y = \alpha + \beta_1 X + \beta_2 X^2 + \beta_3 X^3 + \varepsilon$$

is proposed for the house size-price data. In step 1 SIZE is entered and the analysis of the equation $\hat{Y} = a + b_1 X$ is shown. Under VARIABLES NOT IN THE EQUATION we see SIZESQ (X^2) and SIZEQ (X^3) and the statistics associated with them *if* each were to be added at the next step. That is, if the quadratic term were to be added to the linear term, its partial t would be significant at $\alpha = 0.0327$, but if the cubic term were added to the linear term, its partial t would have attained significance level 0.0216. Since 0.0216 is less than 0.0327, the cubic term is added, giving the regression equation

$$\hat{Y} = 3.47457 + 17.26074X - 0.28471X^3$$

in step 2.

Under VARIABLES NOT IN THE EQUATION in step 2 we see statistics that would result if the quadratic term were next added to the linear and cubic terms. The t-statistic for testing $H_0{:}\beta_2 = 0 | \beta_1, \beta_3$, that is,

$$H_0{:} Y = \alpha + \beta_1 X + \beta_3 X^3 + \varepsilon$$

$$H_1{:} Y = \alpha + \beta_1 X + \beta_2 X^2 + \beta_3 X^3 + \varepsilon,$$

would be significant at $\alpha = 0.1692$. The process terminates after step 2 for two reasons: (1) SPSS^X will not add a variable unless it would be significant at $\alpha \leq 0.05$, and (2) it will not add a variable if its tolerance (mentioned above) becomes less than 0.01. Note that there is a message, FOR BLOCK NUMBER 1 TOLERANCE = 0.010 LIMITS REACHED.

MULTIPLE ENTER SUBCOMMANDS

Using the FORWARD method in analyzing a polynomial model, the researcher is at the mercy of the computer. It is not possible to specify that you first want to see the simple linear model, then the quadratic, and then the cubic, even though for some sets of data it may well turn out that way. The FORWARD method enters terms according to their relative strengths and not according to a prespecified order. Thus, it may be advisable to use multiple ENTER subcommands instead of the FORWARD subcommand if a sequential analysis is desired.

Consider the following REGRESSION command:

```
REGRESSION VARIABLES = SIZE PRICE SIZESQ SIZEQ/
DEPENDENT = PRICE/
ENTER SIZE/
ENTER SIZESQ/
ENTER SIZEQ/
```

SPSS^X will first show a simple linear regression of PRICE on SIZE. Then,

FIGURE 10.44 Cubic Regression Using SPSSˣ FORWARD Method

```
EQUATION NUMBER 1 DEPENDENT VARIABLE..PRICE
BEGINNING BLOCK NUMBER 1. METHOD: FORWARD
VARIABLE(S) ENTERED ON STEP NUMBER 1..SIZE

MULTIPLE R           0.97501   ANALYSIS OF VARIANCE
R SQUARE             0.95065                          DF   SUM OF SQUARES   MEAN SQUARE
ADJUSTED R SQUARE    0.94791   REGRESSION              1    2255.90044       2255.90044
STANDARD ERROR       2.55059   RESIDUAL               18     117.09956          6.50553
                               F = 346.76654       SIGNIF F = 0.0000

               VARIABLES IN THE EQUATION                                    VARIABLES NOT IN THE EQUATION
VARIABLE       B           SE B       BETA        T        SIG T    VARIABLE   BETA IN    PARTIAL    MIN TOLER    T        SIG T
SIZE         12.89823    0.69265    0.97501    18.622    0.0000   SIZESQ    -0.58257   -0.49115   0.03507     -2.325    0.0327
(CONSTANT)    8.70354    1.49810                5.810    0.0000   SIZEQ     -0.34964   -0.52296   0.11040     -2.530    0.0216
--------------------------------------------------------------------------------------------------------------------------------
VARIABLE(S) ENTERED ON STEP NUMBER 2..SIZEQ

MULTIPLE R           0.98191   ANALYSIS OF VARIANCE
R SQUARE             0.96415                          DF   SUM OF SQUARES   MEAN SQUARE
ADJUSTED R SQUARE    0.95993   REGRESSION              2    2287.92573       1143.96287
STANDARD ERROR       2.23704   RESIDUAL               17      85.07427          5.00437
                               F = 228.59285       SIGNIF F = 0.0000

               VARIABLES IN THE EQUATION                                    VARIABLES NOT IN THE EQUATION
VARIABLE       B           SE B       BETA        T        SIG T    VARIABLE   BETA IN    PARTIAL    MIN TOLER    T        SIG T
SIZE         17.26074    1.82838    1.30479     9.440    0.0000   SIZESQ     3.39168    0.33866   3.574E-04    1.440    0.1692
SIZEQ        -0.28471    0.11254   -0.34964    -2.530    0.0216
(CONSTANT)    3.47457    2.44928                1.419    0.1741
--------------------------------------------------------------------------------------------------------------------------------
FOR BLOCK NUMBER 1 TOLERANCE = 0.010 LIMITS REACHED.
```

in the next step, SIZESQ is added to SIZE as a predictor, and an analysis of the model containing linear and quadratic terms is performed. At the third step the cubic term is to be added to the linear and quadratic, although for the data that we are using as an example, SPSS^X will not perform the third step. Instead, a message, `TOLERANCE = 0.010 LIMITS REACHED`, is printed. Tolerance will be explained more fully in Chapter 15; suffice it to say here that $\mathbf{X'X}$ for the cubic model is very close to being singular.

By looking at the various steps resulting from using multiple ENTER subcommands (see Figure 10.45), you can pick out $SS(R)(X)$, $SS(R)(X^2|X)$, and $SS(R)(X^3|X, X^2)$ for a sequential analysis. In the first step, shown in Figure 10.45(a), $SS(R)(X) = 2255.90044$. In the second step, shown in Figure 10.45(b), $SS(R)(X, X^2) = 2284.14810$. Thus, $SS(R)(X^2|X)$ can be computed as

$$\begin{aligned} SS(R(X^2|X) &= SS(R)(X, X^2) - SS(R)(X) \\ &= 2284.14810 - 2255.90044 \\ &= 28.24756. \end{aligned}$$

Had the analysis for the third-degree polynomial been performed, we could have found $SS(R)(X, X^2, X^3)$ from the third step and then the additional contribution of the cubic term to the linear and quadratic terms could have been calculated as

$$SS(R)(X^3|X, X^2) = SS(R)(X, X^2, X^3) - SS(R)(X, X^2).$$

EXERCISES

10.7 Refer to the Deprivation Data in Exercise 2.21.

a. Show the vectors $\mathbf{y}$ and $\hat{\boldsymbol{\beta}}$ and the matrix $\mathbf{X}$ necessary to fit a second-degree polynomial equation to these data. Set up and solve the normal equations and state the regression equation.

b. Using an analysis of variance table, test for the significance of the regression using the second-degree polynomial model.

c. Using sequential F-tests, test for the significance of the linear and quadratic terms, entering the linear term first.

d. Use a computer program to fit a second-degree polynomial equation to the data. Compare the results to those in (c).

e. Test for lack of fit to the second-degree polynomial equation. Are your results surprising? (Refer to the plot of the regression equation in your solution to Exercise 2.21.)

10.8 Refer to the Trees Data in Exercise 9.6.

a. Fit a second-degree polynomial equation to these data.

b. Test for the significance of the polynomial regression.

c. Using sequential F-tests, test for the significance of the linear term and of the quadratic term, given the linear term.

FIGURE 10.45(a) Step 1 of SPSS^X Sequential Analysis Using Multiple ENTER Subcommands

```
EQUATION NUMBER 1 DEPENDENT VARIABLE..PRICE
BEGINNING BLOCK NUMBER 1. METHOD: ENTER SIZE
VARIABLE(S) ENTERED ON STEP NUMBER 1..SIZE

MULTIPLE R           0.97501    ANALYSIS OF VARIANCE
R SQUARE             0.95065                          DF   SUM OF SQUARES    MEAN SQUARE
ADJUSTED R SQUARE    0.94791    REGRESSION             1   2255.90044        2255.90044
STANDARD ERROR       2.55059    RESIDUAL              18    117.09956           6.50553
                                F = 346.76654        SIGNIF F = 0.0000

            VARIABLES IN THE EQUATION                                      VARIABLES NOT IN THE EQUATION
VARIABLE       B          SE B       BETA        T       SIG T    VARIABLE   BETA IN    PARTIAL    MIN TOLER    T        SIG T
SIZE        12.89823   0.69265    0.97501   18.622    0.0000   SIZESQ    -0.58257   -0.49115   0.03507   -2.325   0.0327
(CONSTANT)   8.70354   1.49810               5.810    0.0000   SIZEQ     -0.34964   -0.52296   0.11040   -2.530   0.0216

FOR BLOCK NUMBER 1 ALL REQUESTED VARIABLES ENTERED.
```

FIGURE 10.45(b) Step 2 of SPSS^X Sequential Analysis Using Multiple ENTER Subcommands

```
BEGINNING BLOCK NUMBER 2. METHOD: ENTER SIZESQ
VARIABLE(S) ENTERED ON STEP NUMBER 2..SIZESQ

MULTIPLE R           0.98110    ANALYSIS OF VARIANCE
R SQUARE             0.96256                          DF   SUM OF SQUARES    MEAN SQUARE
ADJUSTED R SQUARE    0.95815    REGRESSION             2   2284.14810        1142.07405
STANDARD ERROR       2.28617    RESIDUAL              17     88.85190           5.22658
                                F = 228.59285        SIGNIF F = 0.0000

            VARIABLES IN THE EQUATION                                      VARIABLES NOT IN THE EQUATION
VARIABLE       B          SE B       BETA        T       SIG T    VARIABLE   BETA IN    PARTIAL    MIN TOLER     T        SIG T
SIZE        20.46853   3.31500    1.54728    6.175    0.0000   SIZEQ     -2.25163   -0.39029   3.574E-04   -1.696   0.1093
SIZESQ      -1.75566   0.75519   -0.58257   -2.325    0.0327
(CONSTANT)   1.77590   3.26848               0.543    0.5939

FOR BLOCK NUMBER 2 ALL REQUESTED VARIABLES ENTERED.

BEGINNING BLOCK NUMBER 3. METHOD: ENTER SIZEQ

FOR BLOCK NUMBER 3 TOLERANCE = 0.010 LIMITS REACHED.
NO VARIABLES ENTERED FOR THIS BLOCK.
```

10.9 For the Police Data in Exercise 1.7, fit a second-degree polynomial equation.
- **a.** Test for the significance of the polynomial regression.
- **b.** Use a sequential F-test to test for the significance of first the linear term and then the quadratic term, given the linear term.
- **c.** Could a lack-of-fit test be performed on these data? Would it be advisable?

10.10 Refer to the Census Data in Exercises 10.3 and 10.4.
- **a.** Fit a second-degree polynomial equation to the population figures, letting $X = 0$ in 1910, $X = 1$ in 1920, $X = -1$ in 1900, and so on.
- **b.** Interpret the meaning of the coefficients a, b_1, and b_2.
- **c.** Use the polynomial function to estimate U.S. population in 1970, and compare to the census figure. Does the polynomial model appear to be a better model for population growth during this period than the exponential model?

10.11 Refer to the MPG Data in Exercise 10.6.
- **a.** Fit a quadratic regression relating mpg to horsepower. Plot this equation with the data, and comment on the fit. Compare to the fit of the exponential equation. Is a quadratic term needed in the model?
- **b.** Fit a quadratic regression relating mpg to weight. Graph this equation on the scatter diagram of the data. How well does the quadratic equation describe the relationship in comparison to how well the exponential equation performed? Is a quadratic term needed in the model?

10.12 Rush hour is exasperating. The more cars there are on the highway, the slower traffic goes. How much slower? A simple random sample of counts of traffic volume and readings of average speed past a certain point is shown in Table 10.15.

TABLE 10.15
Rush Hour Data

No. of Vehicles	7481	7628	9905	8361	7099	5468	6281	6057
Average Vehicle Speed (mph)	50	48	2	38	53	63	55	58

No. of Vehicles	5532	9124	7205	7650	9746	7875	7801	7497
Average Vehicle Speed (mph)	60	20	57	43	5	45	40	49

- **a.** Plot these data.
- **b.** Fit a quadratic regression equation. Graph it on the data.
- **c.** When will traffic come to a standstill?
- **d.** What do the partial F- and t-tests tell you?
- **e.** What do the sequential F-tests tell you?

10.13 Refer to the Professionals Data in Exercise 1.11. Fit a quadratic regression. Explain how income is related to years of experience. Is a quadratic term needed in the model?

10.6

SUMMARY

This chapter examined bivariate models that are more complicated than the simple linear one. Exponential models are appropriate when Y increases or decreases with X at a constant rate, and polynomial models describe Y changing with X at a nonconstant rate. In many cases, exponential and polynomial functions both provide adequate descriptions of the data at hand, and the choice between them would be made on the basis of ease of interpretation of the fitted equation and what makes sense from a theoretical point of view.

In the next and the following chapters we shall see how more than one predictor can be incorporated into a regression equation: We turn now to the study of multiple regression.

ELEVEN

MULTIPLE REGRESSION

11.1 INTRODUCTION

Very seldom would one perform a simple regression study if the objective were to predict fairly accurately the value of some variable. In most situations, the value of a dependent variable is affected by several related variables. Therefore, the more information we have, the better our predictions should be.

In addition, the researcher often can think of several independent variables that *might* be related to a dependent variable. Without running an experiment and collecting data, however, one does not know which variables will indeed be strong predictors of the response variable. Thus, data will be collected on all k predictors and the response. Then a multiple regression analysis is used to eliminate those predictors that are not effective.

11.2 THE MULTIPLE LINEAR REGRESSION MODEL

In multiple linear regression our model is

$$Y = \alpha + \beta_1 X_1 + \beta_2 X_2 + \cdots + \beta_k X_k + \varepsilon.$$

That is, Y is a linear function of k predictor variables, $X_1, X_2, \ldots, X_k$, but an imperfect one. That the function is linear can be seen because no regression coefficient (β_i) is raised to a power greater than 1 (refer to the normal equations presented in the next section). Note that the error term ε is still present. Regardless of the number of factors we take into account, individual variation will prevent us from ever predicting Y exactly.

GEOMETRIC INTERPRETATION

What kind of geometric interpretation can we make for a multiple regression problem? Consider the case in which we have two predictors, X_1 and X_2, of Y. This will lead to a regression equation of the form

$$\hat{Y} = a + b_1X_1 + b_2X_2,$$

which is the equation of a two-dimensional plane in three-dimensional space.

Look at the corner made by two walls and the floor of the room you are sitting in. The corner which runs from floor to ceiling, formed by the joining of the left-hand and the right-hand walls, is the Y-axis. The corner made by the right-hand wall and the floor is the X_1-axis, and the corner made by the left-hand wall and the floor is the X_2-axis. The point ($X_1 = 0$, $X_2 = 0$, $Y = 0$) is the corner made by the two walls and the floor. Below the floor, Y is negative; on the other side of the left-hand wall, X_1 is negative; and on the other side of the right-hand wall, X_2 is negative. Now, imagine a thin board floating in space near the corner. The height of the board from the floor is the Y-intercept, a. The slope of the board with respect to the right-hand wall is b_1, and the slope of the board with respect to the left-hand wall is b_2.

When the number of predictors is greater than two, the real-world geometrical interpretation breaks down. For those with good abstract imaginations,

$$\hat{Y} = a + b_1X_1 + b_2X_2 + \cdots + b_kX_k$$

is the equation of a k-dimensional hyperplane in $(k + 1)$-dimensional space. Because of the difficulty in interpreting the general regression problem in geometric terms, we emphasize interpreting the regression constant and regression coefficients in terms specific to the problem.

OBJECTIVES OF THE ANALYSIS

The objectives in a multiple regression problem are essentially the same as for a simple regression. We want to estimate a and $b_1, \ldots, b_k$; test $b_1, \ldots, b_k$ for the significance of the associated predictors; use the regression equation to estimate Y from $X_1, X_2, \ldots, X_k$; and measure the error involved in estimation. While the objectives remain the same, the calculations and interpretations become more complicated, the more predictors we have.

We will use as an illustrative example the problem of estimating the price (Y) of a house from knowing its size (X_1) and also its age (X_2). The sizes and prices will be the same as in the simple regression problem. What we have done is add ages of houses to the existing data. In real life, one would *not* first go out and collect data on sizes and prices, analyze the simple regression problem, and then go out again and find the ages of the 20 houses sampled. Rather, one would collect all data that might be pertinent on all 20 houses at the outset. Then the analysis performed would throw out predictors that turn out not to be needed.

11.3 NORMAL EQUATIONS

NORMAL EQUATIONS FOR THE TWO-PREDICTOR LINEAR MODEL

When using age and size to predict prices of houses, our model is

$$Y = \alpha + \beta_1 X_1 + \beta_2 X_2 + \varepsilon.$$

We shall use the data of Table 11.1 to find a, b_1, and b_2 in the regression equation

$$\hat{Y} = a + b_1 X_1 + b_2 X_2.$$

In order to find a, b_1, and b_2, we will need to solve three normal equations in three unknowns. Recalling the "trick" for writing down normal equations in Chapter 2, we obtain the normal equations

$$\sum Y = na + b_1 \sum X_1 + b_2 \sum X_2$$

$$\sum X_1 Y = a \sum X_1 + b_1 \sum X_1^2 + b_2 \sum X_1 X_2$$

$$\sum X_2 Y = a \sum X_2 + b_1 \sum X_1 X_2 + b_2 \sum X_2^2.$$

TABLE 11.1 Sizes, Ages, and Prices of 20 Houses

House No.	X_1 = Size (thousands of sq ft)	X_2 = Age (years)	Y = Price (thousands of dollars)
1	1.8	30	32
2	1.0	33	24
3	1.7	25	27
4	2.8	12	47
5	2.2	26	35
6	0.8	25	17
7	3.6	28	52
8	1.1	29	20
9	2.0	25	38
10	2.6	2	45
11	2.3	30	44
12	0.9	23	19
13	1.2	12	25
14	3.4	33	50
15	1.7	1	30
16	2.5	12	43
17	1.4	17	27
18	3.3	16	50
19	2.2	22	37
20	1.5	29	28

(The indices of summation have been deleted; for example, $\sum X_1$ really means

$$\sum_{i=1}^{n} X_{1i} = X_{11} + X_{12} + \cdots + X_{1n},$$

or the sum of all the X_1-values (sizes), of which there are n.) Of course, if we had three predictors, then we would have to solve four equations in four unknowns, and so on.

NORMAL EQUATIONS FOR HOUSE SIZE-PRICE-AGE EXAMPLE

The normal equations indicate that the following calculations need to be performed on the data:

$$\sum X_1 = 40, \qquad \sum X_1^2 = 93.56, \qquad \sum X_1 Y = 1554.9,$$

$$\sum X_2 = 430, \qquad \sum X_2^2 = 10{,}990, \qquad \sum X_2 Y = 14{,}540,$$

$$\sum Y = 690, \qquad \sum Y^2 = 26{,}178, \qquad \sum X_1 X_2 = 844.7.$$

Since what we are calling Y now is the same as in simple regression, $\sum Y$ and $\sum Y^2$ have not changed from the earlier example. What was just X in simple regression is now called X_1, so that what was $\sum X$ before is now $\sum X_1$; what was $\sum X^2$ is now $\sum X_1^2$; and what was $\sum XY$ is now $\sum X_1 Y$. The same sorts of calculations are performed using X_2 and in addition, $\sum X_1 X_2$ is also calculated.

For this example, then, the normal equations are

$$690 = 20a + 40b_1 + 430b_2$$

$$1554.9 = 40a + 93.56b_1 + 844.7b_2$$

$$14{,}540 = 430a + 844.7b_1 + 10{,}990b_2.$$

Now the problem is to solve these three equations in three unknowns. While several equivalent alternative methods can be used, one reason the matrix approach of Chapter 3 was formulated was so that we could use matrix methods to obtain a solution.

NORMAL EQUATIONS IN MATRIX FORM

Define

$$\mathbf{X} = \begin{bmatrix} 1 & X_{11} & X_{21} \\ 1 & X_{12} & X_{22} \\ \vdots & \vdots & \vdots \\ 1 & X_{1n} & X_{2n} \end{bmatrix} = \begin{bmatrix} 1 & 1.8 & 30 \\ 1 & 1.0 & 33 \\ \vdots & \vdots & \vdots \\ 1 & 1.5 & 29 \end{bmatrix}.$$

Then

$$\mathbf{X'X} = \begin{bmatrix} n & \sum X_1 & \sum X_2 \\ \sum X_1 & \sum X_1^2 & \sum X_1X_2 \\ \sum X_2 & \sum X_1X_2 & \sum X_2^2 \end{bmatrix} = \begin{bmatrix} 20 & 40 & 430 \\ 40 & 93.56 & 844.7 \\ 430 & 844.7 & 10{,}990 \end{bmatrix}.$$

Note that $\mathbf{X'X}$ is symmetric. If we had more than two predictors, it would be a simple extension to define $\mathbf{X}$ and $\mathbf{X'X}$. If there are k predictors, $\mathbf{X}$ will be $n \times (k + 1)$. The first column of $\mathbf{X}$ will contain ones, the second column will contain X_1-values, the third will contain X_2-values, . . . , and the $(k + 1)$ column will contain values of X_k. Then $\mathbf{X'X}$ will be $(k + 1) \times (k + 1)$ and will be of the form

$$\begin{bmatrix} n & \sum X_1 & \sum X_2 & \cdots & \sum X_k \\ & \sum X_1^2 & \sum X_1X_2 & \cdots & \sum X_1X_k \\ & & \sum X_2^2 & \cdots & \sum X_2X_k \\ & & & & \vdots \\ \text{sym.} & & & & \sum X_k^2 \end{bmatrix}.$$

Returning to our two-predictor example,

$$\mathbf{y} = \begin{bmatrix} Y_1 \\ Y_2 \\ \vdots \\ Y_n \end{bmatrix} = \begin{bmatrix} 32 \\ 24 \\ \vdots \\ 28 \end{bmatrix},$$

so that

$$\mathbf{X}'\mathbf{y} = \begin{bmatrix} \sum Y \\ \sum X_1 Y \\ \sum X_2 Y \end{bmatrix} = \begin{bmatrix} 690 \\ 1554.9 \\ 14{,}540 \end{bmatrix}.$$

$\mathbf{X}'\mathbf{y}$ also extends directly for $k > 2$ and is of the form

$$\begin{bmatrix} \sum Y \\ \sum X_1 Y \\ \sum X_2 Y \\ \vdots \\ \sum X_k Y \end{bmatrix}.$$

Finally,

$$\hat{\boldsymbol{\beta}} = \begin{bmatrix} a \\ b_1 \\ b_2 \end{bmatrix},$$

and this also extends directly to

$$[a \quad b_1 \quad b_2 \quad \cdots \quad b_k]'$$

if more than two predictors are used. Then the normal equations are

$$(\mathbf{X}'\mathbf{X})\hat{\boldsymbol{\beta}} = \mathbf{X}'\mathbf{y},$$

or

$$\begin{bmatrix} 20 & 40 & 430 \\ 40 & 93.56 & 844.7 \\ 430 & 844.7 & 10{,}990 \end{bmatrix} \begin{bmatrix} a \\ b_1 \\ b_2 \end{bmatrix} = \begin{bmatrix} 690 \\ 1554.9 \\ 14{,}540 \end{bmatrix}.$$

SOLVING NORMAL EQUATIONS

In this example, as in simple regression, the normal equations are solved by

$$\hat{\boldsymbol{\beta}} = (\mathbf{X}'\mathbf{X})^{-1}\mathbf{X}'\mathbf{y},$$

so all we have to do is to invert $\mathbf{X}'\mathbf{X}$ and perform a multiplication. The beauty of the matrix approach is that once $\mathbf{X}$, $\mathbf{y}$, and $\hat{\boldsymbol{\beta}}$ are appropriately defined, the normal equations are solved by $\hat{\boldsymbol{\beta}} = (\mathbf{X}'\mathbf{X})^{-1}\mathbf{X}'\mathbf{y}$ regardless of the number of predictors we have in any problem. The tedious arithmetic involved in inverting $\mathbf{X}'\mathbf{X}$ is best done by a computer. (Note that Exercise 4.9 used the computer to find this inverse.) We find

$$(\mathbf{X}'\mathbf{X})^{-1} = \begin{bmatrix} 0.67164255 & -0.16300717 & -0.01375015 \\ -0.16300717 & 0.07448317 & 0.00065306 \\ -0.01375015 & 0.00065306 & 0.00057879 \end{bmatrix}.$$

(Compare this to the answer for Exercise 4.9.) Notice that eight decimal places are presented in $(\mathbf{X}'\mathbf{X})^{-1}$. It is necessary to carry many decimal places because of the considerable rounding error incurred in inversion. A partial check of the calculations is given by

$$(\mathbf{X}'\mathbf{X})^{-1}\mathbf{X}'\mathbf{X} = \begin{bmatrix} 0.99999970 & -0.00000054 & -0.00000850 \\ -0.00000054 & 0.99999837 & -0.00002000 \\ -0.00000850 & -0.00002000 & 0.99997738 \end{bmatrix}.$$

$$\approx \mathbf{I}.$$

Carrying fewer decimal places results in $(\mathbf{X}'\mathbf{X})^{-1}\mathbf{X}'\mathbf{X}$ deviating further from $\mathbf{I}$; carrying more decimal places will bring the product closer to $\mathbf{I}$.

Proceeding with the solution, we obtain

$$\hat{\boldsymbol{\beta}} = \begin{bmatrix} 0.67164255 & -0.16300717 & -0.01375015 \\ -0.16300717 & 0.07448317 & 0.00065306 \\ -0.01375015 & 0.00065306 & 0.00057879 \end{bmatrix} \begin{bmatrix} 690 \\ 1554.9 \\ 14{,}540 \end{bmatrix}$$

$$= \begin{bmatrix} 10.04632987 \\ 12.83442613 \\ -0.05655391 \end{bmatrix} = \begin{bmatrix} a \\ b_1 \\ b_2 \end{bmatrix}.$$

Substituting back into the normal equations, we see that the check is pretty good:

$$20(10.04632987) + 40(12.83442613) + 430(-0.05655391) = 689.98546130$$

$$40(10.04632987) + 93.56(12.83442613) + 844.7(-0.05655391) = 1554.87101574$$

$$430(10.04632987) + 844.7(12.83442613) + 10{,}990(-0.05655391) = 14{,}539.63499511.$$

If even more decimal places were carried in $(\mathbf{X}'\mathbf{X})^{-1}$, the check would come out even better. Why so many decimal places need be carried will become apparent later. Suffice it to say for now that any element of $(\mathbf{X}'\mathbf{X})^{-1}$ is determined, in part at least, by *every* element of the original matrix, $\mathbf{X}'\mathbf{X}$. It is the relationships among all the elements of $\mathbf{X}'\mathbf{X}$ that cause the difficulty in $(\mathbf{X}'\mathbf{X})^{-1}$. Had only six decimal places been carried, the first equation would have given 691.3173, the second 1557.474, and the third 14,573.631, showing only a very rough check.

CORRECTING X FOR ITS MEAN

Some of this rounding error can be reduced if we correct X_1 and X_2 for their means. That is, instead of using the original X_1- and X_2-values, use

$$U_{1i} = X_{1i} - \overline{X}_1$$
$$U_{2i} = X_{2i} - \overline{X}_2, \quad \text{for } i = 1, 2, \ldots, n.$$

The vector $\mathbf{y}$ will be unaffected, but

$$\hat{\boldsymbol{\beta}} = \begin{bmatrix} a' \\ b_1 \\ b_2 \end{bmatrix} \quad \text{and} \quad \mathbf{X} = \begin{bmatrix} 1 & X_{11} - \overline{X}_1 & X_{21} - \overline{X}_2 \\ 1 & X_{12} - \overline{X}_1 & X_{22} - \overline{X}_2 \\ \vdots & \vdots & \vdots \\ 1 & X_{1n} - \overline{X}_1 & X_{2n} - \overline{X}_2 \end{bmatrix},$$

where $\overline{X}_1$ is the mean of the X_1-values and $\overline{X}_2$ is the mean of the X_2-values.

Then

$$\mathbf{X}'\mathbf{X} = \begin{bmatrix} n & \sum (X_{1i} - \overline{X}_1) & \sum (X_{2i} - \overline{X}_2) \\ \sum (X_{1i} - \overline{X}_1) & \sum (X_{1i} - \overline{X}_1)^2 & \sum (X_{1i} - \overline{X}_1)(X_{2i} - \overline{X}_2) \\ \sum (X_{2i} - \overline{X}_2) & \sum (X_{1i} - \overline{X}_1)(X_{2i} - \overline{X}_2) & \sum (X_{2i} - \overline{X}_2)^2 \end{bmatrix}$$
$$= \begin{bmatrix} n & 0 & 0 \\ 0 & S_{11} & S_{12} \\ 0 & S_{12} & S_{22} \end{bmatrix},$$

since

$$\sum (X_{1i} - \overline{X}_1) = \sum (X_{2i} - \overline{X}_2) = 0,$$
$$\sum (X_{1i} - \overline{X}_1)^2 = S_{X_1X_1}, \quad \text{or } S_{11} \text{ for convenience,}$$
$$\sum (X_{2i} - \overline{X}_2)^2 = S_{X_2X_2}, \quad \text{or } S_{22},$$
$$\sum (X_{1i} - \overline{X}_1)(X_{2i} - \overline{X}_2) = S_{X_1X_2}, \quad \text{or } S_{12}.$$

That is, S_{11} is an S_{XX} using X_1-values; S_{22} is an S_{XX} using X_2-values; and S_{12} is like an S_{XY}, except instead of using an X and a Y, we use an X_1 and an X_2. Calculating formulas for these three quantities are

$$S_{11} = \sum X_1^2 - \frac{(\sum X_1)^2}{n} = 93.56 - \frac{(40)^2}{20} = 13.56$$

(which is what S_{XX} was in the simple regression problem),

$$S_{22} = \sum X_2^2 - \frac{(\sum X_2)^2}{n} = 10{,}990 - \frac{(430)^2}{20} = 1745,$$

$$S_{12} = \sum X_1X_2 - \frac{(\sum X_1)(\sum X_2)}{n} = 844.7 - \frac{(40)(430)}{20} = -15.3.$$

So in our example, if X_1 and X_2 are corrected for their means,

$$\mathbf{X'X} = \begin{bmatrix} 20 & 0 & 0 \\ 0 & 13.56 & -15.3 \\ 0 & -15.3 & 1745 \end{bmatrix},$$

which is close to being diagonal and thus is relatively simple to invert. Think of $\mathbf{X'X}$ partitioned as

$$\mathbf{X'X} = \left[\begin{array}{c|cc} 20 & 0 & 0 \\ \hline 0 & 13.56 & -15.3 \\ 0 & -15.3 & 1745 \end{array}\right].$$

Then

$$(\mathbf{X'X})^{-1} = \left[\begin{array}{c|cc} \frac{1}{20} & 0 & 0 \\ \hline 0 & \multicolumn{2}{c}{\mathbf{S}^{-1}} \\ 0 & & \end{array}\right],$$

where

$$\mathbf{S} = \begin{bmatrix} 13.56 & -15.3 \\ -15.3 & 1745 \end{bmatrix}.$$

Then

$$\begin{aligned} \mathbf{S}^{-1} &= \frac{1}{(13.56)(1745) - (-15.3)^2} \begin{bmatrix} 1745 & 15.3 \\ 15.3 & 13.56 \end{bmatrix} \\ &= \frac{1}{23{,}428.11} \begin{bmatrix} 1745 & 15.3 \\ 15.3 & 13.56 \end{bmatrix} \\ &= \begin{bmatrix} 0.074483 & 0.000653 \\ 0.000653 & 0.000579 \end{bmatrix}, \end{aligned}$$

and

$$(\mathbf{X'X})^{-1} = \begin{bmatrix} 0.05 & 0 & 0 \\ 0 & 0.074483 & 0.000653 \\ 0 & 0.000653 & 0.000579 \end{bmatrix}.$$

Partitioning $\mathbf{X'X}$ makes the inversion relatively easy.

Proceeding to solve the normal equations, we note that if X_1 and X_2 are corrected for their means, then

$$\mathbf{X}'\mathbf{y} = \begin{bmatrix} \sum Y_i \\ \sum (X_{1i} - \overline{X}_1)Y_i \\ \sum (X_{2i} - \overline{X}_2)Y_i \end{bmatrix} = \begin{bmatrix} \sum Y_i \\ S_{1Y} \\ S_{2Y} \end{bmatrix},$$

where

$$S_{1Y} = S_{X_1Y} = \sum (X_{1i} - \overline{X}_1)(Y_i)$$

$$= \sum (X_{1i} - \overline{X}_1)(Y_i - \overline{Y}) = \sum X_1 Y - \frac{(\sum X_1)(\sum Y)}{n}$$

$$= 1554.9 - \frac{(40)(690)}{20} = 174.9$$

(which is what S_{XY} was when our only predictor was $X = X_1 = \text{size}$), and

$$S_{2Y} = S_{X_2Y} = \sum (X_{2i} - \overline{X}_2)(Y_i) = \sum (X_{2i} - \overline{X}_2)(Y_i - \overline{Y})$$

$$= \sum X_2 Y - \frac{(\sum X_2)(\sum Y)}{n}$$

$$= 14{,}540 - \frac{(430)(690)}{20} = -295.$$

Then if

$$\mathbf{X}'\mathbf{y} = \begin{bmatrix} 690 \\ 174.9 \\ -295 \end{bmatrix},$$

the solution to the normal equations is

$$\hat{\boldsymbol{\beta}} = \begin{bmatrix} a' \\ b_1 \\ b_2 \end{bmatrix} = \begin{bmatrix} 0.05 & 0 & 0 \\ 0 & 0.074483 & 0.000653 \\ 0 & 0.000653 & 0.000579 \end{bmatrix} \begin{bmatrix} 690 \\ 174.9 \\ -295 \end{bmatrix}$$

$$= \begin{bmatrix} 34.5 \\ 12.834442 \\ -0.056595 \end{bmatrix}.$$

Note that $a' = \overline{Y} = 3.45$, and b_1 and b_2 have changed from before only slightly. Using these values to check the normal equations, we find

$$20(34.5) + 0(12.834442) + 0(-0.056595) = 690$$

$$0(34.5) + 13.56(12.834442) - 15.3(-0.056595) = 174.900938$$

$$0(34.5) - 15.3(12.834442) + 1745(-0.056595) = -295.125238.$$

When X_1 and X_2 are corrected for their means, we only need to carry six decimal places to obtain as good a check on the solution to the normal equations as we got by carrying eight decimal places when the predictors were not corrected for their means. Thus, a considerable savings in effort and rounding error can be accomplished in multiple regression if the predictors are corrected for their means. As you can imagine, the savings become even more pronounced when more than two predictors are involved, and the calculations must be done by hand.

For the sake of simplicity, we shall use the following values for a, b_1, and b_2, expect when they are needed in other calculations, and by rounding the results to two decimal places, we have:

$$a = 10.05,$$

$$b_1 = 12.83,$$

$$b_2 = -0.06,$$

so that our regression equation is

$$\hat{Y} = 10.05 + 12.83X_1 - 0.06X_2.$$

In the next section we shall examine the interpretation and use of this equation.

EXERCISES

11.1 Referring to the Sales Data in Exercise 1.6, which gives data about the weekly retail sales of stores and their numbers of employees, suppose that we had also the information in Table 11.2 on the sizes of the stores.

TABLE 11.2
Sales Data: II

Y = Average Weekly Retail Sales (thousands of dollars)	X_1 = No. of Employees	X_2 = Size (thousands of sq ft)
7	17	7
17	39	9
10	32	8
5	17	4
7	25	5
15	43	9
11	25	8
13	32	10
19	48	12
3	10	5
17	48	12
15	42	10
14	36	10
12	30	10
8	19	8

a. Find the simple linear regression equation relating Y to X_2.
b. Find the multiple linear regression equation

$$\hat{Y} = a + b_1X_1 + b_2X_2.$$

(You may want to use a computer program to invert $\mathbf{X'X}$.)

11.2 In an effort to decide whether or not to apply for a liquor license, a restaurant owner sampled ten comparable restaurants in his town and recorded the total profits (Y), the food sales (X_1), and the liquor sales (X_2) for the past month. His sample yielded the results in Table 11.3. (All figures are in thousands of dollars.) Find the multiple linear regression equation relating profits to sales of food and liquor.

TABLE 11.3
Restaurant Data

Restaurant	Y	X_1	X_2
1	4.5	3.3	4.0
2	2.0	1.7	1.5
3	1.7	1.5	2.0
4	2.3	1.7	3.0
5	4.0	3.0	3.0
6	4.1	3.2	3.7
7	2.5	1.6	2.0
8	3.7	3.5	4.0
9	3.6	3.0	3.5
10	3.0	3.0	3.5

11.3 A pharmaceutical company testing a new pain-killing drug tests it on 20 people suffering from arthritis. The time elapsed, in minutes, from taking the drug until a noticeable relief in pain is detected is to be predicted from the dosage (in grams) and the age of the patient (in years). The results are given in Table 11.4. Find the regression equation relating time to relief to dosage and age.

TABLE 11.4
Drug Data

Patient	Time	Dosage	Age	Patient	Time	Dosage	Age
1	11	2	59	11	35	7	33
2	3	2	57	12	28	7	49
3	20	2	22	13	23	7	29
4	25	2	12	14	21	7	32
5	27	2	18	15	33	7	20
6	15	5	40	16	27	10	43
7	10	5	64	17	8	10	61
8	34	5	27	18	3	10	69
9	14	5	54	19	12	10	62
10	34	5	22	20	14	10	61

11.4 A company's motor pool collected the information in Table 11.5 concerning the cost of maintenance (in hundreds of dollars) and repairs of the vehicles during the last year. Find the multiple linear regression equation relating maintenance and repair costs to age and mileage of the vehicles.

TABLE 11.5
Auto Data

Vehicle No.	Age	Mileage During Year (thousands)	Cost
1	3	12	1.2
2	3	36	1.5
3	2	10	0.9
4	4	14	1.5
5	3	16	1.0
6	5	22	4.0
7	2	14	0.8
8	2	10	0.8
9	2	24	1.0
10	5	14	2.0
11	5	18	2.0
12	5	21	3.0
13	5	28	3.2
14	0†	27	0.5
15	5	14	1.9

† Age 0 denotes a car less than 1 year old.

11.5 A large discount department store chain advertises on television (X_1), on the radio (X_2), and in newspapers (X_3). A sample of 12 of its stores in a certain area yielded advertising expenditures and revenues during a given month as shown in Table 11.6. (All figures are in thousands of dollars.) Find the regression equation relating revenues to the three kinds of advertising expenditures.

TABLE 11.6
Ads Data

Store	Revenues	TV	Radio	Paper
1	84	13	5	2
2	84	13	7	1
3	80	8	6	3
4	50	9	5	3
5	20	9	3	1
6	68	13	5	1
7	34	12	7	2
8	30	10	3	2
9	54	8	5	2
10	40	10	5	3
11	57	5	6	2
12	46	5	7	2

11.6 In a study of grade school children, ages, heights (in inches), weights, and scores on a physical fitness test were recorded, as given in Table 11.7. Find the multiple linear regression equation relating the scores to the ages, heights, and weights of the children.

TABLE 11.7
Children Data

Child	Score	Age	Height	Weight
1	58	7	47.5	53
2	54	7	45.0	50
3	55	9	52.5	85
4	74	7	48.0	52
5	86	9	55.0	76
6	98	8	51.0	64
7	96	9	53.0	75
8	70	7	46.0	75
9	40	7	48.0	68
10	67	9	50.5	74
11	41	6	45.0	40
12	41	7	48.5	66
13	47	8	50.5	65
14	45	8	49.0	70
15	92	9	51.5	70
16	50	7	46.5	60
17	98	9	53.5	77
18	42	8	45.0	65
19	64	8	52.5	65
20	70	8	51.5	67

11.7 An experiment was run to test the adhesive strength of a new glue (in pounds) under different conditions of temperature (°F) and humidity (percent). The results are given in Table 11.8. Find the multiple linear regression equation relating strength to temperature and humidity. (*Hint:* Code the X-values.)

TABLE 11.8
Glue Data

Strength	Temperature	Humidity
190	80	40
189	80	40
192	90	40
190	90	40
196	80	60
193	80	60
195	90	60
196	90	60
201	80	80
200	80	80
203	90	80
205	90	80

11.8 Returnable soft drink bottles must be sterilized before they are refilled. An experiment was run by a bottler to study the effects of three different temperature levels (coded low, -1; medium, 0; and high, $+1$) and four different sterilizing times (coded short, -2; medium short, -1; medium long, $+1$; and long, $+2$) on the number of impurities remaining. The results are given in Table 11.9. Find the multiple linear regression equation relating the number of impurities remaining to sterilization time and temperature.

TABLE 11.9
Bottles Data

Time	Temperature	Impurities
−2	−1	4
−2	−1	5
−2	0	6
−2	0	4
−2	1	5
−2	1	5
−1	−1	3
−1	−1	5
−1	0	4
−1	0	4
−1	1	2
−1	1	3
1	−1	4
1	−1	3
1	0	3
1	0	3
1	1	2
1	1	1
2	−1	2
2	−1	1
2	0	1
2	0	0
2	1	0
2	1	0

11.9 To determine the effects of nitrogen, potassium, and phosphorus fertilizers on a certain crop, high, medium, and low concentrations (coded −1, 0, 1) of the three elements were tested. The yields are shown in Table 11.10.

a. Find the multiple linear regression equation relating yield (in pounds per acre) to concentrations of the three elements.

b. Find each of the three simple linear regression equations.

c. Find all three two-predictor linear regression equations.

d. Compare the results in (a), (b), and (c).

TABLE 11.10
Yield Data

Yield	Nitrogen	Potassium	Phosphorus
539	−1	−1	−1
319	−1	−1	0
164	−1	−1	1
228	−1	0	−1
207	−1	0	0
178	−1	0	1
180	−1	1	−1
250	−1	1	0
340	−1	1	1
491	0	−1	−1
305	0	−1	0
514	0	−1	1
380	0	0	−1
364	0	0	0

TABLE 11.10 (Continued)

Yield	Nitrogen	Potassium	Phosphorus
366	0	0	1
192	0	1	−1
171	0	1	0
187	0	1	1
354	1	−1	−1
415	1	−1	0
326	1	−1	1
239	1	0	−1
292	1	0	0
446	1	0	1
250	1	1	−1
527	1	1	0
477	1	1	1

11.4 USING THE COMPUTER TO PERFORM MULTIPLE REGRESSION

The calculations involved in a multiple regression problem calculations are usually quite laborious, even when $k = 2$. Rounding error is always a problem, and in large problems human error can creep in, too. When we search for the *best* set of predictors, the labor involved is almost prohibitive without the computer. However, several computer programs can perform the multiple regression analysis quickly and accurately.

MULTIPLE LINEAR REGRESSION USING BMDP1R

The worksheet in Figure 11.1 shows commands for performing a multiple linear regression analysis on the house size-price-age data using the BMDP1R

FIGURE 11.1 Worksheet for Using BMDP1R to Perform a Multiple Regression

```
  {SYSTEM   CARDS
PROBLEM   TITLE='BMDP1R MULTIPLE REGRESSION FOR HOUSES DATA'./
INPUT     VARIABLES=3.
          FORMAT=FREE.
          CASES=2Ø./
VARIABLES NAMES=SIZE,AGE,PRICE./
REGRES    DEPENDENT=PRICE./
PRINT     DATA.
          CORREL.
          COVAR.
          RREG./
PLOT      RESIDUALS./
END/
1.8  3Ø 32
1.Ø  33 24
     : {DATA CARDS
1.5  29 28
FINISH/
```

program. In one of the system cards will appear ...EXEC BIMED,PROG=BMDP1R..., which directs the computer to the appropriate BMD package for multiple linear regression. Very little is different in this program from the program to run a simple linear regression analysis; however, we have asked for some additional information to be printed and for a different kind of plot to be made.

The PROBLEM, INPUT, and VARIABLES paragraphs should be clear to the reader who has used BMD programs before. The same is true for the REGRESSION paragraph, but note that since no INDEPENDENT sentence is included, the computer assumes that both size and age are to be used to predict price. Output resulting from the REGRESSION paragraph is shown in Figure 11.2. Measures of correlation, the analysis of variance, estimates of the regression coefficients, and statistics relevant to them are easy to locate.

Three new sentences have been added in the PRINT paragraph. The CORREL (correlation) sentence specifies that the matrix of bivariate correlation coefficients for all three variables is to be printed. The COVAR (covariance) sentence causes the covariances between pairs of variables to be printed (also in matrix form), and the RREG sentence calls for correlations (R) between the regression coefficients (REG) to be displayed. Printout resulting from these sentences of the PRINT paragraph is shown in Figure 11.3.

The BMDP1R regression programs we have run so far have asked for Y, $\hat{Y}$, and $Y - \hat{Y}$ to be plotted against X. In the case of multiple regression, however, we have more than one X. We could have requested Y, $\hat{Y}$, or $Y - \hat{Y}$ to be plotted first against X_1 and then against X_2, but the RESIDUALS sentence in the PLOT paragraph will produce the graphs in Figure 11.4. The left-hand graph plots $Y - \hat{Y}$ against $\hat{Y}$, and the right-hand graph plots $(Y - \hat{Y})^2$ against $\hat{Y}$. These graphs are useful in examining the appropriateness of the hypothesized model and checking the validity of the assumptions of the model.

FIGURE 11.2 Multiple Regression Analysis From BMDP1R

```
REGRESSION TITLE IS
BMDP1R MULTIPLE REGRESSION FOR HOUSES DATA

DEPENDENT VARIABLE. . . . . . . . . . . . . . . . 3 PRICE
TOLERANCE . . . . . . . . . . . . . . . . . . . . 0.0100
ALL DATA CONSIDERED AS A SINGLE GROUP

MULTIPLE R                         0.9762          STD ERROR OF EST          2.5619
MULTIPLE R-SQUARE                  0.9530

ANALYSIS OF VARIANCE
                   SUM OF SQUARES       DF      MEAN SQUARE       F RATIO      P(TAIL)
        REGRESSION    2261.4187          2       1130.7092        172.271      0.0000
        RESIDUAL       111.5803         17          6.5635

                                                   STD REG
VARIABLE        COEFFICIENT          STD ERROR     COEFF           T             P(2 TAIL)      TOLERANCE
INTERCEPT       10.04634
SIZE  1         12.83445             0.69919        0.970          18.356        0.0            0.99011
AGE   2         -0.05652             0.06164       -0.048          -0.917        0.3719         0.99011
```

FIGURE 11.3 Covariance and Correlation Matrices From BMDP1R

```
CORRELATION MATRIX
                  SIZE  1          AGE  2          PRICE  3
SIZE   1           1.0000
AGE    2          -0.0995           1.0000
PRICE  3           0.9750          -0.1450          1.0000

COVARIANCE MATRIX
                  SIZE  1          AGE  2          PRICE  3
SIZE   1           0.7137
AGE    2          -0.8053           91.84
PRICE  3           9.205           -15.53           124.9

CORRELATION MATRIX OF REGRESSION COEFFICIENTS
                  SIZE  1          AGE  2
SIZE  1           1.0000
AGE   2           0.0995           1.0000
```

MULTIPLE REGRESSION USING MINITAB

The command to Minitab to perform a multiple regression analysis is essentially the same as for a single predictor regression. The dependent variable (or its column number) is specified, and then the number of dependent variables and their names (or column numbers) are given. Standardized residuals, predicted values, and a vector of the regression coefficients resulting from the analysis may also be saved by specifying column numbers in which they are to be stored.

FIGURE 11.4 Residual Analysis Plots From BMDP1R

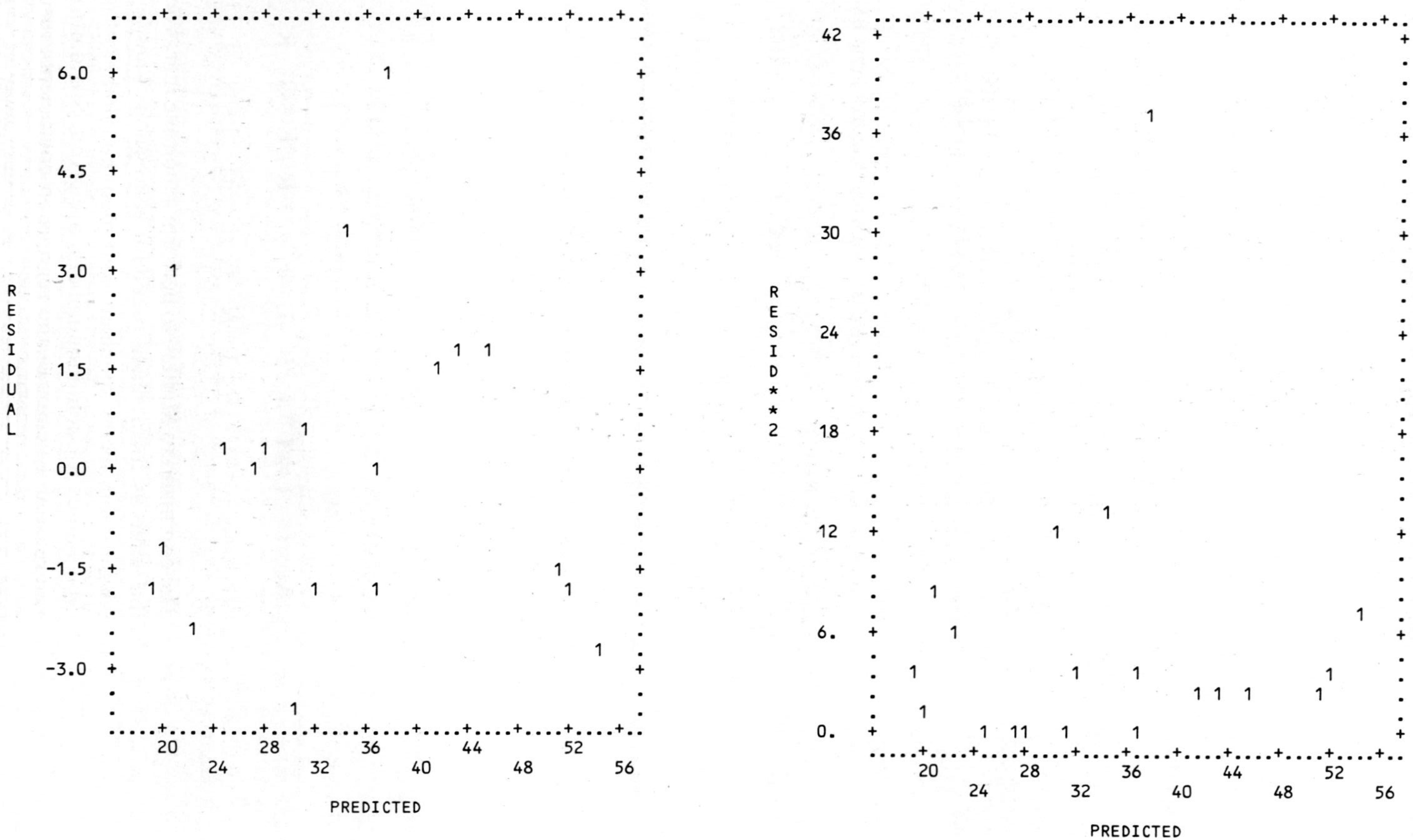

Figure 11.5 shows the output from the Minitab multiple regression of price on size and age for the house size-price-age data. Values of size were entered in column 1, those for age in column 2, and those for price in column 3, under the names SIZE, AGE, and PRICE, respectively. The NOBRIEF command was specified. The command

```
REGRES C3 ON 2 VAR C1 C2/C4/C5/C6
```

causes PRICE to be regressed on the two predictors, SIZE and AGE, the standardized residuals to be stored in column 4, the predicted values to be stored in column 5, and values of a, b_1, and b_2 to be stored in column 6.

Since the NOBRIEF command was used, Minitab produced sequential sums of squares under the headings, FURTHER ANALYSIS OF VARIANCE: SS EXPLAINED BY EACH VARIABLE WHEN ENTERED IN THE ORDER GIVEN. Since in the REGRESSION command the first predictor listed was C1 = SIZE, the first sequential sum of squares is the familiar value

$$\begin{aligned} SS(R)(X_1) &= SS(R)(\text{SIZE}) \\ &= 2255.900, \end{aligned}$$

obtained from fitting just the simple linear model predicting PRICE from SIZE alone. Then the sequential sum of squares for AGE is

$$\begin{aligned} SS(R)(X_2|X_1) &= SS(R)(\text{AGE}|\text{SIZE}) \\ &= 5.520. \end{aligned}$$

Had the order of the predictors in the REGRESSION command been switched, the sequential sums of squares would have been

$$\begin{aligned} SS(R)\ (\text{AGE}) &= 49.871 \\ SS(R)(\text{SIZE}|\text{AGE}) &= 2211.549, \end{aligned}$$

as shown in Figure 11.6. Unless there is some reason to specify the order of the predictor variables in a multiple regression, these sequential sums of squares do not give very useful information, however.

FIGURE 11.6 Minitab Sequential Sums of Squares

```
FURTHER ANALYSIS OF VARIANCE
SS EXPLAINED BY EACH VARIABLE WHEN ENTERED IN THE ORDER GIVEN
DUE TO              DF          SS
REGRESSION          2           2261.420
AGE                 1             49.871
SIZE                1           2211.549
```

FIGURE 11.5
Multiple Regression With Minitab

```
—REGRES C3 ON 2 VAR C1 C2\C4\C5\C6

THE REGRESSION EQUATION IS
Y =     10.0 +  12.8 X1 - 0.0565 X2

                                            ST DEV          T-RATIO =
           COLUMN         COEFFICIENT       OF COEF         COEF/S.D.
                          10.046            2.100            4.78
X1         SIZE           12.8345           0.6992          18.36
X2         AGE            -0.05652          0.06164         -0.92

THE ST DEV OF Y ABOUT REGRESSION LINE IS
S = 2.562
WITH (20 - 3) - 17 DEGREES OF FREEDOM

R-SQUARED = 95.3 PERCENT
R-SQUARED = 94.7 PERCENT, ADJUSTED FOR DF

ANALYSIS OF VARIANCE

DUE TO              DF        SS              MS=SS/DF
REGRESSION           2        2261.420        1130.710
RESIDUAL            17         111.580           6.564
TOTAL               19        2373.000

FURTHER ANALYSIS OF VARIANCE
SS EXPLAINED BY EACH VARIABLE WHEN ENTERED IN THE ORDER GIVEN

DUE TO              DF        SS
REGRESSION          2         2261.420
SIZE                1         2255.900
AGE                 1            5.520

           X1         Y           PRED Y      ST DEV
ROW        SIZE       PRICE       VALUE       PRED Y        RESIDUAL       ST RES
 1         1.80       32.000      31.453      0.780          0.547          0.22
 2         1.00       24.000      21.016      1.105          2.984          1.29
 3         1.70       27.000      30.452      0.640         -3.452         -1.39
 4         2.80       47.000      45.305      0.959          1.695          0.71
 5         2.20       35.000      36.813      0.658         -1.813         -0.73
 6         0.80       17.000      18.901      1.021         -1.901         -0.81
 7         3.60       52.000      54.668      1.353         -2.668         -1.23
 8         1.10       20.000      22.525      0.938         -2.525         -1.06
 9         2.00       38.000      34.302      0.612          3.698          1.49
10         2.60       45.000      43.303      1.360          1.697          0.78
11         2.30       44.000      37.870      0.818          6.130          2.52R
12         0.90       19.000      20.297      0.956         -1.297         -0.55
13         1.20       25.000      24.769      1.024          0.231          0.10
14         3.40       50.000      51.818      1.388         -1.818         -0.84
15         1.70       30.000      31.808      1.422         -1.808         -0.85 X
16         2.50       43.000      41.454      0.867          1.546          0.64
17         1.40       27.000      27.054      0.777         -0.054         -0.02
18         3.30       50.000      51.496      1.099         -1.496         -0.65
19         2.20       37.000      37.039      0.591         -0.039         -0.02
20         1.50       28.000      27.659      0.795          0.341          0.14

R DENOTES AN OBS WITH A LARGE ST RES.
X DENOTES AN OBS WHOSE X VALUE GIVES IT LARGE INFLUENCE.

DURBIN-WATSON STATISTIC = 2.05

(X-PRIME X)INVERSE

           0                 1                 2

0          0.671643
1         -0.163007          0.074483
2         -0.013750          0.000653          0.000579
```

In order to examine the correlation structure between the response variable and the predictors, and among the predictors themselves, the CORRELATION command may be used. Figure 11.7 shows the command to produce a matrix of simple linear bivariate correlation coefficients. Note that in the command CORR C3 C1 C2, the column containing the dependent variable was specified first. It is not necessary to do this, but it was done here so that the first column of the correlation matrix would show correlations between Y = PRICE and each of the predictors.

FIGURE 11.7 Correlation Matrix From Minitab

```
—CORR C3 C1 C2
                 PRICE          SIZE
SIZE             0.975
AGE             -0.145         -0.099
```

If summary statistics, such as the total, sum of squares, mean, standard deviation, maximum and minimum values are required for each of the variables, they may be obtained by using the appropriate commands, as in Figure 4.7. For the variable AGE, these summary statistics are shown in Figure 11.8.

FIGURE 11.8 Summary Statistics for the Variable AGE

```
—COUNT C2
  COUNT    = 20.000
—SUM C2
  SUM      = 430.00
—AVER C2
  AVERAGE  = 21.500
—STAN C2
  ST DEV   = 9.5834
—MAXI C2
  MAXIMUM  = 33.000
—MINI C2
  MINIMUM  = 1.0000
—SSQ C2
  SSQ      = 10990.0
```

Finally, Figure 11.9 shows the standardized residuals plotted against predicted values. Note that the standardized residuals, which were stored in column 4, have been named RESID, and predicted values from column 5 have been named YHAT. Other plots, such as Y versus X_1, and Y versus X_2, could also have been made.

FIGURE 11.9
Minitab Plot of Residuals From Multiple Regression Analysis

```
-- NAME C4 'RESID' C5 'YHAT'

-- PLOT C4 VS C5
```

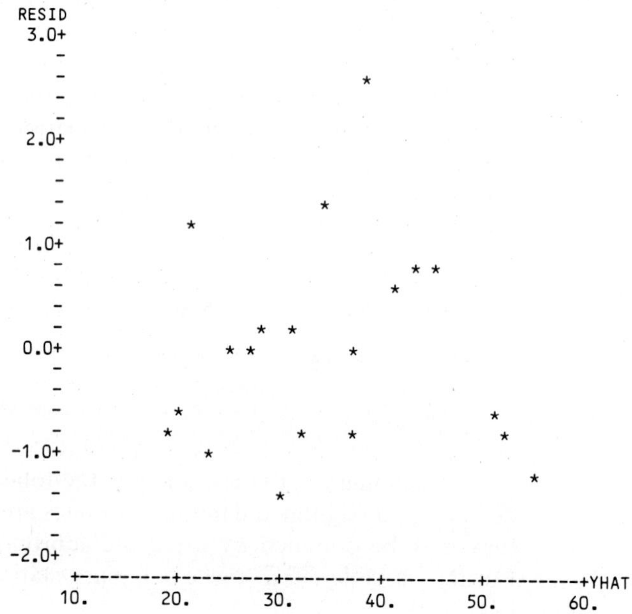

MULTIPLE REGRESSION USING SAS

Figure 11.10 shows a worksheet for using the SAS procedure PROC REG to analyze the house size-price-age data. The only change from previous SAS programs we have demonstrated is that a new procedure has been added:

FIGURE 11.10
Worksheet for Using SAS PROC REG to Regress PRICE on SIZE and AGE

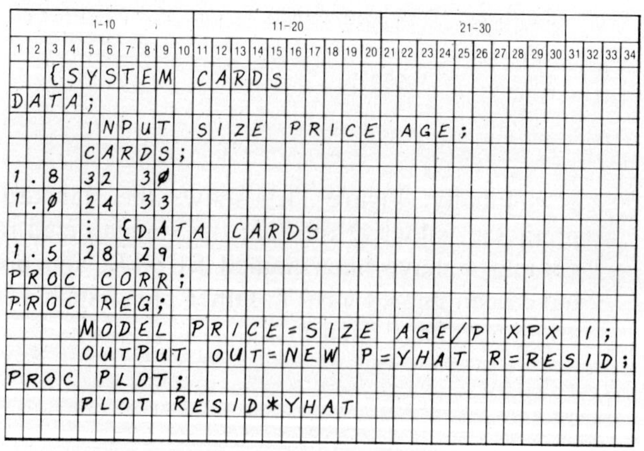

PROC CORR will cause a matrix of bivariate correlation coefficients for all pairs of variables to be printed. This matrix will help us evaluate the separate influences of the individual predictors on Y and also enable us to examine the degree of correlation between pairs of predictors. Figure 11.11 shows the correlation matrix. Note that in the correlation matrix there are two entries in each cell. The top value is the correlation coefficient. Under the correlation coefficient is the significance level attained when testing $H_0{:}\rho = 0$ against the two-sided alternative. Thus, not only does the correlation matrix printed by SAS enable us to see the extent of correlation in the sample, but it also tells whether this correlation is likely to exist in the population.

FIGURE 11.11 Summary Statistics and Correlation Matrix for House Size-Price-Age Data Using SAS

VARIABLE	N	MEAN	STD DEV	SUM	MINIMUM	MAXIMUM
SIZE	20	2.00000000	0.84479833	40.00000000	0.80000000	3.60000000
AGE	20	21.50000000	9.58342868	430.00000000	1.00000000	33.00000000
PRICE	20	34.50000000	11.17563138	690.00000000	17.00000000	52.00000000

CORRELATION COEFFICIENTS / PROB > |R| UNDER H0:RHO=0 / N = 20

	SIZE	AGE	PRICE
SIZE	1.00000 0.0000	-0.09946 0.6765	0.97501 0.0001
AGE	-0.09946 0.6765	1.00000 0.0000	-0.14497 0.5420
PRICE	0.97501 0.0001	-0.14497 0.5420	1.00000 0.0000

There should be no surprises in the printout from a multiple regression analysis using the SAS regression procedure (see Figure 11.12). The analysis of variance table for the full model is shown first, followed by various descriptive statistics. Estimates of the regression coefficients and their standard errors are then shown, followed by partial t-statistics and their associated probabilities.

FIGURE 11.12 Multiple Regression Analysis of House Size-Price-Age Data Using SAS PROC REG

DEP VARIABLE: PRICE

SOURCE	DF	SUM OF SQUARES	MEAN SQUARE	F VALUE	PROB > F
MODEL	2	2261.420	1130.710	172.272	0.0001
ERROR	17	111.580	6.563510		
C TOTAL	19	2373.000			

ROOT MSE	2.561935	R-SQUARE	0.9530
DEP MEAN	34.500000	ADJ R-SQ	0.9474
C.V.	7.425898		

VARIABLE	DF	PARAMETER ESTIMATE	STANDARD ERROR	T FOR H0: PARAMETER=0	PROB > |T|
INTERCEP	1	10.046339	2.099603	4.785	0.0002
SIZE	1	12.834454	0.699193	18.356	0.0001
AGE	1	-0.056523	0.061635	-0.917	0.3719

FIGURE 11.12 (Continued)

OBS	ACTUAL	PREDICT VALUE	RESIDUAL
1	32.000	31.453	0.547337
2	24.000	21.016	2.984
3	27.000	30.452	-3.452
4	47.000	45.305	1.695
5	35.000	36.813	-1.813
6	17.000	18.901	-1.901
7	52.000	54.668	-2.668
8	20.000	22.525	-2.525
9	38.000	34.302	3.698
10	45.000	43.303	1.697
11	44.000	37.870	6.130
12	19.000	20.297	-1.297
13	25.000	24.769	0.230594
14	50.000	51.818	-1.818
15	30.000	31.808	-1.808
16	43.000	41.454	1.546
17	27.000	27.054	-0.053682
18	50.000	51.496	-1.496
19	37.000	37.039	-0.038629
20	28.000	27.659	0.341150

```
SUM OF RESIDUALS            -1.17240E-13
SUM OF SQUARED RESIDUALS    111.5797
```

If the PRINT option is specified, predicted and residual values are printed and can be used to perform a residual analysis. Figure 11.13 shows the residuals plotted against $\hat{Y}$. Other residual analyses that could have been performed are plots of $\hat{Y}$ against SIZE and of $\hat{Y}$ against AGE. For future reference, $\mathbf{X'X}$ and $(\mathbf{X'X})^{-1}$ are shown in Figure 11.14.

SPSS^X MULTIPLE REGRESSION

The worksheet in Figure 11.15 shows how to use SPSS^X to perform a multiple regression analysis of the house size-price-age problem. Most of the program is very much like the SPSS^X programs considered for simple, exponential, or polynomial regressions. Note, however, that in the DESCRIPTIVES subcommand, we have requested that a correlation matrix (CORR) be printed in addition to covariances and cross-products. Figure 11.16 shows the printout from the DESCRIPTIVES subcommand. In each cell of the matrix are three entries: the top one is the simple linear bivariate correlation coefficient between the pair of variables; the second entry is the covariance, $S_{ij}/(n - 1)$, between the pair of variables; and the bottom entry is the sum of squares or cross-products, S_{ij}.

Also new in Figure 11.15 is the STATISTICS subcommand. The defaults on the STATISTICS subcommand are the measures of correlation,

FIGURE 11.13 SAS Plot of Residuals Versus Predicted Values

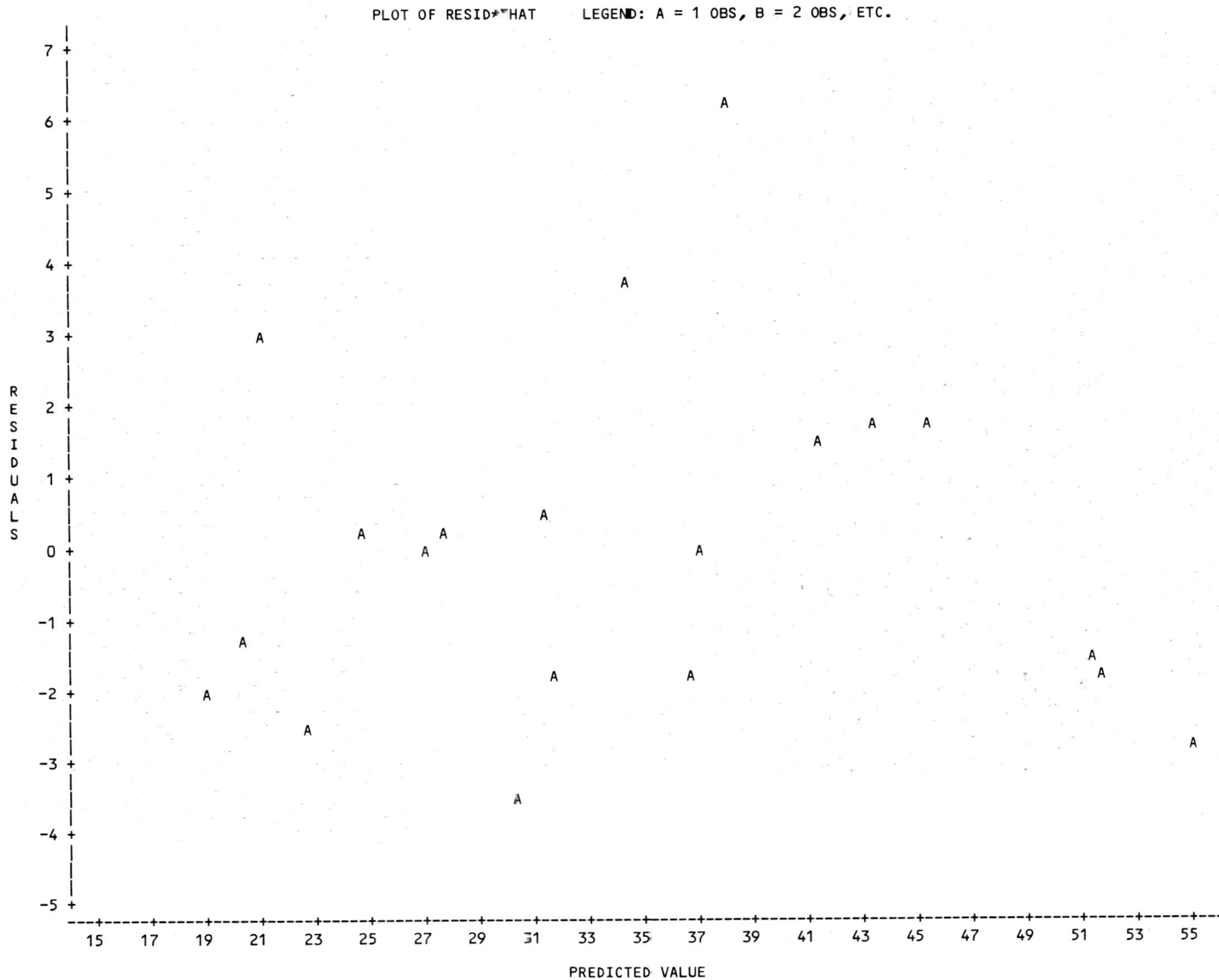

FIGURE 11.14
X′X and **(X′X)**$^{-1}$ Matrices From SAS PROC REG

MODEL CROSSPRODUCTS X'X X'Y Y'Y

X'X	INTERCEP	SIZE	AGE	PRICE
INTERCEP	20	40	430	690
SIZE	40	93.56	844.7	1554.9
AGE	430	844.7	10990	14540
PRICE	690	1554.9	14540	26178

X'X INVERSE, B, SSE

INVERSE	INTERCEP	SIZE	AGE	PRICE
INTERCEP	0.6716425	-0.163007	-0.0137501	10.04634
SIZE	-0.163007	0.07448317	0.0006530616	12.83445
AGE	-0.0137501	0.0006530616	0.0005787919	-0.0565231
PRICE	10.04634	12.83445	-0.0565231	111.5797

FIGURE 11.15 Worksheet for a Multiple Regression Using SPSS^X

```
  {SYSTEM CARDS
UNNUMBERED
TITLE SPSS-X MULT REGR HOUSES DATA
DATA LIST LIST/
          SIZE AGE PRICE
BEGIN DATA
1.8 3Ø 32
1.Ø 33 24
     :
1.5 29 28
END DATA
LIST
REGRESSION DESCRIPTIVES=MEAN STDDEV COV XPROD CORR/
           VARIABLES=SIZE AGE PRICE/
           STATISTICS=DEFAULTS BCOV/
           DEPENDENT=PRICE/
           ENTER/
           CASEWISE=ALL/
           SAVE PRED(YHAT) RESID(R)/
SCATTERGRAM PRICE YHAT WITH SIZE/
           R WITH YHAT
STATISTICS ALL
```

the analysis of variance table, and the regression coefficients with their standard errors, *t*-values, and so on. Had we not even mentioned STATISTICS, this information would have appeared by default (and did, as we saw in earlier examples of SPSS^X programs). However, in this case we have also requested to see correlations between the estimates of the regression coefficients by inserting BCOV. (Note that had we written only STATISTICS-BCOV, SPSS^X would *not* give any of the default statistics but would *only* give the matrix of the correlations and covariances between the regression coefficients.) This matrix is shown in Figure 11.17 below the ANOVA table. On the diagonal in Figure 11.17 are the variances of the regression coefficients b_1 and b_2. Note, for example, that the square root of the variance for AGE is $(0.00380)^{1/2} = 0.06164$, which is the standard error of the regression coefficient for AGE, and similarly for SIZE. The value 0.00429 in the row labeled SIZE and the column labeled AGE (in the cell below the diagonal) gives the covariance

FIGURE 11.16 Descriptive Statistics in SPSS^X Multiple Regression

```
VARIABLE LIST NUMBER 1   LISTWISE DELETION OF MISSING DATA
                  MEAN            STD DEV          LABEL
SIZE              2.000           0.845
AGE              21.500           9.583
PRICE            34.500          11.176

N OF CASES = 20

CORRELATION, COVARIANCE, CROSS-PRODUCT:
                  SIZE            AGE              PRICE
SIZE              1.000          -0.099              0.975
                  0.714          -0.805              9.205
                 13.560         -15.300            174.900
AGE              -0.099           1.000             -0.145
                 -0.805          91.842            -15.526
                -15.300        1745.000           -295.000
PRICE             0.975          -0.145              1.000
                  9.205         -15.526            124.895
                174.900        -295.000           2373.000
```

FIGURE 11.17 SPSS^X Multiple Regression Analysis

```
VARIABLE LIST NUMBER 1   LISTWISE DELETION OF MISSING DATA
EQUATION NUMBER 1    DEPENDENT VARIABLE..PRICE
DESCRIPTIVE STATISTICS ARE PRINTED ON PAGE   4
BEGINNING BLOCK NUMBER  1.  METHOD:  ENTER
VARIABLE(S) ENTERED ON STEP NUMBER    1.. AGE
                                      2.. SIZE

MULTIPLE R           0.97621   ANALYSIS OF VARIANCE
R SQUARE             0.95298                 DF    SUM OF SQUARES    MEAN SQUARE
ADJUSTED R SQUARE    0.94745   REGRESSION     2    2261.42033        1130.71016
STANDARD ERROR       2.56193   RESIDUAL      17     111.57967           6.56351
                     F = 172.27217         SIGNIF F= 0.0000

VAR-COVAR MATRIX OF REGRESSION COEFFICIENTS (B)
BELOW DIAGONAL:  COVARIANCE    ABOVE:  CORRELATION
              AGE             SIZE
AGE           0.00380         0.09946
SIZE          0.00429         0.48887

VARIABLES IN THE EQUATION
VARIABLE              B           SE B         BETA            T          SIG T
AGE            -0.05652        0.06164     -0.04847       -0.917         0.3719
SIZE           12.83445        0.69919      0.97019       18.356         0.0000
(CONSTANT)     10.04634        2.09960                     4.785         0.0002

FOR BLOCK NUMBER  1   ALL REQUESTED VARIABLES ENTERED.
```

between b_1 and b_2, while the value above the diagonal gives the correlation between b_1 and b_2.

The rest of the analysis given in Figure 11.17 should be familiar to the reader, as it contains information on correlations, the analysis of variance for the model, and the regression coefficients, their standard errors, and partial t-values.

Figure 11.18 shows the plot of standardized residuals, along with predicted and residual values. An accompanying table of statistics pertaining to further analysis of the residuals has been omitted here, although it is important in examining the appropriateness of the model.

FIGURE 11.18 SPSS^X Plot of Standardized Residuals in Multiple Regression

```
*: SELECTED   M: MISSING

       -3.0            0.0            3.0
CASE #  0:..............:..............:0     PRICE     *PRED    *RESID
     1   .              .*             .      32.00   31.4527     .5473
     2   .              .      *       .      24.00   21.0155    2.9845
     3   .       *      .              .      27.00   30.4518   -3.4518
     4   .              .   *          .      47.00   45.3045    1.6955
     5   .          *   .              .      35.00   36.8125   -1.8125
     6   .          *   .              .      17.00   18.9008   -1.9008
     7   .         *    .              .      52.00   54.6677   -2.6677
     8   .         *    .              .      20.00   22.5251   -2.5251
     9   .              .       *      .      38.00   34.3022    3.6978
    10   .              .   *          .      45.00   43.3029    1.6971
    11   .              .           *  .      44.00   37.8699    6.1301
    12   .           *  .              .      19.00   20.2973   -1.2973
    13   .              *              .      25.00   24.7694     .2306
    14   .          *   .              .      50.00   51.8182   -1.8182
    15   .          *   .              .      30.00   31.8084   -1.8084
    16   .              .   *          .      43.00   41.4542    1.5458
    17   .              *              .      27.00   27.0537    -.0537
    18   .           *  .              .      50.00   51.4957   -1.4957
    19   .              *              .      37.00   37.0386    -.0386
    20   .              .*             .      28.00   27.6588     .3412
CASE #  0:..............:..............:0     PRICE     *PRED    *RESID
       -3.0            0.0            3.0
```

In order to obtain other plots of predicted values and residuals, we have the option of using either the SCATTERPLOT subcommand of the REGRESSION command or the SCATTERGRAM command. In either case, since there are three variables being used, a choice must be made as to which pair or pairs to plot. It might be important to check that PRICE is linearly related to SIZE and to AGE. Also, it is customary in a multiple regression to plot residuals against predicted values to check for random scatter and thus the appropriateness of the model. These three graphs, with all variables standardized, could be obtained by the use of the subcommand

```
SCATTERPLOT (PRICE,SIZE) (PRICE,AGE) (*RESID,*PRED)
```

under the REGRESSION command. Recall that the asterisks indicate "temporary" variables — those created by the analysis. Although the program in Figure 11.15 does not use the SCATTERPLOT subcommand, the graph of the residuals versus predicted values is shown in Figure 11.19 for your information.

The program in Figure 11.15 used the SCATTERGRAM command to obtain graphs of PRICE versus SIZE, predicted values versus SIZE, and the residuals versus the predicted values. Before this could be done, however, the temporary variables *PRED and *RESID had to be "saved" so that the computer could work with them in a new command. Thus, the SAVE subcommand was included in the REGRESSION command. Note that in saving these variables, the predicted values have been named YHAT and the residuals R. When SPSS^X encounters this subcommand, it prints the following message:

```
FROM EQUATION #1: 2 NEW VARIABLES HAVE BEEN CREATED.
NAME                CONTENTS
YHAT                PREDICTOR
R                   RESIDUAL
```

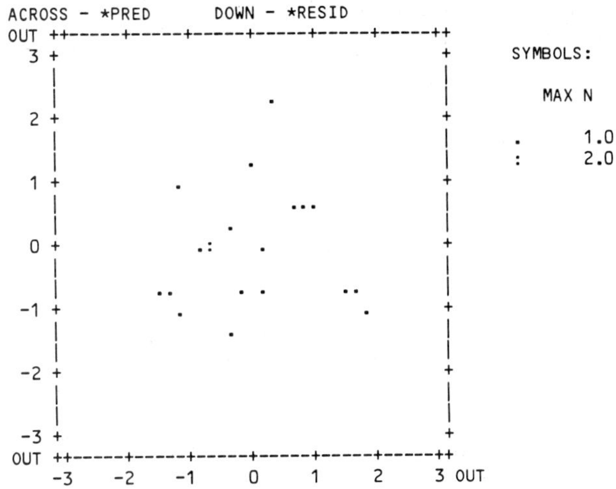

FIGURE 11.19 SPSS^X Scatterplot of Residuals Versus Predicted Values

(One supposes that "predicted" instead of "predictor" was intended.) The graphs of Y versus SIZE, and YHAT versus SIZE, are not shown here. Note, however, in the SCATTERGRAM command that PRICE YHAT WITH SIZE requests two graphs; the first of PRICE versus SIZE and the second of YHAT versus SIZE. A slant separates subcommands within the SCATTERGRAM command when a new horizontal variable is requested. The graph produced by the subcommand R WITH YHAT is shown in Figure 11.20. The statistics associated with this scatter diagram are also shown in Figure 11.20.

EXERCISE

11.10 Perform a regression analysis using an available computer program for one or more of the following data sets, and compare the results to the results of the previous analyses:

a. Sales Data: II (Exercise 11.1);
b. Restaurant Data (Exercise 11.2);
c. Drug Data (Exercise 11.3);
d. Auto Data (Exercise 11.4);
e. Ads Data (Exercise 11.5);
f. Children Data (Exercise 11.6);
g. Glue Data (Exercise 11.7);
h. Bottles Data (Exercise 11.8);
i. Yield Data (Exercise 11.9).

11.5 INTERPRETATION AND USE OF THE MULTIPLE REGRESSION EQUATION

ESTIMATING AN AVERAGE

The same considerations for the use of the regression equation with a single predictor apply for each predictor in multiple regression. First, as before, the regression equation is intended to give only an estimated *average* Y-value for the corresponding X-values. For example, we could estimate that the price of a 26-year-old house having 2,200 sq ft would be

$$\begin{aligned} Y &= 10.04632987 + 12.83442613(2.2) - 0.05655391(26) \\ &= 36.81166570, \end{aligned}$$

or about \$36,812. The interpretation is that the *average* price of houses 2,200 sq ft in size and 26 years old is estimated at \$36,812. House 5 in our sample, of this size and age, sells for \$35,000, which is close to, but not equal to, the predicted price. Factors other than age and size, as well as random variation, lowered the price of house 5 below the typical price of houses that age and size. After we calculate the standard error of estimate for this problem, we will be able to give a confidence interval for these prices and can then tell whether or not house 5 is unusually low-priced.

RELEVANT RANGE

As in simple regression, it is valid to make estimates of price only for houses in the range of sizes *and* the range of ages sampled. For example, since we have information only on houses between 1 and 33 years old, we cannot use the regression equation to predict the price of a 50-year-old house, even if it is in our size range. We do not know whether prices will be more or less expensive for older houses, or if they will behave in the same way as prices of houses between 1 and 33 years old. In addition, if any or all of the predictors

FIGURE 11.20 SPSS[X] Scattergram of Residuals Plotted Against Predicted Values

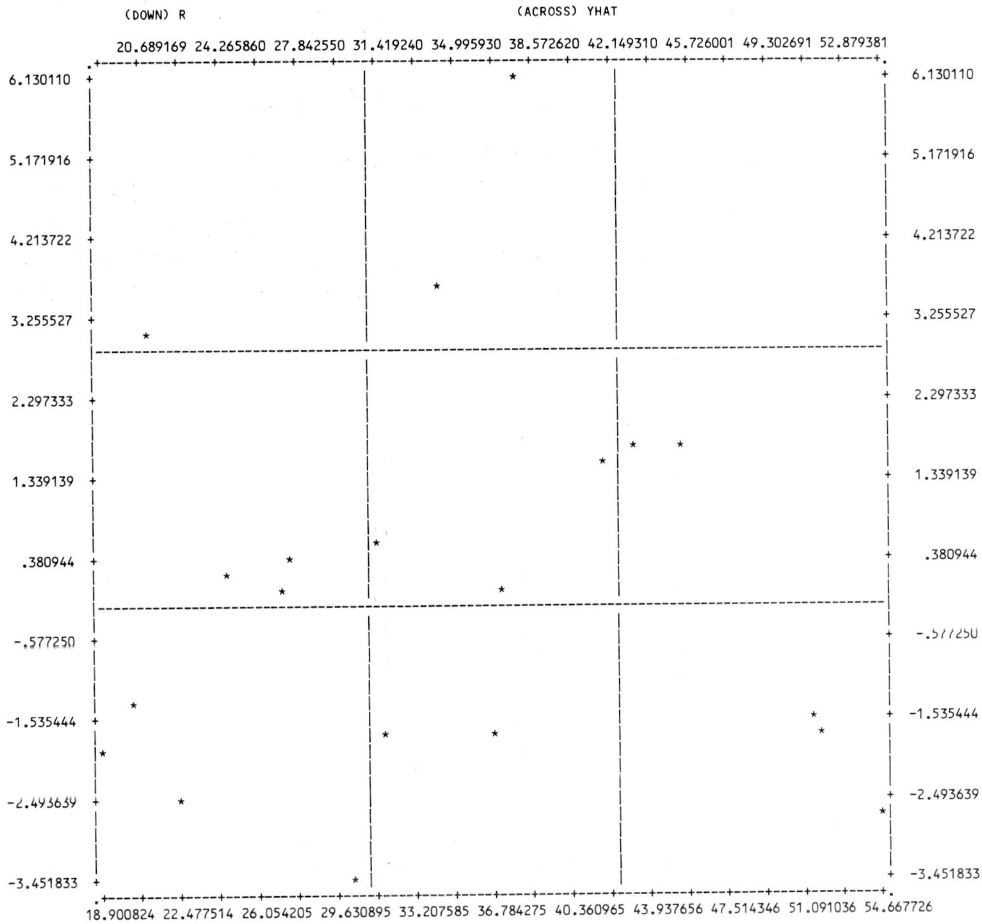

STATISTICS..

CORRELATION (R)-	.00000	R SQUARED -	.00000	SIGNIFICANCE -	.50000
STD ERR OF EST -	2.48975	INTERCEPT (A) -	.00000	SLOPE (B) -	.00000
PLOTTED VALUES -	20	EXCLUDED VALUES-	0	MISSING VALUES -	0

are controlled experimentally, then valid inferences can be made only at the experimental levels chosen for those variables.

THE REGRESSION CONSTANT

For the regression constant, a, to have meaning, the two conditions specified in Chapter 5 must hold for *all* predictors. Thus, for a to be the average Y-value when X_1 and X_2 are zero, (1) It must be physically possible for *both* X_1

and X_2 to be zero; and (2) One must have data around $X_1 = 0$ *and* around $X_2 = 0$. Naturally, if we have $k > 2$ predictors, the two conditions must hold for *all k* of them. For the house size-price-age example, even though a brand-new house would probably be considered to have age zero, it is impossible for a house to have no size. Thus, $a = 10.05$ does *not* mean that the average house of age zero and size zero will sell for \$10,050. Rather, some other factors besides age and size determine this constant, which is needed in order to predict price accurately from size and age.

NET REGRESSION COEFFICIENTS

Net Regression Coefficients

When there is more than one predictor present, $b_1, b_2, \ldots, b_k$ are called **net regression coefficients.** Their interpretation is similar to that in simple regression. For instance, b_1 measures the effect of X_1 on Y by "netting out," or controlling for, the effect of X_2 on Y. Thus, in our example, $b_1 = 12.83$ means that for two houses the same age (holding X_2 constant), if one house is 1,000 sq ft larger than the other, then on the average it will cost \$12,830 more than the smaller house. And $b_2 = -0.06$ means that if two houses are the same size but one is one year older than the other, then on the average the older house will sell for \$60 less than the newer house.

EXERCISES

11.11 Interpret the meaning of a, b_1, and b_2 in the following data sets:

a. Sales Data: II (Exercise 11.1); **b.** Restaurant Data (Exercise 11.2);
c. Drug Data (Exercise 11.3); **d.** Auto Data (Exercise 11.4);
e. Glue Data (Exercise 11.7); **f.** Bottles Data (Exercise 11.8).

11.12 Interpret the meaning of a, b_1, b_2, and b_3 in

a. The Ads Data (Exercise 11.5);
b. The Children Data (Exercise 11.6);
c. The Yield Data (Exercise 11.9).

11.13 a. Using the regression equation found for the Sales Data in Exercise 11.1, estimate the sales of a store 7,000 sq ft in size employing 17 people. Compare your answer to the actual sales in the first store listed in the sample, which had sales of \$7,000, and explain any differences.

b. Using the regression equation found for the Restaurant Data in Exercise 11.2, estimate the total profits in a restaurant with \$1,700 food sales and \$3,000 liquor sales. Compare this estimate to the actual profits of restaurant 4, which had revenues of \$2,300, and explain any differences.

c. Can you use the regression equation found for the Drug Data in Exercise 11.3 to estimate how long it will take a 29-year-old patient to experience relief from pain if he is given 3 gm of drug? Explain why, or why not.
d. Referring to the Auto Data in Exercise 11.4, suppose another car was one year old and was driven 15,000 miles. Estimate the maintenance and repair costs, and interpret the meaning of the estimate.
e. Referring to the Ads Data in Exercise 11.5, can you use the regression equation to estimate the sales of a store spending \$10,000 on television advertising, \$5,000 on radio ads, and \$6,000 on newspaper ads? Explain why, or why not.
f. Use the regression equation found for the Children Data in Exercise 11.6 to estimate the physical fitness score of a nine-year-old child who is 50.5 in. tall and weighs 74 lb. Compare this estimate to child 10, who scored 67, and explain any differences.
g. Use the regression equation from the Glue Data in Exercise 11.7 to estimate the strength of the glue at 80° temperature and 60% humidity. Compare this estimate to the actual strengths found in the sample, 193 and 196 lb, and explain any differences.
h. Referring to the Bottles Data in Exercise 11.8, what is the average number of impurities you would expect to find in bottles sterilized at high temperatures for a medium short time? Compare this estimate to comparable cases in the experiment in which there were 3 and 4 impurities left, and explain any differences.
i. What is the estimated average yield of the crop when fertilized with high concentrations of all three elements in the Yield Data in Exercise 11.9?

11.6 MULTICOLLINEARITY

THE COEFFICIENTS HAVE CHANGED

In the foregoing interpretation of the regression coefficients, the value of b_1 is different from the value obtained for b in simple regression, even though the same data were used and b and b_1 were both supposed to measure the effect of size on price. In this case $b_1 = 12.83$, while in the simple regression problem, $b = 12.9$. Granted, this is not much of a difference, but it is a real one; it is not caused by rounding error. Why has this value changed? The reason is that b_1 is trying to "net out," or control for, the effect of age on price and measure only the effect of size, while b in the simple regression problem ignored the fact that age as well as size might affect price.

One could readily imagine that if a simple regression were run relating price to age alone, the regression coefficient would not exactly equal -0.06.

Further, if a third variable were included, we expect b_1 and b_2 to be different from 12.83 and -0.06, since both b_1 and b_2 would be holding the effect of X_3 constant, which they are not doing at present.

CAN YOU HOLD ONE VARIABLE CONSTANT WHILE VARYING THE OTHER?

This is all fine, *provided* it is possible to hold one variable constant while allowing the other to vary. Of course it can be done *mathematically*, but can it be done in the real world? For example, suppose that in the last 40 years, the trend in housing has been to build smaller and smaller houses. In that case, generally the older the house, the larger it is, and newer houses tend to be smaller. In that case, there is a direct relationship between the two predictors size and age. Then it will be impossible to consider houses the same size but of varying ages, or houses the same age with varying sizes. Instead, if you look at new houses, you will see small sizes, and if you look at old houses, you will see large sizes.

If the predictors themselves are related, can b_1 actually net out the effect of X_2, and can b_2 really control for X_1? Obviously not. Even though b_1 is trying to measure only the effect of X_1 on Y and none of the effect of X_2, if both predictors are related, then b_1 will be contaminated to some extent. It will include some of the effect of X_2 on Y. A similar statement can be made for b_2. We see in our example that b_1 must be contaminated with some of the effect of X_2, because its value in the multiple regression equation has changed from what it was in the simple regression equation. Since the X_1-values and Y-values are the same in each instance, the change in b_1 must be caused by some X_2-values entering into its calculation.

CALCULATION OF REGRESSION COEFFICIENTS

To see how this comes about, let us suppose that X_1 and X_2 are corrected for their means, so that

$$\mathbf{X}'\mathbf{X} = \begin{bmatrix} n & 0 & 0 \\ 0 & S_{11} & S_{12} \\ 0 & S_{12} & S_{22} \end{bmatrix}, \quad \mathbf{X}'\mathbf{y} = \begin{bmatrix} \sum Y \\ S_{1Y} \\ S_{2Y} \end{bmatrix}, \quad \text{and} \quad \hat{\boldsymbol{\beta}} = \begin{bmatrix} a' \\ b_1 \\ b_2 \end{bmatrix}.$$

Then

$$\hat{\boldsymbol{\beta}} = (\mathbf{X}'\mathbf{X})^{-1}\mathbf{X}'\mathbf{y}$$

$$= \begin{bmatrix} \frac{1}{n} & 0 & 0 \\ 0 & \frac{S_{22}}{S_{11}S_{22} - S_{12}^2} & \frac{-S_{12}}{S_{11}S_{22} - S_{12}^2} \\ 0 & \frac{-S_{12}}{S_{11}S_{22} - S_{12}^2} & \frac{S_{11}}{S_{11}S_{22} - S_{12}^2} \end{bmatrix} \begin{bmatrix} \sum Y \\ S_{1Y} \\ S_{2Y} \end{bmatrix}$$

$$= \begin{bmatrix} \frac{1}{n}\sum Y \\ \frac{S_{22}}{S_{11}S_{22} - S_{12}^2} S_{1Y} - \frac{S_{12}}{S_{11}S_{22} - S_{12}^2} S_{2Y} \\ \frac{-S_{12}}{S_{11}S_{22} - S_{12}^2} S_{1Y} + \frac{S_{11}}{S_{11}S_{22} - S_{12}^2} S_{2Y} \end{bmatrix} = \begin{bmatrix} a \\ b_1 \\ b_2 \end{bmatrix}$$

Since $b_1 = (S_{22}S_{1Y} - S_{12}S_{2Y})/(S_{11}S_{22} - S_{12}^2)$, we see that X_2-values are entering into the calculation of b_1 (and a similar statement can, of course, be made about b_2). Only if the covariance S_{12} between the predictors is zero will b_1 measure only the effect of X_1 on Y: if $S_{12} = 0$, then

$$b_1 = \frac{S_{22}S_{1Y} - 0S_{2Y}}{S_{11}S_{22} - 0^2} = \frac{S_{1Y}}{S_{11}},$$

which involves only values of X_1 and Y. The stronger the relationship between X_1 and X_2, the greater the contamination of the regression coefficient by predictors whose effect it is not trying to measure. This holds true, of course, regardless of the number of predictors involved; for example, if X_7 is related to X_4, X_5, and X_{10}, then b_7 will try to measure *only* the effect of X_7 but will in fact also measure some of the effects of X_4, X_5, and X_{10}.

Multi-collinearity

The situation of having correlated predictor variables is called **multicollinearity.** This is a nicely descriptive name: "collinear" means related linearly, and the "multi-" indicates that any or all predictors may be related with any or all others. When multicollinearity is present, the net regression coefficients are said to be *unreliable* measures of the effects of their associated predictor variables; they not only measure the effect of the related predictor but are confounded with the effects of other predictors related to it.

MULTICOLLINEARITY IN HOUSE SIZE-PRICE-AGE EXAMPLE

Let us see to what extent X_1 and X_2 are related in our problem by calculating the correlation coefficient between X_1 and X_2. The formula, similar to that presented in Chapter 8, uses X_1 and X_2 instead of X and Y. We subscript r to indicate that we are calculating the correlation between the two predictors:

$$r_{12} = r_{X_1X_2} = \frac{S_{12}}{\sqrt{S_{11}S_{22}}}$$

$$= \frac{-15.3}{\sqrt{(13.56)(1745)}} = -0.0995.$$

From this we see that our sample gives some evidence of an inverse relationship between the size of a house and its age: Older houses tend to be smaller than newer houses. Since $r_{12}^2 = (-0.0995)^2 = 0.0099 \approx 0.01$, we can say that about 1 percent of the variation in sizes can be explained by differences in ages, or equivalently, that about 1 percent of the variation in age can be accounted for by differences in size. This is not a very strong relationship, and a hypothesis test will show that r_{12} is not significantly different from zero. However, since r_{12} is not zero, our sample shows some correlation between predictors. Thus, *in the sample* it is not possible to hold X_1 or X_2 completely constant while allowing the other to vary. Therefore b_1 and b_2 are not completely reliable measures of the effects of the associated predictors.

STANDARDIZED REGRESSION COEFFICIENTS

Recall that in Chapter 5 we mentioned that it is often desirable to report standardized regression coefficients, or "beta coefficients" or "beta weights," which measure the estimated number of standard deviations that Y will change for each 1 standard deviation increase in X. While in a simple regression, standardized regression coefficients are equivalent to correlation coefficients, their greatest value is in use in multiple regressions, where they are computed so that the regression coefficients may be made comparable in terms of their sizes, even though the predictor variables are measured in different units. That is, in our house size-price-age example one cannot compare $b_1 = 12.834$ to $b_2 = -0.057$ in order to see whether X_1 or X_2 is the stronger predictor, because

b_1 measures the effect on price of a 1,000-sq-ft difference in size, but b_2 measures the effect of a 1-year difference in age. Not only is difference of 1,000 sq ft not comparable to a difference of 1 year but we also see that b_1 and b_2 will be affected by the sizes of the units on X_1 and X_2. For example, if age were measured in units of 10 years, then b_2 would be -0.57 instead of -0.057. Of course, all one need do to determine which predictor is stronger is to look at the correlation coefficients. However, some researchers, notably sociologists and psychologists, prefer a regression equation in which the regression coefficients themselves indicate the relative strengths of their associated variables. Besides, although correlations measure the strengths and directions of the predictor variables, they do not tell *how much* Y changes when X_i changes. Thus, these investigators want to remove the units on the regression coefficients in order to make them directly comparable.

Another instance in which beta coefficients are considered quite useful is that in which an arbitrary measurement scale has been assigned to one of the predictor variables. While an engineer, for instance, has ready-made units for measuring height, weight, pressure, or strength, the social scientist must often make up a scale. For example, a sociologist studying the effect on social class of income and degree of political conservatism has the problem of quantifying the response *social class* and the predictor *degree of conservatism*. A scale from 0 to 4 might do for political conservatism, with 0 meaning not at all conservative, 2 meaning middle-of-the-road, and 4 meaning very conservative, but this is not a natural scale in the sense that length or height or income measurements are. Is a person who rates 4 on the conservatism scale twice as conservative as a person who rates 2? Even though the scale is completely arbitrary, the social scientist's results are very much influenced by the choice: The regression coefficients using a 0–4 scale are not at all the same values he would have obtained with a 0–10 scale, for example. Thus, to remove the influence of this arbitrary scaling, the social scientist often employs a standardized regression coefficient, which provides a unitless measure of the effects of the independent variable on Y. How is this done?

Consider $b_2 = -0.057$ and recall that its units are Y-units over X_2-units (price per age). If the standard deviation of the Y-values were calculated,

$$s_Y = \left[\frac{S_{YY}}{n-1}\right]^{1/2},$$

its units would be Y-units (thousands of dollars). Similarly, one could calculate the standard deviation of the X_2-values,

$$s_2 = s_{X_2} = \left[\frac{S_{22}}{n-1}\right]^{1/2},$$

and its units would be the units of X_2 (years). Then multiplying b_2 by the ratio s_2/s_Y would remove the units on b_2. The resulting quantity is what is referred to as a beta coefficient or beta weight or standardized regression coefficient. The terminology is unfortunate because the word *beta* commonly refers to a population parameter β, estimated by a regression coefficient b, which can never be known from sample information only. A beta weight is, of course, something quite different from a beta. To avoid confusion, we will denote the beta weight by BW. Then for the ith beta weight, $i = 1, 2, \ldots, k$,

$$BW_i = b_i \frac{s_i}{s_Y},$$

where s_i is the standard deviation of the ith predictor variable. In our example, we have that

$$s_1 = 0.8448,$$

$$s_2 = 9.5834,$$

$$s_Y = 11.1756.$$

Then

$$BW_1 = 12.834\left(\frac{0.8448}{11.1756}\right) = 0.9702,$$

$$BW_2 = -0.057\left(\frac{9.5834}{11.1756}\right) = -0.0485.$$

Since BW_1 is greater in absolute value, we see that X_1 (size) is a stronger predictor of price than is X_2 (age).

Note that if multicollinearity is present in the data, a beta weight in a multiple regression is generally not equal to a correlation coefficient. Only when there is no multicollinearity present in the data are correlation coefficients and beta coefficients equal: If there is no multicollinearity, then for example, $b_1 = S_{1Y}/S_{11}$, and

$$BW_1 = \frac{S_{1Y}}{S_{11}} \cdot \frac{[S_{11}/(n-1)]^{1/2}}{[S_{YY}/(n-1)]^{1/2}}$$

$$= \frac{S_{1Y}}{(S_{11}S_{YY})^{1/2}} = r_{1Y}.$$

If multicollinearity is present, then $b_1 \neq S_{1Y}/S_{11}$, and BW_1 is not a correlation coefficient.

The interpretation of exactly what a beta weight measures is a bit awkward. Recall that b_1, for example, measures the number of units and direction that Y changes for a one-unit increase in X_1. BW_1, since it has been standardized by the standard deviations of the variables involved, measures the direction and number of *standard deviations* Y changes for each 1 *standard deviation* increase in X_1. The direction and number of standard deviations Y changes is measured with reference to s_Y, and a 1 standard deviation increase in X_1 is measured with reference to s_1. That is, since $s_1 = 0.8448$ and $s_Y = 11.1756$, then $BW_1 = 0.9702$ says that if X_1 increases 0.8448 units, then the amount of Y will increase $0.9702(11.1756) = 10.8426$ units. (Note that $0.8448(12.83446) = 10.8426$ also.)

One must be careful in interpreting beta coefficients in cases in which multicollinearity is present. We saw that when predictors are correlated, the values of the regression coefficients can change, sometimes drastically, when other predictors are added to or deleted from the model. Thus, it may be risky to use beta weights to compare the influences of the predictors on the response. For example, while X_i may appear to have a greater impact on Y than X_j does when a two-predictor equation is fitted, the reverse may be true if X_k is also included as a predictor.

One final comment about beta weights. Although beta weights are hardly ever used when X is a regulated variable, there is no reason they could not be. If X is regulated, however, the proper standardization is

$$BW_i = b_i \frac{s_i}{s_{Y \cdot X}}.$$

This will not be calculated by the computer, however.

THE VARIANCES OF THE REGRESSION COEFFICIENTS

If we have serious multicollinearity between predictor variables, the net regression coefficients are unreliable not only because they don't measure what they are supposed to but also because their magnitudes and even their signs may be quite far off from what they should be. Refer back to $(\mathbf{X}'\mathbf{X})^{-1}$ from the case in which we are using X_1 and X_2 as predictors.

Note the diagonal elements of the submatrix of $(\mathbf{X}'\mathbf{X})^{-1}$ obtained by deleting the first row and column of $(\mathbf{X}'\mathbf{X})^{-1}$. These diagonal elements when multiplied by $\sigma^2_{Y \cdot X}$ are the *variances of the net regression coefficients*. That is, the variance of b_1 is

$$\sigma^2_{Y \cdot X} \frac{S_{22}}{S_{11}S_{22} - S_{12}^2}$$

and the variance of b_2 is

$$\sigma^2_{Y \cdot X} \frac{S_{11}}{S_{11}S_{22} - S_{12}^2}.$$

The off-diagonal elements, $-S_{12}/(S_{11}S_{22} - S_{12}^2)$ are used to find the covariances between the regression coefficients. Thus, $\sigma^2(\mathbf{X}'\mathbf{X})^{-1}$ is called a **variance-covariance matrix**.

Variance-Covariance Matrix

Consider the denominators $S_{11}S_{22} - S_{12}^2$ of these variances. Note that they can be written as

$$1 - \frac{S_{12}^2}{S_{11}S_{22}} = 1 - r_{12}^2$$

if the numerators are also divided by $S_{11}S_{22}$. Now suppose that X_1 and X_2 are very highly correlated, so that r_{12}^2 is close to 1, and $1 - r_{12}^2$ is close to zero. This will make the ratios $S_{11}/(S_{11}S_{22} - S_{12}^2)$ and $S_{22}/(S_{11}S_{22} - S_{12}^2)$ very large regardless of the values of S_{11} and S_{22}.

What does a large variance for a regression coefficient mean? It means that from sample to sample drawn from this same population the value of the regression coefficient can fluctuate wildly. However, in the long run, if many, many samples are drawn from this same population, the various values of the regression coefficient will *average* out to be correct. But in any one sample the value obtained can be so far off as to even have the wrong sign. Thus, if multicollinearity is severe, not only does b_1 measure a lot of the effect of X_2 as well as the effect of X_1, but you cannot trust b_1 to be anywhere near the correct magnitude or even have the correct sign.

REDUNDANCY

Suppose we had calculated $r_{12} = \pm 1$. Then, in the sample at least, X_1 and X_2 would be perfectly related, meaning that they are, in essence, the same predictor. For example, if X_1 were size in thousands of square feet, and X_2 were size in hundreds of square feet, then X_1 and X_2 are actually the same predictor variable. The two would give redundant information—what one explains about Y is exactly what the other explains. In this case, b_1 and b_2 would be completely unreliable, since b_1 would measure the effect of both X_1 and X_2, and the same goes for b_2. Furthermore, $(\mathbf{X'X})^{-1}$ would not exist, because the denominators $1 - r_{12}^2$ would be zero: we would not even be able to find values for b_1 and b_2. For example, suppose that $X_2 = kX_1$, for k some constant. (X_1 might be a measurement in feet and X_2 a measurement in inches.) Then

$$\begin{aligned} S_{12} &= \sum (X_1 - \bar{X}_1)(X_2 - \bar{X}_2) \\ &= \sum (X_1 - \bar{X}_1)(kX_1 - k\bar{X}_1) \\ &= kS_{11}, \\ S_{22} &= \sum (X_2 - \bar{X}_2)^2 = \sum (kX_1 - k\bar{X}_1)^2 = k^2 S_{11}, \end{aligned}$$

and

$$\mathbf{X'X} = \begin{bmatrix} n & 0 & 0 \\ 0 & S_{11} & kS_{11} \\ 0 & kS_{11} & k^2S_{11} \end{bmatrix},$$

where the third row (column) of $\mathbf{X'X}$ is k times the second row (column). Recall from Chapters 3 and 4 that if one row or column of a matrix is a linear combination of other rows or columns, then the inverse cannot be found—the system of equations cannot be solved. Note that the determinant of the above matrix is

$$n(k^2S_{11}^2 - k^2S_{11}^2) = 0.$$

Even if X_1 and X_2 are not perfectly correlated but are very highly correlated, the rows/columns of **X′X** will be very nearly linear combinations of other rows/columns and the determinant will be close to zero. Then the elements of the inverse matrix become quite large and coefficients become quite large. Recall what happened in Section 4.6 when we used the SAS matrix inversion routine to investigate what happened when the matrix to be inverted was singular or nearly so. Inverting the matrix

$$\begin{bmatrix} 1 & 2 & 3 & 4 \\ 2.001 & 4 & 6 & 7.999 \\ 4 & 7.999 & 12 & 16 \\ 8 & 16 & 24.001 & 32 \end{bmatrix}$$

gave rise to the diagonal elements

$$1600.2, \quad -0.0000000000889039, \quad -0.000000000071039, \quad \text{and} \quad -600.$$

Note, however, that **X′X** always will be symmetric, will never have negative diagonal elements, will never have a negative determinant; and the effect of ill-conditioning will be to make the elements all very *large,* as contrasted to the very small elements obtained in this example. The matrix above is not an **X′X** matrix: it is simply a matrix constructed for purposes of showing what happens when a nearly singular matrix is inverted.

A word of caution is in order. Since we are illustrating the effects of multicollinearity in the two-predictor case, it is possible to examine r_{12}^2 to see the extent of the multicollinearity: if there is any present, it is a linear combination between X_1 and X_2. When we have more than two predictors, however, life is not so easy. Suppose we have predictors X_1, X_2, X_3, and X_4. If X_1 and X_3 are highly correlated, this will show up in r_{13}^2. But suppose that X_3 is a linear combination, or nearly so, of X_1 and X_4. This might not show up in r_{13}^2 or r_{34}^2, and it might be quite a chore to determine which variables the multicollinearity involves. More advanced techniques, such as *eigenvalue* or *characteristic root* analysis are needed in order to spot such relationships among predictors.

In summary, the stronger the relationship between predictors, the more unreliable b_1 and b_2 become as measures of the influences of their associated predictors. Only if $r_{12} = 0$ in the sample are b_1 and b_2 completely reliable; only then can we know the separate effects of the various independent variables on Y.

DESIGNED EXPERIMENTS

In social science studies multicollinearity almost always fouls up the works. Thus, the social science researcher can hardly ever know the individual effects of the predictors on the dependent variable. Almost any variables of interest to a social scientist will be bound up with influences from other variables of interest. One just cannot completely divorce from each other such variables as age, attitudes, incomes, spending habits, amount of education, marital status, and so on. The laboratory scientist who is familiar with the techniques of experimental design, however, is more fortunate. This scientist often has control over the values of the independent variables and can choose levels of $X_1, X_2, \ldots, X_k$ in such a way that all correlations among predictors are zero in the sample. For example, suppose a chemist is studying the effect of temperature (X_1) and pressure (X_2) on reaction time (Y). S/he chooses three evenly spaced temperature levels, which s/he calls high, medium, and low and codes to +1, 0, and −1, respectively. Similarly, two pressure levels, high and low, are specified and coded to +1, and −1. The experiment is replicated three times under each combination of temperature level and pressure level. Table 11.11 shows the experimental design.

TABLE 11.11 Experimental Design for Three Temperature Levels and Two Pressure Levels

X_1 Temperature	X_2 Pressure	Y Reaction Time
−1	−1	Y_1
−1	−1	Y_2
−1	−1	Y_3
−1	1	Y_4
−1	1	Y_5
−1	1	Y_6
0	−1	Y_7
0	−1	Y_8
0	−1	Y_9
0	1	Y_{10}
0	1	Y_{11}
0	1	Y_{12}
1	−1	Y_{13}
1	−1	Y_{14}
1	−1	Y_{15}
1	1	Y_{16}
1	1	Y_{17}
1	1	Y_{18}

Note here that Y_1, Y_2, and Y_3 are reaction times for trials run at low temperature and low pressure; Y_4, Y_5, and Y_6 are reaction times for trials run at low temperature and high pressure, and so on. We note that $\sum X_1 = \sum X_2 = 0$, since temperature and pressure levels are coded. Also, $\sum X_1^2 = 12$, $\sum X_2^2 = 18$, and $\sum X_1X_2 = 0$. Then

$$S_{12} = 0 - \frac{(0)(0)}{18} = 0$$

and

$$r_{12} = \frac{0}{\sqrt{S_{11}S_{22}}} = 0,$$

so that X_1 and X_2 are uncorrelated in the sample. Thus, the chemist can know from the regression analysis what the effect of temperature is on reaction time, and independently what the effect of pressure is. In the equation, b_1 will measure *only* the effect of temperature, uncontaminated by the effect of pressure, and b_2 will measure only the effect of pressure and none of the effect of temperature. (Because in this example neither X_1 nor X_2 is a random variable, it is not strictly correct to calculate a correlation coefficient for them. You are encouraged to satisfy yourself, in the design represented in Table 11.11, that there is no tendency for any given temperature level to occur in conjunction with any particular level of pressure.)

Note also that the calculations involved in solving the normal equations are considerably simplified if the predictors are not correlated. If X_1 and X_2 are not correlated, S_{12} will be zero, and if X_1 and X_2 are corrected for their means,

$$\mathbf{X}'\mathbf{X} = \begin{bmatrix} n & 0 & 0 \\ 0 & S_{11} & 0 \\ 0 & 0 & S_{22} \end{bmatrix},$$

which is diagonal. Thus $(\mathbf{X}'\mathbf{X})^{-1}$ is easily found to be

$$(\mathbf{X}'\mathbf{X})^{-1} = \begin{bmatrix} \frac{1}{n} & 0 & 0 \\ 0 & \frac{1}{S_{11}} & 0 \\ 0 & 0 & \frac{1}{S_{22}} \end{bmatrix}$$

and

$$\hat{\boldsymbol{\beta}} = \begin{bmatrix} \frac{1}{n} & 0 & 0 \\ 0 & \frac{1}{S_{11}} & 0 \\ 0 & 0 & \frac{1}{S_{22}} \end{bmatrix} \begin{bmatrix} \sum Y \\ S_{1Y} \\ S_{2Y} \end{bmatrix} = \begin{bmatrix} \sum \frac{Y}{n} \\ \frac{S_{1Y}}{S_{11}} \\ \frac{S_{2Y}}{S_{22}} \end{bmatrix}.$$

This extends easily to $k > 2$. In this case, $b_1 = S_{1Y}/S_{11}$ and $b_2 = S_{2Y}/S_{22}$, which are both of the form $b = S_{XY}/S_{XX}$. Thus, when predictors are uncorrelated, the multiple regression is just the combination of two simple regressions. That is, a simple regression on the same data using X_1 and Y will yield the same value for b as found by $b_1 = S_{1Y}/S_{11}$. If the value of the regression coefficient is unaffected by the other predictors, the predictors are uncorrelated.

When the regression coefficient changes in the presence of other predictors, multicollinearity is indicated.

EXERCISES

11.14 **a.** In Exercise 2.9 we found $\hat{Y} = -0.3256 + 0.3842X_1$, where X_1 = number of employees, and Y = sales. In Exercise 11.1a we found $\hat{Y} = -3.6183 + 1.7896X_2$, where X_2 = size of store, and Y = sales. In Exercise 11.1b we found $\hat{Y} = -2.447907 + 0.269825X_1 + 0.667393X_2$. Is there multicollinearity present in the data? Explain how you can tell.

b. Recall the following results:

Exercise 11.9a: $\hat{Y} = 322.259259 + 51.166667X_1 - 47.388889X_2 + 8.055556X_3$;

Exercise 11.9b: $\hat{Y} = 322.259259 + 51.166667X_1$
$\hat{Y} = 322.259259 + 47.388889X_2$
$\hat{Y} = 322.259259 + 8.055556X_3$;

Exercise 11.9c: $\hat{Y} = 322.259259 + 51.166667X_1 - 47.388889X_2$
$\hat{Y} = 322.259259 + 51.166667X_1 + 8.055556X_3$
$\hat{Y} = 322.259259 - 47.388889X_2 + 8.055556X_3$.

Is there multicollinearity present in the data? Explain how you can tell.

11.15 **a.** For the Restaurant Data in Exercise 11.2, calculate the correlation coefficient between food and liquor sales. Interpret its meaning with reference to multicollinearity.

b. For the Drug Data in Exercise 11.3, calculate the correlation coefficient between drug dosage and patient age. Interpret its meaning with reference to multicollinearity.

c. Calculate the correlation coefficient between age and mileage in the Auto Data in Exercise 11.4. Interpret its meaning with reference to multicollinearity.

d. Calculate the correlation coefficient between temperature and humidity in the Glue Data in Exercise 11.7. Interpret its meaning with reference to multicollinearity. Is it really appropriate to calculate a correlation coefficient in this problem?

11.16 **a.** Calculate the correlation coefficients between TV and radio advertising, between TV and newspaper advertising, and between radio and newspaper advertising in the Ads Data in Exercise 11.5. Interpret their meanings with reference to multicollinearity.

b. Refer to the Children Data in Exercise 11.6, and calculate correlations between age and height, age and weight, and height and weight. Interpret their meanings with reference to multicollinearity.

11.17 Find standardized regression coefficients for each of the following data sets, interpret their meanings, and compare relative strengths of the predictors:

a. Restaurant Data (Exercise 11.2); **b.** Drug Data (Exercise 11.3);
c. Auto Data (Exercise 11.4); **d.** Ads Data (Exercise 11.5);
e. Children Data (Exercise 11.6).

11.7

THE STANDARD ERROR OF ESTIMATE

Just as in simple regression, we want to find a measure of the variability of the actual values around the average predicted by the regression equation. Recall that in simple regression, we started with the total sum of squares S_{YY} and subtracted the sum of squares that measured variation in Y due to X. We follow the same procedure here, except that from S_{YY} we must subtract variability due to both X_1 and X_2. For the case of two independent variables,

$$s^2_{Y \cdot X} = \frac{S_{YY} - (b_1 S_{1Y} + b_2 S_{2Y})}{n - 3} = \frac{S_{YY} - b_1 S_{1Y} - b_2 S_{2Y}}{n - 3}$$

is the calculating formula for the error variance.

This formula can be obtained by considering the error sum of squares with X_1 and X_2 corrected for their means:

$$\begin{aligned} SS(E) &= \mathbf{y}'\mathbf{y} - \hat{\boldsymbol{\beta}}'\mathbf{X}'\mathbf{y} \\ &= \sum Y^2 - [a' \quad b_1 \quad b_2] \cdot \begin{bmatrix} \sum Y \\ S_{1Y} \\ S_{2Y} \end{bmatrix} \\ &= \sum Y^2 - \left(a' \sum Y + b_1 S_{1Y} + b_2 S_{2Y}\right) \\ &= \sum Y^2 - \overline{Y} \sum Y - b_1 S_{1Y} - b_2 S_{2Y} \\ &= S_{YY} - b_1 S_{1Y} - b_2 S_{2Y}. \end{aligned}$$

We are still measuring the average squared amount that the actual Y-values differ from the predicted:

$$s^2_{Y \cdot X} = \frac{\sum_{i=1}^{n} (Y_i - \hat{Y}_i)^2}{n - 3},$$

where now

$$\hat{Y}_i = a + b_1 X_{1i} + b_2 X_{2i}.$$

Note that the symbol $s^2_{Y \cdot X}$ is intended to be read as the variance in Y, given *all* independent variables X in the model. Alternative notation is $s^2_{Y \cdot X_1, X_2}$, but this becomes more cumbersome as the number of predictor variables increases.

In the denominator of $s^2_{Y \cdot X}$ the $n - 3$ conforms to the general rule that the denominator is $n - (k + 1) = n - p$. The calculating formula extends easily to any number of predictors; the general formula is

$$s^2_{Y \cdot X} = s^2_{Y \cdot X_1, X_2, \cdots, X_k} = \frac{S_{YY} - b_1 S_{1Y} - b_2 S_{2Y} - \cdots - b_k S_{kY}}{n - (k + 1)}$$

$$= \frac{S_{YY} - \sum_{i=1}^{k} b_i S_{iY}}{n - (k + 1)} = \frac{SS(E)}{n - (k + 1)} = MS(E).$$

The standard error of estimate, $s_{Y \cdot X}$, is the square root of $s^2_{Y \cdot X}$. It measures the average amount that the actual values differ from the estimated values. For our house size-price-age example,

$$s^2_{Y \cdot X} = \frac{2373 - 12.83442613(174.9) - (-0.05655391)(-295)}{20 - 3}$$

$$= \frac{111.57546641}{17} = 6.56326273;$$

$$s_{Y \cdot X} = 2.56188655.$$

We interpret the standard error of estimate to mean that when we use the regression equation to estimate the price of a house from its size and age, our estimates will be off from the actual prices by approximately \$2,600, on the average.

C O N F I D E N C E I N T E R V A L S O N M E A N R E S P O N S E

The standard error of estimate can be used to form confidence intervals on the mean price of houses of a given size and age and also confidence or prediction intervals on the price of a particular house of a given size and age. Let $\mu_{Y \cdot X_0}$ denote the true mean Y-value at $X_1 = X_{10}, X_2 = X_{20}, \ldots, X_k = X_{k0}$, where $X_{10}, X_{20}, \ldots, X_{k0}$ denote specified values of $X_1, X_2, \ldots, X_k$, respectively. We can denote these specified X-values as

$$\mathbf{x}_0 = \begin{bmatrix} 1 \\ X_{10} \\ X_{20} \\ \vdots \\ X_{k0} \end{bmatrix}$$

The $\mu_{Y \cdot X_0}$ is estimated by $\hat{Y}_0 = a + b_1X_{10} + b_2X_{20} + \cdots + b_kX_{k0}$. Then the general form of a $100(1 - \alpha)\%$ confidence interval on $\mu_{Y \cdot X_0}$ is

$$\hat{Y}_0 + t_{n-p,\alpha/2}s_{Y \cdot X}\sqrt{\mathbf{x}_0'(\mathbf{X}'\mathbf{X})^{-1}\mathbf{x}_0}.$$

In our example, we estimated the price of a house 2,200 sq ft in size and 26 years old to be $\hat{Y}_0 = 36.81166570$. Then

$$\mathbf{x}_0 = \begin{bmatrix} 1 \\ 2.2 \\ 26 \end{bmatrix} \quad \text{and}$$

$$\mathbf{x}_0'(\mathbf{X}'\mathbf{X})^{-1}\mathbf{x}_0$$

$$= [1 \quad 2.2 \quad 26] \begin{bmatrix} 0.67164255 & -0.16300717 & -0.01375015 \\ -0.16300717 & 0.07448317 & 0.00065306 \\ -0.01375015 & 0.00065306 & 0.00057879 \end{bmatrix} \begin{bmatrix} 1 \\ 2.2 \\ 26 \end{bmatrix}$$

$$= [-0.04447712 \quad 0.017065140 \quad 0.00273512] \begin{bmatrix} 1 \\ 2.2 \\ 26 \end{bmatrix} = 0.06417931.$$

Then a 95% confidence interval on the mean price of houses 2,200 sq ft in size and 26 years old is

$$36.81166570 \pm 2.110(2.56188655)\sqrt{0.06417931},$$

or

$$36.81166570 \pm 1.36943010,$$

or

$$35.44223560 < \mu_{Y \cdot X_0} > 38.18109580.$$

We are 95 percent sure that the mean price of houses 2,200 sq ft in size and 26 years old is between \$35,442 and \$38,181.

CONFIDENCE INTERVALS ON INDIVIDUAL RESPONSE

The confidence interval for an individual house is formed similarly. Let Y_0^* denote the actual price of a house with $X_1 = X_{10}, \ldots, X_k = X_{k0}$, so that

$$\mathbf{x}_0 = \begin{bmatrix} 1 \\ X_{10} \\ X_{20} \\ \vdots \\ X_{k0} \end{bmatrix}.$$

Then a $100(1 - \alpha)\%$ confidence interval on Y_0^* is given by

$$\hat{Y}_0 - t_{n-p,\alpha/2}s_{Y \cdot X}\sqrt{1 + \mathbf{x}_0'(\mathbf{X}'\mathbf{X})^{-1}\mathbf{x}_0} < Y_0^* < \hat{Y}_0 + t_{n-p,\alpha/2}s_{Y \cdot X}\sqrt{1 + \mathbf{x}'(\mathbf{X}'\mathbf{X})^{-1}\mathbf{x}}.$$

Thus, a 95% prediction interval for the price of a house 2,200 sq ft in size and 26 years old is

$$36.81166570 \pm 2.110(2.56188655)\sqrt{1 + 0.06417931},$$

or $$36.81166570 \pm 5.57634653,$$

or $$31.23531917 < Y_0^* < 42.38801223.$$

The interpretation is that 95 percent of all houses 26 years old and 2,200 sq ft in size will cost between $31,235 and $42,388. We note that the price of $35,000 for house 5, which is this size and age, lies in this interval. Thus, house 5 is not considered unusual in price for houses its size and age.

COEFFICIENT OF VARIATION

A coefficient of variation can be formed in exactly the same way as for simple regression studies:

$$CV = \frac{s_{Y \cdot X}}{\bar{Y}}.$$

As before, CV measures the size of the error relative to the size of the response variable being measured. Is the error large because the numbers we are working with are large or because the estimates are inaccurate? In our example,

$$CV = \frac{2.56188655}{34.5} = 0.0742 = 7.42\%$$

says that our estimates will be off, on the average, 7.42 percent.

EFFECT OF MULTICOLLINEARITY ON STANDARD ERROR OF ESTIMATE

Let us now pause and examine the calculating formula for $s^2_{Y \cdot X}$ a bit more closely. One objective in a multiple regression study is to make reasonably accurate estimates, and one feels that more predictors mean more accurate estimates. Thus, if X_1 and X_2 are both effective predictors of Y, then the estimates made using

$$\hat{Y} = a + b_1X_1 + b_2X_2$$

should be closer to the actual values than those made by using only one predictor. To say that the estimates are more accurate means that the error in estimation should be smaller using two predictors than when using only one, or that

$$s^2_{Y \cdot X_1, X_2} < s^2_{Y \cdot X}.$$

This should also be apparent from examination of the numerator of the calculating formula for $s^2_{Y \cdot X}$.

$$S_{YY} - b_1S_{1Y} - b_2S_{2Y}.$$

In simple regression, we would subtract bS_{XY} from S_{YY}, but with two predictors we subtract two terms. Now bS_{XY} in simple regression will not equal b_1S_{1Y} in multiple regression if multicollinearity is present because b will not equal b_1. However, it seems reasonable that even if X_1 and X_2 are correlated, $b_1S_{1Y} + b_2S_{2Y}$ should be larger than bS_{XY}, so that the numerator of $s^2_{Y \cdot X}$ becomes smaller the more predictors we have. Thus, we expect that if X_1 and X_2 are both significant predictors of Y, the standard error should be smaller when two predictors are used than when only one is considered. Similarly, the coefficient of variation should be smaller in a multiple regression.

You may have noticed that the house size-price-age example seems to contradict what we have just said. There, the standard error of estimate using two predictors is *greater* than the standard error using only one. Recall that $s_{Y \cdot X} = \$2550$ when we were predicting price from size alone, but $s_{Y \cdot X} = \$2562$ using age and size. Let us examine what happened.

In Chapter 6 we found

$$s^2_{Y \cdot X} = \frac{117.0996}{18} = \frac{SS(E)(X_1)}{18} = MS(E)(X_1),$$

while for the multiple regression problem, we got

$$s^2_{Y \cdot X} = \frac{111.57546641}{17} = \frac{SS(E)(X_1, X_2)}{17} = MS(E)(X_1, X_2).$$

We note that, as advertised, the numerator of $MS(E)(X_1, X_2)$ is smaller than the numerator of $MS(E)(X_1)$. However, the denominator of $MS(E)(X_1, X_2)$ is *also* smaller, and this is what accounts for $MS(E)(X_1, X_2)$ being larger than $MS(E)(X_1)$. Thus, the reduction in total variability (S_{YY}) obtained by including age (X_2) as a predictor of price is not great enough to overcome the concurrent reduction in degrees of freedom. The question immediately arises: Is X_2 an effective predictor of Y? Obviously not. We cannot predict price more accurately using age and size than we could by using size alone. Only if the effect of X_2 is so strong that $b_2 S_{2Y}$ reduces the numerator of $s^2_{Y \cdot X}$ enough to overcome the reduction in degrees of freedom will it be worthwhile to retain X_2 as a predictor. One might suspect that X_2 is not a significant predictor by simply noting the size of $b_2 = -0.06$. This is not a very large value, especially when compared to the size of the error (\$60 versus \$2600). We will confirm that X_2 is not a significant predictor later, when we perform the multiple regression hypothesis test for the significance of the predictor.

Other strange things can happen to the standard error in "pathological" problems that arise with startling frequency in many actual studies. Especially when the degree of multicollinearity among the predictors is great, many seemingly contradictory results can arise. The most frequent of these is for b_i and S_{iY} to have different signs. In such cases, the quantity $-b_i S_{iY}$ will be positive, so that addition of the predictor X_i may increase the error mean square, $MS(E) = \left(S_{YY} - \sum b_i S_{iY}\right)/(n - p)$. Thus, it is a pretty sure bet that if inclusion of a predictor results in an increase in $s^2_{Y \cdot X}$, the predictor has no business being in the regression equation. Note that the converse does not hold, however: Just because inclusion of a predictor results in a reduction in error does not necessarily mean that that predictor is significant. The final determinant of whether or not a predictor belongs in the regression equation is the hypothesis test, which we shall discuss later.

E X E R C I S E S

11.18 Calculate the standard error of estimate for the following data sets. State its units and interpret its meaning.

a. Sales Data: II (Exercise 11.1); **b.** Restaurant Data (Exercise 11.2);
c. Drug Data (Exercise 11.3); **d.** Auto Data (Exercise 11.4);
e. Ads Data (Exercise 11.5); **f.** Children Data (Exercise 11.6);
g. Glue Data (Exercise 11.7); **h.** Bottles Data (Exercise 11.8);
i. Yield Data (Exercise 11.9).

11.19 For the Sales Data in Exercise 11.1,

a. Calculate the standard error of estimate for part (a), where X_2, size of store, alone was used as a predictor of sales.

b. If sales are predicted from number of employees alone, the standard error of estimate is found to be \$1439.70. Does the size of the store, or the number of employees, when taken alone, provide the better estimates of sales?

c. Compare the two standard errors for the two simple regressions to the standard error from the multiple regression. Are predictions of sales better using two predictors than either one alone?

11.8

MORE COMPLICATED MULTIPLE REGRESSION MODELS

We have considered so far multiple regression models in which Y is strictly *linearly* related to $X_1, X_2, \ldots, X_k$, that is,

$$Y = \alpha + \beta_1 X_1 + \beta_2 X_2 + \cdots + \beta_k X_k + \varepsilon.$$

This was done for the sake of simplicity in introducing the topic of multiple regression. However, the same general linear model techniques can be applied to models involving quadratic or higher-order terms in some or all of the predictors.

Consider, for example, a situation in which it is hypothesized that Y is linearly related to X_1 but is related in a quadratic fashion to X_2. Then our model is

$$Y = \alpha + \beta_1 X_1 + \beta_2 X_2 + \beta_3 X_2^2 + \varepsilon.$$

Define

$$\mathbf{y} = \begin{bmatrix} Y_1 \\ Y_2 \\ \vdots \\ Y_n \end{bmatrix}, \quad \hat{\boldsymbol{\beta}} = \begin{bmatrix} a \\ b_1 \\ b_2 \\ b_3 \end{bmatrix}, \quad \text{and} \quad \mathbf{X} = \begin{bmatrix} 1 & X_{11} & X_{21} & X_{21}^2 \\ 1 & X_{12} & X_{22} & X_{22}^2 \\ \vdots & \vdots & \vdots & \vdots \\ 1 & X_{1n} & X_{2n} & X_{2n}^2 \end{bmatrix}.$$

Then the normal equations are given by

$$\mathbf{X}'\mathbf{X}\hat{\boldsymbol{\beta}} = \mathbf{X}'\mathbf{y},$$

and their solution is

$$\hat{\boldsymbol{\beta}} = (\mathbf{X}'\mathbf{X})^{-1}\mathbf{X}'\mathbf{y}.$$

The error sum of squares is found as

$$\mathbf{y}'\mathbf{y} - \hat{\boldsymbol{\beta}}\mathbf{X}'\mathbf{y},$$

the variances of the regression coefficients a, b_1, b_2, and b_3 are found with reference to the appropriate diagonal elements of $(\mathbf{X}'\mathbf{X})^{-1}$, and to form confidence intervals on the mean response or on an individual response,

$$\mathbf{x}_0' = [1 \quad X_{10} \quad X_{20} \quad X_{20}^2].$$

EXERCISES

11.20 Consider the model

$$Y = \alpha + \beta_1 X_1 + \beta_2 X_2 + \beta_3 X_2^2 + \beta_4 X_1 X_2 + \varepsilon,$$

in which Y is linearly related to X_1, Y is related to X_2 in a quadratic fashion, and X_1 and X_2 *interact* in their effect on Y. (That X_1 and X_2 interact means that the effect of X_1 on Y is not the same for all values of X_2. This concept will be explained more fully in Chapter 14.)

a. Define $\mathbf{y}$, $\hat{\boldsymbol{\beta}}$, and $\mathbf{X}$ for this model.

b. How will $\hat{\boldsymbol{\beta}}$ be obtained?

11.21 For the MPG Data of Exercise 10.6, it was observed that mpg is related to both horsepower and weight in a curvilinear fashion.

a. Use a computer program to fit a model

$$Y = \alpha + \beta_1 X_1 + \beta_2 X_1^2 + \beta_3 X_2 + \beta_4 X_2^2 + \varepsilon,$$

where Y = mpg, X_1 = horsepower, and X_2 = weight.

b. Use a computer program to fit a model

$$Y = \alpha \beta_1^{X_1} \beta_2^{X_2} \varepsilon,$$

or

$$\log Y = \log \alpha + X_1 \log \beta_1 + X_2 \log \beta_2 + \log \varepsilon.$$

11.9 SUMMARY

In this chapter we extended techniques applied in simple regression to the analysis of a multiple regression problem. While most of the earlier results extended to the present problem in a relatively straightforward manner, the analysis became considerably more complicated when we were dealing with several predictors rather than with one.

In order to examine more carefully the effects and implications of multicollinearity in a multiple regression problem, we now turn to the study of how variables can be correlated in multiple regression.

TWELVE

CORRELATION IN MULTIPLE REGRESSION

12.1

INTRODUCTION

In the last chapter we saw how the simple correlation coefficient between two predictor variables can be used as a measure of the extent of multicollinearity in a multiple regression problem. Predictors, however, are not the only variables that might be correlated. One might ask, for instance, if one predictor is correlated with the response, or what percentage of the variation in the response is explained by both predictors together. These questions are addressed here.

12.2

CORRELATIONS BETWEEN PAIRS OF VARIABLES

We saw in Chapter 11 that we could calculate the correlation coefficient or coefficient of determination between the two predictor variables, X_1 and X_2, in our two-predictor multiple regression. This correlation coefficient,

$$r_{12} = \frac{S_{12}}{\sqrt{S_{11}S_{22}}} = -0.0995,$$

is a measure of the degree of multicollinearity among the predictors. In multiple regression studies with $k > 2$, a similar correlation coefficient can be calculated for all pairs of independent variables, to measure the degree of multicollinearity for each pair.

What is perhaps of greater interest is to calculate the linear correlation between each predictor variable and the response variable, Y. This allows us to measure the strength of each predictor separately. The correlation coefficient between X_1 and Y is

$$r_{1Y} = r_{X_1Y} = \frac{S_{1Y}}{\sqrt{S_{11}S_{YY}}} = \frac{174.9}{\sqrt{(13.56)(2373)}} = 0.975.$$

From this we see (as in the simple regression problem) that X_1, size, is a very strong predictor of price, Y, since r_{1Y} is very close to 1. Since r_{1Y} is positive, we also know that larger houses tend to cost more than smaller houses. The coefficient of determination, $r_{1Y}^2 = 0.9507$ says that approximately 95 percent of the variation in prices can be explained entirely by differences in the sizes of the houses. (Recall that the correlation coefficient measures the extent of *linear* association between two variables. The coefficient of determination, if it is calculated as $SS(R)/SS(T)$, can measure variation in Y explained in any fashion by X. If r^2 is calculated as $S_{XY}^2/S_{XX}S_{YY}$, however, then it is also measuring the degree of *linear* association.)

Similarly,

$$r_{2Y} = \frac{S_{2Y}}{\sqrt{S_{22}S_{YY}}} = \frac{-295}{\sqrt{(1745)(2373)}} = -0.1449$$

indicates a weak inverse relationship—older houses tend to cost less than newer houses, but not with a very pronounced effect. Only $r_{2Y}^2 = 0.0210$, or 2 percent of the variation in prices, can be accounted for by the differences in ages of the houses. This result supports our suspicion that age is not a significant predictor of price.

EXERCISE

12.1 Calculate the simple coefficients of determination between Y and each of the predictors, and explain their meanings, for the following data sets:

a. Sales Data: II (Exercise 11.1);
b. Restaurant Data (Exercise 11.2);
c. Drug Data (Exercise 11.3);
d. Auto Data (Exercise 11.4);
e. Ads Data (Exercise 11.5);
f. Children Data (Exercise 11.6);
g. Glue Data (Exercise 11.7);
h. Bottles Data (Exercise 11.8);
i. Yield Data (Exercise 11.9).

1 2 . 3

MULTIPLE COEFFICIENT OF DETERMINATION

The question arises as to what percentage of variation in price is explained by both predictors together. Since the predictors themselves are correlated, we cannot just add r_{1Y}^2 and r_{2Y}^2 to get the percentage of variation in Y explained by X_1 and X_2 together. Since $r_{12} \neq 0$, X_1 and X_2 act alike to some extent; some of the variation in Y explained by X_1 is also explained by X_2, and vice versa, so that X_1 and X_2 overlap in what they tell about Y. The amoeba-like shape in Figure 12.1 represents total amount of variation in Y. Percentages of variation explained by X_1 alone and X_2 alone are indicated, and the total amount of variation explained by both X_1 and X_2 together is shaded with a dark tone. Since the variation explained by X_1 alone and that explained by X_2 alone are not disjoint, their sum is greater than the variation explained by both predictors together. Of course, if X_1 and X_2 are not correlated, these pieces will be disjoint and can be simply added together to determine the total amount of variation explained.

Note in the figure that the variation in Y explained by X_1 and X_2 at the same time (redundantly) is *not* r_{12}^2, since r_{12}^2 does not measure variation in Y.

How then, can we measure the percentage of variation in Y explained by the regression—by both X_1 and X_2? Recall from Chapter 8 that the coefficient

FIGURE 12.1 Percentage of Variation in Y Explained by X_1 and X_2, Where X_1 and X_2 Are Correlated

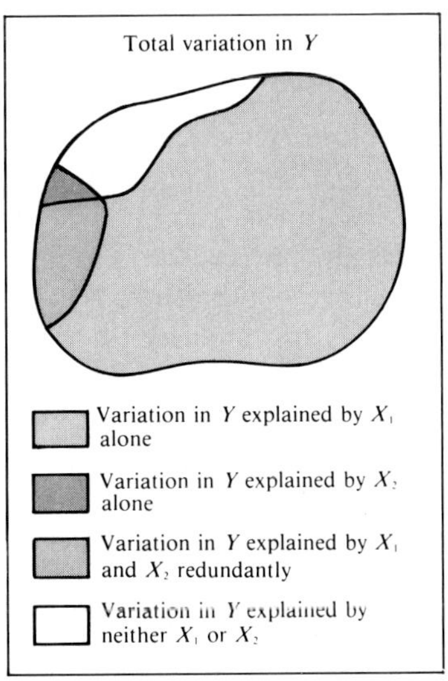

Multiple Coefficient of Determination

of determination is defined as the ratio of variation in Y explained by X to total variation. If we have two predictors, then, the **multiple coefficient of determination** will be the ratio of the variation in Y explained by X_1 and X_2 to the total variation in Y:

$$R^2 = \frac{\text{variation in } Y \text{ explained by } X_1 \text{ and } X_2}{\text{total variation in } Y}.$$

(The notation R^2 is adopted to denote that we are now considering more than one X.) Then

$$R^2 = \frac{SS(R)}{SS(T)} = \frac{SS(\text{regression on } X_1 \text{ and } X_2)}{SS(T)}$$

$$= \frac{b_1 S_{1Y} + b_2 S_{2Y}}{S_{YY}},$$

or in general, for k predictors,

$$R^2 = \frac{b_1 S_{1Y} + b_2 S_{2Y} + \cdots + b_k S_{kY}}{S_{YY}} = \frac{\sum_{i=1}^{k} b_i S_{iY}}{S_{YY}}.$$

Recalling that we partition variation in Y into two parts, one part due to regression and the other due to error, we can equivalently write R^2 as

$$R^2 = \frac{SS(T) - SS(E)}{SS(T)} = 1 - \frac{SS(E)}{SS(T)}.$$

This latter form is convenient because we earlier found $SS(E)$ when calculating the standard error. For our example,

$$R^2 = 1 - \frac{111.57546641}{2373} = 1 - 0.0470 = 0.9530.$$

Differences in the sizes and ages together explain 95.30 percent of the variation in house prices. Note that size alone explained 95.07 percent of the variation in price, so addition of age as a predictor increased the percentage of explained variation by only 0.23 percent. This constitutes further evidence that age is not a significant predictor of price.

Note that while a multiple correlation coefficient, R, is often reported, and indeed is a standard piece of information provided by many computer programs, its interpretation is difficult if not impossible. While R^2 measures the proportion of variation in Y explained by the model—whether it be a simple or multiple regression model, strictly linear or containing higher-order terms—what does R measure? An advantage to calculating r in a simple linear model was that its sign would indicate the direction of the relationship. In multiple regression, however, no meaningful sign can be attached to R, since (as is the case in our house size-price-age example) Y may be directly related to some predictors but inversely related to others. Thus, R is always calculated as the positive square root of R^2. R will also range from 0 to +1, and it can perhaps be interpreted in regard to its strength by mere reference to how close it is to the extremes of its range. Perhaps that is the reason for its popularity among some researchers, although the same information, in interpretable form, is given by R^2. One meaningful interpretation of multiple R is that it is the correlation between the observed and predicted values of Y. Using this interpretation, it makes sense that multiple R should be non-negative.

EXERCISES

12.2 For the following data sets, calculate the multiple coefficient of determination and interpret its meaning with reference to the simple coefficients of determination found in Exercise 12.1.

a. Sales Data: II (Exercise 11.1); **b.** Restaurant Data (Exercise 11.2); **c.** Drug Data (Exercise 11.3); **d.** Auto Data (Exercise 11.4); **e.** Ads Data (Exercise 11.5); **f.** Children Data (Exercise 11.6); **g.** Glue Data (Exercise 11.7); **h.** Bottles Data (Exercise 11.8); **i.** Yield Data (Exercise 11.9).

12.3 Explain why $r_{1Y}^2 + r_{2Y}^2 \neq R^2$ for the following data sets:

a. Sales Data: II (Exercise 11.1); **b.** Restaurant Data (Exercise 11.2); **c.** Drug Data (Exercise 11.3); **d.** Auto Data (Exercise 11.4).

12.4 Explain why $r_{1Y}^2 + r_{2Y}^2 + r_{3Y}^2 \neq R^2$ for the following data sets:

a. Ads Data (Exercise 11.5); **b.** Children Data (Exercise 11.6).

12.5 Explain why $r_{1Y}^2 + r_{2Y}^2 = R^2$ for the following data sets:

a. Glue Data (Exercise 11.7); **b.** Bottles Data (Exercise 11.8); **c.** and why $r_{1Y}^2 + r_{2Y}^2 + r_{3Y}^2 = R^2$ for the Yield Data (Exercise 11.9).

12.4 PARTIAL CORRELATION

The correlation coefficient r_{1Y} does not take into account the fact that the predictor X_2 is included in the regression, and similarly, r_{2Y} does not take the presence of X_1 into account. Since r_{12} is nonzero, we know that some of the variation in Y explained by one variable is also explained by the other. We might ask the question: Once X_1 has explained all it can about Y, what percentage of the left-over variation is explained by X_2? To answer this, we calculate a **partial correlation coefficient**.

Partial Correlation Coefficient

DERIVATION OF THE FORMULA

Suppose we want to correlate X_1 and Y, controlling for X_2. If we run a simple regression using X_2 to predict Y, we would obtain a regression equation of the form

$$\hat{Y} = \overline{Y} + b_1(X_2 - \overline{X}_2),$$

if X_2 is corrected for its mean. The residuals,

$$Y - \hat{Y} = Y - \overline{Y} - b_1(X_2 - \overline{X}_2),$$

then do not depend on X_2 but rather on variables other than X_2 affecting Y. Denote these residuals as W_1:

$$W_1 = Y - \hat{Y}.$$

Similarly, if we were to run a simple regression to predict X_1 from X_2, we would obtain an equation of the form

$$\hat{X}_1 = \overline{X}_1 + b_2(X_2 - \overline{X}_2),$$

and the residuals

$$W_2 = X_1 - \hat{X}_1 = X_1 - \overline{X}_1 - b_2(X_2 - \overline{X}_2)$$

would not depend on X_2.

Since neither W_1 nor W_2 depends on X_2, the correlation between W_1 and W_2 will be the correlation between X_1 and Y, controlling for X_2, or having taken out the effect of X_2 on both X_1 and Y. To calculate the correlation between W_1 and W_2, we first note that

$$\sum W_1 = \sum W_2 = 0.$$

This can be seen by

$$\begin{aligned}\sum W_1 &= \sum [Y_i - \overline{Y} - b_1(X_{2i} - \overline{X}_2)]\\ &= \sum [(Y_i - \overline{Y}) - b_1(X_{2i} - \overline{X}_2)]\\ &= \sum (Y_i - \overline{Y}) - b_1 \sum (X_{2i} - \overline{X}_2)\\ &= 0 - b_1(0) = 0,\end{aligned}$$

and similarly for $\sum W_2$. Therefore,

$$r_{W_1W_2} = \frac{S_{W_1W_2}}{\sqrt{S_{W_1W_1}S_{W_2W_2}}} = \frac{\sum W_1W_2}{\sqrt{\sum W_1^2 \sum W_2^2}}.$$

To find $\sum W_1^2$,

$$\begin{aligned}\sum W_1^2 &= \sum [(Y_i - \bar{Y}) - b_1(X_{2i} - \bar{X}_2)]^2 \\ &= \sum [(Y_i - \bar{Y})^2 - 2b_1(Y_i - \bar{Y})(X_{2i} - \bar{X}_2) + b_1^2(X_{2i} - \bar{X}_2)^2] \\ &= S_{YY} - 2b_1 S_{2Y} + b_1^2 S_{22} \\ &= S_{YY} - 2\left(\frac{S_{2Y}}{S_{22}}\right) S_{2Y} + \left(\frac{S_{2Y}^2}{S_{22}^2}\right) S_{22} \\ &= S_{YY} - \frac{S_{2Y}^2}{S_{22}}.\end{aligned}$$

Similarly,

$$\sum W_2^2 = S_{11} - \frac{S_{12}^2}{S_{22}}.$$

(You can verify this for yourself, following the steps used to find $\sum W_1^2$.) Finally,

$$\begin{aligned}\sum W_1 W_2 &= \sum [(Y_i - \bar{Y}) - b_1(X_{2i} - \bar{X}_2)]\,[(X_{1i} - \bar{X}_1) - b_2(X_{2i} - \bar{X}_2)] \\ &= S_{1Y} - \frac{S_{12}S_{2Y}}{S_{22}}.\end{aligned}$$

Then

$$\begin{aligned} r_{W_1W_2} &= \frac{S_{1Y} - (S_{12}S_{2Y}/S_{22})}{\sqrt{S_{YY} - (S_{2Y}^2/S_{22})}\sqrt{S_{11} - (S_{12}^2/S_{22})}} \\ &= \frac{S_{22}}{S_{22}} \frac{S_{1Y} - (S_{12}S_{2Y}/S_{22})}{\sqrt{(S_{YY}/S_{YY})[S_{YY} - (S_{2Y}^2/S_{22})]}\sqrt{(S_{11}/S_{11})[S_{11} - (S_{12}^2/S_{22})]}} \\ &= \frac{S_{1Y}S_{22} - S_{12}S_{2Y}}{S_{22}\sqrt{S_{YY}(1 - r_{2Y}^2)}\sqrt{S_{11}(1 - r_{12}^2)}} \\ &= \frac{S_{1Y}S_{22} - S_{12}S_{2Y}}{\sqrt{S_{YY}S_{11}S_{22}^2}\sqrt{(1 - r_{2Y}^2)(1 - r_{12}^2)}} \\ &= \frac{\dfrac{S_{1Y}S_{22}}{\sqrt{S_{YY}S_{11}S_{22}^2}} - \dfrac{S_{12}S_{2Y}}{\sqrt{S_{YY}S_{11}S_{22}^2}}}{\sqrt{(1 - r_{2Y}^2)(1 - r_{12}^2)}} \\ &= \frac{r_{1Y} - r_{12}r_{2Y}}{\sqrt{(1 - r_{12}^2)(1 - r_{2Y}^2)}}.\end{aligned}$$

Thus, the partial correlation between X_1 and Y, controlling for X_2, is calculated by using the formula

$$r_{1Y\cdot 2} = \frac{r_{1Y} - r_{12}r_{2Y}}{\sqrt{1 - r_{12}^2}\sqrt{1 - r_{2Y}^2}}.$$

Note that this partial correlation begins with the simple correlation r_{1Y} and then subtracts a piece measuring the correlation between the two predictors, r_{12}, and the correlation, r_{2Y}, between the other predictor and Y. Then a scaling is performed in the denominator. In a similar fashion, the partial correlation between X_2 and Y, controlling for X_1, is

$$r_{2Y\cdot 1} = \frac{r_{2Y} - r_{12}r_{1Y}}{\sqrt{1 - r_{12}^2}\sqrt{1 - r_{1Y}^2}}.$$

PARTIAL CORRELATIONS FOR HOUSE SIZE-PRICE-AGE EXAMPLE

In our example,

$$r_{1Y\cdot 2} = \frac{0.975 - (-0.0995)(-0.1449)}{\sqrt{1 - 0.0099}\sqrt{1 - 0.0210}} = 0.9757,$$

$$r_{2Y\cdot 1} = \frac{-0.1449 - (-0.0995)(0.975)}{\sqrt{1 - 0.0099}\sqrt{1 - 0.9507}} = -0.2168.$$

Notice that taking X_2 into account, the relationship between X_1 and Y is direct, and taking X_1 into account, the relationship between X_2 and Y is inverse. Since $r_{1Y\cdot 2}^2 = (0.9757)^2 = 0.9520$, once X_2(age) has explained all the variation in Y(price) that it can, then of the unexplained variation X_1(size) explains 95.2 percent. And since $r_{2Y\cdot 1}^2 = (-0.2168)^2 = 0.0470$, once X_1(size) has explained all it can about Y, X_2 explains 4.7 percent of what is left. These results again support our suspicion that X_2 is not a significant predictor.

Note that it may be possible for both r_{1Y} and r_{2Y} to be relatively large, but if r_{12} is also large, then X_1 will not improve much over what X_2 can do, and vice versa, because of the large degree of redundancy. In this case, the less important predictor will be discarded: Even though by itself it is a strong predictor, in conjunction with the other independent variable with which it is highly correlated it will not add very much information about Y. Neither $r_{1Y\cdot 2}$ nor $r_{2Y\cdot 1}$ will be very large, even though r_{1Y} and r_{2Y} are reasonably close to ± 1.

An alternative way of interpreting a partial correlation coefficient can be illustrated by the following example. Suppose X_1 = age of child, X_2 = number of hours spent watching television, and Y = score on a manual dexterity test. We might expect to see the following relationships among these three variables: X_1 and Y will probably be directly related, X_1 and X_2 will probably also be directly related (the older the child, the later his bedtime), and perhaps because of the relationship between X_1 and X_2, we will also see X_2 and Y directly related. Suppose, however, that we narrowed down the list and looked at a subset of the children in the sample who are all the same age. When we look at the correlation between X_2 and Y for just these children, we would not be surprised to see a negative correlation: For children the same age, the more time spent in a passive activity such as television viewing, the lower the score on a manual dexterity test. This latter correlation is a partial correlation—it holds X_1 constant and looks at the correlation between X_2 and Y; as such, it measures something quite different from that which is measured by the simple correlation between X_2 and Y.

HIGHER-ORDER PARTIAL CORRELATIONS

When only one variable is controlled for, as in the above example, we have a *first-order* partial correlation. Given more than two predictors, all but one must be controlled for when correlating one predictor and Y. The *order* of the partial correlation is the number of controlled variables. (Thus, simple bivariate unconditional correlations are sometimes referred to as zero-order correlations.) For example, if X_1, X_2, and X_3 predict Y, then the second-order partial correlations are $r_{1Y\cdot 23}$, $r_{2Y\cdot 13}$, and $r_{3Y\cdot 12}$. The formulas for second-order and higher partial correlations appear complicated and require calculation of all lower-order partial correlations. However, their general form is not hard to grasp. For example,

$$r_{1Y\cdot 23} = \frac{r_{1Y\cdot 2} - (r_{13\cdot 2})(r_{3Y\cdot 2})}{\sqrt{1 - r_{13\cdot 2}^2}\sqrt{1 - r_{3Y\cdot 2}^2}},$$

or

$$r_{1Y\cdot 23} = \frac{r_{1Y\cdot 3} - (r_{12\cdot 3})(r_{2Y\cdot 3})}{\sqrt{1 - r_{12\cdot 3}^2}\sqrt{1 - r_{2Y\cdot 3}^2}},$$

is the formula for one possible second-order partial correlation—the correlation between X_1 and Y, correcting for X_2 and X_3. Note that the two formulas are equivalent; you can start off with $r_{1Y\cdot 2}$ or with $r_{1Y\cdot 3}$ and you condition on the same variable in each term; both give the same result. One third-order partial correlation, between X_1 and Y controlling for X_2, X_3, and X_4, is given by the formula

$$r_{1Y\cdot 234} = \frac{r_{1Y\cdot 23} - (r_{14\cdot 23})(r_{4Y\cdot 23})}{\sqrt{1 - r_{14\cdot 23}^2}\sqrt{1 - r_{4Y\cdot 23}^2}}.$$

There are two formulas equivalent to this, one starting with $r_{1Y\cdot 24}$ and another beginning with $r_{1Y\cdot 34}$. Other second- and third-order as well as higher-order partial correlations are calculated similarly.

E X E R C I S E S

12.6 Calculate the partial correlation coefficients $r_{1Y\cdot 2}$ and $r_{2Y\cdot 1}$ and interpret their meanings for the following data sets:
a. Sales Data: II (Exercise 11.1); **b.** Restaurant Data (Exercise 11.2);
c. Drug Data (Exercise 11.3); **d.** Auto Data (Exercise 11.4).

12.7 Find the partial correlations $r_{1Y\cdot 23}$, $r_{2Y\cdot 13}$, and $r_{3Y\cdot 12}$ and interpret their meanings for the
a. Ads Data (Exercise (11.5); **b.** Children Data (Exercise 11.6).

12.8 For the Children Data (Exercise 11.6),
a. One might expect that overweight children would tend to score low on a physical fitness test. If we consider the partial correlation $r_{3Y\cdot 12}$ to be the correlation between weight and score, holding age and height constant, can we conclude that for children the same age and height, the score is lower for heavier children than for lighter-weight children? Recall that $r_{3Y\cdot 12} = -0.2544$.
b. Why does the inverse relationship between weight and score for children the same age and height not show up on the simple correlation, $r_{3Y} = 0.3681$?

12.9 Write the two formulas equivalent to the formula for $r_{1Y\cdot 234}$ given in the text.

12.10 Using the Ads Data (Exercise 11.5), show that the two formulas for $r_{3Y\cdot 12}$ give the same result.

1 2 . 5
S U M M A R Y

There are many correlations that can be calculated in multiple regression problems. Simple correlations between pairs of predictors measure pairwise multicollinearity in the problem. Simple correlations between each predictor and the response tell how strong each predictor is by itself. Partial correlations between a predictor and the response measure the strength of the predictor in the presence of other predictors. Finally, a multiple coefficient of determination measures what percentage of the variation in the response is explained by the entire set of predictors.

We will see that all three kinds of correlations—simple, partial, and multiple—are important in testing the significance of the various predictors and are essential to the understanding of the entire multiple regression problem. Chapter 13 takes up the study of significance tests in multiple regression.

THIRTEEN

TESTS OF SIGNIFICANCE IN MULTIPLE REGRESSION

13.1

INTRODUCTION

The researcher, having collected data on several predictor variables and one response variable, has found a regression equation, a standard error of estimate, and several measures of correlation. Just because such measures can be calculated, however, does not mean that they are appropriate. Before including a predictor in the regression equation, one should be reasonably certain that it is an effective predictor and not redundant with any other predictors.

We are now ready to test our predictors for significance. From the sizes of the net regression coefficients, the standard error of estimate, and the various correlations in our house size-price-age example, we suspect that size, but not age, is an important predictor of price. These suspicions will be supported by the tests of hypotheses.

13.2

THE GENERAL TEST OF SIGNIFICANCE OF THE REGRESSION

Before testing each predictor separately, we will test for the *significance of the regression*. This test will tell us if *any* of the predictors is significantly different from zero. If we find that at least one predictor is significant, then we will turn to the question of which one or ones. If the test of the significance of the regression cannot detect any significant predictors, however, then further testing is generally unnecessary. If your model is at all reasonable, the overall test for significant predictors can be expected to reject the null hypothesis that no predictors are significant. (Of course, there are exceptions.)

THE HYPOTHESIS

We want to test

$$H_0{:}\beta_i = 0 \quad \text{for all } \beta_i,\ i = 1, 2, \ldots, k$$

$$H_1{:}\text{some } \beta_i \neq 0.$$

Note that the alternative hypothesis is two-sided. We are simultaneously testing all k predictors in order to detect those inversely related to Y as well as those directly related. The null and alternative hypotheses can alternatively be written in vector notation as

$$H_0{:}\boldsymbol{\beta}^* = \mathbf{0}$$

$$H_1{:}\boldsymbol{\beta}^* \neq \mathbf{0},$$

where $\boldsymbol{\beta}^*$ is the $k \times 1$ subvector of $\boldsymbol{\beta}$ obtained by deleting the first element, α. That is,

$$\boldsymbol{\beta} = [\alpha \quad \beta_1 \quad \beta_2 \quad \cdots \quad \beta_k]', \qquad \boldsymbol{\beta}^* = [\beta_1 \quad \beta_2 \quad \cdots \quad \beta_k]',$$

and

$$\mathbf{0}_k = [0 \quad 0 \quad \cdots \quad 0]'.$$

THE TEST

The test is run using the analysis of variance shown in Table 13.1, where F has k and $n - (k + 1)$ degrees of freedom. We compare F to $F_{k,n-(k+1),\alpha}$ and

TABLE 13.1 Test of Significance of the Regression

Source	SS	df	MS	F
Regression ($X_1, X_2, \ldots, X_k$)	$SS(R) = \sum_{i=1}^{k} b_i S_{iY}$	k	$MS(R) = \dfrac{\sum_{i=1}^{k} b_i S_{iY}}{k}$	$\dfrac{MS(R)}{MS(E)}$
Error	$SS(E) = S_{YY} - \sum_{i=1}^{k} b_i S_{iY}$	$n - (k + 1)$	$MS(E) = \dfrac{S_{YY} - \sum_{i=1}^{k} b_i S_{iY}}{n - (k + 1)} = s^2_{Y \cdot X}$	
Total	$SS(T) = S_{YY}$	$n - 1$		

reject H_0 if F is greater than this critical value. If we can reject H_0, we are at least $100(1 - \alpha)$ percent sure that at least one of the k predictors is significant, and then we might proceed to look for it. If H_0 cannot be rejected, then usually none of the predictors is strong enough to bother with.

APPLYING THE TEST TO THE HOUSE SIZE-PRICE-AGE EXAMPLE

For the house size-price-age example, the analysis of variance is as shown in Table 13.2. If $\alpha = 0.05$, we will reject $H_0{:}\boldsymbol{\beta}^* = \mathbf{0}$ in favor of $H_1{:}\boldsymbol{\beta}^* \neq \mathbf{0}$ if $F > F_{2,17,0.05} = 3.59$. Since $172.2780 > 3.59$, we can be more than 95 percent sure that either size or age, or both, is a significant predictor of price.

TABLE 13.2
General Regression Significance Test for House Size-Price-Age Example

Source	*SS*	*df*	*MS*	*F*
Regression	2261.4245	2	1130.7122	172.2780
Error	111.5755	$20 - 3 = 17$	6.5633	
Total	2373	$20 - 1 = 19$		

THE TEST IN TERMS OF MATRICES

Before proceeding to find which predictors are significant, it will help to formulate the general test in terms of matrices and vectors. Recall from Chapter 7 that the test for significance can be formulated as shown in Table 13.3 if $H_0{:}\boldsymbol{\beta} = [\alpha \quad \beta_1 \quad \cdots \quad \beta_k]' = \mathbf{0}$ is to be tested, or alternatively, as shown in Table 13.4 if $H_0{:}\boldsymbol{\beta}^* = [\beta_1 \quad \cdots \quad \beta_k] = \mathbf{0}$ is to be tested. We now want to

TABLE 13.3
ANOVA for Testing $H_0{:}\boldsymbol{\beta} = \mathbf{0}$

Source	*SS*	*df*
Regression	$\hat{\boldsymbol{\beta}}'\mathbf{X}'\mathbf{y}$	$k + 1$
Error	$\mathbf{y}'\mathbf{y} - \hat{\boldsymbol{\beta}}'\mathbf{X}'\mathbf{y}$	$n - k - 1$
Total	$\mathbf{y}'\mathbf{y}'$	n

TABLE 13.4 General Regression Significance Test in Matrix Form

Source	SS	df
Regression	$\hat{\boldsymbol{\beta}}'\mathbf{X}'\mathbf{y} - \frac{(\sum Y)^2}{n}$	k
Error	$\mathbf{y}'\mathbf{y} - \hat{\boldsymbol{\beta}}'\mathbf{X}'\mathbf{y}$	$n - k - 1$
Total	$\mathbf{y}'\mathbf{y} - \frac{(\sum Y_i)^2}{n} = S_{YY}$	$n - 1$

show that this latter form is completely equivalent to what we have already done. We do this by showing that

$$\hat{\boldsymbol{\beta}}'\mathbf{X}'\mathbf{y} - \frac{(\sum Y)^2}{n} = b_1 S_{1Y} + b_2 S_{2Y}$$

in the two-predictor case. First,

$$\begin{aligned}
\hat{\boldsymbol{\beta}}'\mathbf{X}'\mathbf{y} &= [a \quad b_1 \quad b_2] \begin{bmatrix} \sum Y \\ \sum X_1 Y \\ \sum X_2 Y \end{bmatrix} \\
&= a \sum Y + b_1 \sum X_1 Y + b_2 \sum X_2 Y \\
&= a \sum Y + b_1 \sum X_1 Y - \frac{b_1 (\sum X_1)(\sum Y)}{n} + \frac{b_1 (\sum X_1)(\sum Y)}{n} \\
&\qquad + b_2 \sum X_2 Y - \frac{b_2 (\sum X_2)(\sum Y)}{n} + \frac{b_2 (\sum X_2)(\sum Y)}{n} \\
&= a \sum Y + b_1 S_{1Y} + b_2 S_{2Y} + \frac{b_1 (\sum X_1)(\sum Y)}{n} + \frac{b_2 (\sum X_2)(\sum Y)}{n}.
\end{aligned}$$

Then

$$\begin{aligned}
\hat{\boldsymbol{\beta}}'\mathbf{X}'\mathbf{y} - \frac{(\sum Y)^2}{n} &= b_1 S_{1Y} + b_2 S_{2Y} \\
&\qquad + \left[a \sum Y - \frac{(\sum Y)^2}{n} + \frac{b_1 (\sum X_1)(\sum Y)}{n} + \frac{b_2 (\sum X_2)(\sum Y)}{n} \right] \\
&= b_1 S_{1Y} + b_2 S_{2Y} + \left(\sum Y\right) \left[a - \frac{\sum Y}{n} + \frac{b_1 (\sum X_1)}{n} + \frac{b_2 (\sum X_2)}{n} \right].
\end{aligned}$$

Recall that the first normal equation for the two-predictor model is

$$\sum Y = na + b_1 \sum X_1 + b_2 \sum X_2,$$

or

$$a = \frac{\sum Y}{n} - \frac{b_1 \sum X_1}{n} - \frac{b_2 \sum X_2}{n},$$

so that

$$a - \left[\frac{\sum Y}{n} - \frac{b_1 \sum X_1}{n} - \frac{b_2 \sum X_2}{n} \right] = a - \frac{\sum Y}{n} + \frac{b_1 \sum X_1}{n} + \frac{b_2 \sum X_2}{n} = 0.$$

Thus,

$$\hat{\boldsymbol{\beta}}'\mathbf{X}'\mathbf{y} - \frac{(\sum Y)^2}{n} = b_1 S_{1Y} + b_2 S_{2Y}.$$

as it was to be shown.

To use the matrix formulation in our house size-price-age problem,

$$\hat{\boldsymbol{\beta}}'\mathbf{X}'\mathbf{y} = [10.04632987 \quad 12.83442613 \quad -0.05655391] \begin{bmatrix} 690 \\ 1554.9 \\ 14{,}540 \end{bmatrix}$$
$$= 26{,}065.9229;$$

and

$$\hat{\boldsymbol{\beta}}'\mathbf{X}'\mathbf{y} - \frac{(\sum Y)^2}{n} = 26{,}065.9229 - \frac{(690)^2}{20}$$
$$= 2260.9229.$$

Except for rounding error this is the same as we found before.

EXERCISE

13.1 Test to see if there are any significant predictors for the following data sets:

a. Sales Data: II (Exercise 11.1);
b. Restaurant Data (Exercise 11.2);
c. Drug Data (Exercise 11.3);
d. Auto Data (Exercise 11.4);
e. Ads Data (Exercise 11.5);
f. Children Data (Exercise 11.6);
g. Glue Data (Exercise 11.7);
h. Bottles Data (Exercise 11.8);
i. Yield Data (Exercise 11.9).

13.3 INDIVIDUAL TESTS: GENERAL

We now turn to the question of which predictors are significant. That is, we want to test the set of k hypotheses

$$H_{01}:\beta_1 = 0,\quad H_{02}:\beta_2 = 0, \cdots, \quad H_{0k}:\beta_k = 0$$
$$H_{11}:\beta_1 \neq 0,\quad H_{12}:\beta_2 \neq 0, \cdots, \quad H_{1k}:\beta_k \neq 0.$$

We have indicated two-sided alternatives here in order to consider the most general case, although one might also be interested in one-sided alternatives. Provided that all alternatives are two-sided, $H_{0i}:\beta_i = 0$, $i = 1, 2, \ldots, k$ can all be tested using

$$F_i = \frac{b_1^2}{c_{ii}s_{Y \cdot X}^2}$$

with 1 and $n - (k + 1)$ degrees of freedom. The c_{ii} in the denominator of F_i is the (i, i) element of the inverse of the matrix **S** defined in Chapter 11. Recall that **S** is the $k \times k$ submatrix obtained from $\mathbf{X'X}$ by deleting the first row and column of $\mathbf{X'X}$. (In Chapter 11, **S** was defined when X_1 and X_2 had been corrected for their means. While it can make life easier if the X's are corrected for their means, it is not absolutely necessary.) That is,

$$\mathbf{X'X} = \begin{bmatrix} n & \sum X_1 & \sum X_2 & \cdots & \sum X_k \\ \sum X_1 & \sum X_1^2 & \sum X_1X_2 & \cdots & \sum X_1X_k \\ \sum X_2 & \sum X_1X_2 & \sum X_2^2 & \cdots & \sum X_2X_k \\ \vdots & \vdots & \vdots & & \vdots \\ \sum X_k & \sum X_1X_k & \sum X_2X_k & \cdots & \sum X_k^2 \end{bmatrix};$$

so

$$\mathbf{S} = \begin{bmatrix} \sum X_1^2 & \sum X_1X_2 & \ldots & \sum X_1X_k \\ \sum X_1X_2 & \sum X_2^2 & \ldots & \sum X_2X_k \\ \vdots & \vdots & & \vdots \\ \sum X_1X_k & \sum X_2X_k & \ldots & \sum X_k^2 \end{bmatrix};$$

and

$$\mathbf{S}^{-1} = \begin{bmatrix} c_{11} & c_{12} & \cdots & c_{1k} \\ c_{21} & c_{22} & \cdots & c_{2k} \\ \vdots & \vdots & & \vdots \\ c_{k1} & c_{k2} & \cdots & c_{kk} \end{bmatrix}.$$

The individual significance tests are much easier and neater when the predictors are uncorrelated. Thus, we will consider separately two cases: (1) predictors uncorrelated, and (2) predictors correlated.

13.4 INDIVIDUAL TESTS: UNCORRELATED PREDICTORS

THE VARIANCE-COVARIANCE MATRIX

We will consider the simpler case first: that in which $X_1, X_2, \ldots, X_k$ are all mutually uncorrelated. If all predictors are uncorrelated, then let us assume that they have all been corrected for their means, so that

$$\mathbf{X'X} = \begin{bmatrix} n & 0 & 0 & \cdots & 0 \\ 0 & S_{11} & S_{12} & \cdots & S_{1k} \\ 0 & S_{21} & S_{22} & \cdots & S_{2k} \\ \vdots & \vdots & \vdots & & \vdots \\ 0 & S_{k1} & S_{k2} & \cdots & S_{kk} \end{bmatrix}$$

and

$$\mathbf{S} = \begin{bmatrix} S_{11} & S_{12} & \cdots & S_{1k} \\ S_{21} & S_{22} & \cdots & S_{2k} \\ \vdots & \vdots & & \vdots \\ S_{k1} & S_{k2} & \cdots & S_{kk} \end{bmatrix}.$$

If all predictors are uncorrelated, then

$$S_{12} = S_{13} = \cdots = S_{1k} = S_{23} = S_{24} = \cdots = S_{2k} = \cdots = S_{(k-1)k} = 0;$$

that is, the only nonzero elements of **S** are $S_{11}, S_{22}, \ldots, S_{kk}$ and

$$\mathbf{X'X} = \text{diag}(n, S_{11}, S_{22}, \ldots, S_{kk}).$$

Thus,

$$\hat{\boldsymbol{\beta}} = (\mathbf{X}'\mathbf{X})^{-1}\mathbf{X}'\mathbf{y} = \begin{bmatrix} 1/n & 0 & 0 & \cdots & 0 \\ 0 & 1/S_{11} & 0 & \cdots & 0 \\ 0 & 0 & 1/S_{22} & \cdots & 0 \\ \vdots & \vdots & \vdots & & \vdots \\ 0 & 0 & 0 & \cdots & 1/S_{kk} \end{bmatrix} \begin{bmatrix} \sum Y \\ S_{1Y} \\ S_{2Y} \\ \vdots \\ S_{kY} \end{bmatrix}$$

$$= \begin{bmatrix} \sum Y/n \\ S_{1Y}/S_{11} \\ S_{2Y}/S_{22} \\ \vdots \\ S_{kY}/S_{kk} \end{bmatrix}$$

so that $b_i = S_{iY}/S_{ii}$ for $i = 1, 2, \ldots, k$, and b_i depends only on X_i. Further, $c_{ii} = 1/S_{ii}$ for $i = 1, 2, \ldots, k$, since $\mathbf{S}^{-1} = \text{diag}(1/S_{11}, 1/S_{22}, \ldots, 1/S_{kk})$.

THE *F*-TEST

Then the F-statistic for testing $H_{0i}:\beta_i = 0$ is

$$F_i = \frac{b_i^2}{c_{ii}s_{Y\cdot X}^2} = \frac{b_i^2}{(1/S_{ii})s_{Y\cdot X}^2} = \frac{b_i^2 S_{ii}}{s_{Y\cdot X}^2}.$$

Since

$$b_i = S_{iY}/S_{ii},$$

$$b_i^2 S_{ii} = \frac{S_{iY}^2}{S_{ii}^2} S_{ii} = \frac{S_{iY}}{S_{ii}} S_{iY} = b_i S_{iY}.$$

Thus

$$F_i = \frac{b_i S_{iY}}{s_{Y\cdot X}^2}.$$

In the analysis of variance shown in Table 13.5, $F_i = b_i S_{iY}/s_{Y\cdot X}^2$ has 1 and $n - (k + 1)$ degrees of freedom. We compare each F to $F_{1,n-(k+1),\alpha}$ and reject H_{0i} for those F_i's which are greater than this critical value, retaining the associated predictors in the regression equation. Those predictors which are not significant are dropped from the model.

Note that the form of F_i is the same as the form of F in a simple regression problem and that the regression sum of squares for the entire regression, $SS(R)$, breaks down into k independent pieces, each measuring the effect of

TABLE 13.5 ANOVA for Mutually Uncorrelated Predictors

Source	SS	df	MS	F
Regression	$\sum_{i=1}^{k} b_i S_{iY}$	k	$\sum_{i=1}^{k} b_i S_{iY}/k$	
on X_1	$b_1 S_{1Y}$	1	$b_1 S_{1Y}$	F_1
on X_2	$b_2 S_{2Y}$	1	$b_2 S_{2Y}$	F_2
⋮	⋮	⋮	⋮	⋮
on X_k	$b_k S_{kY}$	1	$b_k S_{kY}$	F_k
Error	$S_{YY} - \sum_{i=1}^{k} b_i S_{iY}$	$n - (k + 1)$	$s^2_{Y \cdot X}$	
Total	S_{YY}	$n - 1$		

only one predictor. That is,

$$SS(R) = \sum_{i=1}^{k} b_i S_{iY} = b_1 S_{1Y} + b_2 S_{2Y} + \cdots + b_k S_{kY}$$

$$= SS(R)(X_1) + SS(R)(X_2) + \cdots + SS(R)(X_k).$$

Since predictors are uncorrelated, we know the effect of each X_i on Y separately, as measured by b_i, and this measure is not affected by the inclusion or exclusion of any other predictors. The only way in which the multiple regression with k significant predictors differs from k significant simple regressions is in the size of the error mean square: $s^2_{Y \cdot X}$ will typically be smaller than any of the $s^2_{Y \cdot X}$-values in the simple regressions, which means simply that we can predict more accurately using k variables than using any one of them alone.

Any of the k variables found not to be significant are dropped from the regression equation. This will not affect the values for a' and the remaining b_i's, since $a' = \bar{Y}$ and $b_i = S_{iY}/S_{ii}$. However, the standard error of estimate should be recalculated using only those X_i's that remain in the model.

THE t-TEST

In case any one-sided hypotheses are to be tested, a t-test must be employed, since F can be used only for two-sided alternatives. Since $t_\nu^2 = F_{1,\nu}$, we can see that

$$t_i = \frac{b_i}{\sqrt{c_{ii} s^2_{Y \cdot X}}}$$

with $n - (k + 1)$ degrees of freedom. If predictors are uncorrelated, $c_{ii} = 1/S_{ii}$; so

$$t_i = \frac{b_i}{\sqrt{s^2_{Y \cdot X}/S_{ii}}}.$$

Note that this is exactly the same form as the t-test in simple regression, and that t_i depends only on X_i. All t-tests, just like all F-tests, will be independent when predictors are uncorrelated.

A NUMERICAL EXAMPLE

As an example, suppose we have the following data in Table 13.6 from an experiment to determine the effects of altitude and cooking temperature on the baking time required for a new cake mix that is to be marketed. Altitudes are below sea level (coded to -1), normal (0), and high ($+1$), at equally spaced intervals. Temperatures are 350°, 400°, and 450°, also coded to -1, 0, and 1, respectively. Two cakes were baked at each combination of altitude and temperature, yielding the results shown in Table 13.6. From the calculations in the table, we have

$$\mathbf{X'X} = \begin{bmatrix} 18 & 0 & 0 \\ 0 & 12 & 0 \\ 0 & 0 & 12 \end{bmatrix}, \qquad (\mathbf{X'X})^{-1} = \begin{bmatrix} \frac{1}{18} & 0 & 0 \\ 0 & \frac{1}{12} & 0 \\ 0 & 0 & \frac{1}{12} \end{bmatrix},$$

$$\mathbf{X'y} = \begin{bmatrix} 1196 \\ 138 \\ -70 \end{bmatrix}, \qquad \hat{\boldsymbol{\beta}} = \begin{bmatrix} 66.4444 \\ 11.5000 \\ -5.8333 \end{bmatrix}.$$

To test

$$H_0: \beta_1 = \beta_2 = 0$$

$$H_1: \text{some } \beta_i \neq 0, \quad i = 1, 2,$$

we will reject H_0 if $F > F_{0.05,2,15} = 3.68$. Table 13.7 shows the analysis of variance for this example. Since $89.5498 > 3.68$, conclude that either altitude or temperature, or both, are significant predictors of baking time. In order to determine which predictor(s) is significant, if we want to test the two-sided alternatives then we break the regression sum of squares,

$$SS(R) = 1995.3310,$$

into

$$SS(R)(X_1) = b_1 S_{1Y} = 11.5(138) = 1587$$

$$= b_1^2 S_{11} = (11.5)^2(12) = 1587$$

and

$$SS(R)(X_2) = b_2 S_{2Y} = (-5.8333)(-70) = 408.3310$$

$$= b_2^2 S_{22} = (-5.8333)^2(12) = 408.3287.$$

TABLE 13.6 Baking Times at Three Altitudes and Three Temperatures

Altitude	Temperature	Baking Time (minutes)
−1	−1	60
−1	−1	58
−1	0	56
−1	0	50
−1	1	50
−1	1	45
0	−1	78
0	−1	76
0	0	70
0	0	68
0	1	65
0	1	63
1	−1	83
1	−1	80
1	0	77
1	0	75
1	1	70
1	1	72
$\sum(\cdot)$ 0	0	1196
$\sum(\cdot)^2$ 12	12	81,630
$\sum X_1Y = 138$	$S_{YY} = 2162.4444$	$S_{12} = 0$
$\sum X_2Y = -70$	$S_{1Y} = 138$	$S_{11} = 12$
$\sum X_1X_2 = 0$	$S_{2Y} = -70$	$S_{22} = 12$
$n = 18$		

TABLE 13.7 ANOVA Baking Times Example

Source	*SS*	*df*	*MS*	*F*
Regression	1995.3310	2	997.6655	89.5498
Error	167.1134	15	11.1409	
Total	2162.4444	17		

Note that 1587 + 408.3310 = 1995.3310, so that (except for possible rounding error)

$$SS(R)(X_1) + SS(R)(X_2) = SS(R)(X_1, X_2).$$

Then to test

$$H_{01}:\beta_1 = 0 \qquad H_{02}:\beta_2 = 0$$

and

$$H_{11}:\beta_1 \neq 0 \qquad H_{12}:\beta_2 \neq 0,$$

reject H_0 if $F_i > F_{0.05,1,15} = 4.54$, for $i = 1, 2$. Since both F_1 and F_2 are greater than 4.54 (see the analysis of variance in Table 13.8), conclude that both temperature and altitude are significant predictors of baking time.

TABLE 13.8
Tests on Individual Terms for Baking Times Example

Source	*SS*	*df*	*MS*	*F*
Regression	1995.3310	2		
on X_1	1587.0000	1	1587.0000	142.4481
on X_2	408.3310	1	408.3310	36.6515
Error	167.1134	15	11.1409	
Total	2162.4444	17		

If we wished to test

$$H_{01}:\beta_1 = 0 \qquad \text{and} \qquad H_{02}:\beta_2 = 0$$
$$H_{11}:\beta_1 > 0 \qquad\qquad H_{12}:\beta_2 < 0,$$

we would reject H_{01} if $t_1 > t_{15,0.05} = 1.753$ and reject H_{02} if $t_2 < -t_{15,0.05} = -1.753$. We obtain

$$t_1 = \frac{11.5}{\sqrt{11.1409/12}} = 11.5\sqrt{12/11.1409} = 11.9352,$$

$$t_2 = \frac{-5.8333}{\sqrt{11.1409/12}} = -6.0540.$$

Since $t_1 > 1.753$ and $t_2 < -1.753$, reject H_{01} and H_{02} and conclude that the higher the altitude, the longer the baking time, and the higher the temperature, the shorter the baking time. The point is that since the data are from a carefully designed experiment, it is possible to know the effect of temperature completely separate from the effect of altitude. Had two separate simple regressions been run, $SS(R)$ (X_1) and $SS(R)$ (X_2) would not have changed at all. The values of the F-statistics would be a bit smaller, however, since $s^2_{Y \cdot X}$ for either simple regression would be larger than $s^2_{Y \cdot X}$ in the multiple regression.

EXERCISE

13.2 Regardless of the results of the overall regression test, test for the significance of the individual predictors in the following data sets. Comment on any unusual results. Show the regression equation you would use in each problem.

a. Glue Data (Exercise 11.7); **b.** Bottles Data (Exercise 11.8);
c. Yield Data (Exercise 11.9).

13.5

INDIVIDUAL TESTS: CORRELATED PREDICTORS

THE VARIANCE-COVARIANCE MATRIX

If $X_1, X_2, \ldots, X_k$ are correlated, as is usually the case in a social science study, then the analysis becomes more complex. In this case, $\mathbf{X'X}$ will not be diagonal, even if $X_1, X_2, \ldots, X_k$ are all corrected for their means. Then $\mathbf{S}$ is not diagonal, and neither is $\mathbf{S}^{-1}$. Let us first consider the $k = 2$ case when X_1 and X_2 are corrected for their means, and then see how matters become even more complicated if $k > 2$. It was shown earlier that

$$\mathbf{X'X} = \begin{bmatrix} n & 0 & 0 \\ 0 & S_{11} & S_{12} \\ 0 & S_{21} & S_{22} \end{bmatrix} = \left[\begin{array}{c|cc} n & 0 & 0 \\ \hline 0 & \multicolumn{2}{c}{\multirow{2}{*}{$\mathbf{S}$}} \\ 0 & & \end{array}\right]$$

and that

$$(\mathbf{X'X})^{-1} = \left[\begin{array}{c|cc} \frac{1}{n} & 0 & 0 \\ \hline 0 & \multicolumn{2}{c}{\multirow{2}{*}{$\mathbf{S}^{-1}$}} \\ 0 & & \end{array}\right]$$

$$= \begin{bmatrix} \frac{1}{n} & 0 & 0 \\ 0 & \dfrac{S_{22}}{S_{11}S_{22} - S_{12}^2} & \dfrac{-S_{12}}{S_{11}S_{22} - S_{12}^2} \\ 0 & \dfrac{-S_{12}}{S_{11}S_{22} - S_{12}^2} & \dfrac{S_{11}}{S_{11}S_{22} - S_{12}^2} \end{bmatrix}.$$

Then

$$\hat{\boldsymbol{\beta}} = (\mathbf{X}'\mathbf{X})^{-1}\mathbf{X}'\mathbf{y} = \begin{bmatrix} \frac{1}{n} & 0 & 0 \\ 0 & \frac{S_{22}}{|\mathbf{S}|} & \frac{-S_{12}}{|\mathbf{S}|} \\ 0 & \frac{-S_{12}}{|\mathbf{S}|} & \frac{S_{11}}{|\mathbf{S}|} \end{bmatrix} \begin{bmatrix} \sum Y \\ S_{1Y} \\ S_{2Y} \end{bmatrix}$$

$$= \begin{bmatrix} \sum Y/n \\ \frac{S_{22}S_{1Y} - S_{12}S_{2Y}}{S_{11}S_{22} - S_{12}^2} \\ \frac{S_{11}S_{2Y} - S_{12}S_{1Y}}{S_{11}S_{22} - S_{12}^2} \end{bmatrix} = \begin{bmatrix} a' \\ b_1 \\ b_2 \end{bmatrix}.$$

We see that b_1 involves not only the effect of X_1 on Y as measured by S_{1Y} and S_{11} but also the effect of X_2 on X_1 (measured by S_{12}) and the effect of X_2 on Y (measured by S_{2Y} and S_{22}). Of course, this is the effect of multicollinearity—that S_{12} is nonzero brings into b_1 terms involving X_2 and Y. Obviously, if predictors are correlated,

$$b_i \neq \frac{S_{iY}}{S_{ii}},$$

as was the case when predictors were not correlated. In addition, c_{ii} will not involve just X_i: for example,

$$c_{11} = \frac{S_{22}}{S_{11}S_{22} - S_{12}^2}$$

involves something other than just S_{11}. Dividing numerator and denominator of c_{11} by $S_{11}S_{22}$,

$$c_{11} = \frac{S_{22}/S_{11}S_{22}}{(S_{11}S_{22} - S_{12}^2)/S_{11}S_{22}} = \frac{1/S_{11}}{1 - r_{12}^2} = \frac{1}{S_{11}(1 - r_{12}^2)},$$

and similarly,

$$c_{22} = \frac{1}{S_{22}(1 - r_{12}^2)}.$$

Thus, c_{ii} involves the correlation between the two predictor variables, as well as S_{ii}, for $i = 1,2$.

THE F- AND t-TESTS

The F-statistic for testing H_{0i}:$\beta_i = 0$ for $i = 1,2$ is then

$$F_i = \frac{b_i^2}{c_{ii}s_{Y \cdot X}^2} = \frac{b_i^2 S_{ii}(1 - r_{12}^2)}{s_{Y \cdot X}^2}.$$

The t-statistic is

$$t_i = \frac{b_i}{\sqrt{s^2_{Y \cdot X}/S_{ii}(1 - r^2_{12})}} = \frac{b_i}{\sqrt{c_{ii}s^2_{Y \cdot X}}}.$$

THE EFFECT OF MULTICOLLINEARITY

Let us examine the effect of multicollinearity on the tests of significance. The greater the degree of correlation between the predictors, the closer r^2_{12} will be to 1. In the numerator of F_i, then, if X_1 and X_2 are very highly correlated, the quantity $(1 - r^2_{12})$ will be close to zero, thus making the F-statistic close to zero. Also recall that $c_{ii}s^2_{Y \cdot X}$ is an estimate of the variance of b_i. If X_1 and X_2 are highly correlated, c_{ii} will be quite large, making the F-statistic very small. The same thing will happen to the t-statistic. Now, it is only for F-values considerably larger than zero (or t-values larger than zero in absolute value) that we can reject H_{0i}, $\beta_i = 0$. Thus, if predictors are correlated, the hypothesis tests become very conservative in that they will not allow H_{0i} to be rejected unless a predictor affects Y strongly enough to overcome its correlation with the other predictor. That is, b_i must be large enough, taking units of measurement into account, to overcome the effect of r^2_{12} on the F- or the t-statistic. Thus, it is possible that one predictor which is quite strong by itself will not be significant when used in conjunction with another strong predictor with which it is highly correlated. The fact that two predictors are highly correlated means that they are redundant—they are both doing the same thing—so that one is superfluous when used in conjunction with the other and should be discarded.

The denominator of the t-statistic,

$$\left[\frac{s^2_{Y \cdot X}}{S_{ii}(1 - r^2_{12})}\right]^{1/2},$$

is the standard error of the regression coefficient. It measures how much b_i is expected to vary from sample to sample of the same size from the same population. Recall that if this denominator is small, the value of b_i can be trusted pretty well. On the other hand, if the standard error of the regression coefficient is large, then we cannot believe that β_i is close to the value we have obtained for b_i. Severe multicollinearity, which is manifested by a large value of r^2_{12}, will result in a large standard error. The same argument holds when we have more than two predictors. If multicollinearity is severe, the determinant of $\mathbf{X'X}$ will be small, resulting in the elements of $(\mathbf{X'X})^{-1}$ being quite large. These large elements c_{ii} appear in the standard errors of the regression coefficients and make these standard errors large.

Thus, when multicollinearity is present in the data, one should look carefully at the sizes of the standard errors of the regresssion coefficients. If these errors are too large, the whole regression equation is untrustworthy, and another sample would give quite different values for b_i, $i = 1, 2, \ldots, k$.

(Techniques such as *ridge regression* have been recently introduced in attempts to handle this difficulty. Ridge regression is beyond the scope of this text.)

Further, if predictors are correlated, the regression sum of squares does not break down into k independent tests on the individual predictors, as it did in the previous case. The analysis of variance for correlated predictors, where $F_i = b_i^2/c_{ii}s_{Y \cdot X}^2$ with 1 and $n - (k + 1)$ degrees of freedom, is given in Table 13.9.

TABLE 13.9 ANOVA for Correlated Predictors

Source	*SS*	*df*	*MS*	*F*
Regression	$\sum_{i=1}^{k} b_i S_{iY}$	k	$\sum_{i=1}^{k} b_i S_{iY}/k$	
on X_1	b_1^2/c_{11}	1	b_1^2/c_{11}	F_1
on X_2	b_2^2/c_{22}	1	b_2^2/c_{22}	F_2
⋮	⋮	⋮	⋮	⋮
on X_k	b_k^2/c_{kk}	1	b_k^2/c_{kk}	F_k
Error	$S_{YY} - \sum_{i=1}^{k} b_i S_{iY}$	$n - (k + 1)$	$s_{Y \cdot X}^2$	
Total	S_{YY}	$n - 1$		

INDIVIDUAL TESTS FOR THE HOUSE SIZE-PRICE-AGE EXAMPLE

In the house size-price-age example, recall that

$$c_{11} = 0.07448317 \quad \text{and} \quad c_{22} = 0.00057879.$$

Thus

$$SS(R)(X_1) = \frac{b_1^2}{c_{11}} = \frac{(12.83442613)^2}{0.07448317} = 2211.5398$$

$$= b_1^2 S_{11}(1 - r_{12}^2) = (12.83442613)^2(13.56)(1 - 0.0099)$$

$$= 2211.5240$$

$$SS(R)(X_2) = \frac{b_2^2}{c_{22}} = \frac{(-0.05655391)^2}{0.00057879} = 5.5253$$

$$= b_2^2 S_{22}(1 - r_{12}^2) = (-0.05655391)^2(1745)(1 - 0.0099)$$

$$= 5.5253.$$

The analysis of variance is shown in Table 13.10. We will reject $H_{01}:\beta_1 = 0$ and $H_{02}:\beta_2 = 0$ in favor of $H_{11}:\beta_1 \neq 0$ and $H_{12}:\beta_2 \neq 0$ if $F_i > F_{1,17,0.05} = 4.45$. Since $F_1 > 4.45$, conclude that size is a significant predictor of price. But since

TABLE 13.10 Test on Individual Terms for House Size-Price-Age Example

Source	SS	df	MS	F
Regression	2261.4245	2		
on X_1	2211.5398	1	2211.5398	336.9555
on X_2	5.5253	1	5.5253	0.8418
Error	111.5755	17	6.5633	
Total	2373	19		

$F_2 \not> 4.45$, we cannot be 95 percent sure that the age of a house is related to its price. Note that because of multicollinearity, $SS(R)(X_1) + SS(R)(X_2) \neq SS(R)(X_1, X_2)$:

$$2211.5398 + 5.5253 = 2217.0651 \neq 2261.4245.$$

Thus, when the predictors are correlated, the regression sum of squares cannot be partitioned into two disjoint pieces which sum to $SS(R)$.

If one-sided alternatives are to be tested, then t-tests can be used. For example, if

$$H_{01}:\beta_1 = 0 \qquad H_{02}:\beta_2 = 0$$
$$\text{and}$$
$$H_{11}:\beta_1 > 0 \qquad H_{12}:\beta_2 < 0,$$

we will reject H_{01} if $t_1 > t_{17,0.05} = -1.740$, and we will reject H_{02} if $t_2 < -t_{17,0.05} = -1.740$. The test statistics give

$$t_1 = \frac{12.83442613}{\sqrt{6.56326273/13.56(1 - 0.0099)}} = 18.3563$$

and

$$t_2 = \frac{-0.05655391}{\sqrt{6.56326273/1745(1 - 0.0099)}} = -0.9176.$$

Then since $t_1 > 1.740$, we are more than 95 percent sure that the larger a house, generally the higher its price, but since $t_2 \not< -1.740$, we cannot be 95 percent sure that older houses cost less than newer houses.

THE REDUCED MODEL

Since we have only very slight evidence that age is related to price, we will drop X_2 = age from the model $Y = \alpha + \beta_1 X_1 + \beta_2 X_2 + \varepsilon$ and work with the *reduced model*

$$Y = \alpha + \beta X + \varepsilon,$$

where Y = price and X = size. Since we cannot be reasonably certain that the slight relationship we saw between age and size is not just due to chance, there is no reason to retain age as a predictor. Another sample of size 20 from the same population would not be expected to show any relationship between age and price. The regression equation we will use to predict price from size is the equation found in the simple regression problem:

$$\hat{Y} = 8.7 + 12.9X,$$

and the standard error is $s_{Y \cdot X} = 2.55$ (=\$2,550). Note that had we not already performed the simple regression of price on size, but had only run the multiple regression, it would now be necessary to rework the problem using only X = size and Y. Since predictors are correlated, the term involving X_2 in the multiple regression equation cannot simply be dropped, as it was in the case of independent predictors. The regression coefficient on size changes depending upon whether or not X_2 is included in the equation.

EXTENSION TO LARGER PROBLEMS

The reader can appreciate how the situation becomes even more complicated when $k > 2$. For example, if $k = 3$, then

$$S = \begin{bmatrix} S_{11} & S_{12} & S_{13} \\ S_{12} & S_{22} & S_{23} \\ S_{13} & S_{23} & S_{33} \end{bmatrix},$$

and it can be shown that

$$c_{11} = \frac{1}{S_{11}} \cdot \frac{1 - r_{23}^2}{1 - r_{23}^2 - r_{13}^2 - r_{12}^2 + 2S_{12}S_{13}S_{23}/S_{11}S_{22}S_{23}},$$

and similarly for c_{22} and c_{33}. Note that c_{11} involves more than X_1. It involves all pairwise correlations among all three predictor variables and a term involving all three predictor variables at once. Thus a test on β_1 will depend not only on the effect of X_1 on Y but also on the effects of X_2 and X_3 on Y, and on how X_1 is correlated with X_2 and X_3 and how X_2 is correlated with X_3. Obviously, the test on β_1 will not be independent of the tests on β_2 and β_3. Thus, the analysis of the multiple regression problem with correlated predictors is best done with the aid of a computer, using one of the several programs available. The interpretation of the output of some representative programs is discussed in the next chapter.

The reader should note that we have considered here a reasonably well-behaved problem. Strange things can happen when several predictors are highly correlated, so it will be very instructive to study carefully the results of the sample problems in the exercises for Chapters 11–13.

PARTIAL SUMS OF SQUARES AND t- AND F-STATISTICS

The t- and F-tests presented in this chapter are partial tests. They test for the significance of the predictor, conditional on all the other predictors in the model. These partial tests let us see if a predictor adds something to what has already been explained by the other predictors, or whether the predictor is redundant once the other predictors have been included. When a partial t- or F-test tells us to reject $H_0{:}\beta_i = 0$, we conclude that, given that all other predictors are being used, X_i can still yield significant amount of information about Y. When H_0 cannot be rejected, we conclude that either X_i is a worthless predictor or it is giving the same information as some other predictor. In our house size-price-age example, since the partial F-test rejected $H_{01}{:}\beta_1 = 0$, the conclusion is that even if X_2 = age has explained all it can about Y = price, X_1 = size can still improve the percentage of explained variation. In other words, even if age is used as a predictor, we still need size. Since $H_{02}{:}\beta_2 = 0$ was not rejected, either age is a worthless predictor or it adds nothing to what size can tell alone. If size is used as a predictor, age is not needed. The null hypotheses can be written as

$$H_{01} : Y = \alpha + \beta_2 X_2 + \varepsilon \qquad\qquad H_{02} : Y = \alpha + \beta_1 X_1 + \varepsilon$$

and

$$H_{11} : Y = \alpha + \beta_1 X_1 + \beta_2 X_2 + \varepsilon \qquad\qquad H_{12} : Y = \alpha + \beta_1 X_1 + \beta_2 X_2 + \varepsilon,$$

as a reminder that the t- and F-tests are partial, or conditional, tests. We can abbreviate H_{01} and H_{02} as

$$H_{01} : \beta_1 = 0 \mid \beta_2 \quad \text{and} \quad H_{02} : \beta_2 = 0 \mid \beta_1.$$

We discussed partial sums of squares and their associated t- and F-tests in Chapter 10 in connection with polynomial regressions. But there they were less appropriate than were sequential sums of squares and tests. One often sees these partial sums of squares calculated (for the two-predictor case) as

$$SS(R)(X_2 \mid X_1) = SS(R)(X_1, X_2) - SS(R)(X_1)$$

where $SS(R)(X_1)$ is obtained by performing a simple linear regression analysis of Y on X_1, and

$$SS(R)(X_1 \mid X_2) = SS(R)(X_1, X_2) - SS(R)(X_2)$$

where $SS(R)(X_2)$ is obtained by fitting $Y = a + bX_2$. In the house size-price-age example,

$$SS(R)(X_2 \mid X_1) = 2261.42453359 - 2255.90044248 = 5.52409111,$$

$$SS(R)(X_1 \mid X_2) = 2261.42453359 - 49.87106017 - 2211.55347342.$$

Except for rounding error, these are the same values for the individual regression sums of squares as were found previously.

Suppose we had a three-predictor model. Then the partial sums of squares could be calculated as

$$SS(R)(X_1 \mid X_2, X_3) = SS(R)(X_1, X_2, X_3) - SS(R)(X_2, X_3),$$

$$SS(R)(X_2 \mid X_1, X_3) = SS(R)(X_1, X_2, X_3) - SS(R)(X_1, X_3),$$

$$SS(R)(X_3 \mid X_1, X_2) = SS(R)(X_1, X_2, X_3) - SS(R)(X_1, X_2).$$

To find these partial sums of squares in this manner, four different models would need to be fitted. The extension to $k = 4$ and greater is clear. All these partial sums of squares measure the increase in the regression sum of squares, or equivalently, the reduction in error, achieved by adding a predictor to those already in the model. They are sometimes referred to as **extra sums of squares,** or **sums of squares if last;** that is, if this variable were the last one added to the model.

Extra Sums of Squares

Sums of Squares If Last

SEQUENTIAL SUMS OF SQUARES

In some instances sequential sums of squares are useful in multiple regression analysis. Suppose the researcher has an order of preference for the variables in a problem. For example, X_4 seems intuitively important as a predictor of Y. Next in importance comes X_2, because it is very convenient to measure. Then comes X_3 because it is inexpensive. Last in importance is X_1. The researcher is then interested first in what X_4 can do by itself, then in what X_2 adds to X_4, then what X_3 adds to X_2 and X_4, and finally in what X_1 can explain, given X_2, X_3, and X_4. S(he) would then enter the variables in this predetermined order of importance and would calculate sequential sums of squares:

$$SS(R)(X_4),$$

$$SS(R)(X_2 \mid X_4) = SS(R)(X_2, X_4) - SS(R)(X_4),$$

$$SS(R)(X_3 \mid X_2, X_4) = SS(R)(X_2, X_3, X_4) - SS(R)(X_2, X_4),$$

$$SS(R)(X_1 \mid X_2, X_3, X_4) = SS(R)(X_1, X_2, X_3, X_4) - SS(R)(X_2, X_3, X_4).$$

Note that

$$SS(R)(X_4) + SS(R)(X_2 \mid X_4) + SS(R)(X_3 \mid X_2, X_4) + SS(R)(X_1 \mid X_2, X_3, X_4) = SS(R)(X_1, X_2, X_3, X_4),$$

since each sequential sum of squares represents an additional piece of the regression sum of squares.

The question is, in a set of k interrelated predictors, how can we find a subset of predictors that are all effective, given the others? We might like to find a subset of predictors such that they explain the largest possible percentage of variation in Y. Or we might want to find first the single strongest predictor, then add the next strongest predictor, given the first, then the third strongest predictor, given the first two, and so on, until the next strongest predictor is no longer needed, given the others. There are many different ways to define the "best" multiple regression model that can be formed using a set of k possible predictors. In Chapter 15 we shall investigate some of these techniques. The work required is so involved as to be almost prohibitive without the aid of the computer.

EXERCISE

13.3 Regardless of the results of the overall regression significance test, test for the significance of the individual predictors using partial F-tests for the following data sets. Comment on any unusual results. Show the regression equation you would use in each problem.

a. Sales Data: II (Exercise 11.1);
b. Restaurant Data (Exercise 11.2);
c. Drug Data (Exercise 11.3);
d. Auto Data (Exercise 11.4);
e. Ads Data (Exercise 11.5);
f. Children Data (Exercise 11.6).

13.6 CONFIDENCE INTERVALS ON NET REGRESSION COEFFICIENTS

Confidence intervals on $\beta_1, \beta_2, \ldots, \beta_k$ can be formed in just the same way as we found a confidence interval on β in simple regression. Recall that the general form of a $100(1 - \alpha)\%$ confidence interval on any parameter is

$$\text{Point estimate} \pm t_{\nu,\alpha/2} \text{ (standard error of the point estimate).}$$

The point estimates of $\beta_1, \beta_2, \ldots, \beta_k$ are $b_1, b_2, \ldots, b_k$, respectively. The standard error of any b_i is

$$\sqrt{c_{ii} s^2_{Y \cdot X}}.$$

Thus, a $100(1 - \alpha)\%$ confidence interval on β_i will look like

$$b_i - t_{\alpha/2,n-p} s_{Y \cdot X} \sqrt{c_{ii}} < \beta_i < b_i + t_{\alpha/2,n-p} s_{Y \cdot X} \sqrt{c_{ii}}.$$

In our house size-price-age example, suppose we wanted to construct a 95% confidence interval on β_2. Then

$$b_2 = -0.05655391,$$

$$s^2_{Y \cdot X} = 6.5633,$$

$$c_{22} = 0.0005789,$$

$$t_{17,0.025} = 2.110;$$

and the 95% confidence interval on β_2 is

$$-0.05655391 \pm 2.110\sqrt{6.5633(0.0005789)},$$

or

$$-0.05655391 \pm 0.13006,$$

or

$$-0.0735 < \beta_2 < 0.1866.$$

Thus we can be 95 percent sure that for houses the same size, if one house is one year older than another, the older house will sell for between \$73.50 less and \$186.60 more than the newer house. Obviously, this does not give us very much information and can be taken as evidence equivalent to a hypothesis test that age is not a predictor of price. A similar confidence interval could be constructed for β_1.

Such confidence intervals are, of course, conditional on the other parameters in the model. The interval above is a confidence interval on β_2, holding X_1 constant.

Of course, once a predictor is dropped from the model, both $s^2_{Y \cdot X}$ and c_{ii} will change for the predictors left in the model. Thus, unless one is using confidence intervals to determine which predictors to delete and which to retain, it is probably best to wait until the final model is found before constructing confidence intervals on the net regression coefficients. If we used the reduced model from the house size-price-age problem, our confidence interval on β_1 would of course be the same as we found in simple regression.

EXERCISE

13.4 Give 95% confidence intervals on the regression coefficients for the following data sets. Interpret the meaning of these confidence intervals.

a. Sales Data: II (Exercise 11.1); **b.** Drug Data (Exercise 11.3);
c. Glue Data (Exercise 11.7); **d.** Bottles Data (Exercise 11.8).

13.7

SUMMARY

The first test of significance to be performed in a multiple regression problem is intended to detect the presence of any significant predictors in the set under consideration. Provided we have evidence of a relationship for the response with at least one of the predictors, we then perform individual tests to tell which predictors are significant. The procedure is quite straightforward if there is no multicollinearity in the problem; but the results can be confusing if predictors are correlated. For example, the general regression significance test can indicate the presence of at least one significant predictor, but each individual test might indicate that its predictor is not effective.

In the examples we have considered so far, $X_1, X_2, \ldots, X_k$ have been interval- or ratio-level variables. The next chapter examines situations in which some of the predictors are categorical variables.

FOURTEEN

DUMMY VARIABLES

14.1

INTRODUCTION

In Chapter 1 a distinction was made between nominal-, ordinal-, and interval-level variables. Recall that nominal-level variables are those whose values are only categories and that there is no inherent ordering of the categories with regard to size or importance. Variables such as sex (values = "male," "female"), marital status (values = "never married," "married," "formerly married"), job classification (values = "white collar," "blue collar"), political party affiliation ("Republican," "Democrat," "Independent"), and so on can be considered to be nominal-level variables.

In many instances such variables are important predictors in a regression study. Sex may be a very important factor in explaining variation in earnings, for example, as may job classification. In the regression models we have studied so far, the dependent and independent variables have all been interval- or ratio-level. In this chapter we will see how to incorporate nominal-level variables as predictors in a regression equation.

Keep in mind that while the independent variables may be of any measurement level, the dependent variable Y must be at least interval-level for the assumption to be tenable that for a given value of X (or $X_1, X_2, \ldots, X_k$), Y is normally distributed around the regression line. Otherwise, the significance probabilities and any inferences made are untrustworthy. If the *response* is to be a categorical variable, specialized techniques, not covered in this text, are required.

14.2

THE USE OF DUMMY VARIABLES

TWO CATEGORIES

Recall from Chapter 9 our example of a regression relating age and earnings for a group of people. Figure 9.7, showing the relationship, is reproduced as Figure 14.1. Here, it becomes apparent that an important predictor, namely sex, has been omitted from the model. In order to make accurate predictions of earnings from age, we should take sex into account.

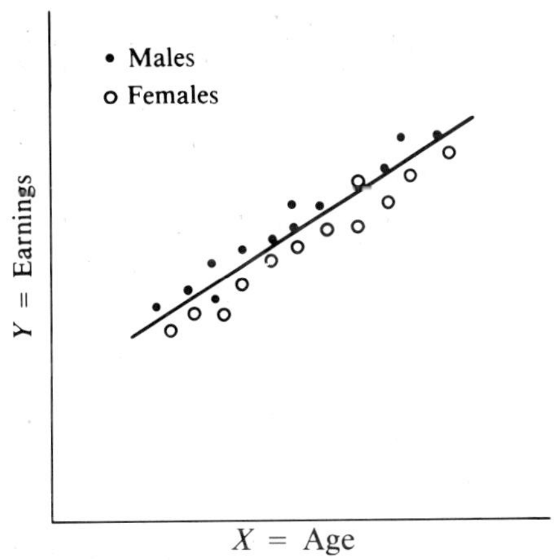

FIGURE 14.1 Scatter Diagram and Regression Line Relating Earnings to Age

If the omitted predictor were, say, years on the job, our task would be simple, because this variable is easily measured. But how does one assign a meaningful numerical measurement to the variable "sex"?

Perhaps we could divide the group of people according to sex and perform a separate simple regression analysis for each group. This would solve our problem but might not produce as good a regression equation as the combined group. If you consider just the dots (males) in Figure 14.1, you can draw a regression line through the dots. Another regression line can be drawn for the circles (females). Both lines have approximately the same slope; but the line for the males is higher. Since the two groups together would give us a more accurate estimate of the common slope, we would like to avoid running separate analyses if possible.

This brings us to the question of how to assign a meaningful numerical measurement to the values "male" and "female" of the variable "sex." For

Dummy Variable

this purpose, we employ a **dummy variable,** or indicator variable. For example, we could call male = 1 and female = 0. Note that there is no inherent ordering in "male-female," as there is in "high-low," for example. Further, we could use male = 3, female = 5 if we wanted to, but it is conventional—because it is easier—to stick with 0 and 1.

What does this do for us? We have a model

$$Y = \alpha + \beta_1 X_1 + \beta_2 X_2 + \varepsilon,$$

where Y = earnings; X_1 = age; and X_2 = sex, where

$$X_2 = \begin{cases} 0 & \text{if female} \\ 1 & \text{if male.} \end{cases}$$

Then, if we are predicting earnings for a female ($X_2 = 0$), our regression equation will be

$$\begin{aligned} \hat{Y} &= a + b_1 X_1 + b_2(0) \\ &= a + b_1 X_1, \end{aligned}$$

which is essentially a simple regression equation for the females; and to predict earnings for a male ($X_2 = 1$), the regression equation is

$$\begin{aligned} \hat{Y} &= a + b_1 X_1 + b_2(1) \\ &= (a + b_2) + b_1 X_1. \end{aligned}$$

The regression coefficient b_1 relating income to age is the same for both males and females. This says that the rate of increase of earnings with age is the same for both males and females. The intercept for the males, however, which is b_2 units greater than that for the females, says that the regression line for males is b_2 units higher than that for females. That is, at all ages, males earn b_2 dollars more than females (if $b_2 > 0$). Figure 14.2 illustrates this example.

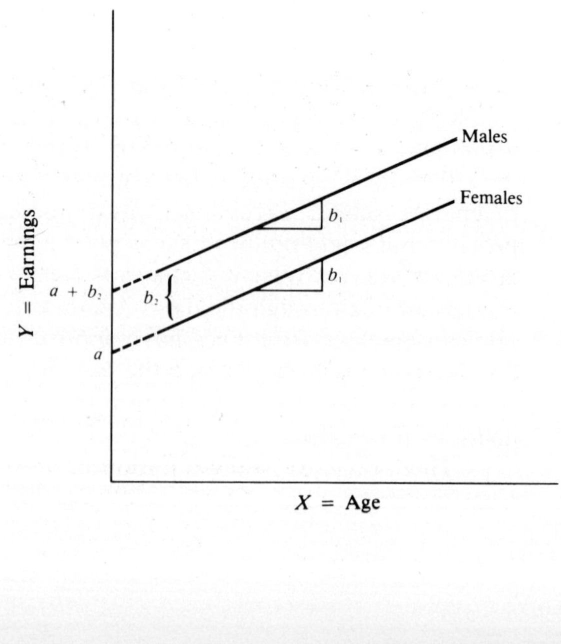

FIGURE 14.2 Interpretation of Regression Coefficients When Dummy Variables Are Used

We have incorporated all features of two separate simple regressions into one multiple regression, and an F-test on the regression coefficient β_2 tells us whether there is any significant difference between the two groups "male" and "female." How to make this test would have been less clear had we used two separate simple regressions.

MORE THAN TWO CATEGORIES

Suppose we have a variable that can be divided into more than two categories. For example, suppose that the strength of a manufactured part is to be related to the percentage of hardening agent used (X_1) as well as to which one of four machines turned out the part. In this case, we need *three* dummy variables. Let the model be

$$Y = \alpha + \beta_1 X_1 + \beta_2 X_2 + \beta_3 X_3 + \beta_4 X_4 + \varepsilon,$$

where

$$Y = \text{strength},$$

$$X_1 = \text{percent of hardening agent},$$

$$X_2 = \begin{cases} 1 & \text{if made on machine 1} \\ 0 & \text{if not made on machine 1,} \end{cases}$$

$$X_3 = \begin{cases} 1 & \text{if made on machine 2} \\ 0 & \text{if not made on machine 2,} \end{cases}$$

$$X_4 = \begin{cases} 1 & \text{if made on machine 3} \\ 0 & \text{if not made on machine 3.} \end{cases}$$

What about machine 4? It has been taken care of automatically. If the part were made on machine 1, then

$$(X_2, X_3, X_4) = (1, 0, 0)$$

indicates this. If the part were made on machine 2, we have

$$(X_2, X_3, X_4) = (0, 1, 0).$$

If the part were made on machine 3, then

$$(X_2, X_3, X_4) = (0, 0, 1).$$

If the part were made on machine 4, then

$$(X_1, X_2, X_3) = (0, 0, 0)$$

says it was not made on machine 1 nor on machine 2 nor on machine 3; thus, it must have been made on machine 4.

Why could we not just add on

$$X_5 = \begin{cases} 1 & \text{if made on machine 4} \\ 0 & \text{if not made on machine 4?} \end{cases}$$

Suppose that two parts are to be made from each machine, so that $n = 8$. Consider the **X** matrix for this situation.

$$\mathbf{X} = \begin{array}{c} \begin{matrix} (a) & (b_1) & (b_2) & (b_3) & (b_4) & (b_5) \end{matrix} \\ \begin{bmatrix} 1 & X_{11} & 1 & 0 & 0 & 0 \\ 1 & X_{21} & 1 & 0 & 0 & 0 \\ 1 & X_{31} & 0 & 1 & 0 & 0 \\ 1 & X_{41} & 0 & 1 & 0 & 0 \\ 1 & X_{51} & 0 & 0 & 1 & 0 \\ 1 & X_{61} & 0 & 0 & 1 & 0 \\ 1 & X_{71} & 0 & 0 & 0 & 1 \\ 1 & X_{81} & 0 & 0 & 0 & 1 \end{bmatrix} \end{array}$$

Note that the last four columns of **X** sum to the first column. We have thus introduced a linear dependency in our data, and **X′X** will be singular. Remember: *If there are c categories to be made, c − 1 dummy variables are required.*

For our model with four categories and three indicator variables, if the part is made by machine 1, then the regression equation is

$$\begin{aligned} \hat{Y} &= a + b_1X_1 + b_2(1) + b_3(0) + b_4(0) \\ &= (a + b_2) + b_2X_1. \end{aligned}$$

For a part made by machine 2, the regression equation is

$$\begin{aligned} \hat{Y} &= a + b_1X_1 + b_2(0) + b_3(1) + b_4(0) \\ &= (a + b_3) + b_1X_1. \end{aligned}$$

If it is made by machine 3,

$$\hat{Y} = (a + b_4) + b_1X_1;$$

and if it is made by machine 4,

$$\begin{aligned} \hat{Y} &= a + b_1X_1 + b_2(0) + b_3(0) + b_4(0) \\ &= a + b_1X_1. \end{aligned}$$

Here, our basis of reference is machine 4. The regression coefficient b_2 shows how much stronger or weaker the parts made by machine 1 were than those made by machine 4, and so on.

SEVERAL CATEGORICAL VARIABLES

If more than one predictor is a categorical variable, all can be handled in the same way. For example, consider

$$Y = \text{degree of political conservatism},$$
$$X_1 = \text{income},$$
$$X_2 = \text{age},$$
$$X_3 = \text{race} \begin{cases} 1 = \text{white} \\ 0 = \text{nonwhite}, \end{cases}$$
$$X_4 = \text{party affiliation} \begin{cases} 1 = \text{belongs to a political party} \\ 0 = \text{does not belong to a political party}. \end{cases}$$

(Only two classifications of both race and party affiliation are being considered.) Then the model is

$$Y = \alpha + \beta_1 X_1 + \beta_2 X_2 + \beta_3 X_3 + \beta_4 X_4 + \varepsilon;$$

and if the person considered is nonwhite and does not belong to a political party, the regression equation becomes

$$\hat{Y} = a + b_1 X_1 + b_2 X_2.$$

A nonwhite who belongs to a political party has regression equation

$$\begin{aligned} \hat{Y} &= a + b_1 X_1 + b_2 X_2 + b_4(1) \\ &= (a + b_4) + b_1 X_1 + b_2 X_2, \end{aligned}$$

so that b_4 indicates the effect of political party membership. A white belonging to a political party has regression equation

$$\begin{aligned} \hat{Y} &= a + b_1 X_1 + b_2 X_2 + b_3(1) + b_4(1) \\ &= (a + b_3 + b_4) + b_1 X_1 + b_2 X_2, \end{aligned}$$

so that $b_3 + b_4$ indicates the effect of being white and belonging to a political party, where b_3 indicates the effect of being white.

USING DUMMY VARIABLES TO TEST FOR INTERACTIONS

Let us return to the example of how sex and age affect earnings. We have a model

$$Y = \alpha + \beta_1 X_1 + \beta_2 X_2 + \varepsilon$$

where Y = earnings, X_1 = age, and X_2 = {0 if female, 1 if male}. We saw that β_2 measures how much higher the line is for males than for females, assuming that both lines relating income to age have the *same slope*. If this assumption is true, the rate of increase of earnings with age is the same for both males and females. Then β_2 measures how much more males earn than females at every age.

It may not be the case that the earnings differential between males and females remains constant regardless of the ages of the employees, however. That is, the regression line for males may not only have a different intercept from the line for females but also a different slope. If the effect of age on earnings varies according to the age of the employees, then we say that there is an **interaction** between the predictors "sex" and "age."

Interaction

In order to investigate whether or not such an interaction is present, we introduce a term involving both X_1 and X_2 into the model:

$$Y = \alpha + \beta_1 X_1 + \beta_2 X_2 + \beta_3 X_1 X_2 + \varepsilon.$$

Now the regression equation for females ($X_2 = 0$) will be

$$\hat{Y} = a + b_1 X_1 + b_2(0) + b_3 X_1(0) = a + b_1 X_1.$$

For males ($X_2 = 1$) we have

$$\hat{Y} = a + b_1 X_1 + b_2(1) + b_3 X_1(1) = (a + b_2) + (b_1 + b_3) X_1.$$

We see that b_2 measures how much higher the intercept is for the males than for females, that $(b_1 + b_3)$ is the slope of the line for males, and that b_3 measures the difference in the slope of the lines for males and females. If $b_3 > 0$, the slope of the line for males is steeper than that for females, and vice versa. Thus, to test the equality of the slopes of two regression lines, we need only test $H_0\!:\beta_3 = 0$. Only if this hypothesis is accepted does β_2 measure the differential between males and females across all ages. If $H_0\!:\beta_3 = 0$ is rejected, then the earnings differential between males and females will either increase or decrease with age. For example, suppose we conclude that $\beta_3 < 0$. Then the situation is as illustrated in Figure 14.3. Here we see that young men earn considerably more than young women, but at some age their earnings become equal, and past that age, women earn more than men of the same age.

As an example, suppose that a female college professor is rated by her students on several items concerning classroom effectiveness, each on a scale from 0 (awful) to 4 (great). The average rating for all items is computed, giving

FIGURE 14.3 Interaction Between Age and Sex in Effect on Income

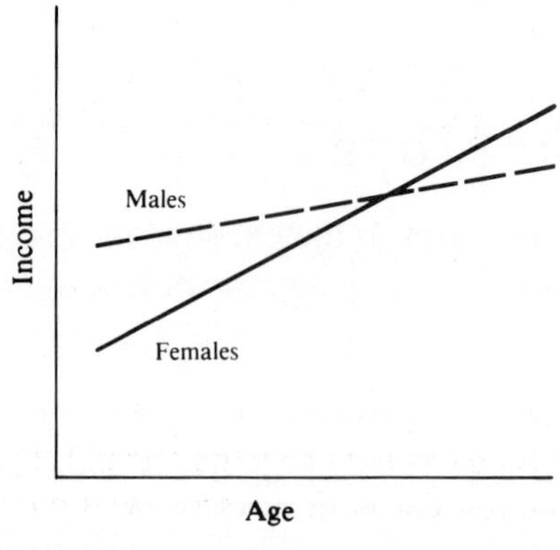

the instructor an overall rating from each student. There is information available concerning the grade in the course earned by each student and the sex of each student (although such information is not shared with the professor). A random sampling of 22 students gives the information in Table 14.1.

TABLE 14.1
Data for Teacher Evaluation Example

Student	Grade	Sex	Rating	Student	Grade	Sex	Rating
1	47	F	1.25	12	75	M	2.90
2	51	F	2.25	13	80	M	3.20
3	50	M	2.80	14	82	M	2.40
4	54	M	2.60	15	88	M	2.10
5	58	F	1.40	16	89	M	2.75
6	59	F	1.90	17	89	F	3.00
7	63	M	3.50	18	91	M	3.75
8	65	M	2.40	19	94	F	2.80
9	69	F	1.80	20	96	M	2.20
10	69	M	3.30	21	96	F	2.75
11	72	F	3.40	22	97	M	3.40

The question arises as to whether or not course grade and/or sex of the student affected the students' ratings of the professor, and also whether or not the effect of grade on rating (if any) was different for male and female students. Accordingly, a model

$$Y = \alpha + \beta_1 X_1 + \beta_2 X_2 + \beta_3 X_1 X_2 + \varepsilon$$

is proposed, where

$$Y = \text{teacher rating},$$

$$X_1 = \text{student's grade},$$

$$X_2 = \text{sex of student} \begin{cases} 0 & \text{if female} \\ 1 & \text{if male.} \end{cases}$$

The data are graphed in Figure 14.4. While it appears that for male students, rating of the teacher is not affected by grade in the course, female students who earned low grades tended to rate the professor low.

Let us find the regression equation relating the teacher's rating to the grade and sex of a student. To fit the model containing an interaction term, we define

$$\mathbf{y} = \begin{bmatrix} 1.25 \\ 2.25 \\ \vdots \\ 3.40 \end{bmatrix}, \quad \hat{\boldsymbol{\beta}} = \begin{bmatrix} a \\ b_1 \\ b_2 \\ b_3 \end{bmatrix}, \quad \mathbf{X} = \begin{bmatrix} 1 & 47 & 0 & 47(0) = 0 \\ 1 & 51 & 0 & 51(0) = 0 \\ \vdots & \vdots & \vdots & \vdots \\ 1 & 97 & 1 & 97(1) = 97 \end{bmatrix}.$$

It is also easy to modify the computer programs previously presented to handle interaction terms. Typically, all that is required is that a new variable, say

FIGURE 14.4 Teacher's Ratings From Students According to Grade and Sex of Student

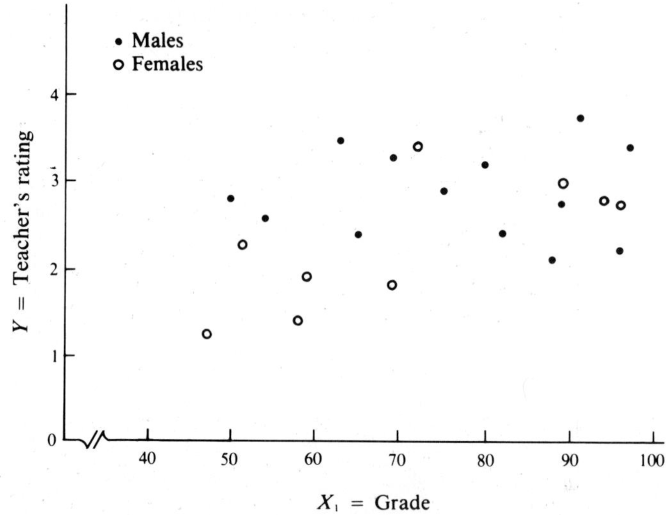

X3=X1*X2, be defined, and that the model specifies Y as a function of X1, X2, and X3.

The regression equation corresponding to the model above is found to be

$$\hat{Y} = 0.243669 + 0.028909X_1 + 2.616660X_2 - 0.028793X_1X_2$$

with significance probabilities of

0.7529, 0.0130, 0.0292, and 0.0648,

respectively, for the regression coefficients. For female students, since $X_2 = 0$, the regression equation relating rating to grade is

$$Y = 0.234669 + 0.028909X_1$$

and for males, since $X_2 = 1$, we have

$$Y = 0.234669 + 0.028909X_1 + 2.616660 - 0.028793X_1$$
$$= 2.860329 + 0.000116X_1.$$

These equations are graphed, along with the data points, in Figure 14.5.

These regression equations confirm what was apparent in the scatter diagram. Male students rate the professor about 2.90 regardless of the grade they earned in the course, but female students with low grades tend to give the professor a low rating. That $b_1 = 0.028909$ is significantly different from zero ($p = 0.0130$) means it is a real effect for female students' ratings to be affected by grade in the course. That $b_2 = 2.616660$ is significantly different from zero ($p = 0.0292$) means there is a real difference in the ratings that male and female students with a course grade of zero (if we may extrapolate) would give the professor. That $b_3 = 0.028973$ is significantly different from zero ($p = 0.0648$) means there is a real differential effect of grade on rating according to sex.

FIGURE 14.5 Regression Equations for Male and Female Students

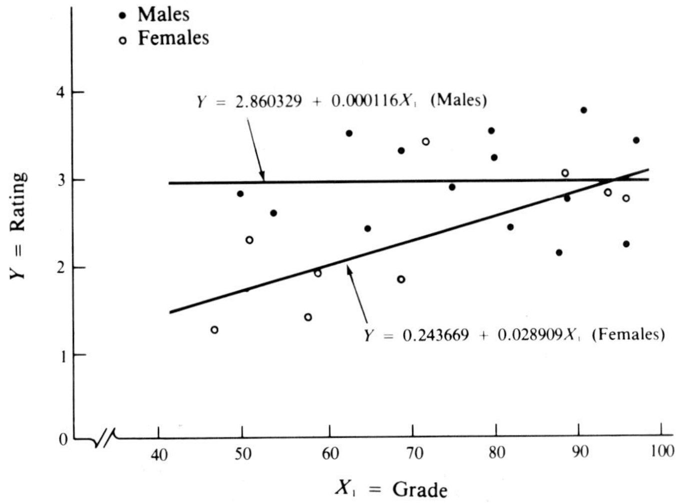

EXERCISES

14.1 In a survey to be taken of undergraduate students at a university, it is desired to indicate which class an interviewed student belongs to (freshman, sophomore, junior, senior). It will be too much trouble, however, to record the number of hours completed in order to indicate class, so dummy variables are to be used. Show how you would define dummy variables in order to take class into account.

14.2 Suppose that in a survey, values of a variable "education" were defined in the following manner:

1 = no formal schooling,
2 = some formal schooling, but no high school diploma,
3 = graduated high school only,
4 = attended college, but no degree,
5 = undergraduate degree only,
6 = schooling past undergraduate degree.

Show an alternative method of indicating education, using dummy variables.

14.3 Suppose that in a study relating the amount spent on medical expenses to the age, sex, and income of a patient, the following regression equation was found:

$$\hat{Y} = 400 + 2X_1 - 20X_2 - 4X_3,$$

where

Y = medical expenses, in dollars, for a given year,

X_1 = age of patient in years,

$$X_2 = \text{sex} \begin{cases} 1 & \text{if female} \\ 0 & \text{if male,} \end{cases}$$

X_3 = annual family income, in thousands of dollars.

What is the meaning of the regression coefficient $b_2 = -20$? Illustrate by comparing the medical expenses of a male and a female patient, both 60 years of age and both having $30,000 annual family income.

14.4 Consider the data in Table 14.2. Suppose it is desired to predict food expenditures from the other three variables. Show the vector **y** and the matrix **X** necessary to set up the problem.

TABLE 14.2
Data for Food Expenditure Exercise

Weekly Food Expenditure (dollars)	Annual Income (thousands of dollars)	Children?	Area
40	14	no	city
72	38	yes	suburb
48	25	no	rural
76	24	yes	suburb
70	19	yes	suburb
59	10	no	city
77	23	yes	suburb
59	12	yes	rural
91	26	yes	city
69	24	yes	suburb

14.5 A sample of 25 employees of a firm was selected, and the information in Table 14.3 was gathered concerning their annual salaries (Y, in thousands of dollars), number of years with firm (X_1), and sex (X_2).

TABLE 14.3
Salary Data

Employee	Salary	Years	Sex	Employee	Salary	Years	Sex
Adams	35	15	M	North	28	6	M
Brown	27	17	F	Owen	29	20	F
Carter	45	25	M	Pitt	19	3	F
Davis	22	13	F	Quincy	29	21	F
Eaves	25	2	M	Roberts	38	19	M
Ford	30	10	M	Smith	19	5	F
Green	37	17	M	Turner	22	1	M
Higgins	25	17	F	Underwood	39	20	M
Irvin	17	1	F	Vance	40	22	M
Jones	28	4	M	Wilson	21	10	F
King	43	25	M	Young	28	7	M
Lane	25	15	F	Ziegler	30	8	M
Martin	22	1	M				

a. Plot salary versus years with the firm on a scatter diagram. Use different symbols to denote data points for males and females.

b. Do the rates of increase in salary with years appear to be different for males and females? If so, describe how.

c. Do starting salaries appear to be different for males and females? If so, describe how.

d. Fit a regression equation describing how salary is related to years with the firm and sex.

e. Test to see if rate of increase in salary with age is the same for males and females. State your conclusion. If there is a difference, give a 95% confidence interval on the difference, and interpret its meaning.

f. Test to see if starting salaries are the same for males and females, and state your conclusion. If there is a difference, give a 95% confidence interval on the difference, and interpret its meaning.

g. Give a 95% confidence interval for the intercept, and interpret its meaning.

h. Give a 95% confidence interval for the coefficient on the number of years with the company, and interpret its meaning.

14.6 Consider the Enrollment Data in Exercise 9.13. In order to get measures of the changes in enrollment from fall quarter through subsequent quarters, fit a regression equation

$$\hat{Y} = a + b_1X_1 + b_2X_2 + b_3X_3 + b_4X_4,$$

where

$$X_1 = \begin{cases} 1 & \text{if winter quarter} \\ 0 & \text{otherwise,} \end{cases}$$

$$X_2 = \begin{cases} 1 & \text{if spring quarter} \\ 0 & \text{otherwise,} \end{cases}$$

$$X_3 - \begin{cases} 1 & \text{if summer quarter} \\ 0 & \text{otherwise,} \end{cases}$$

$$X_4 = \text{quarter number } (1, 2, \ldots, 36).$$

Interpret the meanings of the regression constant and the regression coefficients.

14.7 Refer to the IQ Data in Exercise 9.8, in which score on a test is related to IQ. Information is available on which teacher taught each student in the sample. Fit a regression equation, taking teacher as well as IQ into account when predicting score. Interpret the regression coefficients.

14.8 In order to evaluate the effectiveness of energy-saving electric heating and air conditioning units, a random sample was drawn of single-family residences built two years ago in a certain city. The information in Table 14.4 was obtained. Does

TABLE 14.4
Energy Saver Data

Floor Area (thousands of square feet)	Annual Electricity Cost (hundreds of dollars)	Energy Saver?	Floor Area (thousands of square feet)	Annual Electricity Cost (hundreds of dollars)	Energy Saver?
2.4	13.2	yes			
1.8	10.0	yes	3.0	16.9	yes
3.2	17.8	yes	3.6	20.9	no
3.7	21.6	yes	2.7	16.0	no
1.9	10.6	yes	2.6	15.8	no
3.8	23.0	no	1.1	6.9	no
3.9	22.7	yes	3.6	20.9	no
1.2	7.5	no	3.5	21.7	no
3.9	23.0	yes	1.4	7.2	yes
2.9	16.1	yes	2.8	16.1	no
2.4	14.7	no	2.0	10.8	yes
2.2	13.0	no	2.6	14.2	yes

it make any difference to the relationship between floor space and electricity costs whether or not a heating and air conditioning unit is an energy saver?

14.9 The effectiveness of two weight-reducing diets was to be compared. Twenty people who agreed to participate in the study were randomly assigned to one of the two diets. The subjects were weighed before starting on the diet, and their weight losses at the end of two weeks were calculated. It is suspected that a person's initial weight may be related to weight loss as well. The data are given in Table 14.5.

TABLE 14.5
Diet Data

Subject	DIET A Initial Wt. (pounds)	Pounds Lost	Subject	DIET B Initial Wt. (pounds)	Pounds Lost
1	312	20	11	246	15
2	332	25	12	181	8
3	259	10	13	322	21
4	264	10	14	291	17
5	312	17	15	243	9
6	269	9	16	228	8
7	200	5	17	300	15
8	281	17	18	263	8
9	263	12	19	162	5
10	278	10	20	213	10

a. Plot the data, indicating diet. Does the relationship between weight loss and initial weight appear to be linear?
b. Fit a quadratic regression relating weight loss to initial weights, including a dummy variable indicating diet type.
c. Does type of diet affect the relationship between weight loss and initial weight?

14.10 Referring to the Diet Data in Exercise 4.9, fit a quadratic model relating weight loss to initial weight, including a dummy variable for diet type and an interaction term between initial weight and diet type. Does one type of diet appear to be more effective for heavier people?

14.3 INTERACTIONS WHEN VARIABLES ARE NOT DUMMY VARIABLES

It is possible, and often desirable, to include interaction terms in the model even when one of the variables is not a dummy variable. Recall that an interaction among the variables Y, X_i, and X_j is present when the effect of X_i on Y varies according to the value of X_j. There is nothing in this definition that specifies that X_j must be a dummy variable.

Suppose that a study is run to assess the effects of obesity and age on the blood pressure of adult males. A sample of men between the ages of 40 and 65 is obtained, their blood pressures are taken, and they are weighed and measured. For each man, an "obesity index" is calculated as (weight/height^2) $\times$ 100, a commonly used ratio. Systolic blood pressure, which measures the pressure exerted on arteries when blood is pumped out of the heart, is an indication of the elasticity of the arteries.

It is hypothesized that blood pressure will increase with age and also with obesity. In addition, it is suspected that the rate of increase of blood pressure with age will be faster among obese subjects than among those who are not overweight. Saying it another way, it is thought that the rate of increase of blood pressure with obesity will be faster, the older the subject. That is, age and obesity should interact in their effect on blood pressure. Thus, the model

$$Y = \alpha + \beta_1 X_1 + \beta_2 X_2 + \beta_3 X_1 X_2 + \varepsilon$$

is adopted, where

$$Y = \text{systolic blood pressure } (120 < Y < 170)$$

$$X_1 = \text{obesity index } (2 < X_1 < 5)$$

$$X_2 = \text{age } (40 < X_2 < 65).$$

It is relatively straightforward to define $\mathbf{y}$, $\boldsymbol{\beta}$, and $\mathbf{X}$ for this model. We have values $Y_1, Y_2, \ldots, Y_n$ for the blood pressure readings for each of the n subjects in the sample. Similarly, there are n obesity measures $X_{11}, X_{12}, \ldots, X_{1n}$, and n ages, $X_{21}, X_{22}, \ldots, X_{2n}$. Then

$$\mathbf{y} = \begin{bmatrix} Y_1 \\ Y_2 \\ \vdots \\ Y_n \end{bmatrix}, \quad \hat{\boldsymbol{\beta}} = \begin{bmatrix} a \\ b_1 \\ b_2 \\ b_3 \end{bmatrix}, \quad \mathbf{X} = \begin{bmatrix} 1 & X_{11} & X_{21} & X_{11}X_{21} \\ 1 & X_{12} & X_{22} & X_{12}X_{22} \\ \vdots & \vdots & \vdots & \vdots \\ 1 & X_{1n} & X_{2n} & X_{1n}X_{2n} \end{bmatrix}.$$

A regression equation is found:

$$\hat{Y} = 210 - 35X_1 - 2X_2 + X_1X_2,$$

and the interaction term is significant. What does this mean? First notice that if X_1 increases by 1, then Y increases by

$$-35 + X_2.$$

Thus, the rate of increase of Y with X_1 *depends on the value of* X_2. There is a differential effect on blood pressure of increasing obesity, *depending on the age of the subject*. Similarly, the increase in Y for a one-unit increase in X_2 is

$$-2 + X_1,$$

so that there is a differential effect on blood pressure of increasing age, *depending on the obesity of the subject*.

Although the coefficients b_1 and b_2 on obesity and age, respectively, are negative, this does not imply that blood pressure *decreases* with either an increase in obesity or an increase in age. Consider the rate of change in blood pressure with obesity, for example:

$$-35 + X_2.$$

Since this increase depends on X_2 = age, and there is no age in the sample less than 40, this slope will never be negative. A similar argument holds for the rate of increase of blood pressure with age.

To illustrate the effects of age and obesity on blood pressure, and how they interact, let us choose a few specific values of X_1 = obesity and look at the relationship between Y and X_2 = age graphically. Take X_1 = 2, 3, 4, and 5 (although we could have selected *any* value between 2 and 5).

If $X_1 = 2$, then $\hat{Y} = 210 - 35(2) - 2X_2 + 2X_2 = 140$;

If $X_1 = 3$, then $\hat{Y} = 210 - 35(3) - 2X_2 + 3X_2 = 105 + X_2$;

If $X_1 = 4$, then $\hat{Y} = 210 - 35(4) - 2X_2 + 4X_2 = 70 + 2X_2$;

If $X_1 = 5$, then $\hat{Y} = 210 - 35(5) - 2X_2 + 5X_2 = 35 + 3X_2$.

Figure 14.6 shows these four regression lines of $\hat{Y}$ versus X_2 for the selected values of X_1. We see that at age 40, the more obese the patient, the higher his blood pressure, and that through age 65 the more obese subjects continue to have higher blood pressures than do those who are less obese. Except for the least obese patients, for whom X_1 = 2 and for whom blood pressure remains stable at 140 regardless of their ages, there is a direct

FIGURE 14.6 Plot of Predicted Blood Pressure Versus Age (X_2) for Selected Values of Obesity Index (X_1)

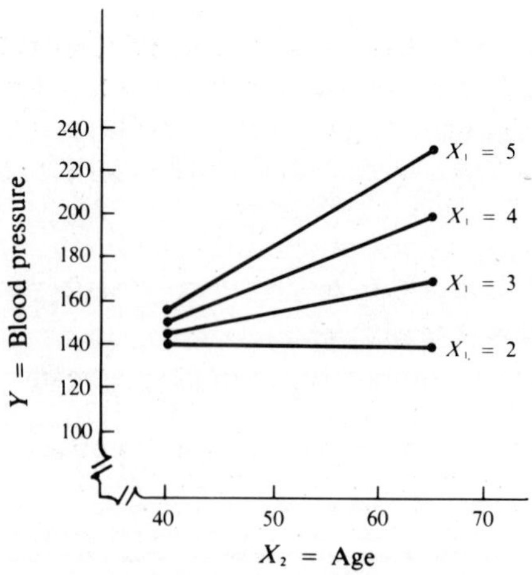

relationship between age and blood pressure, but it is also the case that the more obese a subject is, the faster his blood pressure increases between the ages of 40 and 65.

Looking at the interaction the other way, suppose we plot blood pressure against obesity for several different age groups. While we could look at any or all ages between 40 and 65, let us look at subjects 40, 50, and 60 years old:

$$\text{If } X_2 = 40, \text{ then } \hat{Y} = 210 - 35X_1 - 2(40) + 40X_1 = 130 + 5X_1;$$

$$\text{If } X_2 = 50, \text{ then } \hat{Y} = 210 - 35X_1 - 2(50) + 50X_1 = 110 + 15X_1;$$

$$\text{If } X_2 = 60, \text{ then } \hat{Y} = 210 - 35X_1 - 2(60) + 60X_1 = 90 + 25X_1.$$

These regression equations are graphed in Figure 14.7. We see that for patients who are not obese ($X_1 = 2$), systolic blood pressure is 140 regardless of age. However, blood pressure increases with obesity for all age groups. Furthermore, the rate of increase in blood pressure with obesity is greater, the older the subject.

FIGURE 14.7 Plot of Predicted Blood Pressure Versus Obesity Index (X_1) for Selected Ages (X_2)

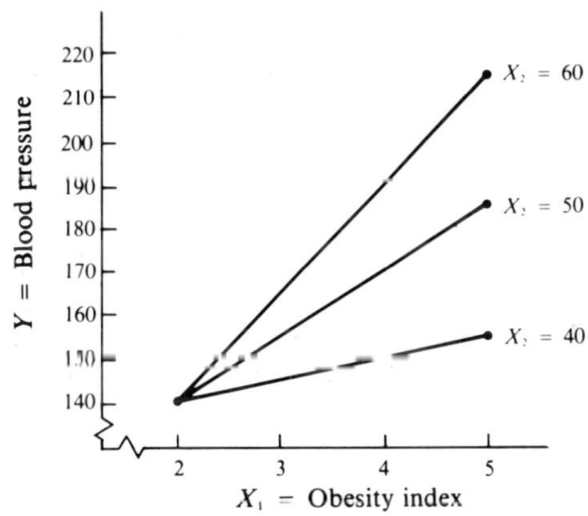

If these were real data and all regression coefficients were significantly different from zero, we would conclude that a combination of advancing age and increasing obesity has a worse effect on blood pressure than does either factor alone. For example, a 65-year-old man with an obesity index of 2 would be expected to have systolic blood pressure of

$$\hat{Y} = 210 - 35(2) - 2(65) + 65(2) = 140.$$

A 40-year-old man with an obesity index of 5 would be predicted to have a blood pressure of

$$\hat{Y} = 210 - 35(5) - 2(40) + 40(5) = 155.$$

But a 50-year-old man with an obesity index of 3.5 might have a blood pressure of

$$\hat{Y} = 210 - 35(3.5) - 2(50) + 3.5(50) = 162.5.$$

In this example the presence of an interaction indicates a compounding, snowballing effect of the predictors on the response. In other cases an interaction may indicate a neutralizing, cancelling-out effect of the predictors, however.

SIGNIFICANCE OF THE REGRESSION COEFFICIENTS

When an interaction term is included in a model, the first partial test that should be looked at is the test of the regression coefficient for the interaction term. If the model is

$$Y = \alpha + \beta_1 X_1 + \beta_2 X_2 + \beta_3 X_1 X_2 + \varepsilon,$$

then we first want to test

$$H_0: Y = \alpha + \beta_1 X_1 + \beta_2 X_2 + \varepsilon$$

versus

$$H_1: Y = \alpha + \beta_1 X_1 + \beta_2 X_2 + \beta_3 X_1 X_2 + \varepsilon,$$

for this tells us if the slope of the line relating Y to X_1 changes depending on the value of X_2 (or equivalently, if the slope of the line relating Y to X_2 changes depending on the value of X_1). If this null hypothesis cannot be rejected, then a model

$$Y = \alpha + \beta_1 X_1 + \beta_2 X_2 + \varepsilon$$

should be fitted.

If the interaction term is included in the model, then how are the tests of

$$H_0: Y = \alpha + \beta_2 X_2 + \beta_3 X_1 X_2 + \varepsilon$$

$$H_1: Y = \alpha + \beta_1 X_1 + \beta_2 X_2 + \beta_3 X_1 X_2 + \varepsilon$$

and

$$H_0: Y = \alpha + \beta_1 X_1 + \beta_3 X_1 X_2 + \varepsilon$$

$$H_1: Y = \alpha + \beta_1 X_1 + \beta_2 X_2 + \beta_3 X_1 X_2$$

to be interpreted? If interaction is present, β_1 no longer tells how Y changes with a one-unit increase in X_1, since this rate of increase depends on the value of X_2; and β_2 does not give the rate of increase in Y with X_2, since this depends on the value of X_1. Instead of being slopes, when interaction is present β_1 and β_2 have to do with *intercepts*.

In our blood pressure example we used a regression equation

$$\hat{Y} = 210 - 35X_1 - 2X_2 + X_1 X_2,$$

where $\hat{Y}$ = estimated systolic blood pressure, X_1 = obesity index, and X_2 = age. Consider a 50-year-old man. The estimating equation for his blood pressure would be

$$\begin{aligned}\hat{Y} &= 210 - 35X_1 - 2(50) + 50X_1 \\ &= [210 - 2(50)] + (-35 + 50)X_1.\end{aligned}$$

We see that b_1, the regression coefficient for X_1 = obesity, is incorporated in the slope of the line relating Y to X_1. However, b_2, the regression coefficient on X_2 = age, becomes involved in the *intercept* of the equation relating Y to X_1. It is the difference in the intercepts of the lines relating Y to X_1 for each one-unit increase on X_2 = age. The test of whether X_2 should be included in the model is thus a test to see if the intercepts of the lines relating Y to X_1 differ according to age.

Arguing the other way, suppose a subject has an obesity index of X_1 = 3. Then to predict his blood pressure, the equation

$$\begin{aligned}\hat{Y} &= 210 - 35(3) - 2X_2 + 3X_2 \\ &= [210 - 35(3)] + (-2 + 3)X_2\end{aligned}$$

is used. In this case, $b_2 = -2$ is involved in measuring how Y = blood pressure changes with X_2 = age, but b_1 tells whether the intercepts of the lines relating blood pressure to age differ according to how obese a subject is.

Suppose in our example we found the significance probabilities indicated in parentheses under the regression coefficients, and that we have specified a significance level of $\alpha = 0.10$:

$$\begin{array}{ccccc}\hat{Y} = & 210 & -35X_1 & -2X_2 & +X_1X_2 \\ & (0.02) & (0.19) & (0.30) & (0.09)\end{array}$$

First, $p = 0.02$ for the regression constant means that it is *not* the case that a man zero years old with obesity index of zero would have zero blood pressure (although this is hardly very enlightening information). The interaction term is significant, so that the effect of age on blood pressure depends on how obese the subject is; or vice versa: The effect of obesity on blood pressure depends on the age of the subject.

That b_1 is not significantly different from zero means that there is no real difference in the intercepts of the lines relating Y to X_2 for different values of X_1, and that b_2 is not significantly different from zero means that the intercepts of the lines relating Y to X_1 do not differ according to the value of X_2. This may seem surprising, because the lines graphed in Figures 14.6 and 14.7 certainly appear to have different intercepts. However, recall that the regression coefficients b_1 and b_2 measure the changes in the intercepts for *one-unit* increases in X_1 and X_2. A one-unit increase in age is a one-year increase, and a difference of 2 in the intercepts of lines for men differing by one year in age is not very much of a difference. The obesity index, (weight/height^2) × 100, can be measured to as many decimal places as desired, and a difference of one unit in the obesity index represents a very pronounced difference. Thus, a difference in intercepts of 35 for a one-unit increase in the obesity index is not so large either.

Note that we have no test of whether, in general, blood pressure increases with either age or obesity. This is because, as we have seen, whether and how blood pressure increases with age depends on how obese the subject is, and vice versa. For example, refer to Figure 14.6, where predicted blood pressure is plotted against age for four selected levels of obesity. While for very obese subjects, blood pressure increases very fast with age, the line for nonobese individuals (index = 2) is horizontal. Thus, the answer to the question, "Does blood pressure increase with age (are the two variables related)?" can only be answered, "That depends on how obese the individuals are."

EXERCISES

14.11 American Package Delivery Service, in order to schedule its delivery trucks, collected the information in Table 14.6 from a simple random sampling of past deliveries.

TABLE 14.6
Package Data

Number of Deliveries	Miles Traveled	Time Required (hours)	Number of Deliveries	Miles Traveled	Time Required (hours)
13	97	5.0	17	66	3.8
11	52	3.0	6	79	3.5
21	106	5.8	8	57	3.1
18	110	5.9	7	78	3.8
11	57	3.0	16	81	4.6
10	72	3.5	19	112	6.3
14	76	4.3	14	94	4.7
13	84	4.5	6	60	2.8
11	92	4.7	12	88	4.5
22	87	5.4	15	100	5.1

a. Fit a model $Y = \alpha + \beta_1 X_1 + \beta_2 X_2 + \beta_3 X_1 X_2 + \varepsilon$.
b. Is there any interaction between number of deliveries and miles traveled in their effects on time required? Explain.
c. Is there any collinearity between number of deliveries and miles traveled? Explain.
d. Explain the difference between interaction and collinearity.

14.12 A survey was performed in order to examine the prescription drug purchasing habits of females who are chronically ill. Among information collected from a sample of 45 such women were their ages and household incomes (in thousands of dollars). The data are given in Table 14.7. It was noted that quarterly drug expenditures (in dollars) tend to increase with age and also with income, and it

TABLE 14.7
Prescription Data

Woman	Age (X_1)	Income (X_2)	Quarterly Drug Expenditures (Y)	Woman	Age (X_1)	Income (X_2)	Quarterly Drug Expenditures (Y)
1	24	10	50	24	49	42	125
2	25	17	70	25	53	35	110
3	29	15	60	26	54	40	130
4	30	12	55	27	60	37	130
5	32	16	75	28	71	38	140
6	35	16	65	29	25	50	105
7	35	20	90	30	33	45	120
8	42	15	90	31	35	55	135
9	48	13	80	32	38	60	150
10	53	12	65	33	42	45	140
11	63	15	90	34	47	44	155
12	70	15	100	35	52	60	175
13	72	12	75	36	55	50	150
14	22	25	75	37	56	60	180
15	25	27	80	38	59	45	170
16	29	30	100	39	60	70	200
17	25	40	100	40	62	65	205
18	32	40	110	41	63	40	180
19	38	30	90	42	69	45	185
20	37	48	120	43	69	60	220
21	44	37	125	44	70	70	240
22	47	30	100	45	72	50	200
23	46	40	130				

was suspected that income level might produce a differential effect of age on drug purchases.

a. Fit a model containing an interaction term relating drug expenditures to age and income.

b. Find and graph regression equations relating drug expenditures to income for women 25, 45, and 65 years of age.

c. Find and graph regression equations relating drug expenditures to age for women with incomes \$15,000, \$40,000, and \$65,000.

d. Interpret the regression coefficients and their significance probabilities. Use $\alpha = 0.10$.

14.4 SUMMARY

This chapter showed how to incorporate categorical variables into a regression analysis. In addition, the concept of an interaction was introduced, and it was noted that while interactions are somewhat easier to understand and analyze when one variable is a dummy, or indicator, variable, the concept may be generalized to include interactions among interval- or ratio-level variables.

What if all the predictor variables are categorical? It is quite possible to analyze such models as regression models, but such models are more properly known as *analysis of variance models*. When analysis of variance was first developed, special techniques for calculation were formulated in order to reduce computational conplexity. With the widespread availability of computer programs simplifying the process of inversion of $\mathbf{X}'\mathbf{X}$ has come an emphasis on treating analysis of variance models as special cases of the general linear model. It has been only recently that textbooks on analysis of variance have begun to break with the traditional approach and to analyze analysis of variance models using the same general techniques as presented here for regression analysis. Thus, what is presented in many older, traditional texts on the analysis of variance may not look very familiar to readers of this text. One text integrating the traditional approach with the linear model approach is Roger E. Kirk's *Experimental Design,* 2d ed. (See Suggestions for Further Reading at the back of the book.) There is a very extensive body of literature on the analysis of variance, so it will not be discussed further here.

Often, when some predictors are categorical and some are interval- or ratio-level, the model is referred to as an *analysis of covariance model*. In such models, the objective is to discover if there is any difference in the response variable for different values, or levels, of the categorical variable(s), adjusting for preexisting differences in the value(s) of some concomitant interval-level variable(s). Covariance models may also be treated as special cases of regression models. The reader is referred to texts on analysis of variance or multivariate analysis for a fuller explanation.

By this time, the reader has probably become rather confused regarding which variables to use in a regression model. Conclusions based on partial t- and F-tests are far from clear-cut in many instances. In Chapter 15 we will examine some computer algorithms for selecting variables to be included in the regression model.

F I F T E E N

SOME VARIABLE-SELECTION PROCEDURES

15.1

ANALYSIS OF A LARGER PROBLEM

The house size-price-age problem has served us well, but it is actually too small to illustrate the many things that can happen in a regression problem. We now want to perform a multiple regression analysis of a larger example to illustrate various procedures for selecting variables.

CRAVENS DATA

For our larger example we will use data kindly provided by Professor David W. Cravens of the Department of Marketing at Texas Christian University. We title them Cravens Data, for short.†

A certain company sells its product all over the United States. The company has designated geographic sales territories, each the responsibility of a single sales representative who services a number of retail outlets within his or her territory. In order to identify factors that influence the number of

† These data were analyzed in "An Analytical Approach for Evaluating Sales Territory Performance," by David W. Cravens, Robert B. Woodruff, and Joe C. Stamper, in *Journal of Marketing,* 36 (January 1972), pp. 31–37.

units of the company's product sold in any territory and to assess the relative importances of the factors, a random sample of 25 sales territories of the firm's national sales organization was drawn. Information was collected on the number of units of the product sold, along with values of eight possible predictor variables, in each territory. The variables are defined briefly in Table 15.1. To elaborate, sales were hypothesized to be possibly a function of the following variables:

1. How long the territory sales representative had been employed by the company;
2. The size of the market in the territory, as measured by total sales of this product and competing products during the last year;
3. The amount of money the company had spent on advertising in that territory during the past year;
4. What percentage this company had of total sales of all similar products in this territory;
5. How much this product's share of the market had increased or decreased in this territory during the past several years;
6. The number of retail outlets the sales representative was responsible for in this territory;
7. The average size of orders per retail outlet;
8. An evaluation of the sales representative's performance by his or her supervisor.

The data are shown in Table 15.2.

TABLE 15.1 Definitions of Variables in Sales Territory Performance Study

(Y = sales territory performance measured by aggregate sales, in units, credited to territory sales representative)†

Predictors	**How Measured**
X_1 = time with company	Length of time employed by company, in months.
X_2 = market potential	Industry sales, in units, of products sold in territory.[a]
X_3 = advertising (or company effort)	Dollar expenditures on advertising in territory.
X_4 = market share	The weighted average of past market share magnitudes for four previous years.
X_5 = market-share change	Change over four years previous to time period analyzed.
X_6 = accounts	Total number of accounts assigned to a sales representative.[a]
X_7 = workload	Average workload per account using a weighted index based on annual purchases of accounts and concentration of accounts.
X_8 = rating	An aggregate rating on a 1–7 scale by applicable field sales manager on eight dimensions of performance.

† These data were coded by use of a linear transformation in order to preserve confidentiality of information for the respondents.

TABLE 15.2 Data for Sales Territory Performance Study

Y Sales	X_1 Time With Company	X_2 Market Potential	X_3 Adver-tising	X_4 Market Share	X_5 Market-Share Change	X_6 Accounts	X_7 Workload	X_8 Rating
3,669.88	43.10	74,065.11	4,582.88	2.51	0.34	74.86	15.05	4.9
3,473.95	108.13	58,117.30	5,539.78	5.51	0.15	107.32	19.97	5.1
2,295.10	13.82	21,118.49	2,950.38	10.91	−0.72	96.75	17.34	2.9
4,675.56	186.18	68,521.27	2,243.07	8.27	0.17	195.12	13.40	3.4
6,125.96	161.79	57,805.11	7,747.08	9.15	0.50	180.44	17.64	4.6
2,134.94	8.94	37,806.94	402.44	5.51	0.15	104.88	16.22	4.5
5,031.66	365.04	50,935.26	3,140.62	8.54	0.55	256.10	18.80	4.6
3,367.45	220.32	35,602.08	2,086.16	7.07	−0.49	126.83	19.86	2.3
6,519.45	127.64	46,176.77	8,846.25	12.54	1.24	203.25	17.42	4.9
4,876.37	105.69	42,053.24	5,673.11	8.85	0.31	119.51	21.41	2.8
2,468.27	57.72	36,829.71	2,761.76	5.38	0.37	116.26	16.32	3.1
2,533.31	23.58	33,612.67	1,991.85	5.43	−0.65	142.28	14.51	4.2
2,408.11	13.82	21,412.79	1,971.52	8.48	0.64	89.43	19.35	4.3
2,337.38	13.82	20,416.87	1,737.38	7.80	1.01	84.55	20.02	4.2
4,586.95	86.99	36,272.00	10,694.20	10.34	0.11	119.51	15.26	5.5
2,729.24	165.85	23,093.26	8,618.61	5.15	0.04	80.49	15.87	3.6
3,289.40	116.26	26,878.59	7,747.89	6.64	0.68	136.58	7.81	3.4
2,800.78	42.28	39,571.96	4,565.81	5.45	0.66	78.86	16.00	4.2
3,264.20	52.84	51,866.15	6,022.70	6.31	−0.10	136.58	17.44	3.6
3,453.62	165.04	58,749.82	3,721.10	6.35	0.03	138.21	17.98	3.1
1,741.45	10.57	23,990.82	860.97	7.37	−1.63	75.61	20.99	1.6
2,035.75	13.82	25,694.86	3,571.51	8.39	−0.43	102.44	21.66	3.4
1,578.00	8.13	23,736.35	2,845.50	5.15	0.04	76.42	21.46	2.7
4,167.44	58.54	34,314.29	5,060.11	12.88	0.22	136.58	24.78	2.8
2,799.97	21.14	22,809.53	3,552.00	9.14	−0.74	88.62	24.96	3.9

The original model for this problem is

$$Y = \alpha + \beta_1 X_1 + \beta_2 X_2 + \cdots + \beta_8 X_8 + \varepsilon.$$

The purpose is to see whether or not any predictors should be eliminated from the model as nonsignificant and to find the appropriate regression equation, standard error, and coefficient of determination. We will run these data through the four multiple regression programs discussed earlier and make some preliminary observations. In the next chapter a more complete analysis is performed.

The proposed model can be criticized as being too naive: Why should sales be a strictly linear function of all eight variables? For example, it could be argued that a quadratic term in X_1 = "time with company" be included in the model, since it could well be that sales increase rapidly with time at first for new sales representatives and then level off after a salesrep has been with the company for a while. Indeed, a scatter diagram of sales against time will show this to be so. Empirical evidence from a scatter diagram or theoretical considerations may well point to the need for inclusion of higher-order terms in the model. If so, such terms certainly should be included. For purposes of illustration here, however, we will consider the strictly linear model and leave as an exercise the formulation and analysis of a more realistic model.

FOUR COMPUTER ANALYSES OF CRAVENS DATA

The BMDP1R, Minitab, SAS, and SPSS^X analyses of the Cravens Data are shown in Figures 15.1–15.5. Except for rounding errors, the results given by all four programs are the same. No residuals or plots are given in Figures 15.1–15.5, although they could have been obtained quite easily. The simple statistics and the correlation matrix are given only for the SAS printout (Figure 15.4).

PRELIMINARY ANALYSIS OF CRAVENS DATA

What can we tell from these preliminary analyses? We should first look at the correlation matrix. From this, we can get an idea of (1) which predictors, by themselves, seem to be good ones, and (2) how highly correlated the predictors

FIGURE 15.1 BMDP1R Analysis of Cravens Data

```
REGRESSION TITLE IS
BMDP1R ANALYSIS OF CRAVENS DATA
DEPENDENT VARIABLE. . . . . . . . . . . . . . . .  1  SALES
TOLERANCE . . . . . . . . . . . . . . . . . . . .  0.0100
ALL DATA CONSIDERED AS A SINGLE GROUP
MULTIPLE R              0.9602        STD ERROR OF EST          449.0256
MULTIPLE R-SQUARE       0.9220


ANALYSIS OF VARIANCE
                  SUM OF SQUARES      DF   MEAN SQUARE       F RATIO    P(TAIL)
    REGRESSION   38153280.0000         8   4769160.0000      23.654     0.0000
    RESIDUAL      3225985.0000        16    201624.0625
                                           STD REG
VARIABLE      COEFFICIENT   STD ERROR   COEFF        T        P(2 TAIL)     TOLERANCE
INTERCEPT    -1507.78711
TIME     2        2.00953     1.93065   0.133      1.041    0.3134        0.29917
POTEN    3        0.03720     0.00820   0.445      4.536    0.0003        0.50566
ADV      4        0.15099     0.04711   0.309      3.205    0.0055        0.52351
SHARE    5      199.02388    67.02754   0.373      2.969    0.0090        0.30905
CHANGE   6      290.85132   186.78220   0.138      1.557    0.1390        0.62433
ACCTS    7        5.55099     4.77552   0.193      1.162    0.2621        0.17733
WORK     8       19.79271    33.67644   0.055      0.588    0.5649        0.54995
RATE     9        8.19148   128.50586   0.006      0.064    0.9500        0.55293
```

FIGURE 15.2 Minitab Multiple Regression for Cravens Data

```
THE REGRESSION EQUATION IS
Y = -1508.  +  2.01 X1 + 0.0372 X2
    +0.151  X3 +   199. X4 + 291. X5
    +5.55   X6 +   19.8 X7 + 8.19 X8
```

	COLUMN	COEFFICIENT	ST DEV OF COEF	T-RATIO = COEF/S.D.
		-1507.8	778.6	-1.94
X1	TIME	2.010	1.931	1.04
X2	POTEN	0.037205	0.008202	4.54
X3	ADV	0.15099	0.04711	3.21
X4	SHARE	199.02	67.03	2.97
X5	CHANGE	290.9	186.8	1.56
X6	ACCTS	5.551	4.776	1.16
X7	WORK	19.79	33.68	0.59
X8	RATING	8.2	128.5	0.06

```
THE ST DEV OF Y ABOUT REGRESSION LINE IS
S = 449.0
WITH (25 - 9) = 16 DEGREES OF FREEDOM

R-SQUARED = 92.2 PERCENT
R-SQUARED = 88.3 PERCENT, ADJUSTED FOR DF

ANALYSIS OF VARIANCE
```

DUE TO	DF	SS	MS=SS/DF
REGRESSION	8	38153564	4769195
RESIDUAL	16	3225985	201624
TOTAL	24	41379548	

themselves are. For example, in Figure 15.4 we see that the number of accounts (ACCTS) is the strongest single predictor of sales, since $r_{6Y} = 0.7540$. Then TIME, POTEN, and ADV are the next strongest predictors, with correlations around 0.6. The variables CHANGE, SHARE, and RATE comprise another group of predictors, all weaker than the preceding group, and finally WORK is the weakest single predictor.

FIGURE 15.3 SAS Analysis of Cravens Data

DEP VARIABLE: SALES

SOURCE	DF	SUM OF SQUARES	MEAN SQUARE	F VALUE	PROB > F
MODEL	8	38153564	4769196	23.654	0.0001
ERROR	16	3225985	201624		
C TOTAL	24	41379549			

ROOT MSE	449.026	R-SQUARE	0.9220
DEP MEAN	3374.568	ADJ R-SQ	0.8831
C.V.	13.30617		

VARIABLE	DF	PARAMETER ESTIMATE	STANDARD ERROR	T FOR H0: PARAMETER=0	PROB > |T|
INTERCEP	1	-1507.814	778.635	-1.936	0.0707
TIME	1	2.009566	1.930654	1.041	0.3134
POTEN	1	0.037205	0.0082023	4.536	0.0003
ADV	1	0.150989	0.047109	3.205	0.0055
SHARE	1	199.024	67.027922	2.969	0.0090
CHANGE	1	290.855	186.782	1.557	0.1390
ACCTS	1	5.550961	4.775550	1.162	0.2621
WORK	1	19.793892	33.676692	0.588	0.5649
RATE	1	8.189284	128.506	0.064	0.9500

FIGURE 15.4 SAS Simple Statistics and Correlation Matrix for Cravens Data

VARIABLE	N	MEAN	STD DEV	SUM	MINIMUM	MAXIMUM
SALES	25	3374.56760000	1313.06811398	84364.1900000	1578.00000000	6519.45000000
TIME	25	87.64200000	86.79694834	2191.0500000	8.13000000	365.04000000
POTEN	25	38858.04960000	15714.54099624	971451.2400000	20416.87000000	74065.11000000
ADV	25	4357.38720000	2689.10018289	108934.6800000	402.44000000	10694.20000000
SHARE	25	7.56480000	2.45978847	189.1200000	2.51000000	12.88000000
CHANGE	25	0.09560000	0.62104804	2.3900000	-1.63000000	1.24000000
ACCTS	25	122.69920000	45.57801891	3067.4800000	74.86000000	256.10000000
WORK	25	18.06080000	3.67007030	451.5200000	7.81000000	24.96000000
RATE	25	3.74400000	0.95920106	93.6000000	1.60000000	5.50000000

CORRELATION COEFFICIENTS / PROB > |R| UNDER H0:RHO = 0 / N = 25

	SALES	TIME	POTEN	ADV	SHARE	CHANGE	ACCTS	WORK	RATE
SALES	1.00000	0.62292	0.59781	0.59618	0.48351	0.48918	0.75399	-0.11722	0.40188
	0.0000	0.0009	0.0016	0.0017	0.0143	0.0131	0.0001	0.5768	0.0464
TIME	0.62292	1.00000	0.45397	0.24919	0.10621	0.25148	0.75782	-0.17932	0.10113
	0.0009	0.0000	0.0226	0.2297	0.6133	0.2253	0.0001	0.3911	0.6305
POTEN	0.59781	0.45397	1.00000	0.17410	-0.21067	0.26829	0.47864	-0.25884	0.35870
	0.0016	0.0226	0.0000	0.4052	0.3121	0.1947	0.0155	0.2115	0.0783
ADV	0.59618	0.24919	0.17410	1.00000	0.26446	0.37652	0.20004	-0.27223	0.41146
	0.0017	0.2297	0.4052	0.0000	0.2014	0.0636	0.3377	0.1880	0.0410
SHARE	0.48351	0.10621	-0.21067	0.26446	1.00000	0.08547	0.40301	0.34935	-0.02356
	0.0143	0.6133	0.3121	0.2014	0.0000	0.6846	0.0458	0.0869	0.9110
CHANGE	0.48918	0.25148	0.26829	0.37652	0.08547	1.00000	0.32742	-0.28766	0.54933
	0.0131	0.2253	0.1947	0.0636	0.6846	0.0000	0.1101	0.1632	0.0045
ACCTS	0.75399	0.75782	0.47864	0.20004	0.40301	0.32742	1.00000	-0.19885	0.22861
	0.0001	0.0001	0.0155	0.3377	0.0458	0.1101	0.0000	0.3406	0.2717
WORK	-0.11722	-0.17932	-0.25884	-0.27223	0.34935	-0.28766	-0.19885	1.00000	-0.27691
	0.5768	0.3911	0.2115	0.1880	0.0869	0.1632	0.3406	0.0000	0.1802
RATE	0.40188	0.10113	0.35870	0.41146	-0.02356	0.54933	0.22861	-0.27691	1.00000
	0.0464	0.6305	0.0783	0.0410	0.9110	0.0045	0.2717	0.1802	0.0000

How are the predictors themselves correlated? We see some large correlations,

$r_{16} = 0.7578$ between time with company and number of accounts (that seems reasonable),

$r_{58} = 0.5493$ between market-share change and rating,

and several other correlations around 0.3 and 0.4. We know that multicollinearity is present and is quite serious in some instances. For example, it would not be surprising to find that, if ACCTS is used as a predictor, TIME will not be able to add very much more information.

Turning to the overall test for significant predictors (Figure 15.3), we find

$$F = 23.65,$$

which is significant at $\alpha = 0.0001$, so we can be quite confident that at least one predictor is significant. All predictors together explain about 92 percent of the variation in sales, and this is a very high coefficient of determination

FIGURE 15.5 SPSS^X Analysis of Cravens Data

```
EQUATION NUMBER 1    DEPENDENT VARIABLE..SALES
DESCRIPTIVE STATISTICS ARE PRINTED ON PAGE 4
BEGINNING BLOCK NUMBER 1.  METHOD: ENTER
VARIABLE(S) ENTERED ON STEP NUMBER 1..    RATE
                                   2..    SHARE
                                   3..    TIME
                                   4..    WORK
                                   5..    ADV
                                   6..    POTEN
                                   7..    CHANGE
                                   8..    ACCTS

                                                  ANALYSIS OF VARIANCE
MULTIPLE R             0.96023                 DF   SUM OF SQUARES   MEAN SQUARE
R SQUARE               0.92204
ADJUSTED R SQUARE      0.88306   REGRESSION     8  38153564.25210   4769195.53151
STANDARD ERROR       449.02566   RESIDUAL      16   3225984.67475    201624.04217
                     F = 23.65390         SIGNIF F = 0.0000

                         VARIABLES IN THE EQUATION
VARIABLE             B             SE B           BETA           T        SIG T
RATE              8.18928      128.50561      5.982E-03      0.064     0.9500
SHARE           199.02354       67.02792      0.37283        2.969     0.0090
TIME              2.00957        1.93065      0.13284        1.041     0.3134
WORK             19.79389       33.67669      0.05532        0.588     0.5649
ADV               0.15099        0.04711      0.30922        3.205     0.0055
POTEN             0.03720      8.2023E-03     0.44526        4.536     0.0003
CHANGE          290.85513      186.78200      0.13757        1.557     0.1390
ACCTS             5.55096        4.77555      0.19268        1.162     0.2621
(CONSTANT)    -1507.81373      770.03404                    -1.936     0.0707

FOR BLOCK NUMBER  1    ALL REQUESTED VARIABLES ENTERED.
```

for real data. If all eight predictors were used, the regression equation would be

$$\hat{Y} = -1507.81 + 2.01X_1 + 0.04X_2 + 0.15X_3 + 199.02X_4 + 290.86X_5 + 5.55X_6 + 19.79X_7 + 8.19X_8,$$

and the standard error of estimate would be 449.03.

Before looking at the partial t- or F-values, let us examine the standard error of the regression coefficients (Figure 15.3). Recall that these measure how much we would expect a, b_1, b_2, . . . , b_8 to vary from sample to sample of size 25 from this same population. If the standard errors are small, this means that we have pretty well zeroed in on the values of α, β_1, β_2, . . . , β_8. However, large standard errors mean that we have not estimated the parameters very well. Perhaps the sample size was too small, but more likely multicollinearity is throwing in a confusion factor. For example, for X_3 = ADV, we have $b_3 = 0.15$ and its standard error is 0.05. Thus, in 68 percent of samples of size 25 drawn from this population, we'd expect to find a value of b_3 between

0.10 and 0.20. This is pretty good. But on the other hand, for X_7 (WORK), $b_7 = 19.79$ with a standard error of 33.68. Thus, approximately 68 percent of all samples of size 25 would yield b_7 between -13.89 and 53.47. One would not consider that we had estimated β_7 very well.

The regression coefficients with relatively small standard errors are b_2(POTEN) and b_3(ADV). Then b_1(TIME) and b_6(ACCTS) have reasonably small standard errors, while those for b_7(WORK) and b_4(SHARE) are relatively large. The standard errors for b_8(RATE) and b_5(CHANGE) are quite large, as is the standard error for a.

Naturally, the sizes of the standard errors of the regression coefficients affect the sizes of the partial t- and F-values. If the standard error is too large relative to the size of the coefficient, then the partial t- and F-statistic will be too small (in absolute value) to show a significant predictor. For the partial t- or F-values, if $\alpha = 0.05$ we see that only X_2(POTEN), X_3(ADV), and X_4(SHARE) turn out to be significant, given all the others. These results are in some ways not surprising and are in other ways quite surprising. These three variables all have reasonably high correlations with Y and reasonably small correlations with each other. However, our single strongest predictor, X_6(ACCTS), did not turn out to be significant—as a matter of fact, it had an extremely small partial t- or F-value. But X_6 was not extremely highly correlated with any of the three variables that ended up being significant. Recall, however, that the partial t- or F-tests answer the question, "Given *all seven* other variables, do I need this one?" Recall, for example, that X_6(ACCTS) is highly correlated with X_1(TIME): $r_{16} = 0.75782$. Thus, given time with company, number of accounts does not add much. But why are we interested in what X_6 can do, given X_1? How do we know that we want X_1 at all?

These partial t- and F-tests are just not that helpful to us if we want to determine which variables should be retained and which variables should be deleted from the model. Only if *all* partial t- and F-values were significant would these partial tests be very useful to us; in this case we would conclude that all predictors add information to what all others provide. This seldom happens, although it is not an impossibility. In most data sets, some variable selection must be performed. The stagewise procedures in the next sections show various ways this can be done.

EXERCISE

15.1 **a.** Use the computer to plot scatter diagrams of Y = sales versus each of the eight predictors in the Cravens Data.

b. If any of the scatter diagrams indicate a quadratic relationship, formulate an appropriate model containing quadratic terms for the Cravens Data.

c. Analyze this model using an appropriate multiple regression computer program.

15.2

SELECTING A SUBSET OF PREDICTORS

In many multiple regression problems the researcher, having identified a large set of possible predictor variables, wants to narrow this set down. This was the case in the two multiple regression examples we have looked at. In the house size-price-age example we started by using sizes and ages of houses to predict prices. Since that was such a small problem we were able to use the partial F-tests to conclude that size is a strong predictor of price, but age is not, especially when used in conjunction with size. Things were less clear-cut for the Cravens Data. There we had eight potential predictors, all correlated with each other to some extent. By using partial F-tests, we detected three variables that are strong predictors, given all the others. But the question is, do we want to use all the others? For example, if X_8 (RATE) is not needed, given the other seven predictors, then perhaps we should throw it out. But if so, the partial F-values we computed are no longer relevant: we have no partial F for, say, X_3 given X_1, X_2, X_4, X_5, X_6, and X_7. The partial F that we have is for X_3 given X_1, X_2, X_4, X_5, X_6, X_7, *and* X_8. We could perform another computer run without X_8 included, but this is not a very efficient method for exploring data. What is needed is a computer program that will look at several different regression models on its own.

Some computer packages contain programs that will perform all possible regressions. However, if there are very large numbers, k, of predictors, this can be a very expensive analysis, since there are 2^k different models. Even if expense is not a consideration, the researcher must then look at 2^k analyses, which is not only time-consuming but confusing. For example, we have eight candidate predictors in the Cravens Data, so there would be $2^8 = 256$ different models to be considered.

What is needed is some procedure by which the computer examines candidate predictors in a systematic fashion, bypasses ineffective predictors, and yields those that will make useful contributions to the model. Such procedures are referred to as **stagewise procedures,** or stepwise procedures, although the term *stepwise* also refers to one *particular* variable selection procedure (see Section 15.3).

Stagewise Procedures

There is no guarantee that any stagewise procedure will find the "best" model. The procedure taken may miss a variable or a combination of variables that should be included in the model. It is quite common for two different variable-selection procedures to produce different "final" models for the same set of data. In addition, even if one were to look at all 2^k possible models, it might be quite difficult to choose among several of them, either on statistical grounds or on the basis of practical or theoretical considerations.

We will use the Cravens Data to illustrate the numerous variable-selection procedures available through BMD, Minitab, SAS, and SPSS^X. The multicollinearity structure in these data makes comparison of these procedures very interesting. Some predictors are redundant with others; some are just ineffective, either alone or in combination with others; some that at first seemed worthless turn out to be very good in combination with others; and predictors that seemed good turn out not to be needed.

We will first discuss the logic underlying some of the most common procedures and then show how BMD, Minitab, SAS, and SPSS^X perform the procedures they offer. For simplicity, we will examine a full model including only linear terms in $X_1, \ldots, X_8$ in the Cravens Data, although selection procedures can also be used to see if quadratic or interaction terms should be retained.

15.3 SELECTION PROCEDURES

FORWARD SELECTION

One of the most straightforward variable-selection procedures is forward selection. The computer first finds the single strongest predictor and performs a simple regression analysis for this model. Let us denote this strongest predictor $X_{(1)}$. Then the next strongest predictor, given $X_{(1)}$, is found, and a two-predictor analysis using $X_{(1)}$ and $X_{(2)}$ is performed. The third strongest predictor $X_{(3)}$, given $X_{(1)}$ and $X_{(2)}$, is then found, and so on. The process terminates either when all variables have been added into the model or when the next strongest predictor is so weak that there is no reason to consider it.

All four computer packages discussed in this text can be made to perform a forward selection. They differ somewhat in the criteria they use to identify the next strongest predictor and in deciding when to stop adding variables to the model.

The problem with a forward selection is that it may terminate too soon and miss a "key set" of variables that might be quite effective. This will be illustrated with the Cravens Data.

BACKWARD ELIMINATION

To try to overcome the problem of missing a "key set" of variables, the backward elimination procedure was invented. This procedure starts with the full model and performs an analysis for it. Then the weakest predictor, given the others, is dropped, and the $(k - 1)$-predictor model is analyzed. In the next step the process continues, dropping the weakest predictor given the others. Finally, no more predictors are judged sufficiently weak to drop from the model, and the process terminates.

Different packages use different criteria for deciding which variables to drop and when to terminate the procedure. The backward elimination procedure is preferred by many statisticians because it starts with the full model, all predictors are given a chance to show how they can perform with other predictors, and no ''key set'' of variables is likely to be missed.

STEPWISE PROCEDURES

One problem with a forward selection procedure is that once a variable has been added to the model, you are stuck with it. The same goes for a backward elimination; once a variable has been dropped from the model, it can never be re-entered. This may or may not matter, depending on the set of data to be analyzed, but it does limit the number of models that would be considered in a selection procedure. Stepwise procedures allow variables entered in one step of a forward selection to be removed at a later step and be eligible for re-entry at any subsequent step. In a backward stepwise procedure any variable removed becomes eligible for re-entry at a later step.

The criteria for removing or entering variables differ from one program to another, but they are commonly based on the size or the significance of the partial F statistic. Suppose that in a forward stepwise procedure variables $X_{(1)}$ and $X_{(2)}$ are entered in the first two steps, as in a forward selection, and both partial F's are significant. Then when $X_{(3)}$ is added, its partial F and the F associated with $X_{(2)}$ are significant, but not the F associated with $X_{(1)}$. If this is the case, $X_{(1)}$ will be dropped, and an analysis using $X_{(2)}$ and $X_{(3)}$ will be shown. Then the computer will look for and add $X_{(4)}$, which has the largest partial F, given $X_{(2)}$ and $X_{(3)}$. In a forward stepwise procedure, $X_{(1)}$ becomes eligible for re-entry at a later step. The procedure continues until either no remaining partial F is large enough to meet the criterion for entry into the model or until the variable that would be entered is the one that was just removed.

A backward stepwise procedure works in a similar fashion. It begins just like a backward elimination procedure, but if at any step an eliminated variable shows a large enough partial F, given the variables in the model at that step, it will be added back in (unless it was the one just removed).

SWAPPING PROCEDURES

In a forward stepwise procedure, if at some step a variable is dropped from the model, then this results in a model with one predictor fewer than at the previous step. For example, if a four-predictor model is under consideration and one variable is dropped, then a three-predictor model is considered at the next step. Just the opposite happens in a backward stepwise procedure, of course.

In a swapping procedure a variable is dropped from the model at some step if it can be exchanged for another variable that makes the resulting model better in some sense. For example, suppose a three-predictor model involving $X_{(1)}$, $X_{(2)}$, and $X_{(3)}$ has been found. If the computer determines that the three predictors $X_{(1)}$, $X_{(3)}$, and $X_{(4)}$ give a "better" three-predictor model than the one involving $X_{(1)}$, $X_{(2)}$, and $X_{(3)}$, then it will replace $X_{(2)}$ by $X_{(4)}$. Both are still three-predictor models, unlike the result of a stepwise procedure.

CRITERIA FOR ENTERING OR REMOVING VARIABLES

Specific criteria for determining which variables will be added or deleted at any step of a stagewise procedure will be discussed in conjunction with the instructions for performing variable selection using BMD, Minitab, SAS, and SPSS^X. In general, there are typically three considerations, which can be used alone or in conjunction with each other:

1. The size of the partial F or its significance;
2. How close $(\mathbf{X}'\mathbf{X})^{-1}$ is to being singular;
3. The size of the coefficient of determination.

Many procedures enter and drop variables based on the size of the associated partial F-statistic or on its significance probability. The philosophy here is that the final model should include predictors whose regression coefficients are all significantly different from zero, given the other variables in the model. The significance level can be selected by either the computer or the researcher.

The standard errors of the estimated regression coefficients should not be too large if the regression equation from the final model is to be used to estimate the values of the regression coefficients. If there is severe multicollinearity among predictors, $\mathbf{X}'\mathbf{X}$ will be nearly singular and the diagonal elements of $(\mathbf{X}'\mathbf{X})^{-1}$ will be quite large, resulting in large standard errors. Thus, many computer programs will not enter a variable if its inclusion would lead to this problem. A measure of **tolerance** is typically calculated, where tolerance is defined as the squared multiple correlation of this variable with those already in the equation.

Tolerance

Some selection procedures just try to find models with large coefficients of determination and will include or swap variables in an effort to maximize R^2. Related to this are those procedures that enter variables according to which one has the largest partial correlation coefficient, given the predictors already in the model, with the dependent variable.

It is easy to see that the variety of variable-selection procedures, together with the variety of criteria for inclusion or deletion of variables, gives a very large combination of methods for searching through a set of candidate predictors to find a "good" subset. There are many other selection procedures that have been proposed, but we are concerned here only with those routinely available through BMD, Minitab, SAS, and SPSS^X. We now turn to the details of the stagewise procedures that are provided by each of these packages.

15.4 BMD SELECTION PROCEDURES

The 2R program, called "Stepwise Regression," is the variable-selection procedure provided in the BMDP series of programs. It contains several options, so that a large variety of selection techniques can be utilized. Forward entry and backward elimination, forward and backward stepwise, and some swapping procedures can all be applied using BMDP2R.

Basically, BMDP2R will perform a forward stepwise selection procedure. This is, variables will be entered one at a time according to their relative strengths. If at any step a variable that has been entered into the model becomes ineffective, given the others, it will be removed. Variations on this theme provide for other selection procedures. We begin with an examination of the default condition of the program. Then we will see how various options can be included to modify this basic procedure.

THE F METHOD

Figure 15.6 shows a worksheet for using the BMDP2R program to perform a forward stepwise selection—its default procedure. Among the system cards will be one containing the statement

```
...EXEC BIMED, PROG=BMDP2R, ...
```

which instructs the computer to run this program. The commands in the program should be familiar to those who have been using BMD programs all along. It is not necessary to specify `CASES=25` in the `INPUT` paragraph. The `PRINT` paragraph is also optional but provides important information relative to the data being analyzed. The sentence `FRAT` (for *F*-ratio) causes a summary table of *F*-values to be printed (we will discuss this table later); `PART` asks for a summary table of partial correlation coefficients (also discussed later). The printout of the data and of the correlation matrix will not be shown here. Other sentences that could have been included in the `PRINT` paragraph

FIGURE 15.6 Worksheet for Analyzing Cravens Data Using BMDP2R F Method

```
 {SYSTEM  CARDS
PROBLEM   TITLE='BMDP2R ANALYSIS OF CRAVENS DATA'./
INPUT     VARIABLES=9.
          FORMAT=FREE.
          CASES=25./
VARIABLES NAMES=SALES,TIME,POTEN,ADV,SHARE,CHANGE,ACCTS,WORK,RATE./
REGRES    DEPENDENT=SALES./
PRINT     DATA.
          CORREL.
          FRAT.
          PART./
PLOT      RESIDUALS./
END/
3669.88  43.1Ø 74Ø65.11  4582.88  2.51  Ø.34  74.86 15.Ø5 4.9
3473.95 1Ø8.13 58117.3Ø  5539.78  5.51  Ø.15 1Ø7.32 19.97 5.1
     :}DATA CARDS
2799.97  21.14 228Ø9.53  3552.ØØ  9.14 -Ø.74  88.62 24.96 3.9
FINISH/
```

but that are not shown in Figure 15.6 are COVAR, for a matrix of covariances among the variables, and RREG, for a matrix of correlations among the regression coefficients. Additionally, the PLOT paragraph in Figure 15.6 is optional. There, a plot of residuals is requested, but other plots are also available. The reader is referred to the BMD manual for the options for the PLOT paragraph, because they are too numerous to consider in detail here.

As standard output from the REGRES paragraph will be shown a list of options requested and a table of univariate descriptive statistics for each variable, but we omit this information here. Of interest to us is the printout shown in Figure 15.7(a). The list at the beginning of Figure 15.7(a) describes the selection procedure. It is instructive to consider the algorithm described, not only to understand what is going on in this procedure but also to see how other selection procedures can be performed by modifying these conditions.

STEPPING ALGORITHM...F specifies that variables will be entered into or removed from the model according to the size of the associated partial *F*-statistics. This is the default algorithm of the BMDP2R program. (We shall see presently what other algorithms are available, and how to specify that they are to be used.)

MAXIMUM NUMBER OF STEPS...18 is a time- and money-saving condition. The default condition on the maximum number of steps is twice the number of variables. Seldom will that many steps be required before a selection procedure terminates because of other conditions, but should more steps be required, a sentence can be added in the REGRES paragraph. For example, to change the maximum to 25, insert a sentence STEP = 25.

The definition of the dependent variable (SALES = variable number 1) and the subscripts of the independent variables (2, 3, . . . , 9) should be self-explanatory.

MINIMUM ACCEPTABLE F TO ENTER...4.000, 4.000 determines that unless a variable has a partial *F*-value of at least 4, it will not be added to the model.

FIGURE 15.7 (a) Step 0 of BMDP2R Analysis of Cravens Data Using **F** Method and Default Conditions F TO ENTER and F TO REMOVE

```
BMDP F METHOD

STEPPING ALGORITHM. . . . . . . . . . . . . . . . .   F
MAXIMUM NUMBER OF STEPS . . . . . . . . . . . . . .   3
DEPENDENT VARIABLE. . . . . . . . . . . . . . . . .     SALES
MINIMUM ACCEPTABLE F TO ENTER . . . . . . . . . . .   4.000,    4.000
MAXIMUM ACCEPTABLE F TO REMOVE. . . . . . . . . . .   3.900,    3.900
MINIMUM ACCEPTABLE TOLERANCE. . . . . . . . . . . .   0.01000
SUBSCRIPTS OF THE INDEPENDENT VARIABLES . . . . . .   2   3   4   5   6   7   8   9

STEP NO. 0

STD ERROR OF EST      1313.0637

ANALYSIS OF VARIANCE
                     SUM OF SQUARES     DF    MEAN SQUARE
         RESIDUAL       41379264.        24    1724136.
```

VARIABLES IN EQUATION							VARIABLES NOT IN EQUATION					
VARIABLE	COEFFICIENT	STD ERROR OF COEFF	STD REG COEFF	TOLERANCE	F TO REMOVE	LEVEL	VARIABLE		PARTIAL CORR	TOLERANCE	F TO ENTER	LEVEL
Y-INTERCEPT	3374.56445						TIME	2	0.62292	1.00000	14.58	1
							POTEN	3	0.59781	1.00000	12.79	1
							ADV	4	0.59618	1.00000	12.68	1
							SHARE	5	0.48351	1.00000	7.02	1
							CHANGE	6	0.48918	1.00000	7.24	1
							ACCTS	7	0.75399	1.00000	30.30	1
							WORK	8	-0.11723	1.00000	0.32	1
							RATE	9	0.40188	1.00000	4.43	1

MAXIMUM ACCEPTABLE F TO REMOVE...3.900, 3.900 says that if at any step a variable already in the equation has a partial F-value of 3.9 or below, it will be dropped from the equation in the next step. F TO ENTER = 4.0 and F TO REMOVE of 3.9 are the default conditions of this program. To change them to, say, 2.0 and 1.0, respectively, add the following sentences in the REGRES paragraph:

```
ENTER = 2.0
REMOVE = 1.0
```

We will explain shortly why you might want to change these conditions and how to decide what new criteria to choose. Note that F TO ENTER must be greater than F TO REMOVE. Otherwise, a stepwise procedure might go into an endless loop of entering a variable at one step, removing it at the next, adding it again at the next, and so on.

Finally, MINIMUM ACCEPTABLE TOLERANCE...0.01000 is a protection against the effects of multicollinearity. If the multicollinearity among predictors is too severe, $\mathbf{X}'\mathbf{X}$ will be nearly singular and the estimates of the regression coefficients will be quite unreliable. Tolerance is the proportion of variation in the independent variable not explained by the variables already in the model. A tolerance of zero means that a predictor under consideration for inclusion is a perfect linear combination of variables already in the model, and a tolerance of 1 means that this predictor is independent of the others already entered. Unless at least 1 percent of the variation in the response variable remains unexplained by predictors already included, the predictor under consideration will not be allowed to enter the model. This default condition on tolerance allows quite a bit of redundancy among predictors. However, one can change the default condition by adding a sentence such as TOL=.05 to the REGRES paragraph. One might want to require higher tolerance than that specified by the default condition, but it is not usually advisable to specify tolerance less than 0.01.

What does all this mean? Let us examine Figure 15.7(a) to see the BMD F Method applied to the Cravens Data. In step 0 no predictors have been entered into the model; that is, step 0 shows an analysis of just the dependent variable, SALES. In the abbreviated analysis of variance table we find $S_{YY} = 41379264$; $df = n - 1 = 24$, and $s_Y^2 = 1724136$. The standard error of estimate in this step is just $s_Y = [\text{MS(E)}]^{1/2} = 1313.0637$. Under VARIABLES IN EQUATION the coefficient given is $\bar{Y} = 3374.56445$.

Things get more interesting when we look under VARIABLES NOT IN EQUATION. Of course, all predictors are listed here. The variable names and numbers are listed. The partial correlations in this case are just the simple correlations of the various predictors with Y. (See the correlation matrix in Figure 15.4.) All tolerances are 1, since no predictors have been entered. In the last column all the listings under LEVEL are 1, to indicate that all eight potential predictors are allowed to compete on an equal footing (we are not forcing any variable into the equation). Under F TO ENTER we see what the F-values in the simple regression analysis would be for each variable, were it allowed to predict Y. Since the largest F to enter is 30.30, corresponding to ACCTS, and since the stepping algorithm is to enter the variable with the largest partial $F \geq 4$ having tolerance at least 0.01, a simple regression using ACCTS to predict SALES will be performed in step 1. (Note that ACCTS also has the largest correlation with SALES.)

So, in step 1 in Figure 15.7(b) we see the simple regression analysis using ACCTS as a predictor. The overall regression significance test is given in the analysis of variance table, and under VARIABLES IN EQUATION we see the regression constant, regression coefficient, standard error of the regression coefficient, standardized regression (beta) coefficient, and partial F-statistic. (Only in step 1 the partial F is just F, since there is only one predictor in the equation.) This partial F is labeled F TO REMOVE. If it were not larger than 3.9, then the predictor ACCTS would be removed from the equation.

Under VARIABLES NOT IN EQUATION are listed the names and numbers of the remaining candidates for entry into the model. Now the partial correlations computed are conditional on ACCTS being used as a predictor. For example, the partial correlation for ADV is 0.69196. This is $r_{4Y\cdot 7}$, where SALES $= Y = X_1$, ADV $= X_2$, and ACCTS $= X_7$. This first-order partial correlation can be calculated from quantities given in the correlation matrix in Figure 15.4:

$$r_{4Y\cdot 7} = 0.69196 = \frac{r_{4Y} - r_{47}r_{7Y}}{\sqrt{(1 - r_{47}^2)(1 - r_{7Y}^2)}}$$

$$= \frac{0.596 - 0.2(0.754)}{\sqrt{[1 - (0.2)^2][1 - (0.754)^2]}} = \frac{0.4452}{0.6436} = 0.6917 \approx 0.69196.$$

The F-levels to enter in step 1 are partial F's, conditional on ACCTS being used as a predictor. Since the largest partial F is that for ADV, ADV enters as a predictor in step 2.

In step 2 we again see the overall test and individual tests on the predictors. Since both F's to remove are greater than 3.9, both ADV and ACCTS will stay in the equation. Under VARIABLES NOT IN EQUATION, POTEN has the largest partial F, given ACCTS and ADV, and this partial F is greater than 4, the minimum F for entry. Thus, POTEN will enter in step 3. The partial correlations given in step 2 are second-order partials. For example, the partial correlation for POTEN (variable 3) is

$$r_{3Y\cdot 47} = 0.48372 = \frac{r_{3Y\cdot 7} - r_{4Y\cdot 7}r_{34\cdot 7}}{\sqrt{(1 - r_{4Y\cdot 7}^2)(1 - r_{34\cdot 7}^2)}}.$$

However, since we are not given a complete list of first-order partial correlations, we cannot compute these from values on the printout.

The procedure continues until we reach step 4, shown in Figure 15.7(c). We see there that four variables have entered, all of which had F to enter values of at least 4 and tolerances of at least 0.01. At no point did any F to remove for any of these four variables fall below 3.9, so no variables were removed. Under VARIABLES NOT IN EQUATION, although all

FIGURE 15.7(b) Steps 1 and 2 of BMDP2R Analysis of Cravens Data Using F Method and Default Conditions on F TO ENTER and F TO REMOVE

```
STEP NO. 1
VARIABLE ENTERED           7  ACCTS

MULTIPLE R                  0.7540
MULTIPLE R-SQUARE           0.5685
ADJUSTED R-SQUARE           0.5497
STD ERROR OF EST          881.0894

ANALYSIS OF VARIANCE
                        SUM OF SQUARES    DF    MEAN SQUARE             F RATIO
REGRESSION              23523920.          1         0.2352392E 08      30.30
RESIDUAL                17855328.         23    776318.6

                          VARIABLES IN EQUATION                                                   VARIABLES NOT IN EQUATION
                           STD ERROR   STD REG                  F TO                                 PARTIAL               F TO
VARIABLE      COEFFICIENT  OF COEFF    COEFF     TOLERANCE    REMOVE   LEVEL    VARIABLE          CORR       TOLERANCE   ENTER   LEVEL
Y-INTERCEPT    709.32837
ACCTS  7        21.72173    3.9460     0.754     1.00000      30.30    1        TIME      2      0.12024    0.42572      0.32   1
                                                                                 POTEN     3      0.41078    0.77090      4.47   1
                                                                                 ADV       4      0.69196    0.95998     20.21   1
                                                                                 SHARE     5      0.29882    0.83758      2.16   1
                                                                                 CHANGE    6      0.39040    0.89280      3.96   1
                                                                                 WORK      8      0.05080    0.96046      0.06   1
                                                                                 RATE      9      0.35889    0.94774      3.25   1
```

```
STEP NO. 2
VARIABLE ENTERED           4  ADV

MULTIPLE R                  0.8804
MULTIPLE R-SQUARE           0.7751
ADJUSTED R-SQUARE           0.7547
STD ERROR OF EST          650.3904

ANALYSIS OF VARIANCE
                        SUM OF SQUARES    DF    MEAN SQUARE             F RATIO
REGRESSION              32073088.          2         0.1603654E 08      37.91
RESIDUAL                 9306166.         22    423007.5

                          VARIABLES IN EQUATION                                                   VARIABLES NOT IN EQUATION
                           STD ERROR   STD REG                  F TO                                 PARTIAL               F TO
VARIABLE      COEFFICIENT  OF COEFF    COEFF     TOLERANCE    REMOVE   LEVEL    VARIABLE          CORR       TOLERANCE   ENTER   LEVEL
Y-INTERCEPT     50.31055
ADV    4         0.22653    0.0504     0.464     0.95999      20.21    1        TIME      2      0.02047    0.41579      0.01   1
ACCTS  7        19.04820    2.9729     0.661     0.95999      41.05    1        POTEN     3      0.48372    0.76451      6.41   1
                                                                                 SHARE     5      0.22213    0.80238      1.09   1
                                                                                 CHANGE    6      0.23226    0.79203      1.20   1
                                                                                 WORK      8      0.31166    0.90417      2.26   1
                                                                                 RATE      9      0.14034    0.80840      0.42   1
```

FIGURE 15.7 (c) Steps 3 and 4 of BMDP2R Analysis of Cravens Data Using F Method and Default Conditions on F TO ENTER and F TO REMOVE

STEP NO. 3
VARIABLE ENTERED 3 POTEN

MULTIPLE R	0.9098
MULTIPLE R-SQUARE	0.8277
ADJUSTED R-SQUARE	0.8031
STD ERROR OF EST	582.6335

ANALYSIS OF VARIANCE

	SUM OF SQUARES	DF	MEAN SQUARE	F RATIO
REGRESSION	34250544.	3	0.1141685E 08	33.63
RESIDUAL	7128700.	21	339461.9	

VARIABLES IN EQUATION

VARIABLE	COEFFICIENT	STD ERROR OF COEFF	STD REG COEFF	TOLERANCE	F TO REMOVE	LEVEL
Y-INTERCEPT	-327.22217					
POTEN 3	0.02192	0.0087	0.262	0.76451	6.41	1
ADV 4	0.21607	0.0453	0.443	0.95212	22.72	1
ACCTS 7	15.55388	2.9994	0.540	0.75638	26.89	1

VARIABLES NOT IN EQUATION

VARIABLE	PARTIAL CORR	TOLERANCE	F TO ENTER	LEVEL
TIME 2	-0.05889	0.40672	0.07	1
SHARE 5	0.64973	0.57320	14.61	1
CHANGE 6	0.20542	0.78232	0.88	1
WORK 8	0.45929	0.87679	5.35	1
RATE 9	0.00633	0.74542	0.00	1

STEP NO. 4
VARIABLE ENTERED 5 SHARE

MULTIPLE R	0.9489
MULTIPLE R-SQUARE	0.9005
ADJUSTED R-SQUARE	0.8805
STD ERROR OF EST	453.8337

ANALYSIS OF VARIANCE

	SUM OF SQUARES	DF	MEAN SQUARE	F RATIO
REGRESSION	37259952.	4	9314988.	45.23
RESIDUAL	4119300.	20	205965.	

VARIABLES IN EQUATION

VARIABLE	COEFFICIENT	STD ERROR OF COEFF	STD REG COEFF	TOLERANCE	F TO REMOVE	LEVEL
Y-INTERCEPT	-1441.91602					
POTEN 3	0.03822	0.0080	0.457	0.54615	22.95	1
ADV 4	0.17499	0.0369	0.358	0.87129	22.48	1
SHARE 5	190.14406	49.7439	0.356	0.57320	14.61	1
ACCTS 7	9.21387	2.8652	0.320	0.50323	10.34	1

VARIABLES NOT IN EQUATION

VARIABLE	PARTIAL CORR	TOLERANCE	F TO ENTER	LEVEL
TIME 2	0.24355	0.35494	1.20	1
CHANGE 6	0.34670	0.77627	2.60	1
WORK 8	0.15438	0.60489	0.46	1
RATE 9	0.09459	0.73791	0.17	1

* * * * * F-LEVELS(4.000, 3.900) OR TOLERANCE INSUFFICIENT FOR FURTHER STEPPING

tolerances were greater than 0.01, none of the remaining four variables had partial F's to enter of 4 or more, so no more predictors were added to the model and the stepping process terminated.

Note that no significance probabilities are calculated by the computer in conjunction with the partial F-values. This is because it is only at the step in which a reasonable model has been found that the calculated F-values are true values of the F-statistic, and thus only at that step does it make sense to calculate their significance probabilities. In any step in which important predictors are omitted, the error mean square will be inflated and thus the F-values will be smaller than they should be to measure the effect of that predictor. In a model containing redundant predictors diagonal elements of $(\mathbf{X}'\mathbf{X})^{-1}$ are inflated, also possibly causing F-values to be reduced. Significance probabilities should be calculated only when a model is found that is reasonable theoretically, has a relatively large multiple R^2, and has small standard errors of the regression coefficients. If all partial F's in that model are conditionally significant, then it is probably a pretty good final model. BMD uses F to enter and F to remove values of approximately 4, however, because (looking at the F-tables in Appendix I), a partial F with 1 degree of freedom in the numerator needs to be around 4 to be significant at the 0.05 level. In the final model produced by the F method, each partial F has 1 and 20 degrees of freedom. Critical values of F at the 0.05 and 0.01 significance levels are 4.35 and 8.10, respectively. Thus, all four variables entered at step 4 are effective predictors with more than 99 percent certainty.

Finally, two summary tables are provided in Figure 15.7(d). The first shows what the regression coefficients for the predictors would be if that predictor had been added at the next step. This can be confusing, so consider step 0. The table says that if, for example, ACCTS had entered in step 1 (which it did), then the regression coefficient for ACCTS would be 21.7217. Looking back to step 1, we see that the regression coefficient for ACCTS was indeed 21.722. On the other hand, if you had wanted to use a simple regression predicting SALES from WORK, then the regression coefficient on WORK would have been -41.9408. Under step 1 we see that the intercept and ACCTS were used (indicated by asterisks). If you wanted to add WORK as a second predictor, then its partial (net) regression coefficient, given ACCTS, would be 12.1821. This does not say, however, that the best regression equation using ACCTS and WORK would be $\hat{Y} = 709.3284 + 21.7217X_7 + 12.1821X_8$, because we know that adding X_8(WORK) to the equation will change the intercept and the coefficient on X_7(ACCTS). We can, however, use this table to pick out the intercept and net regression coefficients for the equation appropriate at each step by looking for the values indicated by asterisks.

The second summary table shows which variables were entered and removed at each step, what the R^2- and F-values were, and how many predictors were used.

These two summary tables are standard pieces of printout from the REGRES paragraph, but the summary tables in Figure 15.8 result from the sentences FRAT and PART in the PRINT paragraph. In the first table of

FIGURE 15.7 (d) Summary Tables of BMDP2R Analysis of Cravens Data Using F Method and Default Conditions on F TO ENTER and F TO REMOVE

STEPWISE REGRESSION COEFFICIENTS

VARIABLES	0 Y-INTERCEPT	2 TIME	3 POTEN	4 ADV	5 SHARE	6 CHANGE	7 ACCTS	8 WORK	9 RATE
STEP 0	3374.5645*	9.4235	0.0500	0.2911	258.1038	1034.2598	21.7217	-41.9408	550.1375
1	709.3284*	1.8314	0.0257	0.2265	114.4932	573.8323	21.7217*	12.1821	331.5022
2	50.3105*	0.2278	0.0219	0.2265*	62.7760	261.6736	19.0482*	55.6116	101.3282
3	-327.2222*	-0.5798	0.0219*	0.2161*	190.1441	203.8129	15.5539*	72.8388	4.1688
4	1441.9160*	1.9512	0.0382*	0.1750*	190.1441*	262.4990	9.2139*	22.4066	47.5583

NOTE-
1) REGRESSION COEFFICIENTS FOR VARIABLES IN THE EQUATION ARE INDICATED BY AN ASTERISK.
2) THE REMAINING COEFFICIENTS ARE THOSE WHICH WOULD BE OBTAINED IF THAT VARIABLE WERE TO ENTER IN THE NEXT STEP.

SUMMARY TABLE

STEP NO.	VARIABLE ENTERED	VARIABLE REMOVED	MULTIPLE R	MULTIPLE RSQ	INCREASE IN RSQ	F TO ENTER	F TO REMOVE	NO. OF INDEPENDENT VARIABLES INCLUDED
1	7 ACCTS		0.7540	0.5685	0.5685	30.3018		1
2	4 ADV		0.8804	0.7751	0.2066	20.2104		2
3	3 POTEN		0.9098	0.8277	0.0526	6.4145		3
4	5 SHARE		0.9489	0.9005	0.0727	14.6112		4

FIGURE 15.8 Summary Tables From BMDP2R PRINT Paragraph

F TO ENTER OR F TO REMOVE OF EACH VARIABLE AT EACH STEP

VARIABLES	2 TIME	3 POTEN	4 ADV	5 SHARE	6 CHANGE	7 ACCTS	8 WORK	9 RATE
STEP 0	14.5834	12.7910	12.6826	7.0176	7.2352	30.3019	0.3205	4.4301
1	0.3227	4.4660	20.2104	2.1571	3.9560	30.3018*	0.0569	3.2526
2	0.0088	6.4145	20.2104*	1.0899	1.1974	41.0530*	2.2593	0.4219
3	0.0696	6.4145*	22.7232*	14.6112	0.8812	26.8918*	5.3468	0.0008
4	1.1980	22.9539*	22.4811*	14.6112*	2.5958	10.3413*	0.4639	0.1715

PARTIAL CORRELATIONS

VARIABLES	2 TIME	3 POTEN	4 ADV	5 SHARE	6 CHANGE	7 ACCTS	8 WORK	9 RATE
STEP 0	0.6229	0.5978	0.5962	0.4835	0.4892	0.7540	-0.1172	0.4019
1	0.1202	0.4108	0.6920	0.2988	0.3904	0.7540*	0.0508	0.3589
2	0.0205	0.4837	0.6920*	0.2221	0.2323	0.8069*	0.3117	0.1403
3	-0.0589	0.4837*	0.7209*	0.6497	0.2054	0.7493*	0.4593	0.0063
4	0.2435	0.7310*	0.7275*	0.6497*	0.3467	0.5838*	0.1544	0.0946

Figure 15.8, *F*-values from steps 0–4 are summarized. Those without asterisks are variables not in the equation at that step and are thus values of *F* to enter. Those with asterisks are *F* to remove values, being associated with variables included in the model at that step. The information in the second table in Figure 15.8, for the partial correlation coefficients, is organized in a similar fashion. Partial correlations for variables included in the model at that step are indicated by asterisks.

Figure 15.9 shows a portion of the analysis of residuals for this problem. A printout of the data, predicted values, and residuals takes two pages, because we entered so many variables, so it is not reproduced here. But we see plots of the predicted values, $\hat{Y}$, from the variables in the final step of the analysis, versus (1) the residuals, and (2) the squares of the residuals (not shown). These plots are still useful in detecting outliers or patterns to the residuals and thus determining whether or not the assumptions of the model have been met. Other plots of residuals might also be helpful in evaluating the appropriateness of this model.

FORWARD SELECTION

By changing the default condition on the maximum *F*-level to remove, we can make the BMDP2R perform a forward selection procedure in which variables are entered one at a time, according to which one gives the largest partial *F*-value. The process will terminate when no variable not yet in the equation has a partial *F* greater than or equal to the *F*-level for entry, or when no variable not in the equation passes the tolerance test. There is no provision for removing a variable should its partial *F* become small when new predictors are added.

To prevent BMDP2R from removing any variable once it has entered into the equation, all we need to do is add the sentence `REMOVE=0` to the `REGRES` paragraph in Figure 15.6. Since *F* cannot be less than zero, no variable can be removed once it has been allowed to enter.

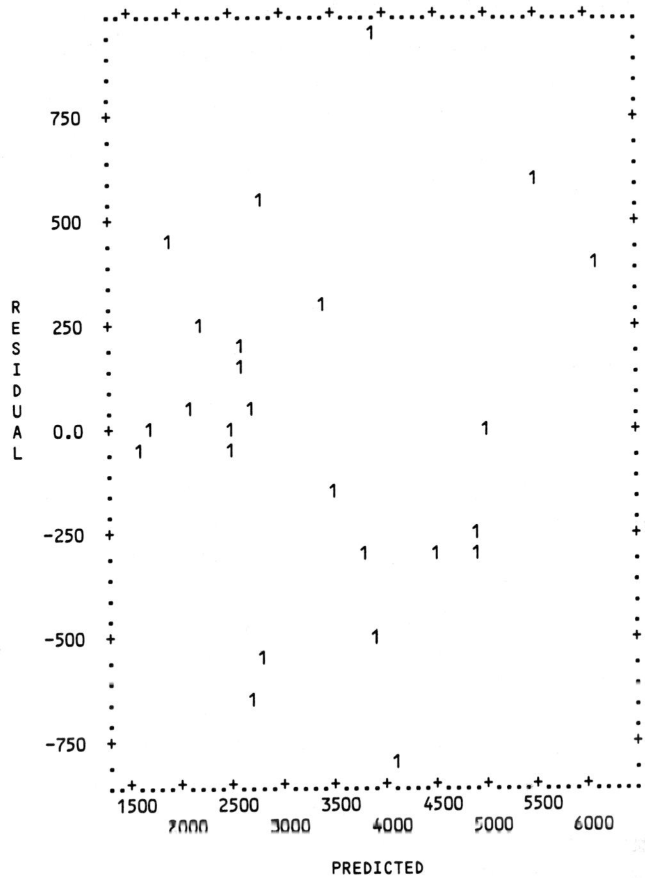

FIGURE 15.9 Partial Analysis of Residuals From BMDP2R

If the Cravens Data were reanalyzed using BMDP2R with the default condition (4.0) on the F to enter and maximum F to remove set at zero, the analysis would look just like that in Figure 15.7, since in Figure 15.7 no variable with F to enter of 4.0 or more subsequently had a partial F less than 3.9. Had the F to enter value been lowered to 2 or less, more variables would have entered, but none would have been removed.

BACKWARD (STEPWISE) ELIMINATION

It is possible to perform a backward elimination procedure with the BMDP2R, but it is rather laborious, at least for the computer. We must first step in every variable. This is accomplished by specifying a very low F to enter and an F

to remove of zero. Once all predictors have entered, the second set of F to enter and F to remove values come into play. If these are set at levels such as 4.0 and 4.0, then the procedure will backward step, removing variables one at a time until no remaining variable has an F to remove less than the second specified value.

Figure 15.10 shows the BMDP2R backward elimination procedure beginning with step 8, in which eight predictors have been entered, because the first F to enter value was set at 0.001 and the first F to remove value specified was zero:

```
ENTER=0.001, 4
REMOVE=0, 4
```

Then in step 9, RATE was removed because its partial F was less than 4, which was the second F to remove specified. In subsequent steps WORK and ACCTS were removed for the same reason. Finally, in step 11 all partial F's are greater than 4, so the process terminates.

Note that whereas in this example three variables were removed from the final model, one at a time, and no variable once removed ever re-entered the equation, this BMD backward selection is actually a stepwise backward procedure. That is, if any variable not in the equation once the program began stepping backward had a partial F greater than 4 (the second F to enter value), then it could be re-entered.

When running a backward stepwise procedure using this program, you may want to change the default condition on the maximum number of steps allowed. Unless otherwise specified, the program will stop after $2k$ steps, where k is the number of variables. This might not be sufficient to fully analyze a problem using the backward procedure, since all variables must first be stepped in before the backward elimination procedure can begin to work. Our example specified 25 as the maximum number of steps instead of 18. This was accomplished by adding a sentence STEPS=25 in the REGRES paragraph.

OTHER SELECTION PROCEDURES AVAILABLE WITH BMDP2R

We have seen that by adjusting the F-values to enter and remove, the standard BMD stepwise selection program can be made to perform both forward selection and backward elimination. There are, however, other variable-selection criteria that can be employed with the BMDP2R.

The default condition on the method of selection is called the *F method* in the BMD manual. It enters the variable with the largest F to enter greater than the specified minimum acceptable. Then, should any F to remove values

FIGURE 15.10 BMDP2R Backward Elimination Procedure

STEP NO. 8
VARIABLE ENTERED 9 RATE

MULTIPLE R	0.9602
MULTIPLE R-SQUARE	0.9220
STD ERROR OF EST	449.0247

ANALYSIS OF VARIANCE

	SUM OF SQUARES	DF	MEAN SQUARE	F RATIO
REGRESSION	38153280.	8	4769160.	23.654
RESIDUAL	3225972.	16	201623.3	

VARIABLES IN EQUATION							VARIABLES NOT IN EQUATION				
VARIABLE		COEFFICIENT	STD ERROR OF COEFF	STD REG COEFF	F TO REMOVE	LEVEL	VARIABLE	PARTIAL CORR	TOLERANCE	F TO ENTER	LEVEL
Y-INTERCEPT		-1507.787									
TIME	2	2.010	1.931	0.133	1.083	1					
POTEN	3	0.037	0.008	0.445	20.575	1					
ADV	4	0.151	0.047	0.309	10.273	1					
SHARE	5	199.024	67.027	0.373	8.817	1					
CHANGE	6	290.851	186.782	0.138	2.425	1					
ACCTS	7	5.551	4.776	0.193	1.351	1					
WORK	8	19.793	33.676	0.055	0.345	1					
RATE	9	8.191	128.506	0.006	0.004	1					

STEP NO. 9
VARIABLE REMOVED 9 RATE

MULTIPLE R	0.9602
MULTIPLE R-SQUARE	0.9220
STD ERROR OF EST	435.6733

ANALYSIS OF VARIANCE

	SUM OF SQUARES	DF	MEAN SQUARE	F RATIO
REGRESSION	38152464.	7	5450352.	28.715
RESIDUAL	3226791.	17	189811.2	

VARIABLES IN EQUATION							VARIABLES NOT IN EQUATION				
VARIABLE		COEFFICIENT	STD ERROR OF COEFF	STD REG COEFF	F TO REMOVE	LEVEL	VARIABLE	PARTIAL CORR	TOLERANCE	F TO ENTER	LEVEL
Y-INTERCEPT		-1485.846									
TIME	2	1.974	1.796	0.131	1.203	1	RATE 9	0.01593	0.55293	0.004	1
POTEN	3	0.037	0.008	0.446	22.560	1					
ADV	4	0.152	0.043	0.31[illegible]	12.343	1					
SHARE	5	198.309	64.117	0.37[illegible]	9.565	1					
CHANGE	6	295.863	164.386	0.140	3.233	1					
ACCTS	7	5.610	4.545	0.195	1.524	1					
WORK	8	19.898	32.636	0.056	0.372	1					

Figure 15.10 (Continued)

```
STEP NO. 10
VARIABLE REMOVED          8  WORK

MULTIPLE R                  0.9593
MULTIPLE R-SQUARE           0.9203
STD ERROR OF EST          428.0024

ANALYSIS OF VARIANCE
```

	SUM OF SQUARES	DF	MEAN SQUARE	F RATIO
REGRESSION	38081904.	6	6346984.	34.648
RESIDUAL	3297348.	18	183186.	

VARIABLES IN EQUATION

VARIABLE		COEFFICIENT	STD ERROR OF COEFF	STD REG COEFF	F TO REMOVE	LEVEL
Y-INTERCEPT		-1165.467				
TIME	2	2.269	1.699	0.150	1.784	1
POTEN	3	0.038	0.008	0.458	25.725	1
ADV	4	0.141	0.038	0.288	13.425	1
SHARE	5	221.603	50.583	0.415	19.193	1
CHANGE	6	285.106	160.559	0.135	3.153	1
ACCTS	7	4.378	3.999	0.152	1.198	1

VARIABLES NOT IN EQUATION

VARIABLE		PARTIAL CORR	TOLERANCE	F TO ENTER	LEVEL
WORK	8	0.14628	0.55128	0.372	1
RATE	9	0.02291	0.55426	0.009	1

```
STEP NO. 11
VARIABLE REMOVED          7  ACCTS

MULTIPLE R                  0.9566
MULTIPLE R-SQUARE           0.9150
STD ERROR OF EST          430.2314

ANALYSIS OF VARIANCE
```

	SUM OF SQUARES	DF	MEAN SQUARE	F RATIO
REGRESSION	37862368.	5	7572473.	40.910
RESIDUAL	3516882.	19	185099.	

VARIABLES IN EQUATION

VARIABLE		COEFFICIENT	STD ERROR OF COEFF	STD REG COEFF	F TO REMOVE	LEVEL
Y-INTERCEPT		-1113.771				
TIME	2	3.612	1.182	0.239	9.343	1
POTEN	3	0.042	0.007	0.504	39.096	1
ADV	4	0.129	0.037	0.264	12.105	1
SHARE	5	256.955	39.136	0.481	43.108	1
CHANGE	6	324.531	157.283	0.153	4.257	1

VARIABLES NOT IN EQUATION

VARIABLE		PARTIAL CORR	TOLERANCE	F TO ENTER	LEVEL
ACCTS	7	0.24985	0.22975	1.198	1
WORK	8	0.01574	0.68721	0.004	1
RATE	9	0.07004	0.57575	0.089	1

***F-LEVELS (4, 4), OR TOLERANCE INSUFFICIENT FOR FURTHER STEPPING fall below the specified maximum F to remove, the variable having the smallest F is removed. If, for example, five variables were entered at step 5 and one of them had an F to remove less than the maximum specified, then it would be removed and there would be only four variables in the equation at step 6.

FSWAP METHOD A variation on this theme can be achieved by specifying METHOD=FSWAP in the REGRES paragraph. The selection procedure will basically be the same as the F method, except that at any step the program looks to see if any variable in the equation could be *exchanged* with any not in the equation to produce a larger multiple R^2. If such an exchange can be made, it is done. Figure 15.11 shows the summary table from the Cravens Data analyzed by the FSWAP method, with default conditions on the F-values to enter and remove. Compare it with the results shown in Figure 15.7. The first three steps in both analyses are identical. However, in the FSWAP procedure, instead of adding a fourth variable, SHARE, in step 4, the program exchanges ACCTS for SHARE. Step 4 then uses three variables, POTEN, ADV, and SHARE, for which $R^2 = 0.8490$. Note that R^2 for these three variables is larger than for the three-predictor model using POTEN, ADV, and ACCTS; $R^2 = 0.8277$. The procedure finds a four-predictor equation in step 5 and then terminates, because no more variables can be added or removed. A summary table is provided for the purpose of comparison with the F method. While the same four-predictor model was obtained by both the F and FSWAP methods for the Cravens Data, one can appreciate the fact that this will not always happen. The three-predictor model using ADV, POTEN, and SHARE was considered by the FSWAP method but not by the F method.

R METHOD A third procedure available through BMDP2R, called the R method, can be used by simply specifying METHOD=R in the REGRES paragraph. Here, variables are entered in the usual way, according to which has the largest F to enter greater than the specified minimum. The removal criterion, however, is a bit different. To quote from the BMD manual (p. 259), "If two or more variables are in the regression equation, the one with the smallest F value is removed if its removal results in a larger multiple R than was obtained for the same number of variables previously." Let's see what happens when the Cravens Data are analyzed by this procedure.

Figure 15.12 starts out looking like the regular stepwise procedure through step 4. Then on step 5, ACCTS is removed, leaving the three-predictor model using ADV, POTEN, and SHARE. Note that for this model $R^2 = 0.8490$, but for the previously found three-predictor model (the one using ACCTS, ADV, and POTEN) the multiple coefficient of determination was 0.8277.

This procedure is similar to the FSWAP, except that no variable will be removed only because its partial F becomes small. It will be removed if its partial F is the smallest of all in the equation *and* if its removal results in a multiple R^2 larger than the one previously found for the same number of predictors.

FIGURE 15.11 Summary Table From BMDP2R FSWAP Method

SUMMARY TABLE

STEP NO.	VARIABLE ENTERED	VARIABLE REMOVED	MULTIPLE R	MULTIPLE RSQ	INCREASE IN RSQ	F TO ENTER	F TO REMOVE	NUMBER OF INDEPENDENT VARIABLES INCLUDED
1	7 ACCTS		0.7540	0.5685	0.5685	30.3018		1
2	4 ADV		0.8804	0.7751	0.2066	20.2104		2
3	3 POTEN		0.9098	0.8277	0.0526	6.4145		3
4	5 SHARE	7 ACCTS	0.9214	0.8490	0.0213	33.6315	10.3413	3
5	7 ACCTS		0.9489	0.9005	0.0515	10.3413		4

FIGURE 15.12 Summary Table From BMDP2R R Method

SUMMARY TABLE

STEP NO.	VARIABLE ENTERED	VARIABLE REMOVED	MULTIPLE R	MULTIPLE RSQ	INCREASE IN RSQ	F TO ENTER	F TO REMOVE	NUMBER OF INDEPENDENT VARIABLES INCLUDED
1	7 ACCTS		0.7540	0.5685	0.5685	30.3018		1
2	4 ADV		0.8804	0.7751	0.2066	20.2104		2
3	3 POTEN		0.9098	0.8277	0.0526	6.4145		3
4	5 SHARE		0.9489	0.9005	0.0727	14.6112		4
5		7 ACCTS	0.9214	0.8490	-0.0515		10.3413	3
6	7 ACCTS		0.9489	0.9005	0.0515	10.3413		4

RSWAP METHOD Finally, the BMDP2R provides the RSWAP method. Again, variables enter according to their F to enter values. As in the R method, they are removed if their removal results in a larger R^2 than previously obtained for the resulting number of predictors. In addition, as in the FSWAP method, instead of removing a variable so that the number of predictors is reduced, any variable removed is exchanged for another. Figure 15.13 shows the RSWAP analysis for the Cravens Data. In step 4 SHARE is exchanged for ACCTS because the model using ADV, POTEN, and SHARE has a larger R^2 than does the model using ADV, POTEN, and ACCTS.

COMPARISON OF THE FOUR METHODS Since all four procedures resulted in the same four-variable model at the final step (using minimum F to enter $= 4.0$ and maximum F to remove $= 3.9$), it may at first be hard to see how these four methods differ. It might help to think of them as combinations of just two principles:

1. Is a variable removed because its partial F becomes small (F method) or because its removal will increase the multiple coefficient of determination over that previously obtained for the resulting number of variables remaining (R method)?
2. When a variable is removed, does this result in fewer predictors being considered (regular method) or the same number (SWAP method)?

One can easily imagine that, depending upon the data being analyzed and the pattern of multicollinearity present, these four methods can give quite different final models.

SUMMARY OF BMDP2R STAGEWISE PROCEDURES

By now we can appreciate that there are just about as many criteria for determining what subset of variables to use as predictors as there are people trying to decide. The typically used selection procedures are the forward selection, backward elimination, and stepwise procedures, and BMDP2R program can perform all three. Variations on these themes can be provided. The criteria for selection can rest either on the significance of the partial F-statistic or on the amount that the multiple coefficient of determination increases when a variable is to be added to the candidate list of predictors. Variables may be added and deleted one at a time or may be exchanged for other variables.

Since the final set of predictors is often determined by the selection criteria, there can be no single universal best model for most sets of data when

FIGURE 15.13 Summary Table From BMDP2R RSWAP Method

SUMMARY TABLE

STEP	VARIABLE		MULTIPLE		INCREASE			NUMBER OF INDEPENDENT
NO.	ENTERED	REMOVED	R	RSQ	IN RSQ	F TO ENTER	F TO REMOVE	VARIABLES INCLUDED
1	7 ACCTS		0.7540	0.5685	0.5685	30.3018		1
2	4 ADV		0.8804	0.7751	0.2066	20.2104		2
3	3 POTEN		0.9098	0.8277	0.0526	6.4145		3
4	5 SHARE	7 ACCTS	0.9214	0.8490	0.0213	33.6315	10.3413	3
5	7 ACCTS		0.9489	0.9005	0.0515	10.3413		4

variables are allowed to stand or fall on their own predictive ability in conjunction with other predictors. Several good models may appear, and it is often not an easy task to choose among them. Generally, all predictors should be significant; the multiple R^2 should be large and the standard errors of the regression coefficients small; and the model should make sense to the researcher who is going to use it.

EXERCISES

15.2 Recall that in the analysis of the Children Data of Exercise 11.6 the overall regression significance test indicated that at least one of the predictors (age, height, or weight) was significant. However, no partial F-test on one predictor, given the other two, indicated a significant predictor. Analyze the data using the BMDP2R stepwise procedure with default conditions on the method and the F-levels for inclusion and removal. See if any significant predictors can be found.

15.3 A simple random sample of 30 of the 95 counties in Tennessee was drawn, and values of eight variables were recorded for each of the sampled counties. Table 15.3 shows the information collected. Run the BMDP2R stepwise regression analysis with F to enter of 1 and F to remove of 1. Use all four methods (F, FSWAP, R, and RSWAP) to predict percent below poverty level from the other eight variables. You might also want to obtain a correlation matrix and other pertinent information. Propose a final model to be used, and justify your choice.

15.4 Consider the MPG Data of Exercise 10.6. Since the relationship of mpg with both horsepower and weight was curvilinear, fit a model

$$Y = \alpha + \beta_1 X_1 + \beta_2 X_2 + \beta_3 X_1^2 + \beta_3 X_2^2 + \beta_5 X_3 + \varepsilon,$$

where Y = mpg, X_1 = HP, X_2 = weight, and X_3 = transmission type.

a. Analyze the full model. Would you use the full model to predict mpg?

b. Use an available BMDP variable-selection procedure to obtain an acceptable reduced model. Examine regression coefficients to see if they make sense.

15.5 In Appendix G is a very large data set containing information on 14 characteristics of residential properties, sold during a given time period. Variables are defined in Appendix G. Use a BMDP variable-selection procedure to determine which characteristics of a residence affect the number of days a residence is on the market, and how.

Note that there is a very high positive correlation between the list price and the selling price, so that it might not be too informative to include both as predictors.

X_{13} = "type" is a dummy variable. X_1 = "zone" is a categorical variable with values as defined in Appendix G. Most new growth in the city has been to the west; you may want to take this into account.

Instead of entering all data for all 389 cases and 14 variables, you might prefer to select a simple random sample to analyze.

TABLE 15.3 Selected Data on a Random Sample of 30 of the 95 Tennessee Counties (Tennessee Data)

County	Population Change 1960–1970	No. Persons Employed in Agriculture	Percent Families Below Poverty Level	Residential and Farm Property Tax Rate	Percent Residences with Telephones	Percent Rural Population	Median Age	Black Population
Benton	13.7	400	19.0	1.09	82	74.8	33.5	360
Cannon	− 0.8	710	26.2	1.01	66	100.0	32.8	193
Carroll	9.6	1,610	18.1	0.40	80	69.7	33.4	3,080
Cheatheam	40.0	500	15.4	0.93	74	100.0	27.8	592
Cumberland	8.4	640	29.0	0.92	65	74.0	27.9	2
DeKalb	3.5	920	21.6	0.59	64	73.1	33.2	230
Dyer	3.0	1,890	21.9	0.63	82	52.3	30.8	3,978
Gibson	7.1	3,040	18.9	0.49	85	49.6	32.4	9,816
Greene	13.0	2,730	21.1	0.71	78	71.2	29.2	1,137
Hawkins	10.7	1,850	23.8	0.93	74	70.6	28.7	992
Haywood	− 16.2	2,920	40.5	0.51	69	64.2	25.1	10,723
Henry	6.6	1,070	21.6	0.80	85	58.3	35.9	3,129
Houston	21.9	160	25.4	0.74	69	100.0	31.4	338
Humphreys	17.8	380	19.7	0.44	83	72.0	30.1	516
Jackson	− 11.8	1,140	38.0	0.81	54	100.0	34.1	12
Johnson	7.5	690	30.1	1.05	65	100.0	30.5	104
Lawrence	3.7	1,170	24.8	0.73	76	69.5	30.0	430
McNairy	1.6	1,280	30.3	0.65	67	81.0	32.4	1,240
Madison	8.4	2,270	19.5	0.48	85	39.1	28.7	20,446
Marshall	2.7	960	15.6	0.72	84	58.4	33.4	1,863
Maury	5.6	1,710	17.2	0.62	84	42.4	29.9	8,035
Montgomery	12.7	1,410	18.4	0.84	86	36.4	23.3	10,620
Morgan	− 4.8	200	27.3	0.73	66	99.8	27.5	211
Sevier	16.5	960	19.2	0.45	74	90.6	29.5	133
Shelby	15.2	11,500	16.8	1.00	87	5.9	25.4	266,159
Sullivan	11.6	1,380	13.2	0.63	85	44.2	28.8	2,432
Trousdale	4.9	530	29.7	0.54	70	100.0	33.1	932
Unicoi	1.1	370	19.8	0.98	75	52.6	30.8	7
Wayne	3.8	440	27.7	0.46	48	100.0	28.4	208
Weakley	19.0	1,630	20.5	0.68	83	72.1	30.4	1,732

Source: Center for Business and Economic Research, The University of Tennessee, Knoxville, *Tennessee Statistical Abstract,* 1974.

15.5 MINITAB SELECTION PROCEDURES

Minitab offers stepwise, forward selection, and backward elimination procedures. Its basic procedure is the stepwise one, and modifications to this procedure are used to produce the other two. We first consider the Minitab stepwise selection procedure.

STEPWISE SELECTION

The stepwise selection procedure first identifies the single predictor with the largest F-value and produces a single-predictor regression analysis using this predictor; that is, provided that this largest F-value is larger than some specified value, called FENTER. Unless you tell it differently, Minitab will assign the value 4 to FENTER; this is the default value. (How to change this default value will be shown later.) Note that an F-value of 4 is, for a single predictor, roughly significant at significance level $\alpha = 0.05$.

At the second step, partial F-values for all variables not in the model, given the one in the model at step 1, are calculated. The variable with the largest partial F is then entered into the model, provided that this partial F is larger than FENTER, and the two-predictor analysis is shown. At this point, Minitab checks the partial F's for variables in the model. If either of them is less than a value FREMOVE (default: FREMOVE = 4), then that variable will be dropped from the model, and (partial) F's recalculated for those variables remaining. When no variables can be removed, partial F's for those predictors not in the model, given those in the model, are calculated and the variable with the largest partial F greater than FENTER is entered. The process continues until no more variables can be entered or removed.

Figure 15.14 shows Minitab's stepwise selection procedure applied to the Cravens Data with FENTER = FREMOVE = 4. The command producing this analysis is

```
STEPWISE C1 on C2-C9
```

where the data have been entered so that Y = sales is in column 1 and the eight predictors are in columns 2 through 9, with appropriate names assigned.

FIGURE 15.14 Minitab Stepwise Selection

```
—STEPWISE C1 ON C2-C9

STEPWISE REGRESSION OF SALES ON 8 PREDICTORS, WITH N = 25

STEP              1           2           3             4
CONSTANT     709.32       50.30     -327.23      -1441.93

ACCTS          21.7        19.0        15.6           9.2
T-RATIO        5.50        6.41        5.19          3.22

ADV                       0.227       0.216         0.175
T-RATIO                    4.50        4.77          4.74

POTEN                                0.0219         0.382
T-RATIO                                2.53          4.79

SHARE                                                 190
T-RATIO                                              3.82

S               881         650         583           454
R-SQ          56.85       77.51       82.77         90.04

MORE? (YES, NO, SUBCOMMAND, OR HELP)
—
NO
```

In Figure 15.14 we see that the variable ACCTS was entered on step 1. Its t-value is 5.50, which is equivalent to an F-value of $(5.50)^2 = 30.25$. The regression equation relating SALES to ACCTS is

$$\hat{Y} = 709.32 + 21.7X.$$

The standard deviation of the points from the regression line is $s_{Y \cdot X} = 881$, and ACCTS explains 56.85 percent of the variation in SALES. Note that in many cases, Minitab rounds off a value quite severely so that printed values can fit into the table.

In step 2 the variable ADV was added to the model, and the regression equation is

$$\widehat{\text{SALES}} = 50.30 + 19.0\ \text{ACCTS} + 0.227\ \text{ADV}.$$

Using both ACCTS and ADV to estimate SALES, estimates will be off, on the average, 650, and these two predictors explain 77.51 percent of the variation in SALES. Since both partial t-values are larger than 2, so that both partial F's are larger than 4, no variables are removed from the model and the procedure continues by adding the variables POTEN in step 3 and SHARE in step 4. The process terminates at step 4, since (1) all partial F's for variables in the model are greater than 4, and (2) no partial F for variables not in the model are greater than 4. The regression arrived at by the stepwise selection process is

$$\widehat{\text{SALES}} = 1441.93 + 9.2\ \text{ACCTS} + 0.175\ \text{ADV} + 0.0382\ \text{POTEN} + 190\ \text{SHARE}.$$

The standard error of estimate is 454, and the multiple R^2 is 90.04 percent. In this data set no variable once entered was ever removed. A more exciting example of the stepwise procedure can be found in Section 15.8, in which hierarchical regression and techniques for forcing variables into the model are discussed.

Since the information provided by the STEPWISE command is so limited, at this point it is a good idea to use the REGRES command to analyze the model containing these four predictors. This is shown in Figure 15.15.

SUBCOMMANDS TO THE STEPWISE COMMAND

Notice in Figure 15.14 that after the summary table for the stepwise selection procedure is printed, there appears the message

```
MORE? (YES, NO, SUBCOMMAND, OR HELP)
```

If you are finished analyzing your data with the STEPWISE command, type NO and Minitab will await other commands. If you type YES, you will most likely get a message that no other variables were entered or removed. In large

FIGURE 15.15 Minitab Multiple Regression Analysis for the Model Selected by Stepwise Regression

```
—REGRES C1 ON 4 VAR 'ACCTS' 'ADV' 'POTEN' 'SHARE'

THE REGRESSION EQUATION IS
Y = -1442. + 9.21 X1 + 0.175 X2
    +0.0382 X3 + 190. X4

                                           ST DEV         T-RATIO =
COLUMN                    COEFFICIENT      OF COEF        COEF/S.D.
X1        ACCTS          -1441.9           423.6          -3.40
X2        ADV                9.214           2.865          3.22
X3        POTEN              0.17499         0.03691        4.74
X4        SHARE              0.038218        0.007977       4.79
                           190.14           49.74           3.82

THE ST DEV OF Y ABOUT REGRESSION LINE IS
S = 453.8
WITH (25 - 5) = 20 DEGREES OF FREEDOM

R-SQUARED = 90.0 PERCENT
R-SQUARED = 88.1 PERCENT, ADJUSTED FOR DF

ANALYSIS OF VARIANCE

DUE TO            DF       SS              MS=SS/DF
REGRESSION         4       37260204        9315051
RESIDUAL          20        4119347         205967
TOTAL             24       41379551

FURTHER ANALYSIS OF VARIANCE
SS EXPLAINED BY EACH VARIABLE WHEN ENTERED IN THE ORDER GIVEN

DUE TO            DF       SS
REGRESSION        4        37260204
ACCTS             1        23524076
ADV               1         8549246
POTEN             1         2177476
SHARE             1         3009407

          X1          Y           PRED Y      ST DEV
ROW       ACCTS       SALES       VALUE       PRED Y      RESIDUAL       ST RES
 1         75         3669.9      3357.6      319.4       312.3          0.97 X
 7        256         5031.7      5037.8      326.3        -6.1         -0.02 X
10        120         4876.4      3941.9      125.3       934.5          2.14R

R DENOTES AN OBS WITH A LARGE ST RES.
X DENOTES AN OBS WHOSE X VALUE GIVES IT LARGE INFLUENCE.

DURBIN-WATSON STATISTIC = 1.60.
```

problems, however, the output will be suspended after available space on the terminal is filled, and typing YES will then cause additional results to be displayed. Typing HELP will give you a message that you are writing subcommands to the STEPWISE command and give you information concerning where to look for instructions. This brings us to the consideration of subcommands, which will enable us to alter the basic STEPWISE procedure.

Subcommands to the STEP command enable us to change FENTER and FREMOVE from the default value of 4, as well as to effect other changes in the selection procedure, which will be discussed later. When subcommands are to be used with the STEP command, follow the STEP command with a semicolon. When the carriage return is pressed, Minitab will await your subcommands before performing the selection procedure and will prompt you with a double dash. Type in subcommands, following them with a semicolon to indicate that more subcommands will follow, or with a period to indicate that you have finished writing subcommands. Figure 15.16 shows the command and subcommands to perform a stepwise regression with FENTER = 2 and FREMOVE = 1. Note that five variables were entered, but none were removed. Note also that FENTER must be greater than or equal to FREMOVE. If you try to specify otherwise, Minitab will print an error message and will automatically assign one of the values to equal the other. For example, if you write FENTER=1 without changing FREMOVE from the default value of 4, then Minitab will automatically set FREMOVE equal to 1. This is done to keep the process from entering a loop in which a variable is entered at one step, removed at the next, entered again on the next step, removed again, and so on.

FIGURE 15.16 Using FENTER and FREMOVE Subcommands in Minitab

```
—STEP C1 ON C2-C9;
—
FENTER=2;
—
FREMOVE=1.
STEPWISE REGRESSION OF SALES ON 8 PREDICTORS, WITH N = 25
STEP             1          2           3            4             5

CONSTANT     709.32      50.30     -327.23     -1441.93      -1285.94

ACCTS          21.7       19.0        15.6          9.2           8.2
T-RATIO        5.50       6.41        5.19         3.22          2.92

ADV                      0.227       0.216        0.175         0.154
T-RATIO                   4.50        4.77         4.74          4.09

POTEN                               0.0219       0.0382        0.0376
T-RATIO                               2.53         4.79          4.90

SHARE                                               190           197
T-RATIO                                            3.82          4.10

CHANGE                                                            263
T-RATIO                                                          1.61

S               881        650         583          454           437
R-SQ          56.85      77.51       82.77        90.04         91.24
```

Figure 15.16 illustrates the use of subcommands at the outset, but as the message

```
MORE? (YES, NO, SUBCOMMAND, OR HELP)
```

indicates, subcommands may be entered after the STEPWISE command has been executed. This is useful when the results of an analysis give ideas for other models to consider. This will be illustrated when we consider the forward selection procedure.

ALTERING THE STEP COMMAND TO PERFORM A FORWARD SELECTION

Although no variables were ever removed from the model in Figures 15.14 and 15.16 the stepwise algorithm does provide for this to be done should any partial *F* for a variable in the model become less than FREMOVE. To perform a forward selection in which, once entered, a variable will not be removed, use the subcommand FREMOVE = 0. Figure 15.17 shows a forward selection process applied to the Cravens Data. Note that in Figure 15.17, FENTER is specified as 1 in order to allow quite a few variables to enter the equation.

FIGURE 15.17 Minitab Forward Selection

```
—STEP C1 ON C2-C9;
 —
FENTER=1;
 —
FREMOVE=0.
STEPWISE REGRESSION OF SALES ON 8 PREDICTORS, WITH N = 25
```

STEP	1	2	3	4	5	6
CONSTANT	709.32	50.30	-327.23	-1441.93	-1285.94	-1165.48
ACCTS	21.7	19.0	15.6	9.2	8.2	4.4
T-RATIO	5.50	6.41	5.19	3.22	2.92	1.09
ADV		0.227	0.216	0.175	0.154	0.141
T-RATIO		4.50	4.77	4.74	4.09	3.66
POTEN			0.0219	0.0382	0.0376	0.0383
T-RATIO			2.53	4.70	4.90	5.07
SHARE				190	197	222
T-RATIO				3.82	4.10	4.38
CHANGE					263	285
T-RATIO					1.61	1.78
TIME						2.3
T-RATIO						1.34
S	881	650	583	454	437	428
R-SQ	56.85	77.51	82.77	90.04	91.24	92.03

The first five steps of Figure 15.17 are exactly the same as the five steps in Figure 15.16, in which FENTER = 2 and FREMOVE = 1. Allowing the variable TIME to enter in step 6 of a forward selection points out an interesting phenomenon: Before, ACCTS was the single strongest predictor; once TIME is entered into the equation the *t*-value for ACCTS falls to 1.09. This is not surprising, since TIME and ACCTS are highly correlated: If one is used, there is not much variation in *Y* left for the other to explain. One wonders what an analysis using TIME, but not ACCTS, along with the other four predictors would look like. To see this analysis, write a subcommand to remove ACCTS when prompted by MORE? The subcommand, as shown in Figure 15.18, is REMOVE 'ACCTS'.

FIGURE 15.18 Using the Minitab REMOVE Subcommand

```
REMOVE 'ACCTS'.
STEP            7              8
CONSTANT    -1114          -1165
ACCTS                        4.4
T-RATIO                     1.09
ADV             0.129        0.141
T-RATIO         3.48         3.66
POTEN           0.0421       0.0383
T-RATIO         6.25         5.07
SHARE         257          222
T-RATIO         6.57         4.38
CHANGE        325          285
T-RATIO         2.06         1.78
TIME            3.6          2.3
T-RATIO         3.06         1.34
S             430          428
R-SQ           91.50        92.03
```

In step 7 in Figure 15.18 we see the regression coefficients and t-values, standard deviation and R^2 for the predictors ADV, POTEN, SHARE, CHANGE, and TIME. All t-values are larger than 2, and $R^2 = 91.50$, so that this five-predictor model might be preferred to the one in step 5 of the forward selection procedure (Figure 15.17). This illustrates the possibility for alternative models arising from the use of subcommands after examining the results of a given selection process.

Note that in step 8 of Figure 15.18 the variable ACCTS was added back in because the subcommand FENTER=1 is still in effect. Also, note that more than one variable can be removed at a time. For example, if after step 6 of the forward selection procedure we had decided to delete both ACCTS and CHANGE, this could have been accomplished by writing REMOVE 'ACCTS' AND 'CHANGE'. Both of these variables would have been eligible for re-entry at subsequent steps, provided that their partial F-values, given those remaining in the model, were sufficiently large.

BACKWARD ELIMINATION

The backward elimination process begins with all predictors in the model and eliminates, in turn, those whose partial F's are the smallest, until all partial F's for variables in the model are greater than FREMOVE. In the classical backward elimination process, once a variable is removed, it may not re-enter. To perform this procedure in Minitab, we first construct the full model using the ENTER subcommand:

```
STEP C1 on C2-C9;
ENTER C2-C9;
```

Then to prevent any eliminated variable from re-entering, assign an impossibly high value for FENTER, such as 100,000. Figure 15.19 shows this analysis, with FREMOVE at the default value, 4. Note that the final model is the same one obtained by intervening in the forward selection process with the subcommand REMOVE 'ACCTS' (Figure 15.18). If, when performing a backward elimination using Minitab, FENTER is not set at some very high value, then a variable once removed may be eligible to re-enter the model, resulting in a backward stepwise process.

FIGURE 15.19
Minitab Backward Elimination

```
—STEP C1 ON C2–C9;
 —
ENTER C2–C9;
 —
FREMOVE=4;
 —
FENTER=100000.
STEPWISE REGRESSION OF SALES ON 8 PREDICTORS, WITH N = 25
```

STEP	1	2	3	4
CONSTANT	-1508	-1486	-1165	-1114
TIME	2.0	2.0	2.3	3.6
T-RATIO	1.04	1.10	1.34	3.06
POTEN	0.0372	0.0373	0.0383	0.0421
T-RATIO	1.54	4.75	5.07	6.25
ADV	0.151	0.152	0.141	0.129
T-RATIO	3.21	3.51	3.66	3.48
SHARE	199	198	222	257
T-RATIO	2.97	3.09	4.38	6.57
CHANGE	291	296	285	325
T-RATIO	1.56	1.80	1.78	2.06
ACCTS	5.6	5.6	4.4	
T-RATIO	1.16	1.23	1.09	
WORK	20	20		
T-RATIO	0.59	0.61		
RATE	8			
T-RATIO	0.06			
S	449	436	428	430
R-SQ	92.20	92.20	92.03	91.50

OTHER SUBCOMMANDS

In addition to FENTER, FREMOVE, ENTER, and REMOVE, Minitab offers the FORCE subcommand (discussed in Section 15.8), the STEPS subcommand, which varies the number of steps to be printed at a time, and the BEST K ALTERNATIVE VARIABLES subcommand. This last subcommand shows the k variables with the *next largest* partial F-values at any step at which a variable enters the equation. Thus, you are able to see variables that are "almost as good" as the variable entered at that step and can consider alternative models. When practical or theoretical considerations dictate that some variables are more important than others, such a procedure can be useful.

Figure 15.20 shows the stepwise analysis of the Cravens Data with the subcommand BEST 2. We see that the second best single predictor of SALES is TIME, with a *t*-value of 3.82. The third best single predictor is POTEN. This agrees with the magnitudes of the correlations of these predictors with SALES. In step 2 ADV was added to ACCTS. The next best choice to add to ACCTS would have been POTEN, with a partial *t*-value of 2.11. One might want to investigate these alternative combinations of predictors.

FIGURE 15.20 Minitab BEST K Subcommand

```
—STEP C1 ON C2-C9;
—
BEST 2.
STEPWISE REGRESSION OF SALES ON 8 PREDICTORS, WITH N = 25

STEP            1           2            3              4
CONSTANT      709.32       50.30      -327.23       -1441.93

ACCTS           21.7        19.0         15.6            9.2
T-RATIO          5.50        6.41         5.19           3.22

ADV                          0.227        0.216          0.175
T-RATIO                      4.50         4.77           4.74

POTEN                                     0.0219         0.0382
T-RATIO                                   2.53           4.79

SHARE                                                  190
T-RATIO                                                  3.82

S             881         650          583             454
R-SQ           56.85       77.51        82.77           90.04

BEST ALT.
VARIABLE       TIME        POTEN        WORK           WORK
T-RATIO        3.82        2.11         1.50           2.31
VARIABLE       POTEN       CHANGE       CHANGE         CHANGE
T-RATIO        3.58        1.99         1.09           0.94
```

A caution is in order: Minitab performs analyses quickly, and it is easy to write subcommands to perform alternative analyses; thus, there is a danger of "over-analyzing" the data. The *Minitab Reference Manual* (p. 79) states this very well:

> Variable-selection procedures can be a valuable tool in data analysis, particularly in the early stages of building a model. At the same time, they present certain dangers. There are several reasons for this. (a) Since the procedures automatically "snoop" through many models, the model selected may fit the data "too well." That is, the procedure can look at many variables and select ones which, by pure chance, happen to fit well. (b) The three automatic procedures are heuristic algorithms, which often work very well but which may not select the model with the highest R-SQ value (for a given number of predictors). (c) Automatic procedures cannot take into account special knowledge the analyst may have about the data. Therefore, the model selected may not be the best from a practical point of view.

EXERCISES

15.6 Recall that in the analysis of the Children Data of Exercise 11.6 the overall regression significance test indicated that at least one of the predictors (age, height, or weight) was significant. However, no partial F-test on one predictor, given the other two, indicated a significant predictor. Analyze the data using the Minitab stepwise procedure with default conditions on `FENTER` and `FREMOVE`, and see if any conditionally significant predictors can be found.

15.7 Refer to the Tennessee Data of Exercise 15.3. Use the Minitab selection procedures of your choice to obtain a model with `FENTER` = `FREMOVE` = 1 to predict percent below poverty level. Obtain a correlation matrix and other pertinent information. Propose a final model, and justify your choice.

15.8 Consider the MPG Data of Exercise 10.6. Since the relationship of mpg with both horsepower and weight was curvillinear, fit a model

$$Y = \alpha + \beta_1 X_1 + \beta_2 X_2 + \beta_3 X_1^2 + \beta_4 X_2^2 + \beta_5 X_3 + \varepsilon,$$

where Y = mpg, X_1 = HP, X_2 = weight, and X_3 = transmission type.

a. Analyze the full model. Would you use the full model to predict mpg?

b. Use an available Minitab variable-selection procedure to obtain an acceptable reduced model. Examine regression coefficients to see if they make sense.

15.9 Work Exercise 15.5 using a Minitab selection procedure.

15.6 SAS SELECTION PROCEDURES

The five selection procedures available through the Statistical Analysis System are called forward selection, stepwise, backward elimination, maximum R^2 improvement, and minimum R^2 improvement. We will discuss them in order, comparing and contrasting what they do.

THE STEPWISE PROCEDURE

Figure 15.21 shows a worksheet for writing an SAS program to perform a forward selection procedure on the Cravens Data. We have not asked for titles, a correlation matrix, a listing of data values, the `REG` procedure, and so on, because these are already shown in Figure 15.3 and 15.4. The `PROC STEPWISE` statement asks the computer to perform a variable selection. (SAS uses the term *stepwise* in the sense that we use *stagewise*, but we shall

FIGURE 15.21 Worksheet for SAS Forward Selection Procedure

```
  {SYSTEM CARDS
DATA CRAVENS;
INPUT SALES TIME POTEN ADV SHARE CHANGE ACCTS WORK RATE;
CARDS;
3669.88   43.10  74065.11  4582.88   2.51   .34   74.86  15.05  4.9
3473.95  108.13  58117.30  5539.78   5.51   .15  107.32  19.97  5.1
          :
2799.97   21.14  22809.53  3552.00   9.14  -.74   88.62  24.96  3.9
PROC STEPWISE;
MODEL SALES=TIME POTEN ADV SHARE CHANGE ACCTS WORK RATE/FORWARD;
```

reserve the term *stepwise* to refer to a *particular* stagewise procedure.) The MODEL line specifies the dependent variable (SALES) and the independent variables to use. The FORWARD (which can be abbreviated as just F) following the slant indicates that the forward selection procedure is to be used. In order to run a backward elimination procedure, replace FORWARD with BACKWARD or B. In order to run the STEPWISE procedure, replace FORWARD with STEPWISE, or simply place a semicolon after the last independent variable and drop the /FORWARD. (This is a default: unless you specify otherwise, the computer will perform a stepwise procedure.) To run the maximum R^2 improvement technique, replace FORWARD with MAXR, and to use the minimum R^2 improvement, replace FORWARD with MINR. Do not forget that each SAS statement must end with a semicolon.

It is often desirable to perform all five selection procedures on a set of data for the sake of comparison. The following model statement will allow all five procedures to be run at once:

```
MODEL SALES=TIME...RATE/FORWARD BACKWARD STEPWISE
MAXR MINR
```

Note that although STEPWISE is the default procedure, if other procedures are also to be performed, it is necessary to specifically request the STEPWISE option. Now that we know how to run the five stagewise procedures using SAS, let us see how these procedures differ from one another.

FORWARD SELECTION

Generally, a forward selection technique begins by running a simple regression analysis using the single strongest predictor, as measured by the *F*-statistic. Then it finds the predictor with the largest partial *F*-value, given the strongest predictor, and denotes this as the second strongest predictor. A two-variable multiple regression analysis is performed using these two strongest predictors. The third strongest predictor, in terms of the variable yielding the largest partial *F*, given the two strongest, is then added and a three-predictor analysis is performed. The procedure continues either until all variables have been added, or until the largest remaining partial F-value is too small to bother with.

SIGNIFICANCE LEVEL FOR ENTRY

What determines when a partial F-value is "too small to bother with" is the significance level for entry. Unless otherwise specified, the process terminates when the largest partial F-value is no longer significant at $\alpha = 0.50$. At this point, no variable will be entered. Why such a high significance level? Why not make the significance level for entry small, say $\alpha = 0.05$ or $\alpha = 0.01$? The reason is that these partial F-values are true F-values *only* when we have found the correct model; that is, only if $s^2_{Y \cdot X} = MS(E)$ estimates true error and not variation due to omitted predictors. If at any stage we do not have the true model, $s^2_{Y \cdot X}$ will be inflated and thus the F-values will be deflated. Setting $\alpha = 0.05$ or $\alpha = 0.01$ would defeat our purposes and would wrongly prevent variables from having a chance to enter into the equation.

One rule of thumb is to set $\alpha = 0.25$. Though not as generous as the default condition of SAS, this rule still allows several variables to enter into the equation. Note that we would actually be using a significance level of 0.25 (if this were specified) *only* at that step in which *all* significant predictors and *only* significant predictors have been included. All the significance level for entry really does is provide a cutoff point: computer time is saved if extremely weak predictors are ignored. Nevertheless, the choice of the significance level for entry will usually have an emphatic effect on the final subset of variables.

To change the significance level for entry from the default condition of 0.50 to $\alpha = 0.25$, add in the MODEL statement in Figure 15.21 SLE = 0.25 before the semicolon. The MODEL statement then reads

```
MODEL SALES=TIME POTEN...RATE/FORWARD SLE=0.25
```

FORWARD SELECTION PROCEDURE FOR CRAVENS EXAMPLE

Figure 15.22 shows the forward selection procedure with $\alpha = 0.50$ applied to the Cravens Data. As we saw in Figure 15.4, ACCTS is the single strongest predictor of SALES. Thus, a simple regression using ACCTS to predict SALES is run as the first step of the forward selection procedure. The F-value of 30.3 shows that ACCTS is a significant predictor of sales, and $R^2 = 0.568$ shows that the variation in the number of accounts explains approximately

FIGURE 15.22 SAS Forward Selection Procedure

```
FORWARD SELECTION PROCEDURE FOR DEPENDENT VARIABLE SALES

STEP 1  VARIABLE ACCTS ENTERED  R SQUARE = 0.56849518   C(P) = 67.55826191

              DF            SUM OF SQUARES          MEAN SQUARE         F        PROB > F
REGRESSION     1         23524074.19189501    23524074.19189501     30.30        0.0001
ERROR         23         17855474.73496109      776324.98847657
TOTAL         24         41379548.92685610

              B VALUE       STD ERROR               TYPE II SS          F        PROB > F
INTERCEPT     709.32383372
ACCTS          21.72176971  3.94603304        23524074.19189501     30.30        0.0001
-------------------------------------------------------------------------------------------
STEP 2  VARIABLE ADV ENTERED  R SQUARE = 0.77510077   C(P) = 27.15634512

              DF            SUM OF SQUARES          MEAN SQUARE         F        PROB > F
REGRESSION     2         32073320.05178234    16036660.02589117     37.91        0.0001
ERROR         22          9306228.87507376      423010.40341244
TOTAL         24         41379548.92685610

              B VALUE       STD ERROR               TYPE II SS          F        PROB > F
INTERCEPT     50.29906160
ADV            0.22652657   0.05038842         8549245.85988733     20.21        0.0002
ACCTS         19.04824598   2.97291380        17365864.05826321     41.05        0.0001
-------------------------------------------------------------------------------------------
STEP 3  VARIABLE POTEN ENTERED  R SQUARE = 0.82772280   C(P) = 18.35666096

              DF            SUM OF SQUARES          MEAN SQUARE         F        PROB > F
REGRESSION     3         34250796.02722474    11416932.00907491     33.63        0.0001
ERROR         21          7128752.89963136      339464.42379197
TOTAL         24         41379548.92685610

              B VALUE       STD ERROR               TYPE II SS          F        PROB > F
INTERCEPT    -327.23338939
POTEN           0.02192192  0.00865564         2177475.97544240      6.41        0.0194
ADV             0.21607079  0.04532744         7713721.79142546     22.72        0.0001
ACCTS          15.55392188  2.99936692         9128825.32465885     26.89        0.0001
-------------------------------------------------------------------------------------------
```

STEP 4 VARIABLE SHARE ENTERED R SQUARE = 0.90044970 C(P) = 5.43082968

	DF	SUM OF SQUARES	MEAN SQUARE	F	PROB > F
REGRESSION	4	37260202.46282121	9315050.61570530	45.23	0.0001
ERROR	20	4119346.46403489	205967.32320174		
TOTAL	24	41379548.92685610			

	B VALUE	STD ERROR	TYPE II SS	F	PROB > F
INTERCEPT	-1441.93182868				
POTEN	0.03821753	0.00797694	4727717.33107410	22.95	0.0001
ADV	0.17499004	0.03690666	4630368.54356448	22.48	0.0001
SHARE	190.14429731	49.74415347	3009406.43559647	14.61	0.0011
ACCTS	9.21389567	2.36521038	2129962.08538966	10.34	0.0043

STEP 5 VARIABLE CHANGE ENTERED R SQUARE = 0.91241574 C(P) = 4.97502565

	DF	SUM OF SQUARES	MEAN SQUARE	F	PROB > F
REGRESSION	5	37755351.59771667	7551070.31954333	39.59	0.0001
ERROR	19	3624197.32913943	190747.22784944		
TOTAL	24	41379548.92685610			

	B VALUE	STD ERROR	TYPE II SS	F	PROB > F
INTERCEPT	1285.94337067				
POTEN	0.03763121	0.00768517	4573489.56079583	23.98	0.0001
ADV	0.15443602	0.03773852	3194373.73424488	16.75	0.0006
SHAPE	196.94952750	48.05692351	3203732.09272627	16.80	0.0006
CHANGE	262.50049338	162.92631286	495149.13489546	2.60	0.1236
ACCTS	8.23411280	2.82357962	1622152.88530014	8.50	0.0089

STEP 6 VARIABLE TIME ENTERED R SQUARE = 0.92031405 C(P) = 5.35404547

	DF	SUM OF SQUARES	MEAN SQUARE	F	PROB > F
REGRESSION	6	38082180.17286742	6347030.02881124	34.65	0.0001
ERROR	18	3297368.75398863	183187.15299937		
TOTAL	24	41379548.92685610			

	B VALUE	STD ERROR	TYPE II SS	F	PROB > F
INTERCEPT	-1165.47855369				
TIME	2.26935112	1.69898362	326828.57515075	1.78	0.1983
POTEN	0.03827800	0.00754688	4712573.84321485	25.73	0.0001
ADV	0.14067029	0.03839221	2459312.09671656	13.43	0.0018
SHARE	221.60469221	50.58309112	3515945.91528346	19.19	0.0004
CHANGE	285.10928426	160.65965553	577623.53145916	3.15	0.0927
ACCTS	4.37770296	3.99903763	219521.27264808	1.20	0.2881

NO OTHER VARIABLES MET THE 0.5000 SIGNIFICANCE LEVEL FOR ENTRY INTO THE MODEL.

57 percent of the variation in SALES. Using the regression equation

$$\hat{Y} = 709.32383372 + 21.72176971X_6,$$

our estimates will be off, on the average, by $[776{,}324.98847657]^{1/2} = 881.093$ units. The standard error of the regression coefficient, 3.946, is not unduly large, so the value $b_6 = 21.72176971$ can be trusted fairly well. (The meaning of the value C_p will be explained in Chapter 16.)

In step 2 the variable X_3 (ADV) is entered, because the partial F-value for ADV given ACCTS is larger than any of the other six partial F-values given ACCTS. (Imagine all the work the computer has done but not printed!) The overall F-value of 37.91 indicates that at least one of the two variables is a good predictor, and the multiple coefficient of determination of 0.775 shows that addition of ADV to ACCTS improves the percentage of explained variation by $77.5 - 56.8 = 20.7\%$. The regression constant and regression coefficients and their standard errors can also be found. Note that the variables are listed according to their order of appearance on the MODEL statement; since ADV $= X_3$ and ACCTS $= X_6$, ADV is listed first.

What is new to us is the Type II sum of squares. By calling it a Type II sum of squares, SAS avoids having to name it. We might refer to it as a "mini-partial" sum of squares or a "sum-of-squares if last." The Type II sum of squares is a partial sum of squares, conditional on all other variables in the model *at that step*. As such, it measures the effect of the variables with which it is associated as if that variable were the last one entered in the model. Thus, the Type II sum of squares for ADV in step 2 is

$$SS(R)(\text{ADV} \mid \text{ACCTS})$$

and that for ACCTS is

$$SS(R)(\text{ACCTS} \mid \text{ADV}).$$

From the probabilities given, we see that if ADV has been included in the model we still need ACCTS, and if ACCTS were already in the model we would still need ADV.

Step 3 continues in much the same way. The variable POTEN (X_2) is entered because

$$F(\text{POTEN} \mid \text{ADV, ACCTS})$$

is larger than any other partial F conditioned on ADV and ACCTS. The Type II sums of squares are

$$SS(R)(\text{POTEN} \mid \text{ADV, ACCTS}) = 2{,}177{,}475.975,$$

$$SS(R)(\text{ADV} \mid \text{POTEN, ACCTS}) = 7{,}713{,}721.791,$$

$$SS(R)(\text{ACCTS} \mid \text{POTEN, ADV}) = 9{,}128{,}825.325.$$

The multiple R^2 has increased to about 83 percent, and all three partial F's are significant at $\alpha < 0.05$.

Step 4 adds SHARE (X_4) and shows all four variables significant at $\alpha = 0.05$. In step 5 CHANGE enters but is not significant at $\alpha = 0.05$, although all other variables are. Recall, however, that the computer is using $\alpha = 0.50$, so it goes on to the next step. Also, the probabilities printed out are valid at *only one step*: the step in which the correct model has been found.

In step 6 a funny thing happens. When the variable TIME (X_1) is entered, it is significant at $\alpha = 0.05$, even though it is supposedly not as strong as the variable entered in step 5 (CHANGE), which was found to be nonsignificant. Also, suddenly our single strongest predictor, ACCTS, is no longer significant (and neither is CHANGE). The intercorrelations among predictors are, of course, to blame for these strange goings-on. Given that you decide to use TIME, POTEN, ADV, SHARE, and CHANGE, you don't need to use ACCTS even though it is strong by itself. Refer to the correlation matrix in Section 15.1. ACCTS is highly correlated with TIME ($r_{16} = 0.75782$), so if TIME is going to be used as a predictor, ACCTS is redundant.

Following step 6, we see the message that no other variable met the 0.50 significance level for entry. That is, neither

$$F(\text{WORK} \mid \text{TIME, POTEN, ADV, SHARE, CHANGE, ACCTS})$$

nor

$$F(\text{RATE} \mid \text{TIME, POTEN, ADV, SHARE, CHANGE, ACCTS})$$

is significant at $\alpha = 0.50$, so the procedure terminates without ever adding these variables to the model. Using the six variables TIME, POTEN, ADV, SHARE, CHANGE, and ACCTS to predict sales, we can explain approximately 92 percent of the variation in SALES. It is interesting to compare the sizes of the R^2-values in each step:

$$0.568,\ 0.775,\ 0.828,\ 0.900,\ 0.912,\ 0.920.$$

The first four steps give relatively large increases in R^2, but the increases in the last two steps are rather small.

Which model do we want to adopt, then? Should we go with the results of step 4 and use POTEN, ADV, SHARE, and ACCTS to predict SALES? All four predictors there are significant at $\alpha = 0.05$. Or should we also use CHANGE and TIME? In step 6 we noticed that once TIME was included in the model, ACCTS became superfluous. Would it be possible to drop ACCTS and replace it with either WORK or RATE and have six predictors which have a multiple R^2 greater than 0.92? What we want to use as a model depends upon our criterion for entry of variables into the model. In this forward selection procedure, setting $\alpha = 0.50$ has allowed us to look at six predictors. We might now drop back to step 4, where all predictors are significant at $\alpha = 0.05$ and say that our model is

$$Y = \alpha + \beta_2 X_2 + \beta_3 X_3 + \beta_4 X_4 + \beta_6 X_6 + \varepsilon.$$

We would then use the analysis associated with these four predictors. Our criterion is that all predictors are significant, given all the others. Not all data sets are as well-behaved as this one, however. It is possible, for instance, to have several predictors become nonsignificant at some step and then all become significant again at a later step.

It is always a good idea to look at the standard errors of the regression coefficients. If these are large, then we are not estimating the regression coefficients very well, and thus our predictions might not be too accurate. In both steps 5 and 7 we see large standard errors for b_5 (CHANGE), and this is evidence that we might not want to go with a six-variable model. It can be further argued that b_4 (SHARE) also has a reasonably large standard error, so that we might want to drop back and use the three-variable model, in which all the standard errors are reasonably small. Again we stress that there is no single answer to the question, "What is the best model?" It all depends upon the criteria you use. Let us look at some other variable-selection procedures that approach the question of which variables to include in different ways.

STEPWISE REGRESSION

One problem with the forward selection procedure is that once a variable is added to the model, you are stuck with it. You can never take it out again. For example, in step 6 of Figure 15.22, once TIME was added, ACCTS became superfluous. We might want to see what a regression with the five variables TIME, POTEN, ADV, SHARE, and CHANGE looks like, but the forward selection would never allow us to do this, because ACCTS was entered on the very first step. Similarly, we might want to replace ACCTS with either WORK or RATE and see if these six variables do a better job.

The stepwise procedure lets us get rid of a previously entered variable that becomes redundant. The stepwise procedure begins just like the forward selection. The default condition for the significance level for entry (SLE) in this procedure is SLE = 0.15. In addition, the procedure looks at all partial F-values. Any variable having a partial F not meeting the significance level for staying in (SLS) is dropped. Then the procedure continues (as in the forward selection) to add more variables, one at a time. Any variable deleted at any step goes back in the hopper and is eligible for reinclusion at a later step. The process terminates either when no new variables can be entered or when the one to be entered was the one dropped at the previous step. Unless otherwise specified, the significance level for staying in is SLS = 0.15. To change the default condition to, say, 0.10, simply add SLS=0.10 before the semicolon on the MODEL statement. Figure 15.23 shows the Cravens Data run through the stepwise procedure with SLS = 0.10. The first five steps are identical to the first five steps of the forward selection procedure. In step 6, however, we see a message that the variable CHANGE was removed from the list of variables in step 5. Thus, step 6 gives the analysis for a four-variable regression using X_2 (POTEN), X_3 (ADV), X_4 (SHARE), and X_6 (ACCTS). We see in step 5 that the F-value for CHANGE was 2.6, which was significant at $\alpha = 0.1236$. Since 0.1236 is greater than 0.10, CHANGE was dropped.

One of two things happened after step 6. Either no partial F-value, given X_2, X_3, X_4, and X_6, was large enough to be significant at $\alpha = 0.50$, or the largest partial F-value was the one for CHANGE, which had just been dropped.

FIGURE 15.23 SAS Stepwise Analysis

STEPWISE REGRESSION PROCEDURE FOR DEPENDENT VARIABLE SALES

STEP 1 VARIABLE ACCTS ENTERED R SQUARE = 0.56849518 C(P) = 67.55826191

	DF	SUM OF SQUARES	MEAN SQUARE	F	PROB > F
REGRESSION	1	23524074.19189501	23524074.19189501	30.30	0.0001
ERROR	23	17855474.73496103	776324.98847657		
TOTAL	24	41379548.92685610			

	B VALUE	STD ERROR	TYPE II SS	F	PROB > F
INTERCEPT	709.32583372				
ACCTS	21.72176971	3.94603304	23524074.19189501	30.30	0.0001

STEP 2 VARIABLE ADV ENTERED R SQUARE = 0.77510077 C(P) = 27.15634512

	DF	SUM OF SQUARES	MEAN SQUARE	F	PROB > F
REGRESSION	2	32073320.05178234	16036660.02589117	37.91	0.0001
ERROR	22	9306228.87507376	423010.40341244		
TOTAL	24	41379548.92685610			

	B VALUE	STD ERROR	TYPE II SS	F	PROB > F
INTERCEPT	50.29906160				
ADV	0.22652657	0.05038842	8549245.85988733	20.21	0.0002
ACCTS	19.04824598	2.97291380	17365864.05826321	41.05	0.0001

STEP 3 VARIABLE POTEN ENTERED R SQUARE = 0.82772280 C(P) = 18.35666096

	DF	SUM OF SQUARES	MEAN SQUARE	F	PROB > F
REGRESSION	3	34250796.02722474	11416932.00907491	33.63	0.0001
ERROR	21	7128752.89963136	339464.42379197		
TOTAL	24	41379548.92685610			

	B VALUE	STD ERROR	TYPE II SS	F	PROB > F
INTERCEPT	-327.23338939	0.00865564	2177475.97544240	6.41	0.0194
POTEN	0.02192192	0.04532744	7713721.79142546	22.72	0.0001
ADV	0.21607079	2.99936692	9128825.32465885	26.89	0.0001
ACCTS	15.55392188				

Figure 15.23 (Continued)

STEP 4 VARIABLE SHARE ENTERED R SQUARE = 0.90044970 C(P) = 5.43082968

	DF	SUM OF SQUARES	MEAN SQUARE	F	PROB > F
REGRESSION	4	37260202.46282121	9315050.61570530	45.23	0.0001
ERROR	20	4119346.46403489	205967.32320174		
TOTAL	24	41379548.92685610			

	B VALUE	STD ERROR	TYPE II SS	F	PROB > F
INTERCEPT	-1441.93182868				
POTEN	0.03821753	0.00797694	4727717.33107410	22.95	0.0001
ADV	0.17499004	0.03690666	4630368.54356448	22.48	0.0001
SHARE	190.14429731	49.74415347	3009406.43559647	14.61	0.0011
ACCTS	9.21389567	2.86521038	2129962.08538966	10.34	0.0043

STEP 5 VARIABLE CHANGE ENTERED R SQUARE = 0.91241574 C(P) = 4.97502565

	DF	SUM OF SQUARES	MEAN SQUARE	F	PROB > F
REGRESSION	5	37755351.59771667	7551070.31954333	39.59	0.0001
ERROR	19	3624197.32913943	190747.22784944		
TOTAL	24	41379548.92685610			

	B VALUE	STD ERROR	TYPE II SS	F	PROB > F
INTERCEPT	-1285.94337067				
POTEN	0.03763121	0.00768517	4573489.56079583	23.98	0.0001
ADV	0.15443602	0.03773852	3194373.73424488	16.75	0.0006
SHARE	196.94952750	48.05692351	3203732.09272627	16.80	0.0006
CHANGE	262.50049338	162.92631286	495149.13489546	2.60	0.1236
ACCTS	8.23411280	2.82357962	1622152.88530014	8.50	0.0089

STEP 6 VARIABLE CHANGE REMOVED R SQUARE = 0.90044970 C(P) = 5.43082968

	DF	SUM OF SQUARES	MEAN SQUARE	F	PROB > F
REGRESSION	4	37260202.46282121	9315050.61570530	45.23	0.0001
ERROR	20	4119346.46403489	205967.32320174		
TOTAL	24	41379548.92685610			

	B VALUE	STD ERROR	TYPE II SS	F	PROB > F
INTERCEPT	-1441.93182868				
POTEN	0.03821753	0.00797694	4727717.33107410	22.95	0.0001
ADV	0.17499004	0.03690666	4630368.54356449	22.48	0.0001
SHARE	190.14429731	49.74415347	3009406.43559647	14.61	0.0011
ACCTS	9.21389567	2.86521038	2129962.08538966	10.34	0.0043

ALL VARIABLES IN THE MODEL ARE SIGNIFICANT AT THE 0.1000 LEVEL.

Given the latter case, it would be useless to add CHANGE at step 7, because it would have just been dropped again at step 8. Therefore, the process terminates after step 6 and gives us the message that all variables are significant at $\alpha = 0.10$. (Inspection shows that as a matter of fact, all variables are significant at $\alpha = 0.05$.) The "best" model arrived at by the stepwise procedure is seen to be different from the "best" forward selection model. We are still not through, however, for these are not the only possible criteria for "best" models.

It would have been interesting to have seen another step, letting TIME enter, then removing ACCTS and going on from there. This might have happened if SLE or SLS had been specified differently, but how were we to know that before the program was run?

BACKWARD ELIMINATION

The problem with both the forward selection and the stepwise procedures is that they can miss a "key set" of variables. For example, suppose that Y (SALES) were a linear combination of X_1 (TIME), X_2 (ADV), and X_8 (RATE). That is, suppose that these three variables could together predict sales almost perfectly, although, for example, X_8 is almost worthless by itself. Neither the forward selection nor the stepwise procedure would discover this magic trio. Both assume that if the variable to be entered at any step is nonsignificant, then any weaker variable cannot be significant.

The backward elimination procedure starts with all variables in the model. It removes variables one at a time according to which one gives the smallest partial F-value, given all the rest, provided that this partial F-value is not significant. The procedure terminates when all partial F-values are significant at the significance level for staying in. A large significance level is usually used in order to give each variable a chance to stay in. A rule of thumb is to use $\alpha = 0.20$ or 0.25, but unless otherwise specified, SAS will use $\alpha = 0.10$.

Figure 15.24 shows the backward elimination procedure for the Cravens Data. Step 0 shows all eight predictors and their partial F-values. The smallest F-value (associated with the largest probability) is that for X_8 (RATE), so in step 1 this variable is dropped. Note that this has hardly any effect on R^2. In step 2 the variable X_7 (WORK) has the smallest F-value, so it is dropped. This hardly affects R^2 either. In step 3, X_6 (ACCTS) is dropped because its F-value is the smallest, having a probability of $0.2881 > 0.10$. Finally, in step 3 the smallest F-value is 4.26, which has probability $0.0530 < 0.10$, so no variables are dropped and the process terminates. A message is given that all variables are significant at $\alpha = 0.10$.

Note that we get quite a different set of predictors with the backward elimination procedure than with either forward selection or stepwise. In the first two procedures we had

X_6 = ACCTS, X_3 = ADV, X_2 = POTEN, X_4 = SHARE

FIGURE 15.24 SAS Backward Elimination Procedure

BACKWARD ELIMINATION PROCEDURE FOR DEPENDENT VARIABLE SALES

STEP 0 ALL VARIABLES ENTERED R SQUARE = 0.92203915 C(P) = 9.00000000

	DF	SUM OF SQUARES	MEAN SQUARE	F	PROB > F
REGRESSION	8	38153564.25210440	4769195.53151305	23.65	0.0001
ERROR	16	3225984.67475170	201624.04217198		
TOTAL	24	41379548.92685610			

	B VALUE	STD ERROR	TYPE II SS	F	PROB > F
INTERCEPT	-1507.81372984				
TIME	2.00956615	1.93065421	218442.90153750	1.08	0.3134
POTEN	0.03720491	0.00820230	4148313.42001816	20.57	0.0003
ADV	0.15098890	0.04710851	2071255.37925263	10.27	0.0055
SHARE	199.02353635	67.02792230	1777623.07698217	8.82	0.0090
CHANGE	290.85513399	186.78199574	488906.44513135	2.42	0.1390
ACCTS	5.55096065	4.77554962	272415.54793787	1.35	0.2621
WORK	19.79389189	33.67669223	69653.96590204	0.35	0.5649
RATE	8.18928366	128.50561301	818.82284513	0.00	0.9500

STEP 1 VARIABLE RATE REMOVED R SQUARE = 0.92201937 C(P) = 7.00406114

	DF	SUM OF SQUARES	MEAN SQUARE	F	PROB > F
REGRESSION	7	38152745.42925927	5450392.20417990	28.71	0.0001
ERROR	17	3226803.49759683	189811.97044687		
TOTAL	24	41379548.92685610			

	B VALUE	STD ERROR	TYPE II SS	F	PROB > F
INTERCEPT	-1485.88075785				
TIME	1.97454330	1.79574963	229490.83764567	1.21	0.2868
POTEN	0.03729049	0.00785101	4282222.67815504	22.56	0.0002
ADV	0.15196094	0.04324545	2343725.58802753	12.35	0.0027
SHARE	198.30848767	64.11718234	1815755.40273993	9.57	0.0066
CHANGE	295.86609393	164.38654930	614867.35490257	3.24	0.0897
ACCTS	5.61018822	4.54495654	289213.78714048	1.52	0.2339
WORK	19.89903131	32.63610128	70565.25639185	0.37	0.5501

STEP 2 VARIABLE WORK REMOVED R SQUARE = 0.92031405 C(P) = 5.35404547

	DF	SUM OF SQUARES	MEAN SQUARE	F	PROB > F
REGRESSION	6	38082180.17286742	6347030.02881124	34.65	0.0001
ERROR	18	3297368.75398868	183187.15299937		
TOTAL	24	41379548.92685610			

	B VALUE	STD ERROR	TYPE II SS	F	PROB > F
INTERCEPT	-1165.47855369				
TIME	2.26935112	1.69898362	326828.57515075	1.78	0.1983
POTEN	0.03827800	0.00754688	4712573.84321485	25.73	0.0001
ADV	0.14067029	0.03839221	2459312.09671656	13.43	0.0018
SHARE	221.60469221	50.58309112	3515945.91528346	19.19	0.0004
CHANGE	285.10928426	160.55965553	577623.53145916	3.15	0.0927
ACCTS	4.37770296	3.99903763	219521.27264808	1.20	0.2881

STEP 3 VARIABLE ACCTS REMOVED R SQUARE = 0.91500898 C(P) = 4.44281083

	DF	SUM OF SQUARES	MEAN SQUARE	F	PROB > F
REGRESSION	5	37862658.90021934	7572531.78004387	40.91	0.0001
ERROR	19	3516890.02663675	185099.47508615		
TOTAL	24	41379548.92685610			

	B VALUE	STD ERROR	TYPE II SS	F	PROB > F
INTERCEPT	-1113.78787943				
TIME	3.61210118	1.18169995	1729460.18780281	9.34	0.0065
POTEN	0.04208812	0.00673122	7236631.11940864	39.10	0.0001
ADV	0.12885675	0.03703609	2240625.11884974	12.10	0.0025
SHARE	256.95554016	39.13606967	7979335.79655743	43.11	0.0001
CHANGE	324.53344995	157.28308421	788061.45599777	4.26	0.0530

ALL VARIABLES IN THE MODEL ARE SIGNIFICANT AT THE 0.1000 LEVEL.

and possibly also

$$X_5 = \texttt{CHANGE} \quad \text{and} \quad X_1 = \texttt{TIME}.$$

Now we have

$$X_1 = \texttt{TIME}, \quad X_2 = \texttt{POTEN}, \quad X_3 = \texttt{ADV}, \quad X_4 = \texttt{SHARE}, \quad \text{and}$$
$$X_5 = \texttt{CHANGE}.$$

If we wanted our predictors all to be significant at $\alpha = 0.05$ instead of $\alpha = 0.10$, we would reduce the set obtained in the backward elimination procedure to just X_1, X_2, X_3, and X_4. This set of variables can also be defended on the basis that b_5 has an extremely large standard error. Notice that the six predictors obtained by forward selection give $R^2 = 0.9203$, the four obtained by stepwise gave $R^2 = 0.9004$, and the five obtained by backward elimination gave $R^2 = 0.9150$. The five variables obtained using backward elimination thus explain almost as much variation as the six selected by forward selection; in addition, all five backward variables are significant at $\alpha = 0.10$. It seems that some relationships in the data detected by the backward elimination procedure escaped the forward selection and stepwise procedures. For this reason, many statisticians prefer the backward elimination procedure, especially if they have no access to the theoretical structure relating to the variables. Only the backward elimination procedure allows "key sets" of variables to act all together.

A potential problem with backward elimination is similar to the problem with forward selection. Here, once a variable is removed, it cannot be reentered. A backward stepwise procedure would seem a good thing to have. (Such a procedure is offered by BMD programs.)

MAXIMUM R^2 IMPROVEMENT

The maximum R^2 improvement technique does not give a final answer, as do the three previous techniques we have studied. Rather, it gives a "best" single-predictor regression, a "best" two-predictor regression, and so on, until it gives the "best" k-predictor regression. The researcher uses the results to decide which regression to use. This technique is designed to look at a large number of regressions, although it falls quite a bit short (purposely) of looking at all possible regressions, since for a large number of predictors, looking at all possible regressions would be prohibitively expensive.

The procedure followed is this. In step 1 the variable is entered that gives the largest R^2 when acting alone. This is the best single-predictor regression. In step 2 the variable that increases R^2 the most, in the presence of the single strongest predictor, is added. This sounds like the forward selection procedure, but there is a difference. At each step the program can perform switches. For example, suppose that the strongest predictor is X_1, and that given X_1, X_2

increases R^2 the most. Variables X_3, X_4, and X_5 are still "in the hopper." The computer looks to see if replacing X_1 by X_3, X_4, or X_5 will yield a larger R^2 than X_1 and X_2 will. If, for example, $R^2_{23,Y} > R^2_{12,Y}$, then the computer will replace X_1 by X_3. It will call this the "best two-predictor model found." Note that this is not guaranteed to be the very best two-predictor model, but only the best one *found*. It may be that X_3 and X_5 give a higher R^2 than do X_2 and X_3, but all the computer has found is that X_2 and X_3 together explain more than do X_1 and X_2.

Suppose that X_2 and X_3 comprise the best two-predictor model found. Now X_1, X_4, and X_5 are in the hopper. For step 3, the computer looks at

$$X_2, X_3, X_1, \qquad X_2, X_3, X_4 \quad \text{and} \quad X_2, X_3, X_5$$

and selects the trio that gives the largest R^2. Suppose this is

$$X_2, X_3, X_4.$$

The computer then holds the last variable to enter, X_4 in this example, and then says, "Can I switch X_2 or X_3 for X_1 or X_5 and get a larger R^2?" In other words, the computer looks at

$$X_4, X_2, X_1, \qquad X_4, X_2, X_5 \quad \text{and} \quad X_4, X_3, X_1, \qquad X_4, X_3, X_5$$

to see if any of these trios will beat X_2, X_3, and X_4. If so, the switch is made that increases the multiple coefficient of determination the most. If not, then the computer proceeds to the next step. The process continues until all k variables have been included.

Note that the computer does not make significance tests in this procedure. The significance levels for inclusion and deletion are not used.

The maximum R^2 improvement technique is applied to the Cravens Data in Figure 15.25. Step 1 enters X_6 (ACCTS), since it has the highest simple correlation with Y (SALES). In step 2, X_3 (ADV) is added, and in step 3, X_2 (POTEN) is added. The variables X_2, X_3, and X_6 together explain 82.77 percent of the variation in Y. But the computer finds that if X_6 (ACCTS) is replaced by X_4 (SHARE), then the trio X_2, X_3, and X_5 explain 84.90 percent of the variation in Y. So it calls the model containing X_2, X_3, and X_5 the "best three-variable model found." Note that X_6 (ACCTS) enters right back in at step 4 and is thrown out again in step 5, only to enter again in step 6.

At the various steps we find the best one-, two-, three-, . . . , eight-predictor models in terms of which ones the computer could find that explained the most variation. Which model do we use? One should consider not only the size of R^2 at each step but also the significance levels given for the F-values. Practical and theoretical considerations also count in many actual problems.

We see in Figure 15.25 that all the variables in the best one-, two-, three-, and four-predictor models are significant at $\alpha = 0.05$. Of course, the multiple coefficient of determination is largest in the four-predictor model, using X_2 (POTEN), X_3 (ADV), X_4 (SHARE), and X_6 (ACCTS). Note that these are the same variables found using the stepwise technique, and the same ones found

FIGURE 15.25 SAS Maximum R^2 Improvement Technique

```
MAXIMUM R-SQUARE IMPROVEMENT FOR DEPENDENT VARIABLE SALES
STEP 1   VARIABLE ACCTS ENTERED   R SQUARE = 0.56849518    C(P) = 67.55826191

                 DF         SUM OF SQUARES         MEAN SQUARE          F         PROB > F
REGRESSION        1      23524074.19189501   23524074.19189501      30.30         0.0001
ERROR            23      17855474.73496109     776324.98847657
TOTAL            24      41379548.92685610

                 B VALUE        STD ERROR          TYPE II SS           F         PROB > F
INTERCEPT      709.32383372
ACCTS           21.72176971     3.94603304   23524074.19189501      30.30         0.0001

THE ABOVE MODEL IS THE BEST 1 VARIABLE MODEL FOUND.
```

```
STEP 2   VARIABLE ADV ENTERED   R SQUARE = 0.77510077    C(P) = 27.15634512

                 DF         SUM OF SQUARES         MEAN SQUARE          F         PROB > F
REGRESSION        2      32073320.05178234   16036660.02589117      37.91         0.0001
ERROR            22       9306228.87507376     423010.40341244
TOTAL            24      41379548.92685610

                 B VALUE        STD ERROR          TYPE II SS           F         PROB > F
INTERCEPT       50.29906160
ADV              0.22652657     0.05038842    8549245.85988733      20.21         0.0002
ACCTS           19.04824598     2.97291380   17365864.05826321      41.05         0.0001

THE ABOVE MODEL IS THE BEST 2 VARIABLE MODEL FOUND.
```

```
STEP 3   VARIABLE POTEN ENTERED   R SQUARE = 0.82772280    C(P) = 18.35666096

                 DF         SUM OF SQUARES         MEAN SQUARE          F         PROB > F
REGRESSION        3      34250796.02722474   11416932.00907491      33.63         0.0001
ERROR            21       7128752.89963136     339464.42379197
TOTAL            24      41379548.92685610

                 B VALUE        STD ERROR          TYPE II SS           F         PROB > F
INTERCEPT     -327.23338939
POTEN            0.02192192     0.00865564    2177475.97544240       6.41         0.0194
ADV              0.21607079     0.04532744    7713721.79142546      22.72         0.0001
ACCTS           15.55392158     2.99936692    9128825.32465885      26.89         0.0001
```

STEP 3 ACCTS REPLACED BY SHARE R SQUARE = 0.84897591 C(P) = 13.99485796

	DF	SUM OF SQUARES	MEAN SQUARE	F	PROB > F
REGRESSION	3	35130240.37743155	11710080.12581052	39.35	0.0001
ERROR	21	6249308.54942455	297586.12140117		
TOTAL	24	41379548.92685610			

	B VALUE	STD ERROR	TYPE II SS	F	PROB > F
INTERCEPT	-1603.58091828				
POTEN	0.05428605	0.00747411	15698916.49197692	52.75	0.0001
ADV	0.16748034	0.04427318	4258521.38246777	14.31	0.0011
SHARE	282.74666591	48.75558026	10008269.67486566	33.63	0.0001

THE ABOVE MODEL IS THE BEST 3 VARIABLE MODEL FOUND.

STEP 4 VARIABLE ACCTS ENTERED R SQUARE = 0.90044970 C(P) = 5.43082968

	DF	SUM OF SQUARES	MEAN SQUARE	F	PROB > F
REGRESSION	4	37260202.46282121	9315050.61570530	45.23	0.0001
ERROR	20	4119346.46403489	205967.32320174		
TOTAL	24	41379548.92685610			

	B VALUE	STD ERROR	TYPE II SS	F	PROB > F
INTERCEPT	-1441.93182868				
POTEN	0.03821753	0.00797694	4727717.33107410	22.95	0.0001
ADV	0.17499004	0.03690666	4630368.54356449	22.48	0.0001
SHARE	190.14429731	49.74415347	3009406.43559647	14.61	0.0011
ACCTS	9.21389567	2.86521038	2129962.08538966	10.34	0.0043

THE ABOVE MODEL IS THE BEST 4 VARIABLE MODEL FOUND.

STEP 5 VARIABLE CHANGE ENTERED R SQUARE = 0.91241574 C(P) = 4.97502565

	DF	SUM OF SQUARES	MEAN SQUARE	F	PROB > F
REGRESSION	5	37755351.59771667	7551070.31954333	39.59	0.0001
ERROR	19	3624197.32913943	190747.22784944		
TOTAL	24	41379548.92685610			

	B VALUE	STD ERROR	TYPE II SS	F	PROB > F
INTERCEPT	-1285.94337067				
POTEN	0.03763121	0.00768517	4573489.56079583	23.98	0.0001
ADV	0.15443602	0.03773852	3194373.73424488	16.75	0.0006
SHARE	196.94952750	48.05692351	3203732.09272627	16.80	0.0006
CHANGE	262.50049338	162.92631286	495149.13489546	2.60	0.1236
ACCTS	8.23411280	2.82357962	1622152.88530014	8.50	0.0089

Figure 15.25 (Continued)

STEP 5 ACCTS REPLACED BY TIME R SQUARE = 0.91500898 C(P) = 4.44281083

	DF	SUM OF SQUARES	MEAN SQUARE	F	PROB > F
REGRESSION	5	37862658.90021934	7572531.78004387	40.91	0.0001
ERROR	19	3516890.02663676	185099.47508615		
TOTAL	24	41379548.92685610			

	B VALUE	STD ERROR	TYPE II SS	F	PROB > F
INTERCEPT	-1113.78787943				
TIME	3.61210118	1.18169995	1729460.18780281	9.34	0.0065
POTEN	0.04208812	0.00673122	7236631.11940864	39.10	0.0001
ADV	0.12885675	0.03703609	2240625.11884974	12.10	0.0025
SHARE	256.95554016	39.13606967	7979335.79655744	43.11	0.0001
CHANGE	324.53344995	157.28308421	788061.45599777	4.26	0.0530

THE ABOVE MODEL IS THE BEST 5 VARIABLE MODEL FOUND.

STEP 6 VARIABLE ACCTS ENTERED R SQUARE = 0.92031405 C(P) = 5.35404547

	DF	SUM OF SQUARES	MEAN SQUARE	F	PROB > F
REGRESSION	6	38082180.17286742	6347030.02881124	34.65	0.0001
ERROR	18	3297368.75398868	183187.15299937		
TOTAL	24	41379548.92685610			

	B VALUE	STD ERROR	TYPE II SS	F	PROB > F
INTERCEPT	-1165.47855369				
TIME	2.26935112	1.69898362	326828.57515075	1.78	0.1983
POTEN	0.03827800	0.00754688	4712573.84321485	25.73	0.0001
ADV	0.14067029	0.03839221	2459312.09671656	13.43	0.0018
SHARE	221.60469221	50.58309112	3515945.91528346	19.19	0.0004
CHANGE	285.10928426	160.55965553	577623.53145916	3.15	0.0927
ACCTS	4.37770296	3.99903763	219521.27264808	1.20	0.2881

THE ABOVE MODEL IS THE BEST 6 VARIABLE MODEL FOUND.

STEP 7 VARIABLE WORK ENTERED R SQUARE = 0.92201937 C(P) = 7.00406114

	DF	SUM OF SQUARES	MEAN SQUARE	F	PROB > F
REGRESSION	7	38152745.42925927	5450392.20417990	28.71	0.0001
ERROR	17	3226803.49759683	189811.97044687		
TOTAL	24	41379548.92685610			

	B VALUE	STD ERROR	TYPE II SS	F	PROB > F
INTERCEPT	-1485.88075785				
TIME	1.97454330	1.79574963	229490.83764567	1.21	0.2868
POTEN	0.03729049	0.00785101	4282222.67815504	22.56	0.0002
ADV	0.15196094	0.04324545	2343725.58802753	12.35	0.0027
SHARE	198.30848767	64.[illegible]1718234	1815755.40273993	9.57	0.0066
CHANGE	295.86609393	164.38654930	614867.35490257	3.24	0.0897
ACCTS	5.61018822	4.54495654	289213.78714048	1.52	0.2339
WORK	19.89903131	32.63610128	70565.25639185	0.37	0.5501

THE ABOVE MODEL IS THE BEST 7 VARIABLE MODEL FOUND.

STEP 8 VARIABLE RATE ENTERED R SQUARE = 0.92203915 C(P) = 9.00000000

	DF	SUM OF SQUARES	MEAN SQUARE	F	PROB > F
REGRESSION	8	38153564.25210440	4769195.53151305	23.65	0.0001
ERROR	16	3225984.67475170	201624.04217198		
TOTAL	24	41379548.92685610			

	B VALUE	STD ERROR	TYPE II SS	F	PROB > F
INTERCEPT	-1507.81372984				
TIME	2.00956615	1.93065421	218442.90153751	1.08	0.3134
POTEN	0.03720491	0.00820230	4148313.4200[illegible]816	20.57	0.0003
ADV	0.15098890	0.04710851	2071255.37925263	10.27	0.0055
SHARE	199.02353635	67.02792230	1777623.07698218	8.82	0.0090
CHANGE	290.85513399	186.78199574	488906.44513135	2.42	0.1390
ACCTS	5.55096065	4.77554962	272415.54793787	1.35	0.2621
WORK	19.79389189	33.57669223	69653.96590204	0.35	0.5649
RATE	8.18928366	128.30561301	818.82284513	0.00	0.9500

THE ABOVE MODEL IS THE BEST 8 VARIABLE MODEL FOUND.

in step 4 of forward selection. In the stepwise procedure the process terminated after four variables had been added, because the next strongest variable, X_5 (CHANGE), was not significant at $\alpha = 0.10$. In the maximum R^2 improvement procedure, we see X_5 entering at step 5 and note that it is not significant, given the others, at $\alpha = 0.10$. However, we see X_6 (ACCTS) replaced by X_1 (TIME), and then all five of these variables are conditionally significant at $\alpha = 0.10$. Thus, while the stepwise procedure found that for significance at $\alpha = 0.10$, the four variables X_2, X_3, X_4, and X_6 together explain $R^2 = 90\%$ of the variation in Y (SALES), the maximum R^2 improvement technique found five variables, X_1, X_2, X_3, X_4, and X_5 that are all significant at $\alpha = 0.10$ and together explain more of the variation ($R^2 = 91.5\%$). Since the maximum R^2 improvement technique can look at more regressions than stepwise, its chances of finding the strongest set of significant predictors are greater.

The disadvantage of the maximum R^2 improvement technique lies in the amount of computer time it takes. It can be quite expensive to run, especially if the original model contains a large number of predictors. There is no automatic cutoff for reducing the amount of output; the program will terminate only when the best k-predictor model, containing all predictors, is found. (However one can instruct the computer to terminate after a certain predetermined number of variables have been included. Refer to the SAS manual.) This procedure does not pay any attention to the significance level of any partial F-value that it computes.

MINIMUM R^2 IMPROVEMENT

The minimum R_2 improvement technique may sound like the dumbest thing anyone would wish to do. It works like the maximum R^2 improvement technique with one exception: It adds variables and performs switches so that the multiple coefficient of determination is increased as *little* as possible at each step.

In step 1 the single *weakest* predictor, X_7 (WORK), is entered first. Subsequently in step 1, X_7 is replaced by the next weakest predictor, X_8 (RATE); X_8 is replaced by the next weakest predictor, and so on. Finally, in the eighth part of step 1 the strongest single predictor, X_6 (ACCTS) is added. The single-predictor model using X_6 has the largest coefficient of determination and is thus called the best model found.

In step 2 the computer looks for the variable that will improve R^2 the *least* when used in conjunction with X_6 (ACCTS). This variable is X_7 (WORK). Switches with these variables are performed in such a way as to increase R^2 by *the smallest amount possible* with each switch until two variables are found (in this case X_2 (POTEN) and X_4 (SHARE)) that form the strongest pair.

This process of increasing R^2 by the smallest amounts possible may seem quite inefficient, and indeed it is. It is designed to look at a very large number of regressions (but still not necessarily all possible regressions). This procedure is useful in that it allows the researcher to see a lot of information and increases the chances that the best pair, triple, or whatever, of variables will be found. It is so time-consuming, however, that in a problem with many predictors the cost of performing this analysis could be prohibitive. In the interest of saving space, Figure 15.26 shows only the *final* 1-variable, 2-variable, . . . , 8-variable models found by the minimum R^2 improvement technique. This is because there are eight parts to step 1, six parts to step 2, six parts to step 3, four parts to step 4, five parts to step 5, three parts to step 6, two parts to step 7, and one to step 8. All this takes up 12 pages of computer printout! The switches performed, along with R^2 values for each model, are summarized in Figure 15.27.

Notice that the maximum R^2 improvement and the minimum R^2 improvement techniques may or may not give the same best models with a designated number of predictors. For example, the best two-predictor model found by the MAXR procedure used the variables X_3 (ADV) and X_6 (ACCTS) and found a multiple R^2 of 77.5 percent. The MINR procedure, however, never considered these two variables together and arrived at a best two-predictor model using X_2 (POTEN) and X_4 (SHARE), which together explain only 74.6 percent of the variation in sales. Both procedures do, however, find the same "best" models for $k = 3, 4, 5, 6, 7$, and of course, for $k = 8$. The MAXR technique finds these best models a lot more quickly than does the MINR procedure, however.

COMPARISON OF THE FIVE SAS PROCEDURES

We can see that the choice of variables to be included in the final model is very much a function of the criteria we apply in selecting variables. Only in extremely clear-cut problems would we expect all five techniques described here to produce the same subset of variables for the reduced model. Usually one wants a model in which all predictors are significant; in which the regression coefficients have small standard errors; and which gives a large multiple coefficient of determination. In addition, the researcher probably has some intuitive idea about which variables to include in the final model, because they make good sense to the problem. Some researcher favoritism might be based on convenience. If one is lucky, one of these five analyses will yield a set of variables that satisfy all these requirements. Then the researcher can be relatively happy with the final model.

In most situations, however, there will be several models to choose from, all of them satisfying some criteria while falling short in others. The final choice of the model then is up to the researcher, rather than the statistician, because the researcher is the one familiar with the theory of the field. While the statistician is useful in explaining what the analysis means, the results of

FIGURE 15.26 SAS Minimum R^2 Improvement Technique

```
MINIMUM R-SQUARE IMPROVEMENT FOR DEPENDENT VARIABLE SALES

STEP 1  TIME REPLACED BY ACCTS  R SQUARE = 0.56849518    C(P) = 67.55826191

                DF            SUM OF SQUARES         MEAN SQUARE          F       PROB > F
REGRESSION       1         23524074.19189501   23524074.19189501      30.30       0.0001
ERROR           23         17855474.73496109     776324.98847657
TOTAL           24         41379548.92685610

                B VALUE       STD ERROR              TYPE II SS           F       PROB > F
INTERCEPT     709.32383372
ACCTS          21.72176971     3.94603304        23524074.19189501    30.30       0.0001

THE ABOVE MODEL IS THE BEST 1 VARIABLE MODEL FOUND.
-----------------------------------------------------------------------------------------
STEP 2  ACCTS REPLACED BY SHARE  R SQUARE = 0.74606224    C(P) = 33.11595710

                DF            SUM OF SQUARES         MEAN SQUARE          F       PROB > F
REGRESSION       2         30871718.99496378   15435859.49748189      32.32       0.0001
ERROR           22         10507829.93189232     477628.63326783
TOTAL           24         41379548.92685610

                B VALUE       STD ERROR              TYPE II SS           F       PROB > F
INTERCEPT   -1578.08881206
POTEN           0.06117826     0.00918324        21197905.35034394    44.38       0.0001
SHARE         340.44371373    58.66779312        16083534.29245808    33.67       0.0001

THE ABOVE MODEL IS THE BEST 2 VARIABLE MODEL FOUND.
-----------------------------------------------------------------------------------------
STEP 3  TIME REPLACED BY ADV  R SQUARE = 0.84897591    C(P) = 13.99485796

                DF            SUM OF SQUARES         MEAN SQUARE          F       PROB > F
REGRESSION       3         35130240.37743154   11710080.12581051      39.35       0.0001
ERROR           21          6249308.54942456     297586.12140117
TOTAL           24         41379548.92685610

                B VALUE       STD ERROR              TYPE II SS           F       PROB > F
INTERCEPT   -1603.58091828
POTEN           0.05428605     0.00747411        15698916.49197694    52.75       0.0001
ADV             0.16748034     0.04427318         4258521.38246776    14.31       0.0011
SHARE         282.74666591    48.75558026        10008269.67486566    33.63       0.0001

THE ABOVE MODEL IS THE BEST 3 VARIABLE MODEL FOUND.
```

STEP 4 WORK REPLACED BY SHARE R SQUARE = 0.90044970 C(P) = 5.43082968

	DF	SUM OF SQUARES	MEAN SQUARE	F	PROB > F
REGRESSION	4	37250202.46282120	9315050.61570530	45.23	0.0001
ERROR	20	4119346.46403490	205967.32320174		
TOTAL	24	41379548.92685610			

	B VALUE	STD ERROR	TYPE II SS	F	PROB > F
INTERCEPT	-1441.93182868				
POTEN	0.03821753	0.00797694	4727717.33107411	22.95	0.0001
ADV	0.17493004	0.03690666	4630368.54356448	22.48	0.0001
SHARE	190.14423731	49.74415347	3009406.43559647	14.61	0.0011
ACCTS	9.21389567	2.86521038	2129962.08538966	10.34	0.0043

THE ABOVE MODEL IS THE BEST 4 VARIABLE MODEL FOUND.

STEP 5 ACCTS REPLACED BY TIME R SQUARE = 0.91500898 C(P) = 4.44281083

	DF	SUM OF SQUARES	MEAN SQUARE	F	PROB > F
REGRESSION	5	37862658.90021933	7572531.78004387	40.91	0.0001
ERROR	19	3516890.02663677	185099.47508615		
TOTAL	24	41379548.92685610			

	B VALUE	STD ERROR	TYPE II SS	F	PROB > F
INTERCEPT	-1113.78787943				
TIME	3.61210118	1.18169995	1729460.18780281	9.34	0.0065
POTEN	0.04208812	0.00673122	7236631.11940865	39.10	0.0001
ADV	0.12885675	0.03703609	2240625.11884974	12.10	0.0025
SHARE	256.95554016	39.13606967	7979335.79655745	43.11	0.0001
CHANGE	324.53344995	157.28308421	788061.45599777	4.26	0.0530

THE ABOVE MODEL IS THE BEST 5 VARIABLE MODEL FOUND.

STEP 6 RATE REPLACED BY ACCTS R SQUARE = 0.92031405 C(P) = 5.35404547

	DF	SUM OF SQUARES	MEAN SQUARE	F	PROB > F
REGRESSION	6	38082180.17286742	6347030.02881124	34.65	0.0001
ERROR	18	3297368.75398868	183187.15299937		
TOTAL	24	41379548.92685610			

	B VALUE	STD ERROR	TYPE II SS	F	PROB > F
INTERCEPT	-1165.47855369				
TIME	2.26935112	1.69898362	326828.57515075	1.78	0.1983
POTEN	0.03827800	0.00754688	4712573.84321486	25.73	0.0001
ADV	0.14067029	0.03839221	2459312.09671656	13.43	0.0018
SHARE	221.60469221	50.58309112	3515945.91528347	19.19	0.0004
CHANGE	285.10928426	160.55965553	577623.53145916	3.15	0.0927
ACCTS	4.37770296	3.99903763	219521.27264808	1.20	0.2881

THE ABOVE MODEL IS THE BEST 6 VARIABLE MODEL FOUND.

Figure 15.26 (Continued)

STEP 7 RATE REPLACED BY WORK R SQUARE = 0.92201937 C(P) = 7.00406114

	DF	SUM OF SQUARES	MEAN SQUARE	F	PROB > F
REGRESSION	7	38152745.42925926	5450392.20417989	28.71	0.0001
ERROR	17	3226803.49759684	189811.97044687		
TOTAL	24	41379548.92685610			

	B VALUE	STD ERROR	TYPE II SS	F	PROB > F
INTERCEPT	-1485.88075785				
TIME	1.97454330	1.79574963	229490.83764567	1.21	0.2868
POTEN	0.03729049	0.00785101	4282222.67815505	22.56	0.0002
ADV	0.15196094	0.04324545	2343725.58802754	12.35	0.0027
SHARE	198.30848767	64.11718234	1815755.40273993	9.57	0.0066
CHANGE	295.86609393	164.38654930	614867.35490257	3.24	0.0897
ACCTS	5.61018822	4.54495654	289213.78714048	1.52	0.2339
WORK	19.89903131	32.63610128	70565.25639185	0.37	0.5501

THE ABOVE MODEL IS THE BEST 7 VARIABLE MODEL FOUND.

STEP 8 VARIABLE RATE ENTERED R SQUARE = 0.92203915 C(P) = 9.00000000

	DF	SUM OF SQUARES	MEAN SQUARE	F	PROB > F
REGRESSION	8	38153564.25210440	4769195.53151305	23.65	0.0001
ERROR	16	3225984.67475170	201624.04217198		
TOTAL	24	41379548.92685610			

	B VALUE	STD ERROR	TYPE II SS	F	PROB > F
INTERCEPT	-1507.81372984				
TIME	2.00956615	1.93065421	218442.90153751	1.08	0.3134
POTEN	0.03720491	0.00820230	4148313.42001817	20.57	0.0003
ADV	0.15098890	0.04710851	2071255.37925263	10.27	0.0055
SHARE	199.02353635	67.02792230	1777623.07698218	8.82	0.0090
CHANGE	290.85513399	186.78199574	488906.44513135	2.42	0.1390
ACCTS	5.55096065	4.77554962	272415.54793787	1.35	0.2621
WORK	19.79389189	33.67669223	69653.96590204	0.35	0.5649
RATE	8.18928366	128.50561301	818.82284513	0.00	0.9500

THE ABOVE MODEL IS THE BEST 8 VARIABLE MODEL FOUND.

FIGURE 15.27 Summary of the Steps in the MINR Procedure Applied to Cravens Data

	R^2
Step 1	
WORK entered	0.0137
WORK replaced by RATE	0.1615
RATE replaced by SHARE	0.2338
SHARE replaced by CHANGE	0.2393
CHANGE replaced by ADV	0.3554
ADV replaced by POTEN	0.3574
POTEN replaced by TIME	0.3880
TIME replaced by ACCTS	0.5685
Step 2	
WORK added to ACCTS	0.5696
WORK replaced by TIME	0.5747
ACCTS replaced by ADV	0.5953
TIME replaced by POTEN	0.6071
ADV replaced by ACCTS	0.6413
ACCTS replaced by SHARE	0.7461
Step 3	
WORK added to POTEN and SHARE	0.7730
WORK replaced by RATE	0.7735
RATE replaced by ACCTS	0.7885
ACCTS replaced by CHANGE	0.8086
CHANGE replaced by TIME	0.8121
TIME replaced by ADV	0.8490
Step 4	
RATE added to POTEN, ADV, SHARE	0.8512
RATE replaced by WORK	0.8513
SHARE replaced by ACCTS	0.8641
WORK replaced by SHARE	0.9004
Step 5	
RATE added to POTEN, ADV, SHARE, ACCTS	0.9013
ACCTS replaced by TIME	0.9025
RATE replaced by ACCTS	0.9064
TIME replaced by CHANGE	0.9124
ACCTS replaced by TIME	0.9150
Step 6	
WORK added to TIME, POTEN, ADV, SHARE, CHANGE	0.9150
WORK replaced by RATE	0.9154
RATE replaced by ACCTS	0.9203
Step 7	
RATE added to TIME, POTEN, ADV, SHARE, CHANGE, ACCTS	0.9204
RATE replaced by WORK	0.9220
Step 8	
RATE added to TIME, POTEN, ADV, SHARE, CHANGE, ACCTS, WORK	0.9220

numerical analyses should never blindly take precedence over expertise in the field of application.

A word of caution is in order. None of these five variable-selection techniques is guaranteed to result in a regression equation that will give good predictions when applied to new sets of data. All results of regression analysis are quite data-dependent: the regression line itself, R^2, the standard error of estimate, and so on. Based on the conclusions of the significance tests, *provided all the assumptions of the model are met,* we can be reasonably certain that significant predictors are those that really should be used. But just because our predictors can explain, say, 92 percent of the variation in Y in this sample gives no assurance that these same predictors would do as well in another sample. If the objective is to make accurate predictions, then we need something beyond variable-selection techniques. The C_p-statistic, which is shown at each step of every SAS stagewise procedure, is designed to help us select the "best" model for use in describing these data and in predicting future values of Y. It will be discussed in Chapter 16.

EXERCISES

15.10 Recall that in the analysis of the Children Data of Exercise 11.6 the overall regression significance test indicated that at least one of the predictors (age, height, or weight) was significant. However, no partial F-test on one predictor, given the other two, indicated a significant predictor.

Run the data through the SAS forward selection or stepwise program, and see if any model can be found in which all predictors are significant at $\alpha = 0.05$. (Use default conditions for the inclusion and deletion criteria.)

15.11 Suppose it is desired to use the Tennessee Data of Exercise 15.3 to estimate the percentage of families in a Tennessee county that have incomes falling below the poverty level. Use the five stagewise procedures available through SAS to analyze these data. Use a significance level for entry of 0.25 and a significance level for staying in of 0.15. You might also want to obtain a correlation matrix and other appropriate statistics. Compare the results of the five stagewise procedures, and propose a final model. Justify your choice.

15.12 Consider the MPG Data of Exercise 10.6. Since the relationship of mpg with both horsepower and weight is curvilinear, fit a model

$$Y = \alpha + \beta_1 X_1 + \beta_2 X_2 + \beta_3 X_1^2 + \beta_4 X_2^2 + \beta_5 X_3 + \varepsilon,$$

where Y = mpg, X_1 = HP, X_2 = weight, and X_3 = transmission type.

a. Analyze the full model. Would you use the full model to predict mpg?

b. Use an available SAS variable-selection procedure to obtain an acceptable reduced model. Examine regression coefficients to see if they make sense.

15.13 Work Exercise 15.5 using an SAS selection procedure.

15.7

SPSS^X SELECTION PROCEDURES

Several variable-selection procedures are available through SPSS^X. In order to perform them, an equation method must be specified by including a method keyword in the `REGRESSION` command. We have already used the `ENTER` method to perform simple and multiple linear regression and the `FORWARD` method in polynomial regression. Let us consider these equation methods, and the others available, in more detail.

In any of the methods used, variables included in the model at any step must pass a tolerance test. This is a protection against the effects of severe multicollinearity. If severely redundant variables are used in a regression, estimates of the regression coefficients are extremely unreliable and partial *F*-values become small. The tolerance of an independent variable is 1 minus the R^2 of that variable with the predictors already included in the model. SPSS^X will not enter a variable if either its tolerance is too small or if entering it would cause the tolerance of a predictor already in the equation to become too small. The minimum acceptable tolerance in SPSS^X is 0.01. It is possible to change this criterion to a larger or smaller value, but it is generally not advisable to do the latter.

THE ENTER METHOD

The `ENTER` method enters all variables (or a block of variables) at once. If just the keyword `ENTER` is used, then a model using all independent variables in the `VARIABLES` subcommand is fitted. (Refer to Section 15.1, in which all eight predictors were used to predict Y = sales.) A subset of predictors could have been specified, however. For example, to use just `TIME`, `POTEN`, and `ADV` to predict `SALES`, when all nine variables have been named in the `VARIABLES` subcommand, write

```
ENTER TIME POTEN ADV
```

in the `REGRESSION` command. The `ENTER` subcommand is most useful in sequential analyses (as we saw in Section 10.5, where a sequential analysis of a polynomial model was performed), and not as useful as a variable-selection procedure in which we have no *a priori* reason to specify the order of entry of variables into the model.

THE FORWARD METHOD

The SPSS^X `FORWARD` method performs a forward selection procedure. At each step the strongest predictor, given those already included, is entered into the model. Strength is measured by the size of the partial F-value associated with that predictor. If there are several candidate variables to be entered, the one with greatest partial F is entered, provided that this partial F is large enough to satisfy the entry criterion. SPSS^X will enter a variable into the equation as long as its partial F is significant at $\alpha \leq 0.05$. This is called the probability of F to enter, or PIN. The default condition is PIN $= 0.05$; unless you specify otherwise, the program will use this value. To change this specification to give more variables a chance to enter, add a subcommand such as `CRITERIA=PIN(0.1)` before the `DEPENDENT` subcommand and after the `VARIABLES` subcommand.

It can be argued that it is not a good idea to calculate significance probabilities at each step of a selection procedure, because these probabilities are valid *only* at the step in which a reasonable model has been found. In steps where some important predictors are omitted, the error mean square will be inflated, and thus F-values will be smaller than they should be to indicate the significance of the predictor. If redundant variables are included in the equation at any step, diagonal elements of $\mathbf{X}'\mathbf{X}$ are inflated, also reducing F-values. Thus, for those who object to adding variables into the model on the basis of a possibly meaningless significance probability, SPSS^X provides for another entry criterion based on the size, but not the significance, of the partial F-statistic. Specifying, for example, `CRITERIA=FIN(2)` will allow variables to enter into the equation as long as their partial F-values are at least 2.00. The default condition for FIN (F in) is 3.84, which is roughly equivalent to significance level 0.05. (Note in the F-tables in Appendix I that in order to be significant at $\alpha = 0.05$, an F-value with 1 degree of freedom in the numerator must be around 4.)

Figure 15.28 shows a worksheet for performing a forward selection procedure on the Cravens Data, with the entry criterion specified as `PIN=0.10`. The only lines in Figure 15.28 that are different from previous SPSS^X regression programs we have run are the `CRITERIA` subcommand and the form of thc

FIGURE 15.28 Worksheet for SPSS^X Forward Selection Procedure

```
 {SYSTEM CARDS
UNNUMBERED
TITLE FORWARD SELECTION
DATA LIST  LIST/
           SALES TIME POTEN ADV SHARE CHANGE ACCTS WORK RATE
BEGIN DATA
3669.88  43.1Ø 74Ø65.11  4582.88  2.51   Ø.34  74.86 15.Ø5 4.9
3473.95 1Ø8.13 58117.3Ø  5539.78  5.51   Ø.15 1Ø7.32 19.97 5.1
   . :}DATA CARDS
2799.97  21.14 228Ø9.53  3552.ØØ  9.14  -Ø.74  88.62 24.96 3.9
END DATA
REGRESSION VARIABLES=SALES TO RATE/
           CRITERIA=PIN(Ø.1Ø)/
           DEPENDENT=SALES/
           FORWARD
```

VARIABLES subcommand. Note that when the list of variables is long, as it is in this example, the subcommand

```
VARIABLES=SALES TO RATE
```

instructs the computer to use those variables listed in the DATA LIST command between SALES and RATE, inclusive.

Figure 15.29 shows printout from the forward selection procedure. In step 1 the variable ACCTS is entered, and a simple linear regression analysis is performed. We see that by itself, ACCTS explains 56.58 percent of the variation in SALES and that ACCTS is almost surely an effective predictor. Regression coefficients, their standard errors, and *t*-values are shown under VARIABLES IN THE EQUATION. Under VARIABLES NOT IN THE EQUATION are shown the standardized partial regression coefficients, partial correlations, tolerances, and partial *t*-values for each of the remaining seven predictors. These are conditional for the variable they are associated with, given ACCTS. For example, the partial *t*-value for the variable TIME is the partial *t* for TIME, given ACCTS. Among the significance probabilities of these partial *t*-values (which are the same as the probabilities associated with partial *F*-values), there are four which are less than 0.10, our maximum acceptable probability to enter. The smallest of these is 0.0002, which is associated with the variable ADV. Thus, ADV is entered in step 2.

Step 2 performs a multiple regression using both ACCTS and ADV to predict SALES. Measures of correlation, the analysis of variance, and partial regression coefficients are shown for this model. Note that the partial *t*-value for ADV under VARIABLES IN THE EQUATION in step 2 is the same partial *t*-value as for ADV under VARIABLES NOT IN THE EQUATION in step 1. Under VARIABLES NOT IN THE EQUATION in step 2 partial *t*-values for the six remaining variables, given ACCTS and ADV, are shown. There is one significance probability that is less than 0.10—the one for the variable POTEN. Thus, POTEN is entered in step 3.

This process continues for one more step. In step 4, under VARIABLES NOT IN THE EQUATION, we find no significance probabilities less than 0.01, so the procedure terminates and the message PIN=0.100 LIMITS REACHED is printed.

The final model obtained by the forward selection procedure uses the four variables ACCTS, ADV, POTEN, and SHARE to predict SALES. Together, all four variables explain 90.045 percent of the variation in SALES, and each individual predictor is significant, given the others, at $\alpha \leq 0.10$.

THE BACKWARD METHOD

In a backward elimination procedure the full model is analyzed first. Then variables are dropped from the full model, one at a time, according to the sizes of their associated significance probabilities. At any step, the variable whose significance probability is the largest will be dropped, provided that this

FIGURE 15.29 SPSS[x] Forward Selection Procedure

```
VARIABLE LIST NUMBER 1  LISTWISE DELETION OF MISSING DATA
EQUATION NUMBER 1  DEPENDENT VARIABLE..SALES
BEGINNING BLOCK NUMBER 1.  METHOD: FORWARD
VARIABLE(S) ENTERED ON STEP NUMBER 1..ACCTS

MULTIPLE R                0.75399    ANALYSIS OF VARIANCE
R SQUARE                  0.56850                       DF    SUM OF SQUARES      MEAN SQUARE
ADJUSTED R SQUARE         0.54973    REGRESSION          1    23524074.19189      23524074.19189
STANDARD ERROR          881.09306    RESIDUAL           23    17855474.73496        776324.98848
                                 F = 30.30184      SIGNIF F = 0.0000

          VARIABLES IN THE EQUATION                                  VARIABLES NOT IN THE EQUATION
VARIABLE       B         SE B       BETA       T      SIG T      VARIABLE  BETA IN  PARTIAL  MIN TOLER  T      SIG T
ACCTS         21.72177    3.94603  0.75399   5.505  0.0000      TIME      0.12106  0.12025  0.42572    0.568  0.5757
(CONSTANT)   709.32383  515.24608            1.377  0.1819      POTEN     0.30733  0.41078  0.77090    2.113  0.0462
                                                                ADV       0.46392  0.69196  0.95998    4.496  0.0002
                                                                SHARE     0.21448  0.29882  0.83758    1.469  0.1561
                                                                CHANGE    0.27141  0.39040  0.89280    1.989  0.0593
                                                                WORK      0.03405  0.05080  0.96046    0.239  0.8136
                                                                RATE      0.24216  0.35889  0.94774    1.803  0.0850
--------------------------------------------------------------------------------------------------------------------
VARIABLE(S) ENTERED ON STEP NUMBER 2..ADV

MULTIPLE R                0.88040    ANALYSIS OF VARIANCE
R SQUARE                  0.77510                       DF    SUM OF SQUARES      MEAN SQUARE
ADJUSTED R SQUARE         0.75466    REGRESSION          2    32073320.05178      16036660.02589
STANDARD ERROR          650.39250    RESIDUAL           22     9306228.87507        423010.40341
                                 F = 37.91079      SIGNIF F = 0.0000

          VARIABLES IN THE EQUATION                                  VARIABLES NOT IN THE EQUATION
VARIABLE       B         SE B       BETA       T      SIG T      VARIABLE  BETA IN  PARTIAL  MIN TOLER  T      SIG T
ACCTS         19.04825    2.97291  0.66119   6.407  0.0000      TIME      0.01506  0.02048  0.41579    0.094  0.9261
ADV            0.22653    0.05039  0.46392   4.496  0.0002      POTEN     0.26236  0.48372  0.75686    2.533  0.0194
(CONSTANT)    50.29906  407.60969            0.123  0.9029      SHARE     0.11760  0.22213  0.80238    1.044  0.3084
                                                                CHANGE    0.12377  0.23226  0.79203    1.094  0.2862
                                                                WORK      0.15544  0.31167  0.90372    1.503  0.1477
                                                                RATE      0.07402  0.14034  0.80840    0.650  0.5230
--------------------------------------------------------------------------------------------------------------------
```

VARIABLES(S) ENTERED ON STEP NUMBER 3..POTEN

MULTIPLE R	0.90979	ANALYSIS OF VARIANCE			
R SQUARE	0.82772		DF	SUM OF SQUARES	MEAN SQUARE
ADJUSTED R SQUARE	0.80311	REGRESSION	3	34250796.02722	11416932.00907
STANDARD ERROR	582.63576	RESIDUAL	21	7128752.89963	339464.42379

F = 33.63219 SIGNIF F = 0.0000

VARIABLES IN THE EQUATION

VARIABLE	B	SE B	BETA	T	SIG T
ACCTS	15.55392	2.99937	0.53989	5.186	0.0000
ADV	0.21607	0.04533	0.44250	4.767	0.0001
POTEN	0.02192	8.6556E-03	0.26236	2.533	0.0194
(CONSTANT)	-327.23339	394.40032		-0.830	0.4160

VARIABLES NOT IN THE EQUATION

VARIABLE	BETA IN	PARTIAL	MIN TOLER	T	SIG T
TIME	-0.03833	-0.05889	0.40289	-0.264	0.7946
SHARE	0.35620	0.64973	0.50323	3.822	0.0011
CHANGE	0.09640	0.20542	0.72575	0.939	0.3591
WORK	0.20359	0.45930	0.74135	2.312	0.0315
RATE	3.044E-03	0.00633	0.70495	0.028	0.9777

VARIABLE(S) ENTERED ON STEP NUMBER 4..SHARE

MULTIPLE R	0.94892	ANALYSIS OF VARIANCE			
R SQUARE	0.90045		DF	SUM OF SQUARES	MEAN SQUARE
ADJUSTED R SQUARE	0.88054	REGRESSION	4	37260202.46282	9315050.61571
STANDARD ERROR	453.83623	RESIDUAL	20	4119346.46404	205967.32320

F = 45.22587 SIGNIF F = 0.0000

VARIABLES IN THE EQUATION

VARIABLE	B	SE B	BETA	T	SIG T
ACCTS	9.21390	2.86521	0.31982	3.216	0.0043
ADV	0.17499	0.03691	0.35837	4.741	0.0001
POTEN	0.03822	7.9769E-03	0.45738	4.791	0.0001
SHARE	190.14430	49.74415	0.35620	3.822	0.0011
(CONSTANT)	-1441.93183	423.58170		-3.404	0.0028

VARIABLES NOT IN THE EQUATION

VARIABLE	BETA IN	PARTIAL	MIN TOLER	T	SIG T
TIME	0.12899	0.24355	0.24193	1.095	0.2874
CHANGE	0.12416	0.34670	0.47988	1.611	0.1236
WORK	0.06263	0.15439	0.39545	0.681	0.5040
RATE	0.03474	0.09458	0.50044	0.414	0.6834

FOR BLOCK NUMBER 1 PIN = 0.100 LIMITS REACHED.

probability is greater than POUT (*P* out). The default condition for POUT is 0.10, so that if a predictor has less than a 90 percent chance of being an effective predictor, it will not be retained in the model.

To perform a backward elimination, the only modification necessary in the worksheet in Figure 15.28 is to replace FORWARD with BACKWARD. If it is desired to change the default condition on POUT to, say, 0.05, then the CRITERIA subcommand should read

```
CRITERIA=PIN(0.049) POUT(0.05)
```

Note that here we have specified POUT = 0.05, which is not greater than the default condition on PIN. For reasons to be explained in the discussion of the stepwise method, SPSS^X regression will not allow POUT to be less than PIN. If you try to specify POUT less than PIN, the program will print a warning message and a notice that it has changed POUT to the smaller of 1.00 and 1.10 times PIN. This will happen regardless of the equation method you select, even though PIN does not apply to a backward elimination procedure. As with forward selection, the criterion for dropping a variable can be based on the size of the partial *F*-value rather than on its significance probability. Specifying FOUT(value) in the CRITERIA subcommand will change this criterion. The default condition on FOUT is 2.71, which roughly corresponds to significance of 0.10.

Figure 15.30 shows the SPSS^X backward elimination procedure applied to the Cravens Data, with POUT specified at 0.05. To perform a backward elimination, the full model must be obtained first. This first step is equivalent to specifying the ENTER method; this is shown in step 1 of Figure 15.30. The computer counts this as the equivalent of eight steps, since eight predictors are included in the model. There are five variables in the model at this step (not counting the intercept) with significance probabilities greater than 0.05. The largest of these probabilities is 0.9500, associated with the variable RATE. Thus, as we see in step 2, which SPSS^X calls step 9, RATE has been removed. This leaves seven predictors in the model, of which four have significance probabilities greater than 0.05. WORK, the variable with the largest POUT (0.5501) is then removed in the next step. The process continues until step 12, in which the predictors SHARE, TIME, ADV, and POTEN are all significant at $\alpha \leq 0.05$. Note that had we not changed the default condition on POUT, the backward elimination procedure would have stopped at step 11, in which CHANGE was also in the model and all predictors were significant at $\alpha \leq 0.10$. As with the forward selection procedure, information contained under VARIABLES NOT IN THE EQUATION pertains to the effects of those variables conditional on the others in the equation at that step.

STEPWISE REGRESSION

In a forward stepwise selection procedure, variables are added to the model one at a time as in a forward selection procedure. However, if at any step the significance probability of the partial *F*-value for a variable in the equation

FIGURE 15.30 SPSS^X Backward Elimination

```
EQUATION NUMBER 2   DEPENDENT VARIABLE..SALES
BEGINNING BLOCK NUMBER 1.  METHOD: ENTER

VARIABLE(S) ENTERED ON STEP NUMBER  1..RATE
                                    2..SHARE
                                    3..TIME
                                    4..WORK
                                    5..ADV
                                    6..POTEN
                                    7..CHANGE
                                    8..ACCTS

                                                        ANALYSIS OF VARIANCE
MULTIPLE R              0.96023                         DF    SUM OF SQUARES     MEAN SQUARE
R SQUARE                0.92204
ADJUSTED R SQUARE       0.88306      REGRESSION          8    38153564.25210   4769195.53151
STANDARD ERROR        449.02566      RESIDUAL           16     3225984.67475    201624.04217

                        F =  23.65390       SIGNIF F = 0.0000

                    VARIABLES IN THE EQUATION

VARIABLE         B             SE B           BETA           T        SIG T

RATE             8.18928     128.50561       5.982E-03      0.064    0.9500
SHARE          199.02354      67.02792       0.37283        2.969    0.0090
TIME             2.00957       1.93065       0.13284        1.041    0.3134
WORK            19.79389      33.67669       0.05532        0.588    0.5649
ADV              0.15099       0.04711       0.30922        3.205    0.0055
POTEN            0.03720       8.2023E-03    0.44526        4.536    0.0003
CHANGE         290.85513     186.78200       0.13757        1.557    0.1390
ACCTS            5.55096       4.77555       0.19268        1.162    0.2621
(CONSTANT)   -1507.81373     778.63494                     -1.936    0.0707

FOR BLOCK NUMBER 1   ALL REQUESTED VARIABLES ENTERED.
```

Figure 15.30 (Continued)

```
VARIABLE(S) REMOVED ON STEP NUMBER 9..RATE

MULTIPLE R             0.96022    ANALYSIS OF VARIANCE
R SQUARE               0.92202                              DF    SUM OF SQUARES     MEAN SQUARE
ADJUSTED R SQUARE      0.88991    REGRESSION                 7    38152745.42926     5450392.20418
STANDARD ERROR       435.67416    RESIDUAL                  17     3226803.49760      189811.97045
                                F = 28.71469      SIGNIF F = 0.0000

              VARIABLES IN THE EQUATION                                    VARIABLES NOT IN THE EQUATION
VARIABLE        B            SE B          BETA       T      SIG T    VARIABLE   BETA IN     PARTIAL   MIN TOLER  T      SIG T
SHARE         198.30849     64.11718      0.37149   3.093   0.0066    RATE       5.982E-03   0.01593   0.17733    0.064  0.9500
TIME            1.97454      1.79575      0.13052   1.100   0.2868
WORK           19.89903     32.63610      0.05562   0.610   0.5501
ADV             0.15196      0.04325      0.31121   3.514   0.0027
POTEN           0.03729      7.8510E-03   0.44629   4.750   0.0002
CHANGE        295.86609    164.38655      0.13994   1.800   0.0897
ACCTS           5.61019      4.54496      0.19474   1.234   0.2339
(CONSTANT) -1485.88076    677.67272                -2.193   0.0425
```

```
VARIABLE(S) REMOVED ON STEP NUMBER 10..WORK

MULTIPLE R             0.95933    ANALYSIS OF VARIANCE
R SQUARE               0.92031                              DF    SUM OF SQUARES     MEAN SQUARE
ADJUSTED R SQUARE      0.89375    REGRESSION                 6    38082180.17287     6347030.02881
STANDARD ERROR       428.00368    RESIDUAL                  18     3297368.75399      183187.15300

                                F = 34.64779      SIGNIF F = 0.0000

              VARIABLES IN THE EQUATION                                    VARIABLES NOT IN THE EQUATION
VARIABLE        B            SE B          BETA       T      SIG T    VARIABLE   BETA IN     PARTIAL   MIN TOLER  T      SIG T
SHARE         221.60469     50.58309      0.41514   4.381   0.0004    WORK       0.05562     0.14629   0.18431    0.610  0.5501
TIME            2.26935      1.69898      0.15001   1.336   0.1983    RATE       8.685E-03   0.02291   0.22118    0.094  0.9258
ADV             0.14067      0.03839      0.28809   3.664   0.0018
POTEN           0.03828      7.5469E-03   0.45810   5.072   0.0001
CHANGE        285.10928    160.55966      0.13485   1.776   0.0927
ACCTS           4.37770      3.99904      0.15195   1.095   0.2881
(CONSTANT) -1165.47855    420.37272                -2.772   0.0126
```

VARIABLE(S) REMOVED ON STEP NUMBER 11..ACCTS

MULTIPLE R	0.95656
R SQUARE	0.91501
ADJUSTED R SQUARE	0.89264
STANDARD ERROR	430.23189

ANALYSIS OF VARIANCE

	DF	SUM OF SQUARES	MEAN SQUARE
REGRESSION	5	37862658.90022	7572531.78004
RESIDUAL	19	3516890.02664	185099.47509

F = 40.91061 SIGNIF F = 0.0000

VARIABLES IN THE EQUATION

VARIABLE	B	SE B	BETA	T	SIG T
SHARE	256.95554	39.13607	0.48136	6.566	0.0000
TIME	3.61210	1.18170	0.23877	3.057	0.0065
ADV	0.12886	0.03704	0.26389	3.479	0.0025
POTEN	0.04209	6.7312E-03	0.50370	6.253	0.0000
CHANGE	324.53345	157.28308	0.15350	2.063	0.0530
(CONSTANT)	-1113.78788	419.88690		-2.653	0.0157

VARIABLES NOT IN THE EQUATION

VARIABLE	BETA IN	PARTIAL	MIN TOLER	T	SIG T
ACCTS	0.15195	0.24984	0.22975	1.095	0.2881
WORK	5.540E-03	0.01575	0.66866	0.067	0.9474
RATE	0.02691	0.07004	0.57575	0.298	0.7692

VARIABLE(S) REMOVED ON STEP NUMBER 12..CHANGE

MULTIPLE R	0.94655
R SQUARE	0.89596
ADJUSTED R SQUARE	0.87516
STANDARD ERROR	463.94781

ANALYSIS OF VARIANCE

	DF	SUM OF SQUARES	MEAN SQUARE
REGRESSION	4	37074597.44422	9268649.36106
RESIDUAL	20	4304951.48263	215247.57413

F = 43.06041 SIGNIF F = 0.0000

VARIABLES IN THE EQUATION

VARIABLE	B	SE B	BETA	T	SIG T
SHARE	259.46981	42.18258	0.48607	6.151	0.0000
TIME	3.81647	1.26982	0.25228	3.006	0.0070
ADV	0.15248	0.03798	0.31227	4.014	0.0007
POTEN	0.04440	7.1578E-03	0.53132	6.202	0.0000
(CONSTANT)	-1312.29211	440.74720		-2.977	0.0074

VARIABLES NOT IN THE EQUATION

VARIABLE	BETA IN	PARTIAL	MIN TOLER	T	SIG T
CHANGE	0.15350	0.42785	0.68929	2.063	0.0530
ACCTS	0.20724	0.31603	0.24193	1.452	0.1628
WORK	-0.02692	-0.07059	0.67762	-0.308	0.7611
RATE	0.09487	0.25038	0.63275	1.127	0.2737

FOR BLOCK NUMBER 2 POUT = 0.050 LIMITS REACHED.

becomes too large, that variable will be dropped from the equation. Then the resulting model will be analyzed, and if any more variables can be dropped, they will be, in subsequent steps. When no more variables can be dropped, variables not in the equation are examined for entry. This process continues until no more variables can be entered or removed, or until a maximum number of steps have been performed.

For the STEPWISE method the default condition on the maximum number of steps is twice the number of independent variables in the full model. This condition is necessary because the procedure could go into an infinite loop in which a variable (or set of variables) removed at one set of steps is entered again at subsequent steps, then removed again, then entered again, and so on. To avoid this, the maximum number of steps is specified. To change the default condition on the maximum number of steps to, say, 20, insert a subcommand CRITERIA=MAXSTEPS(20) before the DEPENDENT subcommand but after the VARIABLES subcommand.

The default entry and removal criteria in the STEPWISE method are the same as for forward selection and backward elimination; that is, PIN = 0.05 and POUT = 0.10. As with the other two procedures, these criteria can be changed to different values, or the decision to add and delete variables can be based on the sizes of the *F*-values, FIN and FOUT. All changes to the default criteria are specified in a single CRITERIA subcommand. For example, to change the criteria to enter a variable if it is significant at $\alpha = 0.10$, keep it if it is significant at $\alpha = 0.05$, and perform no more than 25 steps, the CRITERIA subcommand will read

```
CRITERIA=PIN(0.10) POUT(0.05) MAXSTEPS(25)
```

Note that this particular change in entry and removal criteria could cause a variable to cycle in and out of the equation over and over until the maximum number of steps is reached. Thus, in the SPSS^X program, even though the process would terminate after 25 steps, if PIN is specified to be larger than POUT (or FIN smaller than FOUT), the computer will print a message and change the POUT value to be slightly larger than the PIN (or the FOUT slightly smaller than the FIN). This restricts the usefulness of the procedure somewhat, as we would like to give a variable every opportunity to enter into the equation and show what it can do in conjunction with other variables (so PIN should be rather large), but we don't want to keep it unless it proves itself effective (so POUT should be relatively small).

The stepwise regression using default conditions on PIN and POUT produces an analysis identical to the forward selection procedure shown in Figure 15.29 for this set of data. To obtain this analysis (which could end up being quite different from a forward selection for another set of data) the only change necessary to the program in Figure 15.28 is to replace the method subcommand FORWARD by STEPWISE. Of course, if it is not desired to change the default conditions on PIN and/or POUT, simply omit the CRITERIA subcommand from Figure 15.28.

Referring to Figure 15.29, we see that the variable with the smallest PIN less than 0.05 was ACCTS, so ACCTS was entered in step 1. Once ACCTS got into the equation, its POUT was not greater than 0.10, so it stayed in the equation. In step 2, ADV had the smallest PIN less than 0.05, so it entered the equation. Since POUT for ACCTS, given ADV, was 0.000 and POUT for ADV, given ACCTS, was 0.0002, and both of these probabilities were less than 0.10, both variables stayed in the equation. The process continued, adding POTEN at step 3 and SHARE in step 4. At step 4 all POUTs were less than 0.10, so all four variables stayed in the equation, but no PINs for variables not in the equation were less than 0.05, so no more variables could be added and the process terminated. Even though this does not look different from a forward selection when applied to these data, each partial *t*- or *F*-value was examined to see if that variable should be removed before another variable was entered.

OTHER SPSS^X SELECTION PROCEDURES

In addition to the FORWARD, BACKWARD, STEPWISE, and ENTER methods, SPSS^X offers two other procedures for alternative model analysis: REMOVE and TEST. In the first three of the foregoing methods the computer adds or deletes variables according to their relative strengths, without any direction from the researcher as to which variables to include or which variables to consider first. In the ENTER method either all variables are used in the model, or a subset of variables specified by the researcher is used. In the REMOVE method the researcher specifies a variable or a set of variables to be removed from the full model. In the TEST method the researcher specifies a list of different models to be tested. In both of these procedures the researcher must have an *a priori* reason for specifying that certain variables should be deleted from the model or that particular models should be considered — there should be a reason for wanting to look at those particular reduced models. Since this situation differs from the present one, in which we are simply trying to discover which of several variables collectively are effective predictors of the dependent variable, discussion of the REMOVE and TEST methods will be postponed until Section 15.8.

PERFORMING SEVERAL SELECTION PROCEDURES IN ONE RUN

Since it is quite possible that when forward selection, backward elimination, and stepwise procedures are applied, different final models will result, it is often advisable to try more than one method on any set of data. Clearly one does not want to submit three different SPSS^X programs to do this, and indeed, as many selection procedures as desired can be performed in one program by using multiple sets of subcommands in the REGRESSION command.

Suppose, for example, that for a given set of data we wish to perform a forward selection with PIN = 0.10, a backward elimination with POUT = 0.05, and a stepwise regression with PIN = 0.05 and POUT = 0.10 (the default conditions), in that order. Then the REGRESSION command will read

```
REGRESSION VARIABLES=SALES TO RATE/
CRITERIA=PIN(0.10)/
DEP=SALES/
FORWARD/
CRITERIA=PIN(0.049) POUT(0.05)/
DEP=SALES/
BACKWARD/
CRITERIA=DEFAULTS/
DEP=SALES/
STEPWISE/
```

Note that a CRITERIA subcommand remains in effect until overridden by another CRITERIA subcommand.

CHOICE OF A FINAL MODEL

The various stagewise procedures offered by SPSS^X (or any other package) are likely to indicate several models that might be reasonable for a given data set, and it is often not an easy task to choose among them and propose a single "best" model. Among criteria for good final models are (1) that each variable should have a significant partial F-value associated with it, (2) the standard errors of the regression coefficients should be small enough to ensure that the effects of the predictors are adequately measured, and (3) the multiple R^2 for the model should be reasonably large. There may be several models that satisfy these criteria. Then it is up to the researcher to choose among them according to which one makes the best sense, which is cheapest or easiest to use, and other practical and theoretical considerations. In Section 16.3 we will look at some additional criteria for measuring the apropriateness of various candidate models.

EXERCISES

15.14 Recall that in the analysis of the Children Data of Exercise 11.6 the overall regression significance test indicated that at least one of the predictors (age, height, or weight) was significant. However, no partial F-test on one predictor, given the other two, indicated a significant predictor. Analyze the data using the SPSS^X forward selection procedure with default condition on the inclusion criterion. See if any significant predictor(s) can be found.

15.15 Using the Tennessee Data (Exercise 15.3), run an SPSS^X selection analysis with F-level for inclusion of 1 to predict the percentage of families below poverty level. You might also want to obtain a correlation matrix and other pertinent information. Propose a final model to be used, and justify your choice.

15.16 Consider the MPG Data of Exercise 10.6. Since the relationship of mpg with both horsepower and weight is curvilinear, fit a model

$$Y = \alpha + \beta_1 X_1 + \beta_2 X_2 + \beta_3 X_1^2 + \beta_4 X_2^2 + \beta_5 X_3 + \varepsilon$$

where Y = mpg, X_1 = HP, X_2 = weight, X_3 = transmission type.

a. Analyze the full model. Would you use the full model to predict mpg?

b. Use an available SPSS^X variable-selection procedure to obtain an acceptable reduced model. Examine regression coefficients to see if they make sense.

15.17 Work Exercise 15.5 using an SPSS^X selection procedure.

15.8 OTHER TECHNIQUES FOR SELECTING VARIABLES

In all the stagewise procedures considered using BMD, Minitab, SAS, or SPSS^X, the relative strengths of variables as predictors determine whether or not the variables will be included in the model, and if so, in which order. There are instances, however, when the researcher specifies that one or more particular variables must be included in the model. This decision can be based on theoretical considerations (total revenue from sales must be a function of the number of items sold, for example) or on practical considerations (it may be easy and inexpensive to collect information on certain variables, or some authority requires that they must be used). In other instances, the researcher may want to specify the order in which variables are entered into the model. The model itself may be such that a sequential analysis is appropriate, as in first testing the linear, then the quadratic, and then higher-order terms in a polynomial regression. At other times, theoretical or practical considerations would cause the researcher to look at variables in a specified order but not necessarily to analyze each conditional on the inclusion of those preceding. That is, in a sequential analysis, if X_1, X_2, and X_3 are specified, in that order, then the analysis will be first for X_1, then for X_2 given X_1, and then for X_3 given X_1 and X_2. In a hierarchical regression X_1 is allowed to enter first (if it can), followed by X_2, but either X_1 or X_2 can subsequently leave the model.

In this section we will consider first techniques for performing hierarchical regression. Then we will show how the various selection procedures can be modified to force retention of specified variables in the model.

HIERARCHICAL REGRESSION

Sometimes the researcher has reason to specify that a certain variable or group of variables should be tested before some others. As already mentioned, some models naturally lend themselves to a sequential analysis, and at times practical or theoretical considerations lead a researcher to give certain variables the first whack at being included in the model. With regard to the Cravens Data, suppose that company policy has been to promote sales representatives according to seniority, so it would be important for the company to know how sales relate to time with the company. Also, the home office has been spending a great deal of money on advertising and wants to know what effect this has had on sales. Additionally, there is a very strong argument that if a salesrep has increased his or her territory's share of the market during the last year, then s/he should be rewarded for that effort. Because of company priorities, then, the researcher would like to obtain a regression equation that includes TIME as a predictor. Of secondary importance are ADV and CHANGE, and of third importance are the remaining variables.

Let us see how BMD, Minitab, SAS, and SPSS^X programs can be altered to reflect this hierarchy.

B M D Consider the worksheet shown in Figure 15.6 for performing a stepwise regression using the F method. In the VARIABLES paragraph the variables are listed in the following order:

```
SALES, TIME, POTEN, ADV, SHARE, CHANGE, ACCTS, WORK, RATE
```

We insert the following sentence in the REGRES paragraph, after the DEPENDENT sentence:

```
LEVELS=0,1,3,2,3,2,3,3,3
```

That is, "levels for inclusion" are assigned to each variable: 0 for SALES, 1 for TIME, 3 for POTEN, and so on. The dependent variable SALES has been assigned inclusion level zero, which will prevent it from being used as an independent variable. Any other variables listed in the VARIABLES paragraph that you do not want to use as predictors should also be assigned level zero. Level 1 is assigned to TIME, so TIME will be entered before any other variable (provided its F to enter is at least 4). Then, since ADV and CHANGE are assigned level 2, they will be allowed to compete for entry in the model according to the selection method specified (the F method, in this case). After the level 2 variables have been considered, then all level 3 variables are allowed to compete for entry. Note that if the equation method being used permits variables once entered to be removed subsequently, then it is possible for a level 1 variable to be dropped after a level 2 variable has been added, and so on. At least those variables assigned the lower levels get first chance at being included in the model.

Figure 15.31 shows what happens with this procedure. TIME was the first variable to be included in the model. By itself, TIME is a good predictor, with *F*-value of 14.583. Then ADV and CHANGE, the level 2 variables compete for entry. In step 1 the *F* to enter for ADV is 11.27 and the *F* to enter for CHANGE is 5.26. Since 11.27 is greater than 5.26, ADV enters in step 2. CHANGE is now given a chance to enter, but in step 2 its *F* to enter is only 2.38, which is less than the default criterion of 4.00. Thus, the level 3 variables are allowed to compete for entry, and ACCTS enters at step 3. Then the partial *F* for TIME falls to 0.0088, so TIME is removed in step 4. The partial *F* for CHANGE is only 1.20, so a level 3 variable, POTEN, enters at step 5. There, the *F* to enter for TIME is only 0.07, and the *F* to enter for CHANGE is only 0.88, so another level 3 variable, SHARE, is entered in step 6. The process then terminates because no *F* to enter values are 4 or more in step 6.

MINITAB Using the ENTER subcommand to the STEPWISE command in Minitab, it is possible to begin a selection procedure with a model containing specified variables. Figure 15.32 shows the stepwise selection process with the variable TIME entered at step 1. This differs from the usual stepwise process, in which the variable entered at step 1 would be determined by the predictor that gave the largest *F*-value. After step 1, however, the selection process proceeds as usual, adding variables according to which have the largest partial *F*'s greater than FENTER and deleting variables according to which have the smallest partial *F*'s less than FREMOVE.

In step 1 of Figure 15.32 the variable TIME is entered by the subcommand ENTER 'TIME'. In step 2, ADV is added to TIME, and in step 3, ACCTS is added to TIME and ADV. Once this is accomplished, however, the partial *t*-value for TIME, given ADV and ACCTS, drops to only 0.09. Thus, in step 4, TIME is removed and a two-predictor analysis relating SALES to ACCTS and ADV is produced. Step 5 adds POTEN to these two predictors, and step 6 adds in SHARE.

Thus, while TIME, as a variable of practical importance, was given every opportunity to be included in the model by being entered first, it did not prove to be an effective predictor when subjected to a stepwise selection process. Of course, had a forward selection algorithm been used, TIME would not have been allowed to leave the model once it had been entered. The selection process would have found other variables to use in combination with TIME. Thus, if the practical benefits of including TIME as a predictor outweigh considerations of a statistical nature, such as trying to find a model with as large an R^2-value as possible or having all predictors conditionally

FIGURE 15.31 BMDP2R F Method With LEVELS Specifications

```
STEP NO. 1
VARIABLE ENTERED            2  TIME

MULTIPLE R                     0.6229
MULTIPLE R-SQUARE              0.3880
ADJUSTED R-SQUARE              0.3614
STD ERROR OF EST            1049.2839

ANALYSIS OF VARIANCE
                   SUM OF SQUARES    DF    MEAN SQUARE             F RATIO
         REGRESSION  16056312.        1          0.1605631E 08     14.58
         RESIDUAL    25322944.       23    1100997.

                                VARIABLES IN EQUATION                                          VARIABLES NOT IN EQUATION
                                STD ERROR   STD REG                F TO                                  PARTIAL                F TO
VARIABLE        COEFFICIENT     OF COEFF    COEFF      TOLERANCE   REMOVE   LEVEL     VARIABLE           CORR       TOLERANCE   ENTER   LEVEL
Y-INTERCEPT     2548.66895
TIME  2            9.42353      2.4677      0.623      1.00000     14.58    1         POTEN    3    0.45195    0.79391       5.65   3
                                                                                      ADV      4    0.58203    0.93791      11.27   2
                                                                                      SHARE    5    0.53654    0.98872       8.89   3
                                                                                      CHANGE   6    0.43918    0.93676       5.26   2
                                                                                      ACCTS    7    0.55235    0.42572       9.66   3
                                                                                      WORK     8   -0.00718    0.96784       0.00   3
                                                                                      RATE     9    0.43543    0.98977       5.15   3
```

```
STEP NO. 2
VARIABLE ENTERED            4  ADV

MULTIPLE R                    0.7716
MULTIPLE R-SQUARE             0.5953
ADJUSTED R-SQUARE             0.5586
STD ERROR OF EST            872.4167

ANALYSIS OF VARIANCE
                   SUM OF SQUARES    DF    MEAN SQUARE             F RATIO
         REGRESSION  24634816.        2          0.1231741E 08     16.18
         RESIDUAL    16744438.       22    761110.8

                                VARIABLES IN EQUATION                                          VARIABLES NOT IN EQUATION
                                STD ERROR   STD REG                F TO                                  PARTIAL                F TO
VARIABLE        COEFFICIENT     OF COEFF    COEFF      TOLERANCE   REMOVE   LEVEL     VARIABLE           CORR       TOLERANCE   ENTER   LEVEL
Y-INTERCEPT     1703.67871
TIME  2            7.65122      2.1185      0.506      0.93791     13.04    1         POTEN    3    0.50649    0.78995       7.25   3
ADV   4            0.22957      0.0684      0.470      0.93791     11.27    2         SHARE    5    0.49838    0.92833       6.94   3
                                                                                      CHANGE   6    0.31884    0.83173       2.38   2
                                                                                      ACCTS    7    0.66668    0.42558      16.80   3
                                                                                      WORK     8    0.16695    0.91264       0.60   3
                                                                                      RATE     9    0.27128    0.83070       1.67   3
```

STEP NO. 3
VARIABLE ENTERED 7 ACCTS

MULTIPLE R	0.8805
MULTIPLE R-SQUARE	0.7752
ADJUSTED R-SQUARE	0.7431
STD ERROR OF EST	665.5549

ANALYSIS OF VARIANCE

	SUM OF SQUARES	DF	MEAN SQUARE	F RATIO
REGRESSION	32077008.	3	0.1069234E 08	24.14
RESIDUAL	9302233.	21	442963.4	

VARIABLES IN EQUATION

VARIABLE	COEFFICIENT	STD ERROR OF COEFF	STD REG COEFF	TOLERANCE	F TO REMOVE	LEVEL
Y-INTERCEPT	72.85547					
TIME 2	0.22779	2.4274	0.015	0.41579	0.01	1
ADV 4	0.22578	0.0522	0.462	0.93761	18.73	2
ACCTS 7	18.72830	4.5691	0.650	0.42558	16.80	3

VARIABLES NOT IN EQUATION

VARIABLE	PARTIAL CORR	TOLERANCE	F TO ENTER	LEVEL
POTEN 3	0.48613	0.74782	6.19	3
SHARE 5	0.24824	0.68820	1.31	3
CHANGE 6	0.23360	0.79011	1.15	2
WORK 8	0.31191	0.90411	2.16	3
RATE 9	0.14686	0.77967	0.44	3

STEP NO. 4
VARIABLE REMOVED 2 TIME

MULTIPLE R	0.8804
MULTIPLE R-SQUARE	0.7751
ADJUSTED R-SQUARE	0.7547
STD ERROR OF EST	650.3892

ANALYSIS OF VARIANCE

	SUM OF SQUARES	DF	MEAN SQUARE	F RATIO
REGRESSION	32073120.	2	0.1603656E 08	37.91
RESIDUAL	9306133.	22	423006.0	

VARIABLES IN EQUATION

VARIABLE	COEFFICIENT	STD ERROR OF COEFF	STD REG COEFF	TOLERANCE	F TO REMOVE	LEVEL
Y-INTERCEPT	50.31079					
ADV 4	0.22653	0.0504	0.464	0.95599	20.21	2
ACCTS 7	19.04820	2.9729	0.661	0.95598	41.05	3

VARIABLES NOT IN EQUATION

VARIABLE	PARTIAL CORR	TOLERANCE	F TO ENTER	LEVEL
TIME 2	0.02047	0.41579	0.01	1
POTEN 3	0.48372	0.76451	6.41	3
SHARE 5	0.22213	0.80238	1.09	3
CHANGE 6	0.23226	0.79203	1.20	2
WORK 8	0.31166	0.90417	2.26	3
RATE 9	0.14034	0.80840	0.42	3

Figure 15.31 (Continued)

```
STEP NO. 5
VARIABLE ENTERED          3  POTEN

MULTIPLE R                   0.9098
MULTIPLE R-SQUARE            0.8277
ADJUSTED R-SQUARE            0.8031
STD ERROR OF EST           582.6323

ANALYSIS OF VARIANCE
                 SUM OF SQUARES   DF    MEAN SQUARE          F RATIO
      REGRESSION   34250592.       3        0.1141686E 08     33.63
      RESIDUAL      7128667.      21    339460.3

                              VARIABLES IN EQUATION                                       VARIABLES NOT IN EQUATION
                              STD ERROR  STD REG                 F TO                              PARTIAL              F TO
VARIABLE       COEFFICIENT    OF COEFF   COEFF     TOLERANCE    REMOVE   LEVEL     VARIABLE      CORR       TOLERANCE  ENTER   LEVEL
Y-INTERCEPT   -327.22168
POTEN   3         0.02192      0.0087     0.262     0.76451      6.41     3        TIME     2   -0.05889    0.40672     0.07    1
ADV     4         0.21607      0.0453     0.443     0.95202     22.72     2        SHARE    5    0.64973    0.57320    14.61    3
ACCTS   7        15.55388      2.9994     0.540     0.75685     26.89     3        CHANGE   6    0.20542    0.78232     0.88    2
                                                                                   WORK     8    0.45929    0.87679     5.35    3
                                                                                   RATE     9    0.00633    0.74542     0.00    3
```

```
STEP NO. 6
VARIABLE ENTERED          5  SHARE

MULTIPLE R                   0.9489
MULTIPLE R-SQUARE            0.9005
ADJUSTED R-SQUARE            0.8805
STD ERROR OF EST           453.8318

ANALYSIS OF VARIANCE
                 SUM OF SQUARES   DF    MEAN SQUARE          F RATIO
      REGRESSION   37259968.       4    9314992.              45.23
      RESIDUAL      4119268.      20     205963.4

                              VARIABLES IN EQUATION                                       VARIABLES NOT IN EQUATION
                              STD ERROR  STD REG                 F TO                              PARTIAL              F TO
VARIABLE       COEFFICIENT    OF COEFF   COEFF     TOLERANCE    REMOVE   LEVEL     VARIABLE      CORR       TOLERANCE  ENTER   LEVEL
Y-INTERCEPT  -1441.91602
POTEN   3         0.03822      0.0080     0.457     0.54615     22.95     3        TIME     2    0.24355    0.35494     1.20    1
ADV     4         0.17499      0.0369     0.358     0.87129     22.48     2        CHANGE   6    0.34670    0.77627     2.60    2
SHARE   5       190.14400     49.7437     0.356     0.57320     14.61     3        WORK     8    0.15438    0.60489     0.46    3
ACCTS   7         9.21386      2.8652     0.320     0.50323     10.34     3        RATE     9    0.09459    0.73791     0.17    3

* * * * * F-LEVELS(4.000, 3.900) OR TOLERANCE INSUFFICIENT FOR FURTHER STEPPING
```

FIGURE 15.32 Entering TIME at the First Step of a Stepwise Procedure Using Minitab

```
—STEP C1 ON C2-C9;
 —
ENTER 'TIME'.
STEPWISE REGRESSION OF SALES ON 8 PREDICTORS, WITH N = 25
STEP             1          2          3          4          5          6
CONSTANT   2548.67    1703.67      72.85      50.30    -327.23   -1441.93
TIME           9.4        7.7        0.2
T-RATIO       3.82       3.61       0.09
ADV                     0.230      0.226      0.227      0.216      0.175
T-RATIO                  3.36       4.33       4.50       4.77       4.74
ACCTS                              18.7       19.0       15.6        9.2
T-RATIO                             4.10       6.41       5.19       3.22
POTEN                                                  0.0219     0.0382
T-RATIO                                                  2.53       4.79
SHARE                                                              190
T-RATIO                                                             3.82
S             1049        872        666        650        583        454
R-SQ         38.80      59.53      77.52      77.51      82.77      90.04
```

significant, it would have been better to have performed a forward selection, entering TIME at the first step.

Figure 15.33 shows the three predictors TIME, ADV, and CHANGE entered on step 1 of a stepwise procedure. The difference between this analysis and the previous one is that this stepwise procedure begins with a three-predictor model at step 1 instead of a single-predictor model. Interestingly enough, the same final model is produced by both stepwise processes. The variables chosen are ADV, ACCTS, POTEN, and SHARE. This final model contains one of the three variables, ADV, which were deemed important for practical reasons.

FIGURE 15.33 Entering TIME, ADV, and CHANGE on the First Step of a Stepwise Procedure Using Minitab

```
—STEP C1 ON C2-C9;
 —
ENTER 'TIME' 'ADV' 'CHANGE'.
STEPWISE REGRESSION OF SALES ON 8 PREDICTORS, WITH N = 25
STEP             1          2          3          4          5          6
CONSTANT   1866.63    1703.67      72.85      50.30    -327.23   -1441.93
TIME           7.1        7.7        0.2
T-RATIO       3.39       3.61       0.09
ADV          0.193      0.230      0.226      0.227      0.216      0.175
T-RATIO       2.74       3.36       4.33       4.50       4.77       4.74
CHANGE         470
T-RATIO       1.54
ACCTS                              18.7       19.0       15.6        9.2
T-RATIO                             4.10       6.41       5.19       3.22
POTEN                                                  0.0219     0.0382
T-RATIO                                                  2.53       4.79
SHARE                                                              190
T-RATIO                                                             3.82
S              846        872        666        650        583        454
R-SQ         63.65      59.53      77.52      77.51      82.77      90.04
```

At this point, the researcher might intervene with a subcommand to re-enter the other two predictors of interest, TIME and CHANGE, and see what happens. Figure 15.34 shows the results. While step 6 of Figure 15.33 used four predictors, step 7 of Figure 15.34 contains six, as specified by the ENTER 'TIME' 'CHANGE' subcommand. The stepwise process then continues to step 8, deleting the variable ACCTS, and terminates with a five-predictor model that contains all three desired predictors. (Note that this is the same five-predictor model selected by the backward elimination process. This happens to be so for this data set but could hardly be expected to be true in general. Neither should it be inferred that it will always be possible to find a model containing all desired variables in which all are significant, given the others.)

FIGURE 15.34 Re-Entering TIME and CHANGE Using Minitab

```
ENTER 'TIME' 'CHANGE'.
STEP             7               8
CONSTANT     -1165           -1114
TIME             2.3             3.6
T-RATIO          1.34            3.06
ADV              0.141           0.129
T-RATIO          3.66            3.48
CHANGE         285             325
T-RATIO          1.78            2.06
ACCTS            4.4
T-RATIO          1.09
POTEN            0.0383          0.0421
T-RATIO          5.07            6.25
SHARE          222             257
T-RATIO          4.38            6.57
S              428             430
R-SQ            92.03           91.50
```

SAS Using either the MAXR or the MINR procedures, SAS can be instructed to begin the procedure by looking at models containing a subset of s variables. The s variables to be used must be the first s variables listed in the MODEL statement. Suppose we modify the MODEL statement in the worksheet of Figure 15.21 to read

```
MODEL SALES=TIME ADV CHANGE POTEN SHARE ACCTS WORK RATE/MAXR START=1
```

Then the MAXR procedure will first show a one-variable regression using TIME and then will proceed with its switching process from there. Adding the model statement

```
MODEL SALES=TIME ADV CHANGE POTEN SHARE ACCTS WORK RATE/MAXR START=3
```

will first fit the model using TIME, ADV, and CHANGE and will continue from there with its swapping procedure.

If the START=1 option is used, the SAS MAXR procedure will first show a single-predictor analysis using TIME. Then TIME is exchanged for ACCTS, and the rest of the analysis is just like the MAXR procedure in Figure 15.25. If the START=3 option is used, MAXR will begin at step 3 and use TIME, ADV, and CHANGE as predictors. Figure 15.35 shows the switches performed in step 3, leading to the same "best" three-predictor model found by the regular MAXR procedure. MAXR then finds "best" four-predictor, . . . , eight-predictor models, as before.

FIGURE 15.35 Summary of Switches Made in Step 3 of the SAS MAXR Procedure With the START=3 Option

	R^2
Starting variables entered:	
TIME, ADV, CHANGE	0.6365
Time replaced by ACCTS:	
ADV, CHANGE, ACCTS	0.7872
CHANGE replaced by POTEN:	
ADV, POTEN, ACCTS	0.8277
ACCTS replaced by SHARE:	
ADV, POTEN, SHARE	0.8490

SPSS^X Using multiple ENTER subcommands in an SPSS^X REGRESSION command, the researcher can specify the order of entry of either individual variables or blocks of variables. First, to see a simple linear regression of SALES on TIME, followed by a regression of SALES on TIME, ADV, and CHANGE, and then finally the full model with all predictors, write, after the DEPENDENT=SALES subcommand,

```
ENTER TIME/
ENTER ADV,CHANGE/
ENTER/
```

This last ENTER subcommand could list all the remaining variables, but it is not necessary.

Use of multiple ENTER subcommands produces what is more properly termed a sequential analysis, because the order of entry of variables is completely specified, and once a variable or group of variables is added to the model none are removed in subsequent steps. A combination of subcommands specifying different equation methods might be used, allowing for selection to take place. Consider, for example, the sequence of subcommands

```
ENTER TIME/
ENTER ADV,CHANGE/
STEPWISE/
```

TIME will be entered in the first step. Then ADV and CHANGE will be added to TIME in the second step and a three-predictor model analyzed. Thereafter, variables can be dropped from, or added to, the three-predictor model using the regular STEPWISE algorithm. Figure 15.36 summarizes the results of this sequence of subcommands when applied to the Cravens Data. While of the

FIGURE 15.36
Summary of Steps Using Multiple ENTER and STEPWISE Subcommands in SPSS[X]

	R^2	(Variable)	POUT
Step 1 (ENTER TIME)			
TIME entered	0.62292	(TIME)	0.0009
Steps 2/3 (ENTER ADV, CHANGE)			
ADV and CHANGE added to TIME	0.79780	(TIME)	0.0027
		(ADV)	0.0121
		(CHANGE)	0.1381
Step 4 (STEPWISE)			
CHANGE removed	0.77159	(TIME)	0.0015
		(ADV)	0.0028
Step 5			
ACCTS added to TIME and ADV	0.88045	(TIME)	0.9261
		(ADV)	0.0003
		(ACCTS)	0.0005
Step 6			
TIME removed	0.88040	(ADV)	0.0002
		(ACCTS)	0.0000
Step 7			
POTEN added to ADV and ACCTS	0.90979	(ADV)	0.0001
		(ACCTS)	0.0000
		(POTEN)	0.0194
Step 8			
SHARE added to ADV, ACCTS, POTEN	0.94892	(ADV)	0.0001
		(ACCTS)	0.0043
		(POTEN)	0.0001
		(SHARE)	0.0011

variables deemed *a priori* to be important, only ADV remained in the final model, this combination of methods allowed TIME and CHANGE to be examined in combination with other variables. Different combinations of equation method subcommands could produce many different hierarchical procedures and might result in some very interesting models being considered that might not have been discovered otherwise.

The SPSS[X] TEST method allows the researcher to specify what models s/he would like to see analyzed. However, there is no subsequent computer-performed selection based on the sizes or significances of the partial *F*-values. That is, the researcher has total control over the models to be considered.

To use the TEST method, insert the TEST subcommand under the REGRESSION command (after the DEPENDENT subcommand) and follow it by lists of predictors, enclosed in parentheses and separated by commas. For example, suppose we want to see first how TIME performs as a predictor, then TIME and ADV, then TIME and CHANGE, then TIME, ADV, and CHANGE, and finally ADV and CHANGE. Write the subcommand

```
TEST (TIME) (TIME,ADV) (TIME,CHANGE) (TIME,ADV,CHANGE) (ADV,CHANGE)
```

(If we had not already seen the full model, we would probably want to follow the TEST subcommand by an ENTER subcommand.)

Figure 15.37 shows output from the TEST method applied to the Cravens Data. The regression sums of squares for each of the models specified are shown, along with degrees of freedom, F-values, and significance probabilities. The quantity labeled RSQ CHG (change in R^2) is simply the R^2 for the model being considered. After looking at each individual model specified, the computer collects all independent variables mentioned in the TEST subcommand and performs a "full model" analysis using all these predictors. (Thus, we could have omitted the set (TIME, ADV, CHANGE) from the TEST subcommand and still have seen the analysis using these three variables.)

FIGURE 15.37 The SPSS^X TEST Method

```
EQUATION NUMBER 1      DEPENDENT VARIABLE..SALES
BEGINNING BLOCK NUMBER 1.  METHOD:  TEST    TIME    ADV    CHANGE
VARIABLE(S) ENTERED ON STEP NUMBER  1..CHANGE
                                    2..TIME
                                    3..ADV

HYPOTHESIS TESTS
      SUM OF
DF    SQUARES          RSQ CHG    F           SIG F     SOURCE
 1     8251321.52275   0.19941    11.51935    0.0027    TIME
 2    16435189.1248    0.39718    11.47226    0.0004    TIME       ADV
 2    11629769.7672    0.28105     8.11793    0.0024    TIME       CHANGE
 3    26337225.7608    0.63648    12.25612    0.0001    TIME       ADV       CHANGE
 2    10280750.6685    0.24845     7.17628    0.0042    ADV        CHANGE
 3    26337225.7608               12.25612    0.0001    REGRESSION
21    15042323.1661                                     RESIDUAL
24    41379548.9269                                     TOTAL

                                  ANALYSIS OF VARIANCE
MULTIPLE R            0.79780                        DF    SUM OF SQUARES     MEAN SQUARE
R SQUARE              0.63648
ADJUSTED R SQUARE     0.58455     REGRESSION          3    26337225.76077     8779075.25359
STANDARD ERROR      846.34573     RESIDUAL           21    15042323.16609      716301.10315
                               F =  12.25612      SIGNIF F = 0.0001

                      VARIABLES IN THE EQUATION
VARIABLE          B               SE B          BETA         T         SIG T
CHANGE         470.19772       305.01788      0.22239      1.542      0.1381
TIME             7.08571         2.08771      0.46838      3.394      0.0027
ADV              0.19323         0.07040      0.39573      2.745      0.0121
(CONSTANT)    1866.62601       360.38301                   5.180      0.0000
```

The REMOVE method subcommand allows the researcher to specify that a subset of variables is to be removed, as a block, from the equation. Again, no further selection is performed. Figure 15.38 shows printout from the subcommand

```
REMOVE (TIME,ADV,CHANGE)
```

which we might want to use to see what happens with the "leftover" variables when the variables of primary interest are removed from the model.

FIGURE 15.38 The SPSS^X REMOVE Subcommand

```
EQUATION NUMBER 1    DEPENDENT VARIABLE..SALES
BEGINNING BLOCK NUMBER  2.  METHOD:  REMOVE    TIME     ADV     CHANGE
VARIABLE(S) REMOVED ON STEP NUMBER    9..TIME
                                     10..CHANGE
                                     11..ADV

MULTIPLE R                0.90387    ANALYSIS OF VARIANCE
R SQUARE                  0.81698                           DF    SUM OF SQUARES      MEAN SQUARE
ADJUSTED R SQUARE         0.76881    REGRESSION              5    33806062.51387    6761212.50277
STANDARD ERROR          631.35137    RESIDUAL               19     7573486.41298     398604.54805
                                   F = 16.96221       SIGNIF F = 0.0000

             VARIABLES IN THE EQUATION                             VARIABLES NOT IN THE EQUATION
VARIABLE     B             SE B        BETA       T       SIG T     VARIABLE  BETA IN   PARTIAL  MIN TOLER  T       SIG T
RATE          208.00406   148.41582   0.15195   1.401   0.1772     TIME      0.32311   0.45518  0.22261    2.169   0.0437
SHARE         278.90969    74.79096   0.52249   3.729   0.0014     ADV       0.36889   0.68948  0.39491    4.039   0.0008
WORK          -26.86147    42.06684  -0.07508  -0.639   0.5307     CHANGE    0.18869   0.35210  0.43505    1.596   0.1279
POTEN           0.04310     0.01129   0.51578   3.817   0.0012
ACCTS           7.11235     4.25577   0.24688   1.671   0.1111
(CONSTANT) -1576.31785   1035.09077            -1.523   0.1443
```

FORCING VARIABLES INTO THE MODEL

Suppose that the vice president for research says that the independent variable TIME *must* be included as a predictor in the model regardless of whether or not it is a good predictor of SALES. How can the selection procedures be altered to accommodate this requirement?

BMD Specify inclusion levels as with hierarchical regression. In our example, level 0 was assigned to SALES, level 1 to TIME, level 2 to ADV and CHANGE, and level 3 to all other variables. Adding a sentence FORCE=1 will force all level 1 variables (only one, in this example) into the equation, regardless of the sizes of their partial *F*-values. If we had written FORCE=2, all level 2 variables *and* all level 1 variables would have been forced in.

Figure 15.39 shows the summary table for the *F* method analysis forcing TIME into the model and assigning inclusion level 2 to ADV and CHANGE. It turns out, in this example, that when TIME is forced into the model the *F* method finally comes up with an acceptable model using TIME and both of the level 2 variables. TIME is entered in step 1, followed by ADV in step 2 and ACCTS in step 3. Left to its own devices, the procedure would have removed TIME in step 4 because its partial *F* was only 0.01, but since TIME was forced into the model, it could not do that. Instead, it added POTEN in step 4 (*F* for TIME there was only 0.07) and SHARE in step 5. In step 5 the partial *F* for TIME was 1.20, and also the partial *F* for ACCTS fell to 2.11, so that ACCTS is removed in step 6. The partial *F* for TIME in step 6 rose to 9.03, and stayed at about that value when CHANGE was added in step 7. The final model found by the *F* method uses TIME, POTEN, ADV, SHARE, and CHANGE.

FIGURE 15.39 Summary Table of the BMDP2R F Method With TIME Forced Into the Model, and ADV and CHANGE Assigned Inclusion Level 2

STEP NO.	VARIABLE ENTERED	VARIABLE REMOVED	MULTIPLE R	MULTIPLE RSQ	INCREASE IN RSQ	F TO ENTER	F TO REMOVE	NO. OF INDEPENDENT VARIABLES INCLUDED
1	2 TIME		0.6229	0.3880	0.3880	14.5834		1
2	4 ADV		0.7716	0.5953	0.2073	11.2710		2
3	7 ACCTS		0.8805	0.7752	0.1799	16.8009		3
4	3 POTEN		0.9101	0.8283	0.0531	6.1889		4
5	5 SHARE		0.9520	0.9064	0.0780	15.8328		5
6		7 ACCTS	0.9466	0.8960	-0.0104		2.1083	4
7	6 CHANGE		0.9566	0.9150	0.0190	4.2575		5

MINITAB The FORCE subcommand to the REGRESSION command in Minitab allows one to force a variable or variables into each step of any selection process. The FORCE subcommand differs from the

ENTER subcommand in that a variable put into the model with the ENTER subcommand may be selected out of the model at a later step (if a stepwise or backward elimination procedure is used), but a variable forced in will never be allowed to leave, regardless of how small its partial F becomes.

Figure 15.40 shows the Minitab stepwise selection procedure with TIME forced into the model. Note that even though the final model is the same as the one arrived at by the backward elimination procedure and by using ENTER 'TIME' 'ADV' 'CHANGE' followed by an additional ENTER 'TIME' 'CHANGE' subcommand (Figures 15.33 and 15.34), the sequence of steps in those procedures and in Figure 15.40 are quite different. Using the FORCE subcommand, the variable TIME never leaves the model, even though its partial t-value becomes quite low at steps 3, 4, and 5. Other variables enter or leave the model according to their partial t-values, which are calculated conditionally on TIME (and other variables) being included in the model.

FIGURE 15.40 Using Minitab to Force TIME Into the Model

```
—STEP C1 ON C2-C9;
—
FORCE 'TIME'.
```

STEPWISE REGRESSION OF SALES ON 8 PREDICTORS, WITH N = 25

STEP	1	2	3	4	5	6	7
CONSTANT	2548.67	1703.67	72.85	-390.63	-1349.90	-1312.29	-1113.79
TIME	9.4	7.7	0.2	-0.6	2.0	3.8	3.6
T-RATIO	3.82	3.61	0.09	-0.26	1.09	3.01	3.06
ADV		0.230	0.226	0.218	0.165	0.152	0.129
T-RATIO		3.36	4.33	4.65	4.34	4.01	3.48
ACCTS			18.7	16.3	6.0		
T-RATIO			4.10	3.88	1.45		
POTEN				0.0223	0.0388	0.0444	0.0421
T-RATIO				2.49	4.88	6.20	6.25
SHARE					211	259	257
T-RATIO					3.98	6.15	6.57
CHANGE							325
T-RATIO							2.06
S	1049	872	666	596	452	464	430
R-SQ	38.80	59.53	77.52	82.83	90.64	89.60	91.50

SAS Adding the option INCLUDE=S in the MODEL statement will cause the first s variables listed in the MODEL statement to be included at every step of any selection procedure, whether or not their partial F-values are significantly different from zero. Note that while the START option for performing hierarchical regression can only be used with the MAXR or MINR procedures, the INCLUDE option can be used with any of the stagewise procedures offered by SAS. Figure 15.41 shows a summary of the analysis with TIME forced into the model using the stepwise procedure. Note in step 2 that although TIME becomes nonsignificant, it is not dropped from the

FIGURE 15.41 Summary of Steps Forcing `TIME` Into the Model Using SAS Stepwise Procedure

```
DEPENDENT VARIABLE = SALES
THE FIRST VARIABLE IN EACH MODEL IS AN INCLUDED VARIABLE

STEP 0  INCLUDED VARIABLE (TIME)
ENTERED
  R² = 0.3880     C(P) = 104.5955
                F           PROB > F
TIME         14.58          0.0009

STEP 1  ADV ENTERED
  R² = 0.5953     C(P) = 64.0482
                F           PROB > F
TIME         13.04          0.0016
ADV          11.27          0.0029

STEP 2  ACCTS ENTERED
  R² = 0.7752     C(P) = 29.1370
                F           PROB > F
TIME          0.01          0.9261
ADV          18.73          0.0003
ACCTS        16.80          0.0005

STEP 3  POTEN ENTERED
  R² = 0.8283     C(P) = 20.2341
                F           PROB > F
TIME          0.07          0.7946
POTEN         6.19          0.0218
ADV          21.63          0.0002
ACCTS        15.05          0.0009

STEP 4  SHARE ENTERED
  R² = 0.9064     C(P) = 6.2189
                F           PROB > F
TIME          1.20          0.2874
POTEN        23.80          0.0001
ADV          18.86          0.0004
SHARE        15.83          0.0008
ACCTS         2.11          0.1628

STEP 5  ACCTS REMOVED
  R² = 0.8960     C(P) = 6.3514
                F           PROB > F
TIME          9.03          0.0070
POTEN        38.47          0.0001
ADV          16.12          0.0007
SHARE        37.84          0.0001

STEP 6  CHANGE ENTERED
  R² = 0.9150     C(P) = 4.4428
                F           PROB > F
TIME          9.34          0.0085
POTEN        39.10          0.0001
ADV          12.10          0.0025
SHARE        43.11          0.0001
CHANGE        4.26          0.0530
NO OTHER VARIABLES MET THE 0.1500 SIGNIFICANCE LEVEL FOR ENTRY INTO THE MODEL.
```

model as would ordinarily happen in a stepwise regression. Also, in step 5 the variable `ACCTS` is removed, yielding a four-predictor model in which all four predictors, including `TIME`, are significant at $\alpha = 0.05$. The five predictors included at step 6 are also all significant at the 0.15 level (the default condition), and the process terminates at that point.

SPSS^X Using the `TEST` subcommand, the researcher can specify what model or models s/he would like to see analyzed. No further selection procedures are performed, however: The researcher sees the model containing those variables specified, and no more variables are added to, or deleted from, the model. Refer to instructions given under "Hierarchical Regression" for using the `TEST` subcommand. Various models using `TIME` as a predictor in conjunction with other variables can easily be specified. The effectiveness of this procedure is limited by the insight or luck of the researcher in specifying what other variables should be used in conjunction with the ones forced into the model.

EXERCISES

15.18 Refer to the Tennessee Data of Exercise 15.3. Suppose that the percentage of families below the poverty level is to be predicted, and the client whose problem it is requires that the black population be used as a predictor. Use an available selection procedure to come up with an acceptable model.

15.19 Refer to the Tennessee Data of Exercise 15.3. Suppose that the percentage of families below the poverty level is to be predicted. Perform a hierarchical regression, with black population and percent rural population as the most important predictors, and median age, number of persons employed in agriculture, and percent population change being of secondary importance.

15.20 In Appendix G is a large data set containing information on residences sold in a certain city. (Note that "zone" is a categorical variable, as is "type.") The objective is to find out which variables affect the number of days a house is on the market.

Propose a predictor or set of predictors that you think should be forced into the model. Define at least one other class of predictors that should probably be included but that you may not necessarily want to force into the model. State your proposed model and use available techniques to analyze a random sample from this data set.

15.9 SUMMARY

In this chapter we considered several different variable-selection techniques. Our objective has been to eliminate from a set of potential predictors any variables that are not effective either alone or in conjunction with other predictors. Ideally, we seek a set of predictors that is as small as possible while still explaining a large percentage of the variation in the response.

The choice of predictors for the final reduced model is seldom clear-cut. We typically want all of them to be statistically significant. Only then can we be reasonably certain that a different independent sample drawn from the same population would yield the same predictors.

We would also like the multiple coefficient of determination for the final model to be near 100 percent, although in many practical applications this is not possible. We realize that just because our final set of k predictors can explain, for example, 92 percent of the variation in the response in *this* sample, these same predictors might not explain 92 percent of the variation in response when applied to a new sample. Any predictor included in the regression equation should have an estimated regression coefficient with a small standard error. Then, we would not expect the value of the regression coefficient to change very much if another analysis were run on another sample. In consequence, $\hat{Y}$ should also be a pretty good estimate of the true mean response value. On the other hand, if $\hat{Y}$ is based on a regression equation for which the standard errors of the regression coefficients are large, then $\hat{Y}$ may also become untrustworthy. A final consideration is that theoretical or practical matters may dictate that some predictors should be included in the model.

Very seldom in an actual regression problem are the four criteria—significant predictors, large R^2, small standard errors, and theoretically or practically appealing variables—all met at some step of a stagewise regression procedure. Ideally, one should look at the results of several stagewise techniques and select several alternative possible final models. Then the final model should not be selected until an analysis of residuals shows that the model meets as closely as possible the assumptions made in multiple regression analysis: The errors should be independent, normally distributed random variables with mean zero and common variance in order for the results of the significance tests to be valid.

The final chapter looks briefly at residual analysis for multiple regression. Then two measures of the "goodness" of candidate models are considered.

SIXTEEN

ANALYSIS OF RESIDUALS IN MULTIPLE REGRESSION

16.1 INTRODUCTION

In Chapter 15 we looked at some methods for determining which variables to include in a regression model. The various stagewise procedures proved to be helpful in indicating effective predictors when the researcher has no theoretical grounds for including one predictor rather than another. Although these selection procedures produce candidates for a final model, they do not necessarily come up with a final appropriate model. Before deciding on a final model, one should look at the residuals to make sure that the relationship is truly linear, that no important predictors have been omitted, and that none of the assumptions of the model have been too seriously violated. If the researcher intends to use the final regression equation to make estimates based on data not included in the sample, then it is important to examine how well various models might be expected to perform.

We begin with the analysis of residuals in multiple regression, applying the same kinds of checks on the assumptions as were made in Chapter 9 for simple linear regression. Some adjustments must be made when using these checks for multiple regression problems. We consider, also, two methods to help one choose among candidate models once these have been checked to make sure that no assumptions have been violated.

16.2

EXAMINATION OF RESIDUALS IN MULTIPLE REGRESSION

As in simple linear regression, we have several assumptions to check out before using the regression equation. Just because the computer will perform an analysis on a set of data and come up with a model containing all significant predictors does not mean that we have necessarily done the right thing. For example, $s^2_{Y \cdot X}$ can always be calculated, but it is not meaningful unless we can assume that regardless of the values of $X_1, X_2, \ldots, X_k$, the scatter of the data points around the regression line is uniform. We also need to be reasonably certain that the relationship of Y to $X_1, X_2, \ldots, X_k$ is really linear and not curvilinear. The presence of outliers or an unusual pattern in the residuals also might indicate that we have omitted a predictor that should be taken into account. Finally, the assumption that the errors are normally distributed must be satisfied if we are going to apply significance tests.

Most of the assumptions can be checked out in multiple regression in pretty much the same way that they were in simple regression. We will briefly consider these tests of assumptions, noting the changes to be made for a multiple regression model.

UNIFORM SCATTER

Recall that in simple regression, to see if the scatter were uniform, we either plotted Y against X and looked at the pattern of the points around the regression line, or we plotted the residuals against X. However, in multiple regression, we have many X-variables; and we cannot make a two-dimensional plot with more than one predictor. We could plot residuals versus each of the X's, but perhaps the easiest thing to do is to plot the residuals versus $\hat{Y}$.

Suppose we consider the model using the predictors POTEN, ADV, SHARE, and ACCTS, which was found by the forward selection and stepwise procedures. There, all predictors were significant at $\alpha = 0.05$. The regression equation was

$$Y = -1441.916 + 0.038X_2 + 0.175X_3 + 190.144X_4 + 9.214X_6,$$

where X_2 = POTEN, X_3 = ADV, X_4 = SHARE, and X_6 = ACCTS. Table 16.1 shows the residuals from this model, and Figure 16.1 shows a plot of the residuals versus $\hat{Y}$.

TABLE 16.1 Calculation of $\hat{Y}$ and Residuals Using $\hat{Y} = -1441.916 + 0.038X_2 + 0.175X_3 + 190.144X_4 + 9.214X_6$

Y	$\hat{Y}$	$Y - \hat{Y}$
3669.88	3357.661156	312.218844
3473.95	3785.131639	−311.181639
2295.10	2847.380336	−552.280336
4675.56	4939.637611	−264.077611
6125.96	5525.301106	600.658894
2134.94	2087.445268	47.494732
5031.66	5037.800094	− 6.140094
3367.45	2796.684212	570.765788
6519.45	6127.991101	391.458899
4876.37	3941.926155	934.443845
2468.27	2543.090277	− 74.820277
2533.31	2534.667681	− 1.357681
2408.11	2157.840822	250.269178
2337.38	1904.544659	432.835341
4586.95	4882.934267	−295.984267
2729.24	2669.686565	59.553435
3289.40	3462.109443	−172.799443
2800.78	2632.294688	168.485312
3264.20	4052.445395	−788.245395
3453.62	3935.392933	−481.772933
1741.45	1723.636616	17.813384
2035.75	2704.235023	−668.485023
1578.00	1646.527102	− 68.527102
4167.44	4462.452816	−295.012816
2799.97	2605.821608	194.148392

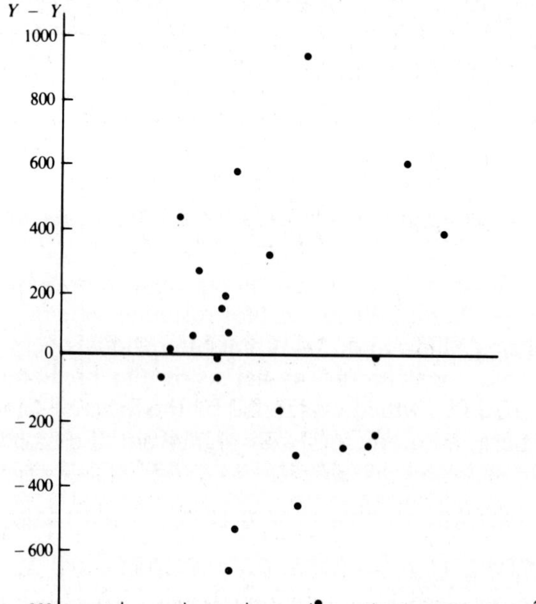

FIGURE 16.1 Plot of Residuals Versus $\hat{Y}$ for a Multiple Regression Model (Cravens Data)

The plot of residuals versus $\hat{Y}$ shows that there appears to be slightly larger scatter toward the center than at either end. If the data are split into two groups, one containing the six largest and the six smallest Y-values and the other containing the middle 13 Y-values, and s^2 is calculated for the residuals in each group, then

$$F = \frac{s_2^2}{s_1^2} = 2.91.$$

This F-value is significant at $\alpha = 0.05$ but not at $\alpha = 0.01$. If one must be 99 percent sure that the scatter is not homogeneous in order to apply remedial measures, then the evidence falls short.

On the other hand, if one uses $\alpha = 0.05$, then the evidence is strong enough to indicate nonhomogeneous scatter and some remedial measures must be taken. One could perhaps transform the data so that the scatter in the transformed data is uniform. In Chapter 9, in cases in which $\sigma^2_{Y \cdot X}$ was proportional to X, we used the transformation

$$Y' = \alpha' + \beta' X' + \varepsilon',$$

where

$$Y' = \frac{Y}{X}, \qquad X' = \frac{1}{X}, \qquad \alpha' = \beta, \qquad \beta' = \frac{\alpha}{X}, \quad \text{and} \quad \varepsilon' = \frac{\varepsilon}{X}.$$

In the multiple regression case the fact that we have more than one X, however, makes this transformation inappropriate. The technique of *weighted least squares,* of which the above transformation is a special case, must be used in order to make the error variances equal. This involves finding a matrix of weights **W**, related to the matrix of variances and covariances of the errors, so that when the normal equations

$$\mathbf{X}'\mathbf{W}^{-1}\mathbf{X}\hat{\boldsymbol{\beta}} = \mathbf{X}'\mathbf{W}^{-1}\mathbf{y}$$

are solved, the residuals resulting from the regression equation found will have homogeneous variance. This is not an easy task, however, and we leave it to more advanced texts.

OUTLIERS

Further examination of the residuals plot shows the possibility of an outlier in our data. All but one of the residuals are between -800 and $+800$. We could test to see if this residual indicates an outlier in the same way we did in Chapter 9: Hold out this observation, rerun the analysis using the remaining observations, form a confidence interval using the remaining observations. If the withheld observation does not lie in the confidence interval, we have some justification for calling it an outlier. Then we can either throw it out as an erroneous measurement or we can search for an omitted predictor that should have been taken into account. (Note that throwing out this extreme residual as an error in measurement will probably help resolve the question of whether or not the scatter is uniform.)

Instead of going to the trouble of constructing a confidence interval based on the $n - 1$ remaining observations, we could apply the Dixon test. Let us test to see if the largest residual in Figure 16.1 is an outlier, using the Dixon test. Table 16.2 shows the residuals ordered from smallest to largest. Since $n = 25$ and since we are testing to the largest value, we calculate

$$r_{22} = \frac{V_{(n)} - V_{(n-2)}}{V_{(n)} - V_{(3)}} = \frac{934.44 - 570.77}{934.44 - (-552.28)} = \frac{363.67}{1486.72} = 0.245.$$

If α is chosen to be 0.01, then we conclude that this residual is an outlier if $r_{22} > 0.516$. Since $0.245 \not> 0.516$, we cannot be 99 percent sure that we have an outlier.

TABLE 16.2 Residuals From Table 16.1 in Order From Smallest to Largest

$Y - \hat{Y}$
$-788.25 = V_{(1)}$
$-668.49 = V_{(2)}$
$-552.28 = V_{(3)}$
⋮
$570.77 = V_{(n-2)}$
$600.66 = V_{(n-1)}$
$934.44 = V_{(n)}$

LACK OF FIT

Sometimes a plot of residuals indicates an unusual pattern. As in simple regression, if we have repeated measurements at any set of $(X_1, X_2, \ldots, X_k)$-values, we can test lack of fit of the model. This can often be done in designed experiments, but seldom if $X_1, X_2, \ldots, X_k$ as well as Y are random variables. Very seldom does chance let us observe the same values on $X_1, X_2, \ldots, X_k$ so that we can see "pure" error by comparing their corresponding Y-values. If such a test can be made, the only changes from the procedure outlined in Section 9.3 is that the degrees of freedom for error will be $n - p$ instead of $n - 2$. Thus the degrees of freedom for lack of fit will be the number of different $(X_1, X_2, \ldots, X_k)$-values minus p.

Even without repeated measurements, however, plots of residuals can often provide a clue to what is wrong with the fitted model, especially if the true relationship is not linear with respect to the variables in the model. For example, suppose that the true model is

$$Y = 4 + X_1 + 2X_1^2 + X_2,$$

and that this is a deterministic relationship (no error is involved). Suppose that 10 observations give the results in Table 16.3. The reader can verify that each Y-value is found by substituting the given values of X_1 and X_2 into the above equation.

TABLE 16.3
$Y = 4 + X_1 + X_1^2 + X^2$

X_1	2	2	3	6	4	6	5	1	1	6
X_2	9	0	3	7	4	6	9	7	4	2
Y	23	14	28	89	44	88	68	14	11	84

Now, if we are ignorant of the presence of the quadratic term in X_1 and fit a linear model

$$Y = \alpha + \beta_1 X_1 + \beta_2 X_2 + \varepsilon,$$

then we will find the regression equation

$$\hat{Y} = 15.2026 + 15.4122X_1 + 1.1802X_2.$$

Predicted values and residuals from fitting this model are shown in Table 16.4. Figure 16.2 shows the residuals plotted against the predicted values. There is

TABLE 16.4
Predicted Values and Residuals from Mis-Specified Model

X_1	X_2	Y	$\hat{Y}$	$Y - \hat{Y}$
2	9	9	26.2436	−3.2436
2	0	0	15.6218	−1.6218
3	3	3	34.5746	−6.5746
6	7	7	85.5320	3.4680
4	4	4	51.1670	−7.1670
6	6	6	84.3518	3.6482
5	9	9	72.4802	−4.4802
1	7	7	8.4710	5.5290
1	4	4	4.9304	6.0696
6	2	2	79.6310	4.3690

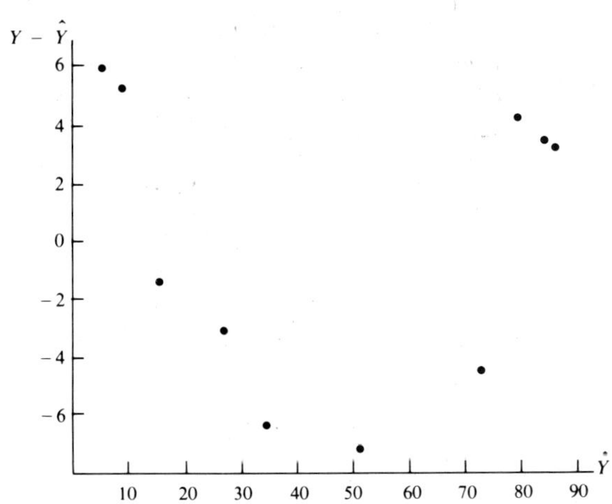

FIGURE 16.2
Plot of Residuals Versus $\hat{Y}$ for Incorrectly Specified Model

obviously a pattern to these residuals—one indicating that a quadratic term has been omitted from the model. But how would we know whether the quadratic term involves X_1 or X_2? Figure 16.3 shows the residuals plotted against X_1, and Figure 16.4 shows the residuals plotted against X_2. Since the pattern emerges again in the plot of the residuals against X_1, we know that an X_1^2 term has been omitted from the model. This example is admittedly oversimplified, but it illustrates how one can detect quadratic terms omitted from the model.

FIGURE 16.3
Plot of Residuals Versus X_1

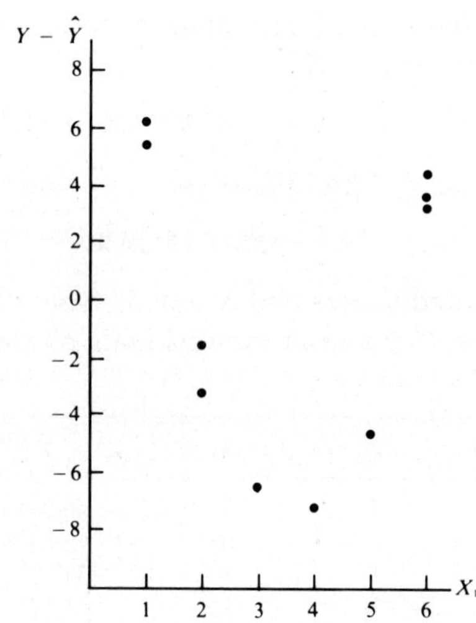

FIGURE 16.4
Plot of Residuals Versus X_2

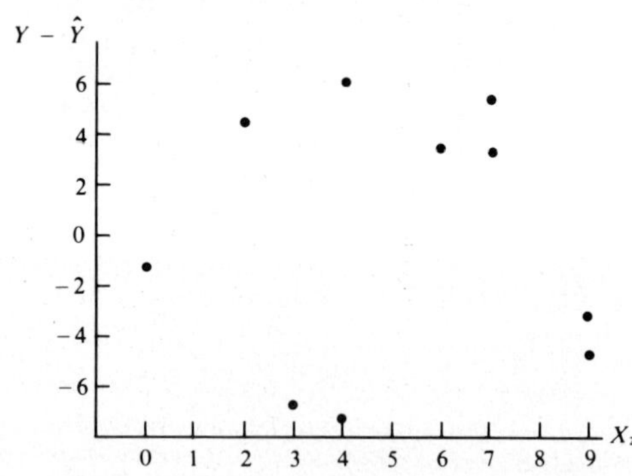

TIME DEPENDENCIES

If the data were collected over time, making it possible that a time effect is causing the errors not to be independent, then a test for serial correlation can be performed. The procedure is exactly the same as that for simple regression in Section 9.4. Most computer programs give serial correlation coefficients, and/or with Durbin-Watson statistics. One need only compare these to the appropriate critical values from the tables in order to discover any serial correlation present.

NORMALITY

Finally, a quick check on the assumption that the errors are normally distributed can be made by computing the standardized residuals, to see if 68 percent of them lie between ± 1, and so on.

EXERCISES

16.1 A diving coach at a large university wants to determine some variables that affect his athletes' performances. An Olympic hopeful is measured, or rated, on six variables over 192 consecutive practice sessions and his performance at each session is rated by the coach. The variables are

Y = performance rating
X_1 = weight (pounds)
X_2 = number of hours sleep the night before
X_3 = number of dives
X_4 = mental rating
X_5 = physical rating
X_6 = coach's attitude

The rating scale for X_4, X_5, and X_6 is:

1 = excellent, 2 = above average, 3 = average, 4 = below average, 5 = poor.

Performance rating is a numerical score between 0 and 10, where 0 = not performed and 10 = perfect, with scores such as 9.65 commonly given. Because of the likelihood of a practice effect, a Durbin-Watson statistic is calculated from the model containing all six predictors. Its value is found to be 1.4863. What does this tell you about any practice effect?

16.2 Refer to the Tennessee Data of Exercise 15.3. Suppose we consider the model found by the backward elimination procedure which uses population change, agricultural employment, percent rural population, and median age to predict the percentage of families below the poverty level. Plot the residuals against $\hat{Y}$ and analyze the plot.

16.3 Using the same model as in Exercise 16.2, find the standardized residuals and check to see whether the normality assumption can be justified.

16.4 For any or all of the models you arrived at using variable-selection techniques in Chapter 15, perform an analysis of residuals. Take corrective action where indicated.

16.3 TWO MODEL SELECTION CRITERIA

The various stagewise procedures discussed in Chapter 15 can all give "best" models according to some criteria. If several selection procedures are applied to the same set of data, it is quite possible—as we have seen—that several different "best" models may be found. We saw with the Cravens Data that several different models were suitable in terms of all predictors being conditionally significant, a large R^2 for the model, and reasonably small standard errors for the regression coefficients. The researcher may be able to choose from among candidate models on theoretical grounds ("This set of predictors makes the most sense in the light of our understanding of the relationship.") or based on practical considerations ("It's easier and cheaper to use two predictors, instead of three, if using three doesn't improve all that much over using just two.").

In this section we will look at two measures of the appropriateness of a proposed model. The first measure, the C_p-statistic, measures the "goodness" of a model in terms of whether or not the error mean square contains only random variation. The second, the PRESS-statistic, measures how well a model will perform when used to predict future values of the dependent variable.

ACCOUNTING FOR NONRANDOM VARIATION

The various stagewise procedures applied to the Cravens Data yielded several models that might reasonably be used. The first columns of Table 16.5 list nine models found by the various SAS selection procedures (as well as by some of the SPSS^X and BMD procedures). All are reasonable models by some combination of relevant criteria. In addition to considering the significance probabilities associated with the included predictors and the sizes of R^2 for each model, we want to look at the error mean square, $s^2_{Y \cdot X}$, for each model.

Recall that for a model

$$Y = \alpha + \beta_1 X_1 + \cdots + \beta_k X_k + \varepsilon,$$

the error term ε can consist of

1. Purely random error,

TABLE 16.5
C_p-Statistics for Nine Models Under Consideration

Model No.	Selection Technique(s)	Variable(s)	Prob F	R^2	$s^2_{Y \cdot X}$	No. Predictors k	$p = k + 1$	C_p
1	FORWARD, STEPWISE, MAXR, MINR	ACCTS	0.0001	0.57	776,325	1	2	67.6
2	MAXR	ADV	0.0002	0.76	423,010	2	3	27.2
		ACCTS	0.0001					
3	MINR	POTEN	0.0001	0.75	477,629	2	3	33.1
		SHARE	0.0001					
4	MAXR	POTEN	0.0194	0.83	339,464	3	4	18.4
		ADV	0.0001					
		ACCTS	0.0001					
5	MAXR, MINR	POTEN	0.0001	0.85	297,586	3	4	14.0
		ADV	0.0011					
		SHARE	0.0001					
6	FORWARD, MAXR, MINR	POTEN	0.0001	0.90	205,967	4	5	5.4
		ADV	0.0001					
		SHARE	0.0011					
		ACCTS	0.0043					
7	BACKWARD, MAXR, MINR	TIME	0.0065	0.92	185,099	5	6	4.4
		POTEN	0.0001					
		ADV	0.0025					
		SHARE	0.0001					
		CHANGE	0.0530					
8	MAXR, MINR	POTEN	0.0001	0.91	190,747	5	6	5.0
		ADV	0.0006					
		SHARE	0.0006					
		CHANGE	0.1236					
		ACCTS	0.0089					
9	MAXR, MINR	TIME	0.1983	0.92	183,187	6	7	5.3
		POTEN	0.0001					
		ADV	0.0018					
		SHARE	0.0004					
		CHANGE	0.0927					
		ACCTS	0.2881					

2. Error due to omitted predictors,
3. Error due to wrong specification of the form of the model.

If the correct model has been found, then the error term contains only random error. Hence, any model with an error mean square too large to be measuring only random variation must not be the "best" model. The C_p-statistic is a measure designed to evaluate candidate models on the basis of how much their MS(E)'s differ from what they "should" be.

THE C_p-STATISTIC

The C_p-statistic is defined as

$$C_p = p + \frac{s_p^2 - \sigma^2}{\sigma^2}(n - p).$$

Here, p is the number of parameters in a candidate model ($p = k + 1$), and s_p^2 is the error mean square associated with the candidate model. The variance σ^2 is the error mean square associated with the true best model, however many predictors it might contain. Now, if the candidate model under consideration is indeed the true best model, then s_p^2 for this model should be very close to σ^2, so that $s_p^2 - \sigma^2$ is very close to zero and C_p is approximately equal to p. Of course, even if we find the true best model, s_p^2 may not exactly equal σ^2, because s_p^2 is based on a sample and is therefore subject to random variability. It may be that s_p^2 is even less than σ^2, so that $C_p < p$ by a little bit. On the other hand, if our candidate model is not the correct one, then s_p^2 will be quite a bit larger than σ^2, since s_p^2 will contain error due to an incorrectly specified model, and C_p will be quite a bit larger than p.

In order to choose from among several candidate models, then, we calculate the C_p-statistic for each and pick the model for which C_p is closest to p.

All this supposes that σ^2, the error mean square for the true best model, is known. But in order to know σ^2, we must already know what the true best model is, and we don't; this is precisely what we are trying to find out. Thus, in order to use the C_p-statistic, we must come up with some method by which σ^2 can be estimated independently. There are generally three methods available for obtaining an independent estimate of σ^2.

The first method can be used if the researcher is quite familiar with the process being studied and can come up with an estimate of the random variability inherent in the process based on past experience. Recall that if the model is correct, $\sigma^2 = \sigma^2_{Y \cdot X}$ does not contain any error due to omitted predictors or to an incorrectly specified model: it contains only pure random error. More often than one might expect, especially in engineering problems, the researcher can give a reasonably accurate estimate of the "noise" in the system. This estimate can then be substituted for σ^2 and the C_p-statistic can be calculated.

The second method for estimating σ^2 is to replicate the experiment or some parts of it, in order to break $s^2_{Y \cdot X}$ into pure error and lack of fit, as we did in Chapter 9. Then the pure error component, which estimates only random variability, is used as an estimate of σ^2. The researcher who anticipates a problem in model selection should take care to have replicated observations, if possible, in order to obtain such an estimate of σ^2.

The least preferred but most common method of estimating σ^2 uses $s_p^2 = s^2_{Y \cdot X}$ from the model containing all candidate predictors. Of course, this means that C_p will always equal p for the candidate model using all predictors. If many of the candidate predictors are not significant, $s^2_{Y \cdot X}$ for the full model is likely to be larger than σ^2. This will lead to C_p being less than p for the "good" candidate models.

Let us illustrate the calculation of C_p for the Cravens Data. Recall from the maximum R^2 improvement technique from SAS, one model found in which all predictors were significant at $\alpha = 0.01$ used ADV and ACCTS to predict SALES. For this model, the error mean square was found to be 423.010. From step 0 of the backward elimination procedure, in which all eight predictors were used to predict SALES, the error mean square was 201,624. Thus, we have

$$n = 25, \qquad \sigma^2 = 201{,}624, \qquad p = k + 1 = 3, \qquad s_p^2 = 423{,}010.$$

Then

$$\begin{aligned} C_p &= 3 + \frac{423{,}010 - 201{,}624}{201{,}624}(25 - 3) \\ &= 3 + 1.0980(22) \\ &= 27.1560. \end{aligned}$$

Since C_p is considerably greater than p, we have evidence that the model using just ADV and ACCTS is not the best.

Table 16.5 shows C_p-statistics for the nine models we previously considered. In all nine calculations, σ^2 is estimated by the error mean square for the full model: $\sigma^2 = 201.624$. Comparing C_p to p for each of the nine, it seems that we can pretty quickly eliminate models 1–5 from further consideration, since for each C_p is quite a bit larger than p. In model 6, C_p is only slightly larger than p, and in models 7–9, C_p is less than p.

COMPUTER CALCULATION OF C_p

BMD The C_p-statistic can be calculated if one performs an "all possible subsets regression" using the BMDP9R program. Since it can be quite expensive to look at all 2^k possible subsets of k predictors if k is large, we have not discussed the P9R program in this text. The interested reader is referred to the BMDP manual for instructions for running this program and requesting that the C_p-statistic be calculated for the models under consideration.

MINITAB The C_p-statistic is not available as standard output from the Minitab REGRESSION or STEPWISE commands.

SAS The C_p-statistic appears as standard output in each step of any SAS variable-selection procedure. Note, for example, Figure 15.22, which shows the SAS forward selection procedure applied to the Cravens Data. At each step the value of C_p is printed on the same line as the multiple R^2. SAS uses s_p^2 from the full model as an estimate of σ^2. For instance, step 1 of the backward elimination procedure (Figure 15.24) has $C_p = 9 = p$ for the initial nine-predictor model. This certainly does not mean that the nine-predictor model is the correct one to use.

SPSS^X SPSS^X provides no means for instructing the computer to calculate C_p automatically for any model.

HYPOTHESIS TESTS BASED ON C_p

There is also available a hypothesis test to eliminate models from consideration. Let

k = total number of predictors under consideration = 8 in our example,

p = number of terms in candidate model (may or may not include intercept term),

q = number of predictors omitted in candidate model (may or may not include intercept),

so that

$$p + q = k + 1.$$

Let $\boldsymbol{\beta}_p$ = vector of parameters for candidate model, with β_i's (and perhaps α) replaced by zeros for omitted terms. For example, the parameter vector for the full model is

$$\boldsymbol{\beta}_k = \begin{bmatrix} \alpha \\ \beta_1 \\ \beta_2 \\ \beta_3 \\ \beta_4 \\ \beta_5 \\ \beta_6 \\ \beta_7 \\ \beta_8 \end{bmatrix}.$$

In model 2, using just ADV $= X_3$ and ACCTS $= X_6$, and including an intercept,

$$\boldsymbol{\beta}_p = \begin{bmatrix} \alpha \\ 0 \\ 0 \\ \beta_3 \\ 0 \\ 0 \\ \beta_6 \\ 0 \\ 0 \end{bmatrix},$$

and $p = 3$, $q = 6$, and $3 + 6 = 8 + 1$.

Then to test

$$H_0: \boldsymbol{\beta} = \boldsymbol{\beta}_p$$
$$H_1: \boldsymbol{\beta} \neq \boldsymbol{\beta}_p$$

where $\boldsymbol{\beta}$ is the parameter vector for the correct model, calculate the test statistic

$$F_p = C_p - p,$$

and reject H_0 if $F_p > q[F_{q,\nu,\alpha} - 1]$ where ν = number of degrees of freedom for error in the candidate model: $\nu = n - p = 25 - p$ in our example, and $F_{q,\nu,\alpha}$ is read from F-tables.

Table 16.6 shows the results of these significance tests for each of the nine candidate models. We see that models 1–5 can be eliminated from further consideration, because in each case $C_p - p$ is greater than the critical value, so that we can be more than 95 percent sure that these models are not appropriate. However, we cannot eliminate any of the remaining four models. While we reached the same conclusion without aid of the hypothesis test in this example, it will not always be so easy to eliminate models just by looking.

TABLE 16.6 Tests of Significance for Nine C_p-Statistics

Model No.	p	$\nu = 25 - p$	$q = 9 - p$	$F_{q,\nu,0.05}$	$q[F_{q,\nu,0.05} - 1]$	$F_p = C_p - p$
1	2	23	7	2.44	10.08	65.6
2	3	22	6	2.55	9.30	24.2
3	3	22	6	2.55	9.30	30.1
4	4	21	5	2.68	8.40	14.4
5	4	21	5	2.68	8.40	10.0
6	5	20	4	2.87	7.48	0.4
7	6	19	3	3.13	6.39	−1.6
8	6	19	3	3.13	6.39	−1.0
9	7	18	2	3.55	5.10	−1.7

Final choice of a model is narrowed down to a choice among models 6, 7, 8, and 9. Selection should be based on which model makes the best sense to the researcher, whether all variables in the model are significant at the desired significance level, the multiple coefficient of determination for the model, which variables are cheaper or easier to measure, and possibly which one has smallest predictor error.

THE MODEL THAT PREDICTS BEST

The various stagewise regression procedures discussed in Chapter 15 can all give best models according to some criteria. One can find several different models in which all predictors are significant, depending upon which variable selection procedure is used. Thus, we sometimes have to choose among various candidate models.

If the objective is to determine which predictors are effective in describing a real-world relationship, then the researcher can probably choose from among various candidate models the one that makes the best sense in terms of the theoretical framework of the problem. On the other hand, if the researcher wants to use a regression equation to make estimates based on future observed values of X_1, X_2, . . . , X_k, then the question of which alternative model to use still remains.

One might think that the model (containing all significant predictors) having the largest R^2 is best, but this is not necessarily the case. The problem is that b_1, b_2, . . . , b_k, and thus R^2, are very data-dependent because the least squares criterion found the line that best fits *these* data. A different sample will give different values for the regression coefficients and the multiple coefficient of determination. If any of the regression coefficients have large standard errors, then we could expect their values to change quite a bit from sample to sample. We are still $100(1 - \alpha)$ percent sure that these predictors are important, but we do not completely trust our estimates of their effects. The regression equation obtained from one set of data may make rather poor predictions when applied to a new set of data.

For example, let us take the Cravens Data and split it in half, randomly choosing 13 of the 25 observations for one-half of the split. We will use as possible predictors X_2, X_3, X_4, and X_6 and run the 13 observations and four predictors through the SAS GLM regression program. The regression equation thus found will then be made to predict values of Y for the remaining 12 of the 25 observations.

Table 16.7 shows the 13 randomly selected observations to which a model is to be fitted. Figure 16.5 shows the results of the multiple regression analysis. We note that for this analysis all predictors are significant at $\alpha = 0.0625$, and

TABLE 16.7
Thirteen Randomly Selected Observations From Cravens Data

Y SALES	X_2 POTEN	X_3 ADV	X_4 SHARE	X_6 ACCTS
3473.95	58,117.30	5539.78	5.51	107.32
6125.96	57,805.11	7747.08	9.15	180.44
5031.66	50,935.26	3140.62	8.54	256.10
6519.45	46,176.77	8846.25	12.54	203.25
4876.37	42,053.24	5673.11	8.85	119.51
2468.27	36,829.71	2761.76	5.38	116.26
2533.31	33,612.67	1991.85	5.43	142.28
2337.38	20,416.87	1737.38	7.80	84.55
2729.24	23,093.26	8618.61	5.15	80.49
2800.78	39,571.96	4565.81	5.45	78.86
3453.62	58,749.82	3721.10	6.35	138.21
1741.45	23,990.82	860.97	7.37	75.61
2035.75	25,694.86	3571.51	8.39	102.44

FIGURE 16.5 SAS Multiple Regression Analysis (GLM Procedure) Using 13 Randomly Selected Observations From Cravens Data

```
DEP VARIABLE: SALES
SOURCE           DF   SUM OF SQUARES      MEAN SQUARE        F VALUE   PR > F          R-SQUARE   C.V.
MODEL             4   28131377.74179306   7032844.43544826   31.89     0.0001          0.940980   13.2357
ERROR             8    1764461.26333002    220557.65791625             STD DEV                    SALES MEAN
CORRECTED TOTAL  12   29895839.00512308                                469.63566508               3548.24538462

SOURCE           DF   TYPE I SS           F VALUE   PR > F   DF        TYPE IV SS         F VALUE   PR > F
POTEN             1   13608399.51206864   61.70     0.0001   1         1539529.83659616    6.98     0.0296
ADV               1    6341582.49673924   28.75     0.0007   1         2865518.01855294   12.99     0.0069
SHARE             1    7149617.26921522   32.42     0.0005   1         2241123.78596977   10.16     0.0128
ACCTS             1    1031778.46376996    4.68     0.0625   1         1031778.46376996    4.68     0.0625

                                    T FOR H0:                   STD ERROR OF
PARAMETER          ESTIMATE         PARAMETER=0     PR > |T|    ESTIMATE
INTERCEPT        -1935.67589269    -3.24            0.0119      598.08515819
POTEN                0.03545828     2.64            0.0296        0.01342099
ADV                  0.21495099     3.60            0.0069        0.05963469
SHARE              272.00492231     3.19            0.0128       85.33062203
ACCTS                8.44672961     2.16            0.0625        3.90531827
```

that $R^2 = 0.94$. The regression equation is

$$\hat{Y} = -1935.675893 + 0.035458X_2 + 0.214951X_3 + 272.004922X_4 + 8.446730X_6,$$

with $s^2_{Y \cdot X} = 469.635665$. Compare the regression coefficients in the above equation to those found using the same predictors for all 25 observations:

$$\hat{Y} = -1{,}441.931829 + 0.038218X_2 + 0.174990X_3 + 190.144297X_4 + 9.213896X_6.$$

We see some considerable differences in the regression coefficients, especially b_4, illustrating that these coefficients are very data-dependent.

Using the equation found in Figure 16.5, let us calculate $\hat{Y}$-values for the remaining 12 observations and see how well this regression equation predicts for a set of data not used in constructing the equation itself. Table 16.8 shows values of $\hat{Y}$ and $Y - \hat{Y}$ for the 12 other observations. Taking the square root of $\sum(Y_i - \hat{Y}_i)^2 = 327{,}684.818949$ found there, we have 572.437611. Comparing this to $s_{Y \cdot X} = 469.635665$ from Figure 16.5, we see that the regression equation found for 13 of the observations does not predict so accurately for the remaining 12 observations.

TABLE 16.8 Calculation of Residuals for 12 Observations, Using $\hat{Y} = -1935.675893 + 0.035458X_2 + 0.214951X_3 + 272.004922X_4 + 8.446730X_6$

Y_i	$\hat{Y}_i$	$Y_i - \hat{Y}_i$
3669.88	2990.673978	679.206022
2295.10	3232.125483	−937.025483
4675.56	4873.708102	−198.148102
2134.94	1876.027628	258.912372
3367.45	2769.498403	597.951597
2408.11	2309.351814	98.758186
4586.95	5471.185262	−884.235262
3289.40	3642.568919	−353.168919
3264.20	4067.984883	−803.784883
1578.00	1563.935130	14.064870
4167.44	5025.793685	−858.353685
2799.97	2871.284574	−71.314574
$\sum(\cdot)$		−2457.137861
$\sum(\cdot)^2$		4,107,660.214107

$$\sum_{i=1}^{12}(Y_i - \hat{Y}_i)^2 = \frac{12(4{,}107{,}660.214107) - (-2457.13861)^2}{12(11)} = 327{,}684.818949$$

Just because a regression equation having a large R^2 is found for a given set of data, we cannot be sure that it will predict as well for a new set of data. We have found several models that could be used. If our interest is in making accurate predictions of Y for new observed values of $X_1, X_2, \ldots, X_k$, then we not only need to look for significant predictors but also to find out which model predicts best.

THE PRESS-STATISTIC

In order to see how well a model predicts, consider the following procedure. Suppose we have some number, say m, candidate models. For one model, hold out one of the observations. Fit the model to the remaining $n - 1$ observations and then see how well the regression equation predicts the withheld observation by looking at its residual $Y_i - \hat{Y}_i$. Do this for each observation in turn. Thus, n regressions, each based on $n - 1$ observations will be run. By summing the n residuals thus obtained, we then see how this model does on the whole set of data. Let $\hat{Y}_i^{(m)}$ denote the prediction of the ith observation using the mth model. Then the PRESS-statistic for the mth model,

$$\text{PRESS}_m = \sum_{i=1}^{n} [\hat{Y}_i^{(m)} - Y_i]^2,$$

is a measure of how well this mth model predicts the set of data on which it is based. PRESS stands for *p*rediction *e*rror *s*um of *s*quares. If we did this procedure for all m models under consideration, then we could find the model giving the smallest value of the PRESS-statistic. This model would then be considered best, in that it gives the most accurate prediction.

While this is a reasonable enough notion, it is obviously too much work, even using a computer; each of the m models would need to have n different regressions run. It would be prohibitive to do this except that it can be shown that nowhere near this much work is actually needed. The computational form of the PRESS-statistic for the mth model is

$$\text{PRESS}_m = \sum_{i=1}^{n} \left[\frac{\delta_i^{(m)}}{Q_i}\right]^2,$$

where

$$\delta_i^{(m)} = Y_i - \hat{Y}_i,$$

the ordinary residual, using all n observations, associated with the ith observation and the mth model and

$$Q_i = 1 - \mathbf{x}_i'(\mathbf{X}'\mathbf{X})^{-1}\mathbf{x}_i,$$

where $\mathbf{x}_i$ is the ith set of $(X_1, X_2, \ldots, X_k)$-values.

It turns out that this statistic is reasonably easy to calculate. For one thing, the $\delta_i^{(m)}$-values are given from the usual regression programs and thus need not be calculated by hand. Also, note that Q_i is related to the width of the confidence interval on the mean Y-value at the point $\mathbf{x}_i = (X_{1i}, X_{2i}, \ldots, X_{ki})$. Recall that the width of the confidence interval at x_i is

$$2t_{n-p,\alpha/2}s_{Y\cdot X}\sqrt{\mathbf{x}_i'(\mathbf{X}'\mathbf{X})^{-1}\mathbf{x}_i} = w_i.$$

Then

$$\sqrt{\mathbf{x}_i'(\mathbf{X}'\mathbf{X})^{-1}\mathbf{x}_i} = \frac{w_i}{2t_{n-p,\alpha/2}s_{Y\cdot X}},$$

or

$$\mathbf{x}_i'(\mathbf{X}'\mathbf{X})^{-1}\mathbf{x}_i = \left[\frac{w_i/2}{t_{n-p,\alpha/2}s_{Y\cdot X}}\right]^2.$$

Thus,

$$Q_i = 1 - \left[\frac{(\text{width of } CI \text{ at } \mathbf{x}_i)/2}{t_{n-p,\alpha/2}s_{Y\cdot X}}\right]^2.$$

COMPUTER CALCULATION OF PRESS

BMD The BMDP9R program for all possible subsets regression can be instructed to calculate the PRESS-statistic for each candidate model. The P9R program has not been discussed in this text, so the reader is referred to the BMD manual for instructions in using it.

MINITAB Minitab does not provide for automatic calculation of the PRESS-statistic.

SAS The SAS GLM procedure will print the value of PRESS automatically if the option requesting confidence limits on the mean response is used. For example,

```
PROC GLM;
MODEL SALES=ACCTS/P CLM;
```

will produce an analysis of `SALES` as a function of `ACCTS`, along with the predicted and residual values and the upper and lower 95% confidence limits for the mean value of `SALES` at that value of `ACCTS`. (This could also be done, of course, for a multiple-predictor model.) After the list of residuals, summary statistics pertinent to the residuals are given. Among them is the value of the PRESS-statistic. Figure 16.6 shows this analysis.

The PRESS-statistic is not an available option under `PROC STEPWISE`, however, and a moment's reflection will reveal why this is so. Even with the "shortcut" method for calculating PRESS, considerable labor is involved. Many of the models considered at the various steps of any stagewise procedure are unacceptable for a variety of reasons. It would waste computer time to calculate PRESS for models that would be eliminated from further consideration. Thus, after using `STEPWISE` to identify several models of interest, these models must be reanalyzed using GLM in order to see values of the PRESS-statistic.

FIGURE 16.6 SAS Analysis of One-Predictor Model Giving Confidence Intervals on the Mean Response

DEP VARIABLE:SALES

SOURCE	DF	SUM OF SQUARES	MEAN SQUARE	F VALUE	PR > F	R-SQUARE	C.V.
MODEL	1	23524074.19189502	23524074.19189502	30.30	0.0001	0.568495	26.1098
ERROR	23	17855474.73496108	776324.98847657		STD DEV		SALES MEAN
CORRECTED TOTAL	24	41379548.92685610			881.09306459		3374.56760000

SOURCE	DF	TYPE I SS	F VALUE	PR > F	DF	TYPE IV SS	F VALUE	PR > F
ACCTS	1	23524074.19189502	30.30	0.0001	1	23524074.19189502	30.30	0.0001

PARAMETER	ESTIMATE	T FOR H0: PARAMETER=0	PR > |T|	STD ERROR OF ESTIMATE
INTERCEPT	709.32383372	1.38	0.1819	515.24608170
ACCTS	21.72176971	5.50	0.0001	3.94603304

OBS	ACTUAL	PREDICT VALUE	RESIDUAL	LOWER 95% CL FOR MEAN	UPPER 95% CL FOR MEAN
1	3669.88000000	2335.41551438	1334.46448562	1801.20540422	2869.62562453
2	3473.95000000	3040.50415924	433.44584076	2654.95998680	3426.04833169
3	2295.10000000	2810.90505338	-515.80505338	2389.29803748	3232.51206928
4	4675.56000000	4947.67553998	-272.11553998	4253.15364349	5642.19743647
5	6125.96000000	4628.79996060	1497.16003940	4032.94795327	5224.65196794
6	2134.94000000	2987.50304114	-852.56304114	2595.02134911	3379.98473318
7	5031.66000000	6272.26905704	-1240.60905704	5123.93321540	7420.60489868
8	3367.45000000	3464.29588633	-96.84588633	3098.20680701	3830.38496564
9	6519.45000000	5124.27352774	1395.17647226	4372.45588585	5876.09116964
10	4876.37000000	3305.29253203	1571.07746797	2939.83125745	3670.75380662
11	2468.27000000	3234.69678047	-766.42678047	2866.39384507	3602.99971587
12	2533.31000000	3799.89722838	-1266.58722838	3401.86213469	4197.93232207
13	2408.11000000	2651.90169909	-243.79169909	2197.32872545	3106.47467272
14	2337.38000000	2545.89946289	-208.51946289	2066.46256078	3025.33636500
15	4586.95000000	3305.29253203	1281.65746797	2939.83125745	3670.75380662
16	2729.24000000	2457.70907786	271.53092214	1956.11243204	2959.30572368
17	3289.40000000	3676.08314102	-386.68314102	3294.34646061	4057.81982144
18	2800.78000000	2422.30259323	378.47740677	1911.47472890	2933.13045755
19	3264.20000000	3676.08314102	-411.88314102	3294.34646061	4057.81982144
20	3453.62000000	3711.48962565	-257.86962565	3325.59432702	4097.38492429
21	1741.45000000	2351.70684166	-610.25684166	1821.95558129	2881.45810203
22	2035.75000000	2934.50192304	-898.75192304	2534.21094862	3334.79289747
23	1578.00000000	2369.30147513	-791.30147513	1844.32811557	2894.27483469
24	4167.44000000	3676.08314102	491.35685898	3294.34646061	4057.81982144
25	2799.97000000	2634.30706562	165.66293438	2175.75327230	3092.86085894

SUM OF RESIDUALS	0.00000000
SUM OF SQUARED RESIDUALS	17855474.73496115
SUM OF SQUARED RESIDUALS - ERROR SS	0.00000007
PRESS STATISTIC	23454156.16210017
FIRST ORDER AUTOCORRELATION	0.18085477
DURBIN-WATSON D	1.55579444

SPSS^X SPSS^X does not provide for automatic calculation of the value of the PRESS-statistic.

PRESS-STATISTICS FOR NINE CANDIDATE MODELS Looking back at the steps of the various stagewise procedures applied to the Cravens Data, several models present themselves for consideration. Let us look at nine models, all with significant or almost significant predictors at $\alpha = 0.05$, and compare them by using the PRESS-statistic. The models we will consider are shown in Table 16.9.

TABLE 16.9

k	Predictors
1	ACCTS
2	ACCTS, ADV
	POTEN, SHARE
3	POTEN, ADV, ACCTS
	POTEN, ADV, SHARE
4	POTEN, ADV, SHARE, ACCTS
5	POTEN, ADV, SHARE, CHANGE, ACCTS
	TIME, POTEN, ADV, SHARE, CHANGE
6	TIME, POTEN, ADV, SHARE, CHANGE, ACCTS

Figure 16.6 shows the analysis from the model using ACCTS to predict SALES. For each of the 25 observations we have residual, $\delta_i^{(1)}$, $i = 1, 2, \ldots, 25$, resulting from this model. We take the difference between the upper 95% confidence limit for the mean response and the lower 95% confidence limit for the mean response to obtain the length of the confidence interval. Also, $s_{Y \cdot X} = 881.093065$ and $t_{0.025,23} = 2.069$. With this information we can find the value of PRESS$_1$-statistic, since

$$Q_i = 1 - \left[\frac{\text{width}}{2(2.069)(881.093065)}\right]^2$$

$$= 1 - \left[\frac{\text{width}}{3645.9633103}\right]^2.$$

Calculations are shown in Table 16.10. The value for PRESS in Table 16.10 differs from the value given by the SAS printout in Figure 16.6 because of rounding error, since we did not carry many decimal places in Table 16.10.

PRESS-statistics can be calculated for each of the other eight models, as well. Of course, the value for $s_{Y \cdot X}$ will be different for each different candidate model, and $t_{n-p,0,025}$ will depend upon the number of predictors in the model under consideration. Table 16.11 shows PRESS-values for all nine candidate models. From Table 16.11 we see that model 7 has the smallest PRESS-statistic. The analysis for this five-predictor model using independent variables POTEN, ADV, SHARE, CHANGE, and ACCTS is shown in Figure 15.25

TABLE 16.10 Calculation of $PRESS_1$

Obs	$\delta_i^{(1)}$	Width of *CI*	Q_i	$[\delta_i^{(1)}/Q_i]^2$
1	1334.46	1068.42	0.91	2,150,446.27
2	433.45	771.09	0.96	230,861.28
3	−515.81	843.21	0.95	294,805.56
4	−272.12	1389.04	0.85	102,489.62
5	1497.16	1191.70	0.89	2,829,796.84
6	−852.56	784.96	0.95	805,380.60
7	−1240.61	2296.67	0.60	4,275,300.58
8	−96.85	732.17	0.96	10,178.79
9	1395.18	1503.63	0.83	2,825,559.28
10	1571.08	730.92	0.96	2,678,263.17
11	−766.43	736.61	0.96	637,378.69
12	−1266.59	796.07	0.95	1,777,555.56
13	−243.79	909.14	0.94	67,262.42
14	−208.52	958.88	0.93	50,274.61
15	1281.66	730.92	0.96	1,782,385.20
16	271.53	1003.20	0.92	87,107.62
17	−386.68	763.47	0.96	162,239.78
18	378.48	1021.66	0.92	169,241.73
19	−411.88	763.47	0.96	184,075.32
20	257.87	771.79	0.96	72,151.33
21	−610.26	1059.50	0.92	440,006.69
22	−898.74	800.58	0.95	895,010.60
23	−791.30	1049.94	0.92	739,789.21
24	491.36	763.47	0.96	261,969.95
25	165.66	917.11	0.94	31,057.01
			$PRESS_1$ =	23,533,587.71

TABLE 16.11 Press-Statistics for Nine Models Under Consideration

Model No.	Predictors	PRESS
1	ACCTS	23,533,587.71
2	ACCTS, ADV	12,437,016.44
3	POTEN, SHARE	14,199,601.98
4	POTEN, ADV, ACCTS	9,990,021.32
5	POTEN, ADV, SHARE	8,601,908.91
6	POTEN, ADV, SHARE, ACCTS	5,794,837.54
7	POTEN, ADV, SHARE, CHANGE, ACCTS	5,338,063.76
8	TIME, POTEN, ADV, SHARE, CHANGE	5,440,757.72
9	TIME, POTEN, ADV, SHARE, CHANGE, ACCTS	6,089,047.96

(step 5 of the MAXR procedure). There we see that all predictors are significant at $\alpha = 0.1236$, and that this model has $R^2 = 0.912416$. The error sum of squares in Figure 15.25 is 3,624,197.33. We can compare this to the value of PRESS for this model: $PRESS_7 = 5{,}338{,}063.76$. This comparison says that by fitting this model to these 25 observations, we get an $SS(E)$ of 3,624,197.33; but if we took another sample of 25 observations and fit this same model to them, we would expect to find an $SS(E)$ of approximately 5,338,000.

Note that when calculating a PRESS-statistic we are weighting each residual before squaring and summing it. The weights have the effect of increasing the absolute values of the residuals, so that PRESS will be larger than the residual sum of squares. How much larger any residual becomes is determined by its weight, which is in turn determined by the width of the associated confidence interval. Recall that the farther $(X_{1i}, X_{2i}, \ldots, X_{ki})$ is from $(\bar{X}_1, \bar{X}_2, \ldots, \bar{X}_k)$, the wider the confidence interval. The wider the confidence interval, the smaller the weight (Q_i) and thus the more the residual is increased. Thus, PRESS takes into account observations that cannot be predicted very well by the model under consideration and assesses a "penalty" for each such observation.

Note that the other five-predictor model we considered—the one using TIME, POTEN, ADV, SHARE, and CHANGE arrived at through the backward elimination procedure—has all predictors significant at $\alpha = 0.0530$ and also has $R^2 = 0.915009$, which is greater than the R^2 for the model selected by PRESS. Usually PRESS will pick the same model as arrived at by the backward elimination procedure. That it did not in this case is probably due to the fact that $n = 25$ is relatively small, and this affects the results of the backward elimination procedure. It could be argued that since PRESS_7 and PRESS_8 are so close, one would prefer to use the model in which all predictors are significant at $\alpha = 0.0530$, that is, the one arrived at by the backward elimination procedure. In any case, if your interest is in choosing a model that will make the most accurate possible predictions for future observations, the PRESS-statistic provides a method for choosing among various models suggested by the different stagewise analyses.

EXERCISES

16.5 Some candidate models that came out of the various stagewise analyses of the Tennessee Data of Exercise 15.3 are shown in Table 16.12. In all four models all predictors are significant at $\alpha = 0.05$. Use the PRESS-statistic to determine which model will make the best predictions when applied to different data.

TABLE 16.12

Model	Predictors
1	Percent of households with telephones
2	Percent of households with telephones
	Population change
3	Population change
	Agricultural employment
	Percent rural population
4	Population change
	Agricultural employment
	Percent rural population
	Median age

16.6 Calculate the C_p-statistic for each of the four candidate models in Exercise 16.5. Use the C_p-statistic to select an appropriate model.

16.4 SUMMARY

In this chapter we considered alternative final models arrived at by several variable-selection procedures in terms of how well they satisfy the assumptions for a regression analysis, whether the error mean square measures only random variation, and how well they will predict the response based on future values of the predictors. Even with these considerations, it is possible that no uniformly best model will be found; rather it is likely that two or three candidate models will be judged about equally appropriate from a statistical point of view. Choice of a final model is considerably facilitated if the researcher has a theoretical or practical framework within which to operate and can choose the model that makes the best sense to the researcher.

There is much more to be said that is not touched upon in this text. In an attempt to obtain more stable estimates of regression coefficients when predictors are highly correlated, least squares estimation procedures are sometimes abandoned in favor of *ridge regression*. Other advanced regression techniques include *regression on principal components* and techniques based on finding the line that minimizes absolute deviations rather than squared deviations. *Nonlinear regression* involves fitting models that cannot be made linear by any sort of transformation on the variables.

Specialized techniques for analyzing relationships between pairs of nominal-level variables and pairs of ordinal-level variables are widely available but are not considered here. Similarly, techniques for regression in cases in which the dependent variable is categorical also exist.

If all predictors are nominal-level variables, but the response is interval- or ratio-level, the regression model is called an *analysis of variance model*. There is a very old and extensive body of literature and techniques concerning these models; the modern approach is to treat them as just another case of the general linear model. If some predictors are categorical and some are interval or ratio, this hybrid of a regression model and an analysis of variance is an *analysis of covariance*.

These topics are covered in other texts. Many of the techniques are quite new, and statisticians are still working on theory and applications regarding them. It is hoped that the reader who has completed study of this text will be sufficiently grounded in the basic techniques of regression to be able to proceed to these other, often more advanced, topics, as well as to a more theoretical treatment of the regression model and of linear models in general. The reader is referred to the *Selected References and Suggestions for Further Reading* at the end of this text.

SELECTED REFERENCES AND SUGGESTIONS FOR FURTHER READING

Here is a short list of some commonly available texts that might be useful to the student who wishes to pursue topics in regression analysis. While this list is far from a complete set of references for regression analysis, it includes:

1. Elementary-level statistics survey texts and books of readings that might help supplement and illustrate concepts, especially simple regression;
2. Texts aimed toward a specialized audience, like sociologists or economists, that show how regression techniques are applied in these fields;
3. Regression texts that explain more advanced topics, such as weighted regression or ridge regression, that are alluded to in this text;
4. Manuals for running regression programs;
5. Theoretical statistics texts for the rigorous mathematical and statistical basis for regression.

A short description of the reference is provided as an aid to directing the reader to potentially useful references for his or her particular interests.

Agresti, Alan, and Barbara F. Agresti. 1979. *Statistical Methods for the Social Sciences*. Dellen, San Francisco.

Sequential chapters on techniques for analyzing relationships among nominal, ordinal, and interval/ratio-level variables are especially well presented, with very good examples.

Blalock, Hubert M. 1979. *Social Statistics: Revised Second Edition*. McGraw-Hill, New York.

A statistics survey text designed for use by sociologists. Uses many examples taken from sociology and includes detailed discussions. Of interest to the reader might be rank correlations as well as interpretations and use of correlations and beta coefficients in sociological studies.

Chatterjee, Samprit, and Bertram Price. 1977. *Regression Analysis by Example*. John Wiley & Sons, New York.

Many actual data sets are analyzed to introduce concepts and also to illustrate failure of the assumptions of the linear model. More advanced topics, such as ridge regression and regression on principal components, are also discussed and illustrated. Relatively sophisticated mathematically.

Daniel, Cuthbert, and Fred S. Wood. 1971. *Fitting Equations to Data: Computer Analysis of Multifactor Data for Scientists and Engineers*. Wiley-Interscience, New York.

Use of computer to analyze data.

Dixon, W. J., ed. 1981. *BMDP Statistical Software 1981*. University of California Press. Berkeley.

Instructions for using BMDP programs.

Draper, Norman, R., and Harry Smith. 1981. *Applied Regression Analysis: Second Edition*. John Wiley & Sons, New York.

A useful text to those with good training in mathematics and statistics. Covers several advanced concepts and contains many solved examples. Provides a more rigorous and thorough treatment of some concepts presented here and is considered as a classic text in applied regression analysis.

Dutta, M. 1975. *Econometric Methods*. South-Western Publishing Co., Cincinnati.

An introductory regression text geared toward the student of economics. Considers some statistical properties and covers several more advanced topics, mostly dealing with time-dependent data.

Edwards, Allen L. 1976. *An Introduction to Linear Regression and Correlation*. W. H. Freeman and Co., San Francisco.

Written for students of the behavioral sciences, this text includes such topics as nonparametric measures of association, regression with standardized variables, and other specialized topics.

Freedman, David, Robert Pisani, and Roger Purves. 1978. *Statistics*. W. W. Norton, New York.

A statistics survey text written for nonmathematically oriented students. Chapters on correlation and regression are well illustrated with historical notes and actual studies.

Graybill, Franklin A. 1961. *An Introduction to Linear Statistical Models, Volume I*. McGraw-Hill, New York.

The bible for statisticians, this very mathematical text provides all the basic statistical theory of regression. Not recommended for the nonstatistician.
See also:

———. 1976. *Theory and Application of the Linear Model*. Duxbury Press, North Scituate, MA.

Kerlinger, Fred N., and Elazar J. Pedhazur. 1973. *Multiple Regression in Behavioral Research*. Holt, Rinehart & Winston, New York.

Elements of simple and multiple regression are covered in the first two chapters, with the rest of the book considering more advanced topics. Examples are taken from the behavioral sciences. Quite useful for advanced work in social sciences.

Kirk, Roger E. 1982. *Experimental Design: Procedures for the Behavioral Sciences* (2d ed.). Brooks/Cole, Belmont, CA.

A very complete treatment of the philosophy and procedures of experimental design and analysis of variance. Integrates the traditional approach with the general linear model formulation.

Kleinbaum, David G., and Lawrence L. Kupper. 1978. *Applied Regression Analysis and Other Multivariable Methods*. Duxbury Press, North Scituate, MA.

A relatively complete treatment of simple and multiple regression and variable-selection techniques, including extensions to experimental design models as well as factor and discriminant analysis. Extensive references to output from computer programs.

Mendenhall, William, and James T. McClave. 1981. *A Second Course in Business Statistics: Regression Analysis*. Dellen, San Francisco.

Good elementary treatment of linear regression, with an introduction to analysis of variance, illustrated with extensive case studies.

Montgomery, Douglas C., and Elizabeth A. Peck. 1982. *Introduction to Linear Regression Analysis*. John Wiley & Sons, New York.

A very complete and up-to-date survey of regression techniques. A more advanced text, but readable to those with a little background. Good for engineering applications.

Mosteller, Fredrick, and John W. Tukey. 1977. *Data Analysis and Regression*. Addison-Wesley Publishing Co., Reading, MA.

Emphasis on appropriate techniques for data sets displaying particular properties. More advanced topics, such as weighted least squares, are well illustrated.

Neter, John, and William Wasserman. 1974. *Applied Linear Statistical Models*. Richard D. Irwin, Homewood, IL.

A very complete text in the use of regression techniques. Especially useful in relating regression to various experimental design models.

Neter, John, William Wasserman, and Michael H. Kutner. 1983. *Applied Linear Regression Models*. Richard D. Irwin, Homewood, IL.

Regression topics from *Applied Linear Statistical Models* (see above) are extracted and somewhat expanded.

Olson, Charles L., and Mario Picconi. 1983. *Statistics for Business Decision Making*. Scott, Foresman and Company, Glenview, IL.

A survey text for business students containing exceptionally interesting examples of the use of regression in business applications.

Ryan, Thomas A., Brian L. Joiner, and Barbara F. Ryan. 1976. *MINITAB Student Handbook*. Duxbury Press, North Scituate, MA.

A good introduction to regression using the Minitab computing system.

SAS User's Guide, 1979 Edition. SAS Institute, Inc., Cary, NC.

SAS User's Guide: Statistics, 1982. SAS Institute, Inc., Cary, NC.

Instructions for using SAS programs.

SPSS, Inc. 1983. *SPSS^x User's Guide*. McGraw-Hill, New York.

Instructions for using SPSS^x programs.

Tanur, G. M., ed. 1972. *Statistics: A Guide to the Unknown*. Holden-Day, San Francisco.

A delightful book of nontechnical essays on the use of statistics in actual applications from astronomy to zoology.

APPENDIX A

THE ALGEBRA OF SUMMATION NOTATION

The three basic rules for manipulating the $\sum$ sign that follow are important to the arguments throughout the text.

RULE 1 The summation sign distributes across the terms of a sum or difference.

$$\sum_{i=1}^{n} (X_i + Y_i - Z_i) = \sum_{i=1}^{n} X_i + \sum_{i=1}^{n} Y_i - \sum_{i=1}^{n} Z_i.$$

This equality is easily verified by letting $n = 2$. Then

$$\sum_{i=1}^{2} (X_i + Y_i - Z_i) = (X_1 + Y_1 - Z_1) + (X_2 + Y_2 - Z_2)$$

$$= (X_1 + X_2) + (Y_1 + Y_2) - (Z_1 + Z_2)$$

$$= \sum_{i=1}^{2} X_i + \sum_{i=1}^{2} Y_i - \sum_{i=1}^{2} Z_i.$$

RULE 2 A constant factors out of a sum.

$$\sum_{i=1}^{n} kX_i = k \sum_{i=1}^{n} X_i, \quad \text{for } k \text{ a constant.}$$

To verify rule 2, let $n = 3$ and $k = 4$. Then

$$\sum_{i=1}^{3} 4X_i = 4X_1 + 4X_2 + 4X_3$$

$$= 4(X_1 + X_2 + X_3)$$

$$= 4 \sum_{i=1}^{3} X_i.$$

RULE 3 The sum of a constant over n terms is n times the constant.

$$\sum_{i=1}^{n} k = kn, \quad \text{for } k \text{ a constant.}$$

Verification: Let $n = 4$ and $k = 5$. Then

$$\sum_{i=1}^{4} 5 = \underbrace{5 + 5 + 5 + 5}_{4 \text{ terms}} = 4(5).$$

APPENDIX B

MINIMIZATION OF $\sum_{i=1}^{n} (Y_i - \hat{Y}_i)^2$

In order to find a minimum value for a function, one takes a first derivative, equates this to zero, and solves for the unknown quantity. Since this gives either the maximum or the minimum, a second derivative check is necessary in order to determine which has been found. Our house size-price example, however, has *two* unknown quantities, a and b, so that we can determine the values in the equation $\hat{Y} = a + bX$. Note that X and Y are *known variables* where respective values are the observed sizes and prices of houses given in Table 1.1. The quantities a and b are *unknown constants*. Our example has *one* unknown a-value and *one* unknown b-value; there are 20 X-values that we do know, and 20 Y-values that are also known.

Since we have two unknowns, we must take partial derivatives. A partial derivative with respect to a will simply ignore the fact that b is also unknown, treating it as a constant, and similarly for a partial derivative with respect to b. Thus, to minimize

$$Q = \sum_{i=1}^{n} (Y_i - \hat{Y}_i)^2 = \sum_{i=1}^{n} (Y_i - a - bX_i)^2$$

we take the first partial derivative of Q with respect to a:

$$\frac{\partial Q}{\partial a} = 2\sum_{i=1}^{n} (Y_i - a - bX_i)(-1) = -2\sum_{i=1}^{n} (Y_i - a - bX_i).$$

Equating to zero, we have

$$-2\sum_{i=1}^{n} (Y_i - a - bX_i) = 0,$$

$$\sum_{i=1}^{n} Y_i - \sum_{i=1}^{n} a - \sum_{i=1}^{n} bX_i = 0,$$

$$\sum_{i=1}^{n} Y_i - na - b\sum_{i=1}^{n} X_i = 0,$$

$$\sum_{i=1}^{n} Y_i = na + b\sum_{i=1}^{n} X_i.$$

Note that this is one equation in two unknowns, a and b. In order to solve for a and b, we need another equation involving these two variables, so we take a first partial derivative of Q with respect to b:

$$\frac{\partial Q}{\partial b} = 2\sum_{i=1}^{n} (Y_i - a - bX_i)(-X_i) = -2\sum_{i=1}^{n} (X_iY_i - aX_i - bX_i^2).$$

Equating to zero,

$$-2\sum_{i=1}^{n}(X_iY_i - aX_i - bX_i^2) = 0,$$

$$\sum_{i=1}^{n} X_iY_i - a\sum_{i=1}^{n} X_i - b\sum_{i=1}^{n} X_i^2 = 0,$$

$$\sum_{i=1}^{n} X_iY_i = a\sum_{i=1}^{n} X_i + b\sum_{i=1}^{n} X_i^2,$$

and this is the second equation in two unknowns. Solving these two equations,

$$\sum_{i=1}^{n} Y_i = na + b\sum_{i=1}^{n} X_i$$

and

$$\sum_{i=1}^{n} X_iY_i = a\sum_{i=1}^{n} X_i + b\sum_{i=1}^{n} X_i^2$$

simultaneously for a and b will give either a maximum or a minimum. Thus, the line $\hat{Y} = a + bX$ will be the line that is, on the average, either the farthest from or the closest to all the (X, Y) points, in terms of squared deviations. The second derivative check will tell us which. Since there are two unknowns, we must see if the determinant of the matrix of second derivations is positive in order for a minimum to have been found:

$$\begin{vmatrix} 2n & 2\sum_{i=1}^{n} X_i \\ 2\sum_{i=1}^{n} X_i & 2\sum_{i=1}^{n} X_i^2 \end{vmatrix} = 4n\sum_{i=1}^{n} X_i^2 - \left(2\sum_{i=1}^{n} X_i\right)^2.$$

(Those unfamiliar with matrices and determinants may want to look back to this after completion of Chapter 3.) You will recall that this determinant is four times the numerator of the calculating formula for the sum of squares,

$$\sum_{i=1}^{n}(X_i - \overline{X})^2 = \frac{n\sum_{i=1}^{n} X_i^2 - \left(\sum_{i=1}^{n} X_i\right)^2}{n}.$$

Since a sum of squares can never be negative, being the sum of squared values, the quantity

$$n\sum_{i=1}^{n} X_i^2 - \left(\sum_{i=1}^{n} X_i\right)^2$$

must always be positive or zero, regardless of what the X-values are. The only way this quantity can be zero is for all X_i-values to be equal to $\overline{X}$, and thus all equal to each other. Thus, we have shown that the solution to the normal equations will yield values a and b so that the line $\hat{Y} = a + bX$ is the one that is, on the average, closest to all data points, in terms of squared deviations.

APPENDIX C

FORMATTING FOR THE COMPUTER

Format Statement

A **format statement** tells the computer how to read the data (numbers) you have entered on cards for a particular problem. Consider, for example, the format statement (which is always enclosed in parentheses)

```
(F6.2, F3.0)
```

This says that

1. There are values of *two* variables entered on each card, since two F's (formats) are specified;
2. The first six columns of each line contain values of the first variable, and the next three columns contain values of the second variable;
3. Thus, all numbers are typed in columns 1–9;
4. Values of the first variable can have as many as two digits to the right of the decimal point; values of the second variable are whole numbers.

Suppose we have a pair of values $X = 223.64$, and $Y = 65.1$. How do we write a format statement to handle these values? First, there are eight numbers to be entered, if we do not type the decimal point. The first value has five numbers, two of which are to the right of the decimal; and the second has three numbers with one decimal place. A format statement such as

```
(F5.2, F3.1)
```

will handle this pair of values, and we would enter, beginning in column 1 on the data card and ending in column 8:

Column	1	2	3	4	5	6	7	8
Punch	2	2	3	6	4	6	5	1

Proofreading the entered data is easier if the decimal point is indicated. For this, we can come up with other format statements. To enter the decimal points in both values requires 10 spaces, so we need this format statement:

```
(F6.2, F4.1)
```

We would enter the data, beginning in column 1 and ending in column 10:

Column	1	2	3	4	5	6	7	8	9	10
Punch	2	2	3	.	6	4	6	5	.	1

To leave a space between the first and second values and still type the decimal points would require

```
(F6.2, F5.1)
```

and we would type

Column	1	2	3	4	5	6	7	8	9	10	11
Punch	2	2	3	.	6	4		6	5	.	1

Note that column 7 is left blank. Finally, we could space between the two values and *not* type the decimal point by using

```
(F5.2, F4.1)
```

and entering

Column	1	2	3	4	5	6	7	8	9
Punch	2	2	3	6	4		6	5	1

You are free to set up the format to your own liking, *as long as your data are entered to conform to the format statement*. For example, suppose the format statement were

```
(2F10.4),
```

indicating that ten spaces have been allotted to both values, with the last four of each ten being decimal places. How would one enter in $X = 123.456$ and $Y = 7.8$? To type the decimals, the number 1 would be placed in column 3, the decimal point in column 6, and the number 6 in column 9; then 7 would appear in column 15, the decimal point in column 16, and the 8 in column 17:

Column	1	2	3	4	5	6	7	8	9	10	11	12	13	14	15	16	17	18	19	20
Punch			1	2	3	.	4	5	6						7	.	8			

If the decimal points are not entered, then the data would be entered as

Column	1	2	3	4	5	6	7	8	9	10	11	12	13	14	15	16	17	18	19	20
Punch				1	2	3	4	5	6							7	8			

Note that if the value 123456 began in column 1 and ended in column 6, the computer, following the format statement `(2F10.4)`, would read the number as 123,456.0000 instead of 123.456. Thus, extreme care must be taken to make the data conform to the format statement.

Right Justify

This last example leads us to the consideration of what it means to **right justify.** Right justifying is simply positioning a value so that the decimal point is in its proper place. If you were to add up the three numbers 10, 243, and 7 by hand, you might write them in a column thus:

```
 10
243
  7
___
```

You have lined up the three numbers so that the understood decimal points are one above the other. Suppose that in columns 9–13 of a card, you are to enter the number 20, with a corresponding format of F5.2, so that the decimal point is understood to stand after column 11. Then the 2 must be typed in column 10 and the zero in column 11. The following table shows how the computer will read the digits 2 and 0 when entered in different columns:

Typed in Column							**Computer Reads**
7	8	9	10	11	12	13	
			2	0			20
				2	0		2
		2	0				200
	2	0					0
			2				20

APPENDIX D

THE BINARY NUMBER SYSTEM

Because human beings have five fingers on each of two hands, most cultures use the decimal (base 10) system. There is really nothing sacred about it, and indeed there exist and have existed cultures that use different systems. Cultures that counted on their toes as well as their fingers developed base 20 number systems, and time is measured in a base 60 system. Whatever the system, there are three main features:

1. The number of symbols required is the same as the base figure.
2. The fewer the symbols, the more of them required to represent any given value.
3. The value is calculated, starting at the decimal point and moving to the left, by multiplying each digit by the base number raised to zero, first, second, . . . power.

In the decimal system, then, we have the ten symbols 0, 1, 2, 3, 4, 5, 6, 7, 8, 9. The numeral 123 represents

$$
\begin{aligned}
1(10)^2 + 2(10)^1 + 3(10)^0 &= 100 + 2(10) + 3(1) \\
&= 100 + 20 + 3,
\end{aligned}
$$

or the value one hundred twenty-three. If we were working in base 5 (suppose people counted on only one hand), then the numeral 123 would represent

$$1(5)^2 + 2(5)^1 + 3(5)^0 = 25 + 10 + 3,$$

or the value thirty-eight.

To represent the value one hundred twenty-three in base 5, we would need to find three symbols s_1, s_2, s_3 in the set 0, 1, 2, 3, 4 such that

$$s_1(5)^2 + s_2(5)^1 + s_3(5)^0 = 100 + 20 + 3.$$

We find that in base 5, $s_1 = 4$, $s_2 = 4$, and $s_3 = 3$. Therefore, the numeral 443 represents

$$4(5)^2 + 4(5)^1 + 3(5)^0 = 100 + 20 + 3,$$

or the value 123.

Computers depend not on ten fingers but on such things as switches that are open or closed, positions on cards that are or are not punched. Thus, the computer recognizes only two symbols—off/on or no/yes—which we usually denote by 0 and 1, respectively. In the base 2 system, then, when a number is moved one place to the left, the value it represents is doubled; moving a digit one place to the right signifies division by 2. Thus,

$$1 = 1(2)^0 = 1,$$

$$10 = 1(2)^1 + 0(2)^0 = 2,$$

$$100 = 1(2)^2 + 0(2)^1 + 0(2)^0 = 4,$$

and similarly,

$$0.1 = 1(\tfrac{1}{2})^1 = \tfrac{1}{2},$$

$$0.01 = 0(\tfrac{1}{2})^1 + 1(\tfrac{1}{2})^2 = \tfrac{1}{4},$$

$$0.001 = 0(\tfrac{1}{2})^1 + 0(\tfrac{1}{2})^2 + 1(\tfrac{1}{2})^3 = \tfrac{1}{8},$$

and so on.

Thus, the binary number 101 represents the decimal number 5, since

$$1(2)^2 + 0(2)^1 + 1(2)^0 = 4 + 0 + 1 = 5.$$

To convert a decimal number to a binary number, repeatedly divide the decimal number by 2 until the quotient is less than 1. Reading the remainders in reverse order gives the binary representation. For example, to convert 553 to the binary system:

$$553 \div 2 = 276, \quad \text{remainder} = 1,$$

$$276 \div 2 = 138, \quad \text{remainder} = 0,$$

$$138 \div 2 = 69, \quad \text{remainder} = 0,$$

$$69 \div 2 = 34, \quad \text{remainder} = 1,$$

$$34 \div 2 = 17, \quad \text{remainder} = 0,$$

$$17 \div 2 = 8, \quad \text{remainder} = 1,$$

$$8 \div 2 = 4, \quad \text{remainder} = 0,$$

$$4 \div 2 = 2, \quad \text{remainder} = 0,$$

$$2 \div 2 = 1, \quad \text{remainder} = 0,$$

$$1 \div 2 = 0, \quad \text{remainder} = 1.$$

Thus, the binary representation of the decimal number 553 is

$$1000101001.$$

To convert fractions to the binary system, instead of repeatedly dividing by 2, we need to multiply by 2. To convert the decimal number 0.625 to a binary number,

$$0.625 \times 2 = 1.250,$$

$$0.250 \times 2 = 0.500,$$

$$0.500 \times 2 = 1.000.$$

Note that at each step, we took only the fractional part of the number and multiplied by 2. We stopped when the fractional part became zero. The binary equivalent of the decimal number 0.625 is then

$$0.101,$$

reading the integral parts of the products in order down the column. The decimal part of the products will not always turn out to be zero. In this case, the decimal part will sooner or later repeat, as in the case below:

$$0.55 \times 2 = 1.10,$$
$$0.10 \times 2 = 0.20,$$
$$0.20 \times 2 = 0.40,$$
$$0.40 \times 2 = 0.80,$$
$$0.80 \times 2 = 1.60,$$
$$0.60 \times 2 = 1.20.$$

Notice that the decimal portion is beginning to repeat. Thus, the binary representation of the decimal number 0.55 is

$$0.10001100110011\ldots = 0.1\overline{0001}1.$$

Binary arithmetic simplifies the circuitry necessary to perform calculations in a computer. The following tables give the rules for multiplication and addition:

Addition

	0	1
0	0	1
1	1	10

Multiplication

	0	1
0	0	0
1	0	1

Note that if any value converted to a binary number results in a repeating decimal, we have a source of rounding error when the value is reconverted to the decimal system.

APPENDIX E

ORTHOGONAL POLYNOMIALS

Suppose one desires to fit a polynomial equation of the form

$$Y = a + b_1X + b_2X^2 + \cdots + b_kX^k$$

to a set of data. Even if $k = 2$, solution of the normal equations

$$\mathbf{X}'\mathbf{X}\hat{\boldsymbol{\beta}} = \mathbf{X}'\mathbf{y}$$

can be quite difficult if $\mathbf{X}'\mathbf{X}$ must be inverted by hand. If, however, X is a controlled variable with values at evenly spaced intervals, one can use orthogonal polynomials and rewrite the equation so that $\mathbf{X}'\mathbf{X}$ will be a diagonal matrix and therefore easy to invert regardless of its dimension. Suppose, for example, that an experiment has been run using the following X-values:

60, 80, 100, 120, 140, 160, 180, 200, 210

For these values, we can show that if our model is $Y = \alpha + \beta X + \varepsilon$, then

$$\mathbf{X}'\mathbf{X} = \begin{bmatrix} 9 & 1{,}250 \\ 1{,}250 & 196{,}100 \end{bmatrix}.$$

While this matrix is not difficult to invert, we saw in Section 2.4 that we can reduce labor if we scale the X-values by

$$Z = \frac{X - \bar{X}}{d},$$

where d is the spacing between levels of X. Since $\bar{X} = 140$ and $d = 20$, we code the X-values to

$$Z = -4, -3, -2, -1, 0, 1, 2, 3, 4$$

and then

$$\mathbf{Z}'\mathbf{Z} = \begin{bmatrix} 9 & 0 \\ 0 & 60 \end{bmatrix},$$

which is diagonal and quite simple to invert.

Let us call this Z-value the *first orthogonal polynomial* and give it a subscript to denote that it is the first one:

$$Z_1 = \frac{X - \bar{X}}{d}.$$

We have seen that the simple linear regression equation

$$\hat{Y} = a + bX$$

can be equivalently written as

$$\hat{Y} = a' + b'Z_1.$$

What if we have a quadratic regression equation

$$\hat{Y} = a + b_1X + b_2X^2?$$

Can we find a second orthogonal polynomial so that the regression equation can be written

$$\hat{Y} = a' + b_1'Z_1 + b_2'Z_2$$

and such that **Z′Z** is a diagonal? The reader has probably guessed that this appendix would not be here unless the answer to this question were yes.

If we wanted to fit a quadratic equation using the original data, then

$$\mathbf{X} = \begin{bmatrix} 1 & 60 & 3{,}600 \\ 1 & 80 & 6{,}400 \\ 1 & 100 & 10{,}000 \\ 1 & 120 & 14{,}400 \\ 1 & 140 & 19{,}600 \\ 1 & 160 & 25{,}600 \\ 1 & 180 & 32{,}400 \\ 1 & 200 & 40{,}000 \\ 1 & 210 & 44{,}100 \end{bmatrix}$$

and **X′X** will not be diagonal. Coding the X-values using the first orthogonal polynomial transforms the second column of **X** into $-4, -3, \ldots, 3, 4$. Define the *second orthogonal polynomial* as

$$Z_2 = \left(\frac{X - \bar{X}}{d}\right)^2 - \frac{n^2 - 1}{12}.$$

In our example, since $n = 9$, $(n^2 - 1)/12 = 80/12 = 6.67$, then

X	Z_1	Z_2
60	−4	$(-4)^2 - 6.67 = 9.33$
80	−3	$(-3)^2 - 6.67 = 2.33$
100	−2	$(-2)^2 - 6.67 = -2.67$
120	−1	$(-1)^2 - 6.67 = -5.67$
140	0	$0^2 - 6.67 = -6.67$
160	1	$(1)^2 - 6.67 = -5.67$
180	2	$(2)^2 - 6.67 = -2.67$
200	3	$(3)^2 - 6.67 = 2.33$
210	4	$(4)^2 - 6.67 = 9.33$
	0	≈ 0

and

$$\mathbf{Z'Z} = \begin{bmatrix} 9 & 0 & 0 \\ 0 & 60 & 0 \\ 0 & 0 & 308 \end{bmatrix},$$

since it is easy to see that $\sum Z_1Z_2 = 0$; the reader can verify that $\sum Z_2^2 = 308$. Thus it is an easy matter to find a', b_1', and b_2' for the regression equation

$$Y = a' + b_1'Z_1 + b_2'Z_2.$$

One can then transform back to find a, b_1, and b_2 for the original equation

$$Y = a + b_1X + b_2X.$$

The third, fourth, . . . orthogonal polynomials, used in fitting cubic or higher-degree polynomial equations can be found using a recursive formula:

$$Z_{i+1} = Z_iZ_1 - \frac{i^2(n^2 - i^2)}{4(4i^2 - 1)}Z_{i-1},$$

where $Z_0 = 1$. For the third orthogonal polynomial, let $i + 1 = 3$, so $i = 2$. Then

$$\begin{aligned} Z_3 &= Z_2Z_1 - \frac{2^2(n^2 - 2^2)}{4[4(2^2) - 1]}Z_1 \\ &= \left[\left(\frac{X - \bar{X}}{d}\right)^2 - \frac{n^2 - 1}{12}\right]\left(\frac{X - \bar{X}}{d}\right) - \frac{n^2 - 4}{15}\left(\frac{X - \bar{X}}{d}\right) \\ &= \left(\frac{X - \bar{X}}{d}\right)^3 - \left(\frac{n^2 - 1}{12} + \frac{n^2 - 4}{15}\right)\left(\frac{X - \bar{X}}{d}\right) \\ &= \left(\frac{X - \bar{X}}{d}\right)^3 - \frac{3n^2 - 7}{20}\left(\frac{X - \bar{X}}{d}\right), \end{aligned}$$

While it can be time-consuming to calculate orthogonal polynomials, a great saving in labor can be achieved with them if one has no access to a computer to solve polynomial equations. They are of most use, however, in designed experiments in which X-values are at evenly spaced intervals; otherwise they are usually more trouble than they are worth.

APPENDIX F

FITTING THE REGRESSION LINE THROUGH THE ORIGIN

We want to find a value for b^* such that

$$Q_1 = \sum_{i=1}^{n} (Y_i - \hat{Y}_i)^2 = \sum_{i=1}^{n} (Y_i - b^*X_i)^2$$

is minimized. Taking the first derivative with respect to b^*, we obtain

$$\frac{dQ_1}{db^*} = 2\sum (Y_i - b^*X_i)(-X_i)$$

$$= -2\sum (X_iY_i - b^*X_i^2).$$

Equating to zero,

$$-2\sum (X_iY_i - b^*X_i^2) = 0,$$

$$\sum X_iY_i - b^*\sum X_i^2 = 0,$$

$$\frac{\sum X_iY_i}{\sum X_i^2} = b^*.$$

For a second-derivative check for a minimum, note that the first derivative can be written as

$$-2\sum X_iY_i + 2b^*\sum X_i^2.$$

Then

$$\frac{dQ_2}{db^{*2}} = 2\sum X_i^2 > 0,$$

and a minimum has been found.

APPENDIX G

A LARGE DATA SET

These data are taken with permission from the *Comparable Book, January 1, 1983–March 31, 1983* of the Knoxville, Tennessee, Board of Realtors. This book is a record of real estate sold by member realtors during this time period. Data extracted are for residential properties in Knoxville and Knox County. $N = 339$.

TABLE G-1 Definitions of Variables

Variable	Code Name	Notes
Selling price (dollars)	PRICE	In some cases, greater than list price
Zone	ZONE	See Table G-2
List price (dollars)	LIST	
Number of bedrooms	BEDRMS	
Number of baths	BATHS	
Number of rooms	ROOMS	10 = 10 or more
Floor space (sq ft)	SQFT	
Age (years)	AGE	
Garage size	CARS	Number of vehicles accommodated in garage or carport
City tax (dollars)	CITYTAX	Zero if residence is in county
County tax (dollars)	COUNTY	
Lot size† (sq yd)	LOT	See Type
Type	TYPE	1 = condo; some have lots, others do not
Days on the market	DOM	

† Some reported values are of questionable accuracy.

TABLE G-2
Zones

RESIDENTIAL NORTH
3 = Mechanicsville
4 = North Knoxville
11 = Fountain City
12 = Halls
13 = Rutledge Pike
24 = Powell†

RESIDENTIAL EAST
1 = Center City
5 = East Knoxville
6 = Holston Hills
14 = East Knoxville, Outside of City

RESIDENTIAL SOUTH
7 = Inside City, South
8 = Alcoa Highway
15 = South Knoxville, Outside of City

RESIDENTIAL WEST
2 = University, Fort Sanders
9 = Sutherland Avenue, Lonsdale, Middlebrook Industrial
10 = Northwest Knoxville
16 = Westmoreland, Riverbend, Sequoyah, West Hills[a]
17 = Cedar Bluff, South of Interstate[a]
18 = Cedar Bluff, North of Interstate[a]
19 = Northshore Drive, Choto[a]
20 = Farragut, South of Interstate
21 = Farragut, North of Interstate
22 = Hardin Valley
23 = Karns

† Indicates that the zone is partly in the city, partly in the county.

Column Count Row	Price 339	Zone 339	List 339	Bedrms 339	Baths 339	Rooms 339	Sq. ft. 339	Age 339	Cars 339	City Tax 339	County 339	Lot 339	Type 339	DOM 339
1	19950.	11.	20000.	4.	1.50000	7.	1587.00	75.	1.	269.	202.00	2333.	0.	33.
2	24900.	4.	24900.	3.	1.00000	8.	1500.00	40.	0.	150.	100.00	2014.	0.	79.
3	26500.	11.	27000.	2.	1.00000	6.	1052.00	27.	0.	192.	159.00	2833.	0.	155.
4	26900.	4.	27500.	2.	1.00000	5.	1400.00	52.	2.	127.	95.00	806.	0.	95.
5	27900.	4.	27900.	3.	1.00000	5.	950.00	12.	1.	227.	170.00	756.	0.	384.
6	25700.	12.	29900.	2.	1.00000	5.	976.00	23.	0.	0.	1141.00	9680.	0.	164.
7	29900.	4.	29900.	2.	1.00000	4.	850.00	17.	1.	155.	113.00	1523.	0.	112.
8	28900.	4.	30000.	2.	1.00000	4.	750.00	25.	1.	140.	104.00	1000.	0.	70.
9	28000.	4.	30900.	2.	1.00000	4.	834.00	30.	0.	114.	94.00	1237.	0.	155.
10	30900.	11.	30900.	3.	1.00000	6.	980.00	25.	1.	147.	110.00	778.	0.	73.
11	29000.	4.	32500.	2.	1.00000	6.	1400.00	30.	1.	148.	123.00	967.	0.	370.
12	28500.	4.	32500.	4.	1.00000	7.	1653.00	51.	0.	201.	150.00	2083.	0.	149.
13	32900.	11.	32900.	2.	1.00000	5.	1123.00	65.	0.	151.	113.00	778.	0.	127.
14	30000.	4.	32900.	2.	1.00000	5.	900.00	30.	1.	164.	123.00	833.	0.	59.
15	25000.	11.	33000.	2.	1.00000	6.	1281.00	50.	1.	182.	137.00	706.	0.	116.
16	32485.	4.	33500.	2.	1.00000	5.	899.00	25.	0.	182.	136.00	934.	0.	77.
17	33900.	11.	33900.	2.	1.00000	4.	768.00	23.	0.	145.	120.00	2696.	0.	167.
18	32500.	4.	36000.	2.	1.00000	5.	918.00	25.	0.	150.	124.00	833.	0.	61.
19	36500.	13.	36500.	3.	1.50000	5.	960.00	10.	1.	0.	1198.00	5324.	0.	59.
20	36900.	4.	37900.	3.	1.00000	7.	1800.00	30.	1.	143.	172.00	1376.	0.	46.
21	39770.	11.	38500.	3.	1.00000	5.	950.00	25.	0.	187.	140.00	2239.	0.	17.
22	37000.	4.	39500.	4.	1.50000	8.	2480.00	40.	2.	301.	225.00	3333.	0.	101.

Column Count Row	Price 339	Zone 339	List 339	Bedrms 339	Baths 339	Rooms 339	Sq. ft. 339	Age 339	Cars 339	City Tax 339	County 339	Lot 339	Type 339	DOM 339
23	35000.	4.	39900.	5.	2.00000	10.	1803.00	27.	1.	269.	202.00	1167.	0.	91.
24	38000.	12.	39900.	3.	1.00000	5.	1008.00	7.	1.	0.	181.00	1889.	0.	67.
25	35000.	24.	39900.	3.	1.00000	7.	1445.00	40.	1.	183.	137.00	1389.	0.	45.
26	39800.	11.	41900.	2.	1.00000	5.	1920.00	21.	0.	198.	148.00	1921.	0.	37.
27	40500.	4.	42500.	2.	1.00000	5.	1220.00	30.	1.	254.	190.00	874.	0.	165.
28	42110.	11.	43200.	3.	1.00000	6.	1136.00	25.	1.	247.	478.00	1889.	0.	55.
29	46500.	12.	44900.	3.	2.50000	7.	1700.00	9.	2.	0.	231.00	1941.	0.	315.
30	45500.	24.	45500.	4.	1.50000	7.	1980.00	11.	2.	293.	240.00	1689.	0.	93.
31	46500.	24.	46500.	3.	1.00000	5.	1221.00	0.	0.	0.	190.00	21296.	0.	245.
32	46500.	4.	46500.	3.	2.00000	6.	1550.00	17.	2.	355.	294.00	2233.	0.	38.
33	46500.	12.	47500.	4.	2.50000	8.	1819.00	18.	1.	0.	254.00	1813.	0.	61.
34	42000.	11.	47900.	2.	1.00000	5.	984.00	25.	1.	200.	150.00	1667.	0.	212.
35	48000.	24.	47950.	3.	2.00000	6.	1086.00	15.	1.	0.	250.00	2111.	0.	43.
36	45950.	24.	47950.	3.	1.50000	7.	1800.00	7.	1.	0.	361.00	4867.	0.	50.
37	45000.	4.	49500.	3.	2.00000	6.	1300.00	20.	2.	179.	174.00	3000.	0.	43.
38	48000.	12.	49900.	3.	1.50000	6.	1196.00	2.	1.	0.	340.00	1583.	0.	234.
39	49000.	4.	49900.	3.	2.00000	7.	1624.00	26.	1.	250.	200.00	1379.	0.	91.
40	49900.	12.	49900.	3.	1.50000	6.	1724.00	10.	2.	0.	252.00	1533.	0.	116.
41	45000.	11.	49900.	4.	2.00000	7.	1560.00	37.	0.	200.	150.00	3000.	0.	120.
42	43000.	24.	49900.	3.	1.00000	6.	1404.00	25.	1.	0.	228.00	2222.	0.	2.
43	47840.	11.	52500.	4.	2.50000	7.	1962.00	6.	2.	345.	286.00	2256.	0.	88.
44	52500.	12.	53000.	4.	2.50000	7.	1694.00	4.	2.	0.	331.00	1667.	0.	162.
45	51650.	24.	53900.	3.	2.00000	6.	1265.00	7.	2.	0.	280.00	2567.	0.	52.
46	52601.	4.	54900.	3.	2.00000	7.	2050.00	24.	2.	260.	211.00	2211.	0.	158.
47	53000.	24.	54900.	3.	2.00000	7.	2020.00	7.	2.	0.	300.00	2460.	0.	59.
48	54000.	12.	54900.	3.	2.00000	7.	1836.00	5.	1.	0.	352.00	3150.	0.	80.
49	54900.	12.	54900.	3.	2.00000	7.	1700.00	5.	2.	0.	350.00	1733.	0.	63.
50	52000.	12.	55900.	3.	2.50000	7.	1900.00	5.	2.	0.	336.00	3727.	0.	102.
51	52900.	24.	55900.	4.	2.00000	8.	1820.00	7.	2.	0.	314.00	948.	0.	147.
52	54442.	12.	55900.	3.	2.50000	7.	1794.00	7.	2.	0.	339.00	1725.	0.	48.
53	53400.	12.	56000.	3.	2.50000	6.	1680.00	3.	2.	0.	358.00	2811.	0.	423.
54	56900.	4.	56900.	2.	1.50000	5.	1360.00	0.	0.	292.	250.00	156.	1.	352.
55	56900.	4.	56900.	2.	1.50000	5.	1360.00	0.	0.	350.	300.00	156.	1.	125.
56	52900.	4.	58900.	3.	1.50000	8.	1959.00	57.	2.	310.	256.00	1290.	0.	62.
57	58000.	11.	59900.	3.	1.00000	7.	1468.00	36.	2.	323.	212.00	8336.	0.	133.
58	52000.	24.	59900.	3.	2.00000	5.	1144.00	5.	2.	0.	350.00	1533.	0.	119.
59	59900.	4.	59900.	3.	2.00000	6.	1500.00	0.	0.	350.	300.00	156.	1.	124.
60	61900.	4.	59900.	3.	2.00000	6.	1500.00	0.	0.	350.	300.00	156.	1.	174.
61	62400.	4.	59900.	3.	2.00000	9.	1360.00	0.	0.	350.	300.00	156.	1.	170.
62	58800.	11.	59900.	4.	2.00000	7.	2231.00	52.	2.	253.	190.00	7869.	0.	90.
63	59900.	13.	59900.	4.	2.00000	8.	2500.00	15.	1.	0.	250.00	24200.	0.	2.
64	61650.	11.	60000.	3.	2.00000	8.	2059.00	23.	1.	339.	409.00	2722.	0.	301.
65	50000.	24.	60900.	3.	2.50000	6.	1600.00	3.	2.	0.	463.00	1889.	0.	6.
66	59900.	24.	61900.	3.	2.00000	6.	1430.00	14.	2.	0.	365.00	4232.	0.	353.
67	59560.	12.	62000.	3.	2.50000	7.	1600.00	11.	1.	0.	250.00	5653.	0.	130.
68	60052.	12.	62500.	3.	2.50000	8.	1845.00	7.	2.	0.	348.00	2333.	0.	159.
69	63900.	12.	63900.	4.	2.50000	8.	2222.00	6.	2.	0.	381.00	2022.	0.	314.
70	59900.	11.	64900.	4.	2.50000	10.	2778.00	14.	1.	364.	301.00	1778.	0.	146.
71	65900.	24.	65900.	3.	2.00000	6.	1557.00	12.	1.	0.	343.00	1556.	0.	88.
72	62000.	12.	66900.	5.	2.50000	10.	2774.00	5.	2.	0.	387.00	2166.	0.	135.

Column Count Row	Price 339	Zone 339	List 339	Bedrms 339	Baths 339	Rooms 339	Sq. ft. 339	Age 339	Cars 339	City Tax 339	County 339	Lot 339	Type 339	DOM 339
73	67000.	24.	67000.	4.	2.00000	9.	2415.00	13.	2.	0.	400.00	1756.	0.	95.
74	66700.	24.	69900.	4.	2.50000	8.	2479.00	4.	2.	0.	563.00	1933.	0.	54.
75	69500.	12.	72500.	3.	2.00000	7.	2200.00	15.	2.	0.	325.00	1836.	0.	71.
76	71500.	11.	74500.	3.	2.50000	8.	2141.00	20.	1.	280.	210.00	1783.	0.	27.
77	75000.	24.	79900.	4.	2.50000	8.	2098.00	6.	1.	0.	466.00	2211.	0.	67.
78	82254.	12.	82254.	4.	2.50000	8.	2249.00	8.	2.	0.	400.00	5066.	0.	338.
79	81900.	12.	82900.	3.	2.00000	7.	1929.00	0.	2.	0.	310.00	2333.	0.	147.
80	68000.	24.	87900.	4.	2.00000	9.	2250.00	6.	1.	0.	500.00	2917.	0.	252.
81	87000.	13.	89000.	3.	3.00000	9.	2694.00	4.	2.	0.	492.00	9680.	0.	146.
82	85000.	24.	91000.	4.	3.00000	8.	2500.00	2.	2.	0.	570.00	2093.	0.	84.
83	85000.	13.	97500.	3.	2.00000	6.	1607.00	22.	1.	0.	447.00	68728.	0.	50.
84	70000.	12.	99900.	3.	2.00000	7.	1675.00	4.	2.	0.	415.00	29040.	0.	115.
85	112500.	12.	117000.	4.	2.50000	10.	2664.00	1.	2.	0.	617.00	2930.	0.	11.
86	127000.	12.	139000.	4.	3.00000	10.	2746.00	9.	2.	0.	1060.00	290400.	0.	36.
87	18000.	14.	21000.	2.	2.00000	4.	720.00	10.	1.	0.	82.00	2368.	0.	66.
88	18000.	5.	22000.	2.	1.00000	5.	1064.00	30.	0.	110.	82.00	956.	0.	171.
89	29800.	14.	32000.	2.	1.00000	5.	850.00	25.	1.	0.	160.00	12584.	0.	38.
90	34000.	5.	34900.	3.	1.00000	6.	1466.00	48.	0.	93.	126.00	806.	0.	154.
91	33000.	14.	35000.	4.	1.00000	7.	2092.00	12.	0.	0.	238.00	16940.	0.	53.
92	52000.	6.	59900.	3.	2.00000	7.	2125.00	21.	2.	373.	309.00	1830.	0.	189.
93	65000.	14.	68900.	3.	2.00000	8.	2187.00	14.	2.	0.	354.00	2600.	0.	69.
94	66800.	6.	74900.	4.	2.00000	8.	2650.00	25.	2.	459.	380.00	4400.	0.	48.
95	60500.	6.	76900.	4.	2.00000	7.	2250.00	31.	1.	410.	340.00	3819.	0.	190.
96	15000.	15.	19000.	2.	1.00000	4.	600.00	4.	0.	0.	101.00	30686.	0.	68.
97	21950.	7.	21950.	3.	1.00000	6.	1260.00	50.	1.	184.	136.00	883.	0.	85.
98	20000.	7.	24950.	3.	1.00000	5.	855.00	8.	2.	249.	185.00	1837.	0.	105.
99	30000.	7.	30000.	2.	1.00000	5.	796.00	26.	0.	124.	150.00	1111.	0.	106.
100	29500.	7.	32000.	2.	1.00000	5.	970.00	4.	0.	184.	152.00	2178.	0.	131.
101	27500.	7.	32500.	2.	1.00000	4.	792.00	23.	1.	156.	130.00	800.	0.	92.
102	34500.	15.	34500.	3.	1.00000	5.	1285.00	8.	1.	0.	241.00	1689.	0.	100.
103	34900.	7.	36900.	2.	1.00000	6.	1050.00	30.	1.	232.	174.00	2000.	0.	148.
104	37500.	7.	37900.	2.	1.00000	6.	1456.00	50.	1.	275.	207.00	4840.	0.	124.
105	35000.	7.	38500.	3.	1.00000	5.	1014.00	27.	1.	219.	164.00	2222.	0.	89.
106	36800.	15.	42500.	3.	1.00000	5.	1245.00	14.	0.	0.	210.00	2483.	0.	34.
107	37500.	7.	42900.	3.	1.00000	4.	1100.00	30.	1.	288.	171.00	1740.	0.	112.
108	42000.	15.	43500.	3.	2.00000	6.	1550.00	25.	2.	220.	165.00	2211.	0.	61.
109	41000.	7.	43500.	3.	1.50000	8.	2016.00	11.	0.	256.	188.00	910.	0.	199.
110	47109.	7.	48000.	4.	2.00000	8.	1806.00	8.	1.	308.	236.00	1528.	0.	7.
111	45000.	7.	49000.	3.	1.00000	6.	1365.00	42.	2.	228.	189.00	2667.	0.	95.
112	49000.	15.	50900.	3.	2.50000	6.	1991.00	13.	2.	0.	302.00	1711.	0.	126.
113	54900.	15.	54900.	2.	1.00000	4.	885.00	35.	1.	0.	245.00	33880.	0.	71.
114	56000.	15.	59900.	3.	2.50000	7.	1814.00	2.	2.	0.	350.00	2222.	0.	148.
115	59900.	15.	59900.	3.	2.50000	7.	2038.00	16.	2.	0.	308.00	1778.	0.	230.
116	55000.	7.	59950.	3.	1.50000	7.	1677.00	29.	2.	430.	322.00	4449.	0.	78.
117	60000.	7.	64900.	3.	3.00000	7.	2520.00	20.	2.	532.	399.00	2222.	0.	62.
118	55024.	15.	64900.	3.	2.50000	7.	2291.00	16.	2.	0.	420.00	2366.	0.	91.
119	76500.	15.	72900.	4.	2.00000	9.	2705.00	12.	2.	0.	310.00	3484.	0.	130.
120	71500.	8.	76500.	3.	2.00000	6.	1822.00	22.	1.	0.	439.00	3133.	0.	105.
121	186500.	8.	209000.	4.	3.00000	9.	3800.00	3.	2.	0.	806.00	3306.	0.	213.
122	10000.	16.	14500.	1.	1.00000	4.	720.00	15.	0.	21.	18.00	694.	0.	102.

Column Count Row	Price 339	Zone 339	List 339	Bedrms 339	Baths 339	Rooms 339	Sq. ft. 339	Age 339	Cars 339	City Tax 339	County 339	Lot 339	Type 339	DOM 339
123	17500.	10.	17500.	2.	1.00000	4.	788.00	40.	0.	55.	41.00	861.	0.	86.
124	17000.	10.	20000.	3.	1.00000	6.	1190.00	25.	0.	119.	89.00	2000.	0.	30.
125	22000.	9.	22500.	3.	1.00000	6.	1200.00	50.	1.	104.	76.00	744.	0.	140.
126	22500.	10.	23900.	1.	1.00000	4.	810.00	35.	1.	123.	82.00	778.	0.	112.
127	25000.	18.	25000.	3.	2.00000	7.	1700.00	40.	0.	0.	168.00	353.	0.	119.
128	23500.	9.	29000.	2.	1.00000	4.	800.00	30.	0.	125.	108.00	833.	0.	118.
129	29130.	18.	29900.	2.	1.00000	4.	840.00	24.	1.	0.	104.00	6355.	0.	21.
130	31500.	19.	31500.	3.	1.00000	5.	1100.00	40.	0.	0.	108.00	1667.	0.	133.
131	33000.	9.	33000.	2.	1.00000	5.	1062.00	30.	0.	248.	186.00	2500.	0.	144.
132	32500.	9.	34500.	2.	1.00000	4.	1080.00	25.	2.	116.	96.00	833.	0.	140.
133	29096.	16.	34900.	3.	1.00000	5.	900.00	18.	1.	213.	173.00	1888.	0.	47.
134	32500.	9.	35900.	2.	1.50000	6.	1150.00	30.	0.	122.	101.00	678.	0.	199.
135	36000.	18.	37900.	3.	1.50000	6.	1300.00	11.	0.	0.	229.00	1167.	0.	48.
136	38450.	18.	38450.	3.	1.00000	6.	1050.00	11.	1.	0.	226.00	1950.	0.	162.
137	38000.	9.	39500.	3.	1.00000	6.	1020.00	25.	1.	227.	169.00	1667.	0.	94.
138	38500.	16.	39900.	3.	1.50000	5.	1040.00	2.	0.	270.	203.00	2911.	0.	78.
139	35000.	16.	39900.	3.	1.50000	7.	2000.00	10.	0.	0.	164.00	1833.	0.	29.
140	38000.	18.	39900.	3.	1.00000	6.	1146.00	40.	0.	0.	132.00	6825.	0.	73.
141	39815.	19.	40000.	3.	1.00000	6.	1420.00	5.	1.	0.	180.00	6050.	0.	10.
142	39394.	10.	41900.	4.	2.00000	7.	1500.00	10.	1.	283.	212.00	1040.	0.	59.
143	39500.	10.	41900.	2.	2.00000	4.	986.00	5.	1.	258.	214.00	948.	0.	57.
144	38000.	10.	41900.	4.	1.50000	8.	1465.00	11.	1.	262.	200.00	708.	0.	34.
145	40000.	16.	41900.	3.	1.00000	5.	1103.00	26.	1.	0.	175.00	2244.	0.	19.
146	42000.	23.	42000.	3.	1.00000	7.	1328.00	0.	0.	0.	208.00	5019.	0.	145.
147	40500.	10.	42000.	3.	1.50000	6.	1067.00	5.	0.	0.	238.00	917.	0.	89.
148	35801.	10.	42500.	3.	1.50000	6.	1244.00	17.	1.	180.	160.00	2243.	0.	36.
149	42500.	9.	42500.	3.	1.00000	6.	1040.00	25.	1.	233.	175.00	1800.	0.	77.
150	40000.	10.	42900.	3.	1.00000	7.	1437.00	23.	0.	244.	202.00	1797.	0.	293.
151	40000.	10.	43000.	3.	1.00000	6.	1480.00	28.	1.	312.	234.00	1667.	0.	94.
152	43000.	18.	43500.	2.	1.50000	5.	1093.00	0.	0.	0.	300.00	336.	1.	350.
153	43500.	18.	44500.	2.	1.50000	5.	1093.00	0.	0.	0.	275.00	336.	0.	335.
154	40000.	16.	44500.	2.	1.00000	4.	1000.00	25.	1.	180.	150.00	833.	0.	95.
155	42000.	10.	44900.	3.	1.00000	7.	931.00	0.	1.	288.	215.00	2560.	0.	148.
156	38730.	18.	44900.	2.	1.00000	4.	900.00	5.	1.	312.	256.00	1170.	0.	66.
157	43321.	16.	44900.	3.	1.00000	5.	912.00	16.	1.	0.	230.00	3940.	0.	44.
158	43500.	18.	44950.	2.	2.00000	5.	1053.00	1.	1.	305.	230.00	516.	0.	119.
159	42291.	22.	44950.	2.	2.00000	6.	1400.00	2.	1.	0.	222.00	1042.	0.	106.
160	39000.	9.	45900.	3.	1.00000	6.	1377.00	25.	0.	141.	140.00	1433.	0.	49.
161	46750.	20.	46750.	3.	1.50000	6.	1008.00	0.	1.	0.	400.00	1500.	0.	117.
162	44000.	10.	46950.	3.	1.50000	6.	1736.00	26.	0.	264.	216.00	1667.	0.	55.
163	47500.	16.	47500.	3.	2.00000	6.	1300.00	28.	1.	222.	167.00	1274.	0.	84.
164	47900.	10.	47900.	3.	2.50000	7.	1942.00	5.	1.	286.	237.00	984.	0.	161.
165	49000.	16.	48500.	3.	1.50000	6.	1539.00	26.	0.	315.	305.00	2053.	0.	307.
166	48500.	10.	48900.	3.	1.50000	7.	1333.00	20.	1.	285.	209.00	1667.	0.	130.
167	49500.	20.	49500.	1.	2.00000	4.	1035.00	5.	1.	0.	223.00	0.	1.	125.
168	42800.	16.	49900.	3.	2.00000	6.	1095.00	26.	1.	270.	250.00	1417.	0.	159.
169	47500.	10.	49900.	2.	1.50000	6.	1550.00	25.	1.	267.	200.00	1667.	0.	96.
170	47000.	16.	49900.	3.	1.50000	7.	1535.00	10.	0.	0.	260.00	2222.	0.	91.
171	48000.	23.	49900.	4.	1.50000	7.	1364.00	16.	1.	0.	183.00	8034.	0.	115.
172	49000.	10.	49900.	3.	1.50000	6.	1182.00	20.	1.	219.	206.00	1691.	0.	82.

Column Count Row	Price 339	Zone 339	List 339	Bedrms 339	Baths 339	Rooms 339	Sq. ft. 339	Age 339	Cars 339	City Tax 339	County 339	Lot 339	Type 339	DOM 339
173	47900.	18.	49900.	2.	1.00000	6.	1470.00	32.	0.	0.	219.00	2862.	0.	64.
174	42000.	10.	49950.	3.	1.00000	6.	1250.00	16.	1.	231.	160.00	1991.	0.	252.
175	49950.	10.	49950.	4.	2.50000	8.	1628.00	4.	2.	347.	260.00	1143.	0.	23.
176	50000.	20.	50000.	2.	2.00000	8.	1151.00	10.	1.	0.	285.00	4444.	0.	61.
177	45600.	18.	50500.	4.	1.50000	7.	1503.00	15.	1.	304.	298.00	1667.	0.	35.
178	44935.	10.	50900.	3.	1.50000	7.	1618.00	5.	1.	310.	216.00	1433.	0.	93.
179	50258.	10.	51000.	3.	1.50000	6.	1461.00	2.	1.	395.	295.00	1444.	0.	146.
180	48500.	16.	51000.	2.	1.50000	5.	1182.00	8.	2.	435.	326.00	0.	1.	153.
181	47000.	16.	51900.	3.	2.50000	7.	1885.00	17.	2.	0.	355.00	1667.	0.	169.
182	50603.	10.	51900.	3.	3.00000	7.	1495.00	3.	1.	360.	320.00	1213.	0.	44.
183	52350.	18.	52350.	3.	2.50000	6.	1690.00	3.	2.	0.	122.00	3250.	0.	193.
184	47988.	18.	52500.	3.	2.00000	6.	1577.00	8.	2.	0.	247.00	2570.	0.	59.
185	51400.	10.	52900.	3.	1.50000	7.	1488.00	9.	2.	341.	283.00	1756.	0.	284.
186	50500.	19.	52900.	3.	1.50000	6.	1200.00	5.	2.	0.	310.00	1870.	0.	83.
187	51000.	17.	53400.	4.	3.00000	9.	1872.00	6.	2.	0.	237.00	1689.	0.	187.
188	51000.	20.	53900.	3.	2.00000	7.	1872.00	10.	0.	0.	304.00	2506.	0.	69.
189	51000.	16.	54000.	2.	2.00000	8.	1954.00	35.	0.	191.	158.00	1783.	0.	184.
190	52500.	16.	54000.	3.	1.00000	7.	1332.00	23.	0.	275.	206.00	1667.	0.	76.
191	48671.	16.	54500.	3.	2.00000	6.	1675.00	22.	0.	0.	199.00	3583.	0.	105.
192	54900.	17.	54900.	2.	1.50000	5.	1396.00	0.	2.	0.	450.00	173.	1.	317.
193	49500.	17.	54900.	3.	2.00000	7.	1250.00	12.	2.	0.	226.00	1660.	0.	183.
194	52000.	20.	54900.	3.	2.00000	6.	1411.00	15.	1.	0.	210.00	1614.	0.	90.
195	56485.	17.	55400.	2.	1.50000	5.	1344.00	0.	1.	0.	450.00	160.	1.	315.
196	56270.	17.	55400.	2.	2.00000	5.	1344.00	0.	1.	0.	450.00	160.	1.	254.
197	46000.	18.	55900.	3.	2.50000	7.	2323.00	18.	2.	0.	298.00	3691.	0.	206.
198	55500.	20.	55900.	3.	2.00000	7.	1500.00	7.	2.	0.	450.00	9413.	0.	49.
199	50000.	22.	56300.	3.	1.00000	6.	1248.00	12.	1.	0.	250.00	11374.	0.	543.
200	56000.	20.	57500.	3.	2.50000	7.	1900.00	5.	2.	0.	360.00	3090.	0.	228.
201	56000.	10.	57900.	3.	2.00000	7.	1596.00	24.	2.	365	288.00	2678.	0.	183.
202	55000.	18.	57900.	3.	2.00000	7.	1750.00	3.	2.	0.	343.00	706.	0.	161.
203	54900.	18.	57900.	3.	2.50000	7.	1912.00	12.	2.	0.	400.00	2833.	0.	95.
204	53100.	17.	57900.	2.	1.50000	5.	1274.00	5.	1.	0.	450.00	156.	1.	134.
205	54500.	18.	57900.	4.	2.00000	8.	1887.00	10.	2.	0.	300.00	1638.	0.	93.
206	57000.	18.	57900.	3.	2.00000	7.	1970.00	7.	2.	0.	360.00	1667.	0.	65.
207	58000.	23.	58000.	4.	2.50000	8.	2200.00	4.	2.	0.	467.00	2185.	0.	123.
208	58700.	17.	58700.	2.	1.50000	5.	1064.00	0.	1.	0.	450.00	267.	1.	253.
209	64000.	17.	58700.	2.	1.50000	5.	1064.00	0.	1.	0.	450.00	267.	1.	258.
210	58790.	10.	58900.	3.	2.50000	7.	1850.00	2.	2.	456.	312.00	1250.	0.	335.
211	58900.	18.	58900.	4.	2.00000	7.	2188.00	11.	0.	0.	323.00	1696.	0.	89.
212	55000.	16.	58900.	3.	2.00000	6.	1463.00	20.	2.	0.	329.00	4308.	0.	44.
213	54500.	18.	59500.	4.	2.00000	7.	1634.00	25.	1.	0.	280.00	21369.	0.	51.
214	55400.	17.	59500.	2.	1.50000	5.	1344.00	2.	1.	0.	450.00	133.	1.	46.
215	58000.	18.	59900.	4.	3.00000	7.	1848.00	4.	2.	0.	331.00	3162.	0.	251.
216	54500.	18.	59500.	4.	2.00000	7.	1634.00	25.	1.	0.	280.00	21369.	0.	51.
217	55400.	17.	59500.	2.	1.50000	5.	1344.00	2.	1.	0.	450.00	133.	1.	46.
218	58000.	18.	59900.	4.	3.00000	7.	1848.00	4.	2.	0.	331.00	3162.	0.	251.
219	55000.	18.	59900.	4.	2.50000	8.	1837.00	6.	2.	0.	329.00	1500.	0.	168.
220	58900.	18.	59900.	4.	2.00000	7.	1853.00	9.	1.	0.	351.00	2652.	0.	64.
221	59600.	2.	59900.	4.	3.00000	8.	2412.00	56.	1.	416.	312.00	619.	0.	100.
222	60000.	20.	60000.	3.	2.00000	7.	1684.00	3.	2.	0.	315.00	1680.	0.	118.

Column Count Row	Price 339	Zone 339	List 339	Bedrms 339	Baths 339	Rooms 339	Sq. ft. 339	Age 339	Cars 339	City Tax 339	County 339	Lot 339	Type 339	DOM 339
223	57800.	20.	60900.	3.	2.00000	7.	1484.00	0.	2.	0.	318.00	1989.	0.	162.
224	58600.	10.	61500.	4.	2.50000	8.	2151.00	3.	0.	382.	280.00	1370.	0.	468.
225	61500.	18.	61500.	5.	2.00000	10.	2150.00	4.	2.	0.	399.00	1776.	0.	111.
226	59900.	17.	61900.	3.	1.50000	8.	1789.00	5.	1.	0.	300.00	1458.	0.	188.
227	61000.	17.	61900.	3.	2.50000	6.	1900.00	3.	1.	0.	429.00	393.	1.	51.
228	60500.	16.	62300.	3.	2.50000	8.	1900.00	16.	1.	0.	342.00	2750.	0.	67.
229	55000.	16.	62500.	3.	3.00000	7.	1903.00	20.	1.	0.	326.00	4103.	0.	120.
230	58000.	16.	62500.	3.	2.50000	7.	2190.00	7.	2.	0.	397.00	2173.	0.	76.
231	58093.	16.	62900.	3.	2.00000	9.	1906.00	9.	0.	0.	247.00	1291.	0.	164.
232	56500.	20.	62900.	2.	2.00000	4.	1250.00	5.	1.	0.	356.00	0.	1.	73.
233	61500.	18.	62900.	3.	3.00000	10.	1900.00	3.	2.	0.	325.00	1328.	0.	49.
234	63350.	23.	63350.	4.	2.50000	9.	2326.00	15.	1.	0.	373.00	3811.	0.	90.
235	61000.	10.	63500.	3.	2.00000	8.	1977.00	23.	2.	346.	259.00	1933.	0.	180.
236	63500.	16.	63500.	3.	2.00000	7.	1760.00	27.	1.	455.	341.00	3093.	0.	101.
237	61225.	18.	63500.	4.	2.50000	8.	2024.00	5.	2.	0.	400.00	1278.	0.	83.
238	62500.	19.	64500.	3.	2.50000	7.	1924.00	4.	2.	0.	380.00	1822.	0.	26.
239	64900.	21.	64900.	4.	2.50000	8.	2238.00	5.	2.	0.	369.00	2000.	0.	124.
240	63900.	21.	64900.	3.	2.50000	8.	1945.00	8.	2.	0.	377.00	1667.	0.	121.
241	63700.	17.	64900.	3.	2.50000	7.	1620.00	4.	2.	0.	350.00	1400.	0.	120.
242	66750.	16.	65900.	3.	2.00000	7.	1596.00	16.	2.	395.	350.00	1833.	0.	274.
243	65000.	20.	65900.	2.	2.00000	4.	1250.00	5.	1.	0.	356.00	0.	1.	78.
244	65333.	20.	65900.	3.	2.50000	10.	1800.00	2.	2.	0.	327.00	3578.	0.	59.
245	66300.	20.	66000.	3.	2.00000	7.	1658.00	4.	2.	0.	358.00	2800.	0.	98.
246	66750.	17.	66750.	3.	2.50000	6.	1900.00	0.	1.	0.	450.00	267.	1.	239.
247	63500.	16.	67000.	3.	2.00000	7.	1800.00	21.	0.	0.	296.00	5808.	0.	75.
248	65000.	16.	67500.	3.	2.00000	6.	1800.00	23.	2.	437.	362.00	2667.	0.	130.
249	67900.	18.	67900.	4.	2.50000	8.	2351.00	6.	2.	0.	429.00	1556.	0.	128.
250	66250.	17.	67900.	3.	2.50000	7.	1588.00	3.	2.	0.	373.00	1885.	0.	168.
251	66900.	20.	67900.	4.	2.00000	8.	2010.00	0.	2.	0.	394.00	1882.	0.	120.
252	64500.	17.	67900.	3.	2.50000	5.	1848.00	4.	2.	0.	400.00	33.	1.	47.
253	55625.	17.	68900.	4.	2.50000	8.	2050.00	3.	2.	0.	358.00	1833.	0.	100.
254	67000.	16.	68900.	3.	2.00000	6.	2100.00	19.	1.	0.	338.00	2222.	0.	63.
255	63000.	18.	69500.	3.	2.00000	6.	1600.00	13.	2.	0.	387.00	2333.	0.	274.
256	69000.	18.	69500.	4.	2.50000	8.	2000.00	6.	2.	0.	390.00	1667.	0.	63.
257	69900.	16.	69900.	3.	2.50000	8.	2363.00	43.	1.	425.	318.00	4033.	0.	240.
258	69900.	22.	69900.	4.	2.00000	7.	1900.00	47.	0.	0.	220.00	24200.	0.	118.
259	71500.	17.	69900.	4.	2.50000	10.	3334.00	15.	2.	0.	390.00	1833.	0.	116.
260	67900.	18.	69900.	3.	2.00000	7.	2000.00	14.	0.	0.	388.00	2083.	0.	103.
261	65000.	17.	69900.	4.	2.50000	8.	2050.00	3.	2.	0.	358.00	1833.	0.	14.
262	64000.	16.	70000.	5.	3.00000	10.	2888.00	18.	1.	453.	375.00	3513.	0.	129.
263	68700.	18.	71900.	4.	2.00000	8.	2440.00	16.	2.	0.	378.00	2722.	0.	179.
264	66900.	18.	71900.	4.	2.50000	8.	2147.00	9.	2.	0.	427.00	2400.	0.	147.
265	71900.	19.	71900.	3.	2.50000	7.	1632.00	1.	2.	0.	563.00	1860.	0.	86.
266	72000.	17.	72000.	3.	2.50000	7.	1848.00	1.	2.	0.	450.00	167.	1.	96.
267	65000.	16.	72000.	3.	1.50000	5.	1350.00	35.	2.	0.	174.00	4156.	0.	55.
268	70000.	16.	72500.	2.	2.00000	5.	1200.00	20.	0.	614.	460.00	0.	1.	79.
269	68000.	21.	72900.	4.	3.00000	8.	2172.00	9.	2.	0.	393.00	3630.	0.	145.
270	70950.	16.	72900.	4.	3.00000	8.	2452.00	21.	2.	409.	316.00	2133.	0.	43.
271	74800.	16.	74800.	5.	3.00000	9.	2348.00	20.	2.	0.	420.00	1878.	0.	91.
272	68000.	18.	74900.	3.	2.00000	7.	1840.00	14.	2.	0.	421.00	2053.	0.	257.

Column Count Row	Price 339	Zone 339	List 339	Bedrms 339	Baths 339	Rooms 339	Sq. ft. 339	Age 339	Cars 339	City Tax 339	County 339	Lot 339	Type 339	DOM 339
273	73000.	18.	74900.	4.	2.50000	8.	2164.00	14.	2.	0.	433.00	1956.	0.	178.
274	73900.	21.	74900.	3.	2.50000	7.	1954.00	3.	2.	0.	372.00	1894.	0.	54.
275	70000.	21.	75900.	4.	2.50000	8.	1976.00	3.	2.	0.	424.00	1567.	0.	105.
276	76150.	17.	76150.	2.	2.00000	5.	1550.00	0.	2.	0.	700.00	0.	1.	376.
277	76500.	17.	76500.	2.	2.00000	5.	1550.00	0.	2.	0.	634.00	0.	1.	152.
278	72000.	18.	76900.	4.	2.50000	8.	2096.00	15.	2.	0.	388.00	2200.	0.	176.
279	73500.	18.	76900.	3.	2.50000	7.	2000.00	10.	2.	0.	450.00	1833.	0.	67.
280	72000.	18.	77500.	4.	2.50000	8.	2500.00	7.	2.	0.	526.00	1750.	0.	104.
281	75900.	21.	77900.	3.	2.00000	5.	2038.00	3.	2.	0.	348.00	1989.	0.	203.
282	72000.	9.	77900.	4.	2.50000	8.	2447.00	8.	2.	570.	421.00	1708.	0.	94.
283	69500.	21.	78900.	3.	2.50000	8.	2550.00	5.	2.	0.	390.00	2213.	0.	58.
284	78000.	21.	79400.	4.	3.00000	8.	2383.00	5.	2.	0.	410.00	2100.	0.	107.
285	70000.	20.	79900.	2.	2.00000	6.	1430.00	5.	1.	0.	394.00	0.	1.	185.
286	68000.	19.	79900.	3.	2.50000	8.	2740.00	6.	2.	0.	437.00	1667.	0.	187.
287	77500.	17.	79900.	4.	3.00000	8.	2280.00	7.	2.	0.	441.00	2124.	0.	76.
288	78000.	18.	79900.	4.	2.50000	8.	2160.00	3.	2.	0.	450.00	4764.	0.	114.
289	79500.	21.	81900.	4.	3.00000	8.	2392.00	3.	2.	0.	417.00	2605.	0.	131.
290	82243.	17.	82243.	3.	2.50000	6.	1850.00	0.	2.	0.	700.00	0.	1.	132.
291	79900.	16.	82500.	3.	2.00000	8.	2173.00	44.	1.	340.	281.00	2108.	0.	191.
292	76000.	21.	82900.	4.	3.00000	8.	2264.00	2.	2.	0.	368.00	1944.	0.	482.
293	84150.	17.	82900.	3.	3.00000	6.	2000.00	0.	2.	0.	634.00	0.	1.	169.
294	80000.	17.	82900.	4.	2.50000	8.	2340.00	5.	2.	0.	515.00	2038.	0.	139.
295	85941.	17.	84900.	3.	2.00000	7.	1772.00	1.	2.	0.	400.00	2562.	0.	91.
296	80000.	18.	85000.	4.	2.50000	8.	2700.00	5.	2.	0.	456.00	2147.	0.	314.
297	83500.	17.	85500.	3.	2.00000	7.	1600.00	5.	2.	0.	380.00	875.	0.	107.
298	85650.	17.	85650.	2.	2.00000	6.	1532.00	0.	1.	0.	600.00	763.	0.	349.
299	89500.	16.	89500.	3.	3.00000	6.	2000.00	3.	2.	772.	550.00	873.	0.	76.
300	83500.	20.	89900.	3.	2.00000	6.	1950.00	5.	1.	0.	561.00	0.	1.	106.
301	84500.	16.	89900.	3.	2.00000	9.	1801.00	2.	2.	432.	358.00	0.	1.	159.
302	86925.	16.	89900.	4.	3.00000	9.	2937.00	25.	3.	457.	598.00	2933.	0.	130.
303	85000.	19.	89900.	3.	2.50000	7.	2310.00	4.	2.	0.	526.00	1833.	0.	66.
304	89000.	20.	93900.	4.	2.50000	8.	2431.00	6.	2.	0.	514.00	3737.	0.	144.
305	94500.	17.	94500.	3.	2.50000	7.	2208.00	2.	2.	0.	468.00	2897.	0.	90.
306	99000.	16.	98500.	3.	2.00000	8.	2440.00	50.	1.	330.	273.00	1007.	0.	152.
307	90000.	17.	99500.	4.	2.50000	10.	2806.00	7.	2.	0.	444.00	2557.	0.	186.
308	99500.	17.	99500.	4.	2.50000	9.	2855.00	4.	2.	0.	399.00	1917.	0.	118.
309	99750.	20.	99750.	4.	2.50000	8.	2518.00	3.	2.	0.	542.00	2227.	0.	196.
310	98000.	20.	99900.	4.	2.50000	9.	2350.00	3.	2.	0.	511.00	2542.	0.	143.
311	98000.	20.	102000.	5.	2.50000	10.	2558.00	3.	2.	0.	472.00	2658.	0.	234.
312	97000.	21.	102000.	4.	2.50000	6.	2500.00	1.	2.	0.	558.00	1750.	0.	71.
313	98750.	16.	102500.	4.	2.50000	10.	2504.00	18.	2.	0.	482.00	7051.	0.	45.
314	99500.	20.	103900.	4.	2.50000	8.	2376.00	3.	2.	0.	577.00	2078.	0.	49.
315	96500.	20.	106900.	4.	3.00000	9.	2951.00	6.	2.	0.	514.00	2191.	0.	226.
316	100500.	23.	107900.	4.	2.50000	9.	3099.00	0.	2.	0.	680.00	3717.	0.	166.
317	108000.	19.	109500.	4.	3.00000	9.	3064.00	6.	2.	0.	473.00	1956.	0.	59.
318	86900.	18.	110000.	5.	2.50000	8.	2950.00	5.	2.	0.	630.00	2167.	0.	267.
319	105000.	16.	110000.	3.	3.00000	8.	2829.00	20.	1.	581.	481.00	1956.	0.	1.
320	108000.	19.	111900.	4.	2.50000	8.	2715.00	1.	0.	0.	613.00	6050.	0.	155.
321	110000.	20.	114500.	4.	2.50000	9.	3184.00	10.	1.	0.	796.00	4833.	0.	177.
322	112000.	16.	117500.	3.	2.50000	8.	2270.00	15.	2.	0.	579.00	3914.	0.	287.

Column Count Row	Price 339	Zone 339	List 339	Bedrms 339	Baths 339	Rooms 339	Sq. ft. 339	Age 339	Cars 339	City Tax 339	County 339	Lot 339	Type 339	DOM 339
323	53300.	20.	122900.	4.	2.50000	9.	3005.00	1.	2.	0.	551.00	2280.	0.	48.
324	110000.	20.	124500.	4.	2.50000	8.	2137.00	3.	2.	0.	573.00	5556.	0.	185.
325	108500.	18.	126000.	3.	2.00000	8.	2300.00	11.	2.	684.	566.00	3911.	0.	303.
326	123000.	20.	129000.	4.	3.00000	8.	2361.00	1.	2.	0.	324.00	2613.	0.	195.
327	110000.	20.	129500.	5.	3.00000	10.	3500.00	4.	2.	0.	682.00	15488.	0.	167.
328	100000.	16.	130000.	4.	3.00000	8.	2954.00	10.	2.	0.	513.00	5266.	0.	520.
329	130500.	20.	135900.	4.	2.50000	8.	3321.00	0.	2.	0.	626.00	2722.	0.	182.
330	150000.	16.	158500.	4.	2.50000	9.	3258.00	16.	2.	863.	647.00	2420.	0.	55.
331	167500.	20.	169900.	5.	3.00000	10.	3706.00	3.	2.	0.	733.00	2874.	0.	43.
332	166900.	20.	179500.	4.	3.00000	10.	1355.00	1.	2.	0.	716.00	4779.	0.	125.
333	157000.	16.	179500.	4.	2.50000	9.	2850.00	50.	2.	615.	509.00	2260.	0.	270.
334	171900.	20.	179900.	4.	3.00000	10.	3289.00	1.	2.	0.	912.00	3646.	0.	59.
335	175000.	20.	184900.	4.	3.00000	9.	3166.00	1.	2.	0.	1302.00	2532.	0.	187.
336	215000.	16.	225000.	4.	3.00000	9.	2550.00	19.	2.	620.	465.00	2478.	0.	116.
337	242850.	20.	295000.	5.	3.00000	10.	5994.00	11.	3.	0.	1500.00	7260.	0.	73.
338	217000.	20.	300000.	3.	1.00000	5.	1200.00	30.	0.	0.	188.00	145200.	0.	13.
339	400000.	17.	450000.	5.	3.00000	10.	4600.00	16.	2.	0.	1120.00	99567.	0.	121.

APPENDIX H

TABLE OF CRITICAL VALUES OF THE STUDENT'S *t* DISTRIBUTION

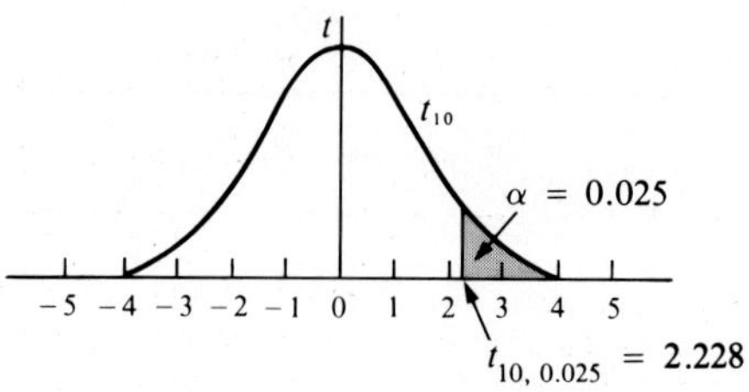

Table gives values of t that cut off $100\alpha\%$ of area under upper tail.

Example: $t_{10,0.025} = 2.228$.

For a t with 10 degrees of freedom, the area under the curve over the interval from $t = 2.228$ to infinity is $2\frac{1}{2}\%$ of the total area.

υ	α				
	0.10	**0.05**	**0.025**	**0.01**	**0.005**
1	3.078	6.314	12.706	31.821	63.657
2	1.886	2.920	4.303	6.965	9.925
3	1.638	2.353	3.182	4.541	5.841
4	1.533	2.132	2.776	3.747	4.604
5	1.476	2.015	2.571	3.365	4.032
6	1.440	1.943	2.447	3.143	3.707
7	1.415	1.895	2.365	2.998	3.499
8	1.397	1.860	2.306	2.896	3.355
9	1.383	1.833	2.262	2.821	3.250
10	1.372	1.812	2.228	2.764	3.169
11	1.363	1.796	2.201	2.718	3.106
12	1.356	1.782	2.179	2.681	3.055
13	1.350	1.771	2.160	2.650	3.012
14	1.345	1.761	2.145	2.624	2.977
15	1.341	1.753	2.131	2.602	2.947

υ	α				
	0.10	**0.05**	**0.025**	**0.01**	**0.005**
16	1.337	1.746	2.120	2.583	2.921
17	1.333	1.740	2.110	2.567	2.898
18	1.330	1.734	2.101	2.552	2.878
19	1.328	1.729	2.093	2.539	2.861
20	1.325	1.725	2.086	2.528	2.845
21	1.323	1.721	2.080	2.518	2.831
22	1.321	1.717	2.074	2.508	2.819
23	1.319	1.714	2.069	2.500	2.807
24	1.318	1.711	2.064	2.492	2.797
25	1.316	1.708	2.060	2.485	2.787
26	1.315	1.706	2.056	2.479	2.779
27	1.314	1.703	2.052	2.473	2.771
28	1.313	1.701	2.048	2.467	2.763
29	1.311	1.699	2.045	2.462	2.756
30	1.310	1.697	2.042	2.457	2.750
40	1.303	1.684	2.021	2.423	2.704
60	1.296	1.671	2.000	2.390	2.660
120	1.289	1.658	1.980	2.358	2.617
∞	1.282	1.645	1.960	2.326	2.576

APPENDIX I

TABLE OF CRITICAL VALUES OF THE *F* DISTRIBUTION

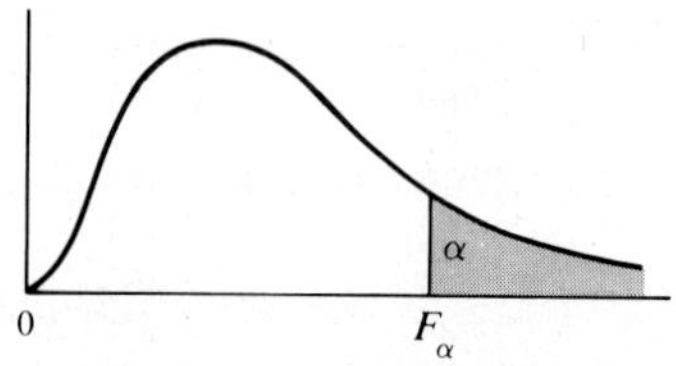

Table gives values of F with ν_1 and ν_2 degrees of freedom that cuts off 100α% of the area under the upper tail of the curve.

$F_{\nu_1,\nu_2,0.05}$

ν_2	ν_1 = 1	2	3	4	5	6	7	8	9
1	161.4	199.5	215.7	224.6	230.2	234.0	236.8	238.9	240.5
2	18.51	19.00	19.16	19.25	19.30	19.33	19.35	19.37	19.38
3	10.13	9.55	9.28	9.12	9.01	8.94	8.89	8.85	8.81
4	7.71	6.94	6.59	6.39	6.26	6.16	6.09	6.04	6.00
5	6.61	5.79	5.41	5.19	5.05	4.95	4.88	4.82	4.77
6	5.99	5.14	4.76	4.53	4.39	4.28	4.21	4.15	4.10
7	5.59	4.74	4.35	4.12	3.97	3.87	3.79	3.73	3.68
8	5.32	4.46	4.07	3.84	3.69	3.58	3.50	3.44	3.39
9	5.12	4.26	3.86	3.63	3.48	3.37	3.29	3.23	3.18
10	4.96	4.10	3.71	3.48	3.33	3.22	3.14	3.07	3.02
11	4.84	3.98	3.59	3.36	3.20	3.09	3.01	2.95	2.90
12	4.75	3.89	3.49	3.26	3.11	3.00	2.91	2.85	2.80
13	4.67	3.81	3.41	3.18	3.03	2.92	2.83	2.77	2.71
14	4.60	3.74	3.34	3.11	2.96	2.85	2.76	2.70	2.65
15	4.54	3.68	3.29	3.06	2.90	2.79	2.71	2.64	2.59
16	4.09	3.63	3.24	3.01	2.85	2.74	2.66	2.59	2.54
17	4.45	3.59	3.20	2.96	2.81	2.70	2.61	2.55	2.49
18	4.41	3.55	3.16	2.93	2.77	2.66	2.58	2.51	2.46
19	4.38	3.52	3.13	2.90	2.74	2.63	2.54	2.48	2.42
20	4.35	3.49	3.10	2.87	2.71	2.60	2.51	2.45	2.39
21	4.32	3.47	3.07	2.84	2.68	2.57	2.49	2.42	2.37
22	4.30	3.44	3.05	2.82	2.66	2.55	2.46	2.40	2.34
23	4.28	3.42	3.03	2.80	2.64	2.53	2.44	2.37	2.32
24	4.26	3.40	3.01	2.78	2.62	2.51	2.42	2.36	2.30
25	4.24	3.39	2.99	2.76	2.60	2.49	2.40	2.34	2.28
26	4.23	3.37	2.98	2.74	2.59	2.47	2.39	2.32	2.27
27	4.21	3.35	2.96	2.73	2.57	2.46	2.37	2.31	2.25
28	4.20	3.34	2.95	2.71	2.56	2.45	2.36	2.29	2.24
29	4.18	3.33	2.93	2.70	2.55	2.43	2.35	2.28	2.22
30	4.17	3.32	2.92	2.69	2.53	2.42	2.33	2.27	2.21
40	4.08	3.23	2.84	2.61	2.45	2.34	2.25	2.18	2.12
60	4.00	3.15	2.76	2.53	2.37	2.25	2.17	2.10	2.04
120	3.92	3.07	2.68	2.45	2.29	2.17	2.09	2.02	1.96
∞	3.84	3.00	2.60	2.37	2.21	2.10	2.01	1.94	1.88

$F_{\nu_1,\nu_2,0.05}$

ν_2	ν_1 = 10	12	15	20	24	30	40	60	120	∞
1	241.9	243.9	245.9	248.0	249.1	250.1	251.1	252.2	253.3	254.3
2	19.40	19.41	19.43	19.45	19.45	19.46	19.47	19.48	19.49	19.50
3	7.79	8.74	8.70	8.66	8.64	8.62	8.59	8.57	8.55	8.53
4	5.96	5.91	5.86	5.80	5.77	5.75	5.72	5.69	5.66	5.63
5	4.74	4.68	4.62	4.56	4.53	4.50	4.46	4.43	4.40	4.36
6	4.06	4.00	3.94	3.87	3.84	3.81	3.77	3.74	3.70	3.67
7	3.64	3.57	3.51	3.44	3.41	3.38	3.34	3.30	3.27	3.23
8	3.35	3.28	3.22	3.15	3.12	3.08	3.04	3.01	2.97	2.93
9	3.14	3.07	3.01	2.94	2.90	2.86	2.83	2.79	2.75	2.71
10	2.98	2.91	2.85	2.77	2.74	2.70	2.66	2.62	2.58	2.54
11	2.85	2.79	2.72	2.65	2.61	2.57	2.53	2.49	2.45	2.40
12	2.75	2.69	2.62	2.54	2.51	2.47	2.43	2.38	2.34	2.30
13	2.67	2.60	2.53	2.46	2.42	2.38	2.34	2.30	2.25	2.21
14	2.60	2.53	2.46	2.39	2.35	2.31	2.27	2.22	2.18	2.13
15	2.54	2.48	2.40	2.33	2.29	2.25	2.20	2.16	2.11	2.07
16	2.49	2.42	2.35	2.28	2.24	2.19	2.15	2.11	2.06	2.01
17	2.45	2.38	2.31	2.23	2.19	2.15	2.10	2.06	2.01	1.96
18	2.41	2.34	2.27	2.19	2.15	2.11	2.06	2.02	1.97	1.92
19	2.38	2.31	2.23	2.16	2.11	2.07	2.03	1.98	1.93	1.88
20	2.35	2.28	2.20	2.12	2.08	2.04	1.99	1.95	1.90	1.84
21	2.32	2.25	2.18	2.10	2.05	2.01	1.96	1.92	1.87	1.81
22	2.30	2.23	2.15	2.07	2.03	1.98	1.94	1.89	1.84	1.78
23	2.27	2.20	2.13	2.05	2.01	1.96	1.91	1.86	.81	1.76
24	2.25	2.18	2.11	2.03	1.98	1.94	1.89	1.84	.79	1.73
25	2.24	2.16	2.09	2.01	1.96	1.92	1.87	1.82	.77	1.71
26	2.22	2.15	2.07	1.99	1.95	1.90	1.85	1.80	.75	1.69
27	2.20	2.13	2.06	1.97	1.93	1.88	1.84	1.79	.73	1.67
28	2.19	2.12	2.04	1.96	1.91	1.87	1.82	1.77	.71	1.65
29	2.18	2.10	2.03	1.94	1.90	1.85	1.81	1.75	.70	1.64
30	2.16	2.09	2.01	1.93	1.89	1.84	1.79	1.74	.68	1.62
40	2.08	2.00	1.92	1.84	1.79	1.74	1.69	1.64	.58	1.51
60	1.99	1.92	1.84	1.75	1.70	1.65	1.59	1.53	.47	1.39
120	1.91	1.83	1.75	1.66	1.61	1.55	1.50	1.43	.35	1.25
∞	1.83	1.75	1.67	1.57	1.52	1.46	1.39	1.32	.22	1.00

$F_{\nu_1,\nu_2,0.01}$

ν_2	ν_1 = 1	2	3	4	5	6	7	8	9
1	4052	4999.5	5403	5625	5764	5859	5928	5981	6022
2	98.50	99.00	99.17	99.25	99.30	99.33	99.36	99.37	99.39
3	34.12	30.82	29.46	28.71	28.24	27.91	27.67	27.49	27.35
4	21.20	18.00	16.69	15.98	15.52	15.21	14.98	14.80	14.66
5	16.26	13.27	12.06	11.39	10.97	10.67	10.46	10.29	10.16
6	13.75	10.92	9.78	9.15	8.75	8.47	8.26	8.10	7.98
7	12.25	9.55	8.45	7.85	7.46	7.19	6.99	6.84	6.72
8	11.26	8.65	7.59	7.01	6.63	6.37	6.18	6.03	5.91
9	10.56	8.02	6.99	6.42	6.06	5.80	5.61	5.47	5.35
10	10.04	7.56	6.55	5.99	5.64	5.39	5.20	5.06	4.94
11	9.65	7.21	6.22	5.67	5.32	5.07	4.89	4.74	4.63
12	9.33	6.93	5.95	5.41	5.06	4.82	4.64	4.50	4.39
13	9.07	6.70	5.74	5.21	4.86	4.62	4.44	4.30	4.19
14	8.86	6.51	5.56	5.04	4.69	4.46	4.28	4.14	4.03
15	8.68	6.36	5.42	4.89	4.56	4.32	4.14	4.00	3.89
16	8.53	6.23	5.29	4.77	4.44	4.20	4.03	3.89	3.78
17	8.40	6.11	5.18	4.57	4.34	4.10	3.93	3.79	3.68
18	8.29	6.01	5.09	4.58	4.25	4.01	3.84	3.71	3.60
19	8.18	5.93	5.01	4.50	4.17	3.94	3.77	3.63	3.52
20	8.10	5.85	4.94	4.43	4.10	3.87	3.70	3.56	3.46
21	8.02	5.78	4.87	4.37	4.04	3.81	3.64	3.51	3.40
22	7.95	5.72	4.82	4.31	3.99	3.76	3.59	3.45	3.35
23	7.88	5.66	4.76	4.26	3.94	3.71	3.54	3.41	3.30
24	7.82	5.61	4.72	4.22	3.90	3.67	3.50	3.36	3.26
25	7.77	5.57	4.68	4.18	3.85	3.63	3.46	3.32	3.22
26	7.72	5.53	4.64	4.14	3.82	3.59	3.42	3.29	3.18
27	7.68	5.49	4.60	4.11	3.78	3.56	3.39	3.26	3.15
28	7.64	5.45	4.57	4.07	3.75	3.53	3.36	3.23	3.12
29	7.60	5.42	4.54	4.04	3.73	3.50	3.33	3.20	3.09
30	7.56	5.39	4.51	4.02	3.70	3.47	3.30	3.17	3.07
40	7.31	5.18	4.31	3.83	3.51	3.29	3.12	2.99	2.89
60	7.08	4.98	4.13	3.65	3.34	3.12	2.95	2.82	2.72
120	6.85	4.79	3.95	3.48	3.17	2.96	2.79	2.66	2.56
∞	6.63	4.61	3.78	3.32	3.02	2.80	2.64	2.51	2.41

$F_{\nu_1,\nu_2,0.01}$

ν_2	ν_1									
	10	12	15	20	24	30	40	60	120	∞
1	6056	6106	6157	6209	6235	6261	6287	6313	6339	6366
2	99.40	99.42	99.43	99.45	99.46	99.47	99.47	99.48	99.49	99.50
3	27.23	27.05	26.87	26.69	26.60	26.50	26.41	26.32	26.22	26.13
4	14.55	14.37	14.20	14.02	13.93	13.84	13.75	13.65	13.56	13.46
5	10.05	9.89	9.72	9.55	9.47	9.38	9.29	9.20	9.11	9.02
6	7.87	7.72	7.56	7.40	7.31	7.23	7.14	7.06	6.97	6.88
7	6.62	6.47	6.31	6.16	6.07	5.99	5.91	5.82	5.74	5.65
8	5.81	5.67	5.52	5.36	5.28	5.20	5.12	5.03	4.95	4.86
9	5.26	5.11	4.96	4.81	4.73	4.65	4.57	4.48	4.40	4.31
10	4.85	4.71	4.56	4.41	4.33	4.25	4.17	4.08	4.00	3.91
11	4.54	4.40	4.25	4.10	4.02	3.94	3.86	3.78	3.69	3.60
12	4.30	4.16	4.01	3.86	3.78	3.70	3.62	3.54	3.45	3.36
13	4.10	3.96	3.82	3.66	3.59	3.51	3.43	3.34	3.25	3.17
14	3.94	3.80	3.66	3.51	3.43	3.35	3.27	3.18	3.09	3.00
15	3.80	3.67	3.52	3.37	3.29	3.21	3.13	3.05	2.96	2.87
16	3.69	3.55	3.41	3.26	3.18	3.10	3.02	2.93	2.84	2.75
17	3.59	3.46	3.31	3.16	3.08	3.00	2.92	2.83	2.75	2.65
18	3.51	3.37	3.23	3.08	3.00	2.92	2.84	2.75	2.66	2.57
19	3.43	3.30	3.15	3.00	2.92	2.84	2.76	2.67	2.58	2.49
20	3.37	3.23	3.09	2.94	2.86	2.78	2.69	2.61	2.52	2.42
21	3.31	3.17	3.03	2.88	2.80	2.72	2.64	2.55	2.46	2.36
22	3.26	3.12	2.98	2.83	2.75	2.67	2.58	2.50	2.40	2.31
23	3.21	3.07	2.93	2.78	2.70	2.62	2.54	2.45	2.35	2.26
24	3.17	3.03	2.89	2.74	2.66	2.58	2.49	2.40	2.31	2.21
25	3.13	2.99	2.85	2.70	2.62	2.54	2.45	2.36	2.27	2.17
26	3.09	2.96	2.81	2.66	2.58	2.50	2.42	2.33	2.23	2.13
27	3.06	2.93	2.78	2.63	2.55	2.47	2.38	2.29	2.20	2.10
28	3.03	2.90	2.75	2.60	2.52	2.44	2.35	2.26	2.17	2.06
29	3.00	2.87	2.73	2.57	2.49	2.41	2.33	2.23	2.14	2.03
30	2.98	2.84	2.70	2.55	2.47	2.39	2.30	2.21	2.11	2.01
40	2.80	2.66	2.52	2.37	2.29	2.20	2.11	2.02	1.92	1.80
60	2.63	2.50	2.35	2.20	2.12	2.03	1.94	1.84	1.73	1.60
120	2.47	2.34	2.19	2.03	1.95	1.86	1.76	1.66	1.53	1.38
∞	2.32	2.18	2.04	1.88	1.79	1.70	1.59	1.47	1.32	1.00

This table is reproduced from Table 18 of *Biometrika Tables for Statisticians,* Vol. I, 1966, by permission of E. S. Pearson and the Biometrika Trustees.

APPENDIX J

TABLE OF DIXON CRITERIA

Criteria for Rejection of Outlying Observations

Statistic	n	α = 0.05	α = 0.02	α = 0.01	α = 0.005
r_{10}	3	0.941	0.976	0.988	0.994
	4	0.765	0.846	0.889	0.926
	5	0.642	0.729	0.780	0.821
	6	0.560	0.644	0.698	0.740
	7	0.507	0.586	0.637	0.680
r_{11}	8	0.554	0.631	0.683	0.725
	9	0.512	0.587	0.635	0.677
	10	0.477	0.551	0.597	0.639
r_{21}	11	0.576	0.638	0.679	0.713
	12	0.546	0.605	0.642	0.675
	13	0.521	0.578	0.615	0.649
r_{22}	14	0.546	0.602	0.641	0.674
	15	0.525	0.579	0.616	0.647
	16	0.507	0.559	0.595	0.624
	17	0.490	0.542	0.577	0.605
	18	0.475	0.527	0.561	0.589
	19	0.462	0.514	0.547	0.575
	20	0.450	0.502	0.535	0.562
	21	0.440	0.491	0.524	0.551
	22	0.430	0.481	0.514	0.541
	23	0.421	0.472	0.505	0.532
	24	0.413	0.464	0.497	0.524
	25	0.406	0.457	0.489	0.516

Adapted from W. J. Dixon, "Processing Data for Outliers," *Biometrics*, 9 (1953), p. 74, with permission from the author and the Biometric Society.

APPENDIX K

TABLE OF DURBIN-WATSON TEST BOUNDS

To test H_0: no positive serial correlation,
if $d < d_L$, reject H_0;
of $d > d_U$, accept H_0;
if $d_L < d < d_U$, the test is inconclusive.
To test H_0: no negative serial correlation, use $d^- = U - d$.

Level of Significance α = 0.05

	$p = 2$		$p = 3$		$p = 4$		$p = 5$		$p = 6$	
n	d_L	d_U	d_L	d_U	d_L	d_U	d_L	d_U	d_L	d_U
15	1.08	1.36	0.95	1.54	0.82	1.75	0.69	1.97	0.56	2.21
16	1.10	1.37	0.98	1.54	0.86	1.73	0.74	1.93	0.62	2.15
17	1.13	1.38	1.02	1.54	0.90	1.71	0.78	1.90	0.67	2.10
18	1.16	1.39	1.05	1.53	0.93	1.69	0.82	1.87	0.71	2.06
19	1.18	1.40	1.08	1.53	0.97	1.68	0.86	1.85	0.75	2.02
20	1.20	1.41	1.10	1.54	1.00	1.68	0.90	1.83	0.79	1.99
21	1.22	1.42	1.13	1.54	1.03	1.67	0.93	1.81	0.83	1.96
22	1.24	1.43	1.15	1.54	1.05	1.66	0.96	1.80	0.86	1.94
23	1.26	1.44	1.17	1.54	1.08	1.66	0.99	1.79	0.90	1.92
24	1.27	1.45	1.19	1.55	1.10	1.66	1.01	1.78	0.93	1.90
25	1.29	1.45	1.21	1.55	1.12	1.66	1.04	1.77	0.95	1.89
26	1.30	1.46	1.22	1.55	1.14	1.65	1.06	1.76	0.98	1.88
27	1.32	1.47	1.24	1.56	1.16	1.65	1.08	1.76	1.01	1.86
28	1.33	1.48	1.26	1.56	1.18	1.65	1.10	1.75	1.03	1.85
29	1.34	1.48	1.27	1.56	1.20	1.65	1.12	1.74	1.05	1.84
30	1.35	1.49	1.28	1.57	1.21	1.65	1.14	1.74	1.07	1.83
31	1.36	1.50	1.30	1.57	1.23	1.65	1.16	1.74	1.09	1.83
32	1.37	1.50	1.31	1.57	1.24	1.65	1.18	1.73	1.11	1.82
33	1.38	1.51	1.32	1.58	1.26	1.65	1.19	1.73	1.13	1.81
34	1.39	1.51	1.33	1.58	1.27	1.65	1.21	1.73	1.15	1.81
35	1.40	1.52	1.34	1.58	1.28	1.65	1.22	1.73	1.16	1.80
36	1.41	1.52	1.35	1.59	1.29	1.65	1.24	1.73	1.18	1.80
37	1.42	1.53	1.36	1.59	1.31	1.66	1.25	1.72	1.19	1.80
38	1.43	1.54	1.37	1.59	1.32	1.66	1.26	1.72	1.21	1.79
39	1.43	1.54	1.38	1.60	1.33	1.66	1.27	1.72	1.22	1.79
40	1.44	1.54	1.39	1.60	1.34	1.66	1.29	1.72	1.23	1.79
45	1.48	1.57	1.43	1.62	1.38	1.67	1.34	1.72	1.29	1.78
50	1.50	1.59	1.46	1.63	1.42	1.67	1.38	1.72	1.34	1.77
55	1.53	1.60	1.49	1.64	1.45	1.68	1.41	1.72	1.38	1.77
60	1.55	1.62	1.51	1.65	1.48	1.69	1.44	1.73	1.41	1.77
65	1.57	1.63	1.54	1.66	1.50	1.70	1.47	1.73	1.44	1.77
70	1.58	1.64	1.55	1.67	1.52	1.70	1.49	1.74	1.46	1.77
75	1.60	1.64	1.57	1.68	1.54	1.71	1.51	1.74	1.49	1.77
80	1.61	1.66	1.59	1.69	1.56	1.72	1.53	1.74	1.51	1.77
85	1.62	1.67	1.60	1.70	1.57	1.72	1.55	1.75	1.52	1.77
90	1.63	1.68	1.61	1.70	1.59	1.73	1.57	1.75	1.54	1.78
95	1.64	1.69	1.62	1.71	1.60	1.73	1.58	1.75	1.56	1.78
100	1.65	1.69	1.63	1.72	1.61	1.74	1.59	1.76	1.57	1.78

Level of Significance α = 0.01

n	$p = 2$ d_L	$p = 2$ d_U	$p = 3$ d_L	$p = 3$ d_U	$p = 4$ d_L	$p = 4$ d_U	$p = 5$ d_L	$p = 5$ d_U	$p = 6$ d_L	$p = 6$ d_U
15	0.81	1.07	0.70	1.25	0.59	1.46	0.49	1.70	0.39	1.96
16	0.84	1.09	0.74	1.25	0.63	1.44	0.53	1.66	0.44	1.90
17	0.87	1.10	0.77	1.25	0.67	1.43	0.57	1.63	0.48	1.85
18	0.90	1.12	0.80	1.26	0.71	1.42	0.61	1.60	0.52	1.80
19	0.93	1.13	0.83	1.26	0.74	1.41	0.65	1.58	0.56	1.77
20	0.95	1.15	0.86	1.27	0.77	1.41	0.68	1.57	0.60	1.74
21	0.97	1.16	0.89	1.27	0.80	1.41	0.72	1.55	0.63	1.71
22	1.00	1.17	0.91	1.28	0.83	1.40	0.75	1.54	0.66	1.69
23	1.02	1.19	0.94	1.29	0.86	1.40	0.77	1.53	0.70	1.67
24	1.04	1.20	0.96	1.30	0.88	1.41	0.80	1.53	0.72	1.66
25	1.05	1.21	0.98	1.30	0.90	1.41	0.83	1.52	0.75	1.65
26	1.07	1.22	1.00	1.31	0.93	1.41	0.85	1.52	0.78	1.64
27	1.09	1.23	1.02	1.32	0.95	1.41	0.88	1.51	0.81	1.63
28	1.10	1.24	1.04	1.32	0.97	1.41	0.90	1.51	0.83	1.62
29	1.12	1.25	1.05	1.33	0.99	1.42	0.92	1.51	0.85	1.61
30	1.13	1.26	1.07	1.34	1.01	1.42	0.94	1.51	0.88	1.61
31	1.15	1.27	1.08	1.34	1.02	1.42	0.96	1.51	0.90	1.60
32	1.16	1.28	1.10	1.35	1.04	1.43	0.98	1.51	0.92	1.60
33	1.17	1.29	1.11	1.36	1.05	1.43	1.00	1.51	0.94	1.59
34	1.18	1.30	1.13	1.36	1.07	1.43	1.01	1.51	0.95	1.59
35	1.19	1.31	1.14	1.37	1.08	1.44	1.03	1.51	0.97	1.59
36	1.21	1.32	1.15	1.38	1.10	1.44	1.04	1.51	0.99	1.59
37	1.22	1.32	1.16	1.38	1.11	1.45	1.06	1.51	1.00	1.59
38	1.23	1.33	1.18	1.39	1.12	1.45	1.07	1.52	1.02	1.58
39	1.24	1.34	1.19	1.39	1.14	1.45	1.09	1.52	1.03	1.58
40	1.25	1.34	1.20	1.40	1.15	1.46	1.10	1.52	1.05	1.58
45	1.29	1.38	1.24	1.42	1.20	1.48	1.16	1.53	1.11	1.58
50	1.32	1.40	1.28	1.45	1.24	1.49	1.20	1.54	1.16	1.59
55	1.36	1.43	1.32	1.47	1.28	1.51	1.25	1.55	1.21	1.59
60	1.38	1.45	1.35	1.48	1.32	1.52	1.28	1.56	1.25	1.60
65	1.41	1.47	1.38	1.50	1.35	1.53	1.31	1.57	1.28	1.61
70	1.43	1.49	1.40	1.52	1.37	1.55	1.34	1.58	1.31	1.61
75	1.45	1.50	1.42	1.53	1.39	1.56	1.37	1.59	1.34	1.62
80	1.47	1.52	1.44	1.54	1.42	1.57	1.39	1.60	1.36	1.62
85	1.48	1.53	1.46	1.55	1.43	1.58	1.41	1.60	1.39	1.63
90	1.50	1.54	1.47	1.56	1.45	1.59	1.43	1.61	1.41	1.64
95	1.51	1.55	1.49	1.57	1.47	1.60	1.45	1.62	1.42	1.64
100	1.52	1.56	1.50	1.58	1.48	1.60	1.46	1.63	1.44	1.65

APPENDIX L

ANALYSES OF DATA SETS

EXERCISE 1.5
CONCESSION DATA

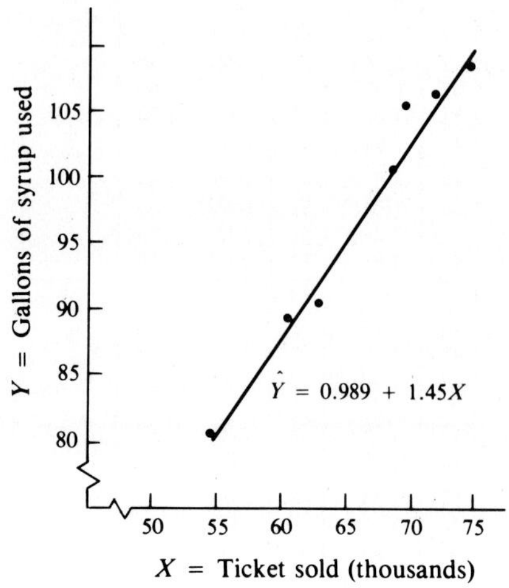

FIGURE L1 Concession Data

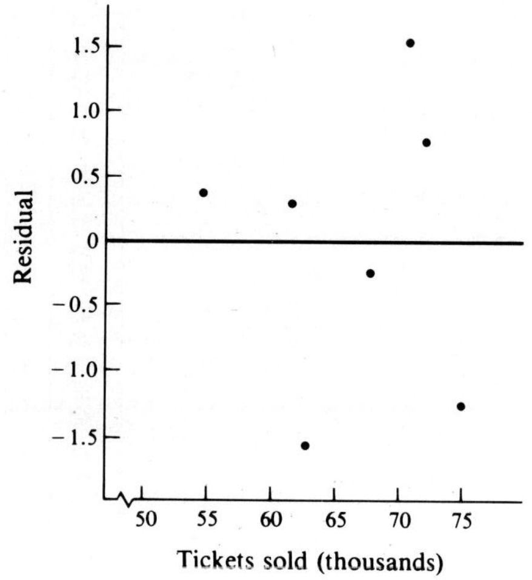

FIGURE L2 Residuals for Concession Data

$n = 7$

$\sum X = 464.4$

$\sum X^2 = 31{,}125.76$

$\sum Y = 681.8$

$\sum Y^2 = 67{,}082.26$

$\sum XY = 45{,}691.98$

$S_{XY} = 459.42$

$S_{XX} = 316.1371$

$S_{YY} = 674.94$

$s_{Y \cdot X} = 1.2081$

C.V. $= 1.2403$

$SS(R) = 667.6430$

$SS(E) = 7.2970$

$SS(T) = 674.9400$

$$\mathbf{X'X} = \begin{bmatrix} 7 & 464.4 \\ 464.4 & 31{,}125.76 \end{bmatrix} \qquad \mathbf{X'y} = \begin{bmatrix} 681.8 \\ 45{,}691.98 \end{bmatrix}$$

$$(\mathbf{X'X})^{-1} = \begin{bmatrix} 14.06522 & -0.209855 \\ -0.209855 & 0.003163184 \end{bmatrix}$$

EXERCISE 1.6
SALES DATA

FIGURE L3 Sales Data

$n = 15$

$\sum X = 463$

$\sum X^2 = 16{,}275$

$\sum Y = 173$

$\sum Y^2 = 2315$

$\sum XY = 6102$

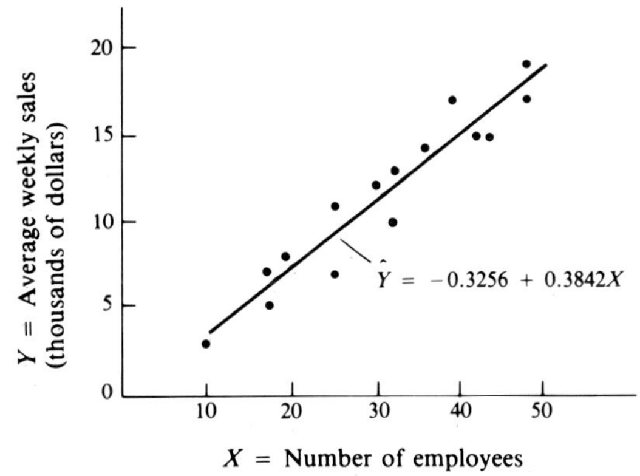

EXERCISE 1.7
POLICE DATA

FIGURE L4 Police Data

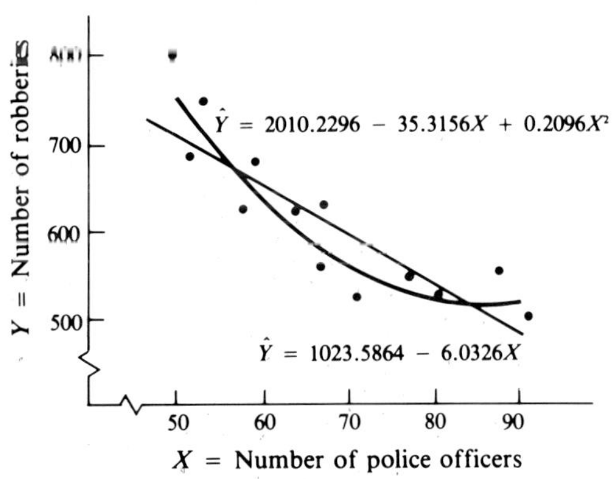

Quadratic:

$$\mathbf{X'X} = \begin{bmatrix} 13 & 878 & 61{,}470 \\ 878 & 61{,}470 & 4{,}454{,}930 \\ 61{,}470 & 4{,}454{,}930 & 333{,}321{,}846 \end{bmatrix} \qquad \mathbf{X'y} = \begin{bmatrix} 8010 \\ 527{,}885 \\ 36{,}104{,}535 \end{bmatrix}$$

$$(\mathbf{X'X})^{-1} = \begin{bmatrix} 80.24475 & -2.34809 & 0.01658431 \\ -2.34809 & 0.06922711 & -0.000492212 \\ 0.01658431 & -0.000492212 & 0.00000352312 \end{bmatrix}$$

$\mathbf{y'y} = 5{,}038{,}700$ $\qquad S_{XY} = -13{,}098.0769$
$s_{Y \cdot X} = 34.3931$ $\qquad S_{XX} = 2171.2308$
$SS(R) = 91{,}478.8660$ $\qquad S_{YY} = 103{,}307.6923$
$SS(E) = 11{,}828.8260$
$SS(T) = 103{,}308.0000$

FIGURE L5 Exponential Function for Police Data

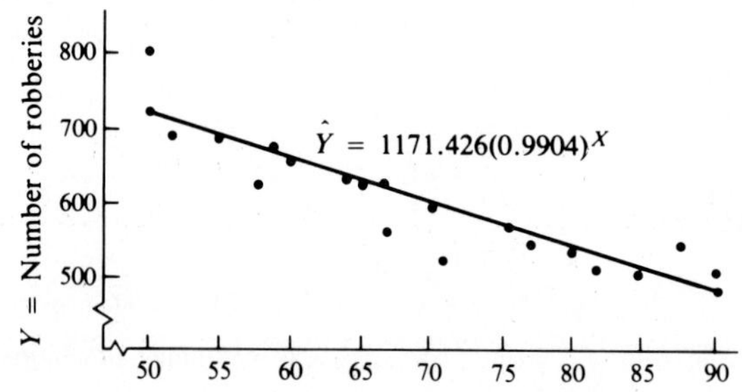

Exponential:

$\sum (\ln Y) = 83.37444$ $\qquad \sum (\log Y) = 36.20906$
$\sum (\ln Y)^2 = 534.9726$ $\qquad \sum (\log Y)^2 = 200.9021$
$\sum X(\ln Y) = 5610.003$ $\qquad \sum X(\log Y) = 2436.393$
$\ln \hat{Y} = 7.065977 - 0.00966202X$ $\qquad \log \hat{Y} = 3.068715 - 0.00419616X$
$\hat{Y} = 1171.4259(0.9904)^X$

EXERCISE 1.8
HEATING DATA

FIGURE L6 Heating Data

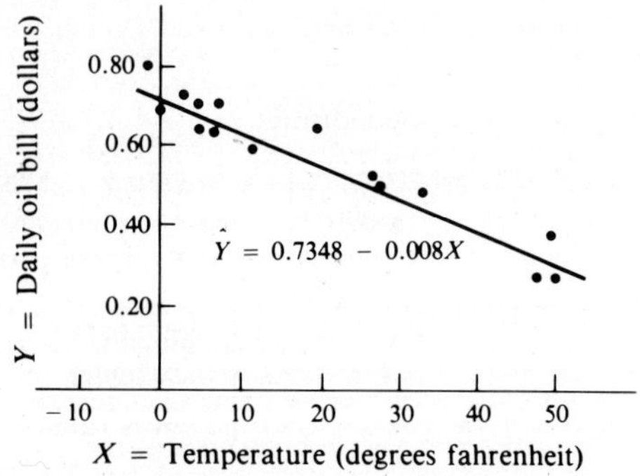

$n = 15$ $\qquad S_{XY} = -39.3667$
$\sum X = 284$
$\sum X^2 = 10{,}302$ $\qquad S_{XX} = 4924.9333$
$\sum Y = 8.75$
$\sum Y^2 = 5.4425$ $\qquad S_{YY} = 0.3383$
$\sum XY = 126.3$

$$\mathbf{X'X} = \begin{bmatrix} 15 & 284 \\ 284 & 10{,}302 \end{bmatrix} \qquad \mathbf{X'y} = \begin{bmatrix} 8.75 \\ 126.3 \end{bmatrix}$$

$$(\mathbf{X'X})^{-1} = \begin{bmatrix} 0.1394537 & -0.00384438 \\ -0.00384438 & 0.0002030484 \end{bmatrix}$$

$s_{Y \cdot X} = 0.0427$
C.V. $= 7.3137$
$SS(R) = 0.3147$
$SS(E) = 0.0237$
$SS(T) = 0.3383$
$\sum e_t^2 = 0.0238$
$\sum e_{t-1}^2 = 0.0228$
$\sum e_t e_{t-1} = 0.0004$
$\sum (e_t - e_{t-1})^2 = 0.0443$

EXERCISE 1.9
SMOKERS DATA

FIGURE L7 Smokers Data

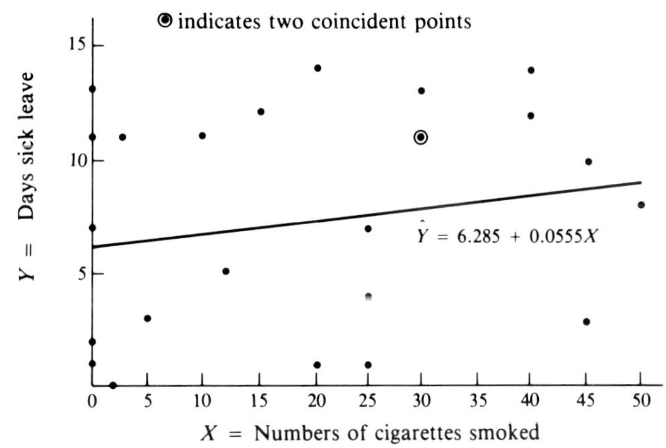

$n = 25$ $S_{XY} = 358.2$
$\sum X = 502$ $S_{XX} = 6451.84$
$\sum X^2 = 16{,}532$
$\sum Y = 185$ $S_{YY} = 582$
$\sum Y^2 = 1951$
$\sum XY = 4073$

FIGURE L8 Residuals for Smokers Data

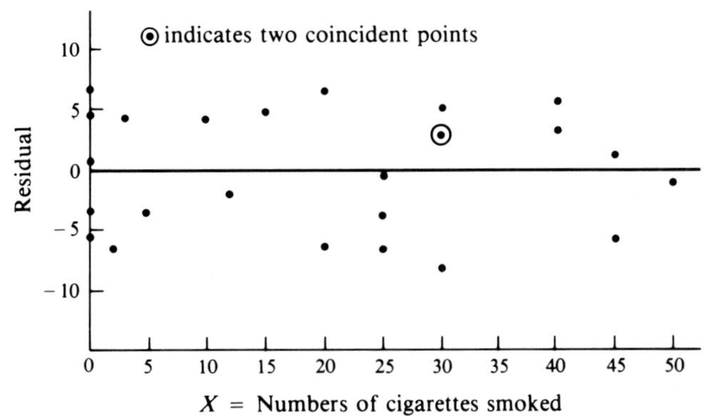

$s_{Y \cdot X} = 4.9437$
C.V. $= 66.8061$
$SS(R) = 19.8869$
$SS(E) = 562.1130$
$SS(T) = 582.0000$

EXERCISE 1.10
TEXTBOOK DATA

FIGURE L9 Textbook Data

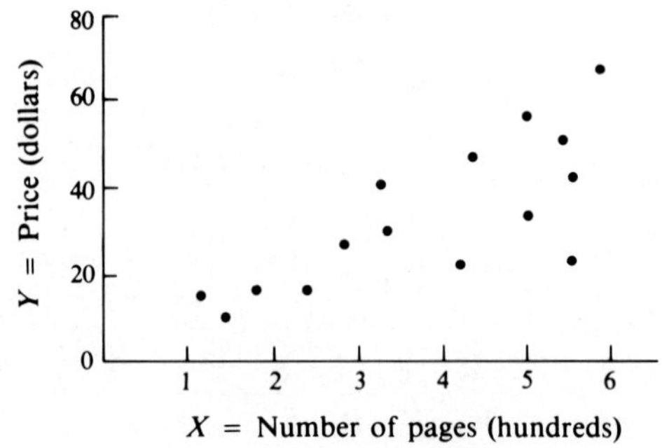

$(n = 15)$:	$(n = 7$ smallest$)$:	$(n = 8$ largest$)$:
$\sum X = 56.5$	$\sum X = 16.0$	$\sum X = 40.5$
$\sum X^2 = 248.41$	$\sum X^2 = 40.8$	$\sum X^2 = 207.61$
$\sum Y = 525$	$\sum Y = 165$	$\sum Y = 360$
$\sum Y^2 = 22{,}617$	$\sum Y^2 = 4603$	$\sum Y^2 = 18{,}014$
$\sum XY = 2274.5$	$\sum XY = 423$	$\sum XY = 1851.5$
$\hat{Y} = 3.5700 + 8.3443X$	$S_{XY} = 45.8571$	$S_{XY} = 29.0000$
	$S_{XX} = 4.2286$	$S_{XX} = 2.5788$
	$S_{YY} = 713.7143$	$S_{YY} = 1814.0000$
	$s^2_{Y \cdot X} = 43.2833$	$s^2_{Y \cdot X} = 247.9799$

FIGURE L10 Residuals for Textbook Data

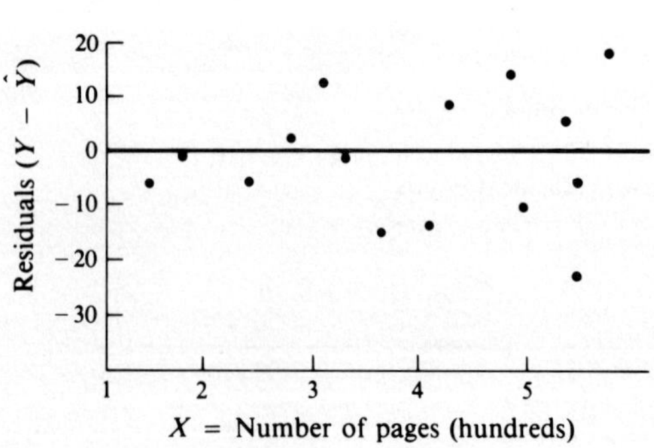

FIGURE L11 Residuals for Transformed Textbook Data

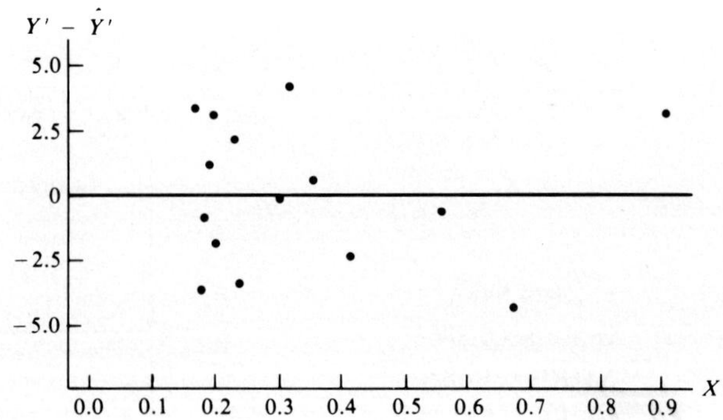

For transformed data,

$(n = 15)$:	$(n = 7$ smallest):	$(n = 8$ largest):
$\sum X' = 5.1326$	$\sum X' = 3.5308$	$\sum X' = 1.6018$
$\sum (X')^2 = 2.4002$	$\sum (X')^2 = 2.0767$	$\sum (X')^2 = 0.3255$
$\sum Y' = 143.7742$	$\sum Y' = 72.9206$	$\sum Y' = 70.8536$
$\sum (Y')^2 = 1503.4735$	$\sum (Y')^2 = 818.6105$	$\sum (Y')^2 = 684.8630$
$\sum X'Y' = 52.2195$	$\sum X'Y' = 38.0913$	$\sum X'Y' = 14.1282$
$S_{X'Y'} = 3.0238$	$S_{X'Y'} = 1.3101$	$S_{X'Y'} = -0.0585$
$S_{X'X'} = 0.6440$	$S_{Y'Y'} = 0.2958$	$S_{X'X'} = 0.0048$
$S_{Y'Y'} = 125.4055$	$S_{Y'Y'} = 58.9799$	$S_{Y'Y'} = 57.3339$
$\hat{Y}' = 7.9831 + 4.6808X'$	$s^2_{Y \cdot X} = 10.6355$	$s^2_{Y \cdot X} = 9.4376$
$s^2_{Y \cdot X} = 8.5578$		

EXERCISE 1.11
PROFESSIONALS DATA

FIGURE L12 Professionals Data

$n = 8$

$\sum X = 180$

$\sum X^2 = 5100$

$\sum X^3 = 162{,}000$

$\sum X^4 = 5{,}482{,}500$

$\sum Y = 369.1$

$\sum Y^2 = 17{,}554.15$

$\sum XY = 9018$

$\sum X^2Y = 265{,}365$

$\hat{Y} = 23.5946 + 1.5504X - 0.0194X^2$

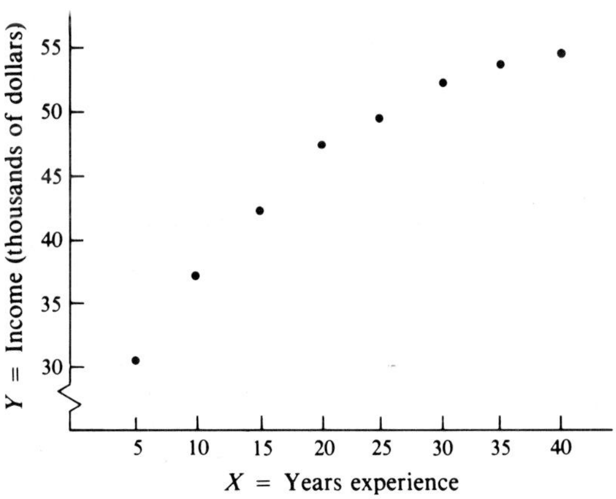

EXERCISE 2.11
REPAIRS DATA

$n = 10$
$\sum X = 500$
$\sum X^2 = 25{,}702$
$\sum Y = 24.7$
$\sum Y^2 = 63.45$
$\sum XY = 1275$
$\hat{Y} = 0.3790 + 0.0570X$

$S_{XY} = 40$
$S_{XX} = 702$
$S_{YY} = 2.441$

$$\mathbf{X'X} = \begin{bmatrix} 10 & 500 \\ 500 & 25{,}702 \end{bmatrix} \qquad \mathbf{X'y} = \begin{bmatrix} 24.7 \\ 1275 \end{bmatrix}$$

$$(\mathbf{X'X})^{-1} = \begin{bmatrix} 3.661254 & -0.0712251 \\ -0.0712251 & 0.001424501 \end{bmatrix}$$

$s_{Y \cdot X} = 0.1422$

EXERCISE 2.12
BARK DATA

Tree	DBH_{ib}
1	26.2
2	20.8
3	34.9
4	26.9
5	23.0
6	18.3
7	33.4
8	21.5
9	27.5
10	33.4
11	25.9
12	33.8
13	29.1
14	22.1
15	27.7

$n = 15$
$\sum X = 449.3$
$\sum X^2 = 13{,}931.17$
$\sum Y = 404.5$
$\sum Y^2 = 11{,}289.97$
$\sum XY = 12{,}540.65$
$\hat{Y} = 0.0908 + 0.8973X$

EXERCISE 2.16
"SUDS" DATA

$n = 18$

$\sum X = 208$ $\qquad S_{XY} = 4320.3000$

$\sum X^2 = 3268$

$\sum Y = 1592.1$ $\qquad S_{XX} = 864.4444$

$\sum Y^2 = 162{,}534$

$\sum XY = 22{,}717.9$ $\qquad S_{YY} = 21{,}712.7550$

$\hat{Y} = 30.6979 + 4.9978X$

EXERCISE 2.17
CLASS SIZE DATA

$n = 22$ $\qquad S_{XY} = -248.2727$

$\sum X = 1086$

$\sum X^2 = 82{,}222$ $\qquad S_{XX} = 28{,}613.0909$

$\sum Y = 70.6$

$\sum Y^2 = 229.66$ $\qquad S_{YY} = 3.0982$

$\sum XY = 3236.8$

$s_{Y \cdot X} = 0.2172$

FIGURE L13 Class Size Data

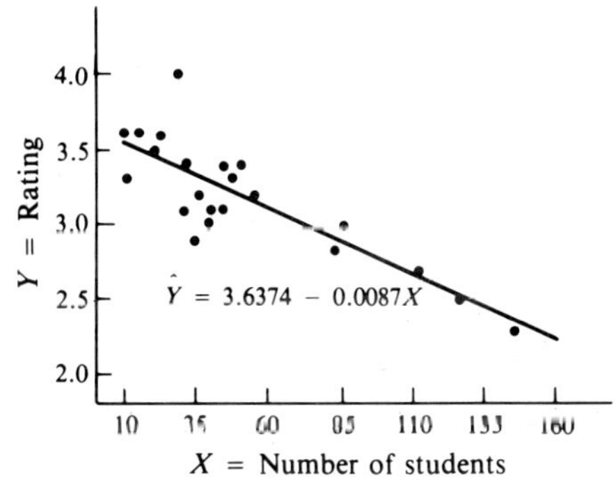

FIGURE L14 Residuals for Class Size Data

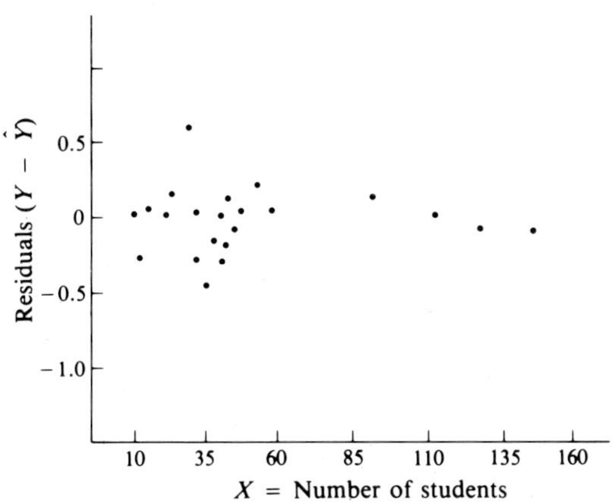

EXERCISE 2.20
FERTILIZER DATA

$n = 9$
$\sum Z = 0$
$\sum Z^2 = 6$
$\sum Y = 542$
$\sum Y^2 = 34,054$
$\sum ZY = 90$

$$\mathbf{X'X} = \begin{bmatrix} 9 & 180 \\ 180 & 3750 \end{bmatrix}$$

$$\mathbf{X'y} = \begin{bmatrix} 542 \\ 11,290 \end{bmatrix}$$

$$(\mathbf{X'X})^{-1} = \begin{bmatrix} 2.777778 & -0.133333 \\ -0.133333 & 0.0066667 \end{bmatrix}$$

$\hat{Y} = 0.2222 + 3X$

FIGURE L15

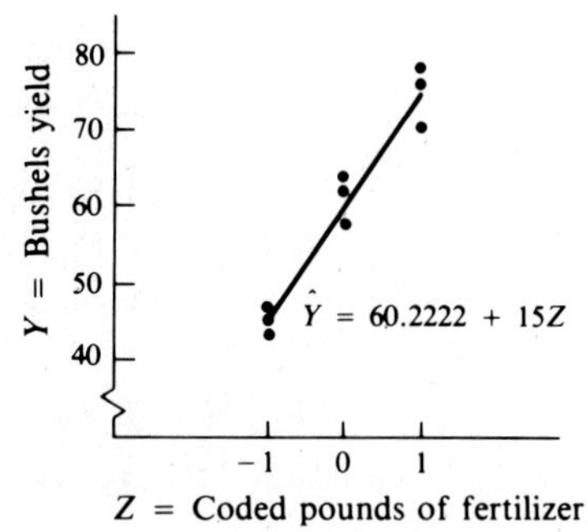

EXERCISE 2.21
DEPRIVATION DATA

$n = 10$
$\sum Z = 0$
$\sum Z^2 = 20$
$\sum Z^3 = 0$
$\sum Z^4 = 68$
$\sum Y = 508$
$\sum Y^2 = 28,108$
$\sum ZY = -53$
$\sum Z^2Y = 1249$

$S_{ZY} = 53$
$S_{ZZ} = 20$
$S_{YY} = 2301.6$

$$\mathbf{Z'Z} = \begin{bmatrix} 10 & 0 & 20 \\ 0 & 20 & 0 \\ 20 & 0 & 68 \end{bmatrix} \qquad \mathbf{Z'y} = \begin{bmatrix} 508 \\ -53 \\ 1249 \end{bmatrix}$$

$$(\mathbf{Z'Z})^{-1} = \begin{bmatrix} 0.242857 & 0 & -0.071429 \\ 0 & 0.05 & 0 \\ -0.071429 & 0 & 0.035714 \end{bmatrix}$$

FIGURE L16 Deprivation Data

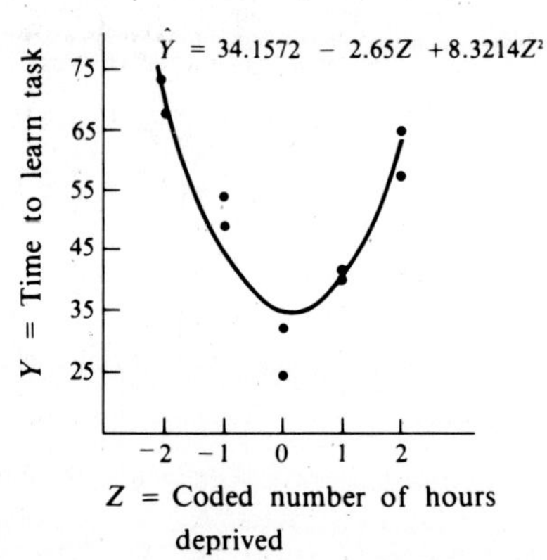

Simple linear regression: $\hat{Y} = 50.8 - 2.65Z$; $s^2_{Y \cdot Z} = 270.1438$
Second-degree polynomial regression: $\hat{Y} = 34.1572 - 2.65X + 8.3214X^2$

FIGURE L17 Residuals from Simple Linear Regression of Deprivation Data

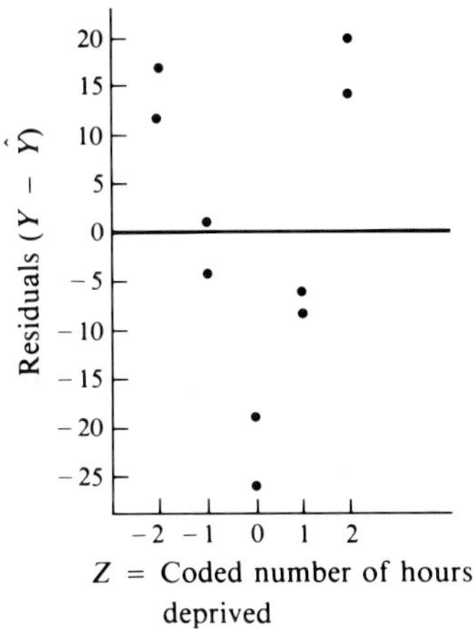

EXERCISE 3.31
MAIL DATA

$n = 15$

$\sum X = 4490$	$S_{XY} = 2933.2$
$\sum X^2 = 1{,}444{,}060$	$S_{XX} = 100{,}053.3333$
$\sum Y = 195.9$	$S_{YY} = 86.776$
$\sum Y^2 = 2645.23$	
$\sum XY = 61{,}572.6$	

$$\mathbf{X'X} = \begin{bmatrix} 15 & 4490 \\ 4490 & 1{,}444{,}060 \end{bmatrix} \qquad \mathbf{X'y} = \begin{bmatrix} 195.9 \\ 61{,}572.6 \end{bmatrix}$$

$$(\mathbf{X'X})^{-1} = \begin{bmatrix} 0.9621935 & -0.00299174 \\ -0.00299174 & 0.00000999467 \end{bmatrix}$$

$\hat{Y} = 4.2846 + 0.0293X$
$s_{Y \cdot X} = 0.2458$
$SS(R) = 85.9908$
$SS(E) = 0.7852$
$SS(T) = 86.7760$

EXERCISE 4.4
APPRAISAL DATA

FIGURE L18 Appraisal Data

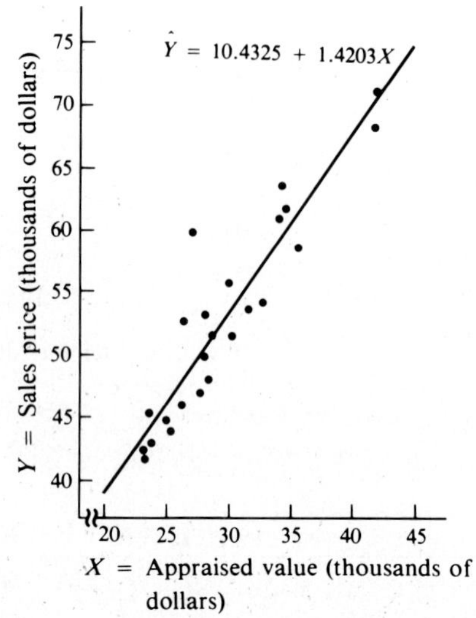

$n = 24$
$\sum X = 716{,}460$
$\sum X^2 = 22{,}043{,}126{,}600$
$\sum Y = 1268$
$\sum Y^2 = 68{,}567.22$
$\sum XY = 38{,}783{,}301$
$s_{Y \cdot X} = 3.3923$

$S_{XY} = 930.331$
$S_{XX} = 655{,}004{,}450$
$S_{YY} = 1574.5533$

FIGURE L19 Residuals for Appraisal Data

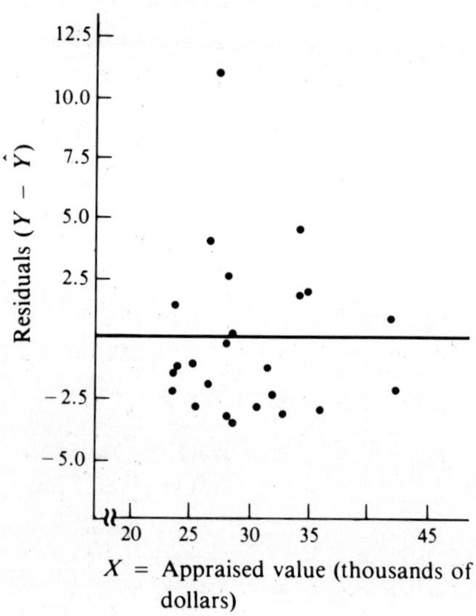

Residuals:

1.	11.119	9.	−0.316	17.	−2.043
2.	4.087	10.	0.927	18.	−2.114
3.	2.571	11.	−1.070	19.	−2.652
4.	1.334	12.	−1.323	20.	−2.852
5.	4.635	13.	−1.367	21.	−3.219
6.	1.835	14.	1.238	22.	−2.694
7.	1.926	15.	−1.745	23.	−2.969
8.	0.118	16.	−2.169	24.	−3.259

EXERCISE 5.12
LOANS DATA

FIGURE L20 Loans Data

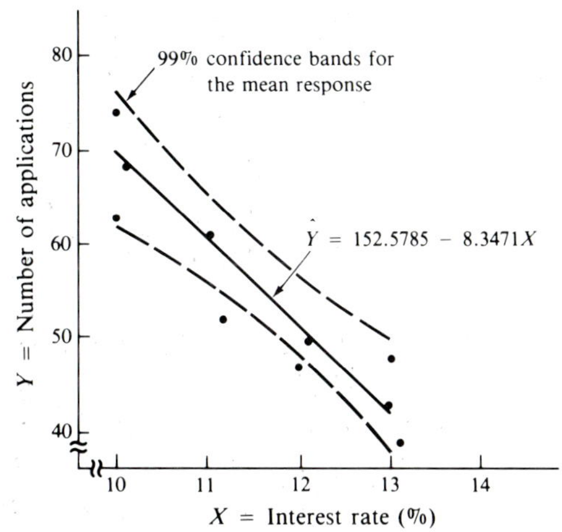

$n = 10$
$\sum X = 117.5$
$\sum X^2 = 1395.75$ $\quad S_{XY} = -126.25$
$\sum Y = 545$ $\quad S_{XX} = 15.125$
$\sum Y^2 = 30{,}877$ $\quad S_{YY} = 1174.5$
$\sum XY = 6277.5$
$\hat{Y} = 152.5785 - 8.3471X$
$s_{Y \cdot X} = 3.8839$

EXERCISE 5.15
HYPERACTIVITY DATA

$n = 11$
$\sum X = 560.7$ $\quad S_{XY} = 293.28$
$\sum X^2 = 33{,}632.53$ $\quad S_{XX} = 5052.1218$
$\sum Y = 40.7$
$\sum Y^2 = 175.87$ $\quad S_{YY} = 25.28$
$\sum XY = 2367.87$
$\hat{Y} = 0.7410 + 0.0581X$
$s_{Y \cdot X} = 0.9577$

EXERCISE 5.16
RESERVATIONS DATA

$n = 24$
$\sum X = 4162$
$\sum X^2 = 762{,}470$ $\quad S_{XY} = 43{,}611.4167$
$\sum Y = 4631$ $\quad S_{XX} = 40{,}709.8333$
$\sum Y^2 = 940{,}587$ $\quad S_{YY} = 46{,}996.9583$
$\sum XY = 846{,}704$
$\hat{Y} = 7.1814 + 1.0713X$
$s_{Y \cdot X} = 3.5493$

FIGURE L21 Reservations Data

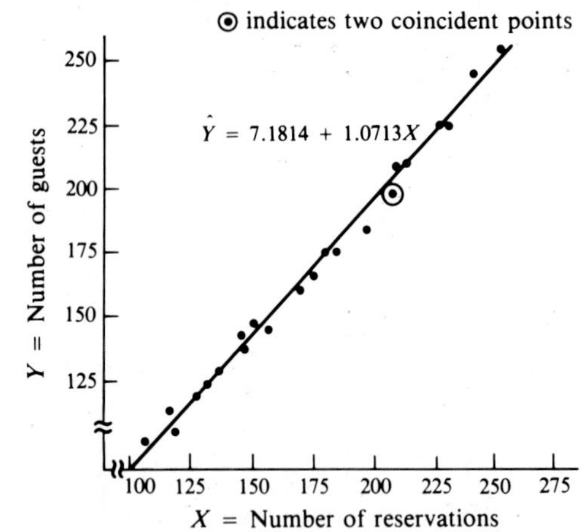

FIGURE L22 Residuals for Reservations Data

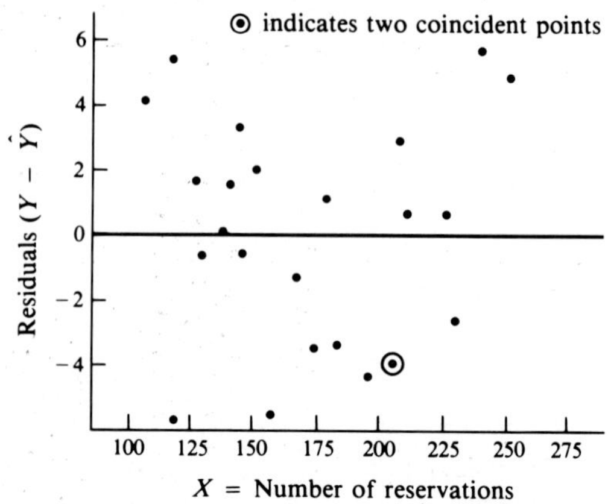

Residuals:

1.	−0.590	9	−3.225	17.	−3.583
2.	0.708	10.	1.678	18.	−5.734
3.	4.192	11.	−5.372	19.	5.479
4.	5.713	12.	−0.588	20.	2.922
5.	−4.151	13.	1.767	21.	1.985
6.	1.060	14.	0.710	22.	−3.864
7.	4.929	15.	−2.503	23.	3.412
8.	−1.156	16.	0.054	24.	−3.864

EXERCISE 5.17
STOCK DATA

$n = 24$
$\sum X = 163$
$\sum X^2 = 329.93$ $\quad S_{XY} = 453.7388$
$\sum Y = -1.5$
$\sum Y^2 = 1086.17$ $\quad S_{XX} = 318.8596$
$\sum XY = 452.72$ $\quad S_{YY} = 1086.0762$
$\hat{Y} = -1.0290 + 1.4230X$
$s_{Y \cdot X} = 4.4742$
$\sum e_t^2 = 440.40$
$\sum e_{t-1}^2 = 438.37$
$\sum e_t e_{t-1} = -16.495$
$\sum (e_t - e_{t-1})^2 = 910.75$

Residuals:

1.	−1.009	9.	−0.333	17.	8.252
2.	2.691	10.	−0.255	18.	−5.897
3.	4.406	11.	−4.102	19.	2.587
4.	5.041	12.	0.310	20.	−2.444
5.	7.441	13.	−2.237	21.	−1.606
6.	1.498	14.	5.879	22.	−1.990
7.	−9.248	15.	−5.294	23.	2.510
8.	−3.952	16.	−3.671	24.	1.425

EXERCISE 5.18
GAUGE DATA

$n = 20$	
$\sum X = 4436$	
$\sum X^2 = 1{,}144{,}090$	$S_{XY} = 160{,}249.6$
$\sum Y = 4488$	$S_{XX} = 160{,}185.2$
$\sum Y^2 = 1{,}167{,}494$	$S_{YY} = 160{,}386.8$
$\sum XY = 1{,}155{,}688$	
$\hat{Y} = 2.5108 + 1.0040X$	
$s_{Y \cdot X} = 2.0107$	

EXERCISE 6.7
MARKETING DATA

FIGURE L23 Marketing Data

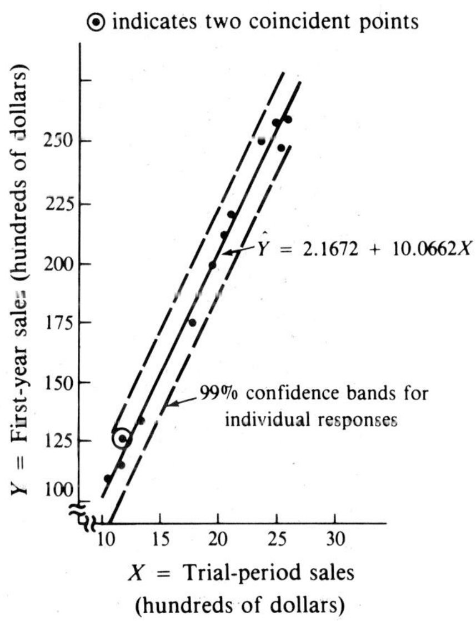

$n = 13$	
$\sum X = 243.7$	$S_{XY} = 4571.9785$
$\sum X^2 = 5022.63$	$S_{XX} = 454.1923$
$\sum Y = 2181.3$	
$\sum Y^2 = 519{,}927.2$	$S_{YY} = 46{,}323.3777$
$\sum XY = 51{,}086.81$	
$\hat{Y} = 2.1672 + 10.0662X$	
$s_{Y \cdot X} = 5.2314$	

EXERCISE 7.10
DIAMETERS DATA

$n = 15$	
$\sum X = 449.3$	
$\sum X^2 = 13{,}931.17$	$S_{XY} = 419.876$
$\sum Y = 449.7$	$S_{XX} = 473.1373$
$\sum Y^2 = 13{,}860.39$	$S_{YY} = 378.384$
$\sum XY = 13{,}889.89$	
$\hat{Y} = 3.3985 + 0.8874X$	
$s_{Y \cdot X} = 0.6664$	

EXERCISE 9.1
SECRETARIES DATA

$n = 20$
$\sum X = 68.7$
$\sum X^2 = 252.01$
$\sum Y = 15{,}040$
$\sum Y^2 = 11{,}345.400$
$\sum XY = 52{,}251$
$\hat{Y} = 625.8359 + 36.7290X$
$s_{Y \cdot X} = 27.589.6$

$S_{XY} = 588.6$
$S_{XX} = 16.0255$
$S_{YY} = 35{,}320$

Residuals:

1.	20.9210	11.	−17.4435
2.	−18.0603	12.	−22.1351
3.	10.3042	13.	−10.3126
4.	14.1920	14.	12.3416
5.	−19.6958	15.	91.9397
6.	−13.9855	16.	−27.6584
7.	−15.8080	17.	−16.0229
8.	11.9397	18.	9.6874
9.	12.5565	19.	−13.3687
10.	13.3603	20.	−22.7519

FIGURE L24 Residuals for Secretaries Data

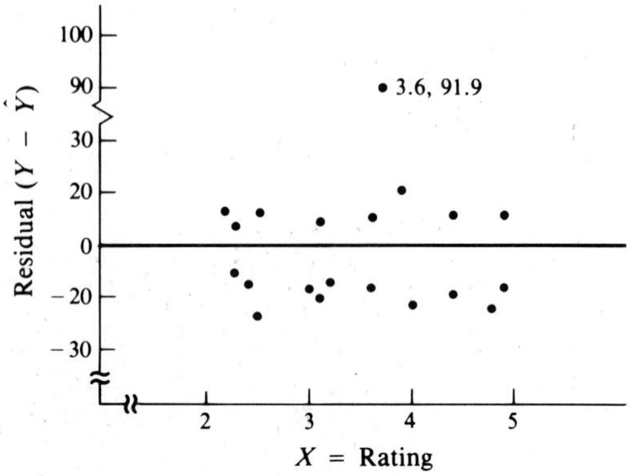

EXERCISE 9.6
TREES DATA

$n = 25$
$\sum X = 97.1$
$\sum X^2 = 411.81$
$\sum Y = 2782$
$\sum Y^2 = 314{,}412$
$\sum XY = 11{,}188.4$
$\hat{Y} = 68.3653 + 11.0491X$
Simple linear regression:
$\sum X^3 = 1888.5350$
$\sum X^4 = 9226.5789$
$\sum X^2Y = 49{,}120.48$
Quadratic regression: $\hat{Y} = 116.4227 - 14.1366X + 3.0210X^2$

$S_{XY} = 383.112$
$S_{XX} = 34.6736$
$S_{YY} = 4831.04$

FIGURE L25 Residuals for Trees Data

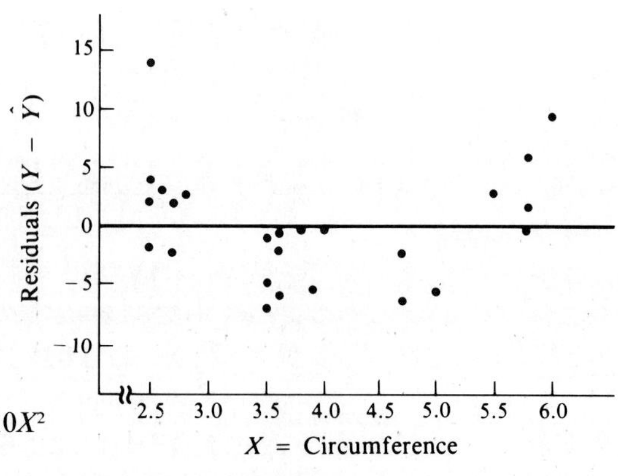

EXERCISE 9.7

TYPIST DATA

$n = 23$
$\sum X = 4779$
$\sum X^2 = 997{,}661$
$\sum Y = 166$
$\sum Y^2 = 1368$
$\sum XY = 35{,}273$
$\hat{Y} = -27.5486 + 0.1673X$

$\sum e_t^2 = 39.2133$
$\sum e_{t-1}^2 = 36.6594$
$\sum e_t e_{t-1} = 28.3715$
$\sum (e_t - e_{t-1}) = 18.7746$

For transformed data,
$n = 22$
$\sum X' = -43$
$\sum (X')^2 = 7939$
$\sum Y' = -5$
$\sum (Y')^2 = 263$
$\sum X'Y' = 1394$
$\hat{Y}' = 0.1120 + 0.1762X'$

$\sum e_t^2 = 17.9308$
$\sum e_{t-1}^2 = 14.8177$
$\sum e_t e_{t-1} = -3.5061$
$\sum (e_t - e_{t-1})^2 = 39.4744$

FIGURE L26 Typist Data

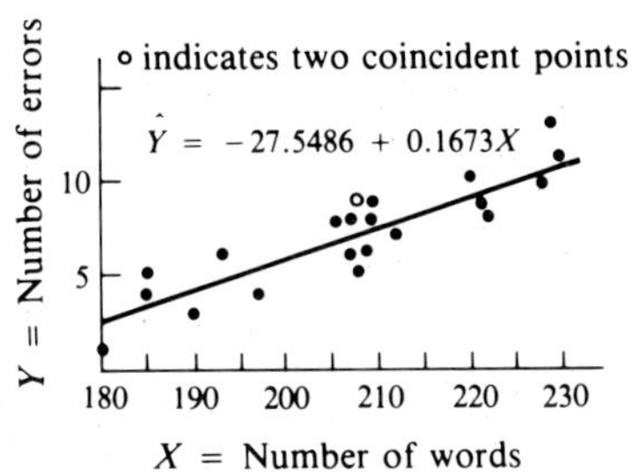

FIGURE L27 Residuals in Time Order for Typist Data

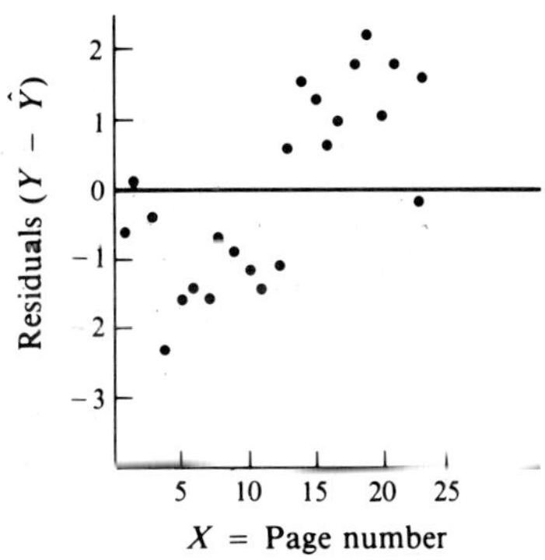

EXERCISE 9.8

IQ DATA

$n = 26$
$\sum Y = 2905$
$\sum Y^2 = 326{,}307$
$\sum X = 2280$
$\sum X^2 = 200{,}680$
$\sum XY = 255{,}790$
$\hat{Y} = 20.2404 + 0.6037X$

For $\hat{Y} = -38.2749 + 1.6885X_1 + 72.2482X_2 - 0.8008X_1X_2$,

$\sum X_1 = 2280$
$\sum X_2 = 13$
$\sum X_1^2 = 200{,}680$
$\sum X_2^2 = 13$
$\sum X_1Y = 255{,}790$
$\sum X_2Y = 1427$
$\sum X_1X_2Y = 121{,}946$

$\sum X_1X_2 = 1110$
$\sum X_1^2X_2 = 94{,}892$
$\sum X_1X_2^2 = 1110$
$\sum X_1^2X_2^2 = 94{,}892$

FIGURE L28 IQ Data

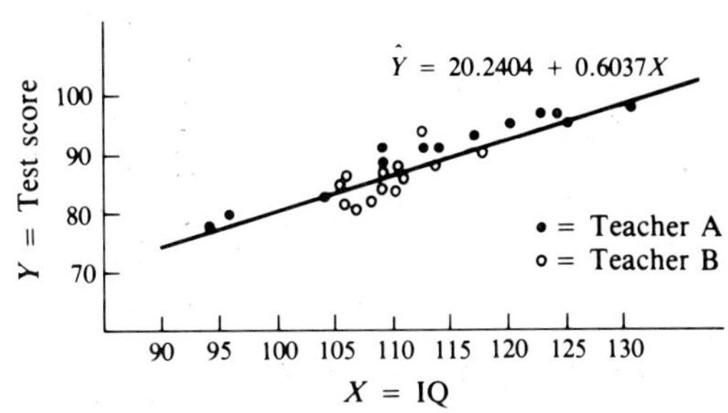

EXERCISE 9.13
ENROLLMENT DATA

$n = 36$
$\sum X = 666$
$\sum X^2 = 16{,}206$
$\sum Y = 768{,}911$
$\sum Y^2 = 17{,}662{,}650{,}431$
$\sum XY = 14{,}813{,}684$
$\hat{Y} = 18{,}555 + 152X$
$\sum e_t^2 = 1{,}150{,}511{,}900$
$\sum e_{t-1}^2 = 1{,}132{,}720{,}600$
$\sum e_t e_{t-1} = -361{,}860{,}810$
$\sum (e_t - e_{t-1})^2 = 3{,}006{,}915{,}900$

Lagged enrollments:
$n = 32$
$\sum X = 656$
$\sum X^2 = 16{,}176$
$\sum Y = 677{,}455$
$\sum Y^2 = 15{,}421{,}694{,}000$
$\sum XY = 14{,}370{,}516$
$\hat{Y} = 1716 - 53.6X$
$\sum e_t^2 = 36{,}507{,}229$
$\sum e_{t-1}^2 = 35{,}985{,}338$
$\sum e_t e_{t-1} = 19{,}283{,}683$
$\sum (e_t - e_{t-1})^2 = 27{,}567{,}467$

For $\hat{Y} = 22{,}327.95 - 2024.1X_1 - 3249.36X_2 - 14{,}241.3X_3 + 157.2635X_4$,

$\sum X_1 = 9$	$\sum X_1^2 = 9$	$\sum X_1X_2 = 0$	$\sum X_1Y = 215{,}796$
$\sum X_2 = 9$	$\sum X_2^2 = 9$	$\sum X_1X_3 = 0$	$\sum X_2Y = 206{,}184$
$\sum X_3 = 9$	$\sum X_3^2 = 9$	$\sum X_1X_4 = 153$	$\sum X_3Y = 108{,}672$
$\sum X_4 = 666$	$\sum X_4^2 = 16{,}206$	$\sum X_2X_3 = 0$	$\sum X_4Y = 14{,}813{,}684$
		$\sum X_2X_4 = 162$	
		$\sum X_3X_4 = 17.1$	

EXERCISE 10.1
CANCER DATA

FIGURE L29 Cancer Data

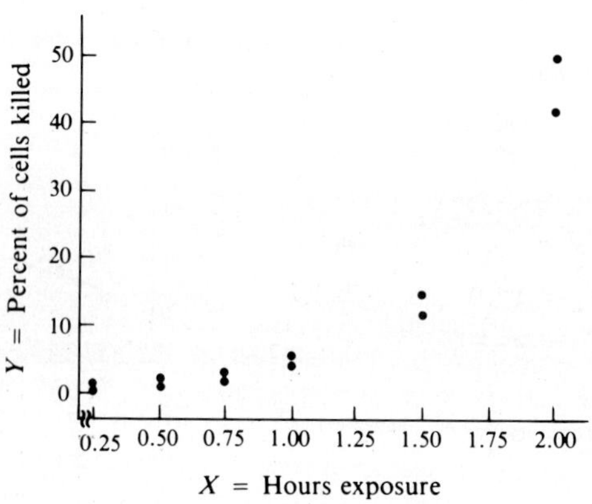

$n = 12$
$\sum X = 12$
$\sum X^2 = 16.25$
$\sum (\ln Y) = 19.42807$ $\sum (\log Y) = 8.437502$
$\sum (\ln Y)^2 = 51.40653$ $\sum (\log Y)^2 = 9.695874$
$\sum X(\ln Y) = 28.6043$ $\sum X(\log Y) = 12.42269$

$\ln \hat{Y} = -0.540109 + 2.159115X$ $\log \hat{Y} = -0.234566 + 0.9376916X$

$$\hat{Y} = (0.5827)(8.6635)^X$$

EXERCISE 10.2
ABRASIVE DATA

FIGURE L30 Abrasive Data

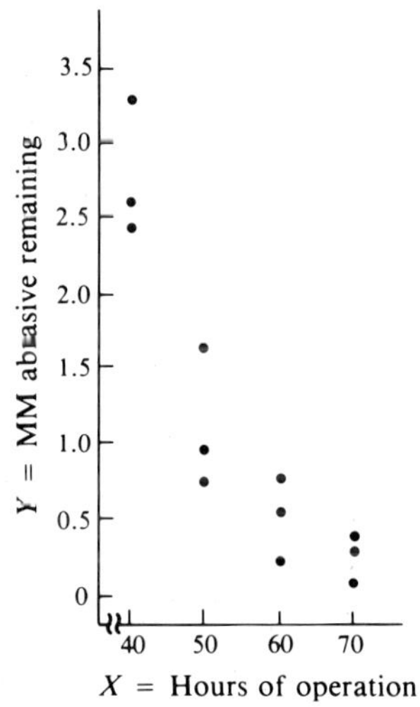

$n = 12$
$\sum X = 660$
$\sum X^2 = 37{,}800$

$\sum (\ln Y) = -3.37529$
$\sum (\ln Y)^2 = 13.46802$
$\sum X(\ln Y) = -309.978$
$\ln \hat{Y} = 4.277731 - 0.82891X$

$\sum (\log Y) = -1.46587$
$\sum (\log Y)^2 = 2.540227$
$\sum X(\log Y) = -134.622$
$\log \hat{Y} = 1.857795 - 0.0359991X$

$$\hat{Y} = 72.0767(0.9205)^X$$

EXERCISE 10.3
CENSUS DATA: I

FIGURE L31 Census Data: I

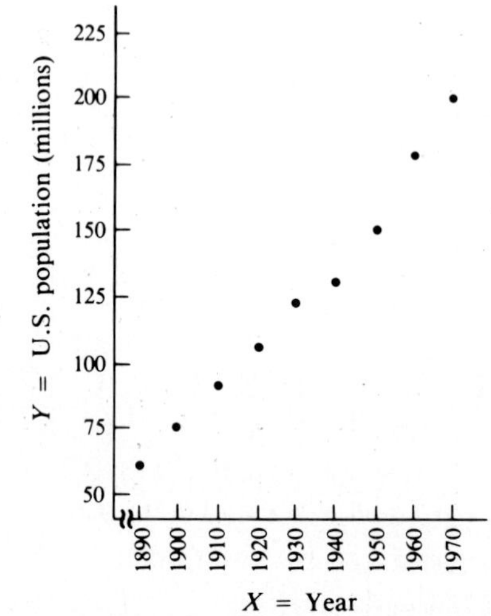

$n = 9$ $\sum(\ln Y) = 42.84983$ $\sum(\log Y) = 18.60944$

$\sum X = 36$ $\sum(\ln Y)^2 = 205.2021$ $\sum(\log Y)^2 = 38.70352$

$\sum X^2 = 204$ $\sum X(\ln Y) = 179.8112$ $\sum X(\log Y) = 78.091$

$\ln \hat{Y} = 4.200301 + 0.1401978X$ $\log \hat{Y} = 1.824167 + 0.06088715X$

$$\hat{Y} = 66.7064(1.1505)^X$$

EXERCISE 10.4
CENSUS DATA: II

$n = 13$ $\sum(\ln Y) = 57.04122$ $\sum(\log Y) = 24.77269$

$\sum X = 0$ $\sum(\ln Y)^2 = 255.8779$ $\sum(\log Y)^2 = 48.26156$

$\sum X^2 = 182$ $\sum X(\ln Y) = 31.52566$ $\sum X(\log Y) = 13.69142$

$\ln \hat{Y} = 4.387786 + 0.173218X$ $\log \hat{Y} = 1.905591 + 0.07522759X$

$$\hat{Y} = 80.4621(1.1891)^X$$

$\sum X^3 = 0$
$\sum X^4 = 4550$
$\sum XY = 2639.19$
$\sum X^2Y = 18{,}922.25$

Quadratic regression: $\hat{Y} = 88.9965 + 14.5010X + 0.5989X^2$

FIGURE L32 Census Data: II

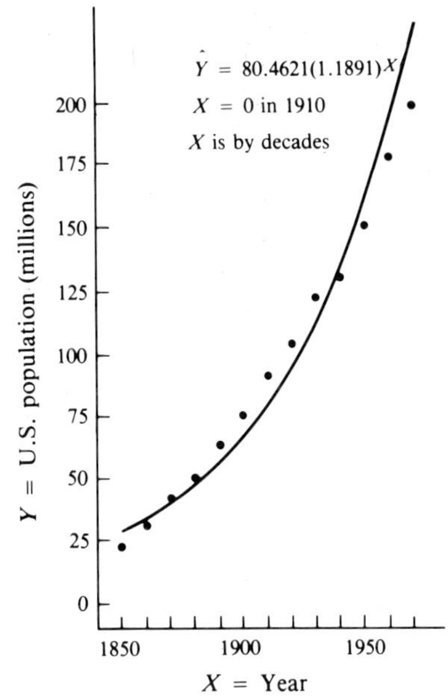

EXERCISE 10.6
MPG DATA

FIGURE L33 MPG Data: MPG Versus Horsepower

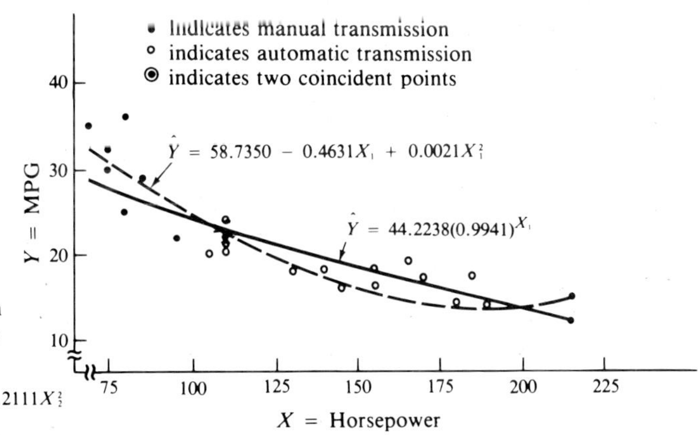

FIGURE L34 MPG Data: MPG Versus Weight

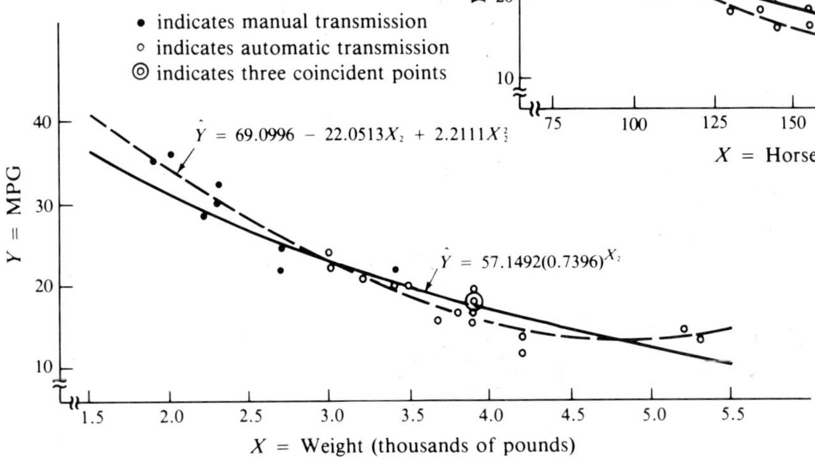

$n = 25$

$\sum Y = 532$	$\sum X_1 = 3260$	$\sum X_2 = 85.4$
$\sum Y^2 = 12{,}364$	$\sum X_1^2 = 473{,}600$	$\sum X_2^2 = 311.02$
$\sum (\ln Y) = 75.37737$	$\sum X_1 Y = 63{,}230$	$\sum X_2 Y = 1688.4$
$\sum (\ln Y)^2 = 229.3417$	$\sum X_1(\ln Y) = 9541.295$	$\sum X_2(\ln Y) = 251.6684$
$\sum (\log Y) = 32.73598$	$\sum X_1(\log Y) = 4143.732$	$\sum X_2(\log Y) = 109.2982$
$\sum (\log Y)^2 = 43.25653$		

$$\left.\begin{aligned}\ln \hat{Y} &= 3.789262 - 0.00593686X_1\\ \log \hat{Y} &= 1.645655 - 0.00257835X_1\end{aligned}\right\} \quad \hat{Y} = 44.2238(0.9941)^{X_1}$$

$$\left.\begin{aligned}\ln \hat{Y} &= 4.045665 - 0.301689X_2\\ \log \hat{Y} &= 1.75701 - 0.131022X_2\end{aligned}\right\} \quad \hat{Y} = 57.1492(0.7396)^{X_2}$$

Quadratic:

$\sum X_1^2 Y = 8{,}465{,}850$	$\sum X_1^3 = 75{,}236{,}750$	$\sum X_2^3 = 1196.5580$
$\sum X_2^2 Y = 5775.16$	$\sum X_1^4 = 12{,}796{,}917{,}500$	$\sum X_2^4 = 4825.4446$

$\hat{Y} = 58.7350 - 0.4631X_1 + 0.0012X_1^2$

$\hat{Y} = 69.0996 - 22.0513X_2 + 2.2111X_2^2$

Multiple quadratic:

$\sum X_1X_2 = 12{,}006.5$

$\sum X_1X_2^2 = 46{,}617.25$

$\sum X_1^2X_2 = 1{,}853{,}96.7$

$\sum X_1^2X_2^2 = 7{,}566{,}227$

$\hat{Y} = 68.3796 - 0.03677X_1 - 0.000004X_1^2 - 20.49278X_2 + 2.2324X_2^2$

Multiple logarithmic:

$\ln \hat{Y} = 3.980636 - 0.0027442X_1 - 0.177897X_2$

or $\log \hat{Y} = 1.728768 - 0.00119179X_1 - 0.077260X_2$

or $\hat{Y} = 53.5511(0.9973)^{X_1}(0.8370)^{X_2}$

FIGURE L35 Multiple Quadratic with Dummy

Full Model: $\text{MPG} = \alpha + \beta_1(\text{HP}) + \beta_2(\text{HP})^2 + \beta_3(\text{WT}) + \beta_4(\text{WT})^2 + \beta_5(\text{TRANS}) + \varepsilon$

Forward Selection/Forward Stepwise

Step 1: WT entered

$p = 0.0001$, $R^2 = 0.8258$

Step 2: $(\text{WT})^2$ entered

$R^2 = 0.9376$

$p(\text{WT}) = 0.0001$, $p(\text{WT})^2 = 0.0001$

Step 3: HP entered

$R^2 = 0.9504$

$p(\text{WT}) = 0.0001$, $p(\text{WT})^2 = 0.0001$, $p(\text{HP}) = 0.0303$

$\widehat{\text{MPG}} = 68.3859 - 0.0380(\text{HP}) - 20.4532(\text{WT}) + 2.2275(\text{WT})$

FIGURE L35 Multiple with Dumnmy (Continued)

Backward Elimination
Step 1: $(HP)^2$ removed
Step 2: TRANS removed gives results as in step 3 of forward selection/stepwise

EXERCISE 10.12
RUSH HOUR DATA

$n = 16$
$\sum X = 120{,}710$
$\sum X^2 = 936{,}945{,}800$
$\sum X^3 = 7.470797 \times 10^{12}$
$\sum X^4 = 6.108289 \times 10^{16}$
$\sum Y = 686$
$\sum Y^2 = 34{,}568$
$\sum XY = 4{,}831{,}747$
$\sum X^2Y = 3.4726629 \times 10^{10}$
$\hat{Y} = -33.83525202 + 0.03407833X - 0.0000030805X^2$

FIGURE L36 Rush Hour Data

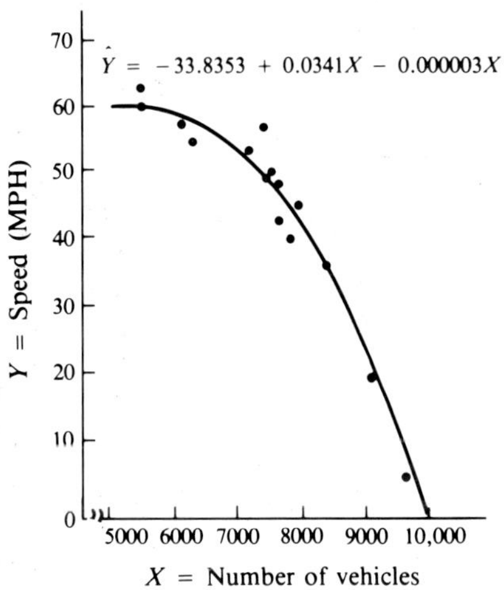

EXERCISE 11.1
SALES DATA: II

$n = 15$ $\quad \sum X_1 = 463$ $\quad \sum X_2 = 127$
$\sum Y = 173$ $\quad \sum X_1^2 = 16{,}275$ $\quad \sum X_2^2 = 1157$
$\sum Y^2 = 2315$ $\quad \sum X_1Y = 6102$ $\quad \sum X_2Y = 1611$
$\sum X_1X_2 = 4260$

$\hat{Y} = -0.3256 + 0.3842X_1$ $\quad s_{Y \cdot X} = 1.4397$
$\hat{Y} = -3.6183 + 1.790X_2$ $\quad s_{Y \cdot X} = 2.1119$
$\hat{Y} = -2.4458 + 0.2697X_1 + 0.6677X_2$ $\quad s_{Y \cdot X} = 1.1730$
$R^2 = 0.9484$
$r_{1Y} = 0.9569$
$r_{2Y} = 0.9048$
$r_{12} = 0.8442$

EXERCISE 11.2
RESTAURANT DATA

$n = 10$ $\quad \sum X_1 = 25.5$ $\quad \sum X_2 = 30.2$ $\quad s_{Y \cdot X} = 0.4485$
$\sum Y = 31.4$ $\quad \sum X_1^2 = 70.97$ $\quad \sum X_2^2 = 98.44$ $\quad R^2 = 0.8352$

$\sum Y^2 = 107.14$ $\sum X_1Y = 86.58$ $\sum X_2Y = 101.37$ $r_{1Y} = 0.9134$
$\sum X_1X_2 = 82.89$ $r_{2Y} = 0.8320$
$\hat{Y} = 0.31148 + 1.0232X_1 + 0.0726X_2$ $r_{12} = 0.8965$

EXERCISE 11.3

DRUG DATA

$n = 20$ $\sum X_1 = 120$ $\sum X_2 = 834$ $s_{Y \cdot X} = 5.8439$
$\sum Y = 397$ $\sum X_1^2 = 890$ $\sum X_2^2 = 41{,}158$ $R^2 = 0.7163$
$\sum Y^2 = 9927$ $\sum X_1Y = 2327$ $\sum X_2Y = 13{,}688$ $r_{1Y} = -0.0932$
$r_{2Y} = -0.7934$
$\sum X_1X_2 = 5472$ $r_{12} = 0.4494$
$\hat{Y} = 35.2210 + 1.1446X_1 - 0.5333X_2$

EXERCISE 11.4

AUTO DATA

$n = 15$ $\sum X_1 = 51$ $\sum X_2 = 280$ $s_{Y \cdot X} = 0.5399$
$\sum Y = 25.3$ $\sum X_1^2 = 209$ $\sum X_2^2 = 6018$ $R^2 = 0.7580$
$\sum Y^2 = 57.13$ $\sum X_1Y = 104.6$ $\sum X_2Y = 502.3$ $r_{1Y} = -0.0179$
$r_{2Y} = 0.8190$
$\sum X_1X_2 = 949$ $r_{12} = 0.2808$
$\hat{Y} = -0.8449 + 0.5253X_1 + 0.0399X_2$

EXERCISE 11.5

ADS DATA

$n = 12$ $\sum X_1 = 115$ $\sum X_2 = 64$ $\sum X_3 = 24$ $s_{Y \cdot X} = 20.7957$
$\sum Y = 647$ $\sum X_1^2 = 1191$ $\sum X_2^2 = 362$ $\sum X_3^2 = 54$ $R^2 = 0.3202$
$\sum Y^2 = 39{,}973$ $\sum X_1Y = 6393$ $\sum X_2Y = 3600$ $\sum X_3Y = 129$ $r_{1Y} = 0.2863$ $r_{12} = -0.0778$
$r_{2Y} = 0.4605$ $r_{13} = -0.3464$
$\sum X_1X_2 = 610$ $\sum X_2X_3 = 129$ $\sum X_1X_3 = 222$ $r_{3Y} = 0.0114$ $r_{23} = 0.0898$
$\hat{Y} = -15.2961 + 2.6202X_1 + 7.5564X_2 + 1.9008X_3$

EXERCISE 11.6

CHILDREN DATA

$n = 20$ $\sum X_1 = 157$ $\sum X_2 = 990$ $\sum X_3 = 1317$ $s_{Y \cdot X} = 16.1822$
$\sum Y = 1288$ $\sum X_1^2 = 1249$ $\sum X_2^2 = 49{,}185.5$ $\sum X_3^2 = 88{,}909$ $R^2 = 0.4738$
$\sum Y^2 = 90{,}910$ $\sum X_1Y = 10{,}329$ $\sum X_2Y = 64{,}537$ $\sum X_3Y = 86{,}350$ $r_{1Y} = 0.6011$ $r_{12} = 0.8233$
$\sum X_1X_2 = 7816.5$ $\sum X_2X_3 = 65{,}612$ $\sum X_1X_3 = 10{,}489$ $r_{2Y} = 0.6514$ $r_{13} = 0.7918$
$r_{3Y} = 0.3681$ $r_{23} = 0.6696$
$\hat{Y} = -136.54 + 9.5298X_1 + 3.3429X_2 - 0.5975X_3$

EXERCISE 11.7

GLUE DATA

$n = 12$ $\sum X_1 = 1020$ $\sum X_2 = 720$ $s_{Y \cdot X} = 1.4011$
$\sum Y = 2350$ $\sum X_1^2 = 87{,}000$ $\sum X_2^2 = 46{,}400$ $R^2 = 0.9444$

$\sum Y^2 = 460{,}526$ $\quad \sum X_1Y = 199{,}810$ $\quad \sum X_2Y = 141{,}960$ $\quad r_{1Y} = 0.1944$

$\sum X_1X_2 = 61{,}200$ $\quad r_{2Y} = 0.9522$

$r_{12} = 0.0000$

$\hat{Y} = 160.8333 + 0.2X_1 + 0.3X_2$

EXERCISE 11.8

BOTTLES DATA

$n = 24$ $\quad \sum X_1 = 0$ $\quad \sum X_2 = 0$ $\quad s_{Y \cdot X} = 0.8825$

$\sum Y = 70$ $\quad \sum X_1^2 = 60$ $\quad \sum X_2^2 = 16$ $\quad R^2 = 0.7723$

$\sum Y^2 = 276$ $\quad \sum X_1Y = -55$ $\quad \sum X_2Y = -9$ $\quad r_{1Y} = -0.8378$

$\sum X_1X_2 = 0$ $\quad r_{2Y} = 0.2655$

$r_{12} = 0.0000$

$\hat{Y} = 2.9167 - 0.9167X_1 - 0.5625X_2$

EXERCISE 11.9

YIELD DATA

$n = 27$ $\quad \sum X_1 = 0$ $\quad \sum X_2 = 0$ $\quad \sum X_3 = 0$

$\sum Y = 8701$ $\quad \sum X_1^2 = 18$ $\quad \sum X_2^2 = 18$ $\quad \sum X_3^2 = 18$

$\sum Y^2 = 3{,}179{,}299$ $\quad \sum X_1Y = 921$ $\quad \sum X_2Y = -853$ $\quad \sum X_3Y = 145$

$\sum X_1X_2 = 0$ $\quad \sum X_2X_3 = 0$ $\quad \sum X_1X_3 = 0$

$\hat{Y} = 322.259 + 51.1667X_1$ $\quad s_{Y \cdot X} = 111.629$

$\hat{Y} = 322.259 - 47.3889X_2$ $\quad R^2 = 0.2364$

$\hat{Y} = 322.259 + 8.0556X_3$ $\quad r_{1Y} = 0.3543$ $\quad r_{12} = 0$

$\hat{Y} = 322.259 + 51.1667X_1 + 47.3889X_2$ $\quad r_{2Y} = -0.3282$ $\quad r_{13} = 0$

$\hat{Y} = 322.259 + 51.1667X_1 + 8.0556X_3$ $\quad r_{3Y} = 0.0558$ $\quad r_{23} = 0$

$\hat{Y} = 322.259 + 47.3889X_2 + 8.0556X_3$

$\hat{Y} = 322.259 + 51.1667X_1 + 47.3889X_2 + 8.0556X_3$

EXERCISE 14.5

SALARY DATA

FIGURE L37 Salary Data

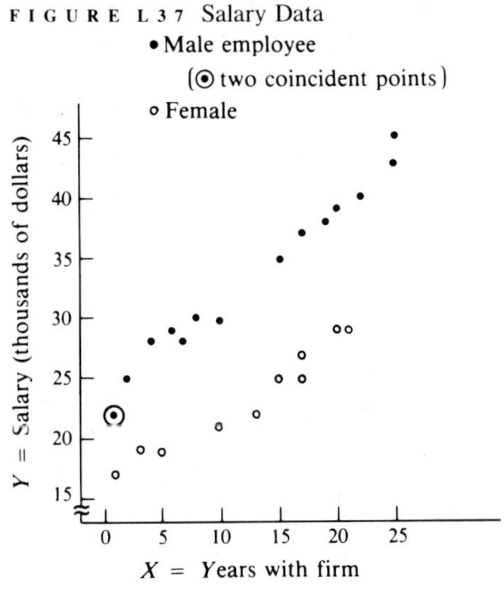

$n = 25$ | $\sum X_1 = 304$ | $\sum X_2 = 15$ | $\sum X_1X_2 = 182$

$\sum Y = 723$ | $\sum X_1^2 = 5228$ | $\sum X_2^2 = 15$ | $\sum X_1^2X_2 = 3280$

$\sum Y^2 = 22.375$ | $\sum X_1Y = 9959$ | $\sum X_2Y = 490$ | $\sum X_1X_2^2 = 182$

$\sum X_1^2X_2^2 = 3280$

$\sum X_1X_2Y = 6846$

$\hat{Y} = 16.1223 + 0.5883X_1 + 6.3477X_2 + 0.2520X_1X_2$

EXERCISE 14.8

ENERGY SAVER DATA

$n = 23$ | $\sum X_1 = 62.2$ | $\sum X_2 = 12$ | $\sum X_1X_2 = 32.7$

$\sum Y = 360.6$ | $\sum X_1^2 = 184.84$ | $\sum X_2^2 = 12$ | $\sum X_1^2X_2 = 97.09$

$\sum Y^2 = 6256.54$ | $\sum X_1Y = 1074.58$ | $\sum X_2Y = 184.1$ | $\sum X_1X_2^2 = 32.7$

$\sum X_1^2X_2^2 = 97.09$

$\sum X_1X_2Y = 550.92$

$\hat{Y} = 0.4201 + 5.8264X_1 - 1.8901X_2 + 0.3430X_1X_2$

EXERCISE 14.9

DIET DATA

FIGURE L 38 Diet Data

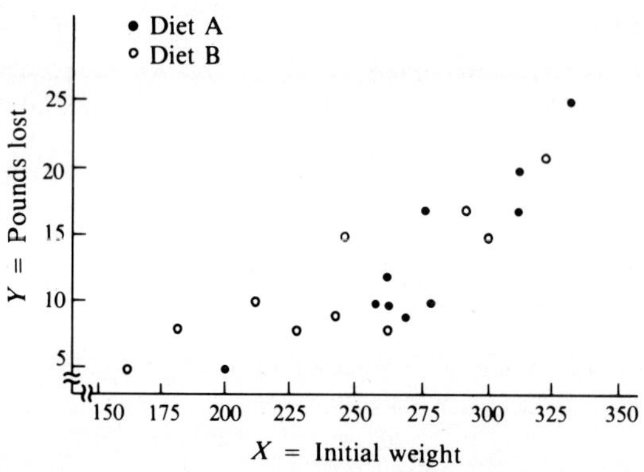

$n = 20$ | $\sum X_1 = 5219$ | $\sum X_2 = 10$ | $\sum X_1X_2 = 2449$

$\sum Y = 251$ | $\sum X_1^2 = 1{,}402{,}921$ | $\sum X_2^2 = 10$ | $\sum X_1^2X_2 = 623{,}457$

$\sum Y^2 = 3731$ | $\sum X_1^3 = 386{,}593{,}223$ | $\sum X_1^3X_2 = 164{,}152{,}927$

$\sum X_1^4 = 108{,}763{,}830{,}793$ | $\sum X_1X_2^2 = 2449$

$\sum X_1Y = 69{,}610$ | $\sum X_1^2X_2^2 = 623{,}457$

$\sum X_1^2Y = 19{,}743{,}896$

$\sum X_2Y = 116$

$\sum X_1X_2Y = 30{,}402$

$\hat{Y} = 28.8141 - 0.2614X_1 + 0.0007X_1^2 + 1.1369X_2$

$\hat{Y} = 14.8384 - 0.1746X_1 + 0.0006X_1^2 + 10.0445X_2 - 0.0332X_1X_2$

EXERCISE 14.11
PACKAGE DATA

$n = 20$
$\sum Y = 87.3$
$\sum Y^2 = 401.47$
$\sum X_1 = 264$
$\sum X_1^2 = 3902$
$\sum X_1 Y = 1226.6$
$\sum X_2 = 1648$
$\sum X_2^2 = 141{,}962$
$\sum X_2 Y = 7529.2$
$\sum X_1 X_2 = 22{,}722$
$\sum X_1^2 X_2 = 347{,}812$
$\sum X_1 X_2^2 = 2{,}037{,}232$
$\sum X_1^2 X_2^2 = 32{,}140{,}614$
$\sum X_1 X_2 Y = 109{,}970.2$

$\hat{Y} = 0.1628 + 0.0580X_1 + 0.0378X_2 + 0.0003X_1X_2$

EXERCISE 14.12
PRESCRIPTION DATA

$n = 45$
$\sum Y = 5540$
$\sum Y^2 = 783{,}550$
$\sum X_1 = 2097$
$\sum X_1^2 = 108{,}507$
$\sum X_1 Y = 279{,}455$
$\sum X_1^2 X_2^2 = 206{,}874{,}389$
$\sum X_1 X_2 Y = 12{,}371{,}240$
$\sum X_2 = 1641$
$\sum X_2^2 = 73{,}317$
$\sum X_2 Y = 234{,}855$
$\sum X_1 X_2 = 80{,}419$
$\sum X_1^2 X_2 = 4{,}309{,}807$
$\sum X_1 X_2^2 = 3{,}746{,}085$

$\hat{Y} = 40.2651 + 0.1852X_1 + 0.5392X_2 + 0.0305X_1X_2$

FIGURE L39 Expenditure Versus Income by Age

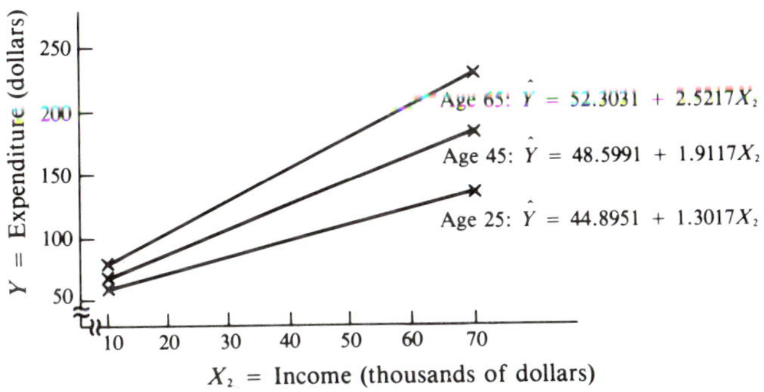

If $X_1 = 45$, $\hat{Y} = 48.5991 + 1.9117X_2$
If $X_1 = 65$, $\hat{Y} = 52.3031 + 2.5217X_2$
If $X_1 = 25$, $\hat{Y} = 44.8951 + 1.3017X_2$

FIGURE L40 Expenditure Versus Age by Income

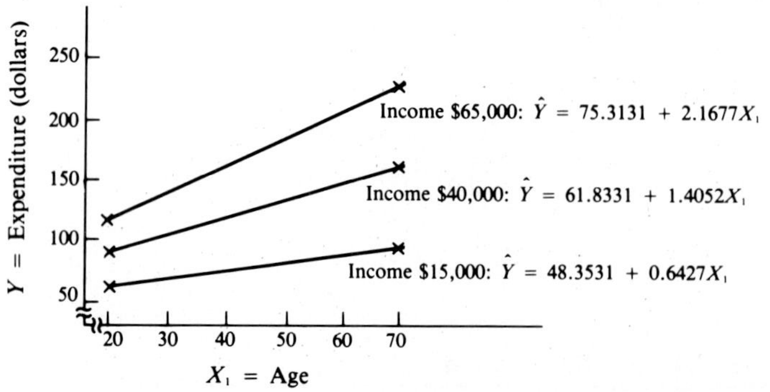

If $X_2 = 15$, $\hat{Y} = 48.3531 + 0.6427X_1$
If $X_2 = 40$, $\hat{Y} = 61.8331 + 1.4052X_1$
If $X_2 = 65$, $\hat{Y} = 75.3131 + 2.1677X_1$

EXERCISE 15.1
CRAVENS DATA

FIGURE L41 SAS Plot of SALES Against TIME

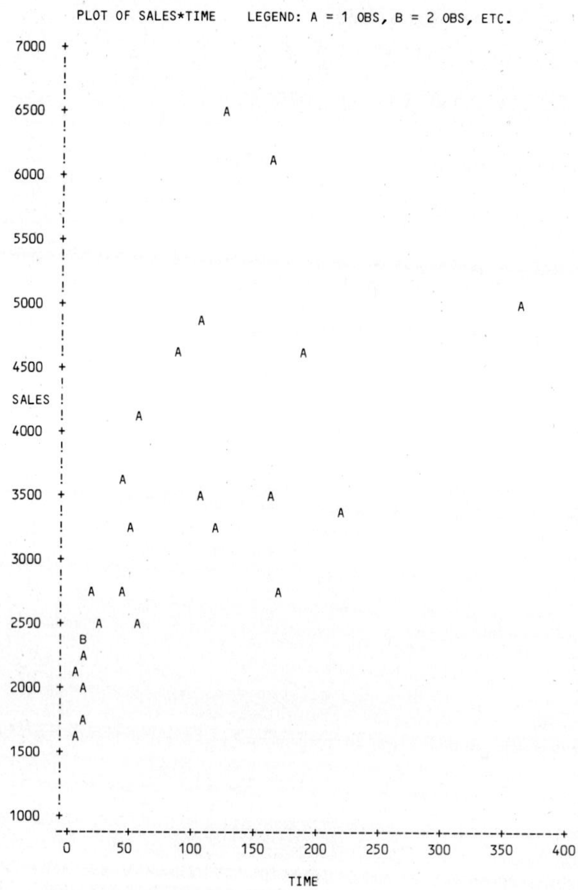

FIGURE L42 SALES Versus POTEN

SAS Plots of SALES Against Each of the Predictors

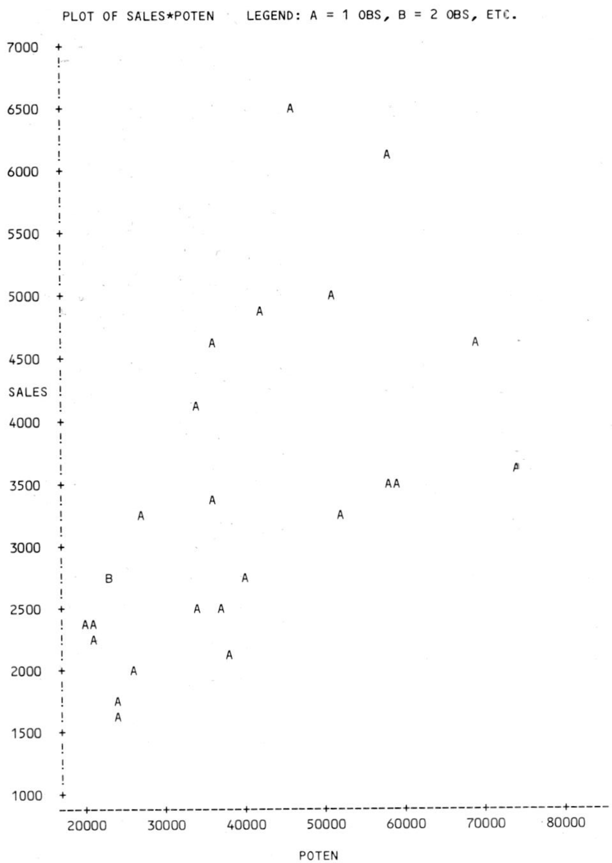

FIGURE L43 SALES Versus ADV

SAS Plots of SALES Against Each of the Predictors

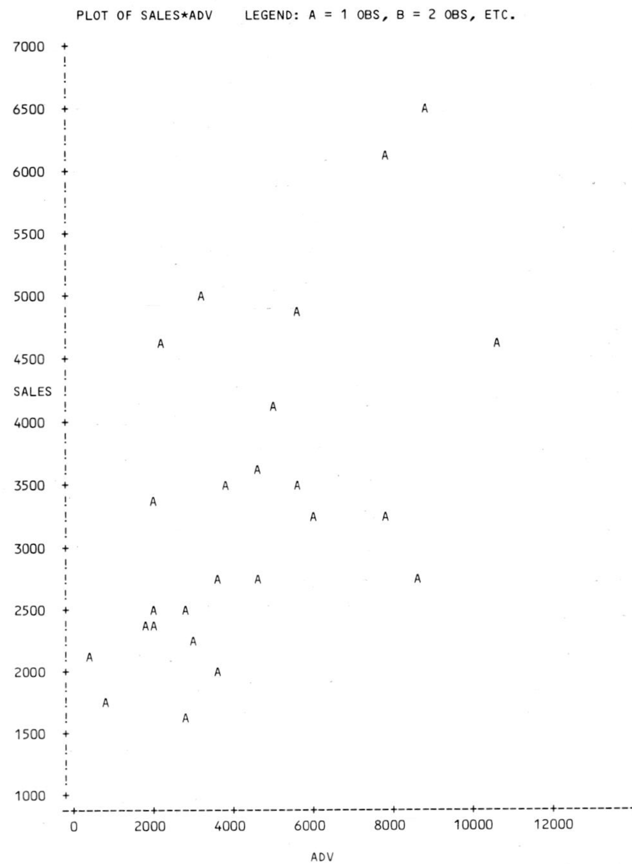

FIGURE L44 SALES Versus SHARE

SAS Plots of SALES Against Each of the Predictors

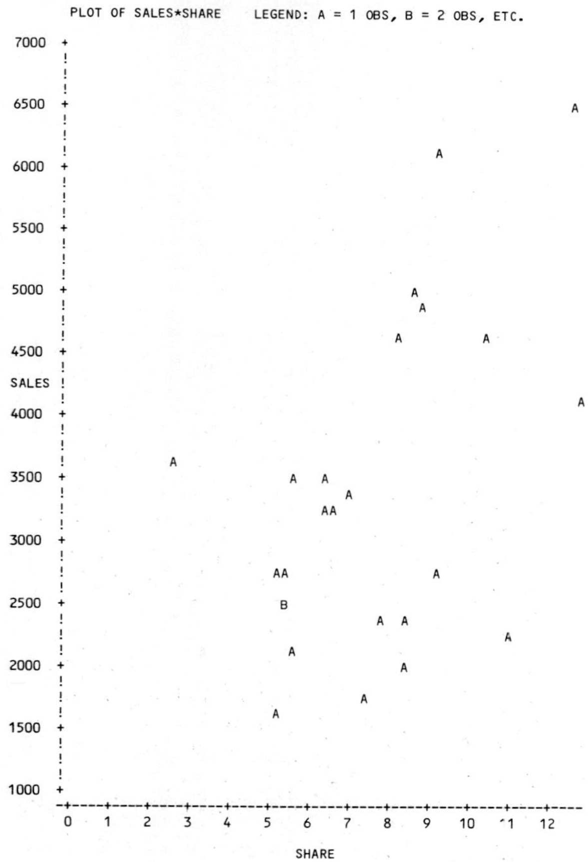

FIGURE L45 SALES Versus CHANGE

SAS Plots of SALES Against Each of the Predictors

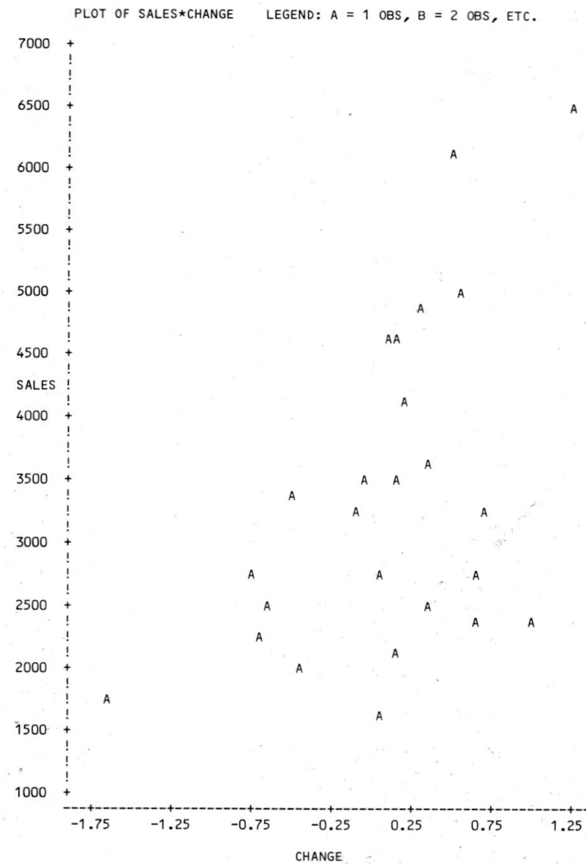

FIGURE L46 SALES Versus ACCTS

SAS Plots of SALES Against Each of the Predictors

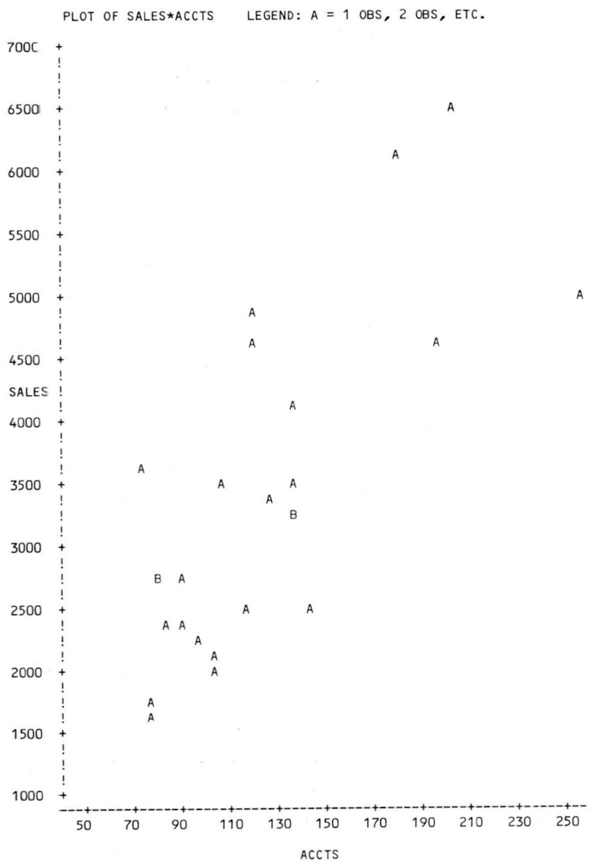

FIGURE L47 SALES Versus WORK

SAS Plots of SALES Against Each of the Predictors

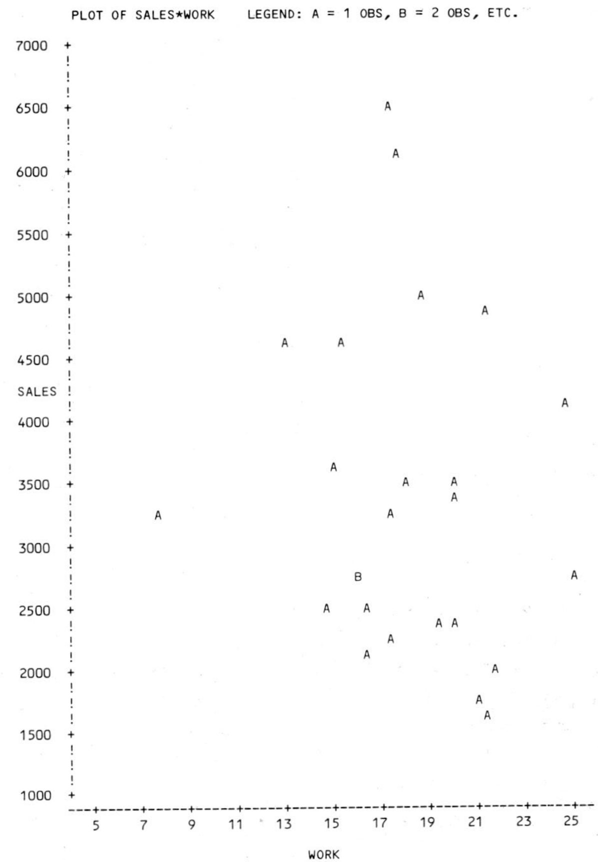

FIGURE L48 SALES Versus RATE

SAS Plots of SALES Against Each of the Predictors

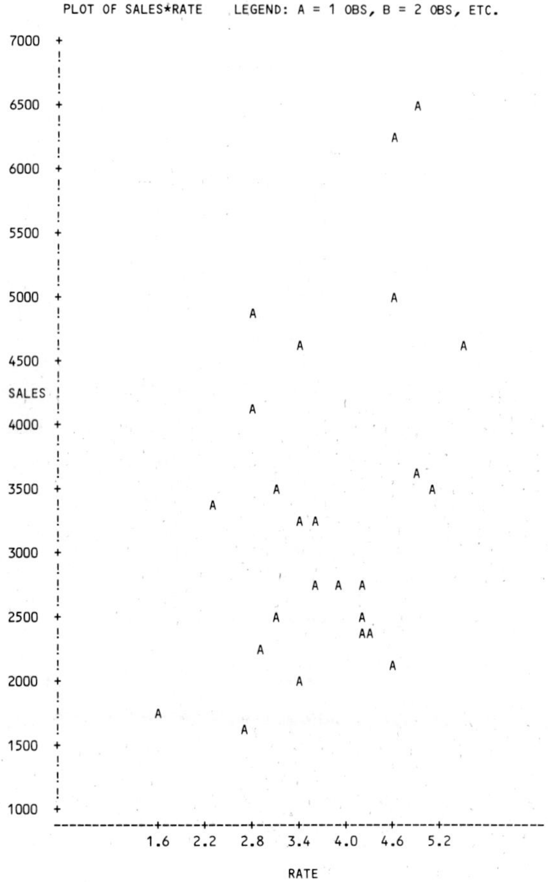

FIGURE L49 SAS Analysis of Cravens Data Including Quadratic Terms

DEP VARIABLE: SALES

SOURCE	DF	SUM OF SQUARES	MEAN SQUARE	F VALUE	PROB > F
MODEL	11	38678025	3516184	16.920	0.0001
ERROR	13	2701524	207810		
C TOTAL	24	41379549			

ROOT MSE	455.861	R-SQUARE	0.9347
DEP MEAN	3374.568	ADJ R-SQ	0.8795
C.V.	13.50873		

VARIABLE	DF	PARAMETER ESTIMATE	STANDARD ERROR	T FOR H0: PARAMETER=0	PROB > |T|
INTERCEP	1	-582.488	1426.101	-0.408	0.6896
TIME	1	12.093113	6.717903	1.800	0.0951
TSQ	1	-0.038148	0.024226	-1.575	0.1394

POTEN	1	0.056129	0.051829	1.083	0.2985
PSQ	1	-3.03145E-07	5.44660E-07	-0.557	0.5873
ADV	1	0.089212	0.063537	1.404	0.1837
SHARE	1	184.911	74.866164	2.470	0.0281
CHANGE	1	171.460	204.448	0.839	0.4168
ACCTS	1	-21.211810	23.235658	-0.913	0.3779
ASQ	1	0.103075	0.081098	1.271	0.2260
WORK	1	25.930707	38.889739	0.667	0.5166
RATE	1	103.541	151.821	0.682	0.5072

OBS	ACTUAL	PREDICT VALUE	RESIDUAL
1	3670	3181	489.135
2	3474	4013	-538.763
3	2295	2447	-152.071
4	4676	5013	-337.287
5	6126	5538	587.787
6	2135	2087	47.458
7	5032	5067	-34.989
8	3367	2975	392.906
9	6519	6512	7.934
10	4876	4072	804.557
11	2468	2620	-152.133
12	2533	2176	357.646
13	2408	2368	39.815
14	2337	2265	72.525
15	4587	4605	-18.196
16	2729	2981	-251.915
17	3289	3213	76.054
18	2801	2953	-152.550
19	3264	3583	-319.117
20	3454	3951	-497.427
21	1741	1569	172.742
22	2036	2438	-402.425
23	1578	1705	-126.812
24	4167	4393	-225.320
25	2800	2640	160.448

SUM OF RESIDUALS 2.38742E-12
SUM OF SQUARED RESIDUALS 2701524

EXERCISE 15.3
TENNESSEE DATA

Some models considered by forward selection, stepwise, maximum R^2 improvement, minimum R^2 improvement:

	B	Std. Error	F	Prob > F
INTERCEPT	58.2588			
PHONE	−0.4710	0.0825	32.60	0.0001

$r^2_{XY} = 0.5380$ $s^2_{Y \cdot X} = 19.7636$
PRESS = 669.66 $C_p = 22.64$

Forward selection, stepwise, maximum R^2 improvement, minimum R^2 improvement:

	B	Std. Error	F	Prob > F
INTERCEPT	52.4945			
PHONE	−0.2698	0.0709	14.47	0.0007
POPCH	−0.3656	0.0732	24.93	0.0001

$r^2_{XY} = 0.6992$ $s^2_{Y \cdot X} = 13.3457$
PRESS = 493.54 $C_p = 7.67$

Maximum R^2 improvement, minimum R^2 improvement:

	B	Std. Error	F	Prob $> F$
INTERCEPT	10.9868			
POPCHG	−0.4037	0.0630	41.07	0.0001
AGRIEMP	0.0010	0.0004	6.05	0.0208
RURAL	0.1921	0.0359	28.56	0.0001
	r^2_{XY} = 0.7358	$s^2_{Y \cdot X}$ = 12.6588		
	PRESS = 514.03	C_p = 6.02		

Backward elimination, maximum R^2 improvement, minimum R^2 improvement:

	B	Std. Error	F	Prob $> F$
INTERCEPT	25.3280			
POPCHG	−0.4215	0.0600	49.32	0.0001
AGRIEMP	0.0009	0.0004	4.42	0.0459
RURAL	0.1998	0.0341	34.34	0.0001
AGE	−0.4777	0.2319	4.24	0.0500
	r^2_{XY} = 0.7725	$s^2_{Y \cdot X}$ = 10.9011		
	PRESS = 482.57	C_p = 3.95		

FIGURE L50 Preliminary Statistics for Tennessee Data

STATISTICAL ANALYSIS SYSTEM

VARIABLE	N	MEAN	STD DEV	SUM	MINIMUM	MAXIMUM
POPCHG	30	7.86666667	10.33225059	236.00000000	-16.20000000	40.00000000
AGRIEMP	30	1548.66666667	2038.38633497	46460.00000000	160.00000000	11500.00000000
POVERTY	30	23.01000000	6.42658004	690.30000000	13.20000000	40.50000000
TAXRT	30	0.71866667	0.20270130	21.56000000	0.40000000	1.09000000
PHONE	30	74.83333333	10.00718133	2245.00000000	48.00000000	87.00000000
RURAL	30	70.72666667	24.02156981	2121.80000000	5.90000000	100.00000000
AGE	30	30.28000000	2.88484745	908.40000000	23.30000000	35.90000000
BLACK	30	11655.33333333	48289.86298017	349660.00000000	2.00000000	266159.00000000

CORRELATION COEFFICIENTS / PROB > |R| UNDER HO:RHO=0 / N = 30

	POPCHG	AGRIEMP	POVERTY	TAXRT	PHONE	RURAL	AGE	BLACK
POPCHG	1.00000	0.04031	-0.64914	0.13095	0.37837	-0.01877	-0.14688	0.12115
	0.0000	0.8325	0.0001	0.4903	0.0392	0.9216	0.4386	0.5237
AGRIEMP	0.04031	1.00000	-0.16769	0.10355	0.35549	-0.65774	-0.36371	0.94176
	0.8325	0.0000	0.3758	0.5861	0.0539	0.0001	0.0482	0.0001
POVERTY	-0.64914	-0.16769	1.00000	0.00898	-0.73347	0.51260	0.02068	-0.19208
	0.0001	0.3758	0.0000	0.9624	0.0001	0.0038	0.9136	0.3092
TAXRT	0.13095	0.10355	0.00898	1.00000	-0.03802	0.02330	-0.04669	0.22795
	0.4903	0.5861	0.9624	0.0000	0.8419	0.9027	0.8065	0.2257
PHONE	0.37837	0.35549	-0.73347	-0.03802	1.00000	-0.74858	-0.07836	0.27032
	0.0392	0.0539	0.0001	0.8419	0.0000	0.0001	0.6807	0.1485
RURAL	-0.01877	-0.65774	0.51260	0.02330	-0.74858	1.00000	0.31442	-0.56116
	0.9216	0.0001	0.0038	0.9027	0.0001	0.0000	0.0906	0.0013
AGE	-0.14688	-0.36371	0.02068	-0.04669	-0.07836	0.31442	1.00000	-0.34873
	0.4386	0.0482	0.9136	0.8065	0.6807	0.0906	0.0000	0.0589
BLACK	0.12115	0.94176	-0.19208	0.22795	0.27032	-0.56116	-0.34873	1.00000
	0.5237	0.0001	0.3092	0.2257	0.1485	0.0013	0.0589	0.0000

Forward selection, forcing in BLACK:

	B	Std. Error	F	Prob $> F$
INTERCEPT	52.6675			
BLACK	0.0000	0.0000	0.02	0.8906
POPCHG	−0.2700	0.0723	13.96	0.0009
PHONE	−0.3683	0.0769	22.91	0.0001

$r^2_{XY} = 0.6994 \quad s^2_{Y \cdot X} = 13.8487$

Stepwise, forcing in BLACK:

	B	Std. Error	F	Prob $> F$
INTERCEPT	13.8034			
BLACK	0.0000	0.0000	3.73	0.0646
POPCHG	−0.4149	0.0660	39.58	0.0001
RURAL	0.1709	0.0340	25.24	0.0001

$r^2_{XY} = 0.7130 \quad s^2_{Y \cdot X} = 13.2205$

	B	Std. Error	F	Prob $> F$
INTERCEPT	28.5599			
BLACK	0.0000	0.0000	2.50	0.1263
POPCHG	−0.4312	0.0624	47.74	0.0001
RURAL	0.1819	0.0324	31.59	0.0001
AGE	−0.5062	0.2391	4.48	0.0444

$r^2_{XY} = 0.7566 \quad s^2_{Y \cdot X} = 11.6598$

Backward elimination, forcing in BLACK:

	B	Std. Error	F	Prob $> F$
INTERCEPT	24.0985			
BLACK	0.0000	0.0000	0.55	0.4651
POPCHG	−0.4100	0.0625	43.00	0.0001
AGRIEMP	0.0016	0.0010	2.26	0.1458
RURAL	0.2059	0.0354	33.86	0.0001
AGE	−0.4794	0.2340	4.20	0.0516

$r^2_{XY} = 0.7776 \quad s^2_{Y \cdot X} = 11.1005$

FIGURE L51 Residuals for Tennessee Data Using Four Predictors

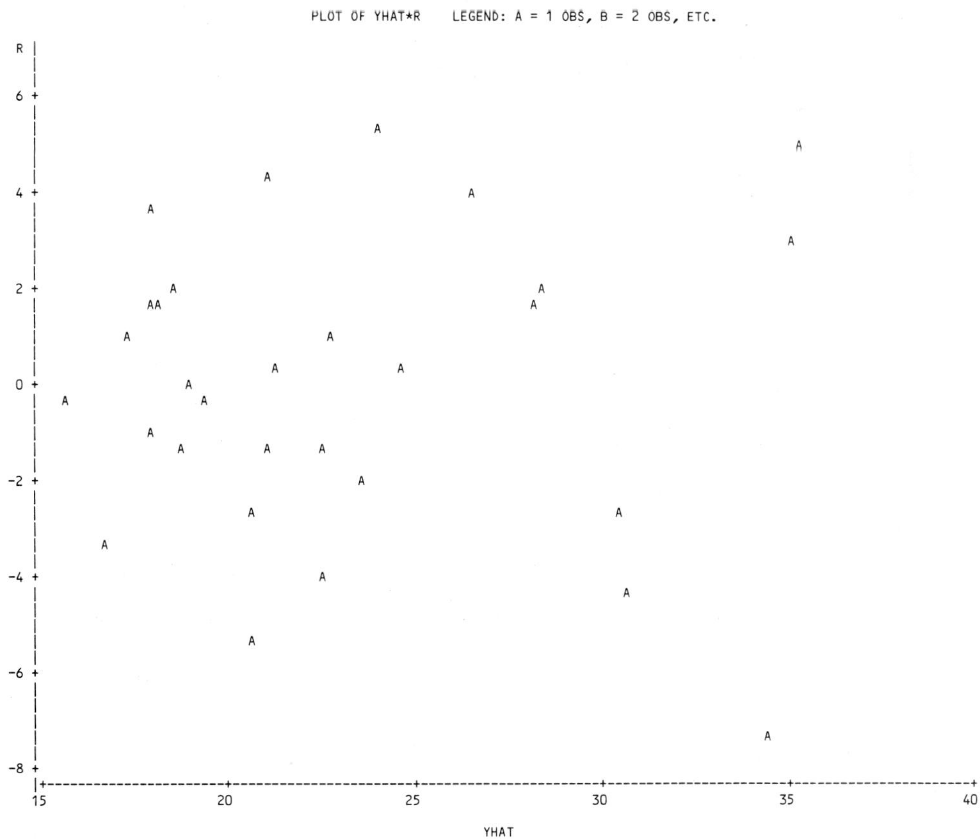

ANSWERS TO SELECTED EXERCISES

ONE

SECTION 1.2, PAGE 7

1.2

	Y	X
a.	Insurance	Income
b.	Sales	Commercials
c.	Time	Number open
d.	Customers	Reservations
e.	Femininity	Size of family
f.	Sessions	Intensity
g.	Mutation	Radiation
h.	House	Dosage
i.	Arbitrary	
j.	Height	Number of trees

1.3 Random: **a, c, d, e, i, j.**
Regulated: **b, f, g, h.**

1.4 **a.** Ratio. **b.** Nominal. **c.** Nominal. **d.** Ratio. **e.** Ordinal. **f.** Ordinal. **g.** Interval. **h.** Ratio. **i.** Ratio. **j.** Nominal.

SECTION 1.3, PAGE 15

1.5 **a.** See Appendix L, Concession Data (p. 632). **b.** Direct. **c.** Linear. **d.** Strong. **e.** Uniform.

1.6 **a.** See Appendix L, Sales Data (p. 633). **b.** Direct. **c.** Linear. **d.** Strong. **e.** Uniform.

1.7 **a.** See Appendix L, Police Data (p. 633). **b.** Inverse. **c.** Curvilinear. **d.** Strong. **e.** Uniform.

1.8 **a.** See Appendix L, Heating Data (p. 634). **b.** Inverse. **c.** Linear. **d.** Strong. **e.** Uniform.

1.9 **a.** See Appendix L, Smokers Data (p. 635). **b.** No apparent relationship. **c.** Uniform.

1.10 **a.** See Appendix L, Textbook Data (p. 636). **b.** Direct. **c.** Linear. **d.** Nonuniform. **e.** Length gives a better indication of price for small books than for large books.

1.11 **a.** See Appendix L, Professionals Data (p. 637). **b.** Direct, curvilinear; increasing at a decreasing rate. **c.** Strong.

1.12 Should look like Figure 1.2, except points will be higher from the horizontal axis.

1.13 **a.** The more fertilizer applied, the greater the yield tends to be. **b.** The longer the time spent playing intramural sports, the lower students' grade averages tend to be. **c.** Regardless of how many jars are sold, the price tends to remain constant. **d.** The longer one cooks the vegetable, typically the lower the vitamin content. **e.** It has been found that families that consume a large amount of electricity also tend to consume a relatively large amount of beer.

1.14

(**a**) Y increasing with X at an increasing rate (**b**) Y decreasing with X at an increasing rate

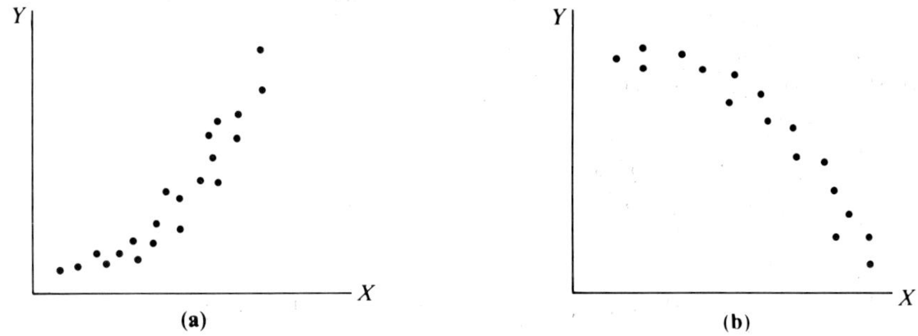

1.15 Y would increase with X at an increasing rate.
1.16 **a.** Linear. Discounting fatigue, each typewriter should require approximately the same amount of time to service. **b.** Possibly curvilinear. Increasing at an increasing rate, because of physical fatigue and increasing distances cases must be transported. **c.** Linear. Ratio of volume to weight should not vary with weight. **d.** Curvilinear. Increasing with age up to around age 21, leveling off between ages 21 and 55, and then decreasing. **e.** Curvilinear. Increasing at an increasing rate.

SECTION 1.4, PAGE 23

1.17 In deterministic models, once X is given, Y is determined without error by applying the rules of arithmetic. In probabilistic models, Y can only be estimated from a given value of X, since probabilistic models take individual differences into account.
1.18 **a.** The number of illustrations and formulas, and the quality of paper, may differ. **b.** One city may have greater unemployment, a larger proportion of poor families, a different ethos. **c.** Customers will have different numbers of items; cashiers vary in speed and efficiency. **d.** If measurements can be made sufficiently accurately, times will vary slightly. **e.** The physicist relates the average time to the degree of incline.
1.19

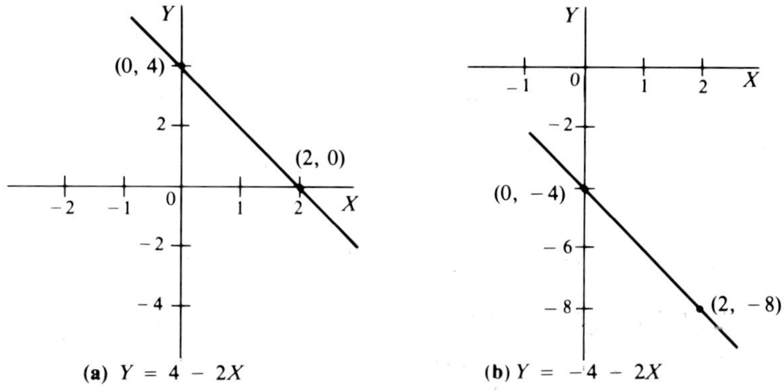

(**a**) $Y = 4 - 2X$ (**b**) $Y = -4 - 2X$

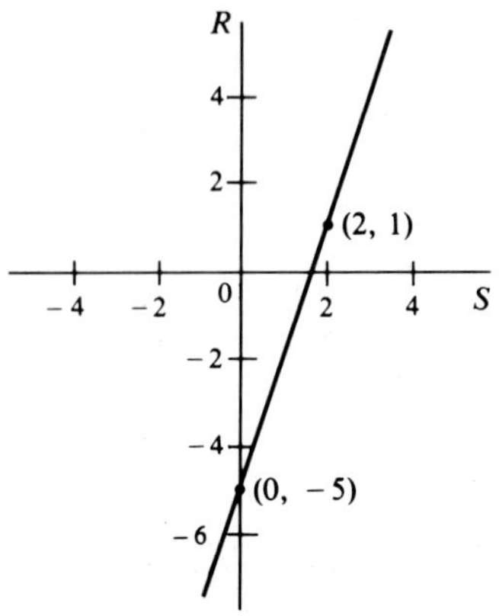

(c) $R = -5 + 3S$

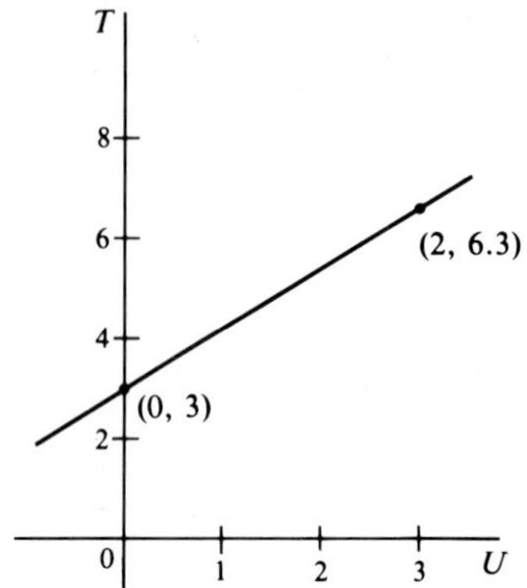

(d) $T = 3 + 1.1U$

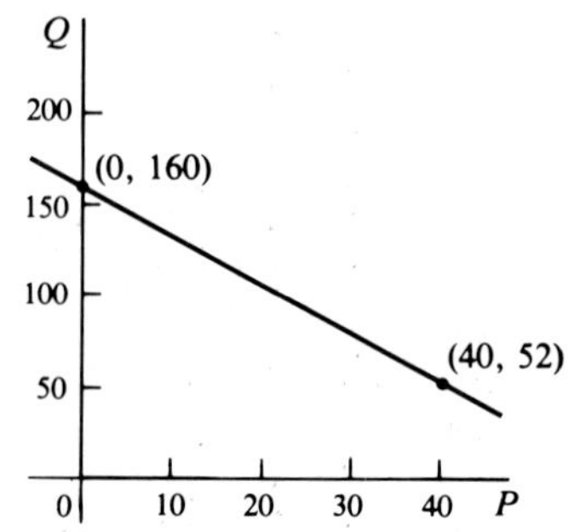

(e) $Q = 160 - 2.7P$

T W O

SECTIONS 2.2–2.3, PAGE 40

2.1 (X_i, Y_i) is a point on the scatter diagram, representing observed values of X and Y. $(X_i, \hat{Y}_i)$ is a point on the regression line, representing an estimated value of Y for an observed value of X.

2.2 **a.** $\sum X_i^2 - \sum X_i + 4n$. **b.** $3\sum X_i + \sum Y_i$. **c.** $\sum X_i^2 - 8\sum X_i + 16n$. **d.** $9\sum X_i^2 - 24\sum X_iY_i + 16\sum Y_i^2$. **e.** $12\sum X_i^2 - 11\sum X_iY_i - 5\sum Y_i^2$. **f.** $\sum Y_i - na - b\sum X_i$. **g.** $\sum X_iY_i - a\sum X_i - b\sum X_i^2$. **h.** $\sum X_iY_i - a\sum X_i - b\sum X_i^2$. **i.** $\sum X_i^2 - 2\bar{X}\sum X_i + n\bar{X}^2$. **j.** $\sum X_iY_i - \bar{X}\sum Y_i - \bar{Y}\sum X_i + n\bar{X}\bar{Y}$.

2.3 **a.** $\sum (X_i - \bar{X}) = \sum X_i - n\bar{X}$. Substitute $(\sum X_i)/n$ for $\bar{X}$. **b.** $\sum (X_i - \bar{X})^2 = \sum X_i^2 - 2\bar{X}\sum X_i + n\bar{X}^2$. Substitute $(\sum X_i)/n$ for $\bar{X}$. **c.** Divide result in (**b**) by $n - 1$. **d.** Refer to Solution 2.2h. Substitute $(\sum X_i)/n = \bar{X}$ and $(\sum Y_i)/n = \bar{Y}$. **e.** Same as (**b**) but with X replaced by Y.

2.4 **a.** $\sum X_i = 15$, $\bar{X} = 3$, $\sum (X_i - \bar{X}) = 0$. **b.** $MD = 6/5$. **c.** $s_X^2 = 5/2$.

2.5 **a.** $(\sum X_i)^2 = 225 \neq \sum X_i^2 = 55$. **b.** $s_X^2 = 5/2$. **c.** False.

2.6 The least squares line is, on the average, closest to all the points in terms of squared deviations.

2.7 **a.**

$$\sum Q_i = na + b\sum P_i,$$
$$\sum P_iQ_i = a\sum P_i + b\sum P_i^2.$$

b.

$$\sum E_i = na + b\sum T_i + c\sum T_i^2,$$
$$\sum T_iE_i = a\sum T_i + b\sum T_i^2 + c\sum T_i^3,$$
$$\sum T_i^2E_i = a\sum T_i^2 + b\sum T_i^3 + c\sum T_i^4.$$

c.

$$\sum Y_i = na + b\sum X_i + c\sum X_i^2 + d\sum X_i^3,$$
$$\sum X_iY_i = a\sum X_i + b\sum X_i^2 + c\sum X_i^3 + d\sum X_i^4,$$
$$\sum X_i^2Y_i = a\sum X_i^2 + b\sum X_i^3 + c\sum X_i^4 + d\sum X_i^5,$$
$$\sum X_i^3Y_i = a\sum X_i^3 + b\sum X_i^4 + c\sum X_i^5 + d\sum X_i^6.$$

d. $\sum Y_i = na + b \sum X_i + c \sum Z_i,$
$\sum X_i Y_i = a \sum X_i + b \sum X_i^2 + c \sum X_i Z_i,$
$\sum Z_i Y_i = a \sum Z_i + b \sum X_i Z_i + c \sum Z_i^2.$

e. $\sum Y_i = nb_0 + b_1 \sum X_{1i} + b_2 \sum X_{2i},$
$\sum X_{1i} Y_i = b_0 \sum X_{1i} + b_1 \sum X_{1i}^2 + b_2 \sum X_{1i} X_{2i},$
$\sum X_{2i} Y_i = b_0 \sum X_{2i} + b_1 \sum X_{1i} X_{2i} + b_2 \sum X_{2i}^2.$

2.8 **a.b.** See Appendix L, Concession Data (p. 632). **c.** $\hat{Y} = 0.3243 + 0.3842(30.8667) = 11.5347$. $\overline{Y} = 11.5333$.

2.9 **a.b.** See Appendix L, Sales Data (p. 633). **c.** $\hat{Y} = 0.3243 + 0.3842(30.8667) = 11.5347$. $\overline{Y} = 11.5333$.

2.10. **a.b.** See Appendix L, Heating Data (p. 634). **c.** $\hat{Y} = 0.7347 - 0.0080(18.9333) = 0.5832$. $\overline{Y} = 0.5833$.

2.11 **a.** See Appendix L, Repairs Data (p. 638). **b.** There is a direct relationship between hours of operation and repair cost, but this does not necessarily mean that large repair costs are incurred because management is unwilling to have routine maintenance performed as often as necessary.

2.12 See Appendix L, Bark Data (p. 638).

SECTION 2.4, PAGE 49

2.13 **a.** $\sum T^2 - (\sum T)/n$. **b.** $\sum R^2 - (\sum R)/n$. **c.** $\sum RT - (\sum R)(\sum T)/n$. **d.** $\sum TR - (\sum T)(\sum R)/n$.

2.14 See Appendix L, Concession Data (p. 632).

2.15 See Appendix L, Heating Data (p. 634).

2.16 See Appendix L, "Suds" Data (p. 639).

2.17 **a.c.** See Appendix L, Class Size Data (p. 639). **b.d.** While an inverse relationship is apparent, there is wider scatter for class sizes 10–60, and more data there. Were the four very large classes omitted, it appears as if the relationship would change considerably.

SECTION 2.5, PAGE 57

2.18 **a.** $\overline{X} = 66.3429$; $U_i = -5.8429, 4.5571, 5.6571, 8.5571, -3.5429, 2.3571, -11.7429$. **b.** $\sum U_i = 0$; $\sum U_i^2 = 316.1371$; $\sum U_i Y_i = 459.3908$; $\sum Y_i = 681.8$; $\sum Y_i^2 = 67{,}082.26$. **c.** $S_{UU} = 316.1371$; $S_{YY} = 674.94$; $S_{UY} = 459.3908$. **d.** $\hat{Y} = 97.4 + 1.4531U$. **e.** The origin has been moved, so the value of Y at the origin must be different. **f.** $\hat{Y} = 97.4 + 1.4531(X - 66.3429) = 0.9971 + 1.4531X$ (differences due to rounding error).

2.19 **a.** $s_X = 7.2588$. **b.** $V_i = 8.3348, 9.7675, 9.9191, 10.3186, 8.6516, 9.4644, 7.5220$. **c.** $\sum V_i = 63.9780$, $\sum V_i^2 = 590.7405$, $\sum Y_i = 681.8$, $\sum Y_i^2 = 67{,}082.26$, $\sum V_i Y_i = 6294.7485$. $S_{VV} = 6$; $S_{YY} = 674.94$; $S_{VY} = 63.2913$. **d.** $\hat{Y} = 0.9888 + 10.5486V$. **e.** Since a one-unit increase in X corresponds to a 7.2588-unit increase in V, the slope must change to reflect how Y increases with V. **f.** $\hat{Y} = 0.9888 + 10.5486(X/7.2588) = 0.9888 + 1.4532X$ (differences due to rounding error).

2.20 **a.** $\mu = 20$; $d = 5$; $Z_i = 1, -1, -1, 0, 0, 0, 1, 1, 1$. **b.** See Appendix L, Fertilizer Data (p. 640). **c.** $\hat{Y} = 60.2222 + 15Z$. **d.** If $X = 15$, $Z = 1$, and $\hat{Y} = 45.2222$. **e.** $\hat{Y} = 60.2222 + 15(X - 20)/5 = 0.2222 + 3X$. **f.** $\hat{Y} = 0.2222 + 3(15) = 45.2222$. **g.** Inferences should be restricted to the three levels at which the experiment was performed unless one is sure that the relationship is linear throughout.

2.21 **a.** $\mu = 72$; $d = 24$; $X_1 = 24$; $X_2 = 48$; $X_3 = 72$; $X_4 = 96$; $X_5 = 120$. **b.** See Appendix L, Deprivation Data (p. 640). Time to learn task is not a linear function of hours of deprivation. **c.** $\sum Z_i = 0$; $\sum Y_i = 508$; $\sum Z_iY_i = -53$; $\sum Z_i^2 = 20$; $\sum Z_i^2Y_i = 1249$; $\sum Z_i^3 = 0$; $\sum Z_i^4 = 68$; $\hat{Y} = 34.1572 - 2.65Z + 8.3214Z^2$. **d.** Lowest point occurs at $Z = 0.1592$, so $\hat{Y} = 33.9458$. **e.** (1) Yes. (2) No.

2.22 $Z_i = -0.2367, -1.1837, -0.3551, 0.9470, 0.2367, -1.4205, 1.8939, -1.0653, 0.0000, 0.7102, 0.3551, -1.3021, -0.9470, 1.6572, -0.3551, 0.5919, -0.7102, 1.5388, 0.2367, -0.5919$. $\sum Z_i = 0$; $\sum Z_i^2 = 19$; $\sum Y_i = 690$; $\sum Y_i^2 = 26.178$; $\sum Z_iY_i = 207.0317$. $S_{ZZ} = 19$; $S_{ZY} = 207.0317$; $S_{YY} = 2373$. $\hat{Y} = 34.5 + 10.8964Z = 34.5 + 10.8964(X - 2)/0.844798 = 8.7035 + 12.8982X$.

T H R E E

SECTION 3.2, PAGE 64

3.1 **a.–d.** Yes.

3.2 **a.** 3×4. **b.** 4×3. **c.** 3×3. **d.** 6×9. **e.** 12×12. **f.** 1×12. **g.** 4×1. **h.** 1×1.

3.3 **a.** Fourth row, seventh column. **b.** Seventh row, fourth column. **c.** Seventh row, seventh column. **d.** ith row, jth column.

3.4 **a.** Yes. **b.–e.** No.

3.5 **a.** $\begin{bmatrix} 1 \\ 2 \\ 3 \\ 4 \\ 5 \end{bmatrix}$ **b.** $[5 \quad 4 \quad 3 \quad 2 \quad 1]$ **c.** $\begin{bmatrix} 6 & 2 \\ 9 & 4 \\ 0 & 6 \\ 1 & 0 \end{bmatrix}$ **d.** $\begin{bmatrix} a & c & e \\ b & d & f \end{bmatrix}$

e. $\begin{bmatrix} a_{11} & a_{21} & a_{31} \\ a_{12} & a_{22} & a_{32} \\ a_{13} & a_{23} & a_{33} \end{bmatrix}$ **f.** $\begin{bmatrix} 1 & 2 \\ 2 & 3 \end{bmatrix}$ **g.** $\begin{bmatrix} 9 & 8 & 7 \\ 8 & 6 & 5 \\ 7 & 5 & 4 \end{bmatrix}$

3.6 **a.** $a_{11}, a_{22}, a_{33}, a_{44}$. **b.** 9, 6, 4. **c.** 1, 1. **d.** 1, 1, 2, 4. **e.** Matrix not square.

3.7 **a.** $\begin{bmatrix} 1 & 0 & 0 & 0 & 0 \\ 0 & 1 & 0 & 0 & 0 \\ 0 & 0 & 2 & 0 & 0 \\ 0 & 0 & 0 & 3 & 0 \\ 0 & 0 & 0 & 0 & 5 \end{bmatrix}$ **b.** $\begin{bmatrix} 7 & 0 & 0 & 0 & 0 & 0 \\ 0 & 0 & 0 & 0 & 0 & 0 \\ 0 & 0 & 3 & 0 & 0 & 0 \\ 0 & 0 & 0 & 4 & 0 & 0 \\ 0 & 0 & 0 & 0 & 1 & 0 \\ 0 & 0 & 0 & 0 & 0 & 0 \end{bmatrix}$ **c.** $\begin{bmatrix} 1 & 0 & 0 & 0 \\ 0 & 1 & 0 & 0 \\ 0 & 0 & 1 & 0 \\ 0 & 0 & 0 & 1 \end{bmatrix}$ **d.** $\begin{bmatrix} 5 & 0 & 0 \\ 0 & 5 & 0 \\ 0 & 0 & 5 \end{bmatrix}$

e. $\begin{bmatrix} a_1 & 0 & \cdots & 0 \\ 0 & a_2 & \cdots & 0 \\ \vdots & \vdots & & \vdots \\ 0 & 0 & \cdots & a_n \end{bmatrix}$

3.8 **a.** $\begin{bmatrix} 5 & 2 & 7 \\ 2 & 1 & 1 \\ 7 & 1 & 2 \end{bmatrix}$ **b.** $\begin{bmatrix} 9 & 9 & 2 & 9 \\ 9 & 2 & 1 & 1 \\ 2 & 1 & 2 & 4 \\ 9 & 1 & 4 & 0 \end{bmatrix}$ **c.** $\begin{bmatrix} a & b & d \\ b & c & e \\ d & e & f \end{bmatrix}$ **d.** $\begin{bmatrix} a_{11} & a_{12} & a_{13} & a_{14} \\ a_{12} & a_{22} & a_{23} & a_{24} \\ a_{13} & a_{23} & a_{33} & a_{34} \\ a_{14} & a_{24} & a_{34} & a_{44} \end{bmatrix}$ **e.** $\begin{bmatrix} 0 & 0 & 9 & 3 \\ 0 & 6 & 2 & 7 \\ 9 & 2 & 4 & 9 \\ 3 & 7 & 9 & 1 \end{bmatrix}$

SECTION 3.3, PAGE 71

3.9 a. $\begin{bmatrix} 7 & 11 \\ 3 & 7 \end{bmatrix}$ **b.** Diag (9, 10, 2, 11). **c.** $\begin{bmatrix} 2 & -8 & 4 & 9 \\ 8 & 5 & 2 & -2 \\ 2 & -3 & 9 & 6 \\ 0 & 14 & 9 & 9 \end{bmatrix}$ **d.** $\begin{bmatrix} 0 & 2 & 0 & 5 \\ -6 & -5 & -10 & 4 \\ 0 & 13 & -5 & -10 \\ -10 & -2 & 3 & -1 \end{bmatrix}$

e. Sum cannot be formed. **f.** Sum cannot be formed. **g.** $\begin{bmatrix} 7 \\ 7 \\ 2 \\ 12 \\ 10 \end{bmatrix}$ **h.** Sum cannot be formed. **i.** $[0 \quad 6 \quad -6 \quad 0 \quad -2]$ **j.** Sum cannot be formed.

3.10 a. $\begin{bmatrix} 16 & -28 & 16 \\ 20 & 32 & 4 \\ -20 & 32 & 8 \end{bmatrix}$ **b.** $\begin{bmatrix} 1 & -7/4 & 1 \\ 5/4 & 2 & 1/4 \\ -5/4 & 2 & 1/2 \end{bmatrix}$ **c.** $\begin{bmatrix} -18 & -15 \\ -21 & -12 \\ -15 & -6 \end{bmatrix}$

d. $[56 \quad 28 \quad 63 \quad 14 \quad 21]$. **e.** $\begin{bmatrix} -35 \\ -10 \\ -5 \\ 10 \\ 0 \\ -35 \end{bmatrix}$

3.11 a. $(2 \times 3)(3 \times 2) = (2 \times 2)$. **b.** $(3 \times 2)(2 \times 3) = (3 \times 3)$. **c.** $(3 \times 3)(3 \times 3) = (3 \times 3)$. **d.** $(3 \times 3)(3 \times 3) = (3 \times 3)$. **e.** No. **f.** $(1 \times 4)(4 \times 1) = (1 \times 1)$. **g.** $(4 \times 1)(1 \times 4) = (4 \times 4)$. **h.** No. **i.** Scalar times a vector (3×1).

3.12 a. $\begin{bmatrix} 131 & 140 \\ 90 & 95 \end{bmatrix}$ **b.** $\begin{bmatrix} 62 & 45 & 117 \\ 46 & 29 & 72 \\ 72 & 52 & 135 \end{bmatrix}$ **c.** $\begin{bmatrix} 42 & 84 & 70 \\ 17 & 35 & 23 \\ 23 & 28 & 27 \end{bmatrix}$ **d.** $\begin{bmatrix} 32 & 14 & 24 \\ 46 & 16 & 23 \\ 90 & 23 & 56 \end{bmatrix}$

e. Cannot be formed. **f.** 17. **g.** $\begin{bmatrix} 0 & 6 & 2 & 3 \\ 0 & 12 & 4 & 6 \\ 0 & 6 & 2 & 3 \\ 0 & 6 & 2 & 3 \end{bmatrix}$ **h.** Cannot be formed. **i.** $\begin{bmatrix} 4 \\ 8 \\ 4 \end{bmatrix}$

3.13 a. The products are not equal—the product matrices are not the same dimensions. **b.** Even though the products are the same dimensions, they are not equal.

3.14 a. $\mathbf{SS'} = \begin{bmatrix} 4 & 18 & 14 \\ 18 & 110 & 81 \\ 14 & 81 & 70 \end{bmatrix}$ $\mathbf{S'S} = \begin{bmatrix} 3 & 9 & 16 & 8 \\ 9 & 41 & 63 & 27 \\ 16 & 63 & 114 & 53 \\ 8 & 27 & 53 & 26 \end{bmatrix}$ **b.** Both are symmetric. **c.** $\mathbf{SS'} \neq \mathbf{S'S}$.

3.15 For any of the given matrices **M**, $\mathbf{IM} = \mathbf{MI} = \mathbf{M}$.

SECTION 3.4, PAGE 79

3.16 a. -25. **b.** 7. **c.** $a_{11}a_{22} - a_{12}a_{21}$. **d.** $n \sum X_i^2 - (\sum X_i)^2$. **e.** $4 - 6 = -2$. **f.** 0. **g.** 42. **h.** Matrix not square. **i.** 8. **j.** $a_1 a_2$.

3.17 a. Matrix in 3.16f is singular. **b.** Matrices in 3.16a, b, e, g, i, are nonsingular. If $a_{11}a_{22} - a_{12}a_{21} \neq 0$, then the matrix in **c** is nonsingular. Matrix in **d** is nonsingular as long as $X_i \neq 0$ for all i.

3.18 **a.** 320. **b.** 12. **c.** −347. **d.** 0. **e.** 0. **f.** Matrix not square. **g.** Matrix not square. **h.** 280. **i.** 105. **j.** $(d_1)(d_2)(d_3)$.

3.19 **a.** Singular: **d** and **e**. **b.** Nonsingular: **a, b, c, h,** and **i** and **j** as long as no d_i, $i = 1, 2, 3$, is zero.

3.20 **a.** −28. **b.** 20. **c.** 0. **d.** 0. **e.** 1179. **f.** 504. **g.** 1960. **h.** $(a_1)(a_2) \ldots (a_n)$. **i.** Matrix not square. **j.** 0. **k.** 12,222.

3.21 **a.** Singular: **c, d, j.** **b.** Nonsingular: **a, b, e, f, g, k,** and **h** as long as $a_i \neq 0$, $i = 1, 2, \ldots, n$.

SECTION 3.5, PAGE 85

3.22 **a.** (a) $\begin{bmatrix} -1/25 & 3/25 \\ 9/25 & -2/25 \end{bmatrix}$ (b) $\begin{bmatrix} 1 & 0 \\ 9/7 & 1/7 \end{bmatrix}$ (c) $\dfrac{1}{a_{11}a_{22} - a_{12}a_{21}} \begin{bmatrix} a_{22} & -a_{12} \\ -a_{21} & a_{11} \end{bmatrix}$

(d) $\dfrac{1}{n\sum X_i^2 - (\sum X_i)^2} \begin{bmatrix} \sum X_i^2 & -\sum X_i \\ -\sum X_i & n \end{bmatrix}$ (e) $\begin{bmatrix} -2 & 1 \\ 3/2 & -1/2 \end{bmatrix}$ (g) $\begin{bmatrix} 1/6 & 0 \\ 0 & 1/7 \end{bmatrix}$ (i) 1/8.

b. (a) $\dfrac{1}{320} \begin{bmatrix} -16 & 42 & -14 \\ -56 & 7 & 51 \\ 48 & -6 & 2 \end{bmatrix}$ (b) $\dfrac{1}{12} \begin{bmatrix} 12 & 0 & 0 \\ 32 & 4 & -4 \\ -43 & -2 & 5 \end{bmatrix}$ (c) $\dfrac{1}{-347} \begin{bmatrix} -40 & -21 & 49 \\ -21 & 15 & -35 \\ 49 & -35 & -34 \end{bmatrix}$

(h) Diag (1/7, 1/5, 1/8). (i) Diag (1/7, 1/5, 1/3). (j) Diag $(1/d_1, 1/d_2, 1/d_3)$.

c. (a) $\dfrac{1}{-28} \begin{bmatrix} -28 & 0 & 0 & 0 \\ 0 & 8 & 4 & -8 \\ 0 & -10 & -12 & 10 \\ 84 & 16 & 36 & -30 \end{bmatrix}$ (b) $\dfrac{1}{20} \begin{bmatrix} 4 & -18 & 10 & 20 \\ 0 & 15 & -15 & -10 \\ 0 & 40 & -20 & -60 \\ 0 & -20 & 20 & 20 \end{bmatrix}$

(e) $\dfrac{1}{1179} \begin{bmatrix} 99 & -171 & -342 & 306 \\ -132 & 228 & 63 & -15 \\ 172 & -47 & -225 & 91 \\ 32 & -91 & 342 & -175 \end{bmatrix}$ (f) Diag (1/3, 1/6, 1/7, 1/4). (g) Diag (1/7, 1/7, 1/8, 1/5). (h) Diag $(1/a_1, 1/a_2, \ldots, 1/a_n)$.

3.23 **a.** $\begin{bmatrix} 1 & 0 \\ 0 & 1/6 \end{bmatrix}$ **b.** If $\mathbf{A} = \left[\begin{array}{c|cc} 1 & 0 & 0 \\ \hline 0 & & \mathbf{M} \\ & & \end{array}\right]$ then $\mathbf{A}^{-1} = \left[\begin{array}{c|cc} 1 & 0 & 0 \\ \hline 0 & & \\ 0 & & \mathbf{M}^{-1} \end{array}\right]$

c. $\begin{bmatrix} 1 & 0 & 0 & 0 \\ 0 & -\frac{16}{320} & \frac{42}{320} & -\frac{14}{320} \\ 0 & -\frac{56}{320} & \frac{7}{320} & \frac{51}{320} \\ 0 & \frac{48}{320} & -\frac{6}{320} & \frac{2}{320} \end{bmatrix}$ **d.** $\left[\begin{array}{c|ccc} 1 & 0 & \cdots & 0 \\ \hline 0 & & & \\ \vdots & & \mathbf{M}^{-1} & \\ 0 & & & \end{array}\right]$ **e.** $\left[\begin{array}{c|ccc} 1/d & 0 & \cdots & 0 \\ \hline 0 & & & \\ \vdots & & \mathbf{M}^{-1} & \\ 0 & & & \end{array}\right]$

SECTION 3.6, PAGE 92

3.24 a. $\mathbf{x} = \begin{bmatrix} 14/25 \\ -1/25 \end{bmatrix}$ **b.** $\mathbf{x} = \begin{bmatrix} 1 \\ 11/7 \end{bmatrix}$ **c.** $\mathbf{x} = \begin{bmatrix} -6 \\ 4 \end{bmatrix}$ **d.** No solution. **e.** $\mathbf{x} = \begin{bmatrix} 7/6 \\ 3/7 \end{bmatrix}$

f. No solution. **g.** $x = 5/8$.

3.25 a. $\mathbf{x} = \frac{1}{320}\begin{bmatrix} -300 \\ 270 \\ 180 \end{bmatrix}$ **b.** $\mathbf{x} = \frac{1}{12}\begin{bmatrix} 36 \\ 100 \\ -113 \end{bmatrix}$ **c.** $\mathbf{x} = \frac{1}{347}\begin{bmatrix} -324 \\ 281 \\ -193 \end{bmatrix}$

d. No solution. **e.** No solution. **f.** $\mathbf{x} = \begin{bmatrix} 5/7 \\ 1 \\ 5/8 \end{bmatrix}$

3.26 a. $\mathbf{x} = \frac{1}{28}\begin{bmatrix} 196 \\ 44 \\ -34 \\ -486 \end{bmatrix}$ **b.** $\mathbf{x} = \frac{1}{20}\begin{bmatrix} -56 \\ 40 \\ 140 \\ -40 \end{bmatrix}$ **c.** No solution. **d.** $\mathbf{x} = \frac{1}{1179}\begin{bmatrix} -5409 \\ 2889 \\ -3300 \\ 2265 \end{bmatrix}$

e. No solution.

SECTION 3.7, PAGE 102

3.27 a. $\mathbf{y} = [89.2 \quad 105.6 \cdots 80.7]'$, $\hat{\boldsymbol{\beta}} = [a \quad b]'$,

$$\mathbf{X} = \begin{bmatrix} 1 & 60.5 \\ 1 & 70.9 \\ \vdots & \vdots \\ 1 & 54.6 \end{bmatrix}$$

b.–d. See Appendix L, Concession Data (p. 632).

3.28 a. $\mathbf{y} = [89.2 \quad 105.6 \cdots 80.7]'$, $\hat{\boldsymbol{\beta}} = [a \quad b]'$,

$$\mathbf{U} = \begin{bmatrix} 1 & -5.8429 \\ 1 & 4.5571 \\ \vdots & \vdots \\ 1 & 2.3571 \end{bmatrix}$$

b. $\mathbf{U}'\mathbf{U} = \begin{bmatrix} 7 & 0 \\ 0 & 316.1371 \end{bmatrix}$, $\mathbf{U}'\mathbf{y} = \begin{bmatrix} 681.8 \\ 459.3908 \end{bmatrix}$

c. $(\mathbf{U}'\mathbf{U})^{-1} = \text{diag}\,(1/7,\ 1/316.1371)$. **d.** $\hat{\boldsymbol{\beta}} = [681.8/7 \quad 459.3908/316.1371]' = [97.4 \quad 1.4531]'$ **e.** $\hat{Y} = 97.4 + 1.4531U = 97.4 + 1.453(X - 66.3429) = 0.9971 + 1.4531X$.

3.29 a. $\mathbf{y} = [0.30 \quad 0.70 \cdots 0.75]'$, $\hat{\boldsymbol{\beta}} = [a \quad b]'$,

$$\mathbf{X} = \begin{bmatrix} 1 & 50 \\ 1 & 4 \\ \vdots & \vdots \\ 1 & 2 \end{bmatrix}$$

b.–d. See Appendix L, Heating Data (p. 634).

3.30 a. $\mathbf{y} = [2.7 \quad 2.0 \cdots 2.3]'$, $\hat{\boldsymbol{\beta}} = [a \quad b]'$,

$$\mathbf{X} = \begin{bmatrix} 1 & 55 \\ 1 & 40 \\ \vdots & \vdots \\ 1 & 45 \end{bmatrix}$$

b.–d. See Appendix L, Repairs Data (p. 638).

3.31 $\mathbf{y} = [10.3 \quad 13.0 \cdots 16.2]'$, $\hat{\boldsymbol{\beta}} = [a \quad b]'$,

$$\mathbf{X} = \begin{bmatrix} 1 & 211 \\ 1 & 310 \\ \vdots & \vdots \\ 1 & 403 \end{bmatrix}$$

b.–d. See Appendix L, Mail Data (p. 641).

3.32 a. $\mathbf{y} = [625 \quad 750 \cdots 625]'$, $\hat{\boldsymbol{\beta}} = [a \quad b \quad c]'$,

$$\mathbf{X} = \begin{bmatrix} 1 & 64 & (64)^2 = 4096 \\ 1 & 53 & (53)^2 = 2809 \\ \vdots & \vdots & \vdots \\ 1 & 58 & (58)^2 = 3364 \end{bmatrix}$$

b.–d. See Appendix L, Police Data (p. 633).

3.33 a. $\mathbf{y} = [45 \quad 43 \cdots 76]'$, $\hat{\boldsymbol{\beta}} = [a \quad b]'$,

$$\mathbf{X} = \begin{bmatrix} 1 & 15 \\ 1 & 15 \\ \vdots & \vdots \\ 1 & 25 \end{bmatrix}$$

b.–d. See Appendix L, Fertilizer Data (p. 640).

3.34 a. $\mathbf{y} = [45 \quad 43 \cdots 76]'$, $\hat{\boldsymbol{\beta}} = [a' \quad b']'$,

$$\mathbf{X} = \begin{bmatrix} 1 & -1 \\ 1 & -1 \\ \vdots & \vdots \\ 1 & 1 \end{bmatrix}$$

b. $\mathbf{X}'\mathbf{X} = \begin{bmatrix} 9 & 0 \\ 0 & 6 \end{bmatrix}$, $\mathbf{X}'\mathbf{y} = \begin{bmatrix} 542 \\ 90 \end{bmatrix}$

c. $(\mathbf{X}'\mathbf{X})^{-1} = \text{diag}(1/9, 1/6)$ **d.** $\hat{\boldsymbol{\beta}} = [542/9 \quad 90/6]' = [60.2222 \quad 15]'$

3.35 a.

$$\mathbf{Z} = \begin{bmatrix} 1 & -2 & 4 \\ 1 & -2 & 4 \\ 1 & -1 & 1 \\ \vdots & \vdots & \vdots \\ 1 & 2 & 4 \end{bmatrix} \quad \hat{\boldsymbol{\beta}} = \begin{bmatrix} a' \\ b' \\ c' \end{bmatrix} \quad \mathbf{y} = \begin{bmatrix} 73 \\ 68 \\ \vdots \\ 65 \end{bmatrix}$$

b.
$$\mathbf{Z'Z} = \begin{bmatrix} 10 & 0 & 20 \\ 0 & 20 & 0 \\ 20 & 0 & 68 \end{bmatrix} \quad \mathbf{Z'y} = \begin{bmatrix} 508 \\ -53 \\ 1249 \end{bmatrix}$$

c.
$$(\mathbf{Z'Z})^{-1} = \frac{1}{5600}\begin{bmatrix} 1360 & 0 & -400 \\ 0 & 280 & 0 \\ -400 & 0 & 200 \end{bmatrix}$$

d.
$$\hat{\boldsymbol{\beta}} = \begin{bmatrix} 34.1571 \\ -2.65 \\ 8.3214 \end{bmatrix}$$

F O U R

SECTION 4.2, PAGE 112

4.1 See Appendix L, Concession Data (p. 632).
4.2 See Appendix L, Smokers Data (p. 635).
4.3 See Appendix L, Class Size Data (p. 639).
4.4 See Appendix L, Appraisal Data (p. 642).

SECTION 4.3, PAGE 122

4.5 See Appendix L, Concession Data (p. 632).
4.6 See Appendix L, Smokers Data (p. 635).
4.7 See Appendix L, Class Size Data (p. 639).
4.8 See Appendix L, Appraisal Data (p. 642).
4.9 See Appendix L, Concession Data (p. 632).
4.10 See Appendix L, Smokers Data (p. 635).
4.11 See Appendix L, Class Size Data (p. 639).
4.12 See Appendix L, Appraisal Data (p. 642).

SECTION 4.5, PAGE 142

4.13 See Appendix L, Concession Data (p. 632).
4.14 See Appendix L, Smokers Data (p. 635).
4.15 See Appendix L, Class Size Data (p. 639).
4.16 See Appendix L, Appraisal Data (p. 642).

SECTION 4.6, PAGE 149

4.17 Refer to the solution to Exercise 3.22a.
4.18 Refer to the solution to Exercise 3.22c.

4.19 $\begin{bmatrix} 0.6716384 & & (\text{sym}) \\ -0.1630060 & 0.0744829 & \\ -0.0137501 & 0.0006531 & 0.005788 \end{bmatrix}$

F I V E

SECTIONS 5.1–5.6, PAGE 169

5.1 **a.** $\hat{Y} = 0.989 + 1.45(60) = 87.989$ gallons. **b.** 80,000 tickets is a bit beyond the range for which the relationship was found. **c.** On the average, 1.45 additional gallons of syrup will be required for each additional 1,000 advance tickets sold. **d.** Factors other than the number of advance tickets combine to require 0.989 gallons of syrup per game, on the average. **e.** While it is true that if no advance tickets were sold, the concessionaire might infer that he will sell no soft drinks, such an occurrence is beyond the scope of this study and probably not likely to happen. **f.** $b = \left[\frac{45{,}691.98}{31{,}125.76}\right] = 1.4680.$ This line is not appreciably different from $\hat{Y} = 0.989 + 1.45X$.

5.2 **a.** $6,205.80. **b.** The typical store employing 17 people has weekly sales of approximately $6,206. **c.** The estimate is only an average and is not intended to coincide with any actual values. **d.–f.** We only have information concerning the relationship for stores employing between 10 and 48 people. **g.** $a = -0.3256$ is a constant affecting sales determined by something other than the number of employees. **h.** It would seem reasonable that a store with no employees might have no sales. **i.** If one store has one more employee than another, then its average weekly sales are $384.20 higher than in the other's. **j.** Probably larger stores tend to have both large numbers of employees and high sales.

5.3 **a.** 584.98 robberies. **b.** The average number of robberies in cities employing 67 police officers is 584.98. **c.** Since the estimate is only an average, we would not expect it to coincide with the actual number of robberies in either city. **d.** We have no evidence that the same relationship holds for cities employing more than 90 policemen. **e.** No, out of range. **f.** Yes. **g.** The existence of a relationship does not imply a cause-effect relationship. **h.** The intercept is a constant affecting the number of robberies determined by something other than the number of police officers.

5.4 **a.** 47¢. **b.** The average daily bill when the temperature is 33 degrees is 47¢. **c.** The estimate is only an average and is not intended to coincide with any actual bill. **d.–f.** Estimates should be made only for days with temperatures between -2 and 50 degrees, since this is the only range for which we have information. **g.** The average bill on zero-degree days is 73¢. **h.** We saw that the average bill on zero-degree days is 73¢, not 0¢. **i.** If one day the temperature was one degree warmer than on another, the bill on the warmer day was 0.8¢ lower, on the average, than on the colder day.

5.5 **a.** 75.2222 bu. **b.** The average yield from plots to which 25 lb of fertilizer was applied is 75.2222 bu. **c.** The estimate is only an average and is not intended to coincide with any actual yield. **d.–e.** The data contain information about yield at only 15, 20, and 25 lb fertilizer applied, so estimates should be restricted to these levels. **f.** If $z = 0$, or 20 lb fertilizer was applied, the average yield was 60.22 bu. **g.** 0.2222 is a constant affecting yield determined outside of the relationship of yield to fertilizer. **h.** It seems unlikely that yield would be zero if no fertilizer were applied. **i.** Plots receiving 5 lb fertilizer more than others have, on the average, 15 bu greater yield. **j.** Plots receiving 5 lb fertilizer more than others have, on the average, $3(5) = 15$ bu greater yield.

5.6 a. With every additional 100 hours a machine runs, on the average the repair costs will increase $5.70. The intercept is a constant determined by factors other than hours of operation, which is needed to estimate repair costs accurately. **b.** While the regression equation indicates that repair costs increase with increasing hours of operation, one must know more about management policies and other factors affecting repair costs before one can establish a cause-effect relationship.

5.7 a. On the average, a "Suds" establishment will have gross sales of $30,698 unrelated to the size of the student body. If one "Suds" is located near a campus with 1,000 more students than another, the former may be expected to have sales $4,998 greater than the latter. **b.** $\hat{Y} = 30.698 + 4.99778(18) = \$120,658$. **c.** $75 = 30.698 + 4.99778X$; $X = 8,860$ students.

5.8 While it was found that teacher ratings tend to be lower in larger classes than in smaller ones, this does not demonstrate that teaching larger sections was the cause of this professor's low ratings. This professor may teach unpopular courses or be a poor teacher.

5.9 a. On the average, the company can expect 428.46 orders per day, plus an additional 2.93 orders per pound of mail. **b.** A cause-effect relationship could be supported, since orders come in the mail.

5.10 a. On the average, employees take 6.285 days sick leave regardless of how many cigarettes they smoke. There is a slight tendency for those who smoke more to take more days sick leave, but only about 1 additional day per additional pack (20 cigarettes). **b.** Given the small slope and the wide variation in sick leave taken for all degrees of smoking, there is not much evidence to support the contention that sick leave taken is related to number of cigarettes smoked.

5.11 a. A tree with an outside diameter 1 cm larger than another is estimated to have an inside diameter 0.897 cm larger than the other. The constant 0.0908 is determined by factors that are unrelated to the outside diameter but that affect inside diameter. **b.** $\hat{Y} = 0.0908 + 0.89726(30) = 27.0086$.

5.12 a. See Appendix L, Loans Data (p. 643). **b.** $\hat{Y} = 152.579 - 8.347107(12.75) = 46.15$ applications. **c.d.** Both 8% and 15% interest rates indicate unusual conditions that were not studied in this exercise. **e.** Since higher interest rates result in higher mortgage payments, it could be argued that increasing interest rates cause a decline in the number of persons applying for mortgage loans.

5.13 Residences sell for $10,433 plus 14¢ for every dollar of the appraised value.

5.14 $15 = 2.167 + 10.0662X$; $X = 1.2486$ or 124.86.

5.15 a. See Appendix L, Hyperactivity Data (p. 643). **b.** The existence of a statistical relationship is not sufficient to establish that this food additive causes hyperactivity, either in this sample or for all children. Physiological effects must be demonstrated. **c.** It represents the combined effects of variables other than the amount of preservative ingested on the hyperactivity level of children.

5.16 a. See Appendix L, Reservations Data (p. 653). **b.** On the average, 1.07 guests arrive for each reservation made, and an additional 7.18 guests arrive regardless of the number of reservations made.

5.17 a. See Appendix L, Stock Data (p. 644). **b.** Since the regression coefficient is 1.423, this indicates that XZQ is an aggressive stock. **c.** If the market return is zero, XZQ's return can be expected to be -1.02%.

5.18 a. Regress readings from the new gauge on readings from the old gauge, since it is assumed that readings from the old gauge will be accurate. **b.** Intercept = 0, slope = 1. **c.** See Appendix L, Gauge Data (p. 645). The new gauge tends to read about 2.5 lb higher than the old gauge regardless of the pressure being measured.

5.19 a. X could cause Y, Y could cause X, or perhaps larger firms (Z) both spend more and have greater sales. **b.** Probably X causes Y. **c.** X and Y are probably related by chance. **d.** Probably size of the store (Z) determines both X and Y. **e.** Probably related by chance.

S I X

SECTION 6.2, PAGE 183

6.1 Scatter not homogeneous; the 15 different variances cannot be estimated from the data in this sample.

6.2 **a.c.** See Appendix L, Concession Data (p. 632). **b.** Using the regression equation to estimate how much syrup will be needed, the concessionaire can expect to be off by 1.208 gallons.

6.3

$$\mathbf{y}'\mathbf{y} = \sum Y_i^2 = 67{,}082.26$$

$$\hat{\boldsymbol{\beta}}'\mathbf{X}'\mathbf{y} = [0.988566 \quad 1.45323] \begin{bmatrix} 681.8 \\ 45{,}691.98 \end{bmatrix}$$

$$= 67{,}074.9604$$

$$\mathbf{y}'\mathbf{y} - \hat{\boldsymbol{\beta}}'\mathbf{X}'\mathbf{y} = 67{,}082.26 - 67{,}074.9604 = 7.2996$$

$$s_{Y \cdot X}^2 = 7.2996/5 = 1.4599$$

$$s_{Y \cdot X} = (1.4599)^{1/2} = 1.2083$$

6.4 **a.c.** See Appendix L, Heating Data (p. 634). **b.** Using temperature to estimate a heating bill, one will be off by \$0.04, on the average.

6.5 **a.b.** See Appendix L, Smokers Data (p. 635). **c.** On the average, the estimate of number of sick days will be off by 4.93 days. Since the average number of sick days is 7.4, these estimates are not very accurate.

6.6 Estimates of the number of orders will be off by only 24.6, on the average.

6.7 On the average, trial period sales can predict first-year sales within \$523.

6.8 **a.** See Appendix L, Police Data (p. 633). **b.** 10 degrees of freedom: sample size is $n = 13$; number of parameters in the model are $p = 3$.

6.9 **a.** $s_{Y \cdot X}^2 = (\mathbf{y}'\mathbf{y} - \hat{\boldsymbol{\beta}}'\mathbf{X}'\mathbf{y})/(n - p)$. **b.** $n - 4$.

SECTION 6.6, PAGE 199

6.10 **a.** 1.725. **b.** −2.473. **c.** −1.356. **d.** 2.2787. **e.** ±2.447.

SECTION 6.7, PAGE 203

6.11 $95.4485 \pm 2.571(1.2081)[1/7 + (65 - 66.3429)^2/316.1371]^{1/2} = 94.25$ to 96.65 gallons.

6.12 $0.5748 \pm 2.650(0.0427)[1/15 + (20 - 18.9333)^2/4924.9333]^{1/2} = \$0.55$ to \$0.60. You are 99% sure that this interval covers the real average heating bill on days for which the temperature is 20 degrees.

6.13 **a.** $2852 \pm 1.708(4.9437)[1/25 + (0 - 20.08)^2/6451.84]^{1/2} = 3.58$ to 9.00 days. **b.** $7.3956 \pm 1.708(4.9437)[1/25 + (20 - 20.08)^2/6451.84]^{1/2} = 5.71$ to 9.08 days. **c.** There is not much difference in the number of sick days taken by nonsmokers and those who smoke a pack a day.

6.14

Day	Lower limit	Upper limit	Day	Lower limit	Upper limit
1	146.459	150.721	13	140.970	145.497
2	232.233	236.350	14	246.853	251.726
3	118.957	124.659	15	249.980	255.027
4	261.431	267.144	16	151.940	155.952
5	215.438	218.865	17	192.081	195.086
6	197.423	200.456	18	133.274	138.195
7	272.867	279.276	19	129.972	135.069
8	185.640	188.671	20	229.092	233.063
9	201.682	204.767	21	168.322	171.709
10	157.413	161.192	22	225.948	229.780
11	173.754	176.989	23	161.783	165.392
12	161.783	165.392	24	225.948	229.780

6.15 See Appendix L, Loans Data (p. 643).

6.16 If X_i is very far away from $\overline{X}$, the confidence bands become so wide as to be useless or even meaningless. Estimation of mean Y-value has very little accuracy and confidence.

SECTION 6.8, PAGE 205

6.17 $95.4485 \pm 2.571(1.2081)[1 + 1/7 + (65 - 3429)^2/316.1371]^{1/2} = 92.12$ to 98.78 gallons. Wider than in Exercise 6.11 because it is easier to pin down an average than to capture all 95% of the individuals.

6.18 $11.6136 \pm 2.602(0.2458)[1 + 1/15 + (250 - 299.3333)^2/100{,}053.3333]^{1/2} = 10.95$ to 12.28 hundred orders. 99% of the time, if 250 lb of mail are received, between 1,095 and 1,228 orders will be processed.

6.19

Day	Lower limit	Upper limit	Day	Lower limit	Upper limit
1	140.927	156.253	13	135.532	150.934
2	226.649	241.935	14	241.536	257.043
3	113.914	129.701	15	244.722	260.285
4	256.392	272.183	16	146.317	161.575
5	209.594	224.709	17	186.071	201.096
6	191.424	206.455	18	127.973	143.495
7	268.044	284.099	19	124.731	140.310
8	179.641	194.671	20	223.454	238.702
9	195.704	210.745	21	162.462	177.568
10	151.703	166.902	22	220.258	235.470
11	167.835	182.908	23	156.009	171.166
12	156.009	171.166	24	220.258	235.470

6.20 See Appendix L, Marketing Data (p. 645).

S E V E N

SECTION 7.3, PAGE 223

7.1 **a.** $H_0{:}\beta = 0$; $H_1{:}\beta > 0$. One would expect that demand for soft drinks would increase with attendance. **b.** $t = 21.389 > t_{5,0.05} = 2.105$, so you are more than 95% sure that the greater the number of tickets sold, the more syrup will be needed. **c.** F cannot test a one-sided alternative.

7.2 **a.** $H_0{:}\beta = 0$; $H_1{:}\beta < 0$. One would expect higher heating bills on colder days. **b.** $t = -13.148 < -t_{13,0.05} = -1.771$, so you are more than 95% sure that the heating bill rises as the temperature drops. **c.** F cannot test a one-sided alternative.

7.3 **a.** $H_0{:}\beta = 0$; $H_1{:}\beta > 0$. **b.** $t = 0.902 \not> t_{23,0.05} = 1.714$, so the evidence is insufficient to conclude with 95% certainty that the more one smokes, the more days sick leave one takes.

7.4 **a.** $t = 10.616 > t_{8,0.01} = 2.896$, so he can be more than 99% sure that the greater the hours of operation, the higher the repair cost. **b.** The foreman has evidence that costs increase with hours of operation. Without knowledge of the operating conditions, the existence of a statistical relationship is supporting evidence—not proof.

7.5 **a.** $t = -6.756 < -t_{20,0.01} = -2.528$, so you are more than 99% sure that the larger the class, the lower the rating. **b.** Existence of a statistical relationship is insufficient evidence to establish cause and effect.

7.6 **a.** $t = 10.716 > t_{22,0.025} = 2.074$. Reject $H_0{:}\beta = 0$ in favor of $H_1{:}\beta \neq 0$. **b.** $F = 114.829 > F_{1,22,0.05} = 4.30$. **c.** You are more than 95% sure that the sale price is related to the appraised value.

7.7 **a.** $F = 18.562 > F_{1,9,0.01} = 10.56$, so reject $H_0{:}\beta = 0$ in favor of $H_1{:}\beta \neq 0$. **b.** $t = 4.308 > t_{9,0.005} = 3.250$. **c.** You are more than 99% sure that hyperactivity level is related to the amount of preservative ingested.

7.8 $t = (1.4230 - 1)/0.2506 = 1.688 > t_{22,0.10} = 1.321$, but $t \not> t_{22,0.05} = 1.717$, so you are more than 90% sure but less than 95 percent sure that XZQ is an aggressive stock.

7.9 $t = (1.0004 - 1)/0.0050 = 0.08 \not> t_{18,0.005} = 2.878$, so there is not sufficient evidence to conclude that the reading by the new gauge varies systematically with readings by the old gauge.

7.10 See Appendix L, Diameters Data (p. 645). $t = (0.8874 - 1)/0.0306 = 3.680$, and $-3.680 < -t_{13,0.005} = -3.012$. Thus, we are more than 99% sure that the accuracy with which students measure varies systematically with the diameters of the trees (as reflected by instructor's measurements).

SECTION 7.4, PAGE 227

7.11 **a.** $t = 0.218 \not> t_{5,0.025} = 2.571$. Evidence is insufficient to conclude that the intercept is different from zero. While one might argue that it is reasonable to expect that if no advance tickets are sold, there will be no demand for soft drinks, such an occurrence on a game week is unlikely if not impossible. Additionally, such an instance was not considered by the sample. **b.** $b^* = \sum XY/\sum X^2 = 45{,}691.98/31{,}125.76 = 1.4680$; $s^2_{Y\cdot X} = (1/6)\left[\sum Y^2 - (\sum XY)^2/\sum X^2\right] = (1/6)[67{,}082.26 - (45{,}691.98)^2/31{,}125.76] = 1.2278$. $t = b^*/(s^2_{Y\cdot X}/\sum X^2)^{1/2} = 1.4680/(1.2278/31{,}125.76)^{1/2} = 233.734 > t_{6,0.05} = 1.943$, so you are more than 95% sure that the greater the number of advance tickets sold, the more gallons of syrup are required.

7.12 **a.** $t = 3.971 > t_{23,0.005} = 2.807$. Reject H_0 in favor of H_1. **b.** Null: those who don't smoke take no sick leave. Alternative: those who don't smoke do take sick leave. **c.** You are more than 99% sure that nonsmokers take sick leave.

7.13 $t = 2.090 > t_{18,0.05} = 1.734$. You are more than 95% sure that the new gauge reads consistently higher than the old gauge.

7.14 $t = 3.640 > t_{13,0.005} = 3.012$, so you are more than 99% sure that the student consistently measures the tree diameters differently from the instructor.

7.15 $t = 2.290 > t_{22,0.05} = 1.717$, so you are more than 95% sure that there will always be guests arriving without reservations.

7.16 $t = 0.995 \not> t_{9,0.10} = 1.383$, so you cannot be even 90% sure that the intercept is not zero. Since the data did not include a control group of children who did not ingest any preservative, one cannot conclude that such children would have a hyperactivity level of zero. Rather, it may not be necessary to include the intercept as a correction factor in the estimating equation.

SECTION 7.5, PAGE 229

7.17 **a.** $p = 0.025$. **b.** $p = 0.05$. **c.** $0.005 < p < 0.01$. **d.** $0.01 < p < 0.02$. **e.** $0.025 < p < 0.05$. **f.** $0.05 < p < 0.10$. **g.** $p < 0.005$. **h.** $p < 0.001$. **i.** $p > 0.10$. **j.** $p > 0.20$.

7.18 **a.** $p = 0.05$. **b.** $0.01 < p < 0.05$. **c.** $p > 0.05$. **d.** $p > 0.01$. **e.** $p > 0.05$.

7.19 **a, b, c, d, e, g, h.**

7.20 **a, b, d.**

7.21 **a.** $t = 21.389 > t_{5,0.005} = 4.032$, so you are more than 99.5% sure that the more tickets sold, the more syrup is required ($p < 0.005$). **b.** $t = -13.148 < -t_{13,0.005} = -3.012$, so you are more than 99.5% sure that the heating bill increases as the temperature decreases ($p < 0.005$). **c.** $t = 0.902 \not> t_{23,0.10} = 1.319$, so you cannot be even 90% sure that days of sick leave increase with number of cigarettes smoked ($p > 0.10$). **d.** $t = 10.716 > t_{22,0.005} = 2.819$, so you are more than 99% sure that the sale price is related to the appraised value ($p < 0.01$). **e.** $t = 0.218 > t_{5,0.10} = 1.476$, so you cannot be even 80% sure that the intercept is different from zero ($p > 0.20$). **f.** $t = 3.971 > t_{23,0.005} = 2.807$, so you are more than 99% sure that nonsmokers take sick leave ($p < 0.01$).

SECTION 7.6, PAGE 237

7.22 **a.** $\mathbf{y} = [Y_1 Y_2 \cdots Y_n]'$, $\hat{\boldsymbol{\beta}} = [a \quad b_1 \quad b_2]'$,

$$\mathbf{X} = \begin{bmatrix} 1 & X_{11} & X_{21} \\ 1 & X_{12} & X_{22} \\ \vdots & \vdots & \vdots \\ 1 & X_{1n} & X_{2n} \end{bmatrix}$$

b.

$$\mathbf{X'X} = \begin{bmatrix} n & \sum X_{1i} & \sum X_{2i} \\ \sum X_{1i} & \sum X_{1i}^2 & \sum X_{1i}X_{2i} \\ \sum X_{2i} & \sum X_{1i}X_{2i} & \sum X_{2i}^2 \end{bmatrix}, \quad \mathbf{X'y} = \begin{bmatrix} \sum Y_i \\ \sum X_{1i}Y_i \\ \sum X_{2i}Y_i \end{bmatrix}$$

c. $\hat{\boldsymbol{\beta}}\mathbf{X'y} = a\sum Y_i + b_1 \sum X_{1i}Y_i + b_2 \sum X_{2i}Y_i$. **d.** 3 (three terms in the sum). **e.** $H_0: \alpha = \beta_1 = \beta_2 = 0$.

f.

ANOVA

Source	SS	df
Regression	$\hat{\boldsymbol{\beta}}'\mathbf{X'y} - \frac{(\sum Y_i)^2}{n}$	2
Error	$\mathbf{y'y} - \hat{\boldsymbol{\beta}}'\mathbf{X'y}$	$n - 3$
Total	$\mathbf{y'y} - \frac{(\sum Y_i)^2}{n}$	$n - 1$

7.23 **a.** $\mathbf{y} = [Y_1 \quad Y_2 \cdots Y_n]'$, $\hat{\boldsymbol{\beta}} = [b_1 \quad b_2]'$,

$$\mathbf{X} = \begin{bmatrix} X_{11} & X_{21} \\ X_{12} & X_{22} \\ \vdots & \vdots \\ X_{1n} & X_{2n} \end{bmatrix}$$

b. $$\mathbf{X}'\mathbf{X} = \begin{bmatrix} \sum X_{1i}^2 & \sum X_{1i}X_{2i} \\ \sum X_{1i}X_{2i} & \sum X_{2i}^2 \end{bmatrix}, \quad \mathbf{X}'\mathbf{y} = \begin{bmatrix} \sum X_{1i}Y_i \\ \sum X_{2i}Y_i \end{bmatrix}$$

c. $\hat{\boldsymbol{\beta}}'\mathbf{X}'\mathbf{y} = b_i \sum X_{1i}Y_i + b_2 \sum X_{2i}Y_i$. **d.** 2 = number of terms in the sum. **e.** $H_0\colon \beta_1 = \beta_2 = 0$.

f. ANOVA

Source	*SS*	*df*
Regression	$\hat{\boldsymbol{\beta}}'\mathbf{X}'\mathbf{y}$	2
Error	$\mathbf{y}'\mathbf{y} - \hat{\boldsymbol{\beta}}'\mathbf{X}'\mathbf{y}$	$n - 2$
Total	$\mathbf{y}'\mathbf{y}$	n

7.24 **a.** $\mathbf{y} = [Y_1 \quad Y_2 \cdots Y_n]'$, $\hat{\boldsymbol{\beta}} = [a' \quad b_1 \quad b_2]'$,

$$\mathbf{X} = \begin{bmatrix} 1 & X_{11} - \overline{X}_1 & X_{21} - \overline{X}_2 \\ 1 & X_{12} - \overline{X}_1 & X_{22} - \overline{X}_2 \\ \vdots & \vdots & \vdots \\ 1 & X_{1n} - \overline{X}_1 & X_{2n} - \overline{X}_2 \end{bmatrix}$$

b. $$\mathbf{X}'\mathbf{X} = \begin{bmatrix} n & 0 & 0 \\ 0 & \sum (X_{1i} - \overline{X}_1)^2 & \sum (X_{2i} - \overline{X}_2) \\ 0 & \sum (X_{1i} - \overline{X}_1)(X_{2i} - \overline{X}_2) & \sum (X_{2i} - \overline{X}_2)^2 \end{bmatrix}$$

$$\mathbf{X}'\mathbf{y} = \begin{bmatrix} \sum Y_i \\ \sum (X_{1i} - \overline{X}_1)Y_i \\ \sum (X_{2i} - \overline{X}_2)Y_i \end{bmatrix}$$

c. $\hat{\boldsymbol{\beta}}'\mathbf{X}'\mathbf{y} = a' \sum Y_i + b_1 \sum (X_{1i} - \overline{X}_1)Y_i + b_2 \sum (X_{2i} - \overline{X}_2)Y_i$. **d.** 3 = number of terms in the sum.
e. $H_0\colon \alpha' = \beta_1 = \beta_2 = 0$.

f. ANOVA

Source	*SS*	*df*
Regression	$\hat{\boldsymbol{\beta}}'\mathbf{X}'\mathbf{y} - \dfrac{(\sum Y_i)^2}{n}$	2
Error	$\mathbf{y}'\mathbf{y} - \hat{\boldsymbol{\beta}}'\mathbf{X}'\mathbf{y}$	$n - 3$
Total	$\mathbf{y}'\mathbf{y} - \dfrac{(\sum Y_i)^2}{n}$	$n - 1$

7.25 **a.** $s_a = (c_{00}s_{Y\cdot X}^2)^{1/2}$. **b.** $s_{b_1} = (c_{11}s_{Y\cdot X}^2)^{1/2}$. **c.** $s_{b_2} = (c_{22}s_{Y\cdot X}^2)^{1/2}$.

SECTION 7.7, PAGE 241

7.26 **a.** $1.4532 \pm 2.571\sqrt{1.4594(0.0032)} = 1.2775$ to 1.6289. You are 95% certain that this interval contains the true average additional number of gallons of syrup required for each additional 1,000 tickets sold. **b.** $0.9886 \pm 4.032\sqrt{1.4594(14.0652)} = -17.2790$ to 19.2562. There is evidence that the intercept term could be dispensed with.

7.27 **a.** $-0.0080 \pm 3.012\sqrt{0.0018(0.0002)} = -0.0098$ to -0.0062. You are 99% confident that if on one day the temperature is one degree higher than on another, the heating bill on the warmer day will be between \$0.0062 and \$0.0098 lower. **b.** $0.7347 \pm 2.160\sqrt{0.0018(0.1395)} = 0.7005$ to 0.7689. You are 95% sure that this interval, \$0.7005 to \$0.7689, covers the true average heating bill on days when the temperature is zero degrees.

7.28 **a.** $0.0555 \pm 1.714\sqrt{24.4397(0.0002)} = -0.0643$ to 0.1753. **b.** $6.2852 \pm 1.714\sqrt{24.4397(0.1025)} = 3.5724$ to 8.9980. **c.** We are 90% sure that nonsmokers take, on the average, between 3.6 and 9 days sick leave, but we cannot be 90% sure that smokers are any different from nonsmokers in this regard.

7.29 **a.** $1.0004 \pm 2.878\sqrt{4.0430(0.000006)} = 0.9862$ to 1.0146. **b.** $2.5108 \pm 2.878\sqrt{4.0430(0.3571)} = -0.9473$ to 5.9689. **c.** You cannot be 99% sure that the new gauge performs differently from the old gauge.

7.30 **a.** $3.3985 \pm 2.160\sqrt{0.4441(1.9629)} = 1.3818$ to 5.4152. **b.** $0.8874 \pm 2.160\sqrt{0.4441(0.0021)} = 0.8214$ to 6.9534. **c.** For each 1 cm increase in diameter, the student's measurement will increase between 0.82 and 0.95 cm; in addition, the student will add between 1.38 and 5.42 cm regardless of the diameter of the tree.

7.31 $1.4230 \pm 1.717\sqrt{20.0184(0.0031)} = 0.9953$ to 1.8507. Since the ends of the interval are not both above 1.00, you cannot be 90% sure that XZQ is an aggressive stock.

SECTION 7.8, PAGE 245

7.32 **a.** $F = 457.475$, $p = 0.0001 < 0.05$, so you are more than 95% sure that syrup used is related to advance tickets sold. **b.** $t = 21.389$, $p = (1/2)(0.0001) = 0.00005 < 0.05$, so you are more than 95% sure that the more tickets sold, the more syrup needed. **c.** $t = 0.218$, $p = 0.8359 > 0.05$. You can be only 16.41% sure that the intercept is different from zero. **d.** $1.4532 \pm 2.571\sqrt{1.4594(0.0032)} = 1.2775$ to 1.6298. You are 95% sure that for each additional 1,000 tickets sold, between 1.28 and 1.63 additional gallons of syrup will be required. **e.** $0.9886 \pm 2.571\sqrt{1.4594(14.0652)} = -10.6597$ to 12.6369. We have some evidence that the intercept term could be dropped from the model.

7.33 **a.** $F = 0.814$, $p = 0.3764$. The evidence is insufficient to conclude that days sick leave are related to smoking habits. **b.** $t = 0.902$, $p = (1/2)(0.3764) = 0.1882 > 0.10$. The evidence is insufficient to be 90% certain that days sick leave increase with the number of cigarettes smoked.

7.34 **a.** $H_0{:}\beta = 0$; $H_1{:}\beta \neq 0$. $F = 45.643$, $p = 0.0001$. Rating is related to class size. **b.** $t = -6.756$, $p = (1/2)(0.0001) = 0.00005$. Larger classes give lower ratings. **c.** $t = 46.327$, $p = (1/2)(0.0001) = 0.00005$. An intercept term is needed in the model. **d.** $-0.0087 \pm 2.845\sqrt{0.0472(0.00003)} = -0.0121$ to -0.0053. You are 99% sure that the teacher in a class having one more student than another class will receive a rating between 0.0121 and 0.0053 points lower. **e.** $3.6374 \pm 2.845\sqrt{0.0472(0.1306)} = 3.4140$ to 3.8608. You are 99% sure that this interval covers the true intercept in the population.

E I G H T

SECTION 8.2, PAGE 255

8.1 Variation in advance ticket sales explains 98.92% of the variation in amount of syrup used.
8.2 Variation in temperature explains 93.01% of the variation in the heating bill.
8.3 Variation in the number of cigarettes smoked explains only 3.42% of the variation in the number of sick days taken.
8.4 Variation in the number of pounds of mail received explains 99.10% of the variation in the number of orders.
8.5 **a.** The quadratic model explains 88.55% of the variation in the number of robberies. **b.** $1 - [1182.883/(103.308/12)] = 0.8626$.
8.6 **a.** $r^2 = 0.6024$; adj $r^2 = 1 - [(330/13)/(830/14)] = 0.5718$. **b.** $r^2 = 0.6024$; adj $r^2 = 1 - [(330/10)/(830/14)] = 0.4434$. **c.** That one predictor could produce $SS(R) = 500$ should indicate a stronger relationship than if four predictors together produce the same $SS(R)$.

SECTION 8.3–8.4, PAGES 260 AND 263

8.7 **a.** $r = 0.9946$; strong direct relationship. **b.** $H_0{:}\rho = 0$; $H_1{:}\rho > 0$; $t = 21.4 > t_{5,0.005} = 4.032$. You are more than 99.5% sure that the more tickets sold, the more syrup is needed. **c.** $r^2 = 0.9892$.
8.8 **a.** $r = -0.9644$; strong inverse relationship. **b.** $H_0{:}\rho = 0$; $H_1{:}\rho < 0$; $t = -13.15 < -t_{13,0.005} = -3.012$. You are more than 99.5% sure that the colder the day, the higher the heating bill will be. **c.** $r^2 = 0.9301$.
8.9 **a.** $r = 0.1849$; weak direct relationship. **b.** $t = 0.902 \not> t_{23,0.10} = 1.319$. You cannot be even 80% sure that sick days are related to the number of cigarettes smoked. **c.** $r^2 = 0.0342$.
8.10 **a.** $r = 0.9955$; strong direct relationship. **b.** $t = 37.835 > t_{13,0.005} = 3.012$. You are more than 99.5% sure that the heavier the mail, the more orders will be received. **c.** $r^2 = 0.9910$.
8.11 **a.** $r = -0.8746$. **b.** $r^2 = 0.7649$. The value for r^2 in Exercise 8.5 measures the proportion of variation in number of robberies explained by a quadratic model; this r^2 measures proportion of variation explained by a linear model.
8.12 Reject H_0 if $r/\sqrt{(1 - r^2)/(n - 2)} > t_{0.05}$. With some algebra, this can be written as reject H_0 if $r > t/\sqrt{t^2 + n - 2}$.

n	$t_{n-2,0.05}$	Critical r	n	$t_{n-2,0.05}$	Critical r
5	2.353	0.8053	30	1.701	0.3060
10	1.860	0.5495	50	1.684	0.2362
15	1.771	0.4409	75	1.671	0.1919
20	1.734	0.3783	100	1.671	0.1664
25	1.714	0.3365			

8.13 $\sum Z_1 = \sum Z_2 = 0$;

$$\sum Z_1Z_2 = \sum \left(\frac{X - \bar{X}}{s_X}\right)\frac{(Y - \bar{Y})}{s_Y} = \frac{1}{s_X s_Y} S_{XY} = \frac{S_{XY}}{\sqrt{[S_{XX}/(n-1)][S_{YY}/(n-1)]}}$$
$$= (n - 1)r.$$

N I N E

SECTION 9.2, PAGE 279

9.1 **a.b.** See Appendix L, Secretaries Data (p. 646). **c.** Ms. Moore's salary does appear to be unusually high. **d.** If Ms. Moore's salary is omitted, a 99% confidence interval on her salary is \$703.10 to \$802.98. **e.** $r_{22} = 0.6816$; $p < 0.005$.

9.2 **a.** See Appendix L, Textbook Data (p. 636). **b.** $F = 0.1745 < 1/F_{6,5,0.05} = 1/4.95 = 0.2020$. Conclude nonuniform scatter. **c.** See Appendix L, Textbook Data. **d.** $F = 1.1269 \not> F_{5,6,0.05} = 4.39$ and $\not< 1/F_{6,5,0.05} = 1/4.95 = 0.2020$. Cannot detect nonuniform scatter in transformed data. **e.** $t = 4.6808/(8.5578/0.6440)^{1/2} = 1.2841 < t_{13,0.10} = 1.350$, so we cannot be 80% sure that any linear relationship exists in the transformed data.

9.3 See Appendix L, Appraisal Data (p. 642). One residual appears to be much larger than the others. Dixon test: $r_{22} = 0.4991$, so you are between 98% and 99% sure that it is an outlier. The reason for this should be investigated.

9.4 See Appendix L, Reservations Data (p. 643). Residuals appear to display uniform scatter.

SECTION 9.3, PAGE 286

9.5 **a.** See Appendix L, Deprivation Data (p. 640). $t = 0.7210 \not> t_{0.10,8} = 1.397$. There is no significant linear relationship. **b.** See Appendix L, Deprivation Data.

c. Z	Y	s^2
−2	73.68	12.5
−1	49.54	12.5
0	32.25	24.5
1	42.40	2.0
2	60.65	12.5
		64.0

SS (pure error) $= 64$ with $10 - 5 = 5$ df.
$SS(E) = 2161.15$ with 8 df.
SS (lack of fit) $= 2161.15 - 64 = 2097.15$ with $5 - 2 = 3$ df.
$F = (2097.15/3)/(64/5) = 54.6133 > F_{3,5,0.01} = 5.41$. A simple linear model is not appropriate.

9.6 **a.** See Appendix L, Trees Data (p. 646).

b. X	Y	s^2
2.5	110, 94, 100, 98	46.3333
2.7	96, 100	8.0000
3.5	100, 102, 106	9.3333
3.6	108, 102, 106	9.3333
4.7	114, 118	8.0000
5.8	134, 138, 132	9.3333

SS (pure error) = 3(46.3333) + 8 + 2(9.3333) + 2(9.3333) + 8 + 2(9.3333) = 211 with 25 − 14 = 11 df.
$SS(E)$ = 597.9972 with 23 df.
SS (lack of fit) = 386.9972 with 14 − 2 = 12 df.
$F = (386.9972/12)/(211/11) = 1.6813 \not> F_{12,11,0.05} = 2.79$. We cannot conclude a significant lack of fit (probably because of the large variance of Y-values at $X = 2.5$).

9.7 **a.** See Appendix L, Typist Data (p. 647). **b.** Residuals increase in a time order.

9.8 **a.b.** See Appendix L, IQ Data (p. 647). **c.** Since scores for students taught by teacher A tend to be above the regression line, while those for students taught by teacher B tend to be below the line, the teacher is apparently a factor in predicting score.

9.9 It appears that the four very large classes have great influence on the slope of the line. If we omit the four largest classes, the regression equation is $\hat{Y} = 3.64 - 0.0090X$, which is practically identical to the regression equation on data including the four largest classes. There is no reason to treat these four classes differently from the others.

SECTION 9.4, PAGE 300

9.10 **a.** $r_{t,t-1} = 0.7739$; $d = 0.4788 < 1.26$. There is a positive serial correlation. **b.** $r_{t,t-1} = -0.2366$; $d = 2.2015$; $d^- = 1.7985 > 1.44$. There is no negative serial correlation in the first differences.

9.11 $r_{t,t-1} = 0.0175$; $d = 1.8613 > 1.36$. There is no significant serial correlation.

9.12 $r_{t,t-1} = -0.0376$; $d^- = 2.0680$; $d^- = 1.932 > 1.45$. There is no significant serial correlation.

9.13 **a.** Because of the seasonal pattern, one would expect to see a negative serial correlation. See Appendix L, Enrollment Data (p. 648). **b.** **c.** $r_{t,t-1} = -0.3195$; $d = 2.6135$; $d^- = 1.3865 < 1.41$. There is a significant negative serial correlation.

9.14 **a.** Lagged enrollments appear to display a positive serial correlation. **b.** $r_{t,t-1} = 0.5359$; $d = 0.7551 < 1.37$. Residuals show positive serial correlation. **c.** $Y_t - Y_{t-4}$ represents growth from one quarter of one year to the same quarter the next year. Such growth is most likely to be affected by periods of economic expansion and contraction.

SECTION 9.5, PAGE 304

9.15

Std. Resid.	Frequency
< −3.000	0
−3.000 to −2.001	0
−2.000 to −1.001	0
−1.000 to 1.000	21
1.001 to 2.000	2
2.001 to 3.000	0
>3.000	1
Total	24

Standardized residuals do not appear to be normally distributed.

9.16

Std. Resid.	Frequency
< −3.000	0
−3.000 to −2.001	0
−2.000 to −1.001	6
−1.000 to 1.000	14
1.001 to 2.000	4
2.001 to 3.000	0
>3.000	
Total	24

58.3% of values lie within 1 standard deviation of the mean; 100% of values lie within 2 standard deviations of the mean. Normality is questionable.

9.17

Std. Resid.	Frequency
< −3.000	0
−3.000 to −2.001	1
−2.000 to −1.001	2
−1.000 to 1.000	16
1.001 to 2.000	5
2.001 to 3.000	0
>3.000	0
Total	24

66.7% of values lie within 1 standard deviation of the mean; 95.8% lie within 2 standard deviations; 100% lie within 3 standard deviations. Residuals appear to be normally distributed.

T E N

SECTIONS 10.2–10.3, PAGE 327

10.1 a. b. See Appendix L, Cancer Data (p. 648). Data do not indicate a linear relationship. **c.** We would expect 0.58% of cells to die not related to amount of exposure; with each additional hour of exposure, 766% more cells are killed.

10.2 a. b. See Appendix L, Abrasive Data (p. 649). The graph indicates that the thickness of the abrasive decreases quickly at first and then levels out as the hours of operation increase. **c.** 72.0767 is a constant determined outside of the relationship between thickness and hours of operation. With each additional hour of operation, the thickness remaining is 92.05% of the thickness the previous hour.

10.3 a. See Appendix L, Census Data: I (p. 650). **b.** Starting at an estimated level of 66.7 million people in 1890, the U.S. population grew, on the average, by 15.05% per decade until 1970. **c.** $\hat{Y} = 66.7064(1.1505)^9 = 235.58$ million. U.S. population in 1980 was approximately 226 million.

10.4 a. b. See Appendix L, Census Data: II (p. 650). Exponential curve appears to be increasing faster than the data for more recent years. **c.** $\hat{Y} = 80.4621(1.1891)^6 = 227.4582$ million, which is quite a bit higher than the actual figure, 220.25 million. The exponential model does not appear to give good forecasts.

10.5 a. c. See Appendix L, Police Data (p. 638). **b.** 1171.4259 is a constant affecting the number of robberies determined by something other than the relationship with the number of police officers. The number of robberies decreases on the average by 0.96% with each additional officer employed.

10.6 a. See Appendix L, MPG Data (p. 651). Relationships are not linear. **b.** See Appendix L, MPG Data (p. 651). 44.2238 is a constant affecting mpg unrelated to horsepower. With each additional foot-pound of horsepower, mpg decreases on the average by 0.59%. **c.** See Appendix L, MPG Data. 57.1492 is a constant affecting mpg unrelated to weight. With each additional 1,000 lb of weight, mpg decreases 26.04%. **d.** Yes—those automobiles with manual transmissions tend to have the lowest horsepower and weight.

SECTIONS 10.4–10.5, PAGE 368

10.7 a. See Appendix L, Deprivation Data (p. 640).

b. ANOVA

Source	*SS*	*df*	*MS*	*F*
Regression	2079.3429	2	1039.6714	32.74
Error	222.2571	7	31.7510	
Total	2301.6000	9		

$F = 32.74 > F_{2,7,0.01} = 9.55$, so reject $H_0: \beta_1 = \beta_2 = 0$ in favor of H_1: either β_1 or β_2 or both are different from zero. **c.** $SS(R)(X) = 140.45$, $F = 4.42$; $SS(R)(X^2 \mid X) = 1938.8929$, $F = 61.07$; $F_{1,7,0.05} = 5.59$, $F_{1,7,0.01} = 12.25$. We cannot be 95% certain that a linear term is needed; addition of a quadratic term to the linear term improves the model. **e.** $F = (159.2836/2)/64.15 = 6.2220 > F_{2,5,0.05} = 5.79$, so H_0: second-degree polynomial model is inadequate at the $\alpha = 0.05$ significance level. The plot reveals that the curve does not come very close to several of the data points.

10.8 **a.** See Appendix L, Trees Data (p. 646). **b.** $F = (4536.2226/2)/(294.8174/22) = 169.25 > F_{2,22,0.01} = 5.72$, so reject $H_0: \beta_1 = \beta_2 = 0$ in favor of $H_1: \beta_1 \neq 0$ and/or $\beta_2 \neq 0$. **c.** For linear term, $F = (4233.0420/1)/(294.8174/22) = 315.88$. For quadratic term, given linear term, $F = (303.1806/1)/(294.8174/22) = 22.62$. Since $F_{1,22,0.01} = 7.95$, we are more than 99% sure that (1) $Y = \alpha + \beta X + \varepsilon$ is a better model than $Y = \alpha + \varepsilon$, and (2) $Y = \alpha + \beta_1 X + \beta_2 X^2 + \varepsilon$ is a better model than $Y = \alpha + \beta X + \varepsilon$.

10.9 **a.** $F = (91{,}478.8664/2)/(11{,}828.8259/10) = 38.67 > F_{2,10,0.01} = 7.56$, so $H_0: \beta_1 = \beta_2 = 0$ is rejected in favor of $H_1: \beta_1 \neq 0$ and/or $\beta_2 \neq 0$. **b.** Linear: $F = (79{,}014.9170/1)/(11{,}828.8259/10) = 66.80 > F_{1,10,0.01} = 10.04$, so $Y = \alpha + \beta X + \varepsilon$ is preferred to $Y = \alpha + \varepsilon$. Quadratic, given linear: $F = (12{,}463.9494/1)/(11{,}828.8259/10) = 10.54 > F_{1,10,0.01} = 10.04$, so $Y = \alpha + \beta_1 X + \beta_2 X^2 + \varepsilon$ is preferred to $Y = \alpha + \beta X + \varepsilon$. **c.** The test could be performed, since there is one instance of repeated observations at a given X-value; however, pure error would not be estimated very well based on only two observations.

10.10 **a.** See Appendix L, Census Data I (p. 650). **b.** Estimated population in 1910 ($X = 0$) is 88.9965 million. In 1910 population was increasing by 14.5010 million per decade. With each subsequent decade, population increased by 2(0.5989) = 1.1978 million more than in the previous decade. **c.** $\hat{Y} = 88.9965 + 14.5010(6) + 0.5989(36) = 197.56$ million. Actual = 200.25 million. Quadratic is closer than exponential, which gave an estimate of 227.5 million for 1970.

10.11 **a.** See Appendix L, MPG Data (p. 651). $SS(R)(X^2 \mid X) = 114.9908$, $F = 16.87 > F_{1,22,0.01} = 7.95$, so the quadratic term adds to the model. **b.** See Appendix L, MPG Data. $SS(R)(X^2 \mid X) = 116.6295$, $F = 39.43 > F_{1,22,0.01} = 7.95$, so the quadratic term adds to the model.

10.12 **a. b.** See Appendix L, Rush Hour Data (p. 653). **c.** $X = 99.61$ vehicles. **d.** $F(X \mid X^2) = 39.67$; $F(X^2 \mid X) = 76.72 > F_{1,13,0.01} = 9.07$. If the quadratic term is used, the linear term is also needed; if the linear term is used, the quadratic term is also needed. **e.** $F(X) = 613.08$, $F(X^2 \mid X) = 76.72 > F_{1,13,0.01} = 9.07$. The model $Y = \alpha + \beta X + \varepsilon$ is preferred to $Y = \alpha + \varepsilon$; the model $Y = \alpha + \beta_1 X + \beta_2 X^2 + \varepsilon$ is preferred to $Y = \alpha + \beta X + \varepsilon$.

10.13 See Appendix L, Professionals Data (p. 637). Income increases with age, but levels off. $F(X^2 \mid X) = 206.02 > F_{1,5,0.01} = 16.26$, so a quadratic term is needed in the model.

E L E V E N

SECTION 11.3, PAGE 382

11.1 See Appendix L, Sales Data (page 653).
11.2 See Appendix L, Restaurant Data (page 653).
11.3 See Appendix L, Drug Data (page 654).
11.4 See Appendix L, Auto Data (page 654).
11.5 See Appendix L, Ads Data (page 654).
11.6 See Appendix L, Children Data (page 654).
11.7 See Appendix L, Glue Data (page 654).
11.8 See Appendix L, Bottles Data (page 655).
11.9 See Appendix L, Yield Data (page 655).

SECTION 11.5, PAGE 404

11.11 **a.** $a = -2.447909$ is a constant determined outside of the relationship. For two stores the same size, if one has one more employee than another, then on the average its sales will be \$269.82 more than in the other. Among stores with the same number of employees, a store 1000 sq ft larger than another will have \$667.39 greater sales on the average. **b.** $a = 0.311472$ is determined outside of the relationship. For restaurants with the same liquor sales, if one has \$1000 greater food sales than another, then it will average \$1023.20 greater total profits than the other. If two restaurants have the same food sales but one has liquor sales \$1000 greater than another, then it will average \$72.63 higher profits than the other. **c.** The intercept is determined by something other than the effect of dosage and age on time to relief. Among patients receiving the same dosage, if one is one year older than another, he will experience relief about half a minute sooner than will the younger. For patients the same age, one receiving one gram more than another will experience relief 1.14 minutes later than the other. **d.** The constant represents an effect on cost determined by something other than age and mileage. If one car is one year older than another but both were driven the same number of miles, then the older car averaged \$52.53 more in maintenance and repair costs. For two cars the same age, if one were driven 1,000 miles farther than another, then its cost of upkeep averaged \$3.99 more. **e.** $a = 195.833333$ is determined outside the relationship. If humidity is held constant, then holding strength is 1 lb greater on the average at the higher temperature level than at the lower. If temperature is held constant, the holding strength is 6 lb greater for each higher level of humidity. **f.** a was determined outside of the relationship. If temperature is held constant, then an average of 0.92 fewer impurities are found at longer sterilization times than at the next shorter times. For equal sterilization times, an average of 0.56 fewer impurities are found at higher temperatures than at the next lower temperature level.

11.12 **a.** The constant is determined outside of the relationship. If two stores spend the same amount on radio and newspaper advertising but one spent \$1000 more on TV ads, then on the average the store spending more on TV ads will have \$2620.38 greater sales. For stores spending the same amount on TV and newspaper, one spending \$1000 more on radio will have \$7556.19 greater sales than another. If one store spends \$1000 more on newspaper ads than another but both spend the same amount on radio and TV, then the higher-spending store will have \$1900.66 greater sales. **b.** The regression constant is determined outside of the relationship. For children the same height and weight, the average score of one child a year older than another is 9.5272 points higher. For two children the same age and weight, if one is one inch taller than another her average score will be 3.349 points higher. For children the same age and height, if one is one pound heavier than another, his score will be 0.5969 points lower than the lighter child's, on the average. **c.** The average yield for plots receiving medium amounts of nitrogen, phosphorus, and potassium was 322.26 lb. For plots receiving the same amounts of potassium and phosphorus, those receiving higher levels of nitrogen averaged 51.17 lb greater yield than those at the next lower levels. For plots receiving the same application of nitrogen and phosphorus, those receiving one level higher application of potassium averaged 47.39 lb less yield than those at the lower level. If two plots received the same amounts of nitrogen and potassium but one received one level higher application of phosphorus than another, the yield from the plot receiving more phosphorus was 8.06 lb greater than from the other.

11.13 **a.** \$6810.90. It is only an average and is not intended to coincide with any actual value. **b.** \$2268.80. Other factors affecting total profits have not been included. **c.** The experiment included no trials on which patients were given 3 gm. **d.** \$27.95. **e.** Sample included no information on expenditures as large as \$6000 on newspaper advertisements. **f.** 74.1715. The estimate is only an average and is not expected to coincide exactly with any particular score. **g.** 194.833333. Factors other than temperature and humidity, including random variation, also affect holding strength. **h.** 3.270834; other factors omitted. **i.** 334.092593 lb.

SECTION 11.6, PAGE 416

11.14 **a.** b_1 and b_2 changed from the simple regressions to the multiple regression, so we know that multicollinearity is present. **b.** No multicollinearity is present.

11.15 **a.** $r_{12} = 0.8965$; regression coefficients will be very unreliable. **b.** $r_{12} = 0.4494$; regression coefficients will be rather unreliable. **c.** $r_{12} = -0.0179$; regression coefficients slightly unreliable. **d.** $r_{12} = 0$; regression coefficients are reliable measures of the effects of the independent variables.

11.16 **a.** $r_{12} = -0.0778$, $r_{13} = -0.3464$, $r_{23} = 0.0898$; regression coefficients unreliable. **b.** $r_{12} = 0.8233$, $r_{13} = 0.7918$, $r_{23} = 0.6696$; regression coefficients unreliable.

11.17 **a.** 1.0232(0.8127/0.9743) = 0.8535; 0.0726(0.8967/0.9743) = 0.0668. For a 1 standard deviation increase in food sales, there is a 0.8535 standard deviation increase in profits; for a 1 standard deviation increase in liquor sales, there is a 0.0668 standard deviation increase in profits. Food sales are a stronger predictor of profits than are liquor sales. **b.** 1.1446(2.9912/10.3785) = 0.3299; −0.5333(18.3248/10.3785) = 0.9416. For a 1 standard deviation increase in dosage, there is a 0.3299 standard deviation increase in time to relief; for a 1 standard deviation increase in age, there is a 0.9416 standard deviation decrease in time to relief. Age is the stronger predictor. **c.** 0.5253(1.5946/1.0162) = 0.8243; 0.0399(7.5182/1.0162) = 0.2952. If one car is 1 standard deviation older than another, then its expected cost of repair is 0.8243 standard deviations greater. If one car was driven 1 standard deviation more miles than another, its expected cost of repair is 0.2952 standard deviations greater. Age is the stronger predictor of cost. **d.** 2.6202(2.8431/21.5088) = 0.3463; 7.5564(1.3707/21.5088) = 0.4815; 1.9008(0.7385/21.5088) = 0.0653. A 1 standard deviation increase in TV, radio, and newspaper advertising result in 0.3463, 0.4815, and 0.0653 standard deviation increases, respectively, in revenue. By this criterion, radio advertising is the most effective. **e.** 9.5298(0.9333/20.4718) = 0.4345; 3.3429(3.0822/20.4718) = 0.5033; −0.5975(10.7227/20.4718) = −0.3130. If a child is 1 standard deviation older than another, then it is expected that her score is 0.4345 standard deviations higher. If she is 1 standard deviation taller, her score is expected to be 0.5033 standard deviations higher. If she is 1 standard deviation heavier, then her score is on the average 0.3130 standard deviations lower.

SECTION 11.7, PAGE 422

11.18 **a.** Estimating sales from numbers of employees and sizes of stores, we will be off, on the average, by $1172.30. **b.** Using food and liquor sales to estimate total profits, we will be off by $448.44 on the average. **c.** Our estimates will be off by an average of 5.844 minutes when estimating time to relief from dosage and age. **d.** Upkeep costs can be estimated from age and mileage within $53.99 on the average. **e.** Revenues can be estimated from TV, radio, and newspaper advertising within $19,606.40 on the average. **f.** Physical fitness score can be estimated from age, height, and weight within 15.6896 points on the average. **g.** When estimating holding strength from temperature and humidity, the estimates will be off by 1.4011 lb on the average. **h.** The number of impurities can be estimated from temperature and time of sterilization within 0.8825 on the average. **i.** Yield can be estimated from amounts of the three elements applied within 109.2790 lb on the average.

11.19 **a.** $2111.80. **b.** Predictions are more accurate when using the number of employees than when using the sizes. **c.** More accurate predictions can be made by using both predictors than by using either predictor alone.

SECTION 11.8, PAGE 424

11.20 **a.** $\mathbf{y} = [Y_1 \quad Y_2 \cdots Y_n]'$, $\hat{\boldsymbol{\beta}} = [a \quad b_1 \quad b_2 \quad b_3 \quad b_4]'$,

$$\mathbf{X} = \begin{bmatrix} 1 & X_{11} & X_{21} & X_{31} & X_{11}X_{21} \\ 1 & X_{12} & X_{22} & X_{32} & X_{12}X_{22} \\ \vdots & \vdots & \vdots & \vdots & \vdots \\ 1 & X_{1n} & X_{2n} & X_{3n} & X_{1n}X_{2n} \end{bmatrix}$$

b. $\hat{\boldsymbol{\beta}} = (\mathbf{X}'\mathbf{X})^{-1}\mathbf{X}'\mathbf{y}$.

11.21 See Appendix L, MPG Data (p. 651).

T W E L V E

SECTION 12.2, PAGE 426

12.1 **a.** $r_{1Y}^2 = 0.9156$; differences in the numbers of employees alone explain 91.56% of the variation in the sales of the stores. $r_{2Y}^2 = 0.8187$; differences in the sizes of the stores explain 81.87% of the variation in sales. **b.** Variation in food sales explains 83.44% $= r_{1Y}^2$ of the variation in total profits, and variation in liquor sales explains $r_{2Y}^2 = 69.22\%$ of the variation in total profits. **c.** Differences in drug dosages explain 0.87% of the variation in time to pain relief, and differences in ages explain 58.8% of the variation in time to pain relief. **d.** 67.07% of the variation in upkeep costs can be explained by differences in the ages of the cars, and 7.88% of the variation in cost can be accounted for by mileage differences. **e.** Differences in amounts spent on television, radio, and newspaper advertising explain 8.2%, 21.2% and 0.01%, respectively, of the variation in revenues. **f.** Variation in ages of the children explains 36.13% of the variation in scores, variation in heights explains 42.44%, and variation in weights explains 13.55%. **g.** Differences in temperature explain 3.78% of the variation in holding strength, and differences in humidity account for 90.66% of the variation in strength. **h.** Differences in sterilization times account for 70.19% of the variation in the number of impurities remaining, and 7.05% of the variation can be explained by differences in sterilization temperatures. **i.** Differences in the amounts of nitrogen, potassium, and the phosphorus applied explain 12.56%, 10.77%, and 0.31% of the variation in yield, respectively.

SECTION 12.3, PAGE 429

12.2 **a.** Together, differences in sizes and numbers of employees explain 94.84% of the variation in sales. **b.** Differences in food and liquor sales together explain 83.50% of the variation in profits. **c.** Differences in dosages and patient ages together explain 71.63% of the variation in time to relief. **d.** Age and mileage differences together explain 75.80% of the variation in cost of upkeep. **e.** Using amounts spent on radio, television, and newspaper advertising, we can explain 52.02% of the variation in revenues. **f.** Differences in ages, heights, and weights together account for 47.45% of the variation in scores. **g.** Using both temperature and humidity, we can explain 94.44% of the variation in strength. **h.** 77.24% of the variation in the number of impurities can be accounted for by differences in sterilization times and temperatures. **i.** Differences in the amounts of nitrogen, potassium, and phosphorus applied account for 23.64% of the variation in yield.

12.3 **a.** 71.27% of the variation in the number of employees can be explained by the differences in sizes of the stores; the two predictors are giving very redundant information. **b.** 80.37% of the variation in food sales can be explained by variation in liquor sales; the two predictors are giving very much the same information. **c.** Since there was a tendency for older patients to receive larger doses, we cannot separate the percentages of variation explained by dosage alone and by ages alone. **d.** There was a slight tendency for newer cars to be driven farther than older cars, so the percentages of variation explained by upkeep and mileage are not disjoint.

12.4 **a.** The amounts spent on advertising in the three media are not independent and cannot explain variation in revenue independently. **b.** Since ages, heights, and weights of children are not independent, they cannot explain independent percentages of variation in scores.

12.5 In all cases the predictor variables are uncorrelated; thus they explain independent percentages of variation in the response and the total percentage of explained variation is equal to the sum of the percentages explained by each predictor separately.

SECTION 12.4, PAGE 434

12.6 **a.** After sizes of the stores have been taken into account, variation in the number of employees explains 71.52% of the variation in sales. For stores with the same number of employees, variation in sizes explains 38.84% of the variation in sales. **b.** For stores with the same liquor sales, variation in food sales explains 46.5% of the variation in profits. After differences in food sales has explained all it can, variation in liquor sales accounts for 0.52% of the variation in profits. **c.** Once variation in ages of patients has explained all it can, variation in dosages explains 19.23% of the remaining variation in time to relief. Once variation in dosage has explained all it can of variation in time to relief, differences in ages of the patients account for 66.42% of the remaining variation. **d.** For cars with the same mileage, differences in ages explain 73.74% of the variation in upkeep costs; for cars the same age, 26.50% of the variation in upkeep costs can be explained by mileage differences.

12.7 **a.** After differences in amount spent on radio and newspaper advertising have explained all they can, variation in amounts spent on TV advertising explain 13.47% of the remaining variation in revenues. After TV and newspaper advertising have explained all they can, radio advertising explains 25.24% of the remaining variation. After radio and TV advertising have told all they can, newspaper advertising explains 0.58% of the remaining variation. **b.** For children the same height and weight, age differences account for 8.76% of the variation in scores. For children the same age and weight, height differences account for 13.4% of the variation in scores. For children the same age and height, weight differences explain 6.47% of the variation in scores.

12.8 **a.** For children the same age and height, there is a slight tendency for heavier children to score lower than do lighter children. **b.** The older, taller, and heavier child tends to score relatively high on the test.

12.9
$$r_{1Y\cdot 234} = \frac{r_{1Y\cdot 24} - r_{13\cdot 24}r_{3Y\cdot 24}}{\sqrt{1 - r_{13\cdot 24}^2}\sqrt{1 - r_{3Y\cdot 24}^2}}$$

$$= \frac{r_{1Y\cdot 34} - r_{12\cdot 34}r_{2Y\cdot 34}}{\sqrt{1 - r_{12\cdot 34}^2}\sqrt{1 - r_{2Y\cdot 34}^2}}$$

12.10 0.0760.

T H I R T E E N

SECTION 13.2, PAGE 439

13.1 a. Since $F = 110.1992 > 3.89$, conclude that either the number of employees, the size of the store, or both are significant predictors of sales. **b.** Since $F = 17.7213 > 4.74$, either food or liquor sales, or both, are significant predictors of profits. **c.** $F = 21.4640 > 3.59$. At least one predictor is significant. **d.** $F = 18.7901 > 3.89$. At least one predictor is significant. **e.** $F = 1.2558 \ngtr 4.07$. We cannot be 95% sure that amounts spent on television, radio, or newspaper advertising effectively predict revenues. **f.** $F = 4.8150 > 3.24$; at least one predictor is significant. **g.** $F = 76.4137 > 4.26$; at least one predictor is significant. **h.** $F = 35.6241 > 3.47$; at least one predictor is significant. **i.** $F = 2.3731 \ngtr 3.03$; we cannot be 95% sure that the amount of nitrogen, potassium, or phosphorus is an effective indicator of yield.

SECTION 13.4, PAGE 446

13.2 a. $F_1 = 6.1131$, $F_2 = 146.7142$; use $\hat{Y} = 195.8333 + X_1 + 6X_2$. **b.** $F_1 = 64.747$, $F_2 = 6.5012$; use $\hat{Y} = 2.1967 - 0.9167X_1 - 0.5625X_2$. **c.** $F_1 = 3.78172$, $F_2 = 3.24391$, $F_3 = 0.09374$; no significant predictors.

SECTION 13.5, PAGE 455

13.3 a. $F_1 = 30.1577$, $F_2 = 7.6022$; use $\hat{Y} = -2.447907 + 0.269825X_1 + 0.667383X_2$. **b.** $F_1 = 6.0686$, $F_2 = 0.0373$; use $\hat{Y} = 0.3477 + 1.0950X_1$. **c.** $F_1 = 5.1992$, $F_2 = 42.403$; use $\hat{Y} = 35.221155 + 1.143994X_1 - 0.533289X_2$. **d.** $F_1 = 33.6725$, $F_2 = 4.3405$; use $\hat{Y} = -0.0878 + 0.5219X_1$. **e.** $F_1 = 1.2393$, $F_2 = 2.6998$, $F_3 = 0.0439$; no significant predictors. **f.** $F_1 = 1.2481$, $F_2 = 2.4867$, $F_3 = 1.1075$; no F's are significant. (Contradicts results of Exercise 12.1f.)

SECTION 13.6, PAGE 456

13.4 a. $0.162766 < \beta_1 < 0.376884$
$0.139968 < \beta_2 < 1.194818.$

b. $0.085392 < \beta_1 < 2.20261$
$-0.706090 < \beta_2 < -0.360488.$

c. $0.084917 < \beta_1 < 1.915083$
$4.879154 < \beta_2 < 7.120746.$

d. $-0.916667 < \beta_1 < -0.679703$
$-1.021369 < \beta_2 < -0.103731.$

F O U R T E E N

SECTION 14.2, PAGE 467

14.1 $X_1 = \begin{cases} 1 \text{ if sophomore,} \\ 0 \text{ if otherwise;} \end{cases}$ $X_2 = \begin{cases} 1 \text{ if junior,} \\ 0 \text{ if otherwise;} \end{cases}$ $X_3 = \begin{cases} 1 \text{ if senior,} \\ 0 \text{ if otherwise.} \end{cases}$

14.2 $X_1 = \begin{cases} 1 \text{ if no high school diploma,} \\ 0 \text{ if otherwise;} \end{cases}$ $X_2 = \begin{cases} 1 \text{ if graduated high school,} \\ 0 \text{ if otherwise;} \end{cases}$ $\ldots;$ $X_5 = \begin{cases} 1 \text{ if graduate work,} \\ 0 \text{ if otherwise.} \end{cases}$

14.3 If two patients have the same income and are the same age, but one is male and the other is female, then on the average the female's medical expenses will be \$20 less than the male's.

14.4

$$\mathbf{y} = \begin{bmatrix} 40 \\ 72 \\ \vdots \\ 69 \end{bmatrix} \qquad \mathbf{X} = \begin{bmatrix} 1 & 14 & 0 & 1 & 0 \\ 1 & 38 & 1 & 0 & 1 \\ 1 & 25 & 0 & 0 & 0 \\ 1 & 24 & 1 & 0 & 1 \\ 1 & 19 & 1 & 0 & 1 \\ 1 & 10 & 0 & 1 & 0 \\ 1 & 23 & 1 & 0 & 1 \\ 1 & 12 & 1 & 0 & 0 \\ 1 & 26 & 1 & 1 & 0 \\ 1 & 24 & 1 & 0 & 1 \end{bmatrix}$$

14.5 **a.** See Appendix L, Salary Data (p. 655). **b. c.** It appears that not only do males start out at a higher salary than females but also that the rate of increase in salary with years on the job is greater for males than for females. **d.** See Appendix L, Salary Data. **e.** For interaction, $t = 4.279$, so rate of increase in salary with years with the firm is different for males and females. Confidence interval: $0.2520 \pm 2.080(0.0589) = 0.1295$ to 0.3745. We are 95% sure that salary increases with years of employment by between \$129.50 and \$374.50 faster per year for males than for females. **f.** $t = 7.582$, so starting salaries differ for males and females. $6.3477 \pm 2.080(0.8372) = 4.6063$ to 8.0891. We are 95% sure that the starting salary for males is between \$4,606 and \$8,089 greater than for females. **g.** $15.1223 \pm 2.080(0.6878) = 13.6917$ to 16.5529. We are 95% sure that the starting salary for females is between \$13,692 and \$16,553. **h.** $0.5883 \pm 2.080(0.0493) = 0.4858$ to 0.6908. We are 95% certain that the average annual raise for females is between \$486 and \$691.

14.6 See Appendix L, Enrollment Data (p. 648). $a = 23.327.951$ is the estimated enrollment in fall 1970 (when $X_1 = X_2 = X_3 = X_4 = 0$). On the average, enrollments increased by 157 students per quarter. Generally, winter quarter enrollments are 2024 less than in the fall, spring enrollments are 3249 less than in the fall, and summer enrollments are 14,241 less than in the fall.

14.7 Regression equation for teacher A ($X_2 = 0$): $\hat{Y} = -38.2749 + 1.6885X_1$. Regression equation for teacher B ($X_2 = 1$): $\hat{Y} = (-38.2749 + 72.2482) + (1.6885 - 0.8008)X_1 = 33.9733 + 0.8877X_1$. $a = -38.2749$ is the intercept of the equation relating score to IQ for teacher A and is not interpretable in this problem. $b_2 = 72.2482$ tells how much higher the intercept for teacher B is than for teacher A. $b_1 = 1.6885$ says that for students taught by teacher A, a one-unit increase in IQ results in a 1.6885-point increase in score. $b_2 = -0.8008$ says that for teacher B, the increase in score for a 1-point increase in IQ is 0.8008 points less than for teacher A's students.

14.8 To test for difference in intercepts, $t = -3.358$; to test for difference in slopes, $t = 1.728$. You can be more than 99% sure that electricity costs for energy savers are lower than for nonsavers but cannot be quite 90% sure that there is any difference in how cost is related to area for energy savers and nonsavers.

14.9 **a.** See Appendix L, Diet Data (p. 656). Relationship appears to be curvilinear. **b.** See Appendix L, Diet Data. **c.** $t = 0.942$, so type of diet does not appear to affect the relationship between initial weight and weight loss.

14.10 See Appendix L, Diet Data (p. 656). For interaction, $t = -1.063$; for diet, $t = 1.187$, so type of diet does not appear to affect the relationship between initial weight and pounds lost.

SECTION 14.3, PAGE 676

14.11 **a.** See Appendix L, Package Data (p. 657). **b.** $t = 0.552$, so there is little evidence to conclude that interaction is present. **c.** $r_{12} = 0.9463$, so there is severe collinearity: typically, the greater the number of deliveries, the greater the number of miles traveled. **d.** Collinearity refers to the extent to which the two predictors are related. Interaction refers to the differential effect of one predictor on the response, depending on the value of the other predictor.

14.12 **a.–c.** See Appendix L, Prescription Data (p. 657). **d.** The intercept, 40.2651, is not interpretable, since the sample does not include individuals zero years of age with zero income. For each additional one year of age, expenditure increases \$0.18; for each additional \$1000 income, expenditure increases \$0.54; and for each combination of one additional year and \$1000 additional income, expenditure increases \$0.03. Significance probabilities indicate that the model needs an intercept ($p = 0.0006$), that expenditure is probably not related to age ($p = 0.4131$), but may be related to income ($p = 0.0904$), and is a function of the interaction between age and income ($p = 0.0001$).

F I F T E E N

SECTION 15.1, PAGE 486

15.1 **a.** See Appendix L, Cravens Data (p. 658). **b.** Plots indicate possible quadratic relationships with TIME, POTEN, and ACCTS. **c.** See Appendix L, Cravens Data.

SECTIONS 15.4–15.7, PAGES 509 AND 519

15.2, 15.6, 15.10, 15.14 Only height is conditionally significant.

15.3, 15.7, 15.11, 15.15 *Forward selection:* Final model with all predictors significant at $\alpha = 0.25$ contains predictors population change, agricultural employment, percent with telephones, percent rural, and median age. If $\alpha = 0.05$, then use only population change and percent with telephones. *Stepwise:* Uses population change and percent with telephones, both significant at $\alpha = 0.05$. *Backward:* Predictors population change, agricultural employment, percent rural, median age. All significant at $\alpha = 0.05$. MAXR: "Best" models with all predictors significant at $\alpha = 0.05$:

k	Predictors
1	Percent with telephones
2	Percent with telephones
	Population change
3	Population change
	Agricultural employment
	Percent rural
4	Population change
	Agricultural employment
	Percent rural
	Median age

MINR: "Best" models with all predictors significant at $\alpha = 0.05$ are the same as found by MAXR.

15.4, 15.8, 15.12, 15.16 See Appendix L, MPG Data (p. 651). **a.** Do not use the full model; it contains redundant predictors. *p*-values for HP, the quadratic term in HP, and the dummy variable indicating transmission type are 0.5127, 0.8013, and 0.4829, respectively. **b.** Almost all procedures come up with the model $\widehat{\text{MPG}} = 68.3859 - 0.0380\text{HP} - 20.4532\ \text{WT} + 2.2275(\text{WT})^2$. mpg decreases with increasing horsepower and weight, but at a decreasing rate with weight.

S I X T E E N

SECTION 16.2, PAGE 581

16.1 $d = 1.4863 < d_L = 1.57$; conclude positive serial correlation.
16.2 Reasonably well-behaved plot.

16.3

Interval	Percent
−1, 1	67
−2, 2	97
−3, 3	100

SECTION 16.3, PAGE 596

16.5

Model No.	PRESS
1	669.66
2	493.54
3	514.03
4	482.57

16.6

Model	C_p	$C_p - p$	$q[F_{p,v,0.05} - 1]$
1	22.64	20.64	8.70
2	7.67	4.67	7.85
3	6.02	2.02	6.96
4	3.95	−1.05	

I N D E X